Differential Optical Absorption Spectroscopy

Physics of Earth and Space Environments

The series *Physics of Earth and Space Environments* is devoted to monograph texts dealing with all aspects of atmospheric, hydrospheric and space science research and advanced teaching. The presentations will be both qualitative as well as quantitative, with strong emphasis on the underlying (geo)physical sciences.
Of particular interest are

- contributions which relate fundamental research in the aforementioned fields to present and developing environmental issues viewed broadly
- concise accounts of newly emerging important topics that are embedded in a broader framework in order to provide quick but readable access of new material to a larger audience

The books forming this collection will be of importance for graduate students and active researchers alike.

U. Platt · J. Stutz

Differential Optical Absorption Spectroscopy

Principles and Applications

With 272 Figures and 55 Tables

Professor Dr. Ulrich Platt
Universität Heidelberg
Inst. Umweltphysik
Im Neuenheimer Feld 229
69120 Heidelberg
Germany
e-mail: ulrich.platt@iup.
uni-heidelberg.de

Professor Dr. Jochen Stutz
University of California Los Angeles
Department of Atmospheric
and Oceanic Sciences
7127 Math Sciences
Los Angeles, CA 90095–1565
USA
e-mail: jochen@atmos.ucla.edu

ISBN: 978-3-540-21193-8 e-ISBN: 978-3-540-75776-4

Physics of Earth and Space Environments ISSN: 1610-1677

Library of Congress Control Number: 2008920395

Cover design: eStudio Calamar S.L., F. Steinen-Broo, Pau/Girona, Spain

Printed on acid-free paper

9 8 7 6 5 4 3 2 1

springer.com

Preface

Light is the essential source of information about the surrounding world for most of us. We see objects of different brightness and colours, or, in scientific terms, we distinguish objects by the way they reflect light of different wavelengths with varying efficiency. It was exciting for both of us to learn that scientific instruments can expand our senses and provide information from this very same radiation that goes far beyond the simple recognition of objects. The ability to "see" the composition of the air surrounding us and above our heads, to us and others, one of the most fascinating aspects of the application of modern technology to study the atmosphere. This fascination drove many of the applications of the Differential Optical Absorption Spectroscopy (DOAS) method presented in this book, and continues to be the motivation for many current developments.

DOAS is an elegant and powerful analytical method to study the atmosphere and is based on the relatively simple principles of classical absorption spectroscopy. It has therefore always been quite surprising to us that this method has not found wider use in atmospheric research and air-quality monitoring.

We have introduced many university students and researchers to DOAS, often wishing that we could improve our efforts with a comprehensive text that describes both the theoretical basis and the practical applications of the method. Through our collaborations with air-quality monitoring agencies, we also realized that few of our partners had a deeper understanding of the requirements and the benefits of this method, explaining why DOAS was not considered as an air-quality monitoring technique by these agencies.

Many of the principles of DOAS and examples of scientific applications have been published in the literature over the past three decades, while some of the basics such as hardware and software design can be found only in unpublished documents or have been communicated orally in the few research groups specializing in this method.

Over the years it has become clear to us that the lack of a comprehensive text was an obstacle to the advancement of DOAS. This book emanated from the desire to provide such a text for students, researchers at universities and in industry, as well as air-monitoring agencies wishing to employ this method.

The development and first application of DOAS dates back more than 30 years. We would like to thank the large number of colleagues who have made contributions to the improvement of the method and the development of new uses over the years. In particular, we want acknowledge Dieter Perner, with whom Ulrich Platt first introduced the idea of DOAS, and Dieter Ehhalt, who supported this new development. Fruitful discussions with John Noxon, who had similar ideas at the time, lead to improvements in this method. John Burrows, who so stubbornly and resourcefully pursued the idea of DOAS in space and trusted us that it would work, deserves much of the credit for the development of satellite DOAS instruments. We are also in debt to Jim Pitts and Paul Crutzen, who immediately saw the potential of the technique. This book would have been impossible to write without the help of all the students and researchers who have worked in our groups. Among them we would like to single out Gerd Hönninger, who played a crucial role in the development of the MAX-DOAS technique, and who tragically passed away during a research visit to Alaska. We would also like to thank Yasmine Stutz, who helped us in the preparation of this book.

Finally, and most of all, we would like to thank our families for their continuing support and great patience with us.

Contents

1

Introduction

The atmosphere is a thin layer of gases separating the earth's surface from space. It protects the surface from radiation, provides a blanket that elevates the surface temperature, is responsible for the redistribution of water and heat, provides us with oxygen, and removes harmful gaseous compounds. These properties are essential to the earth's ability to support life. For example, it is believed that only the ozone layer formed 0.5–1 billion years ago reduced harmful solar UV radiation sufficiently to allow life to leave the oceans and further develop on land. While the atmosphere has undergone natural variations during the past 4.5 billion years, the arrival and expansion of human societies over the recent millennia have led to changes in the chemistry and physics of the atmosphere. Industrialisation, which began approximately 200 years ago, has had a marked impact on the composition of the atmosphere, for example by increasing levels of carbon dioxide and tropospheric ozone. During the same period scientific interest in the chemistry and physics of the atmosphere has expanded. While at the beginning this interest was driven by scientific curiosity, over the past few decades research has been increasingly motivated by the urgency to understand the response of the atmosphere to anthropogenic influence.

As the atmosphere is the medium in which most plants, animals, and humans live, the 'health' of the atmosphere is of great societal concern today. Numerous examples have shown the impact that emission and transformation of gases have on the atmosphere. In particular, urban air quality has been a concern for many years. Beginning with observation of increased numbers of deaths in industrialised areas in 19th century Europe, air pollution was first recognised as a serious threat after the London 'killer-smog' in December 1952. A new kind of air pollution was simultaneously recognised in Los Angeles, which lead to serious plant damage, and lung and eye irritation in humans. Although recognised for half a century, urban and regional air quality problems remain among the most challenging issues of the 21st century. After five decades of air pollution research, political efforts to improve air quality, and an increasing public awareness, more than 70 million people in the United

U. Platt and J. Stutz, *Introduction*. In: U. Platt and J. Stutz, Differential Optical Absorption Spectroscopy, Physics of Earth and Space Environments, pp. 1–4 (2008)
DOI 10.1007/978-3-540-75776-4_1

States still live in or near areas where ozone levels exceed national air quality standards (NRC, 2000). Künzli et al. (2000) found that between 19,000 and 44,000 people per year die prematurely from the effects of air pollution in three European countries. Air pollution is also an increasing problem in developing countries that strive to reach western living standards through rapid industrialisation. Secondary pollutants, such as ozone, particulate matter, and most likely nitrogen dioxide are known to have detrimental effects on human health (Brunekreef and Holgate, 2002; Fenger, 1999) The World Health Organization estimates that 1.4–6 million people die each year because of air pollution (WHO, 1999). Pollution can also be transported over long distances, for example between Asia and the United States, and the United States and Europe, therefore influencing the composition of the atmosphere on a global scale (IPCC, 2002).

Another well-known example of human impact on the composition of the atmosphere is stratospheric ozone depletion events over Antarctica. While scientists had anticipated a thinning of the ozone layer due to the release of man-made chemicals summarised under the term chlorofluorocarbons (CFCs), discovery of the extent and severity of this 'ozone hole' in 1985 was a surprise. The urgent need to protect the ozone layer, which was supported by a sound scientific assessment of the depletion mechanism, led to a number of international treaties to curb the release of CFCs. Although the ozone hole will probably not 'close' until the year 2050, the ban of CFCs is the first successful international effort to abate a severe change in the atmospheric composition on a global scale.

The most publicly discussed topic of atmospheric change in the beginning of the 21st century is the anthropogenic impact on the climate system of our earth. In this context, climate change refers not only to temperature but to all aspects of the atmospheric system, including weather and rain patterns, the fate of global ice sheets, the frequency of severe weather events, etc. The fact that the earth's climate is altered by a change in atmospheric composition is well known (Arrhenius, 1896). The analyses of various paleorecords, such as ice cores, have shown that atmospheric composition and climate have been correlated over the past 100,000 years. Major volcanic eruptions also perturbed the climate on shorter time scales. For example, weather pattern changes and a global temperature decrease of up to 1.2°C were observed for 5 years after the eruption of Krakatau in 1883. These natural variations in the earth climate have been ongoing throughout the entire history of our atmosphere.

It is, however, becoming increasingly clear that human activities have changed the climate of our planet in an unprecedented way during the past two centuries. Observations show a 0.6°C increase in surface temperature over the past 100 years, a shrinking of Arctic sea ice volume, a retreat of glaciers, and many other examples that all support this conclusion (IPCC, 2002). The main instigators of these changes are trace gases that trap infrared radiation. These trace gases are released from human activities or formed as a consequence of these emissions. Emission and formation of small particles suspended in air is also likely to play an important role. Although many uncertainties about the processes influencing our climate system remain, it is clear that the possible consequences of these changes are severe, and that strategies to mediate these changes are needed.

Efforts to understand the fundamental physical and chemical processes that control the atmosphere are thus not only motivated by scientific curiosity, but also more importantly by the desire to assess and mitigate anthropogenic impact. Abatement strategies for air pollution, climate change, and other environmental problems must be based on a sound scientific understanding, as well as an accurate quantification of the changes and the efficiency of remediation efforts.

Measurement techniques are an important source of knowledge about atmospheric chemistry and physics. Scientific interest in the quantitative determination of atmospheric trace constituents dates back more than 150 years, when Christian Schönbein measured the concentration of atmospheric ozone, which at that time was considered beneficial for human health. Since then a wide variety of analytical methods to study trace gas concentrations have been developed. Spectroscopic techniques (initially introduced by Kirchhoff and Bunsen in Heidelberg around 1859) have been on the forefront of scientific research and have made major contributions to many aspects of atmospheric research. Discoveries of the UV absorption of atmospheric ozone by Cornu and Hartley around 1880, the ozone layer by Dobson (1925) and Götz (1926) in the 1920s, and methane and carbon monoxide in the earth's atmosphere by Migeotte (1948, 1949) were some of the early milestones.

Since then spectroscopic methods have played an important role in the study of the chemical and physical processes dominating the atmospheric composition. Among other discoveries, they elucidated the central role of free radicals by making the first unambiguous identification of the hydroxyl radical (OH) in the nightglow (Bates and Nicolet, 1950), and the first detection of tropospheric OH (Perner et al. 1976). The history of the discovery and study of the ozone hole was also driven by spectroscopic techniques. The first report of the ozone hole was based on the long-term spectroscopic monitoring of ozone in Antarctica (Farman et al., 1985). The identification and quantification of various halogen radicals such as ClO, OClO, and BrO shed light on the catalytic processes responsible for the destruction of ozone.

Among the many spectroscopic techniques, differential optical absorption spectroscopy (DOAS) has proved to be one of the most powerful methods to measure a wide variety of trace gases. On the basis of the measurement of absorptions in the open atmosphere, it is particularly useful for the observation of highly reactive trace gases, and allowed the first detection of radicals such as OH, NO_3, OClO, BrO, IO, and OIO both in the troposphere and the stratosphere. The calibration-free absolute measurements, and the unequivocal identification of these and other trace gases, make this method a unique tool for the investigation of atmospheric chemistry and physics. Developments in recent years have introduced new DOAS applications which range from smog chamber studies, monitoring emissions from volcanoes, and atmospheric chemistry process studies to satellite measurements of global trace gas levels.

This book provides an introduction to the DOAS technique and an overview of the current state of this measurement method. It was written for readers who want to become familiar with the DOAS technique, and therefore provides the necessary background to appreciate and understand the breadth of DOAS applications. The physical and technical principles that form the basis of DOAS are conveyed for readers with a background in the natural sciences. The book is intended to present the details of this method for future and established DOAS users. It will provide enough information to make educated use of DOAS instruments, and aid in the construction of new instruments and the improvement of the analytical methods. The book is divided into 12 chapters. A short review of the chemistry of the troposphere and the stratosphere is given in Chap. 2, introducing the trace gases and concepts needed to understand the various applications of DOAS. The third chapter will briefly convey the fundamentals of atomic and molecular spectroscopy that form the basis of all spectroscopic methods. Chapter 4 describes the propagation of radiation in the open atmosphere. A review of various analytical methods to measure atmospheric trace constituents is presented in Chap. 5. Chapter 6 gives a general introduction to DOAS. The technical aspects in the construction of DOAS instruments, the mathematical basics of the analytical techniques, and radiative transfer methods are presented in Chaps. 7–9, respectively. Chapters 10 and 11 then present a number of examples for the application of DOAS in the troposphere and the stratosphere. A look into the future of DOAS is given at the end of the book in Chap. 12.

2

Atmospheric Chemistry

Air quality, the health of the ozone layer, and the earth's global climate are closely tied to the composition of the atmosphere and the chemical transformation of natural and anthropogenic trace gases. Although the purpose of this book is not to cover the full breadth of atmospheric chemistry, this chapter provides a short introduction into this topic, with the intent to motivate and clarify the applications of the differential optical absorption spectroscopy (DOAS) technique. Key elements of the contemporary understanding of atmospheric chemistry and ongoing research efforts are presented. This chapter is mostly restricted to chemical reactions occurring between gas molecules, although a few surface reactions are also presented. It should be noted that much of our knowledge of atmospheric chemistry comes from studies employing sophisticated instrumentation for the detection of atmospheric trace gases, including the DOAS technique.

Much of the motivation to study atmospheric chemistry and the composition of the atmosphere is based on various man-made environmental problems that have emerged over the past two centuries. Human activities have upset the natural balance of the atmosphere by influencing the trace gas and aerosol composition on local, regional, and even global scales. The following list names the most significant atmospheric environmental problems:

- 'London Type' smog was first recognised as an environmental problem in the 19th century (Brimblecombe and Heymann, 1998). The emission of soot particles and the formation of aerosols consisting of sulphuric acid, which were produced from photochemical oxidation of SO_2 emitted from combustion sources, had a severe impact on human health. During the so-called 'London Killer Smog', approximately 4000 excess deaths were counted in London during a 4-day period. It should be noted that the term 'Smog' was coined to describe the interaction of Smoke and Fog (Smog) that formed the strong haze observed in London during winter time.

U. Platt and J. Stutz, *Atmospheric Chemistry*. In: U. Platt and J. Stutz, Differential Optical Absorption Spectroscopy, Physics of Earth and Space Environments, pp. 5–75 (2008)
DOI 10.1007/978-3-540-75776-4_2

- Crop damage and human discomfort were the first indication of high levels of oxidants during sunny and hot days in Los Angeles (Haagen-Smit, 1952; Haagen-Smit and Fox, 1954; Finlayson-Pitts and Pitts, 2000). The formation of this 'Los Angeles Type' smog was found to be primarily due to photochemical formation of large amounts of ozone, carbonyl compounds, and organic aerosol from car exhaust and industrial emissions of nitrogen oxides, carbon monoxide, and volatile organic compounds (VOCs) (Haagen-Smit, 1952; Finlayson-Pitts and Pitts 2000). Despite many decades of research and mitigation activities, Los Angeles type smog remains one of the most common air pollution problems in urban areas today.
- Forest decline and lake acidification were signs of another man-made environmental problem. Increased emissions of sulphur dioxide and nitrogen oxides, followed by their gas and aqueous phase oxidation to sulphuric and nitric acid, lead to 'Acid rain' that lead to the deposition of these acids to various ecosystems (Charlson and Rhode, 1982). While acid rain has successfully been reduced in Europe and the United States, it remains a problem in many developing countries.
- The health of the atmospheric ozone layer has always been a concern of atmospheric scientists. In 1971, Johnston (1971) predicted the loss of stratospheric ozone due to emission of oxides of nitrogen by a planned fleet of supersonic passenger aircraft. Molina and Rowland (1974) warned of a gradual loss of stratospheric ozone due to catalysed ozone destruction processes by halogen species transported to the stratosphere in the form of extremely stable halogen-containing compounds, so called chlorofluorocarbons (CFCs), used as coolants or spray can propellants. Their predictions became reality in 1985, when Farman et al. (1985) discovered the stratospheric 'ozone hole' over Antarctica. This recurring phenomenon reduces the thickness of the stratospheric ozone layer over Antarctica to less than one third of its normal level every Antarctic Spring (e.g. Farman et al., 1985; Solomon, 1999). While steps have been taken to stop the emissions of CFCs, it is estimated that it will take another three decades before the ozone hole will close again.
- One of the most serious environmental problems is the impact of human activities on the climate of our planet. The potential of global warming caused by IR-active 'greenhouse' gases such as CO_2, CH_4, N_2O, O_3, CFCs (e.g. $CFCl_3$ and CF_2Cl_2), and the direct and indirect effects of chemically generated aerosol, have been known for some time (e.g. Arrhenius, 1896; Rhode et al., 1997; IPCC, 2002). Human activities, and thus the emission of the various greenhouse gases, over the past century have reached levels where an impact on global surface temperatures, the global water cycle, ocean levels, etc. has become very likely. Today, many signs point at the beginning of a global climate change which will have a severe impact on our earth. The global increase of tropospheric ozone, which apparently started in the middle of the 19th century, also contributes to this phenomenon. It is thought to be caused by catalytic ozone production due to wide-spread

emission of oxides of nitrogen and hydrocarbons (HC) (Volz and Kley, 1988; Staehelin et al. 1994).

Many of these phenomena are closely related to the atmospheric chemistry presented in this chapter.

Environmental conditions, such as temperature, pressure, and the solar spectrum, change with altitude and several altitude regimes of atmospheric chemistry can be distinguished. These properties of the atmosphere relevant for atmospheric chemistry will be presented in Sect. 2.1. The composition of the atmosphere is influenced by emissions of natural and man-made species. A short overview over these emission and their sources is given in Sect. 2.2. Tropospheric chemistry is dominated by ozone and other oxidants, i.e. hydroxyl/peroxy radicals, nitrogen oxides, and volatile organic carbons. While their chemistry is closely linked, we will discuss the chemistry of each class of species in Sects. 2.4–2.7. Sulphur chemistry (Sect. 2.8) also plays an important role as it aids in the formation of atmospheric acid and is also crucial for the formation of particles in the atmosphere. The chemistry of reactive halogen species (RHS) in the troposphere, which is suspected to influence ozone levels on a global scale, is presented in Sect. 2.9. The various radical species in the troposphere also play a crucial role for the 'self-cleaning' power of the atmosphere (Sect. 2.11). The atmospheric region between 15 and 50 km height, the stratosphere, hosts the ozone layer, which protects the earth surface from solar UV radiation. Consequently, stratospheric chemistry (Sect. 2.10) revolves around the formation and destruction pathways of ozone, as well as the chemical mechanisms leading to the destruction of ozone over Antarctica in spring, the so-called ozone hole.

2.1 Atmospheric Structure and Composition

For practical purposes, air can be viewed as an ideal gas. The relationship between pressure p, absolute temperature T, and volume V for a given number of moles n (1 mole is equal to $\mathrm{N_A} = 6.023 \times 10^{23}$ molecules of air) is given by the ideal gas law:

$$pV = nRT \tag{2.1}$$

where $R = 8.315\,\mathrm{J\,mole^{-1}K^{-1}}$ denotes the universal gas constant.

At standard sea-level pressure, $p_0 = 1.013 \times 10^5\,\mathrm{Nm^{-2}}$ and temperature $T_0 = 273.15\,\mathrm{K}$ (or 0.0° Celsius), each cubic centimetre ($\mathrm{cm^3}$) of air contains 2.69×10^{19} molecules. At higher altitudes, z, the atmospheric pressure, $p(z)$ drops exponentially from its value at sea level:

$$p\,(z) = p_0 e^{-\frac{Mgz}{RT}} = p_0 e^{-\frac{z}{z_S}} \tag{2.2}$$

where $M = 0.02897\,\mathrm{kg\,mole^{-1}}$ is the mean molar mass of air, $\mathrm{g} = 9.81$ is the acceleration of gravity, and $z_S = \frac{RT}{Mg} \approx 7 \pm 1\,\mathrm{km}$ denotes the 'scale height' of the atmosphere.

While there is a monotonic drop in pressure with altitude, the vertical temperature profile shows distinct variations, with several maxima and minima. The variation of temperature, type of chemical processes, and mixing mechanisms at different altitudes leads to a division of the atmosphere into a number of distinct compartments (see Fig. 2.1):

The *troposphere* is the lowest layer of the atmosphere, and is relatively well mixed. It extends from the surface to about 8 km altitude at the poles, and from the surface to about 18 km at the equator. The troposphere can be further subdivided into:

- The atmospheric boundary layer (BL), which roughly covers the lowest 1–2 km of the atmosphere. Air movement and mixing in the BL are influenced by friction on the earth's surface.
- The free troposphere, which extends from the top of the BL to the tropopause.

The *stratosphere* is a poorly mixed layer extending above the troposphere to about 50 km altitude. Common subdivisions of the stratosphere include:

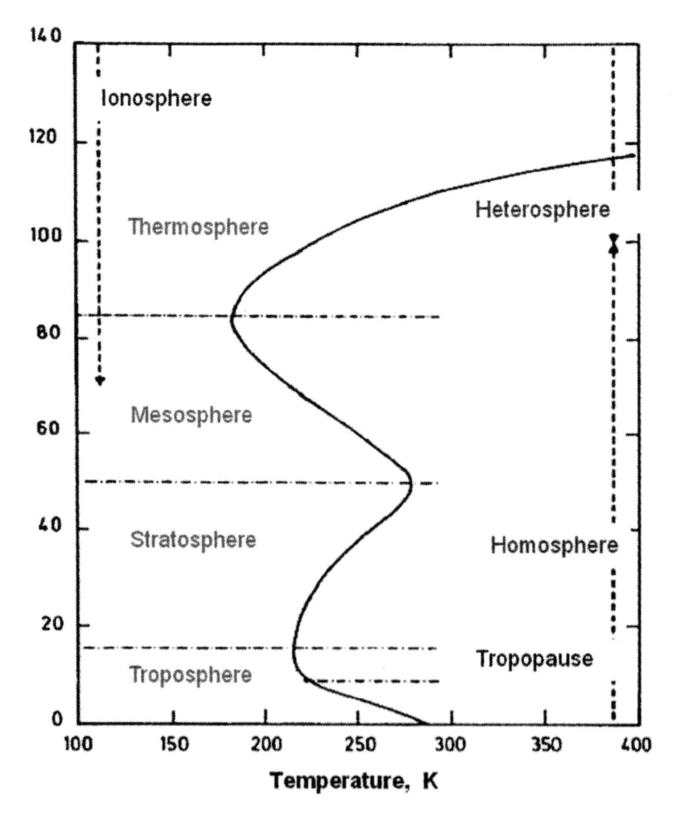

Fig. 2.1. The layers of the atmosphere, divided according to temperature, mixing mechanism, and degree of ionisation (from Brasseur and Solomon, 1986)

- The tropical tropopause layer
- The lower and middle stratosphere
- The upper stratosphere

The *upper atmosphere* extending above the stratopause is subdivided into the mesosphere, thermosphere, and exosphere.

Despite the vertical temperature and pressure variation up to about 100 km altitude, the composition of the atmosphere is fairly constant with respect to its main components. The three most abundant species, which constitute more than 99.9% of dry air (see Table 2.1), are molecular nitrogen and oxygen, as well as the noble gas argon. In addition, the atmosphere contains water vapour between 10^{-4}% and a few % (by volume). Other species with comparable abundances are carbon dioxide, and all remaining nobel gases. With the exception of water vapour and carbon dioxide, the species listed in Table 2.1 have atmospheric residence times in excess of 1000 years, and are therefore referred to as 'permanent' constituents. In addition to the above species, the atmosphere contains a multitude of other species at much lower abundance. They are called 'trace gases' and include methane (CH_4, $\approx$ 0.00017%), molecular hydrogen (H_2, $\approx$ 0.00005%), and nitrous oxide (N_2O, $\approx$ 0.00003%) among others.

2.1.1 Trace Species in the Atmosphere

In addition to the main gaseous constituents (Table 2.1), the atmosphere contains a large number of trace gases. Only a few trace species (e.g. CH_4) exceed a mixing ratio of one molecule in a million air molecules. However, a very large number of species (e.g. ozone, carbonyl sulphide, difluoro–dichloro–methane, methyl chloride, and methyl bromide) are present at mixing ratios around or below one molecule in a billion air molecules. Atmospheric chemistry

Table 2.1. The main constituents of the (unpolluted), dry atmosphere

Gas	Chemical formula	Mixing ratio by volume	Mixing ratio vol. %
Nitrogen	N_2	0.7808	78.08
Oxygen	O_2	0.2095	20.95
Argon	Ar	$9.3 \cdot 10^{-3}$	0.93
Carbon dioxide	CO_2	$0.37 \cdot 10^{-3}$	0.037
Neon	Ne	$18 \cdot 10^{-6}$	0.0018
Helium	He	$5.2 \cdot 10^{-6}$	0.00052
Methane	CH_4	$1.7 \cdot 10^{-6}$	0.00017
Krypton	Kr	$1.1 \cdot 10^{-6}$	0.00011
Xenon	Xe	$0.9 \cdot 10^{-6}$	0.00009
Hydrogen	H_2	$0.5 \cdot 10^{-6}$	0.00005
Dinitrogen oxide	N_2O	$0.3 \cdot 10^{-6}$	0.00003

is predominately concerned with the fate of these trace species, which, despite their low concentration, have a very noticeable impact on our atmosphere.

Another very important class of components in the atmosphere, which is not gaseous, is small particles of liquid or solid matter dispersed and suspended in air, the *aerosol*. The literal translation of the word aerosol is 'solution in air'. Since a solution includes both a solvent and a solute, one might insist that the term aerosol should only be used to denote the entire system, including both the suspended particles and the carrier gas. Usually, however, the term aerosol is used in the literature synonymously with 'suspended particles', excluding the carrier gas.

An important property of a solution is that it is a stable system. Applying a similar criterion to aerosol particles means that only those particles which remain suspended in air for a sufficiently long time to be transported by the wind over reasonably long distances can be part of an aerosol. The atmospheric residence time of large aerosol particles is limited by their settling velocity, which is the cause of sedimentation. The settling velocity varies approximately with the square of the particle radius, and therefore the upper limit for the size of aerosol particles is relatively sharp. Particles with radii much larger than about 10 μm have too large settling velocities to behave like true aerosol particles, and are thus classified as 'coarse dust particles'. (A spherical particle of 10 μm diameter has a settling velocity of about $2.42\,\mathrm{cm\,s^{-1}}$, or $87\,\mathrm{m\,h^{-1}}$, and thus will be rapidly removed from the atmosphere by sedimentation).

Defining the lower limit for the existence range of aerosol particles is more difficult. Clearly, aerosol particles must exceed the size of individual molecules, which have radii in the range of fractions of a nanometre (e.g. the gas kinetic radius of N_2 or O_2 is ≈ 0.2 nm). Nucleation theory teaches us that clusters composed of several condensable molecules (e.g. a mixture of sulphuric acid and water molecules) can only be stable when they exceed a certain critical size. Before reaching that critical size, the clusters tend to evaporate again. It appears attractive to use that critical size as a lower limit for the definition range of aerosol particles in the atmosphere. However, the critical size is not uniquely defined, but depends on atmospheric conditions, in particular on temperature, relative humidity, and the oxidation rate of SO_2, which yields the condensable sulphuric acid molecules. Very often the critical size of these so-called 'secondary aerosols' is in the range of 1 nm.

2.1.2 Quantification of Gas Abundances

The amount of trace gases present in the atmosphere can be quantified by using two different descriptions. First, we define the **concentration** of a trace gas as the amount of trace gas per volume of air (at a given temperature):

$$c = \frac{\text{Amount of trace gas}}{\text{Volume of air}} \tag{2.3}$$

where 'amount' refers to either mass (c_m), number of molecules (c_n), or number of moles (c_M). Examples for units of concentration are micrograms per

m^3 or molecules per cm^3. The latter is also known as 'number density' of a gas. The partial pressure of a species is also a measure of its concentration.

Second, we can define the *mixing ratio* of a trace gas as the ratio of the amount of a trace gas to the amount of air, including the trace gas:

$$x = \frac{\text{Amount of trace gas}}{\text{Amount of air + Trace gas}} \approx \frac{\text{Amount of trace gas}}{\text{Amount of air}} \tag{2.4}$$

At typical atmospheric trace gas mixing ratios of $<10^{-6}$, the distinction between the 'amount of air' and 'amount of air + trace gas' is so small that it can be neglected for practical purposes. It is necessary to distinguish whether the 'amount' is in volume, number of moles, number of molecules, or mass.

An example of a mixing ratio is parts per million (ppm) by volume:

$$x_V = \frac{\text{Unit volume of trace gas}}{10^6 \text{ unit volumes of (air + trace gas)}} \text{ppm}$$

For smaller trace gas mixing ratios, x_V is given as parts per billion (ppb) and parts per trillion (ppt), which are analogously defined as:

$$x_V = \frac{\text{Unit volume of trace gas}}{10^9 \text{ unit volumes of (air + trace gas)}} \text{ppb}$$

$$x_V = \frac{\text{Unit volume of trace gas}}{10^{12} \text{ unit volumes of (air + trace gas)}} \text{ppt}$$

While mixing ratios could also be given by mass, this is rarely done in atmospheric chemistry. Nevertheless, in the literature the terms ppmv, ppbv, and pptv are sometimes used to stress the fact that volume mixing ratios are understood. Similar to volume mixing ratios are molar mixing ratios:

$$x_M = \frac{\text{Moles of trace gas}}{\text{Mole of (air + trace gas)}} \tag{2.5}$$

It should be noted that x_M is the new SI unit for mixing ratios. Since air under ambient conditions can be regarded in good approximation as an ideal gas, for practical purposes a mixing ratio x_M specified in moles per mole equals the mixing ratio x_V. Thus, the terms micromole per mole, nanomole per mole, and picomole per mole are essentially equivalent to ppm, ppb, and ppt by volume, respectively.

A common problem in atmospheric chemistry is the conversion of units of the amounts of trace gases. For example, for trace gas i we obtain the following conversion between the number density c_n (in molecules per cm^3) and mass concentration c_m (in grams per cm^3):

$$(c_m)_i = \frac{c_n \cdot M_i}{N_A} \tag{2.6}$$

where M_i denotes the molecular mass of the species i in g per mole and N_A is Avogadro's number with $N_A = 6.0221420 \times 10^{23}$ molecules $mole^{-1}$. Thus, the conversion between number density and mass concentration is different for species with different molecular (atomic) mass.

The conversion of number density c_n (in molecules per cm^3) into the corresponding (volume) mixing ratio is given by:

$$x_\mathrm{V} = c_\mathrm{n}\frac{V_0}{N_\mathrm{A}} \text{ or } c_\mathrm{n} = x_\mathrm{V}\cdot\frac{N_\mathrm{A}}{V_0} \tag{2.7}$$

where V_0 denotes the molar volume in cm^3 for the pressure p and temperature T at which the number density c_n was measured. For standard conditions ($p_0 = 101325\,\mathrm{Pa} = 1$ Atmosphere, $T = 273.15\mathrm{K}$) the molar volume is $V_0 = 22414.00\,\mathrm{cm^3\,mole^{-1}}$. For arbitrary temperature and pressure conditions, we can use:

$$x_\mathrm{V} = c_\mathrm{n}\frac{1}{N_\mathrm{A}}\frac{T}{p}\frac{p_0}{T_0}V_0 = c_\mathrm{n}\frac{1}{N_\mathrm{A}}\cdot\frac{RT}{p} = c_\mathrm{m}\frac{1}{M_i}\cdot\frac{RT}{p} \tag{2.8}$$

or

$$c_\mathrm{m} = x_\mathrm{V}\cdot M_i\cdot\frac{p}{RT} \tag{2.9}$$

Table 2.2 gives a few examples for the conversion of the different units in which the abundance of trace gases is customarily expressed.

It should be noted that mixing ratios, x_V and x_M, are independent of temperature and pressure. Consequently, they are conserved during vertical transport in the atmosphere. In contrast, the concentration c depends on both parameters and change when air is transported. However, trace gas concentrations are relevant for the calculation of chemical reaction rates and radiative properties (such as UV absorption) of the atmosphere. Spectroscopic measurement techniques (including DOAS) also give results in number density or concentration, not mixing ratios.

Table 2.2. The different units for the abundance of atmospheric trace gases ($T = 293.15\mathrm{K}$, $p = 101325\,\mathrm{Pa}$)

Trace gas	Molecular mass g/mole	Mixing ratio x_V ppb	Number density c_n molec. cm^{-3}	Concentration c_m µg m^3
O_3	48.00	1.000	$2.503\cdot10^{10}$	1.995
		0.501	$1.254\cdot10^{10}$	1.000
SO_2	64.06	1.000	$2.503\cdot10^{10}$	2.662
		0.376	$0.941\cdot10^{10}$	1.000
NO	30.01	1.000	$2.503\cdot10^{10}$	1.251
		0.799	$2.000\cdot10^{10}$	1.000
NO_2	46.01	1.000	$2.503\cdot10^{10}$	1.912
		0.532	$1.33\cdot10^{10}$	1.000
CH_4	16.04	1.000	$2.503\cdot10^{10}$	0.667
		1.500	$3.755\cdot10^{10}$	1.000
CH_2O	30.03	1.000	$2.503\cdot10^{10}$	1.248
		0.801	$2.005\cdot10^{10}$	1.000
CO	28.01	1.000	$2.503\cdot10^{10}$	1.164
		0.859	$2.150\cdot10^{10}$	1.000

Thus both types of units are useful and must be converted into each other as needed. Atmospheric trace gas profiles might look quite different when viewed in terms of mixing ratios or concentrations, as illustrated by the example of the atmospheric ozone profile in Fig. 2.2. The concentration (in molecules per cm^3) has a noticeable, essentially constant level in the troposphere, and a maximum at about 22 km in the lower stratosphere. In contrast, tropospheric mixing ratios (in ppm or µmole per mole) are very low and peak at about 36 km.

Another category for specifying trace gas abundances is the column density S, which is defined as the integral of the number density along a certain path in the atmosphere:

$$S = S_n = \int_{\text{Path}} c_n(s)\, ds \tag{2.10}$$

The unit of S_n, and column densities in general, is molecules per cm^2. In analogy to the mass mixing ratio c_m, the mass column density S_m can be defined as:

$$S_m = \int_{\text{Path}} c_m(s)\, ds \tag{2.11}$$

Typical units for mass column densities are $\mu g/cm^2$. The conversion between S_n and S_m is analogous to the conversion between c_n and c_m.

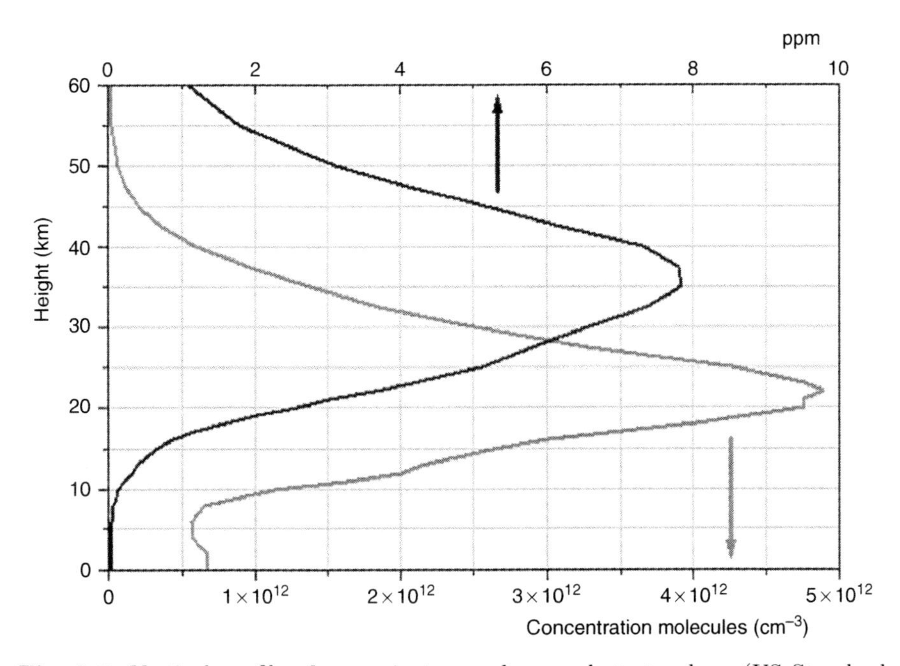

Fig. 2.2. Vertical profile of ozone in troposphere and stratosphere (US Standard Atmosphere), both in mixing ratio (in ppm or µmole per mole), peaking at about 36 km and concentration (molecules per cm^3), peaking at about 22 km

A frequently used path is the total (vertical) atmospheric column density V, where the concentration is integrated vertically from the surface to infinity:

$$V = \int_0^\infty c_\mathrm{n}(z)\,dz \tag{2.12}$$

2.1.3 Lifetime of Trace Gases in the Atmosphere

The atmosphere is not a static system with respect to its components. Rather, gases are released to the atmosphere or formed by chemical transformations, and removed again by chemical degradation or deposition on the ground. A useful quantity in this context is the steady-state (average) lifetime, or residence time, of a species in the atmosphere.

If the concentration of a species has reached a steady-state value, i.e. its rate of production P equals its rate of destruction D, its lifetime can be defined as the ratio:

$$\tau = \frac{c}{D} = \frac{c}{P} \tag{2.13}$$

Assuming the production P to be constant and the destruction D to be proportional to c, i.e. $D = c/\tau$ with the constant of proportionality $1/\tau$, we obtain:

$$c(t) = P \cdot \tau \cdot \left(1 - e^{-\frac{t}{\tau}}\right) \xrightarrow[\tau \to \infty]{} P \cdot \tau \tag{2.14}$$

The constant τ becomes clear by calculating $c(\tau) = 1 - 1/e \approx 63.2\%$ of the final value $c(\infty) = c(t \to \infty) = P \cdot \tau$. In other words, the lifetime τ is the time constant with which the trace gas concentration approaches its steady state.

Figure 2.3 gives approximate lifetimes of atmospheric trace constituents. The abundance of atmospheric constituents frequently is in a stationary state which depends on the lifetime of the particular gas. The variation of the concentration of a gas in space and time also depends on its lifetime. For instance, the long-lived components such as N_2, O_2, or noble gases, with lifetimes of thousands of years, show very little variation, whereas the concentrations of short-lived species may vary considerably with time and location. A quantitative relationship of residence time and variation of the mixing ratio of a gas was first established by Junge (1963).

2.2 Direct Emission of Trace Gases to the Atmosphere

As described above (Sect. 2.1.3), most gaseous (and particulate) components of our atmosphere are in a balance between continuous addition to

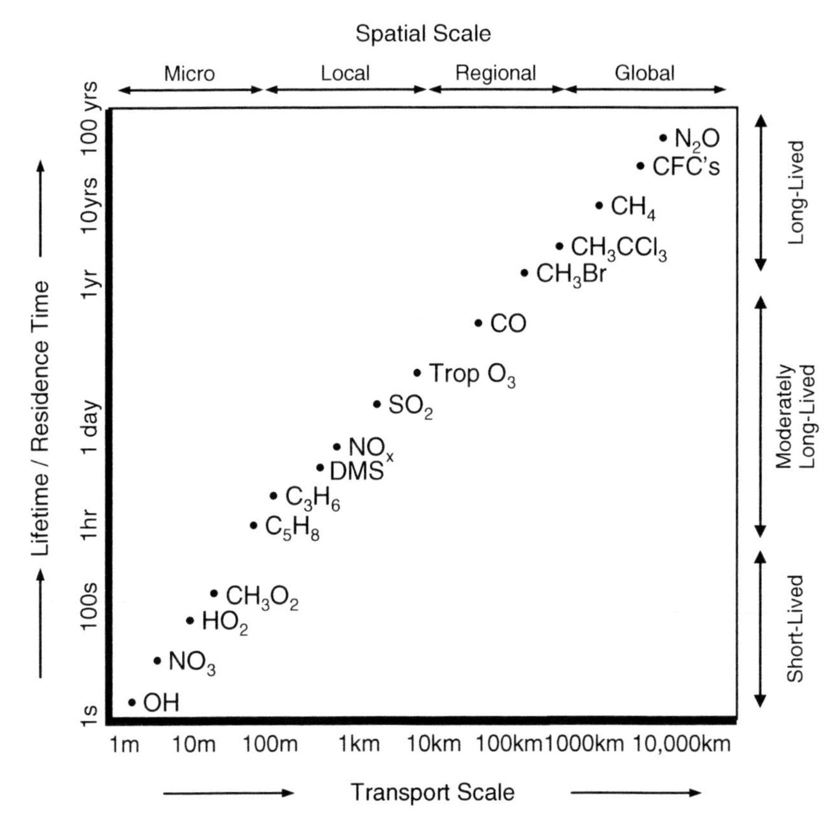

Fig. 2.3. Average lifetimes (residence times) of gases in the atmosphere range from seconds (and below) to millennia. Accordingly, transport can occur on scales reaching from a few meters to the global scale (adapted from Seinfeld and Pandis, 1998)

the atmosphere and removal from it. To understand the levels of trace gases in the atmosphere and their chemistry, it is important to study their sources or, in the case of gases which are formed chemically, the sources of their precursors. In particular with respect to air pollution, this is an important issue since human activities have added new trace gas sources over the past centuries. These sources have to be compared with natural emission sources in order to assess their impact on the composition of the atmosphere. While this section only gives a very brief overview of trace gas sources, which is limited to the gases that are of interest in this book, it should be noted that the accurate determination of emissions is a crucial aspect of atmospheric chemistry.

Emission sources are typically subdivided into natural and anthropogenic processes. The following list gives examples of the types of sources found in the atmosphere:

Natural emission of gases includes the following processes:

- Emission of gases from soil (e.g. methane, oxides of nitrogen NO and N_2O)
- Emission of gases by vegetation (primarily volatile organic compounds (VOC))
- Emission by biomass burning (VOCs, CO, NOx, etc.)
- Marine emissions [Dimethyl sulphide (DMS) and sea salt aerosol]
- Volcanic emissions (CO_2, SO_2, HCl, BrO, etc.)
- Formation in thunderstorms (NO)

Anthropogenic emissions are due to a series of human activities, primarily traffic, heating, and industrial processes, and also agricultural activities. The latter include emission from bare, or fertilised soils, flooded areas (e.g. for rice growing), or biomass burning. Anthropogenic emissions include:

- Emission during combustion processes (e.g. cars, power stations, and industry). The emitted gases include oxides of nitrogen (NO and NO_2), carbon monoxide, and VOCs
- Emission by industrial processes

Some anthropogenic emissions occur through the same processes as natural emission:

- Impact on gas emissions from soil due to agriculture: oxides of nitrogen (NO and N_2O)
- Emission by biomass burning (VOCs, CO, and NO_X)
- Emission of gases, by vegetation (primarily VOC)

We will briefly discuss the source and sink strengths, as well as average mixing ratios and lifetimes (residence times), of a series of important atmospheric trace gases, which have their dominant sources at the surface. The discussion is subdivided into nitrogen, sulphur, and carbon-containing species.

2.2.1 Nitrogen Species

Fixed nitrogen species play an important role in the chemistry and the climate of the atmosphere. We will focus here on the three most nitrogen important species that are emitted into the atmosphere – nitrous oxide (N_2O), nitrogen oxides (NO and NO_2), and ammonia (NH_3).

N_2O

Nitrous oxide is an important greenhouse gas and also the dominant source of reactive nitrogen in the stratosphere. N_2O is naturally emitted by bacteria in the soil and the ocean during the nitrogen fixation process (Table 2.3). Human activities have lead to an increase of these sources, predominantly due to the intensivation and expansion of agriculture. As a consequence, N_2O mixing ratios have increased from ~275 ppb before the year 1800 to over 310 ppb today.

Table 2.3. Tropospheric sources, sinks, and mixing ratios of nitrogen species [Source strength in million tons of fixed nitrogen (Tg N) per year (from Lee et al., 1997, Ehhalt 1999)]

Species	Mixing ratio, ppb	Major sources	Average source strength Tg/year	Major Sinks	Sink strength Tg/year	Atm. life time
Nitrous oxide, N_2O	310	Soil emission	10	Loss to stratosphere	25	110 a
		Ocean emission	6			
		Anthropogenic	9			
Nitrogen oxides, NO, NO_2	0.03–5	Fossil fuel combustion	21	Oxidation by OH and O_3 to HNO_3	44	2d
		Biomass burning	8			
		Soil emission/microbial production	10			
		Thunderstorms	5			
		Aircraft emissions	0.8			
Ammonia, NH_3	0.1 marine 5 continent	Domestic animals	26	Dry deposition	17	5d
		Emission from vegetation	6	Conversion to NH_4+ aerosol	36	
		Ocean emission	9			
		Fertilizer use	8			

h = hour, d = day, a = year.

NO_X ($NO + NO_2$)

The significance of nitrogen oxides will be discussed in detail in Sect. 2.5. Oxides of nitrogen, NO and NO_2, are produced in a large number of natural and anthropogenic processes. Particularly important are those where air is heated to high temperatures, such as in combustion processes (in internal combustion engines), forest fires, or lightning strikes. Natural sources of nitrogen oxides produce globally about 13–31 million tons of nitrogen fixed as NO_X per year (Table 2.3). The largest natural contributions come from brush and forest fires, lightning storms, and emission from the soil. Smaller contributions come from diffusion from the stratosphere and oxidation of ammonia (Ehhalt and Drummond, 1982; Lee et al., 1997).

Anthropogenic NO_X emissions can be grouped into three categories that essentially all originate from combustion processes. Thus, the largest source is the stationary combustion of fossil fuel (power stations, industry, and home heating). A further important – and growing – source is emission by automobiles. This emission additionally occurs in densely populated areas at very low emission height. Finally, part of the NO_X emission from forest fires and from soil is due to anthropogenic influence (intentional forest fires and artificial fertilisation).

In Table 2.3, the contributions of the various NO_X sources are summarised. On a global scale, the contributions of natural and anthropogenic NO_X sources are about equal in magnitude. The spatial distribution of strong NO_X sources, as derived from satellite-based DOAS measurements, is shown in Figs. 11.45 and 11.46 (Leue et al., 2001; Beirle et al., 2004a,c).

It should, however, be considered that the natural NO_X sources are much more equally distributed over the surface of the earth than the anthropogenic sources, which are concentrated on a very small fraction of the earth's surface (see also Fig. 11.45). For instance, in Germany [NO_X emission 2000: 0.9 million tons nitrogen, as N (Schmölling, 1983; Fricke and Beilke, 1993)], the natural contribution to the total NO_X emission is small.

NH_3

Ammonia plays a crucial role in the atmosphere as it is the only significant atmospheric base neutralising acids such as sulphuric and nitric acids. The ammonium ion, which is formed upon the uptake of ammonia on particles, is an important part of the aerosol. Ammonia is emitted predominately by livestock wastes and fertilised soils. Emissions from the ocean and vegetation also play a role (Table 2.3). Modern cars with catalytic converters also emit ammonia.

2.2.2 Sulphur Species

The increase of sulphur emissions since the onset of industrialisation has led to numerous environmental problems, such as London-type smog and acid rain. Naturally sulphur is emitted through biological processes in the soil and the ocean in its reduced form as carbonyl sulphide (COS), hydrogen sulphide (H_2S), and DMS (CH_3SCH_3). Volcano eruptions can also contribute to the natural sulphur emissions, in the form of sulphur dioxide, SO_2 (Table 2.4). Of the natural emissions, those of DMS dominate and are particularly important for the global sulfur budget (Sect. 2.7).

Anthropogenic emissions of SO_2 originate primarily from fossil fuel combustion. Over the past 200 years, sulphur emissions have sharply increased. As a response to the growing acid rain problem, the emission of sulphur has been greatly reduced in industrialised countries since 1975 (see also Fig. 2.4). This is due to stack gas desulphurisation measures, and also to the change of the economies in the eastern European countries. On a global scale, however, there is an expected increase of sulphur emission, due to enhanced coal combustion in Asian countries (Fig. 2.4).

The sulphur species that are most important in the atmosphere, their typical concentrations and atmospheric residence times, and their degradation mechanisms are summarised in Table 2.4.

2.2.3 Carbon-Containing Species

A very large number of carbon-containing trace species are present in the atmosphere. The most important carbon-containing species is carbon dioxide (CO_2). While CO_2 plays a crucial role as a greenhouse gas, it undergoes little chemistry in the atmosphere, and its sources will not be a topic of this section. The second most abundant carbon species is methane (CH_4), which, as will be discussed in Sect. 2.4, is chemically degraded in the atmosphere, directly participating in the formation of tropospheric ozone. Methane is also a greenhouse gas. It is difficult to distinguish natural and anthropogenic sources of methane as the emissions predominately stem from animals, including farm livestock, and wetlands, which includes rice paddies (Table 2.5). Fossil fuel consumption also contributes to today's methane emissions. The expansion of agriculture over the past two centuries has led to an increase of methane sources and a rise of methane mixing ratios from $\sim$750 ppb to over 1700 ppb today. Carbon monoxide (CO) is a product of incomplete combustion of natural material, for example forest fires, and fossil fuels. In addition, CO is a product of the chemical degradation of methane and volatile organic carbon (VOC) species in the atmosphere. Because CO is toxic and participates in the formation of photochemical smog, various measures have been taken to reduce its emissions from combustion sources. The most prominent example is the catalytic converter of automobiles, which was primarily designed to reduce CO emissions.

Hydrocrabons – often the term non-methane hydrocarbons (NMHC) is used to exclude CH_4 from this group – constitute a large class of carbon-containing species in the atmosphere. The number of different species is very large and we will consider only the most important members of this group

Table 2.4. Tropospheric sources, sinks, and mixing ratios of sulphur species [Source strength in million tons of sulphur (Tg S) per year (from Ehhalt 1999)]

Species	Mixing ratio, ppb	Major sources	Source strength Tg/year	Major sinks	Sink strength Tg/year	Atm. life time
Carbonyl sulphide, COS	0.5	Soil emission	0.3	Uptake by vegetation	0.4	7 a
		Ocean emission	0.3	Flux to stratosphere	0.15	
		CS_2 + DMS ox.	0.5			
Hydrogen sulphide, H_2S	0.005–0.09	Soil emission	0.5	Reaction with OH	2.7	3 d
		Vegetation	1.2			
		Volcanoes	1			
Dimethyl sulphide, DMS, CH_3SCH_3	0.005–0.1	Ocean emission	68	Reaction with OH, NO_3, BrO	70	2 d
		Soil emission	2			
Sulphur dioxide, SO_2	0.02–0.09 (marine), 0.1–5 cont.	Fossil fuel burning	160	Dry deposition, reaction with OH, liquid phase	100	4 d
		Volcanoes	14	Oxidation to SO_4 = + Wet deposition	140	
		Sulphide oxidation	70			

h = hour, d = day, a = year.

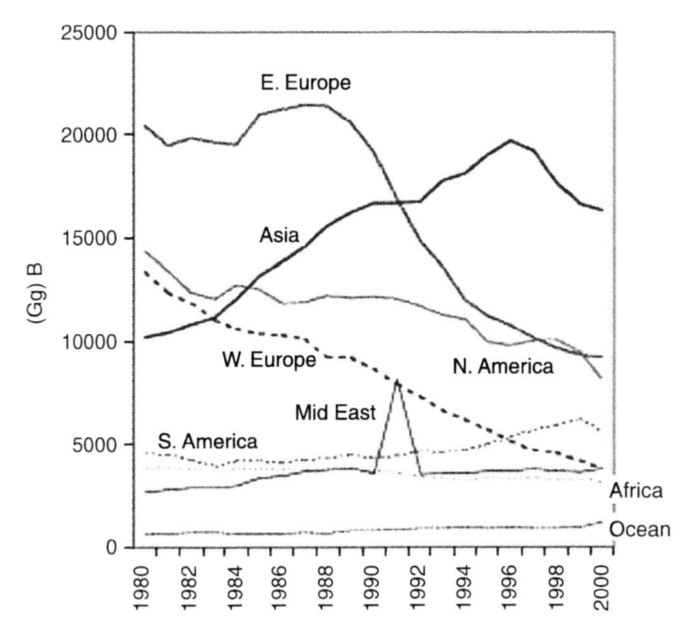

Fig. 2.4. Regional trends of global sulphur emissions (in giga-grams of sulphur per year) during the past two decades (from Stern, 2005, Copyright Elsevier, 2005)

here. On a global scale, the emission by vegetation of isoprene, C_5H_8, and to a lesser extent, terpenes, $C_{10}H_{16}$, dominate the emissions of VOC. These species are highly reactive in the atmosphere and participate in the formation of ozone and particles. Globally, natural emissions of other species are of lesser importance. Anthropogenic (and natural) emissions of VOC are an essential part of Los Angeles-type smog. A complex mixture of different trace gases is emitted by fossil fuel consumption, both through incomplete combustion and the loss of VOC vapours before combustion, industrial processes, refining of oil, solvents, etc. The emitted mixture, which contains alkanes (ethane, propane, etc.), alkenes (ethane, propene, etc.), alkynes and aromatics (benzene, toluene, etc.), and a variety of oxidised VOC (formaldehyde, acetaldehyde, etc.), depends on the emissions signatures of the different sources, as well as the distribution of these sources. For details on urban VOC emissions, we refer the reader to atmospheric chemistry textbooks, such as Finalyson-Pitts and Pitts (2000).

2.3 Ozone in the Troposphere

Ozone is a key compound in the chemistry of the atmosphere. In the troposphere it is a component of smog, poisonous to humans, animals and plants, as well as a precursor to cleansing agents (such as the OH radical, see Sect. 2.4.1). Tropospheric ozone is also an important greenhouse gas.

Table 2.5. Important organic trace gases, approximate mixing ratio, emission to the atmosphere, removal from the atmosphere, and average lifetime (residence time) τ in the atmosphere [see (2.14)]

Species	Mixing ratio, ppb	Major sources	Source strength Tg/year	Major Sinks	Sink strength Tg/year	Atm. life time
Methane, CH_4	1700	Rice fields	75	Reaction with OH	490	8 a.
		Domestic animals	100	Export to stratosphere	40	
		Swamps/marshes	200	Soil uptake	30	
		Biomass burning	40			
		Fossil fuel	100			
Carbon monoxide, CO	200 (NH) 50 (SH)	Anthropogenic emission	440	Reaction with OH	2400	0.2 a
		Biomass burning	770	Flux to stratosphere	100	
		CH_4 oxidation	860	Soil uptake	400	
		O_x. of natural VOC	610			
Isoprene, C_5H_8	0.6–2.5	Emission from deciduous trees	570	Reaction with OH, ozonolysis	570	0.2 d
Terpenes, $C_{10}H_{16}$	0.03–2	Emission from coniferous trees	140	Reaction with OH, ozonolysis	140	0.4 d
Total VOC		Fossil fuel	65	Reaction with OH	850	
		Biomass burning	35			
		Foliar emission	744			
		Ocean emission	6			

The table is based on Ehhalt (1999), but has been extended.
h = hour, d = day, a = year.

Ozone is formed by two distinctly different mechanisms in the troposphere and stratosphere. In the stratosphere, O_2 molecules are split by short-wave UV radiation into O-atoms, which combine with O_2 to form O_3. This process is the core of the 'Chapman Cycle' (Chapman, 1930; see Sect. 2.10.1). Until the late 1960s, it was believed that tropospheric ozone originated from the stratosphere. Today we know that large amounts of O_3 are formed and destroyed in the troposphere. Influx of O_3 from the stratosphere is only a minor contribution to the tropospheric ozone budget. Recent model calculations (IPCC, 2002) put the cross tropopause flux of O_3 at 390–1440 Mt a^{-1} (very recent investigations indicate that values near the lower boundary of the range are more likely), while they derive ozone formation rates in the troposphere at 2830–4320 Mt a^{-1}. The formation is largely balanced by photochemical destruction in the troposphere amounting to 2510–4070 Mt a^{-1}. Another, relatively small contribution to the O_3 loss is deposition to the ground, modelled at 530–900 Mt a^{-1}.

2.3.1 Mechanism of Tropospheric Ozone Formation

In the early 1950s, it became clear that under certain conditions in the atmosphere near the ground high concentrations of ozone are formed (e.g. Haagen-Smit, 1952). In fact, it could be shown in 'smog-chamber' experiments that large amounts of ozone are formed when mixtures of NO_X ($NO + NO_2$) and VOC are exposed to solar UV radiation. While the phenomenon of ozone formation as a function of VOC and NO_X in illuminated mixtures was empirically found in the 1960s, the exact mechanism only became clear much later, due to the work of Weinstock (1969), Crutzen (1970), and Levy et al. (1971). Ozone formation in the troposphere is initiated by the production of $O(^3P)$ from NO_2 photolysis (indicated by the term $h\nu$, see Chap. 3).

Under clear sky conditions at noontime, the average lifetime of the NO_2 molecule is only on the order of 2 min [$j_{NO2} = j_2 \approx 8 \times 10^{-3} s^{-1}$; e.g. Junkermann et al. (1989)]:

$$NO_2 + h\nu \rightarrow NO + O(^3P) \qquad (R2.1)$$

This reaction is followed by the rapid recombination of O with O_2:

$$O(^3P) + O_2 + M \rightarrow O_3 + M \qquad (R2.2)$$

Where M denotes any atmospheric molecule. At high pressure (and thus M and O_2 concentrations) in the troposphere, other reactions of $O(^3P)$, in particular its reaction with O_3, are negligible. Therefore, for each photolysed NO_2 molecule, an ozone molecule is formed. Reactions R2.1 and R2.2 are essentially the only source of ozone in the troposphere. However, ozone is often rapidly oxidised by NO to back NO_2:

$$O_3 + NO \rightarrow NO_2 + O_2 \qquad (R2.3)$$

Reactions R2.1–R2.3 lead to a 'photo-stationary' state between O_3, NO, and NO_2. The relation between the three species can be expressed by the *Leighton Relationship* (Leighton, 1961):

$$\frac{[NO]}{[NO_2]} = \frac{j_{2.1}}{k_{2.3} \cdot [O_3]} \tag{2.15}$$

where $j_{2.1}$ denotes the photolysis frequency of NO_2 and $k_{2.2}$ denotes the rate constant for the reaction of ozone with NO. For typical ozone mixing ratios of 30 ppb (1 ppb $\approx 10^{-9}$ mixing ratio), the $[NO]/[NO_2]$ ratio during daytime near the ground is on the order of unity. The reaction cycle formed by R2.1–R2.3 does not lead to a net formation of ozone. However, any reaction that converts NO into NO_2 without converting an O_3 molecule interferes with this cycle and leads to net ozone production. The key factor in tropospheric O_3 formation is, thus, the chemical conversion of NO to NO_2.

In the troposphere, the conversion of NO to NO_2 without O_3 occurs through a combination of the reaction cycles of hydroxyl HO_X ($OH + HO_2$), peroxy radicals, and NO_X (Fig. 2.5). In these cycles, OH radicals are converted to HO_2 or RO_2 radicals, through their reaction with CO or HC. HO_2 and RO_2, on the other hand, react with NO to reform OH, thus closing the HO_X/RO_X cycle. This reaction also converts NO to NO_2 (see also Sect. 2.5.1), which is then photolysed back to NO (R2.1). The oxygen atom formed in the NO_2 photolysis then reacts with O_2 to form ozone (R2.2). The process, shown in Fig. 2.5, therefore acts like a machine that, in the presence of NOx and sunlight, converts the 'fuel' CO and HC into CO_2, water, and ozone. Because, HO_X and NO_X are recycled, this catalytic ozone formation can

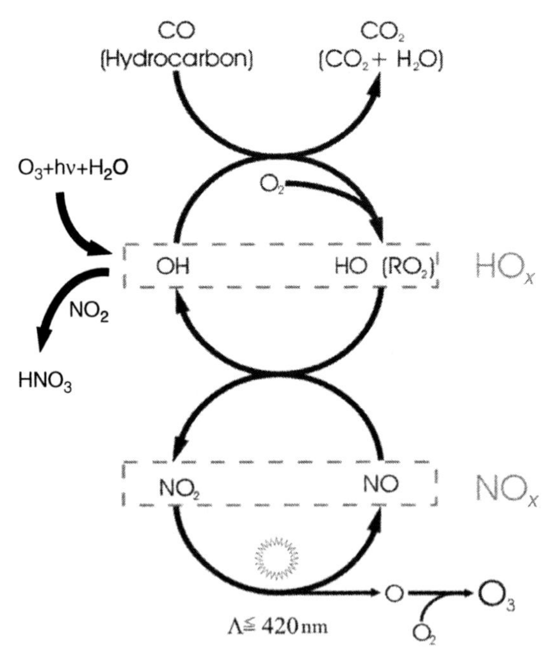

Fig. 2.5. Ozone formation in the troposphere is catalysed by hydrogen radicals ($OH + HO_2{=}HO_X$), peroxy radicals, and NO_X

be quite efficient (Crutzen, 1970). HO_X radicals are always present in the sunlit atmosphere (see Sect. 2.4); they are, for example, formed through the photolysis of ozone in the presence of water vapour. The cycles are only interrupted if either a NO_X or a HO_X is removed from the respective cycles, for example by the reaction of OH with NO_2, or the self-reactions of HO_2 and RO_2.

In background air, fuel for ozone formation is always present in the form of methane (mixing ratio of ~1.7 ppm) and CO, which is formed as a degradation product of CH_4 (Sect. 2.4.1). However, in clean air, the NO_X level might be very low, and thus insufficient to act as catalyst. This effect is outlined in more detail in Sect. 2.5.1.

2.3.2 Ozone Formation in Urban Centres and Downwind

In many urban centres, car exhaust dominates as a source of pollution. Emissions from home heating and industry can also contribute considerably. These sources emit large amounts of NO_X (mostly in the form of NO) and VOC, the main ingredients of ozone formation. As a consequence, ozone formation rates can be orders of magnitude higher than in slightly polluted air outside urban centres. On the other hand, in urban centres there are also effective ozone sinks:

- Reaction of O_3 with freshly emitted NO
- Reaction of O_3 with olefins.

Overall, the ozone lifetime can be quite short (on the order of a few hours) in urban areas. Therefore, there are usually large diurnal variations of the O_3 level in these regions. In fact, in urban centres the ozone level near the ground frequently reaches zero a few hours after sunset and remains at zero for the rest of the night. After sunrise, enhanced vertical mixing replenishes some O_3 from higher layers, thus initiating radical chemistry again (see Fig. 2.8). In addition, photolysis of formaldehyde and nitrous acid (see Sects. 2.4.1 and 2.5.1) can considerably speed up radical processes, and thus growth of the O_3 mixing ratio in the morning.

In a series of experiments where the concentrations of NO_X and VOC were independently varied, the ozone formation after a fixed time was found to vary with the initial levels of both groups of species in a characteristic way: At a given initial level of NO_X, the O_3 increases linearly with the initially present level of VOC. In other experiments with higher initial VOC levels, the amount of O_3 levels off. Increasing initial VOC even further does not lead to higher O_3 production. This regime is called 'NO_X limited'. Conversely, at a given initial level of VOC, the O_3 increases linearly with the initially present level of NO_X. At higher initial NO_X levels, the amount of O_3 levels off. Increasing initial NO_X even further in this 'VOC limited' regime does

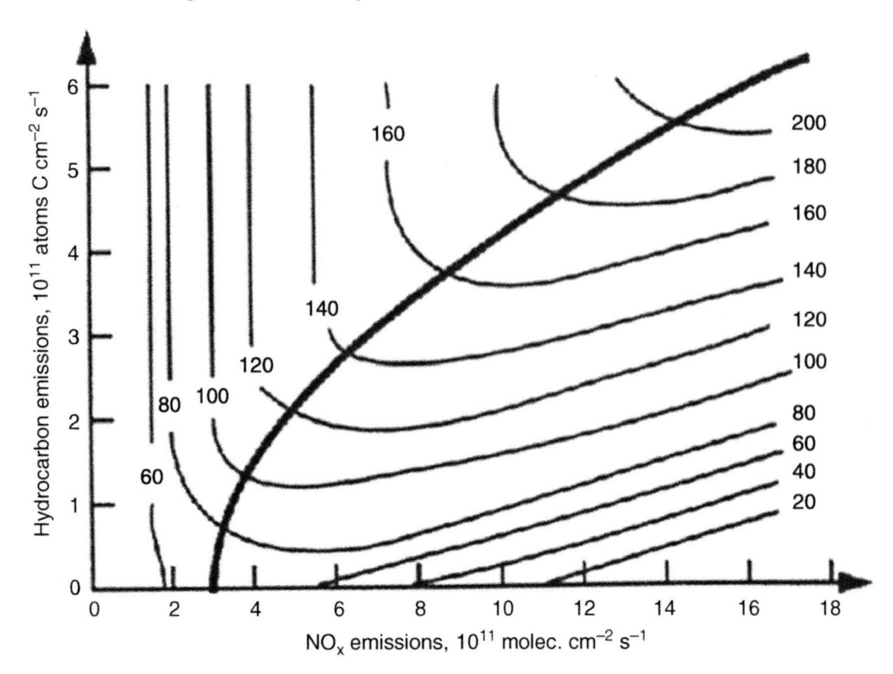

Fig. 2.6. Ozone concentrations simulated by a regional photochemical model as a function of NO_X and hydrocarbon (VOC) emissions. The thick line separates the NO_X limited regime (top left) from the VOC limited regime (bottom right) (from Sillman et al., 1990)

not lead to higher O_3 production. In fact, due to removal of OH by reaction with NO_2 (Fig. 2.5), O_3 formation will be reduced. Figure 2.6 illustrates this relationship. The lines of constant ozone formation as function of the VOC and NO_X composition of the initial mixture (or VOC and NO_X emission) are called ozone isopleths. Measurements in the open atmosphere also illustrate this dependence of the ozone production on the NO_X level (an example is shown in Fig. 2.7).

When considering which 'mixture' of HC (VOC) and NO_X is most effective in ozone production (at a given level of solar UV), it turns out that NO_X and VOC must be present in a certain ratio $R_{X0} = [NO_X]/[VOC]$, indicated by the ridge line in Fig. 2.6. Often this maximum is not reached in urban centres, where the primary emissions occur. Because the atmospheric lifetimes of VOC and NO_X are different, R_{X0} for an air-mass changes with time. In addition, it takes several hours for high ozone levels to be formed. Consequently, the highest O_3 levels are often found downwind of urban centres. This is illustrated in Fig. 2.8 which shows the diurnal variation of ozone levels in different types of air masses at a rural site (Forest, Weltzheimer Wald, Germany), and in a city (Heilbronn, southern Germany).

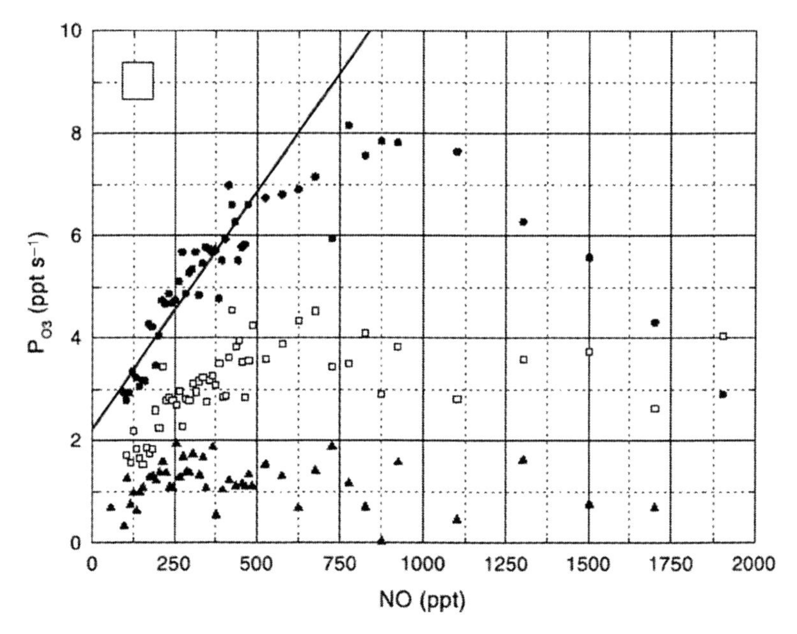

Fig. 2.7. Ozone production rates (P_{O3}) calculated from simultaneous observations of NO, NO_2, O_3, OH, HO_2, H_2CO, actinic flux, and temperature during the 1999 'Southern Oxidant Study' (June 15 to July 15) at Cornelia Fort Airpark, Nashville, Tennessee. Averaged P_{O3} is plotted as a function of the NO mixing ratio. The data were placed into three P_{HOx} bins: high ($0.5 < P_{HOx} < 0.7\,\mathrm{ppt\,s^{-1}}$, circles), moderate ($0.2 < P_{HOx} < 0.3\,\mathrm{ppt\,s^{-1}}$, squares), and low ($0.03 < P_{HOx} < 0.07\,\mathrm{ppt\,s^{-1}}$, triangles), and then averaged as a function of NO. All three P_{HOx} regimes demonstrate the expected generic dependence on NO. P_{O3} increases linearly with NO for low NO (<600 ppt NO), and then P_{O3} becomes independent of NO for high NO (>600 ppt NO). The crossover point between NO_X-limited and NO_X-saturated O_3 production occurs at different levels of NO in the three P_{HOx} regimes (from Thornton et al., 2002, Copyright by American Geophysical Union (AGU), reproduced by permission of AGU).

2.4 Radical Processes in the Atmosphere

Free radicals are the driving force for most chemical processes in the atmosphere. Since the pioneering work of Weinstock (1969) and Levy (1971), photochemically generated HO_X radicals (hydrogen radicals, OH + HO_2) have been recognised to play a key role in tropospheric chemistry. In particular, hydrogen radicals:

- initiate the degradation and thus the removal of most oxidisable trace gases emitted into the atmosphere
- give rise to the formation of strongly oxidising agents (mostly in the troposphere), such as ozone or hydrogen peroxide
- catalytically destroy stratospheric ozone (see Sect. 2.10)

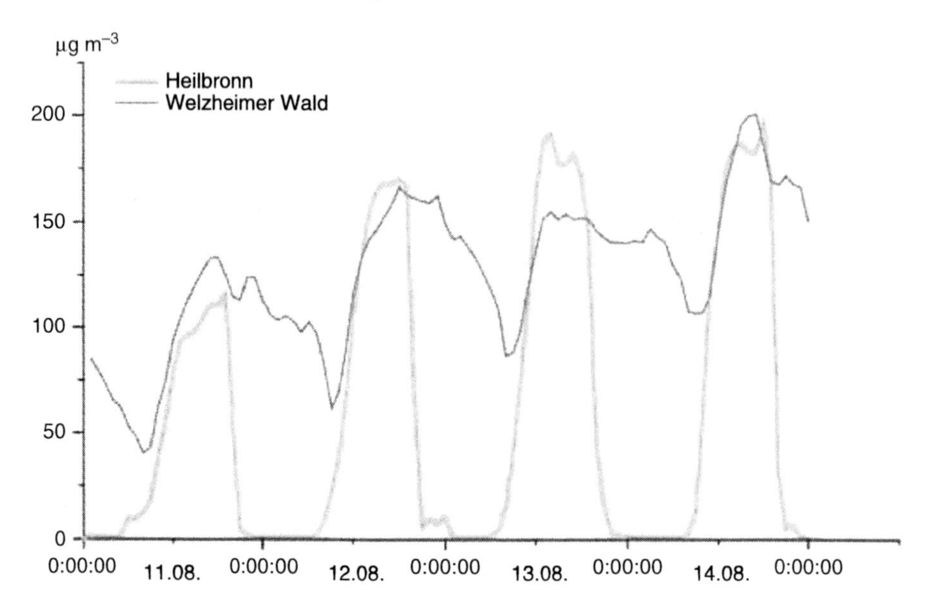

Fig. 2.8. Diurnal variation of ozone levels in different types of air masses during August 11–14, 2000: Forest (Weltzheimer Wald, Germany) and city of Heilbronn (southern Germany).
Source: Landesanstalt für Umweltschutz Baden-Württemberg and UMEG

- are difficult to remove once they are generated, since radical-molecule reactions tend to regenerate radicals.

Today we have an enormous amount of direct and indirect evidence of the presence of HO_X radicals (see, for example, Ehhalt et al., 1991; Wennberg et al. 1998; Platt et al., 2002), and the importance of HO_X for atmospheric chemistry can be assumed to be proved beyond reasonable doubt. Nevertheless, the possible role of other radicals, beginning with the (historical) idea of the impact of oxygen atoms $O(^3P)$ or excited oxygen molecules $O_2(^1\Delta)$ has been the topic of past and current investigations. In particular, the nitrate radical, NO_3, (see Sect. 2.5.2) and the halogen atoms and halogen oxide radicals BrO, IO, and ClO (Sect. 2.8) can make a considerable contribution to the oxidising capacity of the troposphere. For instance, reaction with NO_3 or BrO can be an important sink of DMS in marine environments. In addition, night-time reactions of nitrate radicals with organic species and NOx play an important role for the removal of these species. NO_3 chemistry can also be a source of peroxy radicals (such as HO_2 or CH_3O_2), and even OH radicals (Sect. 2.5.2). Table 2.6 shows an overview of the most important radical species in the troposphere and their significance for atmospheric chemistry. The details of the chemistry of NO_3 and halogen oxides will be discussed in the following sections. Here, we will concentrate on the tropospheric chemistry of hydroxyl radicals.

Table 2.6. Free radical cycles pertinent to tropospheric chemistry, and key processes influenced or driven by reaction of those radicals

	Species	Significance
HO_X cycle	OH	Degradation of most volatile organic compounds (VOC)
		Key intermediate in O_3 formation
		$NO_X \Rightarrow NO_Y$ conversion
	HO_2	Intermediate in O_3 formation
		Intermediate in H_2O_2 formation
	RO_2	Intermediate in ROOR' formation
		Aldehyde precursor
		PAN precursor
		Intermediate in O_3 formation
NO_3 cycle	NO_3	Degradation of certain VOC (olefins, aromatics, DMS, etc.)
		$NO_X \Rightarrow NO_Y$ conversion (via N_2O_5 or DMS-reactions)
		RO_2 precursor (night-time radical formation)
XO_X cycle	XO (X = Cl, Br, I)	Catalytic O_3 destruction (cause of 'Polar Trop. Ozone Hole')
		Degradation of DMS (BrO)
		Change of the NO_2/NO (Leighton) ratio
	X	Degradation of (most) VOC (Cl)
		Initiates O_3 formation
		RO_2 precursor
		Initiates particle formation (IO_X)

2.4.1 Sources of Hydrogen Radicals (OH and HO_2)

Hydroxyl radicals are probably the most important free radicals in the atmosphere. The degradation of most (but not all) oxidisable trace gases (such as HC or CO) is initiated by OH reaction (cf. Platt, 1999). OH radicals are therefore sometimes called the 'cleansing agent' of the atmosphere. Figure 2.9 shows a simplified outline of the HO_X ($OH + HO_2$) and RO_2 cycle.

Several primary OH production mechanisms are known:

(a) Globally, the most important process forming OH is initiated by UV photolysis of ozone to form electronically excited oxygen atoms:

$$O_3 + h\nu(\lambda < 320\,\text{nm}) \rightarrow O(^1D) + O_2(^1\Delta) \quad \text{(R2.4a)}$$

Recent research shows that photolysis of vibrationally excited ozone can considerably enhance the rate of reaction (R2.4a). In addition, some contribution comes from the channel:

$$O_3 + h\nu(\lambda < 420\,\text{nm}) \rightarrow O(^1D) + O_2(^3\Sigma) \quad \text{(R2.4b)}$$

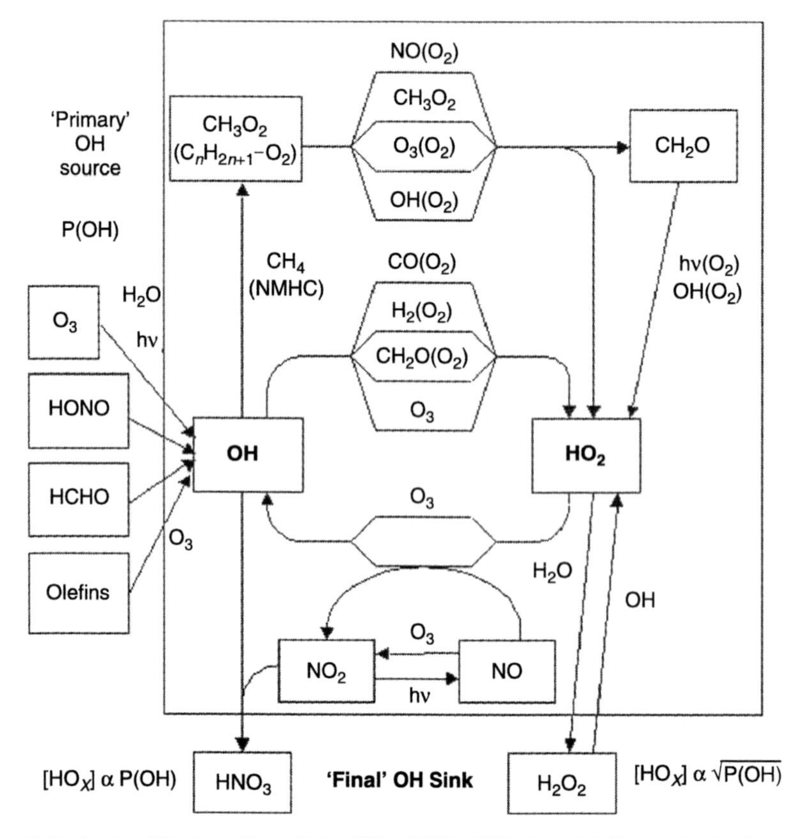

Fig. 2.9. A simplified outline of the $HO_X(OH + HO_2)$ and RO_2 cycles in the troposphere. The very reactive OH attacks most oxidiseable species (e.g. CO, hydrocarbons). Typical OH lifetimes are therefore below 1 s. Consequently, noontime concentrations only reach 10^6–$10^7\,\mathrm{cm}^{-3}$ (0.04–0.4 ppt). These reactions usually lead to the formation of HO_2 or RO_2, which, in turn, can be reconverted to OH by reaction with NO or O_3, thus preserving HO_X. Alternatively HO_X is destroyed by self-reaction of HO_2, producing H_2O_2. The only other 'final' HO_X, sink of importance is the $OH + NO_2$ recombination. In a very simplified picture, one can imagine the HO_X reservoir to be fed by R2.4 followed by R2.6 (rate of OH production = P_{OH}) and drained by either HO_2 self-reaction (if NO_X is low) or $OH + NO_2$ recombination (at high NO_X). All other reactions (inside the box) only interconvert OH and HO_2 (partly via RO_2). Note that $[HO_X] \propto \sqrt{P_{OH}}$ at low NO_X and $[HO_X] \propto P_{OH}$ at high NO_X (i.e. $NO_X > 1$ ppb)

(see e.g. Talukdar et al., 1998). The excited oxygen atoms $O(^1D)$ from (R2.4) can be deactivated by collision with N_2, O_2, or H_2O (i.e. any molecule 'M'):

$$O(^1D) + M \rightarrow O(^3P) + M \qquad \text{(R2.5)}$$

A certain fraction X of the excited oxygen atoms, however, will react according to:

$$O(^1D) + H_2O \rightarrow OH + OH \qquad \text{(R2.6)}$$

Assuming $k_{2.5a}$ and $k_{2.5b}$ denote the rate constants of the reaction of $O(^1D)$ with N_2 or O_2, respectively, and $k_{2.6}$ the rate constant of reaction (R2.6), we obtain for X:

$$X = \frac{k_3 \cdot [H_2O]}{k_{2.5a} \cdot [N_2] + k_{2.5b} \cdot [O_2] + k_{2.6} \cdot [H_2O]} \qquad (2.16)$$

(Chemical symbols in square brackets denote the concentration of the particular species in the atmosphere.) In the lower troposphere, X is typically in the range of 0.05–0.1. At very high relative humidity, up to 20% can be reached. The production rate of OH radicals, $P(OH)$, is accordingly calculated from the O_3-concentration and the photolysis frequency [see (2.18) below] for the above process (R2.4a):

$$\frac{d}{dt}\left[O(^1D)\right] = [O_3] \cdot J_1 \qquad (2.17)$$

$$P(\mathrm{OH}) = \frac{d}{dt}[\mathrm{OH}] = 2 \cdot X \cdot [\mathrm{O_3}] \cdot J_1 \qquad (2.18)$$

(b) Another important source of OH is the reaction of HO_2 radicals with NO:

$$HO_2 + NO \rightarrow OH + NO_2 \qquad (2.19)$$

An analogous reaction of HO_2 with ozone (R2.18) also yields OH. HO_2 radicals are largely formed by OH reactions. The consequences of this 'radical recycling' will be analysed below. In addition, the photolysis of aldehydes forms HO_2. This process is particularly significant in urban areas, where the oxidation of VOC and direct emissions lead to high aldehyde levels. A number of non-photochemical mechanisms also produce HO_2 radicals. Among these, reactions of NO_3 radicals with certain organic species, such as olefins, phenols, or DMS, are particularly important (e.g. Platt et al. 1990, 2002; Geyer et al., 2003a; Geyer and Stutz, 2004; see Sect. 2.5.2).

(c) A further OH source is the photolysis of HONO (nitrous acid, see Sect. 2.5.3):

$$HONO + h\nu \rightarrow OH + NO \qquad \text{(R2.8)}$$

This OH source is only of importance in polluted air, where HONO is formed by heterogeneous processes, i.e. chemical reactions on surfaces e.g. of building walls or aerosol particles (see, e.g. Platt 1986, Alicke et al. 2002, 2003 Sect. 2.5).

(d) The ozonolysis of olefins also produces OH radicals (Atkinson et al., 1992; Atkinson and Aschmann, 1993; Paulson et al., 1997, 1999):

$$\text{Olefin} + O_3 \rightarrow OH + \text{products} \quad \text{(R2.9)}$$

An ozone molecule adds across the C=C double bond to form a primary ozonide, which decomposes to form vibrationally excited carbonyl oxide and carbonyl products. The subsequent chemistry in the gas phase is still not completely understood. For a recent publication, see e.g. (Finlayson-Pitts and Pitts, 2000; Paulson et al., 1999). The OH production yields of these reactions vary from 7% to 100%, depending on the structure and size of the alkene (Paulson et al., 1997; 1999). Note that this source (though usually weak when compared with others) is independent of the light, and thus also provides radicals at night.

(e) Finally, photolysis of hydrogen peroxide produces OH radicals:

$$H_2O_2 + h\nu \rightarrow OH + OH \quad \text{(R2.10)}$$

In contrast to processes (b) and (c), the latter source would only be relevant in air masses with very low NO_X pollution, since only under these conditions can significant amounts of H_2O_2 be formed.

2.4.2 Temporal Variation of the HO_X Source Strength

The photolysis frequency $J_{2.4}$, and thus the strength of the main OH source [P(OH)], is given by the expression (e.g. Junkermann et al., 1989): where $I(\lambda)$ denotes the solar intensity (actinic flux), $\sigma(\lambda)$ the ozone absorption cross-section, and $\Phi(\lambda)$ the quantum efficiency of reaction R2.4. Because the solar intensity varies with the time of day, cloudiness, and seasons, the OH source strength is highly variable in the atmosphere. In addition, other environmental variables, such as the water vapour concentration, influence the formation of OH radicals. Figure 2.10, for example, shows the diurnal variation of OH concentrations. The OH levels follow the diurnal behaviour of the O_3 photolysis rate closely. The seasonal dependence of the OH source strength at mid-latitudes is shown in Fig. 2.11.

In polluted air, the diurnal variation of the ozone concentration (with a maximum in the early afternoon) has an additional influence on OH. Since the rate of R2.4 is proportional to the ozone concentration, P(OH) should also be proportional to $[O_3]$. However, a higher O_3 level will shift the NO/NO_2 stationary state towards higher NO_2 (see Sect. 2.5.1), and thus will reduce the OH concentration. The strength of other OH sources (e.g. HONO – photolysis, see Sect. 2.5.1) frequently peak in the morning.

2.4.3 Chemistry of Hydrogen Radicals (OH and HO_2)

Hydroxyl radicals react with most oxidisable gases in the atmosphere, for instance with carbon monoxide:

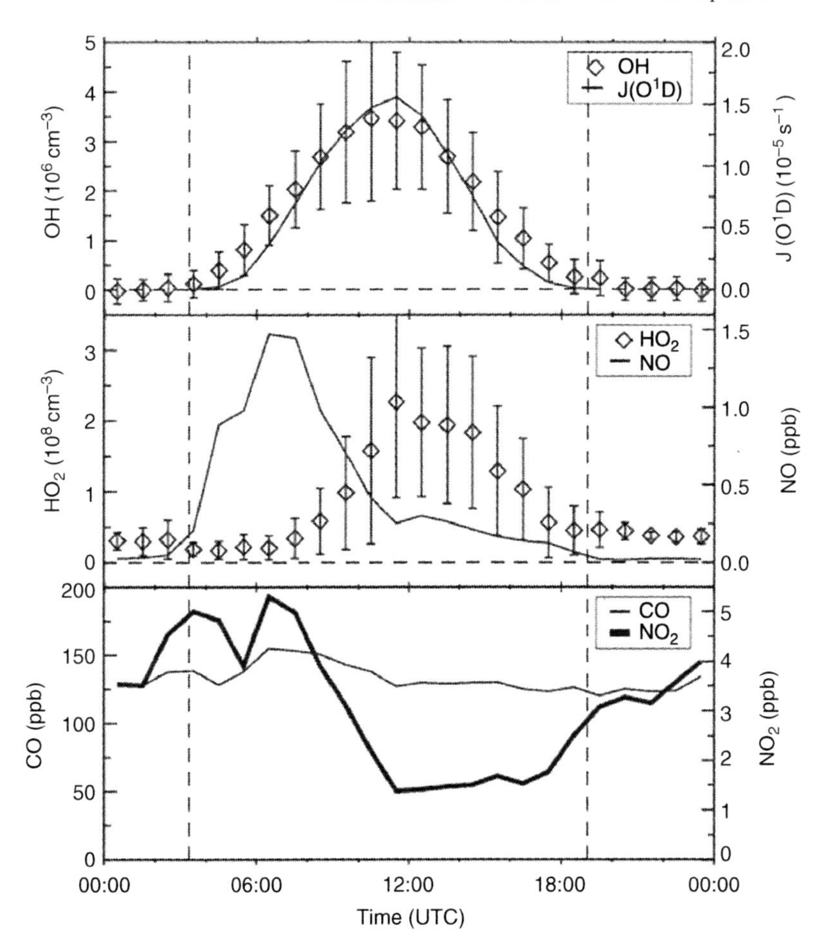

Fig. 2.10. Average diurnal OH, HO_2, and $J(O^1D)$ levels as observed during the BERLIOZ field experiment near Berlin in summer 1998. The OH concentration follows the solar radiation closely, while the levels of HO_2 are also influenced by the concentrations of NO (from Holland et al., 2003, Copyright by American Geophysical Union (AGU), reproduced by permission of AGU).

$$OH + CO \rightarrow CO_2 + H \quad \text{(R2.11)}$$

In this reaction, both CO_2 and hydrogen atoms are produced. In the atmosphere, the hydrogen atoms immediately form hydro-peroxy radicals, HO_2:

$$H + O_2 \xrightarrow{M} HO_2 \quad \text{(R2.12)}$$

As products of the degradation of VOCs, peroxy radicals are also produced. An example is methane degradation:

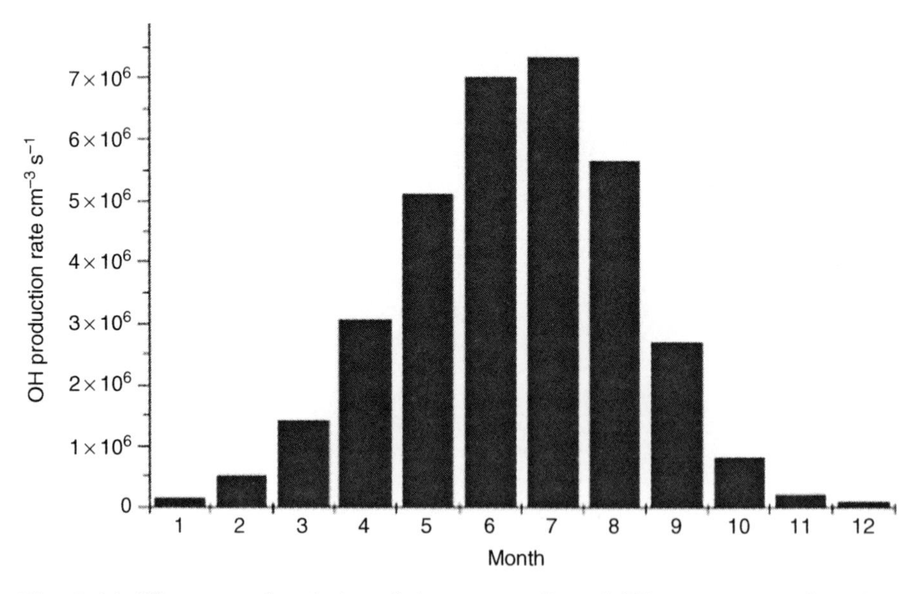

Fig. 2.11. The seasonal variation of the average diurnal OH source strength under cloud-free conditions in mid-latitudes (Germany)

$$OH + CH_4 \rightarrow CH_3 + H_2O \quad \text{(R2.13)}$$

$$CH_3 + O_2 \xrightarrow{M} CH_3O_2 \quad \text{(R2.14)}$$

Further important reactions of OH are the oxidation of molecular hydrogen and formaldehyde (and higher aldehydes):

$$OH + H_2 \rightarrow H_2O + H \quad \text{(R2.15)}$$

$$OH + CH_2O \rightarrow CHO + H_2O \quad \text{(R2.16)}$$

$$CHO + O_2 \rightarrow HO_2 + CO \quad \text{(R2.16a)}$$

In these reactions, the primary product is also HO_2 (R2.15 followed by R2.12, or R2.16 followed by R2.16a, respectively)

Thus, in all of the above reactions, and in fact most OH reactions, free radicals (here HO_2 radicals, which are easily converted to OH) are regenerated. An important exception to this rule is the reaction of OH with oxides of nitrogen:

$$OH + NO_2 \xrightarrow{M} HNO_3 \quad \text{(R2.17)}$$

The nitric acid produced in this reaction will (in the troposphere) be removed from the gas phase by various processes, such as dry deposition to the ground or rainout. Figure 2.9 shows the simplified reaction scheme of the hydrogen radicals $HO_X(OH + HO_2)$, as outlined above.

The balance between OH sources and sinks determines the levels of OH and HO_2 and their respective lifetimes. Figure 2.12 shows the OH and HO_2 budget as a function of the NO_X ($NO + NO_2$) concentration. The OH concentration shows a non-monotonous behaviour as a function of the NO_X level: At very low NO_X (below e.g. 0.1 ppb) essentially each OH radical reacting with CO (or HC) is converted to HO_2 (or an organic peroxy radical, RO_2). These radicals are, in turn, lost by radical–radical interaction. Thus, the OH concentration remains relatively low. At medium NO_X levels (about 1–2 ppb) most HO_2 and RO_2 radicals react with NO to reform OH. Therefore, OH levels are much higher than in the case of low NO_X. At very high NO_X levels (10 ppb or higher) most of the OH radicals react with NO_2 to form HNO_3, and therefore OH levels are low again.

As a consequence, OH levels show a maximum at around 1 ppb of NO_X, while HO_2 levels drop monotonously with increasing NO_X. This behaviour is illustrated in Fig. 2.13.

Although its concentration (in clean air) can be up to two orders of magnitude larger than that of OH (see Fig. 2.13), the reactivity of the HO_2 radical is much lower than that of the OH radical. However, the HO_2 radical is an important OH reservoir, and it is also the precursor for atmospheric H_2O_2. In addition, there is most likely a direct role of HO_2 in liquid-phase chemistry, for instance in cloud droplets (Chameides and Davis, 1982; Chameides, 1984; Lelieveld and Crutzen, 1990, 1991).

Only four HO_2 reactions play a role in the troposphere. Their significance depends on the levels of NO_x:

(1)
$$HO_2 + NO \rightarrow OH + NO_2 \tag{R2.7}$$

(2)
$$HO_2 + O_3 \rightarrow OH + 2O_2 \tag{R2.18}$$

(3)
$$HO_2 + HO_2 \rightarrow H_2O_2 + O_2 \tag{R2.19}$$

(4)
$$HO_2 + OH \rightarrow H_2O + O_2 \tag{R2.20}$$

The first two reactions reduce HO_2 radicals to OH. The third and forth reactions can be regarded as 'final sinks' for the HO_2 radical (and thus for HO_X). While hydrogen peroxide can be photolysed to yield two OH radicals (R2.10), this is a slow process. Thus rainout and washout of H_2O_2 is much more likely, since it is highly water soluble. A further sink of H_2O_2 is reaction with OH:

$$H_2O_2 + OH \rightarrow H_2O + HO_2 \tag{R2.21}$$

Hydrogen Peroxide is an important oxidant in the liquid phase, e.g. in raindrops, or cloud or haze droplets. Reactions of H_2O_2 in cloud or haze droplets contribute significantly to the oxidation of SO_2 (S(IV)) to sulphate (S(VI)).

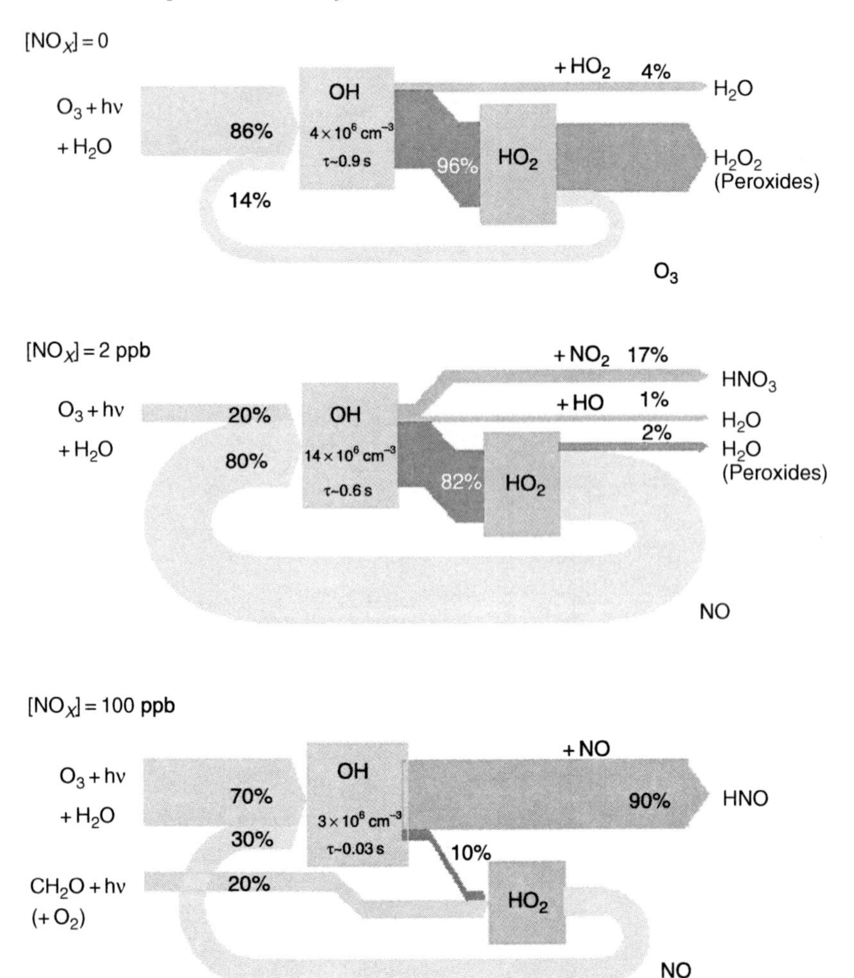

Fig. 2.12. The relative strength of OH sources and sinks, and the resulting OH concentration as a function of the NO_X ($NO + NO_2$) level. Here, three cases are shown: (**a**) At very low NO_X, essentially each OH radical reacting with CO (or hydrocarbons) is converted to HO_2 (or an organic peroxy radical, RO_2). These radicals are, in turn, lost by radical–radical interaction, and thus the OH concentration remains relatively low. (**b**) At medium NO_X levels (about 1 ppb), most HO_2 and RO_2 radicals react with NO to reform OH. Therefore OH levels are much higher than in case (**a**). (**c**) At very high NO_X levels, most of the OH radicals react with NO_2 to form HNO_3, and therefore OH levels are low again

Hydrogen peroxide is essentially only formed via reaction R2.19 (i.e. HO_2 recombination). A small contribution is assumed to come from the reaction of olefines with ozone (Becker et al., 1990). Assuming reaction R2.19 to be the dominating source, high rates of H_2O_2 production (and thus usually also high

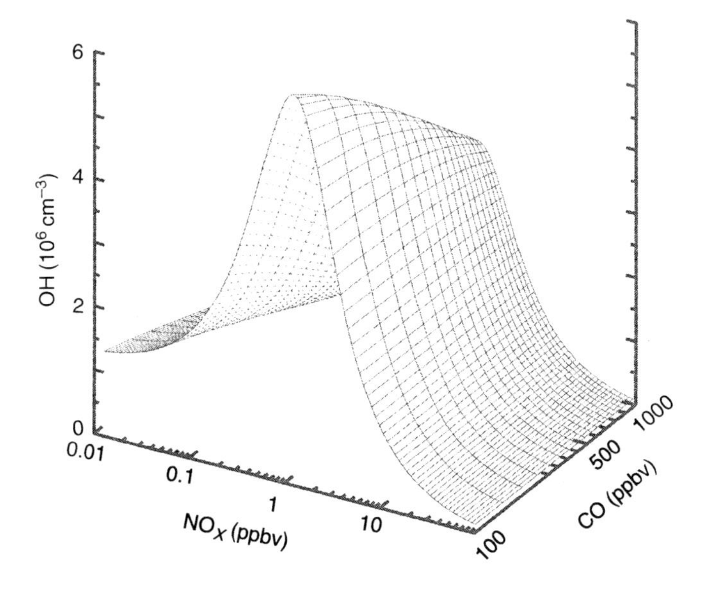

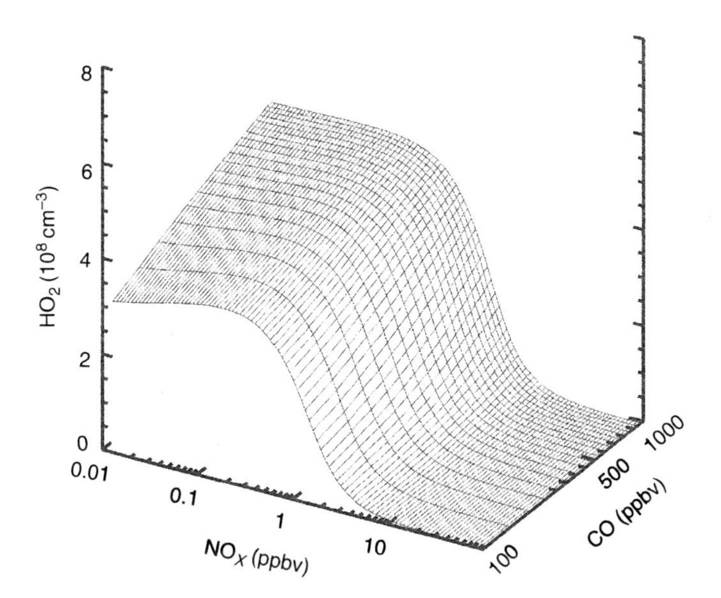

Fig. 2.13. Concentrations of OH (upper panel, units of 10^6 molecules cm$-^3$) and HO_2 (lower panel, units of 10^8 molecules cm^{-3}) as a function of the NO_X ($NO+NO_2$) mixing ratio for typical conditions (37.6 ppb ozone, 88.7 ppb CO, 0.92 ppb CH_2O, $J(O_3 - O^1D) = 3 \times 10^{-5}\,s^{-1}$, $J(NO_2) = 9.1 \times 10^{-3}\,s^{-1}$). The HO_2 concentration, and therefore H_2O_2 production rate, are highest in unpolluted air at NO_X levels below a few 100 ppt (from Enhalt, 1999)

levels) are to be expected under conditions of high HO_2. Thus, in air masses with high humidity and ozone, and low NO, the highest H_2O_2 levels are to be expected.

2.5 Oxides of Nitrogen in the Atmosphere

The oxides of nitrogen, NO and NO_2 (NO_X), are key species in atmospheric chemistry. They regulate many trace gas cycles and influence the degradation of most pollutants in clean air, as well as in polluted regions:

NO_X concentration has a strong influence on the atmospheric level of hydroxyl radicals (OH, see Sect. 2.4.1), which, in turn, are responsible for the oxidation processes of most trace gases. In addition, NO_X is a catalyst for tropospheric ozone production (see Sect. 2.3). Oxides of nitrogen (or acids formed from them) can also react with VOC degradation products to form organic nitrates or nitrites [e.g. peroxy acetyl nitrate (PAN) or methyl nitrite], as well as nitrosamines (Platt et al., 1980b). These species can be much more detrimental to human health than the primary oxides of nitrogen. Finally, the thermodynamically most stable and ultimate degradation product of all atmospheric oxides of nitrogen, nitric acid, is (besides sulphuric acid) the main acidic component in 'acid rain'.

An overview of the most important oxidised nitrogen species in the atmosphere is given in Fig. 2.14. The main reaction pathways between the various species are indicated by arrows. Oxides of nitrogen are primarily emitted in the form of NO (plus some NO_2) and N_2O. While N_2O is a very inert species and therefore plays no role for the chemical processes in the troposphere, NO reacts rapidly with natural ozone to form NO_2. On the other hand, NO_2 will be destroyed (photolysed) by sunlight during the daytime, and therefore a stationary state between these two NO_X species will be established. NO_2 then further reacts with OH radicals or NO_3, ultimately forming nitric acid (or nitrate aerosol).

The 'classic' role of oxides of nitrogen in atmospheric chemistry (see Sect. 2.5.1) has been known since the 1980s. Since then, the importance of this class of species has been underlined by newly obtained results, largely made possible by improved measurement techniques:

- The confirmation of an additional NO_X degradation pathway via heterogeneous and homogeneous hydrolysis of N_2O_5 at night (Heintz et al., 1996; Mentel et al., 1996; Wahner et al., 1998a; Geyer et al., 2001c; Brown et al., 2003; Stutz et al., 2004).
- The non-photochemical production of peroxy radicals (HO_2 and RO_2) via NO_3–VOC reactions as a further consequence of elevated NO_X levels in the atmosphere (Platt et al., 1990; LeBras et al., 1993; Mihelcic et al., 1993; Geyer et al., 1999; Geyer and Stutz, 2004).

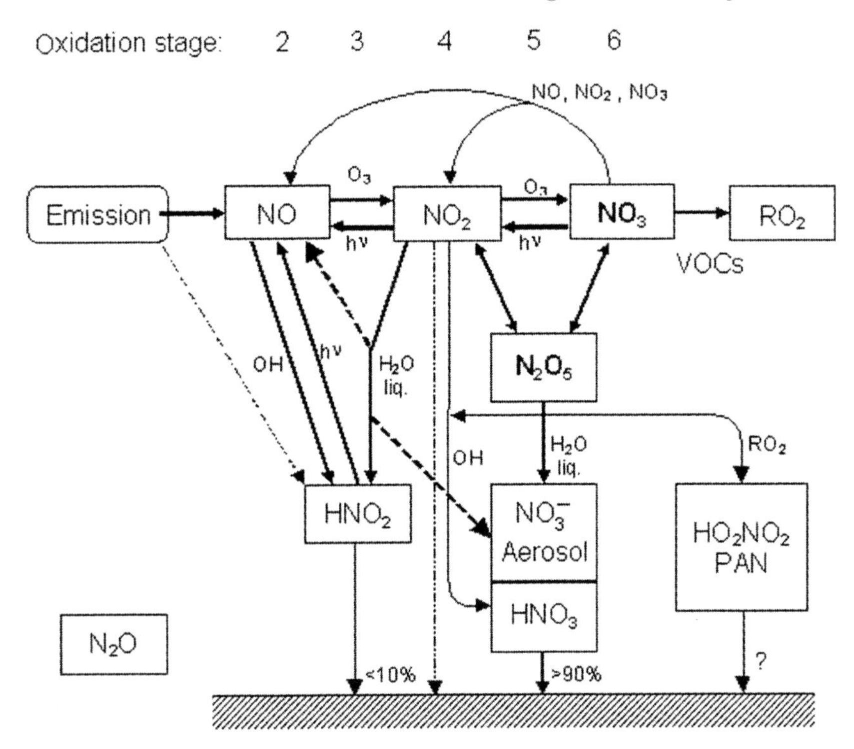

Fig. 2.14. Simplified overview of the NO_X reaction scheme in the atmosphere. Arrows indicate main reaction pathways

- The establishment of HONO photolysis as another significant OH source in polluted air (Perner and Platt, 1979; Harris et al., 1982; Platt, 1986; Harrison et al., 1996; Alicke et al., 2002, 2003).
- Finally, an important role of oxidised nitrogen species is postulated in the mechanisms releasing RHS from sea salt as, for example, observed in the polar BL (Finlayson-Pitts and Johnson, 1988; Finlayson-Pitts et al., 1989, 1990).

The improved knowledge of the cycle of oxidised nitrogen species also has great practical importance: The exact knowledge of the NO_X lifetime allows the estimation of NO_X levels and the health effects of atmospheric NO_X, the amount of long-range transport (and thus, for instance, the amount of acid rain in non-polluted areas), and finally the NO_X level in clean air (and thus the amount of ozone formation there).

2.5.1 Classical Chemistry of Oxides of Nitrogen in the Atmosphere

In this section, we briefly summarise the central aspects of the classical chemistry of oxides of nitrogen in the atmosphere (see also Fig. 2.14 and Table 2.3).

Extensions of our image of the chemical processes related to oxidised nitrogen species are discussed in Sect. 2.5.2.

The transformation pathways between the atmospheric nitrogen reservoir are closely coupled to the reactions of ozone (as discussed in Sect. 2.3) and to the cycles of hydrogen radicals (OH and HO_2). An important question in this context is: Are NO_X reactions sources or sinks for ozone? The simple system described by reactions R2.1–R2.3 (Sect. 2.3) neither destroys nor produces ozone. However, in the atmosphere there are additional reactions oxidising NO to NO_2 without consuming ozone. Thus, the NO_X level ultimately determines whether there is net ozone formation or destruction. In the unpolluted (background) atmosphere, the processes oxidising NO to NO_2 without consuming O_3 (and thus producing O_3 via reactions R2.1 and R2.2) are reactions of NO with peroxy radicals (HO_2, CH_3O_2, and higher RO_2). In the simple case of HO_2, we have:

$$NO + HO_2 \rightarrow NO_2 + OH \qquad \text{(R2.7)}$$

The HO_2 radicals are produced, for example, by reaction of OH radicals with CO (R2.11, followed by R2.12). At the same time, the reactions R2.7, R2.11, and R2.12 facilitate another stationary state in the atmospheric concentration ratio of the hydrogen radicals OH and HO_2, (HO_X). Since ozone also reacts with HO_2 radicals (but not with other peroxy radicals), the HO_X system also destroys ozone:

$$O_3 + HO_2 \rightarrow OH + 2O_2 \qquad \text{(R2.18)}$$

Whether the above-described interplay of the NO_X and HO_X systems leads to net ozone formation (as shown in Fig. 2.7) or O_3 destruction depends on the relative rates of the reactions R2.7 and R2.18, and thus on the NO concentration. At a NO concentration around 1/3000 of the O_3 concentration (i.e. $[NO] \approx 0.01$ ppb at 30 ppb O_3), the rates of both reaction pathways are equal. In other words, ozone formation equals ozone destruction. At higher NO_X (and thus NO) levels net O_3 formation occurs, while at lower NO levels O_3 is photochemically destroyed. In regions influenced by industrial activities, where NO levels can reach many ppb, O_3 formation by far exceeds destruction. This can also be seen from the observation that ozone levels in the (more industrialised) northern hemisphere exceed those in the southern hemisphere by about 50%. Figure 2.7 shows model calculated and observed ozone formation rates as a function of the NO level.

Oxides of nitrogen influence the concentrations of OH and HO_2 radicals in the atmosphere by several mechanisms. The most important removal process for OH radicals is their recombination with NO_2. This reaction (R2.6) leads to final removal of OH from the radical chain. Consequently, the OH concentration is reduced by high levels of NO_X (and thus NO_2). On the other hand, the reaction of NO with HO_2 radicals reduces the stationary state $[HO_2]/[OH]$ ratio (reactions R2.7, R2.11, and R2.12), and consequently leads to higher OH radical concentrations. This relationship is illustrated in Figs. 2.12 and 2.13.

The crucial step initiating the eventual removal of nitrogen oxides from the atmosphere is their conversion to nitric acid (HNO_3), which is subsequently removed from the atmosphere by dry and wet deposition (see also Table 2.3). The direct (dry) deposition of nitrogen oxides to the ground is of minor importance in comparison (<10%). On the other hand, re-conversion of nitric acid to NO_X is negligible in the troposphere. Nitric acid is formed in reactions of NO_2:

- Directly by reaction with OH (R2.17).
- Indirectly by reaction of NO_2 with O_3, leading to NO_3 (R2.22), which can either abstract a hydrogen atom (e.g. from DMS) or further react with NO_2 to form N_2O_5, (R2.26) the anhydride of nitric acid, which in turn is converted to HNO_3 by reaction with gas-phase or liquid water (R2.27).

2.5.2 Tropospheric Chemistry of Nitrate Radicals, NO_3

At night, when levels of OH radicals are low, other oxidants play an important role in the troposphere. Among those, the NO_3 radical plays an important role for the budget of NO_X and for the concentrations of certain organic compounds. Nitrate radicals are formed by reaction of NO_2 with ozone (see also Fig. 2.14)

$$NO_2 + O_3 \rightarrow NO_3 + O_2 \tag{R2.22}$$

which constitutes the only relevant source of NO_3 in the troposphere. The rate of NO_3 production, P_{NO3}, can be calculated from the concentration of NO_2 and O_3:

$$\frac{d\,[NO_3]}{dt} = [NO_2] \cdot [O_3] \cdot k_1 = P_{NO3} \tag{2.20}$$

During the daytime, NO_3 radicals are effectively destroyed by photolysis:

$$NO_3 + h\nu \rightarrow NO + O_2 \tag{J2.23a}$$

$$\rightarrow NO_2 + O \tag{J2.23b}$$

(photolytic lifetime is about 5 s), and rapid reaction with NO ($k_{2.24} = 2.6 \times 10^{-11}\,cm^3s^{-1}$ at 293K):

$$NO_3 + NO \rightarrow NO_2 + NO_2 \tag{R2.24}$$

At night, when photolysis is unimportant and NO levels are usually low, the NO_3 loss rate is much smaller and thus its concentration is larger. Loss of NO_3 due to gas-phase reaction with nitrogen dioxide or the NO_3 self-reaction is slow. Similarly, the unimolecular decomposition of NO_3 is probably of minor importance in the atmosphere (Johnston et al., 1986; Russel et al., 1986). Gas-phase reaction of NO_3 with water vapour is endothermic.

Main destruction mechanisms are thus reaction with organic species (see Table 2.7) or heterogeneous loss (i.e. reaction at the ground or at the surface of aerosol or cloud particles) of either NO_3 or N_2O_5.

Table 2.7. Reaction of OH, NO_3, O_3, and Cl with VOC (at 25°C) (from Atkinson, 1994)

Species	Reaction rate constant 10^{15} cm^3 $molec.^{-1}$ s^{-1}			
	With OH	With NO_3	With Ozone	With Cl
Alkanes				
Methane	6.14	<0.001	<0.00001	100
Ethane	254	0.0014	<0.00001	59000
Propane	1120	0.017	<0.00001	137000
n-Butane	2190	0.046	<0.00001	205000
n-Pentane	4000	0.04	<0.00001	280000
Alkenes				
Ethene	8520	0.21	0.0017	99000
Propene	26300	9.4	0.013	230000
1-Butene	31400	12	0.011	220000
2-Methyl-2-Butene	68900	9300	0.423	
2,3-Dimethyl-2-Butene	110000	57000	1.16	
Alkynes				
Ethyne	5100	0.0082	<0.00001	
Aromatics				
Benzene	1230	<0.03	<0.00001	
Toluene	5960	0.068	<0.00001	
Phenol	26300	3780		
o-Cresol	42000	13700	0.00026	
m-Cresol	64000	9740	0.00019	
p-Cresol	47000	10700	0.00047	
Benzaldehyde	12900	2.6	No data	
Isoprene	101000	678	0.013	460000
Terpenes				
α - Pinene	53700	6160	0.0866	480000
β - Pinene	78900	2510	0.015	
3-Carene	88000	9100	≈0.037	
Limonene	171000	12200	≈0.2	
Sabinene	117000	10000	≈0.086	
α - Terpinene	363000	180000	8.47	
Myrcene	215000	11000	0.47	
Aldehydes				
Formaldehyde	9370	0.6	<0.00001	73000
Acetaldehyde	15800	2.7	<0.00001	79000

NO_3 + VOC Reactions

NO_3 reacts with a variety of organic compounds in the atmosphere. In particular, reactions with biogenic VOC, such as isoprene and terpenes, as well as the reactions with DMS, are of significance.

NO_3 reactions with alkanes and aldehydes proceeds via H abstraction (Atkinson, 1990):

$$NO_3 + RH \rightarrow R + HNO_3 \quad (R2.25)$$

The alkyl radicals produced in reaction R2.25 are transformed to peroxy radicals (RO_2) by reaction with O_2.

Reactions with alkenes (including isoprene and terpenes) proceed via NO_3 addition to a C = C double bond. Further reaction steps are O_2 addition, followed by either reaction with other peroxy radicals or NO_2. Reasonably stable end products are organic nitrates and peroxy radicals (Atkinson, 1990).

Reactions with aromatic compounds are also important. While reaction with benzene is quite slow (rate constant on the order of 10^{-17} cm^{3}molec.$^{-1}$s^{-1}), hydroxy-substituted benzenes react very fast, with reaction rate constants on the order of 10^{-12} cm^3 molec.$^{-1}$ s^{-1} (furane and phenol) to 10^{-11} cm^3 molec.$^{-1}$ s^{-1} (cresols).

Finally, NO_3 radicals react rapidly with DMS (reaction rate constant 10^{-12} cm^3 s^{-1}). This can be significant for the budgets of both DMS and NO_3, in particular in the marine atmosphere. The exact reaction mechanism is still subject to investigations. Recent investigations show that the primary step is probably H abstraction. Thus, nitric acid and peroxy radicals should be the dominant products (Butkovskaya and LeBras, 1994; Platt and LeBras, 1997).

Heterogeneous Loss of NO_3 and N_2O_5

In the atmosphere, NO_3 and NO_2 are in a chemical equilibrium with N_2O_5:

$$NO_2 + NO_3 \overset{M}{\longleftrightarrow} N_2O_5 \quad (R2.26)$$

The equilibrium concentration of N_2O_5 is given by:

$$[N_2O_5] = K \cdot [NO_2] \cdot [NO_3] \quad (2.21)$$

Under 'moderately polluted' conditions (NO_2 mixing ratio of about 1 ppb) typical for rural areas in industrialised countries, NO_3 and N_2O_5 concentrations are roughly equal (293K) (e.g. Wängberg et al., 1997). While homogeneous reactions of N_2O_5 with water vapour and other atmospheric trace gases are believed to be extremely slow (Mentel et al., 1996), heterogeneous reaction with water contained in aerosol or cloud particles will convert N_2O_5 to nitric acid:

$$N_2O_5 + H_2O \xrightarrow{het} HNO_3 \quad (R2.27)$$

Accommodation coefficients γ for N_2O_5 and NO_3 on water (droplet) surfaces are in the range of 0.04–0.1 (the higher values applying to water containing

H_2SO_4) (Van Doren et al., 1990; Mozurkewich and Calvert, 1988; Mentel et al., 1996; Wahner et al., 1998; Hallquist et al., 2003), and >0.0025 (Thomas et al., 1993), respectively. If reaction R2.27 leads to accumulation of nitrate in the liquid phase, the accommodation coefficients can be considerably (by about one order of magnitude) reduced. For instance, Wahner et al. (1998) find γ in the range of 0.002–0.023 at relative humidities of 48–88% for the accommodation of N_2O_5 on sodium nitrate aerosol.

NO_3 Lifetime

From measured concentrations of O_3, NO_2, and NO_3, the atmospheric lifetime of NO_3 (τ_{NO3}, see Sect. 2.1.3), limited by the combination of any first-order loss process, can be calculated (Heintz et al., 1996; Geyer et al., 2001a,b,c, 2002; Brown et al., 2003; Geyer and Stutz, 2004):

$$NO_3 + Z \rightarrow \text{products} \qquad \text{(R2.28)}$$

Assuming pseudo stationary state conditions with respect to NO_3, τ_{NO3} becomes

$$\tau_{NO3} = \frac{[NO_3]}{P_{NO3}} = \frac{[NO_3]}{[NO_2] \cdot [O_3] \cdot k_1} \qquad (2.22)$$

Since $k_{2.22}$ is known from laboratory measurements, and the concentrations of O_3, NO_2, and NO_3 can be simultaneously measured (for instance, by the DOAS technique), τ_{NO3} can readily be calculated for various atmospheric conditions. It must be emphasised that the steady-state assumption is an approximation, which is quite good for most atmospheric situations. The NO_3 lifetime can also be limited by any irreversible loss of N_2O_5 (for instance $k_{2.27} > 0$), because both species are in a rapidly established thermodynamic equilibrium (R2.28, the time constant to reach equilibrium, is on the order of 1 min at ambient temperatures). Under conditions where both NO_3 and N_2O_5 are lost, the observed NO_3 lifetime τ_{NO3} is given by:

$$\tau_{NO3} = \frac{1}{\dfrac{1}{\tau'_{NO3}} + \dfrac{[NO_2]}{K} \cdot \dfrac{1}{\tau_{N2O5}}} \qquad (2.23)$$

Of course, in the case of negligible loss of N_2O_5($\tau_{N2O5} \rightarrow \infty$), $\tau'_{NO3} = \tau_{NO3}$. For conditions where direct loss of NO_3 molecules can be neglected ($\tau'_{NO3} \rightarrow \infty$), (2.23) reduces to:

$$\tau_{NO3} = \frac{K}{[NO_2]} \cdot \tau_{N2O5} \qquad (2.24)$$

The simple stationary state assumption predicts an observed NO_3 lifetime τ_{NO3} inversely proportional to the NO_2 concentration in cases where only

N_2O_5 is lost. Since NO_3 formation is proportional to $[NO_2]$ (R2.20), $[NO_3]$ should become independent of the NO_2 concentration under such conditions.

Formation of Hydrogen Radicals by NO_3 + VOC Reactions

Since most of the above reactions of NO_3 with organic species appear to lead to the formation of peroxy radicals, an interesting consequence of the night-time reaction of NO_3 with organic species is a considerable contribution to the HO_X budget (Platt et al., 1990; Mihelcic et al., 1993; Carslaw et al., 1997; Geyer et al., 2001; Geyer and Stutz, 2004).

In circumstances where NO levels are very low, peroxy radicals undergo reactions with other RO_2 radicals, O_3, or NO_3:

$$RO_2 + NO_3 \rightarrow RO + NO_2 + O_2 \qquad \text{(R2.29)}$$

In urban areas, where ubiquitous NO emissions and vertical transport can lead to an active mixing of NO_3, N_2O_5, and NO, the likely fate of RO_2 radicals is reaction with NO (Geyer and Stutz, 2004):

$$RO_2 + NO \rightarrow RO + NO_2 \qquad \text{(R2.30)}$$

This gives rise to a radical chemistry which is similar to that during the day. However, the region in which NO_3 and NO can coexist is often shallow and it moves vertically throughout depending on the NO emission strength and vertical stability (Geyer and Stutz, 2004).

2.5.3 Nitrous Acid, HONO in the Atmosphere

The photolysis of nitrous acid, HONO, is an important source for OH radicals in the polluted urban atmosphere (Perner and Platt, 1979; Platt et al., 1980b; Kessler et al., 1981; Harris et al., 1982; Alicke et al., 2002, 2003; Zhou et al., 2002; Aumont et al., 2003; Ren et al., 2003; Kleffmann et al., 2005), and the polar BL (Li, 1994; Zhou et al., 2001). In addition, HONO is toxic (Beckett et al., 1995) and its chemistry leads to the formation of carcinogenic nitrosamines, (Shapley, 1976; Famy and Famy, 1976; Pitts, 1983). HONO formation at chamber walls is a candidate for an intrinsic 'chamber dependent' radical formation, which is observed in most smog chamber experiments (e.g. Gleason and Dunker, 1989; Rohrer et al., 2004).

The photolysis of HONO, which absorbs in the wavelength range from 300 to 405 nm, leads to a formation of an OH radical and NO (Stockwell and Calvert, 1978; Stutz et al., 2000):

$$HONO + h\nu \rightarrow OH + NO \quad (300\,\text{nm} < \lambda < 405\,\text{nm}) \qquad \text{(R2.8)}$$

The photolysis of HONO begins early in the morning, when HONO concentrations are typically highest. The combination of elevated HONO concentrations at sunrise and fast photolysis results in a peak in the production of OH which surpasses other HO_X sources, such as O_3 and HCHO photolysis (Alicke et al., 2002).

Nitrous acid was first identified and quantified in the atmosphere by DOAS (Perner and Platt, 1979; Platt and Perner, 1980). Typical HONO mixing ratios are in the range of 0 – 15 ppb (see Calvert et al., 1994; Lammel and Cape, 1996; or Alicke, 2000 for reviews). Generally, nitrous acid concentrations scale with the NO_2 concentration, and therefore approximately with the degree of pollution. The diurnal variation of the mixing ratio of HONO is dominated by its photolysis, which leads to lower levels during the day. During the night concentrations increase, often reaching a constant value in the early morning. The amount of HONO formed is often correlated with the stability of the nocturnal BL, as inferred for example from radon data (Febo et al., 1996).

Several studies have discussed the importance of HONO photolysis as an OH source (e.g. Platt et al., 1980b; Harrison et al., 1996; Stutz et al., 2002; Alicke et al., 2002, 2003). Alicke et al. (2002) showed OH production rates of up to 3×10^7 molecules $cm^{-3} s^{-1}$ in Milan, Italy. Other observations confirm these results (Harris et al., 1982; Sjödin, 1988). Model studies (e.g. Harris et al., 1982) show that HONO photolyis can lead to an increase in the maximum daytime ozone concentration of up to 55%, if 4% of the initial NO_X is present as HONO (Los Angeles case, initial NO_x: 0.24 ppm), compared with a calculation without nitrous acid. Jenkin et al. (1988) showed HONO photolysis can account for a fivefold increase in OH at 6:00 AM, a 14% increase in OH present at the daily maximum (noon), and a 16% increase in net photochemical ozone production. In addition, the formation of ozone begins earlier in the day due to the HONO photolysis in the morning.

HONO is formed through various chemical mechanisms in the atmosphere. The only important gas-phase reaction forming nitrous acid is 2.31 (Stuhl and Niki, 1972; Pagsberg et al., 1997; Zabarnick, 1993; Nguyen et al., 1998):

$$OH + NO(+M) \rightarrow HONO(+M) \quad \text{(R2.31)}$$

Gas-phase nitrous acid producing pathways, such as the reaction of NO_2 with HO_2, have been reported to be of minor importance (e.g. Howard, 1977; Tyndall et al., 1995). The bulk of urban HONO is suspected to be formed from either of the mechanisms summarised in reaction R2.32 or R2.33, involving only NO_X and water vapour (Perner and Platt, 1979; Kessler and Platt, 1984; Platt, 1986).

$$NO_2 + NO + H_2O \Leftrightarrow 2\,HONO \quad \text{(R2.32)}$$

$$2NO_2 + H_2O \Leftrightarrow HONO + HNO_3 \quad \text{(R2.33)}$$

These reactions (and their reverse reactions) appear to proceed heterogeneously on surfaces (Kessler, 1984; Kessler et al., 1981; Sakamaki et al., 1983; Lammel and Perner, 1988; Notholt et al., 1991, 1992; Junkermann and Ibusuki, 1992; Ammann et al., 1998; Calvert et al., 1994; Longfellow et al.,

1998; Goodman et al., 1999; Kalberer et al., 1999; Veitel, 2002; Veitel et al., 2002). A number of laboratory studies (Sakamaki et al., 1983; Pitts et al., 1984b; Jenkin et al., 1988; Svensson et al., 1987; Kleffmann et al., 1998) have shown reaction R2.32 to be insignificant. Several field observations, where the presence of high ozone at night excluded NO, or low NO was documented, confirm this result (e.g. Kessler and Platt, 1984; Harrison and Kitto, 1994).

The exact mechanism of the heterogeneous formation of HONO summarised in reaction R2.33 is unknown, but several studies (Svensson et al., 1987; Jenkin et al., 1988; Kleffmann et al., 1998; Finlayson-Pitts et al., 2003) have shown that it is first order in NO_2 and water. Neither the reaction rate constants nor the nature of the surface is known, which makes calculation of the OH production by models difficult. In addition, reactions on 'urban surfaces' like organic aerosols or soot aerosols were suggested (e.g. Gutzwiller et al., 2002; Bröske et al., 2003). In fact reactions of NO_2 on asphalt or roof tile surfaces appear to be sufficiently rapid to explain observations (Trick, 2004).

The most important gas-phase chemical removal process for HONO, besides photoloysis, is reaction R2.34. A few percent of HONO is expected to be destroyed by OH radicals:

$$\mathrm{OH} + \mathrm{HONO} \rightarrow .\mathrm{H_2O} + \mathrm{NO_2} \qquad \text{(R2.34)}$$

Several studies have been carried out to estimate the strength of direct emission of nitrous acid from combustion processes (Kessler and Platt, 1984; Pitts et al., 1984a; Kirchstetter et al., 1996; Ackermann, 2000; Winer and Biermann, 1994). Up to 1% of the emitted NO_X was found as nitrous acid, making this source important, especially in heavily polluted areas with high amounts of traffic. (Kirchstetter et al., 1996) report a HONO/NO_X emission ratio of 0.35% for a north-American car fleet, while (Ackermann, 2000; Kurtenbach et al., 2002) found 0.65% of NO_X emitted as HONO in a traffic tunnel in Germany.

2.6 Tropospheric Chemistry of VOCs

Organic compounds are an extremely large class of chemicals that play a significant role in the atmosphere. As discussed in Sect. 2.3, they provide the fuel for ozone formation. In addition, chemical reactions with organic carbons can lead to the formation of particles, so called secondary organic aerosol. They also impact radical budgets (see Sect. 2.4) and thus influence radical chemistry.

The chemistry of volatile organic carbon compounds (VOCs) is a complex topic which, by far, exceeds the scope of this book. Here, we will only briefly discuss typical levels of organics in the atmosphere, and in particular in urban areas, and their initial reactions with various radical species. VOCs can be subdivided into a number of different classes encompassing hydrocarbons (HC) and oxidised species (Table 2.7):

- alkanes, with the overall formula C_nH_{2n+2}
- alkenes (or olefins), with the overall formula C_nH_{2n} (mono-olefins)
- alkynes, with the overall formula C_nH_{2n-2}
- aromatics, which contain one or more benzene rings (C_6H_6)
- isoprene and terpenes
- oxidised species: aldehydes, ketones, alcohols, and organic acids

Members of each class often undergo similar chemistry. We will therefore discuss their initial reactions steps in a general way.

Alkanes are molecules containing carbon and hydrogen molecules, whereas the carbon molecules are held together by single molecular bonds. Consequently, the degradation of alkanes by OH proceeds by abstraction of one hydrogen atom, in analogy to the reaction of methane (R2.35 and R2.36):

$$OH + RH \rightarrow R + H_2O \qquad \text{(R2.35)}$$

where R denotes an alkyl radical (C_nH_{2n+1}), e.g. C_2H_5. The rate constants for some alkane–OH reactions are given in Table 2.7.

Analogous to CH_3 (see Sect. 2.4.1), alkyl radicals immediately react with molecular oxygen to form organic peroxy radicals:

$$R + O_2 \xrightarrow{M} RO_2 \qquad \text{(R2.36)}$$

The fate of peroxy radicals in the atmosphere are self-reaction (forming peroxides), reaction with O_3, and, mostly in polluted areas, reaction with NO:

$$RO_2 + NO \rightarrow RO + NO_2 \qquad \text{(R2.37)}$$

Alkoxy radicals RO react with O_2 to form aldehydes and HO_2 radicals:

$$RO + O_2 \rightarrow R'CHO + HO_2 \qquad \text{(R2.38)}$$

where R′ denotes an alkyl radical with one less C atom ($C_{n-1}H_{2n-1}$).

Initial reactions of alkanes with NO_3 radicals and Cl atoms also proceed by hydrogen abstraction, followed by reactions R2.36–R2.38 (Table 2.7). Ozone only reacts very slowly with alkanes and its ractions are unimportant in the atmosphere (Table 2.7).

Alkenes are HC which contain one or more carbon double bonds. While hydrogen abstraction can still occur, the predominant first step in alkene+OH reactions is the addition of the OH radical to the double bond, followed by addition of O_2 to the resulting peroxy radical:

$$OH + RCH_2CH_2 \xrightarrow{M} RCH_2CH_2OH \qquad \text{(R2.39)}$$

$$RCH_2CH_2OH + O_2 \xrightarrow{M} O_2RCH_2CH_2OH \qquad \text{(R2.40)}$$

Alkenes with aliphatic chains can also (though less likely) react by hydrogen abstraction from the aliphatic chain. The rate constants for some of the more abundant alkenes with OH radicals are listed in Table 2.7. Similar to the alkane reaction chain, the $O_2RCH_2CH_2OH$ radical can react with NO forming a β-hydroxyalkoxy radical:

$$O_2RCH_2CH_2OH + NO \xrightarrow{M} ORCH_2CH_2OH \qquad (R2.41)$$

The $ORCH_2CH_2OH$ radical then further decays into hydroxyl or alkoxy radicals and aldehydes.

Reactions of alkenes with NO_3 radicals proceed primarily through the addition of NO_3 to the double bond. The radical formed in the NO_3 addition reacts further with O_2 to form a β-nitratoalkyl peroxy radical, which can then react with NO to form nitrates, hydroxyl radicals, and other products. Alkenes can also react with O_3. Although these reactions are slower than those of OH, they can be important at night, when OH levels are low. Their significance also stems from the fact that OH radicals are formed from the reaction of ozone and larger alkenes (Paulson et al., 1999). Cl atom reactions

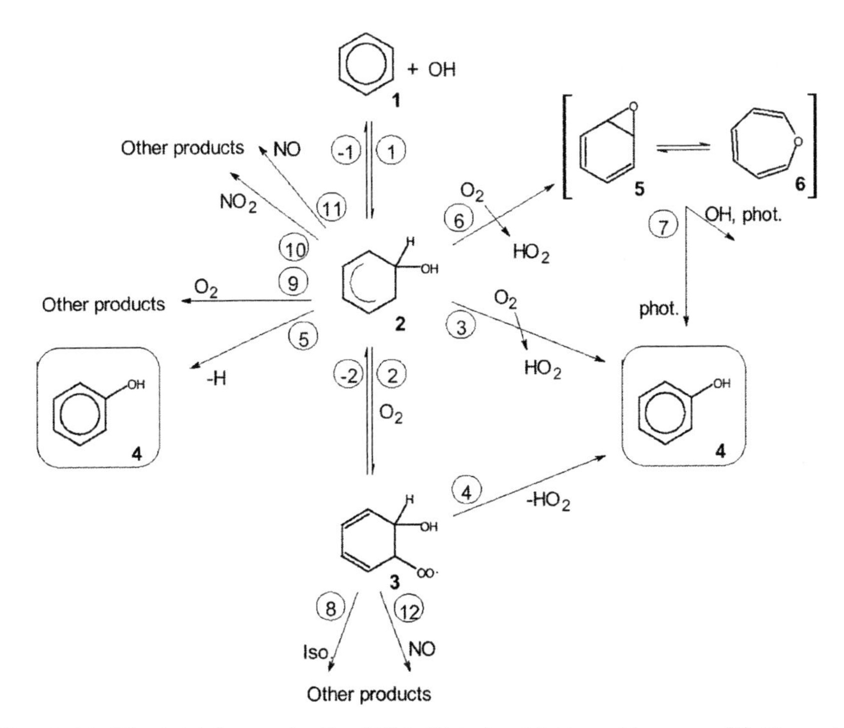

Fig. 2.15. The initial steps in the OH initiated oxidation of benzene (1) Currently proposed loss processes of the aromatic-OH adduct (2) and the peroxy radical (3) (which are in rapidly established equilibrium) are shown that lead in part to the formation of phenol (4) Intermediate species are indicated by bold numbers. Similar schemes for the phenol forming pathways can also be drawn for the alkyl-substituted aromatics (from R. Volkamer, 2001)

also proceed through Cl addition to the double bond. Details of the reaction chains for individual alkenes can, for example, be found in Finlayson-Pitts and Pitts (2000).

Alkynes contain carbon triple bonds. Their reactions proceed in similar ways as those of alkenes. An important product of these reactions is glyoxal (CHOCHO).

For monocyclic *aromatics* the primary reaction pathway is also OH addition. The resulting radical can either lose a hydrogen atom (through HO_2 formation) and form a phenol, or the ring can open leading to rapid degradation of the products (see Fig. 2.15) (Volkamer et al., 2002). In the latter case, glyoxal was found to be an important product (Volkamer et al., 2001).

Isoprene and terpenes are olefins with more than one double bond; they are very reactive towards OH, NO_3, O_3, and Cl, with reaction rate constants towards OH radicals approaching the collision limit. Because these molecules have a complex chemical structure, their reactions mechanisms are quite complex, and beyond the scope of this book. The reader is referred to a review by Atkinson and Arey (1998).

Aldehydes, such as formaldehyde and acetaldehyde, react with OH, NO_3, and Cl by abstraction of the weakly bond aldehydic hydrogen (see for example R2.16). The radical formed in this reaction then reacts with O_2, similar to the reactions described above. Ketones react by abstraction of the H atom from the alkyl chains, similar to the reactions of alkanes.

2.7 Tropospheric Chemistry of Sulphur Species

Sulphur compounds play an important role in the atmosphere, in clean air as well as in polluted areas (e.g. Rotstayn and Lohmann, 2002). In particular, sulphur species with low vapour pressure (such as sulphuric acid or methane sulphonic acid) have a large influence on particle formation in the atmosphere. The legendary London Smog largely consisted of sulphuric acid aerosol. The most important atmospheric sulphur species, their typical mixing ratios, residence times, and degradation pathways are summarised in Table 2.8. In this section we will briefly discuss the chemistry of these species and the impact of sulphur chemistry on the climate.

2.7.1 Sulphur Dioxide – SO_2

Sulphur dioxide is oxidised in the gas phase and, owing to its relatively high solubility in water, by reactions in atmospheric liquid water, such as cloud, fog, or aerosol (see Fig. 2.16)

The most important gas-phase reaction is oxidation, which is initiated by OH addition:

Table 2.8. Sulphur compounds in the atmosphere

Species	Chemical formula	Typ. mixing ratio ppb	Atmospheric life-time	Degraded by reaction with
Sulphur dioxide	SO_2	$10–10^5$	1–9 days	OH, liquid-phase oxidation
Hydrogen sulphide	H_2S	0–300	∼3 days	OH
Methyl mercaptan	CH_3SH		0.2 days	OH
Dimethyl sulphide (DMS)	CH_3SCH_3	0.01–1	0.3–3 days	OH, NO_3, BrO
Dimethyl disulphide	CH_3SSCH_3		∼1day	Like DMS
Dimethyl sulphoxide (DMSO)	CH_3SOCH_3	0–0.02	Days	OH, deposition
Methansulphonic acid (MSA)	CH_3SO_3H	0–0.06		OH, gas–particle conv.
Carbon Disulphide	CS_2	25 ± 10	12 days	OH
Carbonyl sulphide	COS	510 ± 70	≈6 years	OH, O
Sulphurous acid	H_2SO_3			Liquid phase
Sulphuric acid	H_2SO_4	$10^{-4}–10^{-3}$	0.01 day	Gas to particle conversion
Sulphates, e.g. Ammonium sulphate	$(NH_4)SO_4$ Aerosol		2–8 days	Deposition
Sulphur hexafluoride	SF_6	0.002	>1000 years	Ion reactions
Short lived intermediates				
Sulphur trioxide	SO_3	10^{-5}	10^{-6} s	H_2O
SH-radicals	SH		Seconds	O_3, O_2
SO-radicals	SO		<3 ms	O_2

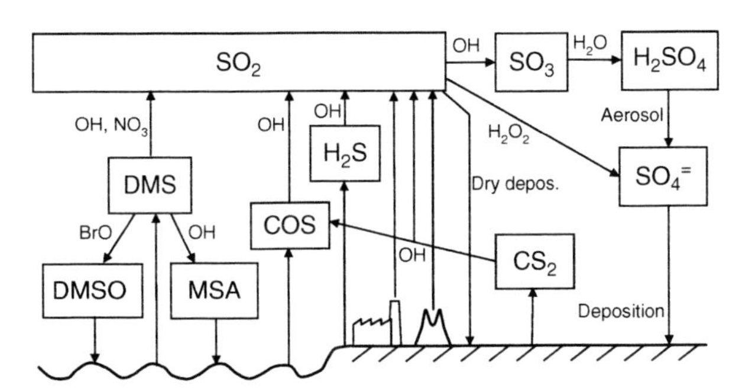

Fig. 2.16. Overview of the cycles of sulphur species in the atmosphere. The surface ocean (lower left corner) is relevant as source of atmospheric sulphur compounds

$$SO_2 + OH \xrightarrow{M} HSO_3 \tag{R2.42}$$

$$HSO_3 + O_2 \rightarrow SO_3 + HO_2 \tag{R2.43}$$

$$SO_3 + H_2O \xrightarrow{M} H_2SO_4 \tag{R2.44}$$

In this reaction sequence, discovered by Stockwell and Calvert (1983), the number of HO_X radicals is conserved, as in many other OH-initiated oxidation processes. Overall, OH oxidation of SO_2 facilitates about one third of the total SO_2 degradation.

The oxidation of SO_2 in the liquid phase also plays an important role (see Sect. 2.4.1). Gaseous SO_2 is in an equilibrium with aqueous SO_3^-:

$$SO_{2(g)} + H_2O_{(liq)} \leftrightarrow SO_2 \cdot H_2O_{(liq)} \tag{R2.45}$$

$$SO_2 \cdot H_2O_{(liq)} \leftrightarrow HSO^-_{3\,(liq)} + H^+_{(liq)} \tag{R2.46}$$

$$HSO_3^- \leftrightarrow SO_3^- + H^+ \tag{R2.47}$$

$HSO_3{}^-$ and $SO_3{}^-$ can be further oxidised to H_2SO_4 or sulphate, for example by hydrogen peroxide. This oxidation mechanism can be summarised as:

$$HSO_3{}^- \rightarrow \ldots \rightarrow SO_4{}^= \tag{R2.48}$$

The sulphate formed is eventually transported to the ground together with the (cloud or fog) droplets, thus contributing to 'acid rain' or 'acid haze'.

Finally, SO_2 can also react directly with the ground. This process is termed dry deposition (in contrast to processes where precipitation events play a role). About 30–50% of the SO_2 is removed from the atmosphere by dry deposition.

2.7.2 Reduced Sulphur Species: DMS, COS, CS_2, H_2S

Most sulphur compounds, with the exception of SO_2, are emitted in a reduced form and are rapidly oxidised in the atmosphere. With the exception of volcanic SO_2 all natural sulphur emissions are in the form of the following species:

Hydrogen Sulphide (H_2S) is present at mixing ratios of up to 300 ppb in the atmosphere. It is a by-product of the degradation of proteins and smells like rotten eggs. The oxidation of H_2S is initiated by reaction with OH:

$$H_2S + OH \rightarrow H_2O + SH \tag{R2.49}$$

The reaction pathways of the SH radical are currently unclear. While an analogy to the OH radical should exist, (cf. sulphur $\leftrightarrow$ oxygen), analogous reactions of SH with methane and CO are endothermic (Becker et al., 1975b). The reaction of SH with molecular oxygen is assumed to be very slow. However, SH reacts with ozone, forming SO radicals. The SO radicals finally react with O_2 or ozone to form SO_2.

Carbonyl Sulphide is, to a large extent, produced by the photolysis of sulphur-containing organic species (e.g. of amino acids) in the surface layer of the ocean (e.g. Ulshöfer and Andreae, 1998). Once released to the water, the majority of COS is degraded by hydrolysis to H_2S (timescale of about 10–12 h). The remaining COS is emitted to the atmosphere. In addition, there is some emission from soils and marshes (Ulshöfer, 1995). In addition to the direct emission, degradation of DMS and CS_2 by OH radicals is a source of COS (Chin and Davis, 1993):

$$CS_2 + OH \rightarrow SCSOH \tag{R2.50}$$

$$SCSOH \rightarrow COS + SO_2 + H \tag{R2.51}$$

Carbonyl sulphide is removed from the atmosphere by plant uptake and reaction with OH. The lifetime of COS in the troposphere is on the order of 2–3 years, and thus only small deviations from its average mixing ratio of 500 ppt are found (Ulshöfer, 1995).

Vertical transport of COS plays an important role as a sulphur source in the stratosphere (much like N_2O for the transport of oxidised nitrogen, CFCs for chlorine, and halons for Bromine, see Sect. 2.10.2). The degradation of COS in the stratosphere partially sustains the stratospheric sulphate – aerosol layer [named after its discoverer Christian Junge, 'Junge-layer' (Junge et al., 1961)].

Carbon Disulphide (CS_2) is naturally released via volcanoes, the ocean, marshes and forests, and industrial activities (Watts, 2000). Atmospheric mixing ratios are on the order of 15–35 ppt. Recently, CS_2 was found by DOAS in the city of Shanghai at levels up to 1.2 ppb (Yu et al., 2004). CS_2 is degraded in the atmosphere by OH oxidation (R2.50).

Dimethyl Sulphide (CH_3SCH_3) is produced by biological processes in the ocean (Andreae et al., 1985). Besides sporadic volcanic eruptions, oceanic DMS emissions are the largest natural source of sulphur in the atmosphere. Due to the important role of sulphur in the formation of aerosol particles and cloud condensation nuclei (CCN) DMS has received considerable attention (see also Sect. 2.7.3).

The degradation mechanisms of this species are not fully elucidated to date. Reactions of free radicals are probably largely responsible for its degradation. A role of OH in the degradation is likely. The first step in OH initiated degradation of DMS is OH abstraction from one of the methyl groups, or OH addition to the sulphur atom (Yin et al., 1990a,b).

$$OH + CH_3SCH_3 \rightarrow H_2O + CH_3SCH_2 \tag{R2.52}$$

$$OH + CH_3SCH_3 + M \rightarrow CH_3S(OH)CH_3 + M \tag{R2.53}$$

Intermediate products in this reaction chain are dimethyl sulphoxide (DMSO), CH_3SOCH_3, and $CH_3SO_2CH_3$. Stable end products are sulphuric acid and methane sulphonic acid (CH_3SO_3H). The involvement of NO_3 radicals has also been suggested (Winer et al., 1984; Platt and Le Bras, 1997), where the H abstraction channel appears to be predominant (Butkovskaya and Le Bras, 1994):

$$NO_3 + CH_3SCH_3 \rightarrow HNO_3 + CH_3SCH_2 \quad (R2.54)$$

In addition, there are several reports of a possible role of halogen oxide radicals, in particular of BrO (Toumi, 1994). The product of the BrO–DMS reaction is DMSO:

$$BrO + CH_3SCH_3 \rightarrow Br + CH_3SOCH_2 \quad (R2.55)$$

While sulphuric acid and methane sulphuric acid form particles, DMSO does not. Thus, the fraction of DMS degraded by BrO may determine the efficiency of particle formation in marine areas, as discussed by von Glasow et al. (2004).

2.7.3 Influence of Sulphur Species on the Climate, the CLAW Hypothesis

The low vapour pressure sulphur compounds, H_2SO_4 and methane sulphonic acid, will quickly condense after their formation in the gas phase (e.g. by reaction R2.44). In this process, small particles are formed, which can act, under appropriate conditions, as cloud condensation nuclei (CCN) and thus impact cloud formation. Therefore, biological emission of sulphur species could affect cloud formation and thus our climate. This theory was proposed in 1987 by Charlson, Lovelock, Andreae and Warren (referred to as the CLAW hypothesis after the authors, initials) (Charlson et al., 1987; Bates et al., 1987). The authors postulate the following mechanism: Due to microbiological activities in the ocean, volatile sulphur species (essentially DMS) are released. DMS is oxidised by photochemical processes in the atmosphere to sulphuric acid or methane sulphonic acid. Due to their low vapour pressure, these species condense to form small particles (mean radius $\sim$0.07 μm), which can act as CCN.

In large areas of the worlds oceans, the supply of CCN is so small that cloud formation can be limited by CCN availability and not by the water vapour supply (about 200 CCN cm^{-3} are required, corresponding to a DMS mixing ratio of less than 100 ppt). The extent of cloud cover now feeds back on the insulation and thus the ocean surface temperature. If the biological DMS formation is positively correlated with the ocean surface temperature, which is likely, this process could form a biological negative feedback loop which effectively stabilises the surface temperature of earth [see also Gaia Hypothesis (Lovelock, 1979)]. Since its publication, the CLAW hypothesis has been under intense debate. For instance, Schwartz (1988) argued that the emission of gaseous sulphur species in the northern hemisphere (but not in the southern hemisphere) largely increased during the last 100 years due to anthropogenic activities (in fact the present anthropogenic S emission exceeds the DMS source in the northern hemisphere), without major change in cloud albedo or temperature (Schwartz, 1988).

2.8 Chemistry of Halogen Radicals in the Troposphere

During the last decade, significant amounts of the halogen oxides BrO, IO, OIO, and ClO were detected in the tropospheric BL by DOAS (Table 2.9). Direct and indirect evidence for Cl and Br atoms, as well as for Br_2 and BrCl, was also found under certain conditions. In addition, there is growing evidence for a BrO 'background' in the free troposphere. Observations were made at a variety of sites (see also Table 2.9 and Fig. 2.17):

The presence of these RHS in the troposphere has many consequences. Elevated RHS levels are associated with ozone destruction, which can lead to complete loss of BL ozone in the Arctic. This 'Polar Tropospheric Ozone Hole' was the first hint for tropospheric halogen chemistry (Oltmans and Komhyr, 1986; Bottenheim et al., 1986, 1990; Barrie et al., 1988; Platt and Lehrer, 1997; Barrie and Platt, 1997; Platt and Hönninger, 2003). In addition, other disturbances of tropospheric chemistry can occur, as detailed in Sect. 2.8.3.

2.8.1 Tropospheric Sources of Inorganic Halogen Species

The sources of RHS (X, X_2, XY, XO, HOX, $XONO_2$, HX, where X = Cl, Br, I) in the troposphere are the degradation of organic halogen compounds and the volatilisation of halogen ions (X^-) from sea salt aerosol or surface salt deposits.

Fully halogenated compounds (such as CF_2Cl_2 or CF_2ClBr), which are the main source of RHS in the stratosphere, are photolytically stable in the troposphere. RHS can therefore only be released from less stable precursors (Cauer, 1939), such as partially halogenated organic compounds like CH_3Br, or polyhalogenated species, such as $CHBr_3$, CH_2Br_2, CH_2I_2, CH_2BrI (Cicerone, 1981; Schall and Heumann, 1993; Khalil et al., 1993; Schauffler et al., 1998; Carpenter et al., 1999, 2001), or even I_2 (Saiz-Lopez et al., 2004b; Peters, 2004; Peters et al., 2005). While some of these species, such as CH_3Br, are emitted by up to 50% from anthropogenic sources, most of them, and in particular polyhalogenated species, are only emitted from biological sources predominately in the ocean or in coastal areas.

The atmospheric lifetime of halogenated organics varies widely. The lifetime of methyl bromide, CH_3Br, is on the order of 1 year, bromoform, $CHBr_3$, has a lifetime of several days, while diiodo methane CH_2I_2 is photodegraded in minutes. (Wayne et al., 1995; Yvon and Butler, 1996; Davis et al., 1996; Carpenter et al., 1999). It is possible that even less stable halocarbon species are emitted, which, probably due to their instability, have escaped detection (Carpenter et al., 1999).

The release of inorganic halogen species from sea salt (e.g. sea salt aerosol or sea salt deposits), appears to proceed by three main pathways:

(1) Strong acids can release HCl (but not HBr) from sea-salt halides. Under certain conditions (see below), HX can be heterogeneously converted to RHS.

Table 2.9. Observation of reactive halogen species in the troposphere and their probable source mechanism

Species	Measurement site	Technique	Conc. level	Reference
Br	Arctic BL	Hydrocarbon Clock	$(1\text{–}10) \cdot 10^7\,cm^{-3}$	Jobson et al. (1994), Ramacher et al. (1997, 1999)
BrO	Arctic and Antarctic BL	DOAS (ground based)	Up to 30 ppt	Hausmann and Platt (1994), Tuckermann et al. (1997), Hegels et al. (1998), Martínez et al. (1999), Hönninger and Platt (2002), Frieß et al. (2004a), Hönninger et al. (2004c)
BrO	Arctic and Antarctic BL	DOAS (satellite)	Around 30 ppt (assuming 1000 m layer)	Wagner and Platt (1998), Richter et al. (1998), Hegels et al. (1998), Martinez et al. (1999), Wagner et al. (2001b), Hollwedel et al. (2004)
BrO	Salt lakes (Dead Sea, Salt Lake City, Caspian Sea, Salar de Uyuni)	Active DOAS, MAX-DOAS	Up to 176 ppt	Hebestreit et al. (1999), Matveev et al. (2001), Stutz et al. (2002), Wagner et al. (2001b), Hönninger et al. (2004b)
BrO	Mid-Lat. Marine BL	MAX-DOAS	Up to 2 ppt	Leser et al. (2002), Saiz-Lopez (2004a)
BrO	Mid-Lat. Free Troposphere	DOAS (difference)	1–2 ppt	Frieß et al. (1999), van Roozendael et al. (2002), Pundt et al. (2002)
BrO	Polar free Troposphere	Airborne DOAS	0–20 ppt	McElroy et al. (1999)

BrO	Volcanic plumes	MAX-DOAS	Up to 1000 ppt	Bobrowski et al. (2003), Bobrowski (2005)
$HOCl/Cl_2$	Marine BL	Mist Chamber	up to 254 ppt	Pszenny et al. (1993)
Cl	Arctic BL	Hydrocarbon Clock	$(1–10) \cdot 10^4\,cm^{-3}$	Jobson et al. (1994), Ramacher et al. (1997, 1999)
Cl	Arctic BL	$^{12}C/^{13}C$ in CO	$(1–10) \cdot 10^4\,cm^{-3}$	Röckmann et al. (1999)
Cl	Remote Marine BL	Hydrocarbon Clock	$(1–15) \cdot 10^3\,cm^{-3}$	Wingenter et al. (1996)
Cl	Extra-tropical S. hemisphere	$^{12}C/^{13}C$ in CH_4	$2.6 - 18 \times 10^3$	Platt et al. (2004)
ClO	Salt Lake City	DOAS	Up to 15 ppt	Stutz et al. (2002)
ClO	Volcanic plumes	MAX-DOAS		
Br_2, BrCl	Arctic BL	APCIMS	Up to 30 ppt	Foster et al. (2001), Spicer et al. (1998, 2002)
I_2	Mid-Lat. Marine BL	Active DOAS	Up to 90 ppt	Saiz-Lopez (2004b), Peters (2004), Peters et al. (2005)
$Cl_2/HOCl$ $Br_2/HOBr$	Arctic BL		$Cl_X \leq 100$ $Br_X \leq 38$ ppt	Impey et al. (1999)
IO, OIO	Coastal Areas	DOAS	Up to 6 ppt (≈ 0.2 ppb h^{-1})	Alicke et al. (1999), Allan et al. (2005a), Wittrock et al. (2000a), Frieß et al. (2001), Saiz-Lopez et al. (2005), Zingler and Platt (2005)

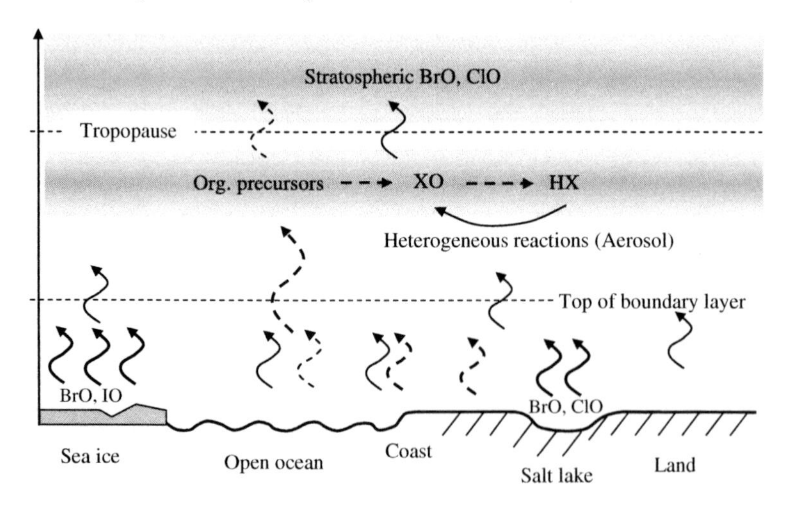

Fig. 2.17. The occurrence of reactive halogen species (XO, X = Cl, Br, I) in the atmosphere: BrO is found above polar sea ice during springtime, IO (and BrO) are present at most coastal areas. It is suspected that there is a layer of BrO in the free troposphere (see von Glasow et al., 2004). *Drawn arrows*: emission of RHS, *dashed arrows*: emission of organohalogens

(2) Oxidising agents may convert Br^- or Cl^- to gaseous Br_2 or BrCl. In particular, HOX (i.e. HOBr and HOCl) is such an oxidant. Direct photochemical (Oum et al., 1998a) or non-photochemical oxidation of halides by O_3 may also occur (Oum et al., 1998b; Hirokawa et al., 1998), though probably quite slowly.
(3) Oxidised nitrogen species, in particular N_2O_5 and NO_3, and perhaps even NO_2, can react with sea-salt bromide or chloride to release HBr, HCl, or photolabile species, in particular $ClNO_2$ and $BrNO_2$ (Finlayson-Pitts and Johnson, 1988; Finlayson-Pitts et al., 1990; Behnke et al., 1993, 1997; Rudich et al., 1996; Schweitzer et al., 1999; Gershenzon et al., 1999).

Because these mechanisms are closely linked to the atmospheric cycling of RHS, we will discuss their details in the next section.

2.8.2 Tropospheric Cycles of Inorganic Halogen Species

In this section, we review the reaction cycles of inorganic halogen species in the troposphere (see Fig. 2.18), (see Wayne et al., 1995; Platt and Janssen, 1996; Platt and Lehrer, 1997; Lary et al., 1996; Platt and Hönninger, 2003; von Glasow Crutzen, 2003).

Following release, inorganic and organic halogen species are photolysed to form halogen atoms which then predominately react with ozone:

$$X + O_3 \rightarrow XO + O_2 \tag{R2.56}$$

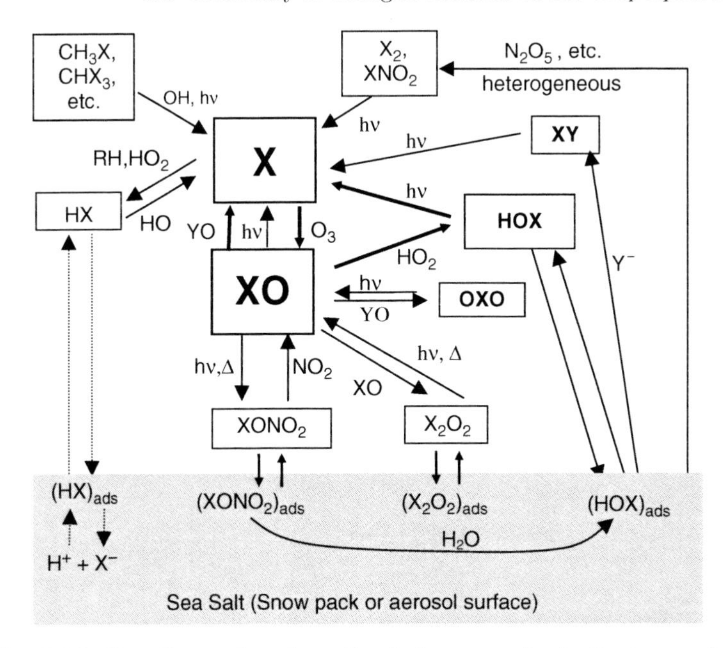

Fig. 2.18. Reaction cycles of reactive halogen species in the troposphere

Typical lifetimes of halogen atoms (X) due to R2.56 at tropospheric background O_3 levels are around 0.1 s for Cl, and on the order of 1 s, for Br and I. Halogen atoms are regenerated in a series of reactions including photolysis of XO, which is of importance for X = I, Br, and, to a minor extent, Cl:

$$XO + h\nu \rightarrow X + O \tag{R2.57}$$

where $J_{2.57} \approx 3 \times 10^{-5}\,s^{-1}$, $4 \times 10^{-2} s^{-1}$, $0.2 s^{-1}$ for X = Cl, Br, and I, respectively. The reaction of XO with NO also recycles halogen atoms:

$$XO + NO \rightarrow X + NO_2 \tag{R2.58}$$

This reaction leads to a shift of the NO/NO_2 photostationary state in the atmosphere, by providing a shortcut in the normal $NO/NO_2/O_3$ cycle (R2.1/R2.2/R2.3). This has consequences for the photochemistry in the troposphere and, in particular, ozone formation (Stutz et al., 1999).

The self-reactions of XO (or reaction with another halogen oxide YO) also play a role:

$$XO + YO \rightarrow X + Y + O_2 \tag{R2.59a}$$

$$XO + YO \rightarrow XY + O_2 \tag{R2.59b}$$

$$XO + YO \rightarrow OXO + O \tag{R2.59c}$$

$$XO + YO \xrightarrow{M} X_2O_2 \tag{R2.59d}$$

Reactions R2.56, R2.59a,b,d, and the photolysis of the halogen molecules formed in R2.59 constitute a catalytic cycle which destroys ozone in the troposphere. The efficiency of this cycle is determined by the amount of RHS present and the rates of R2.59a,b,d. This type of halogen-catalysed ozone destruction has been identified as the prime cause of polar BL ozone destruction (Oltmanns and Komhyr, 1986; Barrie et al., 1988; Barrie and Platt, 1997).

Halogen atoms can also react with saturated (Cl) or unsaturated HC (Cl and Br) to form hydrogen halides, e.g.:

$$RH + Cl \rightarrow R + HCl \qquad (R2.60)$$

Here, R denotes an organic radical. An alternative is the reaction of halogen atoms with formaldehyde (or higher aldehydes) or HO_2, also leading to the conversion of halogen atoms to hydrogen halides:

$$X + HO_2 \rightarrow HX + O_2 \qquad (R2.61)$$

$$X + HCHO \rightarrow HX + CHO \qquad (R2.62)$$

The former reaction occurs for X = Cl, Br, I, the latter only for X = Cl, Br. In particular, Cl is a very strong oxidant and a Cl concentration of 10^4 cm^{-3} can already considerably contribute to the oxidation capacity of the troposphere.

The main loss of RHS is their conversion to hydrogen halides which are highly water soluble, and can thus be easily lost from the atmosphere by wet or dry deposition. The only relevant gas-phase 'reactivation' mechanism of HX is reaction with OH:

$$HX + OH \rightarrow X + H_2O \qquad (R2.63)$$

Only at high HO concentration and in the absence of deposition processes, such as in the free troposphere, does this reaction play a role. HX can also be recycled through the aerosol phase, as we will discuss below.

Other important inorganic halogen species are HOX and $XONO_2$. HOX is formed via the reaction of peroxy radicals with XO (Canosa-Mas et al., 1999):

$$XO + HO_2 \rightarrow HOX + O_2 \qquad (R2.64)$$

Formation of HOX is followed by its photolysis:

$$HOX + h\nu \rightarrow X + OH \qquad (R2.65)$$

Reactions R2.56, R2.64, and R2.65 form another catalytic cycle that destroys ozone during the day. This cycle dominates at low RHS levels, such as those found in the marine BL. The catalytic ozone destruction cycle through R2.59 is important at higher RHS levels, such as those found in polar regions. Reactions R2.64 and R2.65 also lead to a shift in the HO_2/OH ratio in the troposphere, and thus directly influence the oxidising capacity of the troposphere (Bloss et al., 2005).

Another important reaction cycle involving HOBr is the liberation of gaseous bromine species (and to a lesser extent chlorine species) from (sea-salt) halides (Fan and Jacob, 1992; Tang and McConnel, 1996; Vogt et al., 1996):

$$HOBr + (Br^-)_{Surface} + H^+ \rightarrow .Br_2 + H_2O \qquad (R2.66)$$

The required H^+ [the reaction appears to occur at appreciable rates only at $pH < 6.5$ (Fickert et al., 1999)] can be supplied by strong acids, such as H_2SO_4 and HNO_3 from anthropogenic or natural sources. Reaction R2.66, together with R2.64, R2.56, and the photolysis of Br_2, form a cycle where Br^-, for example, in sea ice or the aerosol, is volatilised. Because the uptake of one HOBr molecule leads to the formation of two Br atoms, which, ignoring any Br loss, can form two HOBr molecules (R2.64), this cycle is autocatalytic. The explosion-like behaviour of this cycle has lead to the term 'Bromine Explosion' for this cycle, which is believed to be responsible for the high reactive bromine levels found in the Arctic (Platt and Lehrer, 1995; Platt and Janssen, 1996; Wennberg, 1999).

The reaction of halogen oxides with NO_2 forms halogen nitrate, $XONO_2$:

$$XO + NO_2 \xrightarrow{M} XONO_2 \qquad (R2.67)$$

Bromine nitrate is assumed to be quite stable against thermal decay, but is readily photolysed and may be converted to HOX by heterogeneous hydrolysis (van Glasow et al., 2002):

$$XONO_2 + H_2O \xrightarrow{M} HNO_3 + HOX \qquad (R2.68)$$

or to Br_2 or BrCl by heterogeneous reaction with HY:

$$XONO_2 + HY \xrightarrow{Surface} HNO_3 + XY \qquad (R2.69)$$

Overall, $XONO_2$ is probably of minor importance at the low NO_X levels typically found in the free troposphere, but it can play an important role in polluted air, for instance at polluted coastlines. There $XONO_2$ can be an important source of HOX, thus in particular $BrONO_2$ might contribute to bromine explosion events.

Under conditions of high NO_X, halogen release can also occur via the reactions:

$$N_2O_5(g) + NaX(s) \rightarrow NaNO_3(s) + XNO_2(g) \qquad (R2.70)$$

(e.g. Finlayson-Pitts et al., 1989). The XNO_2 formed in the above reaction may photolyse to release a halogen atom, or possibly further react with sea salt (Schweitzer et al., 1999):

$$XNO_2(g) + NaX(s) \rightarrow NaNO_2(s) + X_2(g) \qquad (R2.71)$$

The above reaction sequence would constitute a dark source of halogen molecules. The direct reaction of NO_3 on salt surface can also release reactive halogens, for example in the form of halogen atoms:

$$NO_3(g) + NaX(s) \rightarrow NaNO_3(s) + X(s) \tag{R2.72}$$

(Seisel et al., 1997; Gershenzon et al., 1999). The uptake coefficient of NO_3 by aqueous solutions of NaBr and NaCl was found by Rudich et al. (1996) to be near 0.01 for sea water, while Seisel et al. (1997) found up to 0.05 for dry NaCl.

2.8.3 Potential Impact of Inorganic Halogen Species on Tropospheric Chemistry

The role of reactive halogens in the troposphere has been a controversial issue and remains a poorly understood aspect of atmospheric chemistry. However, it is becoming clear that the RHS are present in many parts of the troposphere (Table 2.9). At the levels shown in Table 2.9, RHS can have a noticeable effect on several aspects of tropospheric chemistry. These include:

(1) RHS, and in particular reactive bromine and iodine can readily destroy tropospheric ozone through catalytic cycles (Solomon et al., 1994b; Platt and Janssen, 1996; Davis et al., 1996; Sander et al., 2003; von Glasow et al., 2004). This has consequences for atmospheric chemistry and, considering that ozone is a greenhouse gas, also the global climate (e.g. Roscoe et al., 2001).
(2) Reactions R2.64, and J2.65 will lead to the conversion of HO_2 to OH, and thus reduce the HO_2/OH ratio, with consequences for the atmospheric oxidation capacity and thus indirectly for the lifetime of methane, another important greenhouse gas.
(3) The presence of highly reactive chlorine atoms, and to a lesser extent bromine atoms, has a direct impact on the oxidation capacity of the troposphere.
(4) The reaction of BrO with DMS might be important in the unpolluted remote marine BL, where the only other sink for DMS is the reaction with OH radicals (Toumi, 1994; von Glasow and Crutzen, 2003, 2004). It may thus have an impact on cloud formation as explained in Sect. 2.8.3.
(5) Deposition of mercury was found to be enhanced by the presence of reactive bromine species, in particular in polar regions (e.g. Barrie and Platt, 1997; Schroeder et al., 1998). This process appears to be one of the most important pathways for toxic mercury compounds to enter the arctic (Lindberg et al., 2002), and possibly also other ecosystems.
(6) Iodine species might be involved in particle formation in the marine BL (Hoffmann et al., 2001; O'Dowd et al., 2002; Jimenez et al., 2003; Burkholder et al., 2004).

At present, more measurements are needed to ascertain the distribution and levels of RHS in the troposphere and to assess the significance of RHS on local, regional, and global atmospheric chemistry and our climate.

2.9 Oxidation Capacity of the Atmosphere

The term oxidation capacity, sometimes also called oxidative power, refers to the capability of the atmosphere (or rather a part thereof) to oxidise (or otherwise degrade) trace species emitted into it. This ability is crucial for the removal of trace species, such as the greenhouse gas methane, and is thus often also referred to as the 'self-cleaning' capacity of the atmosphere. Although there is no general definition, the oxidation capacity is frequently associated with the abundance of OH. However, as explained above, many other oxidants (including O_2 and O_3), as well as free radicals other than OH, can contribute to the oxidation capacity of the atmosphere. While the oxidation capacity is quite a popular concept, it is nevertheless difficult to define.

One difficulty in defining the term oxidation capacity lies in the fact that a given agent (e.g. OH radicals) might act quite differently on various pollutants. Perhaps the best definition for the oxidation capacity C_i of a radical R_i (e.g. $R_i = NO_3$, OH, O_3, Cl) with the atmospheric abundance (concentration) $[R]$, with respect to the degradation (oxidation) of species X_j (e.g. volatile organic species, CO, and NO_X), oxidised by the molecule (per second and cm^{-3}), might be a sum such as:

$$C_i = \sum_j [X_j] \cdot [R_i] \cdot k_{ij} \qquad \text{(R2.73)}$$

where $[R_i]$ denotes the concentration of radical species i with the reaction rate constant k_{ij} towards species X_j (Geyer et al., 2001; Platt et al., 2002). Frequently 24-h average values of the oxidation capacity are calculated.

Table 2.10 summarises typical and maximum tropospheric concentrations of the radicals discussed, and shows minimum and typical fractional lifetimes of key VOCs against degradation by the various free radicals.

As an example, the relative contribution of the atmospheric oxidants OH, NO_3, and ozone to the 24 h average of the degradation of organic species in the atmosphere, as observed during the BERLIOZ 1998 campaign at Papstthum near Berlin, Germany, is shown in Fig. 2.19. When all VOCs (including CH_4 and CO) are considered, OH reactions constitute about three quarters of the oxidation capacity. For NMHC, NO_3 contributes almost one third of the oxidation capacity (Geyer et al., 2001).

Open questions include the role of heterogeneous reactions in radical cycles. While the contribution of heterogeneous reactions to HO_X chemistry is unclear to date, it is likely that heterogeneous reactions play an important role in the atmospheric cycles of most other radicals. Another interesting question

Table 2.10. Estimated lifetime (hours) for selected VOC

Species	OH (av.)	OH (peak)	NO_3 (av.)	NO_3 (peak)	ClO (av.)	ClO (peak)	Cl (av.)	Cl (peak)	BrO (av.)	Br (av)
Conc. molec. cm^{-3}	6×10^5	1×10^7	1.5×10^8	1×10^{10}	10^6	10^8	10^3	10^5	10^8	5×10^6
DMS	110	6.3	1.7	0.03	3×10^4	280	870	8.7	10^c	26
n-Butane	190	11	2.8×10^4	430	–		1400	14		
C_2H_4	56	3.4	1×10^4	160			2800	28		
Isoprene	4.6	0.3	1.9	0.03	–		<1100	<11		0.9
Toluene	77	4.6	3.1×10^4	460	–					
CH_4	7×10^4	4×10^3	$>5 \times 10^4$	>700	$>7 \times 10^7$	$>7 \times 10^5$	3×10^6	3×10^4		

is whether the effect of several oxidising species is additive. In fact, it is possible that the presence of halogen oxides may reduce the OH concentration due to an increase in the photo-stationary state NO_2/NO ratio. In addition, it is well possible that there are other, yet unrecognised, mechanisms involving the above or other free radicals.

2.10 Stratospheric Ozone Layer

The role of atmospheric ozone as a filter for the Sun's harmful UV radiation has lead to an early interest in ozone chemistry in the investigation of the chemistry of the atmosphere. Spectroscopic measurements by Gordon Miller Bourne Dobson in 1925 showed an ozone column density larger than expected from assuming the O_3 mixing ratio found near the surface to be constant throughout the atmosphere (e.g. Dobson and Harrison, 1926; Dobson, 1968). Subsequent spectroscopic measurements using the so-called 'Umkehr' effect by Paul Götz proved the theory of an ozone layer and determined the altitude of its maximum at 25 km (Götz et al., 1934). This stratospheric ozone layer contains 90% of the atmosphere's ozone and is thus responsible for the majority of the absorption of solar UV radiation.

2.10.1 Stratospheric Ozone Formation: The Chapman Cycle

In the late 1920s, Sidney Chapman proposed a reaction scheme which explained the observed vertical profile of ozone, with relatively low mixing ratios in the troposphere and a maximum around 25 km altitude, the 'Chapman Mechanism' (Chapman, 1930). The initial process is the photolysis of oxygen molecules to form two oxygen atoms in their ground state (indicated by the spectroscopic notation (^3P), see Chap. 3). In the stratosphere, sufficiently energetic UV light (i.e. light with wavelengths below 242 nm) is available to photolyse oxygen molecules:

$$O_2 + h\nu(\lambda < 242\,\text{nm}) \rightarrow O(^3P) + O(^3P) \qquad \text{(R2.74)}$$

The oxygen atoms can react in three ways: (1) Recombine with an oxygen molecule to form ozone. Since two particles (O and O_2) combine to make one (O_3), collision with a third body (M, likely N_2 or O_2,) is required to facilitate simultaneous conservation of energy and momentum. The reaction is therefore pressure dependent:

$$O(^3P) + O_2 + M \rightarrow O_3 + M \qquad \text{(R2.2)}$$

Alternatively, (2) the oxygen atom can react with an existing ozone molecule:

$$O(^3P) + O_3 \rightarrow O_2 + O_2 \qquad \text{(R2.74)}$$

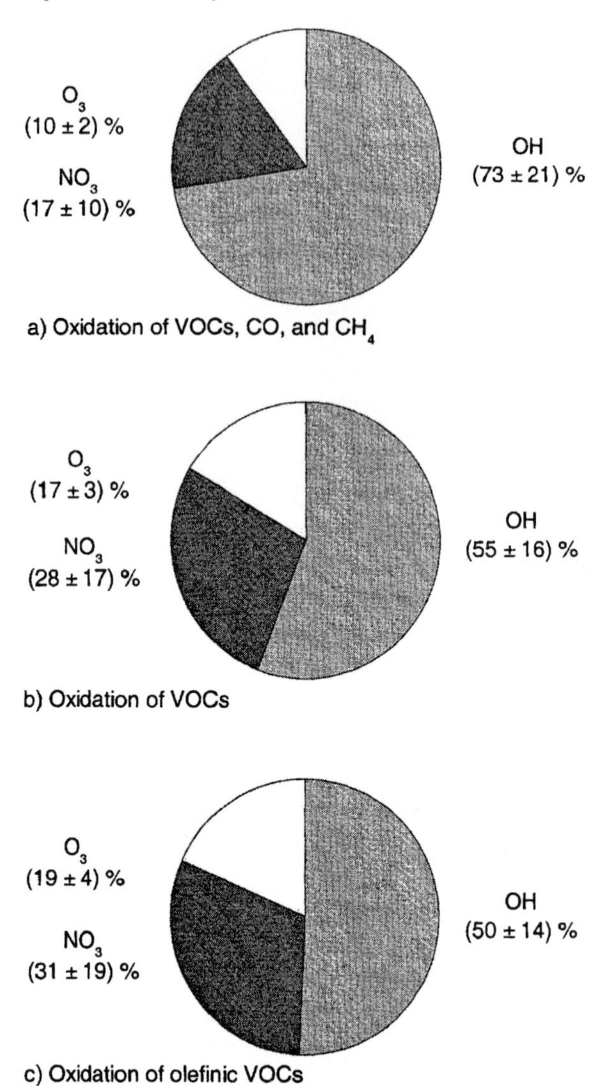

Fig. 2.19. The relative contribution of the atmospheric oxidants OH, NO_3, and ozone to the degradation of organic species in the atmosphere on a 24-h basis. Case study during the BERLIOZ 1998 campaign at Papstthum near Berlin, Germany. (**a**) VOCs including CH_4 and CO, (**b**) non-methane VOCs, (**c**) alkenes only (from Geyer et al., 2001, Copyright by American Geophysical Union (AGU), reproduced by permission of AGU)

Finally, (3) the recombination of two oxygen atoms to form molecular oxygen is possible but largely unimportant in the stratosphere:

$$O(^3P) + O(^3P) + M \rightarrow O_2 + M \qquad \text{(R2.75)}$$

In addition to the 'primary' production of O atoms by the photolysis of O_2, the photolysis of O_3 also provides 'secondary' O atoms. In fact, photolysis of ozone molecules occurs at a much higher rate than that of oxygen molecules:

$$O_3 + h\nu(\lambda < 612\,\text{nm}) \rightarrow O(^3P) + O_2 \qquad \text{(R2.76)}$$

In summary, the above reactions, also known as the 'Chapman Reactions', lead to a steady-state O_3 level in the stratosphere, in which the O atom production via reactions R2.74 and R2.76 is in balance with their destruction via recombination with O_2 and reaction with O_3. The above set of reactions explains the formation of a layer of ozone with a maximum concentration in the lower stratosphere. In the lower stratosphere and troposphere, the rate of O_2 photolysis, and thus the ozone formation rate, becomes extremely low (despite the much higher O_2 concentration there). However, O_3 destruction still occurs via O_3 photolysis, which takes place at much longer wavelengths, and the reaction of $O + O_3$. This explains why the ozone concentration should increase with height (in fact the Chapman mechanism predicts zero O_3 formation in the troposphere). On the other hand, in the upper part of the stratosphere the recombination of $O + O_2$ (reaction R2.2) becomes slower, since the concentration of air molecules necessary as a 'third body' (M) in the recombination of $O + O_2$ (reaction R2.2) reduces proportionally to the atmospheric pressure. Thus, despite increasing levels of UV radiation, the O_3 concentration (and also the mixing ratio) will eventually decrease with altitude. Figure 2.20 depicts the ozone profile predicted by the Chapman cycle.

2.10.2 Stratospheric Ozone Chemistry: Extension of the Chapman Cycle

The Chapman mechanism gave a satisfactory explanation for the occurrence of an ozone layer. However, detailed investigation during the 1960s of the elementary reactions and photolysis processes involved revealed that quantitatively the mechanism overestimates the O_3 levels by about a factor of three, see Figure 2.20. It subsequently became clear that there are many other trace gas cycles affecting stratospheric O_3 levels. In particular, a group of reactions were found to catalyse the elementary reaction of $O + O_3$ (reaction R2.74 above). These reaction sequences follow the general scheme:

$$O_3 + Z \rightarrow ZO + O_2 \qquad \text{(R2.77)}$$

$$ZO + O(^3P) \rightarrow Z + O_2 \qquad \text{(R2.78)}$$

With the net result,

$$O(^3P) + O_3 \xrightarrow{Z} O_2 + O_2 \qquad \text{(R2.79)}$$

where Z (and ZO) denotes a species acting as catalyst Z for reaction R2.75. In the stratosphere, the following species are active (see Table 2.11): Cl (ClO),

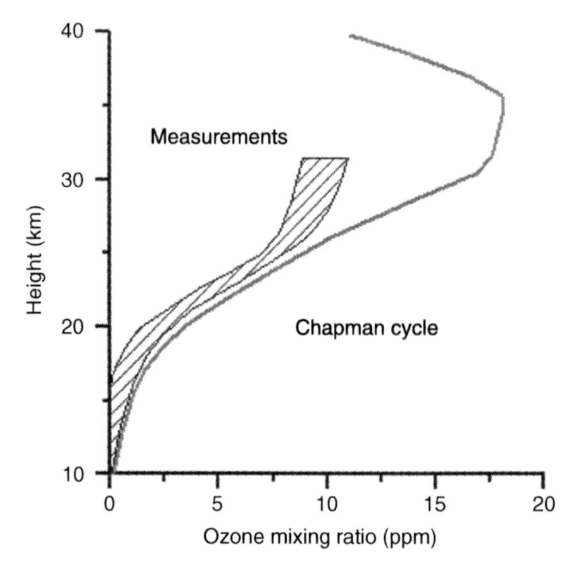

Fig. 2.20. The stratospheric ozone profile according to the Chapman Mechanism, compared to observations (figure adapted from Roth, 1994)

Br (BrO), NO (NO_2), or OH (HO_2). Inclusion of these reactions brings observations and model calculations in very good agreement. Figure 2.21 illustrates the contribution to the ozone destruction of each of the catalysts at different altitudes.

All species listed in Table 2.11 play an important role in stratospheric chemistry. The *oxides of nitrogen* (NO, NO_2, see Sect. 2.5) are formed in the stratosphere by reaction of excited oxygen atoms $O(^1D)$ with N_2O (which can be regarded as a 'transport species' for oxidised nitrogen into the stratosphere):

$$\begin{aligned} N_2O + O(^1D) &\rightarrow NO + NO \\ &\rightarrow N_2 + O_2 \end{aligned} \tag{R2.80}$$

Table 2.11. 'Catalysts' destroying stratospheric ozone

Catalyst Z	ZO	Stratospheric source
NO	NO_2	Degradation of N_2O from the troposphere
OH	HO_2	$O^1D + H_2O$
Cl	ClO	Degradation of (mostly man-made) organo chlorides
Br	BrO	Degradation of CH_3Br, and other organo bromine species

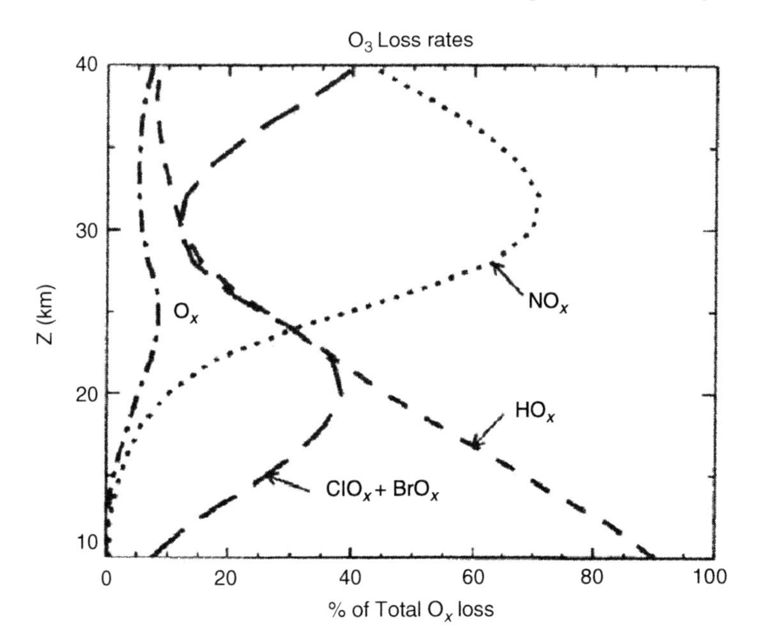

Fig. 2.21. The contribution of the ClO_X, BrO_X, NO_X, and HO_X, catalysed destruction of stratospheric ozone to the original Chapman Mechanism as a function of altitude (from Portmann et al., 1999, Copyright by American Geophysical Union (AGU), reproduced by permission of AGU)

The excited oxygen atoms $O(^1D)$ are formed in the UV photolysis of O_3 by (R2.4). As in the troposphere, a photo-stationary state between NO, NO_2, and O_3 is established and conversion of NO_2 to N_2O_5 will take place. Under 'normal' stratospheric conditions, loss of N_2O_5 occurs due to photolysis, leading to an increase of NO_X during the course of a day (only during polar winter do heterogeneous reactions of N_2O_5, e.g. R2.98 become important).

Halogen atoms are released by the degradation of long-lived halocarbon species (e.g. $CH_3Cl, CFCl_3, CH_3Br$, etc.), which are emitted naturally or anthropogenically in the troposphere and then transported into the stratosphere. Halogen atoms (denoted as X = F, Cl, Br, I), which are released in the degradation processes, react quickly with O_3 to form halogen oxides (see Sect. 2.8). In addition, halogen atoms can react with O_2 by association reactions to form halogen peroxide radicals. These radicals are unstable and dissociate by the reverse of their formation reaction, so that under atmospheric conditions the XO_2 species exists in equilibrium with X and O_2.

$$X + O_2(+M) \rightarrow XO_2(+M) \qquad \text{(R2.81)}$$

The stability of XO_2 decreases in the series F > Cl > Br > I. The equilibrium for atmospheric conditions lies in favour of the atom, and XO_2 reactions do not play a significant role in stratospheric chemistry.

A rapid reaction occurs between XO and NO:

$$XO + NO \rightarrow X + NO_2 \qquad (R2.82)$$

NO_2 is rapidly photolysed to produce atomic oxygen, which reforms ozone, and thus the overall effect of this reaction coupled with the $X + O_3$ reaction on ozone is neutral. On the other hand, this reaction will lead to a higher stationary state X/XO ratio, and thus (at least in the case of chlorine) enhance the rate of $X \rightarrow HX$ conversion by reaction of halogen atoms with HC (i.e. CH_4). Consequently, the presence of nitrogen oxides tends to increase the HX/XO_X ratio, and thus reduce the ozone depletion by halogen radicals.

The chlorine oxide radical undergoes self-reaction to form a dimer, Cl_2O_2 (Cox and Derwent, 1979). The ClO dimer plays a major part in the catalytic destruction of ozone occurring in the polar lower stratosphere during wintertime, which in Antarctica leads to the formation of the ozone hole (see Sect. 2.10.3 below). The relevant cycle is:

$$ClO + ClO + M \rightarrow Cl_2O_2 + M \qquad (R2.83)$$

$$Cl_2O_2 + h\nu \rightarrow Cl + ClO_2 \qquad (R2.84)$$

$$ClO_2 + M \rightarrow Cl + O_2 \qquad (R2.85)$$

$$Cl + O_3 \rightarrow ClO + O_2 \qquad (R2.86)$$

The ClO dimer is relatively unstable, and the cycle is only important at low temperatures ($T \approx 200$K) when the reverse decomposition of Cl_2O_2:

$$Cl_2O_2 + M \rightarrow ClO + ClO + M \qquad (R2.87)$$

is slow compared with its photolysis.

Symmetric chlorine dioxide is formed in the stratosphere from the coupled reaction of BrO with ClO:

$$BrO + ClO \rightarrow OClO + Br \qquad (R2.88a)$$

The discovery of atmospheric OClO using ground-based DOAS spectroscopy at the McMurdo Sound base in Antarctica (Solomon et al., 1987b), was the first positive indication that partitioning of atmospheric chlorine between its active and inactive forms was greatly perturbed in the Antarctic ozone hole. Chlorine dioxide is rapidly photolysed by visible light to yield O atoms

$$OClO + h\nu \rightarrow O + ClO \qquad (R2.89)$$

and thus is not a catalytic agent for O_3 destruction. A consequence of its rapid photolysis are low OClO concentrations during daytime.

Observations of the evolution of stratospheric OClO during sunset allow deduction of information on the BrO concentration in that BrO is an essential 'catalyst' for its formation. On the other hand, BrO is lost by BrCl production via reaction:

$$\mathrm{BrO + ClO \rightarrow BrCl + O_2} \tag{R2.88b}$$

while the third channel of reaction R2.88 does not change the XO reservoir, but leads to the destruction of ozone:

$$\mathrm{BrO + ClO \rightarrow Cl + Br + O_2} \tag{R2.88c}$$

Since $k_{2.88b}/(k_{2.88a}+k_{2.88b}+k_{2.88c}) \approx 0.08$, the late night OClO concentration cannot become larger than $1/0.08 \approx 12$ times the daytime BrO concentration (e.g. Wahner and Schiller, 1992).

The behaviour of halogen oxides in the stratosphere is strongly influenced by the chemistry of temporary reservoir species containing chlorine and bromine. Temporary reservoirs are molecules which are formed and broken down at a time scale on the order of 1 day, contain halogens, and are not active in O_3 destruction. For example, chlorine nitrate is an important reservoir for ClO_x. It is formed by the reaction of ClO with NO_2 (R2.67), and active chlorine is released by photolysis; which occurs mainly by the reaction:

$$\mathrm{ClONO_2 + h\nu \rightarrow Cl + NO_3} \tag{R2.90}$$

or

$$\mathrm{ClONO_2 + h\nu \rightarrow ClO + NO_2} \tag{R2.91}$$

The other important temporary reservoir for chlorine is HOCl, formed by reaction of ClO with HO_2:

$$\mathrm{ClO + HO_2 \rightarrow HOCl + O_2} \tag{R2.92}$$

Chlorine is released from HOCl by photolysis. The major channel of this reaction is:

$$\mathrm{HOCl + h\nu \rightarrow HO + Cl} \tag{R2.93}$$

In an analogous way ClO may react with CH_3O_2, producing methylhypochlorite CH_3OCl, or other products (Helleis et al., 1993, 1994):

$$\mathrm{ClO + CH_3O_2 \rightarrow CH_3OCl + O_2} \tag{R2.94a}$$

$$\mathrm{\rightarrow Cl + products} \tag{R2.94b}$$

Cycles involving photolytic production of Cl from $ClONO_2$ and HOCl can lead to depletion of ozone in the lower stratosphere, and these may be significant under some circumstances [see for instance Toumi and Bekki, 1993]. Similarly, reaction R2.94a, followed by the photolysis of CH_3OCl or reactions of type R2.94b, could cause ozone loss (Crutzen et al., 1992).

In the polar stratosphere and in circumstances of elevated concentrations of volcanic aerosol, heterogeneous reactions of $ClONO_2$ and HOCl on polar stratospheric clouds (PSCs; particles containing nitric acid trihydrate and/or water ice) lead to gross perturbation of the distribution of ClO_x between inactive reservoirs and active ozone depleting forms. The reactions involved are:

$$ClONO_2 + H_2O(s) \xrightarrow{\text{surface}} HOCl + HNO_3(s) \quad \text{(R2.95)}$$

$$ClONO_2 + HCl(s) \rightarrow Cl_2 + HNO_3(s) \quad \text{(R2.96)}$$

$$HOCl + HCl(s) \rightarrow H_2O(s) + Cl_2 \quad \text{(R2.97)}$$

$$N_2O_5 + HCl(s) \rightarrow ClNO_2 + HNO_3(s) \quad \text{(R2.98)}$$

Here, (s) refers to the chemical species adsorbed or absorbed in the solid or liquid phase.

As bromine is less tightly bound than chlorine, only a small fraction of the bromine released from bromocarbons is sequestered in the form of HBr and $BrONO_2$, rendering this atom very effective for ozone loss (e.g. Lary, 1995). In particular, the combined Br–Cl catalytic cycles are very efficient in depleting ozone, and can therefore cause equal or even larger ozone destruction than chlorine alone in the lower stratosphere. Although there are significant human sources of bromine, the contemporary abundance of total stratospheric Br is only about 0.5% of that of Cl (e.g. Schauffler et al., 1998; Wamsley et al., 1998). The organic precursor species of inorganic bromine, $= 50\% CH_3Br$, halons, $CHBr_3$, etc. are still increasing. When the precursor species reach the lower stratosphere, they are photolysed and release inorganic bromine (Br_Y = BrO, HBr, HOBr, $BrONO_2$, BrONO, Br, and Br_2). Due to the lower binding energy of HBr compared with HCl, and to the more rapid photolysis of Br_2, BrCl, HOBr, $BrONO_2$ compared with their Cl analogues, BrO is the most abundant bromine species during the daytime and constitutes between 50% and 70% of total Br_Y. In mid-latitude summer with high stratospheric NO_2 levels, BrO is less abundant ($\approx$50% of the total Br_Y), while BrO can amount up to 70% of Br_Y at high latitudes in a denoxified stratosphere. This is because the Br_Y reservoir species $BrONO_2$ is less abundant in a stratosphere with low NO_X loading.

The possible participation of iodine chemistry in stratospheric processes is still under debate. Owing to the much shorter atmospheric lifetimes of iodine-containing compounds (e.g. CH_3I emitted from the oceans) compared with their Cl- and Br-containing analogues, it has generally been assumed that iodine does not reach the stratosphere in significant quantities, and the major focus of studies of iodine photochemistry has been the troposphere, as discussed in detail earlier. However, should iodine reach the stratosphere, its impact on ozone depletion 'molecule for molecule' is probably even greater than that of Br. The main reason for this is the inherent instability of potential iodine reservoir species (HI, HOI, or $IONO_2$) compared with their Cl- and Br-containing counterparts (Solomon et al., 1994). In addition, the potential coupling of the chemistry of IO with ClO and BrO may be far more efficient than the XO self-reaction cycles referred to above.

2.10.3 Stratospheric Ozone Hole

In 1985, Farman et al. (1985) observed a decrease of the total ozone column over Antarctica after polar sunrise. Satellite observations revealed that this ozone depletion occurred over the entire Antarctic continent. It soon became clear that, during Antarctic winter, special conditions prevail which lead to the release of reactive chlorine from its 'reservoir compounds' (HCl and $ClNO_3$) by heterogeneous reactions, i.e. chemical processes at the surface of particles in the stratosphere (see reactions R2.95–R2.98) (e.g. Solomon, 1999). This massive activation, bringing the fraction of reactive chlorine from around 1% to near 50%, leads to dramatic ozone losses in the lower stratosphere. A few examples are illustrated in Fig. 2.22, which shows normal ozone profiles recorded in August (Antarctic winter) and disturbed profiles in October (Antarctic spring). There is nearly complete ozone loss in the altitude range from about 14 to 24 km.

A number of factors contribute to the formation of the ozone hole: The primary cause is the increase in stratospheric chlorine and bromine levels due to the release of man-made CFCs, such as CF_3Cl and CF_2Cl_2. These compounds, which have no sinks in the troposphere, have been accumulating since their introduction in the 1930s. Because of their chemical stability, CFCs were widely used as propellants, in refrigerators, as solvents, and in

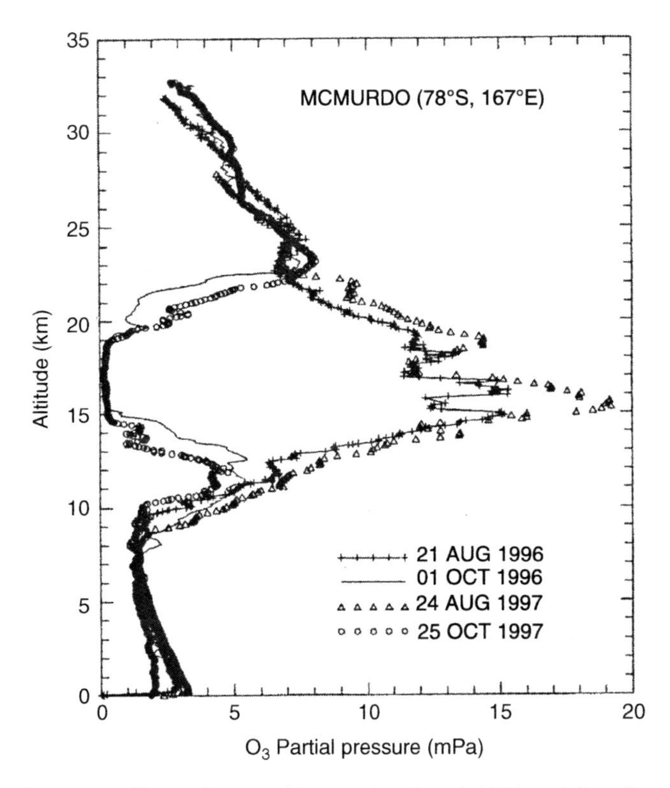

Fig. 2.22. Ozone profiles in August (Antarctic winter) 1996 and October (Antarctic spring) of 1997 (from Nardi et al., 1999, Copyright by American Geophysical Union (AGU), reproduced by permission of AGU)

other industrial applications. Most man-made halocarbons transported to the stratosphere contain more than one halogen atom, and the release of halogen atoms occurs in stages, with intermediate formation of the carbonyl halides (e.g. CFClO). For all halogens, the sink is provided by transport of the most stable (and therefore most abundant) forms (e.g. HX) to the troposphere.

A polar vortex forms in the winter polar stratosphere, which inhibits gas-exchange with lower latitudes. As a consequence, colder polar air cannot mix with warmer lower latitude air, leading to very cold stratospheric temperatures. In addition, ozone from lower latitudes cannot replenish ozone destroyed by halogen catalysed cycles. The polar vortex is very stable over Antarctica, and can exist from June to November. The northern polar vortex is less stable and is typically not stable for more than a few weeks. The more unstable northern vortex is the main reason why the ozone depletion over the north pole has not reached the same extent as that over the south pole.

Very cold stratospheric temperatures develop over Antarctica in winter due to radiative cooling and lack of solar heating. In addition, the lack of air exchange with mid-latitudes also contributes to the low temperature. Once these temperatures fall below a certain value ($\sim$195K), PSCs form. The clouds are made out of nitric acid tri-hydrate and water. PSCs provide a surface on which the chlorine reservoir species, such as HCl and $ClONO_2$ can be converted to Cl_2 (R2.95 and R2.96). In addition, PSC formation removes nitrogen oxides from the gas phase, further increasing the amount of reactive halogens. These heterogeneous processes proceed for several months, as long as PSCs are present.

The rapid photolysis of Cl_2 releases reactive chlorine, in the form of Cl atoms, at polar sunrise. Cl reacts quickly with O_3 forming ClO, which then participates in the catalytic ozone destruction cycles described earlier. ClO is the most important halogen oxide in the stratosphere: its concentration ranges from 10–100 ppt in the undisturbed lower stratosphere to 1–2 ppb under conditions of disturbed stratospheric chemistry, i.e. the ozone hole, where more than 50% of the available chlorine can be present as ClO. Under the conditions of very high ClO levels and low temperatures, the ClO self-reaction to form a dimer, Cl_2O_2 [reaction R2.83 above (Cox and Derwent, 1979; Molina and Molina, 1987)] becomes the dominant catalytic destruction cycle of ozone, leading ultimately to the formation of the ozone hole. It should be noted that cycles involving BrO also contribute to the ozone destruction, although to a lesser extend, since the partitioning between active and reservoir species favours the active forms already in the undisturbed stratosphere.

The break-up of the polar vortex in December marks the disappearance of the ozone hole, as warmer and ozone-rich air masses are transported over Antarctica.

The formation of the ozone hole is a recurring phenomenon, which starts every Antarctic winter.

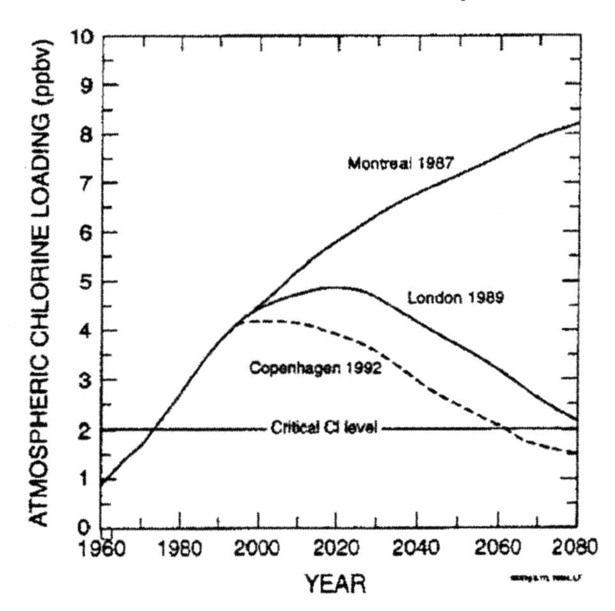

Fig. 2.23. Time series of atmospheric chlorine loading from 1960 – today, with projections to 2080. The expected effect of the treaties for limitation of CFMs is shown (from Brasseur and Solomon 1986)

2.10.4 Recovery of Stratospheric Ozone

Recently, there have been signs of recovery of stratospheric ozone, which is largely due to the successful measures employed to reduce the release of halogen transport species (e.g. CFCs).

Figure 2.23 shows time series of the atmospheric chlorine loading, the total amount of chlorine present as transport, reservoir, and active species from 1960 to 1995 (measured data), and projections to 2080.

The chlorine loading of the troposphere (Montzka et al., 1996) and stratosphere (Rinsland et al., 2003) peaked in 2004, and has begun to decline. The first indication of the expected recovery of stratospheric ozone may already be visible (Newchurch et al., 2003; Bodeker et al., 2005).

3

Interaction of Molecules with Radiation

Atoms and molecules can exist in many states that are different with respect to the electron configuration, angular momentum, parity, and energy. Transitions between these states are frequently associated with absorption or emission of electromagnetic radiation. In fact, historically, most of the insight into these different states of molecules and atoms has been gained by studying the interaction of atoms and molecules with radiation. However, interaction of radiation with matter can also be used to determine the presence and abundance of molecules or atoms in a sample.

3.1 Electromagnetic Radiation and Matter

We know electromagnetic radiation in many forms, which are solely distinguished by their wavelength or frequency (see Fig. 3.1). The wavelength λ is connected to the frequency ν by:

$$c = \nu \cdot \lambda ,$$

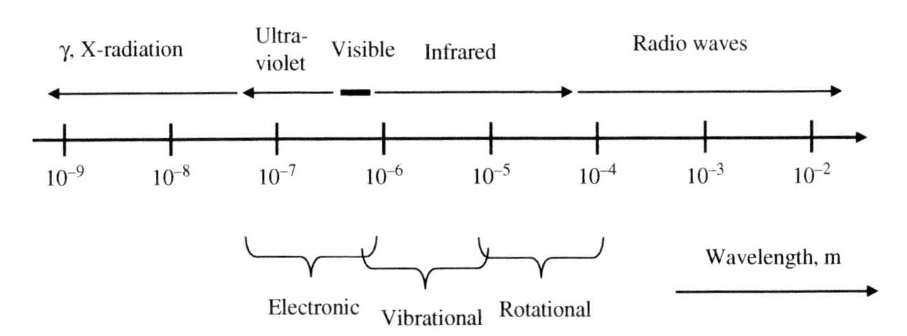

Fig. 3.1. The electromagnetic spectrum – radio waves to X-rays. The types of transition in molecules or atoms (electronic, vibrational, rotational) induced by radiation of different wavelength ranges are indicated

U. Platt and J. Stutz, *Interaction of Molecules with Radiation.* In: U. Platt and J. Stutz, Differential Optical Absorption Spectroscopy, Physics of Earth and Space Environments, pp. 77–90 (2008)
DOI 10.1007/978-3-540-75776-4_3

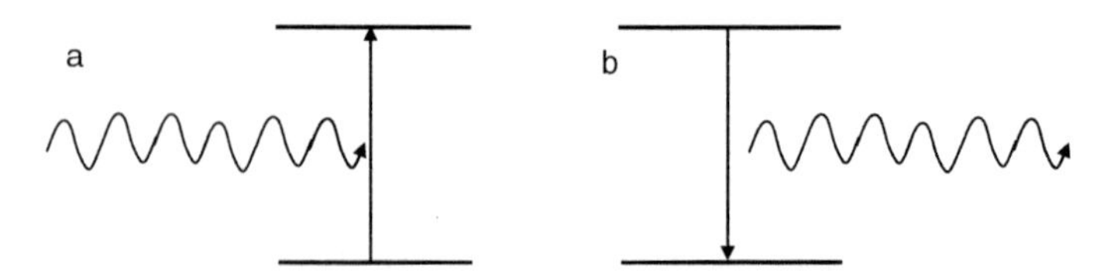

Fig. 3.2. Interaction of radiation with matter: Upon absorption of a photon with appropriate energy, the atom or molecule enters an excited state (**a**). Return to the ground state can occur by collisions with other molecules or atoms, eventually converting the photon energy into heat, or by re-radiating the photon (**b**)

where $c = 2.99792458 \times 10^8$ m/s $\approx 3 \times 10^8$ m/s denotes the speed of light.

Radiation of very short wavelength ($\lambda < 30$ nm) is called γ-radiation or (at somewhat longer wavelengths) X-radiation (X-rays). At longer wavelengths, we speak of the ultraviolet and infrared ranges framing the visible spectral range extending from about 400 to 780 nm. Radiation of even longer wavelength is known as sub-mm wave ($\lambda < 1$ mm), microwave, and radio wave radiation.

It is a well-known result of quantum mechanics that some aspects of electromagnetic radiation have to be described as behaviour of waves, while others can only be understood when describing radiation as a shower of particles, the photons. Depending on the energy E of the photons, which is given by $E = h\nu = hc/\lambda$, with $h \approx 6.626176 \times 10^{-34}$ Js denoting Planck's constant, there are different forms of interaction of radiation with matter: X-rays interact largely by ionising atoms, while UV- and visible radiation will lead to reconfiguration of the outer electron shell of atoms or molecules. This process is frequently called electronic excitation. Radiation with less energy will excite vibrational or rotational states in molecular gases, as indicated in Figs. 3.1 and 3.2 and detailed in Sect. 3.3.1. Excited atoms or molecules can return to the ground state as a result of collisions with other molecules, eventually converting the energy of the absorbed photon into heat, or by re-radiating the photon (see Fig. 3.2b).

3.2 Energy Levels and Transitions in Atoms

Atoms and molecules are quantum mechanical multi-particle systems; their individual states are discrete and described by a set of quantum numbers. The states relevant to chemistry and optical spectroscopy are determined by the configuration of the outer (loosely bound) electrons.

In atoms, the energy levels E_n are determined by the quantum numbers n, l, m, s, m_s of a single outer electron, where l and m enumerate components of the orbital angular momentum of the electron, while s and m_s characterise its intrinsic angular momentum or spin. For small atoms, E is in good approximation given by the "Rydberg formula":

$$E_n = -\frac{R_{\mathrm{Ry}}}{n^2}, \tag{3.1}$$

where R_{Ry} denotes the Rydberg constant for the particular atom. Note that by convention binding energies are negative; they can be thought of the energy that was released when free electrons were attached to the nucleus (or nuclei in the case of a molecule). The energy of a photon exchanged (i.e. absorbed or emitted) by an atom (or molecule) then corresponds to the difference of the energy of two particular states. Using the famous Planck relationship between energy ΔE_{el} and frequency ν (and thus wavelength) of a photon we obtain:

$$\Delta E_{\mathrm{el}} = E_{n_1} - E_{n_2} = R_{\mathrm{Ry}} \left(\frac{1}{n_1^2} - \frac{1}{n_2^2} \right) = h\nu_{1,2} \,, \tag{3.2}$$

where n_1, n_2 denote the lower, upper state of the atom and $\nu_{1,2}$ the frequency of the photon, respectively. The quantity ΔE_{el} signifies the fact that we deal with an "electronic" transition, i.e. one where the configuration of the electron shell changes.

Frequently, the state of an atom (or molecule) is determined by several outer electrons. The relevant quantum numbers are the sum of the individual orbital angular momenta $L = \Sigma l_{\mathrm{i}}$, and the sum of the electron spins $S = \Sigma s_{\mathrm{i}}$. Also, frequently the relevant quantum number is one (j, or J) that characterises the sum of orbital angular momentum and electron spin. States with the same main quantum number n and different quantum numbers L or S (or J) have slightly different energies, the so called "fine structure" of the transitions.

3.3 Energy Levels and Transitions in Molecules

Besides electronic states of different energy, molecules have two excitation schemes not found in atoms: first, the rotation of the entire molecule and thus angular momentum of the entire molecule. The rotation of the molecule should not to be confused with the orbital angular momentum due to the electrons (which can be additionally present in molecules just as in atoms and spin of the constituent elementary particles of the molecules). Second, the vibration of the atoms within the molecule relative to each other.

The typical transition energies are:

- Electronic transitions on the order of 1 eV: corresponding wavelengths for transitions between an electronically excited state and the ground state correspond to the visible- or near UV spectral ranges.

- Vibrational transitions on the order of 0.1 eV: corresponding wavelengths for transitions between a vibrationally excited state and the ground state are in the infra red (IR) spectral range.
- Rotational transitions on the order of 10^{-3}–10^{-2} eV: corresponding wavelengths for transitions between rotationally excited state and the ground state are in the sub-mm or microwave range.

The electronic states of molecules are designated by the following (traditional) designations:

$$[\text{State}]^{2S+1}[\text{AM}]_{\Lambda+\Sigma} ,$$

where [State] is a code for the particular electronic state of the molecule, with [State] $= X$ denoting the lowest energy (ground) state;

- S denotes total electron spin just as in the case of atoms;
- [AM] is the code for the angular momentum. This is either the projection of the total orbital momentum on the molecule's axis or a combination of the total orbital momentum with the rotation of the molecule, depending on the coupling case. [AM] $= 1, 2, 3, \ldots$ is coded as Σ, Π, Δ, ...
- Finally, $\Lambda + \Sigma = |M_L| + M_S$ denotes the projection of the total angular momentum of the electrons on the axis connecting the nuclei of the molecule.

For instance, the ground state of the NO molecule with the angular momentum 2 and $S = 1/2$ is designated as:

$$X^2\Pi .$$

3.3.1 Rotational Energy Levels and Transitions

The angular momenta (total orbital angular momentum, total electron spin, "nuclear" rotation) in a molecule can couple in different ways, which are denoted as "Hund's Coupling Cases" to form a total angular momentum $\vec{J}$ and associated quantum number J.

Rotational energy levels in a molecule are given by:

$$E_j = B \cdot J(J+1) , \tag{3.3}$$

where $B = \frac{\hbar^2}{2\Theta}$ denotes the rotational constant of the particular molecule and rotation mode (axis) with the moment of inertia Θ with respect to this axis. In first approximation, Θ is assumed to be independent of J (rigid rotor model). However, molecules are not rigid, i.e. the atoms within a molecule can move compared to each other (see Sect. 3.3.2), therefore Θ will increase somewhat at higher values of J compared to its value at low rotational levels. As a consequence $B \propto 1/\Theta$ will decrease.

Selection rule: The difference in the angular momentum quantum number of initial and final states is given as $\Delta J = \pm 1$, since the photon exchanged with the atom/molecule has a spin (intrinsic angular momentum) of unity. The transitions are denoted as "P-branch" ($\Delta J = -1$), "Q-branch" ($\Delta J = 0$, which only can occur if electronic transitions take place at the same time), and "R-branch" ($\Delta J = +1$). Consequently, the photon energy $h\nu = \Delta E$ of allowed transitions is given by the energy difference of two consecutive states:

$$\Delta E_j = E_{j+1} - E_j = B \cdot [(J+1) \cdot (J+2) - J \cdot (J+1)] = 2B(J+1) \propto J \ . \quad (3.4)$$

In other words, the energy of absorption or emission lines varies proportionally to the rotational quantum number of the ground state; the spectra consist of lines with equidistant spacing (the energy difference of two neighbouring lines being just $2B$). The energy differences between consecutive states are on the order of $10^{-3} - 10^{-2}$ eV, the corresponding wavelengths of photons exchanged in such "purely rotational" transitions in the sub-mm- or microwave ranges. These energies are on the order of the thermal kinetic energy of molecules at room temperature, thus most molecules are rotationally excited under ambient conditions (see below).

3.3.2 Vibrational Energy Levels

Vibration of molecules can – in first approximation – be treated as harmonic oscillation, where the vibrational energy levels are given by:

$$E_\nu = \left(\nu + \frac{1}{2}\right) \cdot \hbar\omega_0 \ , \quad (3.5)$$

where $\nu = 0, 1, 2, \ldots$ denotes the vibrational quantum number (vibration level) and $1/2\hbar\omega_0$ is the zero-point energy of the molecular oscillator. Thus energies of different vibrational states vary in proportion to the vibrational quantum number ν. In addition to vibrational excitation, a molecule can also be rotationally excited (in fact is very likely to be at ambient temperatures). Thus each vibrational transition splits into a series of rotational lines (see Fig. 3.3). Energies of vibrational transitions are on the order of 0.1 eV, with the corresponding wavelengths in the IR spectral range.

3.3.3 Electronic Energy Levels

Analogous to atoms, molecules can exist in several, distinct configurations of the electronic shell; each configuration has its own set of vibrational and rotational levels, as sketched in Fig. 3.3. The situation is further complicated by the fact that the moment of inertia Θ, and thus the rotational constant B, can be different in the ground and excited electronic states (usually denoted as B'' and B', respectively). This is largely due to different inter-atomic distances in the molecule in its different electronic states. As a consequence, the rotational line spacing is no longer equidistant. Rather, the lines tend to group towards a certain wavelength either below ("blue shaded") or above ("red shaded") the wavelength given by electronic and vibrational states alone. In detail, the line positions are given by the "Fortrat Parabolas" determined by the different rotational constants (B'' and B').

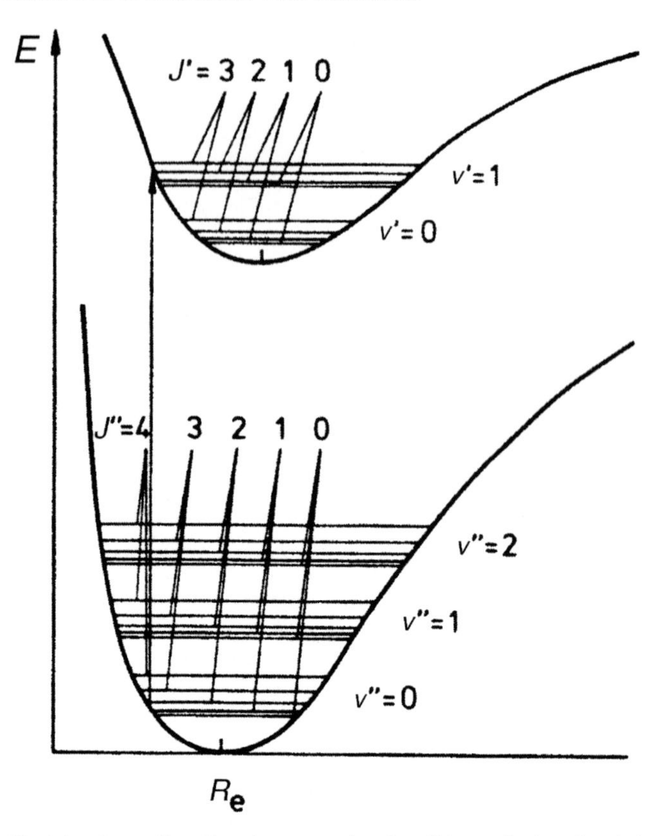

Fig. 3.3. Sketch of ro-vibrational energy levels of two electronic states of a (diatomic) molecule. Electronic energies are given as a function of the distance between the nuclei of the atoms within the molecule. They are minimal at a certain distance R_e, which is usually different in the different electronic states. The equidistant, horizontal groups of lines denote the lowest rotational levels of the vibrational states (ground state: $\nu'' = 0, 1, 2$, excited state: $\nu' = 0, 1$). Note that the energy separation of the rotational states is exaggerated in comparison to the vibrational energies

3.4 Population of States

The relative population $n(E)$ of excited states of an atom or molecule with the energies E_1, E_2 above the ground state (with $E = E_1$) is given by the **Boltzmann distribution**:

$$\frac{n(E_2)}{n(E_1)} = \frac{G_2}{G_1} \cdot \mathrm{e}^{-\frac{E_2 - E_1}{kT}} . \tag{3.6}$$

Here T denotes the absolute temperature and k the Boltzmann constant ($\mathrm{k} = 1.38 \times 10^{-23}$ J/K), while G_1, G_2 are the statistical weights or degeneration factors (i.e. the number of different states with the same energy E) of the respective states.

In the case of an atom, $G_n = 2n^2$, while vibrational excitation in molecules is not degenerated (i.e. $G_v = 1$). Rotational states in molecules are degenerated by $G_J = 2J + 1$, thus the population of rotational states is given by:

$$\frac{n(E_2)}{n(E_1)} = \frac{2J_2+1}{2J_1+1} \cdot \mathrm{e}^{-\frac{E_2-E_1}{kT}} = \frac{2J_2+1}{2J_1+1} \cdot \mathrm{e}^{-\frac{B(J_2(J_2+1)-J_1(J_1+1))}{kT}} . \tag{3.7}$$

With $E(J) = B \cdot J(J+1)$, where B is the rotation constant of the particular molecule and rotation axis (see above).

The kinetic energy of a molecule of around 0.03 eV at room temperature is comparable to or larger than the lowest rotational energy level, but smaller than the lowest vibrational energy level of most molecules. Therefore, several rotational states are usually populated, but only a small fraction of the molecules is vibrationally excited.

3.5 Molecular Spectra

As outlined above, the UV–visible spectra of molecules exhibit vibrational and rotational structures. Frequently, the rotational structure consists of so many lines, the spacing of which is smaller than their line-width (or which cannot be separated by the spectrometer used), thus leading to a quasi-continuous absorption within a vibrational "band". Figure 3.4 shows the cross-section spectrum of the IO molecule as an example. A striking feature of the spectrum

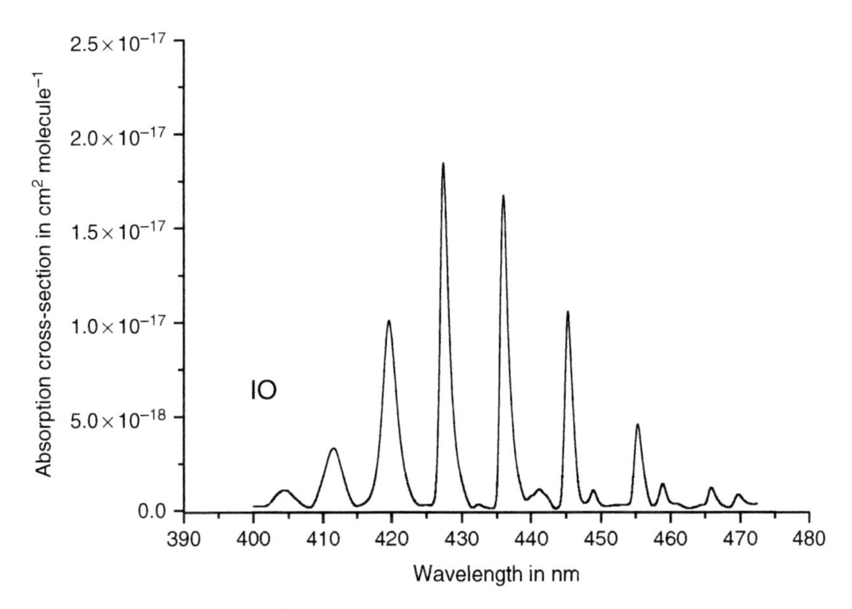

Fig. 3.4. Example of the (ro-) vibrational structure of the electronic spectrum of the iodine oxide (IO) molecule

is the quite different strength of its vibrational bands. This can be explained by the Franck–Condon principle stating that the electronic transition from the ground state to the excited state upon absorption of a photon by the molecule is much faster than the time needed for one vibration. Since the distance between the nuclei is usually different in the ground state and electronically excited state, the molecule will most likely find itself in a certain, excited vibrational state. Transitions to higher and lower vibrational states are less likely, thus having a smaller absorption cross-section, as shown in Fig. 3.4.

3.6 Broadening Mechanisms and Line Width of Absorption Lines

We first consider an excited atom or molecule with the energy E_0 above the ground state (with $E = 0$). According to classical electrodynamics, the energy of the excited electrical dipole will decrease exponentially with time t due to emission of electromagnetic radiation:

$$E(t) = E_0 \mathrm{e}^{-\delta t} , \tag{3.8}$$

where δ denotes some damping constant, and $\tau = 1/\delta$ is the time in which the energy is decayed to 1/e (with e = 2.7182.. denoting the base of the natural logarithm) of its initial value. Accordingly, the corresponding amplitude of the electric field $\widehat{E}(t)$ of the emitted radiation will exponentially decrease with time t:

$$\widehat{E}(t) = \widehat{E}_0 \mathrm{e}^{-\delta t} \cdot \cos(\omega_0 t + \varphi) \ . \tag{3.9}$$

Here ω_0 denotes the frequency of the oscillation, which is related to the energy E_0 of the excited state, $\omega_0 = 2\pi E_0/h$ (with h as Planck's constant), while φ is the phase of the oscillation. Since the amplitude $E(t)$ is damped, the radiation cannot be truly monochromatic. This is quantified by calculating the Fourier analysis of $E(t)$:

$$E(t) = \frac{1}{\sqrt{2\pi}} \int_{-\infty}^{\infty} a(\omega) \mathrm{e}^{i\omega t} d\omega \ , \tag{3.10}$$

with the spectral function $a(\omega)$:

$$a(\omega) = \frac{E_0}{2} \left[\frac{\mathrm{e}^{ij}}{i(\omega - \omega_0) - \frac{\delta}{2}} - \frac{\mathrm{e}^{-ij}}{i(\omega + \omega_0) + \frac{\delta}{2}} \right] . \tag{3.11}$$

Since $\delta \ll \omega_0$ at frequencies ω near the resonance frequency ω_0, we can neglect the second term in the square bracket of (3.11). We obtain the radiation intensity $I(\omega)$ as the square of $a(\omega)$ (and normalising the integral over I to unity):

$$I(\omega) d\omega = I_0 \frac{\frac{\delta^2}{4}}{(\omega - \omega_0)^2 - \frac{\delta^2}{4}} d\omega \ . \tag{3.12}$$

This equation is known as the Lorentz distribution; its maximum is at $\omega = \omega_0$ while $I(\omega) = 0.5I(\omega_0)$ at $\omega = \omega_0 \pm \delta/2$, i.e. δ denotes the full width at half maximum (FWHM) of the distribution.

3.6.1 The Natural Line Width

The above described classical image of an oscillating dipole cannot directly be applied to an excited atom or molecule; however, the correspondence principle suggests that, at least at large quantum numbers, (3.12) should be a good approximation. In practice it turns out that (3.12), can be used to describe the profile of an emission line of an isolated atom when using an empirical δ. In addition, there might be more than one level which an excited atom or molecule can decay to, thus $\delta_{\mathrm{L}} = \Sigma\delta_{\mathrm{i}}$ has to be used instead of δ. The individual δ_{i} describe the transition frequency to the states (including the ground state) to which the excited state under consideration can decay into. Thus, we obtain the description of the **natural line shape** of an isolated atom or molecule, i.e. the intensity distribution centred at ω_0:

$$I_{\mathrm{L}}(\omega)d\omega = I_0 \cdot \frac{\frac{\delta_{\mathrm{L}}^2}{4}}{(\omega - \omega_0)^2 - \frac{\delta_{\mathrm{L}}^2}{4}} d\omega \,, \tag{3.13}$$

with the **natural line width** δ_L (FWHM), which essentially depends on the lifetime of the excited state of the molecule, but *not* directly on the energy difference of the transition and thus the frequency of the emitted (absorbed) radiation. For allowed transitions, the natural lifetime is on the order of 10^{-8} s, corresponding to $\delta_{\mathrm{L}} = 10^8$ Hz. For radiation of $\lambda = 400$ nm wavelength ($\omega = 2\pi c/\lambda \approx 4.7 \times 10^{15}$ Hz), this would amount to $\Delta\omega/\omega \approx 2.1 \times 10^{-8}$ or $\Delta\lambda \approx 8.5 \times 10^{-6}$ nm (≈ 0.01 pm).

3.6.2 Pressure Broadening (Collisional Broadening)

Collisions of molecules (among themselves or with other types of molecules in the gas) will reduce the lifetime of the excited state below the value given by the radiation transition alone, see (3.13), which determines the natural line width. The shape of a pressure broadened line is therefore Lorentzian as given in (3.12), with $\delta = \delta_{\mathrm{P}}$ as the width of the pressure broadened line. In principle, the collisional damping constant δ_{p} is given by the product of the gas kinetic collision frequency, z_{AB}, and the deactivation probability, p_{AB}, per collision (here A refers to the species under consideration, B to the gas providing the pressure). Since p_{AB} can be different for each molecular species, the amount of pressure broadening depends not only on the molecular species under consideration but also on the surrounding gas. Of particular interest are the self broadening (the species itself is the pressure gas, $p_{\mathrm{AB}} = p_{\mathrm{AA}}$) and air broadening (species A is occurring in traces in air, which is species B). In the latter case, the deactivation is almost exclusively occurring in collisions between air molecules and species A molecules. The collision frequency is directly proportional to the product of gas density (and thus to pressure p times T^{-1}) and to the average molecular speed (and thus to $T^{1/2}$). In summary, the pressure-broadened line width δ_{P} is therefore given by:

$$\delta_{\mathrm{P}}(p,T) = \delta_{\mathrm{P}}(p_0,T_0) \cdot \frac{p}{p_0} \cdot \sqrt{\frac{T_0}{T}} = \delta_0 \cdot \frac{p}{p_0} \cdot \sqrt{\frac{T_0}{T}} \,. \tag{3.14}$$

Here $\delta_{\mathrm{P}}(p_0,T_0) = \delta_0$ denotes the pressure broadening at some reference pressure and temperature. Typical values at one atmosphere for pressure broadening of small molecules in the near UV are $\Delta\lambda \approx 1$ pm.

3.6.3 Doppler Broadening

In reality, the atoms or molecules of a gas are not at rest but move about. This **Brownian motion** has two main consequences:

1. The energy E can be not only dissipated by radiation, but also by collisions with other molecules present in the gas. This effect manifests itself in an increased $\delta_{\mathrm{P}} > \delta_{\mathrm{L}}$, as discussed above. Since the effect can be described in terms of a damped oscillation, the line shape is also given by (3.13).
2. The Doppler effect will change the frequency ω_0 of the emitted radiation according to (in first approximation):

$$\omega = \omega_0 \cdot \left(1 + \frac{v_{\mathrm{x}}}{c}\right) \,, \tag{3.15}$$

where v_{x} denotes the velocity component of the emitting molecule with respect to the observer and c the speed of light. The distribution of an individual component of the velocity $N(v_{\mathrm{x}})dv_{\mathrm{x}}$ (denoting the number N of molecules that have the x-component of the velocity in the range of $v_{\mathrm{x}} \ldots v_{\mathrm{x}} + dv_{\mathrm{x}}$) is Gaussian (this should not be confused with the distribution of the absolute value of the velocity, which is given by the Maxwell–Boltzmann distribution).

$$N\left(v_{\mathrm{x}}\right) dv_{\mathrm{x}} = \mathrm{const} \cdot \mathrm{e}^{-\frac{mv_{\mathrm{x}}^2}{2RT}} dv_{\mathrm{x}} \,, \tag{3.16}$$

where m denotes the atomic or molecular weight (kg/mol) and R the universal gas constant. The intensity distribution centred at ω_0 follows as:

$$I_{\mathrm{D}}(\omega)d\omega = I_0 \cdot \mathrm{e}^{-\frac{mc^2}{2RT} \cdot \frac{(\omega_0-\omega)^2}{\omega_0^2}} \cdot d\omega = I_0 \cdot \mathrm{e}^{-\frac{\delta_{\mathrm{D}}}{2} \cdot (\omega_0-\omega)^2} \cdot d\omega \,, \tag{3.17}$$

with the FWHM of a purely Doppler broadened line:

$$\delta_{\mathrm{D}} = \omega \frac{2\sqrt{2R\ln 2}}{c} \cdot \sqrt{\frac{T}{m}} = \omega \cdot K_{\mathrm{D}} \cdot \sqrt{\frac{T}{m}} \,. \tag{3.18}$$

The constant $K_{\mathrm{D}} = 2\sqrt{2R \ln 2}/c$ in the above equation has the value of 2.26×10^{-8} $(\mathrm{kg/mol})^{1/2}/\mathrm{K}^{1/2}$. For radiation with $\lambda = 400$ nm, at $T = 300$ K and a molecular species with the molecular weight of air (0.03 kg/mol), we obtain $\delta_{\mathrm{D}}/\omega_0 = \Delta\omega/\omega = 2.26 \times 10^{-8}$, or a Doppler width of $\Delta\lambda = 9 \times 10^{-4}$ nm (about 1 pm).

3.6.4 Realistic Broadening in the UV- and Visible Spectral Ranges

Of the three broadening mechanisms discussed earlier, the natural line width, i.e. natural lifetime of the excited state, leads to very small broadening (typ. 0.01 pm), while the effect of pressure broadening and Doppler broadening are roughly equal (typ. 1 pm) in the UV–visible spectral region. Note that the two types of line shapes discussed above (Lorentzian and Gaussian) have quite different properties:

(1) The intensity (or absorption cross-section) of a Lorentzian line [e.g. (3.13)] only decays with the square of the deviation from the centre frequency $1/(\omega - \omega_0)^2$, and thus a large fraction of the total emission (or total absorption) is in the "wings" of the line. In comparison, the Gaussian profile [e.g. (3.17)] decays exponentially (proportional to $\exp\left[-(\omega-\omega_0)^2\right]$), and consequently there is very little emission (or absorption) in the wings of the line.

(2) In the case of Lorentzian (i.e. natural- or pressure) broadening, the half-width (3.13) is independent of the frequency ω (or wavelength λ), while Gaussian (i.e. Doppler) broadening (3.18) is proportional to ω (or λ). Therefore, Gaussian broadening is, under atmospheric conditions, usually negligible at low frequencies, i.e. in the far IR or microwave range, becomes noticeable in the near IR and dominates in the short-wavelength UV.

As can be seen from the examples above the natural line width is usually negligible compared to Doppler and/or pressure broadening. In particular, in the visible and UV spectral range Doppler (Gaussian) broadening is comparable to pressure (Lorentzian) broadening. The resulting line shape can be obtained by convoluting the Gaussian and Lorentzian line shapes, which is known as Voigt shape (neglecting the natural line width):

$$I_{\mathrm{V}}(\omega, \delta_{\mathrm{D}}, \delta_{\mathrm{P}}) = \int_{-\infty}^{\infty} I_{\mathrm{D}}(\omega', \delta_{\mathrm{D}}) \cdot I_{\mathrm{P}}(\omega - \omega', \delta_{\mathrm{P}})\, d\omega' \,. \tag{3.19}$$

Since the convolution can only be performed by numerical calculation, approximations of the Voigt shape have been developed for describing line shapes where both broadening mechanisms contribute. Figure 3.5 shows a comparison of the three line shapes.

Since absorption is the inverse process to emission, the formulas derived above also describe the line shapes of atomic- or molecular absorption lines, and thus the absorption cross-section $\sigma(\omega)$ [or $\sigma(\lambda)$] will show the same wavelength dependence as $I(\omega)$ in (3.19).

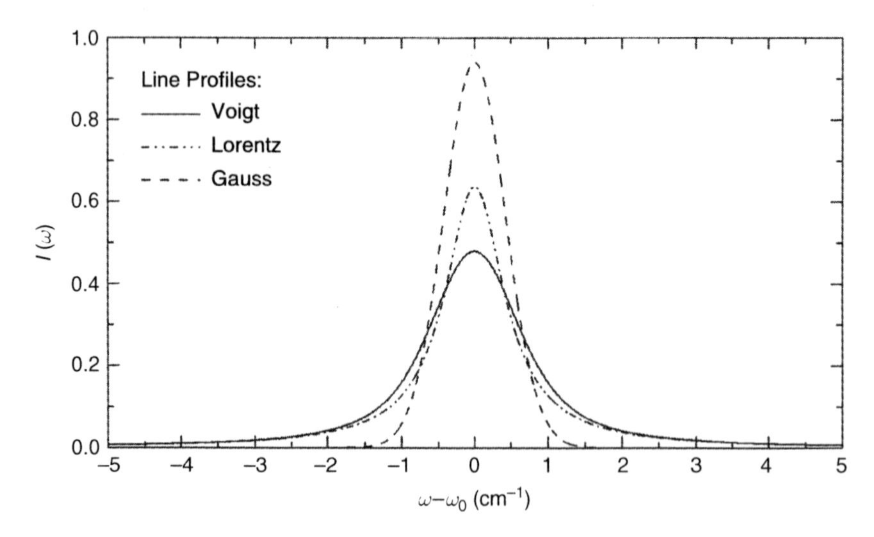

Fig. 3.5. Different line shapes: Lorentz (due to collision broadening), Doppler (due to thermal motion of molecules), and Voigt shape resulting from simultaneous Doppler and collision broadening

3.7 Spectroscopic Techniques for Chemical Analysis

The interaction of radiation with matter provides a powerful tool for a wide variety of investigations, for instance about the structure of atoms and molecules and the chemical composition of complex mixtures like the atmosphere. In principle, we distinguish between two different experimental approaches of spectroscopy:

3.7.1 The Fluorescence Techniques

The emission of radiation by excited atoms or molecules is called fluorescence. If bound states are involved in the excited state and ground state, the radiation has a characteristic set of wavelengths with the photon energy corresponding to the energy difference of the participating states (see Sects. 3.2 and 3.3).

The energy to change the molecule from the ground state to its excited state (i.e. the energy then radiated by the molecule) can be supplied by various mechanisms including thermal excitation (at ambient temperatures in the mid-infrared), electron bombardment, chemical reactions, or absorption of radiation. The latter mechanism, (light)-induced fluorescence, is employed in several measurement instruments for atmospheric trace gases (e.g. OH). A general outline of the principle is given in Fig. 3.6. In the simplest case (Fig. 3.6a), the number of fluorescence photons emitted by the molecules (or atoms) in a sample volume illuminated by the excitation radiation is proportional to the number of exciting photons multiplied by the number of trace

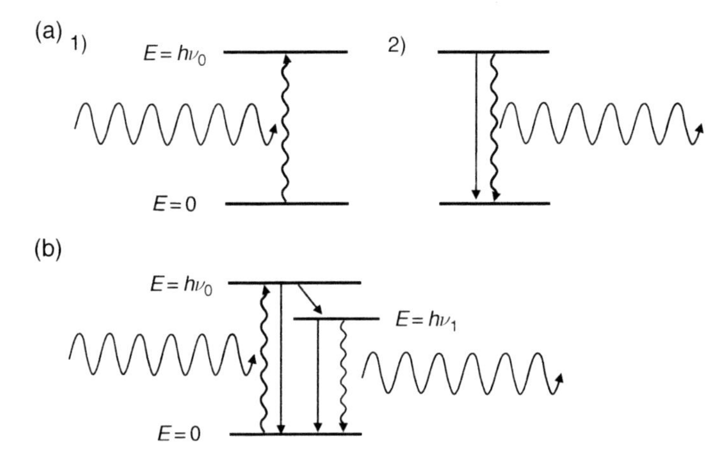

Fig. 3.6. The principle of trace gas detection by induced fluorescence spectroscopy. Scheme (**a**) uses two energy levels within the molecule, the energy difference of which is just $\Delta E = h\nu_0$. After excitation (1) by a photon with energy $E = h\nu_0$, the excited state can lose its energy (be quenched, 2) either by collision with other molecules (*straight line*) or radiating another photon with the energy $E = h\nu_0$ (*undulated line*). Scheme (**b**) uses three energy levels within the molecule. The energy difference between the lowest and the highest level is $\Delta E = h\nu_0$. After excitation by a photon with energy $E = h\nu_0$, the excited state can lose its energy completely or partially by collision with other molecules. The observed photon has a lower energy (longer wavelength) $E = h\nu_1$

gas molecules with matching transitions. In practice, geometrical factors, the absorption cross-section, and quenching also must be taken into account.

3.7.2 Absorption Spectroscopy

This spectroscopic technique makes use of the absorption of electromagnetic radiation by matter (Fig. 3.7). Quantitatively, the absorption of radiation is expressed by Lambert–Beer's law (or Bouguer–Lambert law, see Sect. 6.1):

$$I(\lambda) = I_0(\lambda) \exp[-L\sigma(\lambda)c] \,, \tag{3.20}$$

where $I_0(\lambda)$ denotes the initial intensity emitted by some suitable source of radiation, while $I(\lambda)$ is the radiation intensity after passing through a layer of thickness L, where the species to be measured is present at the concentration (number density) c. The quantity $\sigma(\lambda)$ denotes the absorption cross-section at the wavelength λ. The absorption cross-section is a characteristic property of any species. The absorption cross-section $\sigma(\lambda)$ can be measured in the laboratory, while the determination of the light-path length L is in many cases trivial. Once those quantities are known, the trace gas concentration c can be calculated from the measured ratio $I_0(\lambda)/I(\lambda)$:

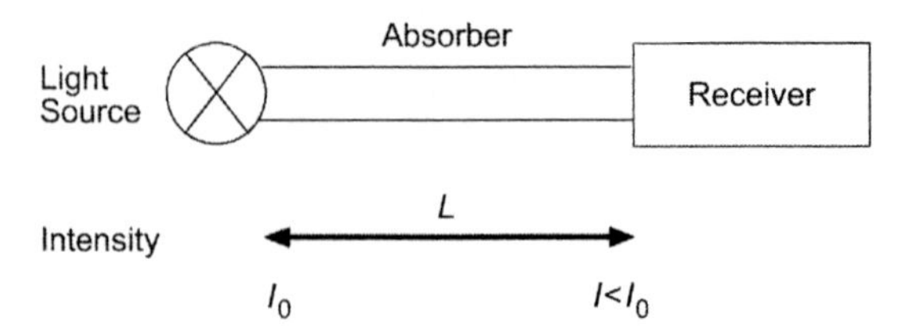

Fig. 3.7. The basic principle of absorption spectroscopic trace gas detection

$$c = \frac{\log\left(\frac{I_0(\lambda)}{I(\lambda)}\right)}{\sigma(\lambda) \cdot L} = \frac{D}{\sigma(\lambda) \cdot L} \,. \tag{3.21}$$

The expression

$$D = \log\left(\frac{I_0(\lambda)}{I(\lambda)}\right) \tag{3.22}$$

is called the **optical density** of a layer of a given species. (Note that in the literature, the decadic as well as the natural logarithm are used in the definition of the optical density. In this book we always use the natural logarithm.)

4

Radiation Transport in the Atmosphere

Absorption based measurements of trace gases in the open atmosphere have considerable advantages, such as avoidance of wall losses, and potential for remote sensing. In order to understand the implication of light-path arrangements, an in-depth knowledge of the transmission of radiation in the atmosphere is necessary. The basic issues are extinction of radiation in the atmosphere and possible generation of scattered light. This chapter describes the basic quantities pertinent to radiation transport and the fundamental laws governing radiation transport in absorbing and scattering media, i.e. the atmosphere.

4.1 Basic Quantities Related to Radiation Transport

A light source will emit a certain amount of energy W in the form of radiation.

1. The radiant flux Φ is defined as radiation energy per unit time (regardless of the direction of its emission):

$$\Phi = \frac{\text{Radiated energy}}{\text{Time interval}} = \frac{dE}{dt} \quad \frac{\text{Ws}}{\text{s}} = \text{W} . \tag{4.1}$$

   The radiant flux is measured in joules/second or Ws/s = watt.

2. The irradiance B is defined as the radiant flux Φ received by an 'illuminated' area A_{e}:

$$B = \frac{\Phi}{A_e} \quad \frac{\text{W}}{\text{m}^2} . \tag{4.2}$$

3. The radiation intensity (Ω = solid angle):

$$F = \frac{\Phi}{\Omega} \quad \frac{\text{W}}{\text{sr}} . \tag{4.1}$$

U. Platt and J. Stutz, *Radiation Transport in the Atmosphere*. In: U. Platt and J. Stutz, Differential Optical Absorption Spectroscopy, Physics of Earth and Space Environments, pp. 91–112 (2008)
DOI 10.1007/978-3-540-75776-4_4

4. The radiance (A_s = radiating area):

$$I = \frac{\Phi}{\Omega \cdot A_s} \quad \frac{\mathrm{W}}{\mathrm{sr} \times \mathrm{m}^2} . \tag{4.2}$$

All areas (A_e and A_s) are assumed to be oriented perpendicular to the direction of propagation of the radiation. In addition, frequently the radiation quantities are given as wavelength or frequency-dependent quantities i.e. the specific radiance $I(\lambda)\,\mathrm{d}\lambda$.

4.2 Interaction Processes of Radiation in the Atmosphere

There are a multitude of processes in which radiation interacts with the atmosphere:

- Absorption, i.e. radiation is removed from the radiation field and converted into some other form of energy, e.g. heat.
- Elastic scattering changes the direction of propagation of an individual photon, but not its energy. Scattering can be due to air molecules (Rayleigh scattering) or aerosol particles (Mie scattering) present in the air.
- Inelastic scattering: As in elastic scattering, the direction of a photon is changed, but inelastic scattering also changes its energy. Inelastic scattering by molecules is called Raman scattering; here, the energy of the scattered photon can be reduced at the amount of energy transferred to the scattering molecule (Stokes scattering). Similarly, energy can be transferred from the (thermally excited) molecules to the photon (anti-Stokes scattering).
- Thermal emission from air molecules and aerosol particles: The emission at any given wavelength cannot exceed the Planck function (or emission from a black body) for the temperature of the atmosphere; thus, noticeable thermal emission only takes place at infrared wavelengths longer than several micrometres. Due to Kirchhoff's law, absorbing gases (such as CO_2, H_2O and O_3, but not the main components of air N_2, O_2 and Ar) can emit radiation.
- Aerosol fluorescence: Excitation of molecules within aerosol particles by radiation can result in (broad band) fluorescence. This process will not be discussed further here.

4.2.1 Absorption Processes

Radiation is absorbed by molecules in the atmosphere (such as ozone, oxygen, nitrogen dioxide, or water vapour) or aerosol (such as soot). The absorption of solar energy in the atmosphere is a key process in the climate system of earth, while absorption of solar UV radiation at wavelengths below about 300 nm by the stratospheric ozone layer (see Chap. 2) is an important prerequisite for life on land.

4.2.2 Rayleigh Scattering

Elastic scattering (i.e. scattering without change of the photon energy) by air molecules is called Rayleigh scattering. Although this is not an absorption process, light scattered out of the probing light beam will normally not reach the detector; thus, for narrow beams (NB) it is justified to treat Rayleigh scattering as an absorption process. The Rayleigh scattering cross-section $\sigma_R(\lambda)$ in cm^2 is given by (Rayleigh, 1899):

$$\sigma_R(\lambda) = \frac{24\pi^3}{\lambda^4 N_{air}^2} \cdot \frac{\left(n_0(\lambda)^2 - 1\right)^2}{\left(n_0(\lambda)^2 + 2\right)^2} \cdot F_K(\lambda) \approx \frac{8\pi^3}{3\lambda^4 N_{air}^2} \cdot \left(n_0(\lambda)^2 - 1\right)^2 \cdot F_K(\lambda) \;, \tag{4.3}$$

where λ denotes the wavelength in cm, $n_0(\lambda)$ is the wavelength-dependent index of refraction of air, N_{air} is the number density of air (e.g. 2.4×10^{19} molec. cm^{-3} at 20°C, 1 atm), and $F_K(\lambda) \approx 1.061$ is a correction for anisotropy (polarisability of air molecules).

Note that $n_0(\lambda)^2 - 1 \approx 2(n_0(\lambda) - 1) \propto N_{air}$, since $n_0 \approx 1$ [in fact $n_0(550\,\text{nm}) = 1.000293$], and $n_0 - 1 \propto N_{air}$; thus, $\sigma_R(\lambda)$ is essentially independent of N_{air}.

For simple estimates, the Rayleigh scattering cross-section can be written as:

$$\sigma_R(\lambda) \approx \sigma_{R0} \cdot \lambda^{-4} (\sigma_{R0} \approx 4.4 \times 10^{-16}\,\text{cm}^2\,\text{nm}^4 \text{ for air}) \tag{4.4}$$

On the basis of (4.4), a better expression of the Rayleigh scattering cross-section (in cm^2) is given by Nicolet (1984), (note that here λ has to be given in μm):

$$\sigma_R(\lambda) \approx \frac{\sigma_{R0}}{\lambda^{4+x}}, \qquad \sigma_{R0} = 4.02 \cdot 10^{-28}, \tag{4.5}$$

with $x = 0.04$ for $\lambda > 0.55\,\mu\text{m}$

$x = 0.389\lambda + 0.09426/\lambda - 0.3328$ for $0.2\,\mu\text{m} < \lambda < 0.55\,\mu\text{m}$.

An even more accurate treatment is found in Penndorf (1957).

The Rayleigh extinction coefficient $\varepsilon_R(\lambda)$ is then given by:

$$\varepsilon_R(\lambda) = \sigma_R(\lambda) \cdot N_{air} \;. \tag{4.6}$$

The Rayleigh phase function (see Fig. 4.1) is given by:

$$\Phi(\cos\vartheta) = \frac{3}{4}\left(1 + \cos^2\vartheta\right) \;. \tag{4.7}$$

Taking the anisotropy of the polarisability into account, (4.7) becomes (Penndorf, 1957):

$$\Phi(\cos\vartheta) = 0.7629 \cdot \left(0.9324 + \cos^2\vartheta\right) \;. \tag{4.8}$$

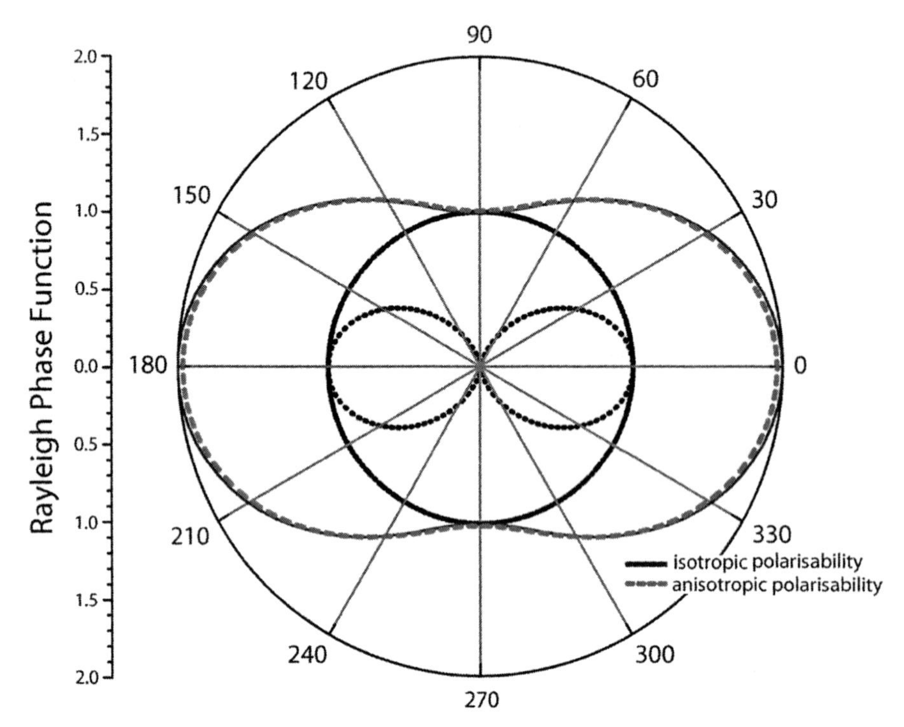

Fig. 4.1. Polar diagram of the Rayleigh scattering phase function $\Phi(\vartheta)$ for non-polarised incident light. The contribution of light polarised parallel to the scattering plane shows the $\sin^2 \vartheta'$ dependence of a Hertz dipole (*dotted line*), with $\vartheta' = \pi/2 - \vartheta$ being the angle between dipole axis and the direction of the incident radiation, while the contribution of light polarised perpendicular to the scattering plane is independent of ϑ (*solid line*) (from Filsinger, 2004)

4.2.3 Raman Scattering

While Rayleigh (and Mie) scattering can be regarded as 'elastic' scattering processes, where no energy is transferred between the scattering particle and the photon, 'inelastic' scattering occurs if the scattering particle (i.e. molecule) changes its state of excitation during the scattering process. A part of the photons energy is then passed from the photon to the molecule (Stokes lines, S-branch) or vice versa (Anti-Stokes, O-branch). The term rotational Raman scattering (RRS) is used if only the rotational excitation is affected ($\Delta v = 0$). If the vibrational state also changes, the term (rotational-) vibrational Raman scattering (VRS) is used ($\Delta v = \pm 1$). Only discrete amounts of energy given by the difference between the discrete excitation states of a molecule can be absorbed/emitted. For air (oxygen and nitrogen), RRS frequency shifts of up to $\pm 200\,\mathrm{cm}^{-1}$ occur; for VRS a vibrational shift of $\pm 2331\,\mathrm{cm}^{-1}$ for nitrogen and $\pm 1555\,\mathrm{cm}^{-1}$ for oxygen must be added. The VRS is one order of magnitude weaker than the RRS, see Table 4.1.

Table 4.1. Comparison of the total cross-section for the different scattering types for 770 nm, 273 K

Scattering type	Cross-section (cm^2)	Ratio (%)
Rayleigh Scattering	1.156×10^{-27}	100
O_2 RRS	7.10×10^{-29}	6.1
N_2 RRS	2.94×10^{-29}	2.5
Air RRS	3.82×10^{-29}	3.3
VRS	–	0.1

In this section, we give a quantitative description of RRS and VRS by O_2 and N_2 (Bussemer, 1993; Burrows et al., 1996; Haug, 1996; Sioris and Evans, 1999). The scattered power density $I_{v,\,J\to v',\,J'}$ given in W/m^2 scattered into the full solid angle 4π involving a transition (v, $J \to v'$, J') is given by (Schrötter and Klöckner, 1979):

$$I_{\nu,\,J\to\nu',\,J'} = I_o \cdot \sigma_{\nu,\,J\to\nu',\,J'} \cdot L \cdot N \cdot g_J \cdot (2J+1) \cdot \frac{1}{Z} \cdot e^{-E(\nu,\,J)/kT} \,, \qquad (4.9)$$

where I_0 is the incident power density, N is the number of molecules in the scattering volume, L is the length of the latter, and g_J is the statistical weight factor of the initial rotational state due to the nuclear spin. J and v are the rotational and vibrational quantum numbers, respectively. The factor $(2J+1)$ accounts for the degeneracy due to the magnetic quantum numbers while exp $(-E(v,\,J)/kT)$ accounts for the population of the initial state of the molecule at temperature T. The state sum Z is given by the product of the rotational state sum Z_{rot} and the vibrational state sum Z_{vib}. The absolute cross-section in (4.11) is given by $\sigma_{v,\,J\to v',\,J'}$ which can be obtained by integration of the differential cross-section $d\sigma_{v,\,J\to v',\,J'}/d\Omega$ over the entire solid angle Ω (note that the term differential refers to the solid angle). The energy of the molecule is characterised by the vibrational (v) and rotational (J) quantum numbers (see Sect. 4.2.3) and is given by

$$E(\nu, J) = E_{vib}(\nu) + E_{rot}(J) = hc\tilde{\nu}\left(\nu + \frac{1}{2}\right) + hcBJ(J+1) \,, \qquad (4.10)$$

assuming no coupling between rotation and vibration. B denotes the rotational constant and $\tilde{\nu} = 1/\lambda = \nu/c$ is the wave number in 1/cm of the ground vibration. Allowed transitions are $\Delta J = 0, \pm 2$, resulting in the Q-, O- and S-branches, and $\Delta v = 0, \pm 1$ for vibrational transitions. Due to the temperatures in the earth's atmosphere, only the ground vibrational state is occupied significantly; thus, leading to only Stokes transitions of the vibrational states. The differential cross-section for an incident light beam with the wave number $\tilde{\nu}$ can be written as:

$$\frac{d\sigma_{\nu,J\to\nu',J'}}{d\Omega} = 8\pi^4 \cdot (\tilde{\nu}_{in} + \tilde{\nu}_{\nu,J\to\nu',J'})^4 \cdot \left\{ \sum_{ij} \overline{[\alpha_{ij}]^2}_{\nu,J\to\nu',J'} \cdot f_{ij}(\Theta) \right\} \,, \qquad (4.11)$$

Table 4.2. Squares of the polarisability tensor (from Long, 1977)

Transition	$\overline{[\alpha_{ii}]^2}_{v,J\to v',J'}$	$\overline{[\alpha_{ij}]^2}_{v,J\to v',J'}$
Cabannes line $\Delta v = 0$, $\Delta J = 0$	$a^2 + \frac{4}{45} b_{J,J} \cdot \gamma^2$	$\frac{1}{15} b_{J,J} \cdot \gamma^2$
O-, S-branch $\Delta v = 0$, $\Delta J = \pm 2$	$\frac{4}{45} b_{J,J\pm 2} \cdot \gamma^2$	$\frac{1}{15} b_{J,J\pm 2} \cdot \gamma^2$
Q-branch $\Delta v = 1$, $\Delta J = 0$	$(a'^2 + \frac{4}{45} b_{J,J} \cdot \gamma'^2) \cdot \frac{h}{8c\pi^2\tilde{\nu}}$	$(\frac{1}{15} b_{J,J} \cdot \gamma'^2) \cdot \frac{h}{8c\pi^2\tilde{\nu}}$
O-, S-branch $\Delta v = 1$, $\Delta J = \pm 2$	$(\frac{4}{45} b_{J,J\pm 2} \cdot \gamma'^2) \cdot \frac{h}{8c\pi^2\tilde{\nu}}$	$(\frac{1}{15} b_{J,J\pm 2} \cdot \gamma'^2) \cdot \frac{h}{8c\pi^2\tilde{\nu}}$

where $\overline{[\alpha_{ij}]^2}_{v,J\to v',J'}$ are the spatial averages of the squares of the polarisability tensor, as given by Long (1977), see Table 4.2. The invariants of this tensor are the average polarisability a, responsible for the isotropic part of the scattered light, and the anisotropy γ for pure rotational transitions, responsible for the anisotropic part, and a' and γ' for vibrational and rotational transitions, respectively (see Young, 1981). The Placzek–Teller coefficients $b_{J\to J'}$ of Table 4.2 can be found in Penney et al. (1974) as:

$$b_{J\to J} = \frac{J(J+1)}{(2J-1)(2J+3)}$$
$$b_{J\to J+2} = \frac{3(J+1)(J+2)}{2(2J+1)(2J+3)}$$
$$b_{J\to J-2} = \frac{3J(J-1)}{2(2J+1)(2J-1)}. \qquad (4.12)$$

The geometric factors $f_{ij}(\Theta)$ in (4.13) are determined by the relative orientations of the incident light polarisation, the polarisability tensor and the direction of observation, given by the scattering angle Θ. For incident unpolarised light, they can be expressed by the normalised phase functions $p^{iso}(\Theta)$ and $p^{aniso}(\Theta)$ according to a, a' and γ, γ', respectively:

$$p^{iso}(\Theta) = \frac{3}{4}(1 + \cos^2\Theta)$$
$$p^{aniso}(\Theta) = \frac{3}{40}(13 + \cos^2\Theta). \qquad (4.13)$$

The total cross-sections of the Q-branch of the vibrational transition can be calculated (Schrötter and Klöckner, 1979) as:

$$\sigma^{Q,total}(T) = \sum_{\nu,J} \sigma_{\nu,J\to\nu',J} \cdot \frac{1}{Z} \cdot g_J \cdot (2J+1) \cdot e^{-E(\nu,J)/kT}$$
$$= \frac{128\pi^5(\tilde{\nu}_{in} + \tilde{\nu})^4}{9(1 - e^{hc\tilde{\nu}/kT})} \left[3a'^2 + \frac{2}{3}\gamma'^2 \cdot S(T)_{\Delta J=0}\right] \cdot \frac{h}{8c\pi^2\tilde{\nu}} \qquad (4.14)$$

Table 4.3. Numerical values of the weighting factors $S(T)$, as used in the calculation of the total cross-section (4.15) of the Q-branch.

Temperature	$S(T)_{\Delta J=0}$		$S(T)_{\Delta J=2}$		$S(T)_{\Delta J=-2}$		$\Sigma_{\Delta J=0,\pm 2} S(T)_{\Delta J}$	
	N_2	O_2	N_2	O_2	N_2	O_2	N_2	O_2
234 K	0.2561	0.2617	0.4450	0.4311	0.2989	0.3068	1.0000	0.9997
300 K	0.2552	0.2592	0.4369	0.4246	0.3078	0.3147	0.9999	0.9985
Relative deviation	0.4%	1.0%	1.8%	1.5%	−2.9%	−2.5%	0.01%	−0.12%

Using the weighting factors $S(T)$:

$$S(T)_{\Delta J} = \sum_J \frac{1}{Z_{rot}} \cdot g_J \cdot (2J+1) \cdot b_{J,J+\Delta J} \cdot e^{-E_{rot}(J)/kT}, \tag{4.15}$$

which depend slightly on the temperature; the numerical values are given in Table 4.3. The observed wavelength shifts in scattered light spectra depend on the involved transitions and are presented in Table 4.4. The exact values for a' and γ' of O_2 and N_2 are not found in the literature, but the total differential cross-sections of the Q-branch for the O_2 and N_2 have been measured by several authors [for references, see Schrötter and Klöckner (1979)]. Thus, either a' or γ' can be eliminated from (4.14) yielding (within a relative error of 5%):

(i) $a' = 0 \Rightarrow \gamma'^2{}_{N2} = 6.79 \times 10^{-5}\,\mathrm{cm}^4\,\mathrm{kg}^{-1}$

$$\gamma'^2{}_{O2} = 4.44 \times 10^{-5}\,\mathrm{cm}^4\,\mathrm{kg}^{-1}$$

(ii) $\gamma' = 0 \Rightarrow a'^2{}_{N2} = 2.7 \times 10^{-6}\,\mathrm{cm}^4\,\mathrm{kg}^{-1}$

$$a'^2{}_{O2} = 1.8 \times 10^{-6}\,\mathrm{cm}^4\,\mathrm{kg}^{-1}$$

Table 4.4. Observed wave number shifts in scattered light spectra as a function of the transition involved

Vibrational transitions	Rotational transitions		Wave number shifts
$\Delta v = 0$	Q-branch	$\Delta J = 0$	$\tilde{\nu} = 0$
	S-branch	$\Delta J = +2$	$\tilde{\nu} = 4B(J+3/2)$
	O-branch	$\Delta J = -2$	$\tilde{\nu} = +4B(J-1/2)$
$\Delta v = 1$	Q-branch	$\Delta J = 0$	$\tilde{\nu}_{N2} = -2331\,\mathrm{cm}^{-1}$
	S-branch	$\Delta J = +2$	$\tilde{\nu}_{O2} = -1555\,\mathrm{cm}^{-1}$
	O-branch	$\Delta J = -2$	$\tilde{\nu} = \tilde{\nu}_{N2/O2} - 4B(J+3/2)$
			$\tilde{\nu} = \tilde{\nu}_{N2/O2} + 4B(J-1/2)$

Table 4.5. The relative differential cross-sections of the elastic component (Cabannes line)

Scenario	Relative differential cross-section $d\sigma/d\Omega$		
	Cabannes ($\Delta v = 0$, $\Delta J = 0$): RRS ($\Delta v = 0$, $\Delta J = \pm 2$): VRS ($\Delta v = 1$, $\Delta J = 0, \pm 2$)		
	N_2	O_2	Air ($0.8 \cdot N_2$–$0.2 \cdot O_2$)
$a' = 0,\ \gamma' \neq 0$	100:1.98:0.27	100:5.31:0.32	100:1.58:0.22–100:1.01:0.06
$a' \neq 0,\ \gamma' = 0$	100:1.94:0.07	100:5.31:0.08	100:1.58:0.06–100:1.01:0.02

In scenario (i), the isotropic part of the vibrational band is omitted. Therefore, only the anisotropic part is present. Since for linear molecules the Q-branch contains a quarter of the entire scattered light of the vibrational band (Weber and Anderson, 1973), the entire scattered radiance of the vibrational band can be calculated from the radiance of the Q-branch (see Table 4.5). In scenario (ii), the anisotropic part is set to zero, and thus there are no O- and S-branches in this vibrational band.

Table 4.5 shows the relative differential cross-sections of the elastic component (Cabannes line), the rotational Raman and vibrational Raman parts, setting the contribution of the elastic component to 100%. The values were derived for a scattering angle of 0° (or 180°) and depend slightly on the incident wavelength (neglecting the weak wavelength dependence of the averaged polarisability a and the anisotropy γ).

4.2.4 Polarisation Properties of Vibrational Raman Scattered Light and Line Filling in

The polarisation properties of the isotropic and anisotropic components of Raman scattered light are described by:

$$\begin{aligned} \frac{I_{\text{parallel}}}{I_{\text{perp}}}(\text{anisotropic}) &= \frac{6+\cos^2\Theta}{7} \\ \frac{I_{\text{parallel}}}{I_{\text{perp}}}(\text{isotropic}) &= \cos^2\Theta \end{aligned}, \tag{4.16}$$

where the terms parallel and perpendicular refer to the plane defined by the sun, the scattering point and the observer. Thus, essentially only the isotropic part leads to enhanced polarisation of scattered light, especially at large scattering angles [solar zenith angle (SZA)]. Since the Q-branch of the vibrational band of the Raman scattered light consists of an isotropic fraction, this will lead to an enhanced degree of polarisation in the centre of a Fraunhofer line. But due to the small cross-section of VRS, this enhancement is very small (e.g. $\cong$0.3% for the Ca-I-line at 422.7 nm and a resolution of 0.01 nm) and occurs only at large scattering angles. Thus, the observed high degrees of polarisation at small SZAs, cannot be attributed to VRS (Clarke and Basurah, 1989).

Due to the relatively small cross-section of vibrational–rotational Raman scattering, the additional filling in of Fraunhofer or terrestrial absorption lines is about 10% of the filling in due to RRS (see Table 4.5).

4.2.5 Scattering and Absorption of Radiation by Particles (Mie Scattering)

Radiation may be absorbed and scattered by particles. These processes are described by the absorption and scattering coefficients $\varepsilon_a(r, \lambda)$ and $\varepsilon_s(r, \lambda)$, respectively:

$$dI = -I(\lambda) \cdot \varepsilon_a(r, \lambda)\, ds. \qquad (4.17)$$

$$dI_s = -I(\lambda) \cdot \varepsilon_s(r, \lambda)\, ds. \qquad (4.18)$$

$I(\lambda)$ and $dI(\lambda)$ are the radiation flux and its change after passing through a layer of the aerosol of thickness ds, respectively. We define the 'single scattering albedo' of the aerosol as:

$$A_S = \frac{\varepsilon_s(r, \lambda)}{\varepsilon_a(r, \lambda) + \varepsilon_s(r, \lambda)}, \qquad (4.19)$$

and the extinction coefficient as:

$$\varepsilon_M(r, \lambda) = \varepsilon_a(r, \lambda) + \varepsilon_s(r, \lambda). \qquad (4.20)$$

Mie scattering was first described in a comprehensive way by Gustav Mie and is defined as the interaction of light with (particulate) matter of dimensions comparable to the wavelength of the incident radiation (Mie, 1908). It can be regarded as the radiation resulting from a large number of coherently excited elementary emitters (i.e. molecules) in a particle. Since the linear dimension of the particle is comparable to the wavelength of the radiation, interference effects occur. The most noticeable difference compared to Rayleigh scattering is the typically much weaker wavelength dependence (typically proportional to $\lambda^{-1.3}$) and a strong dominance of the forward direction in the scattered light. The calculation of the Mie scattering cross-section can be very complicated (involving summing over slowly converging series), even for spherical particles, but even more so for particles of arbitrary shape. Generally, the Mie phase function, $P(\cos \nu)$, depends on the ratio between radius r and wavelength λ of the scattered radiation, the **size parameter**:

$$\alpha_S = \frac{2\pi r}{\lambda}.$$

However, the Mie theory (for spherical particles) is well developed and a number of numerical models exist to calculate scattering phase functions and

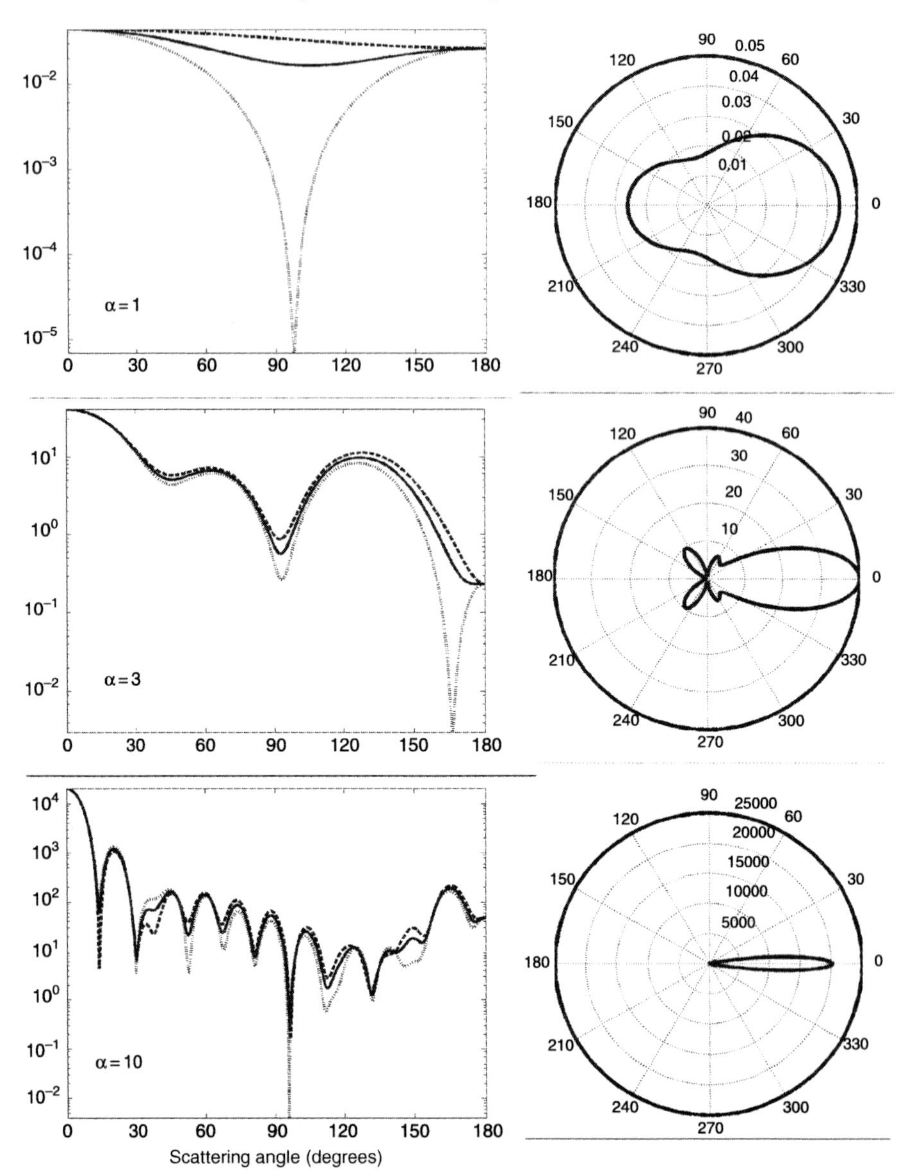

Fig. 4.2. Sample phase function for Mie scattering from individual particles (or mono-disperse aerosol) with size parameters $\alpha_S = 1, 3$, and 10. **Left column**: Phase function on log scale; **Right column**: Polar diagrams of the same phase functions on linear scale. It becomes apparent that for larger values of α_S essentially all energy is scattered in forward direction (from Shangavi, 2003)

extinction coefficients for given aerosol types and particle size distributions (van de Hulst, 1980; Wiscombe, 1980). Figure 4.2 gives a few examples for single particles or mono-disperse aerosols, respectively.

The computational effort is substantially reduced by the introduction of an analytical expression for the scattering phase function, which only depends on a few observable parameters. Most common is the Henyey–Greenstein parameterisation:

$$\Phi\left(\cos\vartheta\right) = \frac{\left(1-g^2\right)}{4\pi\left(1+g^2-2g\cos\vartheta\right)^{3/2}}, \tag{4.21}$$

which depends only on the asymmetry factor g (average cosine of the scattering function):

$$g = \langle\cos\vartheta\rangle = \frac{1}{2}\int_{-1}^{1} P\left(\cos\vartheta\right)\cdot\cos\vartheta\, d\cos\vartheta \tag{4.22}$$

For is isotropic scattering, $\Phi(\cos\vartheta)$=constant, the asymmetry factor. For complete $g = 0$ forward scattering, g would be 1. Typical values for tropospheric aerosol are $g \approx 0.6 - 0.7$.

Tropospheric aerosol is either emitted from the surface (sea salt, mineral dust, and biomass burning) or forms in the gas phase by condensation of chemically formed hygroscopic species (primarily sulphate, nitrates, or oxidised organic material). The aerosol load of the atmosphere, i.e. particle number density and size distribution, depends on the aerosol origin and history.

Parameters for typical aerosol scenarios (urban, rural, maritime, and background) can be found in the data base for the radiative transfer model LOWTRAN (Isaacs et al., 1987), which includes the extinction coefficients and the asymmetry factors, as well as their spectral dependence. Another important aspect is Mie scattering on cloud particles. A radiative transfer model including all cloud effects known to date was e.g. developed by Funk (2000).

Mie scattering is partly an absorption process, but by similar arguments as in the case of Rayleigh scattering, for NB it can be treated as an absorption process with the extinction coefficient:

$$\varepsilon_{\mathrm{M}}(\lambda) = \varepsilon_{\mathrm{M0}}\cdot\lambda^{-\alpha}$$

with the Angström exponent α being inversely related to the mean aerosol particle radius and a constant $\varepsilon_{\mathrm{M0}}$. Typically α is found in the range 0.5–2.5, with an ‘average’ value of $\alpha = 1.3$ (Angström, 1930). For the ideal case of an exponential aerosol size distribution:

$$\frac{\Delta N}{\Delta r} = r^{-(\nu_{\mathrm{J}}+1)}.$$

The Angström exponent is related to the Junge index ν_{J} by $\nu_{\mathrm{J}} = \alpha+2$ (Junge, 1963). Thus, an Angström exponent of 1.3 would correspond to a Junge index of 3.3.

Table 4.6. Typical Mie scattering coefficients, calculated Rayleigh scattering coefficients, corresponding extinction lengths and visibility range

Wavelength (nm)	Rayleigh scattering cross-section σ_R[a] $(10^{-26}\,cm^2)$	Rayleigh scattering coefficient ε_R[b] $(10^{-5}\,m^{-1})$	Typical Mie scattering coefficient ε_M $(10^{-5}\,m^{-1})$	Rayleigh extinction length (km)	Total extinction length L_E (km)	Visibility range[c] $L_V(C = 0.02)$ (km)
300	5.653	13.524	5.8	7.39	5.1	–
400	1.672	4.013	4	24.9	12.5	49
500	0.672	1.612	3	62.1	21.7	85
600	0.317	0.760	2.4	131.6	31.6	123
700	0.170	0.408	1.9	245.4	43.3	169

[a] According to (4.7).
[b] At 1 atm and 20°C.
[c] For definition of visibility range see Sect. 4.7.

In summary, a more comprehensive description of atmospheric extinction (in the presence of a single trace gas species) can be expressed as:

$$I(\lambda) = I_0(\lambda) \exp\left[-L(\sigma(\lambda) \cdot c) + \varepsilon_R(\lambda) + \varepsilon_M(\lambda)\right] . \tag{4.23}$$

Typical extinctions due to Rayleigh and Mie scatterings at 300 nm are 1.3×10^{-6} and $1\text{–}10 \times 10^{-6}\,cm^{-1}$, respectively (see Table 4.6).

4.3 The Radiation Transport Equation

The propagation of radiation (radiation transport) in absorbing and scattering media, such as the atmosphere, is a complex process. This is mostly due to the fact that scattered light can propagate in any direction and can undergo further scattering and absorption processes. In addition, at long wavelengths, thermal emission from gas molecules (and aerosol) can also play a role. Although the contribution of thermal emission is negligible compared to the solar radiation in the near IR or even visible or UV spectral regions, we include it here for completeness. In this section, we present the basic equations describing the radiation transport in absorbing and scattering media as described earlier by the elementary processes. Analytical solutions of the equations can only be derived for greatly simplified cases; for instance for a collimated light beam travelling a finite distance through the atmosphere or a horizontally infinite atmosphere (see Sect. 4.4.1). In general, solutions have to be calculated by numerical methods, which can be divided into two groups. First, numerical solutions for the resulting radiation field can be calculated. Second, in a statistical approach the paths of an ensemble of individual photons (with randomly chosen parameters) through the atmosphere are calculated for the desired conditions. From the photon density the radiation field can then be derived. Examples of the former approach are the 'Discrete Ordinate Method' [e.g. DISORT, (Stamnes et al., 1988)] of the latter Monte Carlo Methods [e.g. TRACY (von Friedeburg, 2003)].

4.3.1 Absorption of Radiation

The radiance $I(\lambda)$ will be reduced by the amount $dI_\mathrm{a}(\lambda)$ after traversing an absorbing layer of the (infinitesimal) thickness ds:

$$dI_\mathrm{a} = \frac{d\Phi}{\Delta\Omega \cdot A_\mathrm{s}} = -I(\lambda) \cdot \varepsilon_\mathrm{a}(\lambda)\, ds = -I(\lambda) \cdot \sigma_\mathrm{a}(\lambda) \cdot N\, ds\ , \tag{4.24}$$

where ε_a is the absorption coefficient, $\sigma_\mathrm{a}(\lambda)$ is the absorption cross-section of the absorber (molecule), and N is the number of absorbing molecules (atoms) per unit volume

Integration of (4.24) yields [with $I_0(\lambda)$ denoting the initial radiance]:

$$\ln\left(\frac{I_0(\lambda)}{I(\lambda)}\right) = \sigma_\mathrm{a}(\lambda) \cdot \int_0^L N ds = \sigma_\mathrm{a}(\lambda) \cdot S = \tau(\lambda)\ , \tag{4.25}$$

where

$$S = \int_0^L N\, ds\ , \tag{4.26}$$

is the column density of the absorber in which L denotes the thickness of the layer penetrated by the radiation; and D the optical density, which can also be written as:

$$D = \int \varepsilon_\mathrm{a}(\lambda)\ ds \text{ or } dD = \varepsilon_\mathrm{a}(\lambda)\ ds\ . \tag{4.27}$$

Thus, the radiance after traversing the absorbing layer becomes:

$$I(\lambda)_L = I_0(\lambda) \cdot e^{-\sigma_\mathrm{a}(\lambda) \cdot S} = I_0(\lambda) \cdot e^{-D}\ . \tag{4.28}$$

This is the well-known Lambert–Beer's law (or Bouguer–Lambert law, see Sect. 6.1).

4.3.2 Scattering of Radiation

The scattered radiance is (in analogy to the absorption) given by:

$$dI_\mathrm{s} = \frac{d\Phi_\mathrm{s}}{\Delta\Omega \cdot A_\mathrm{s}} = I(\lambda) \cdot \varepsilon_\mathrm{s}(\lambda)\, ds = I(\lambda) \cdot \frac{d\sigma_\mathrm{s}(\lambda)}{d\Omega} \cdot N\, d\Omega\, ds\ , \tag{4.29}$$

where ε_s is the scattering coefficient, $\sigma_\mathrm{s}(\lambda)$ is the scattering cross-section of a scattering centre (molecule), and N is the number of scattering centres (e.g. molecules or particles) per unit volume.

The scattered radiant flux is given by:

$$d\Phi_{\mathrm{s}} = I(\lambda) \cdot \Delta\Omega \cdot \frac{d\sigma_{\mathrm{s}}(\lambda)}{d\Omega} \cdot N\, d\Omega dV\ , \tag{4.30}$$

where $d\sigma s(\lambda)/d\Omega$ denotes the differential scattering cross-section (depending on scattering angle ϑ and polar angle ϕ).

The total scattering cross-section is thus given by:

$$\sigma_{\mathrm{s}}(\lambda) = \int_{4\pi} \frac{d\sigma_{\mathrm{s}}(\lambda)}{d\Omega} \cdot d\Omega\ . \tag{4.31}$$

Frequently, a dimensionless scattering function is useful:

$$S(\vartheta, \Phi) = \frac{4\pi}{\sigma_{\mathrm{s}}} \cdot \frac{d\sigma_{\mathrm{s}}(\lambda)}{d\Omega}\ . \tag{4.32}$$

The total extinction experienced by the radiation after traversing a layer of the medium of thickness ds is thus given by the sum of absorption and scattering:

$$dI = -dI_{\mathrm{a}} - dI_{\mathrm{s}} = -I(\lambda) \cdot (\varepsilon_{\mathrm{a}}(\lambda) + \varepsilon_{\mathrm{s}}(\lambda))\, ds\ . \tag{4.33}$$

Radiance added by scattering is given by:

$$dI_{\mathrm{S}}^{*}(\lambda) = \varepsilon_{\mathrm{s}}(\lambda)\, ds \int_{0}^{\pi}\int_{0}^{2\pi} I^{*}(\lambda, \vartheta^{*}, \phi^{*}) \cdot \frac{S(\vartheta^{*}, \phi^{*})}{4\pi} d\phi^{*} \cdot \sin\vartheta^{*} d\vartheta^{*}\ . \tag{4.34}$$

4.3.3 Thermal Emission

The thermal emission $[dI_{th}(\lambda, T)]$ from a volume element $(dV = Asds)$ is given as:

$$dI_{\mathrm{th}}(\lambda, T) = \varepsilon_{\mathrm{a}}(\lambda) \cdot I_{\mathrm{p}}(\lambda, T)\, ds\ , \tag{4.35}$$

where ε_a denotes the absorption coefficient as in (4.27) and $I_p(\lambda, T)$ denotes the Planck function (see also Sect. 7.1):

$$dI_{\mathrm{p}}(\lambda, T) = \frac{2hc^2}{\lambda^5} \cdot \frac{d\lambda}{e^{hc/\lambda kT} - 1}\ . \tag{4.36}$$

The three processes discussed above, i.e. absorption, scattering, and thermal emission are combined in the radiation transport equation (4.37):

$$\frac{dI(\lambda)}{ds} = -(\varepsilon_{\mathrm{a}}(\lambda) + \varepsilon_{\mathrm{s}}(\lambda)) \cdot I(\lambda) + \varepsilon_{\mathrm{a}}(\lambda) \cdot I_{\mathrm{p}}(\lambda, T) + \varepsilon_{\mathrm{s}}(\lambda) \cdot \int_{0}^{\pi}\int_{0}^{2\pi} I^{*}(\lambda, \vartheta^{*}, \phi^{*}) \cdot \frac{S(\vartheta^{*}, \phi^{*})}{4\pi} d\phi^{*} \cdot \sin\vartheta^{*} d\vartheta^{*}. \tag{4.37}$$

4.3.4 Simplification of the Radiation Transport Equation

Frequently, simplifications of the radiation transport equation are possible, if only partial systems are of interest.

For instance, at short wavelengths (UV and visible), usually the Planck term can be neglected and (4.37) reduces to:

$$\frac{dI(\lambda)}{ds} = -\left(\varepsilon_{\mathrm{a}}(\lambda) + \varepsilon_{\mathrm{s}}(\lambda)\right) \cdot I(\lambda) + \varepsilon_{\mathrm{s}}(\lambda) \int_0^{\pi} \int_0^{2\pi} I(\lambda, \vartheta, \phi) \cdot \frac{S(\vartheta, \phi)}{4\pi} d\phi \cdot \sin\vartheta d\vartheta \,. \tag{4.38}$$

If the thermal radiation (from the atmosphere) is of interest due to its long wavelength, frequently Rayleigh scattering and Mie scattering at aerosol particles (but not at cloud droplets) can be neglected and given by:

$$\frac{dI(\lambda)}{ds} = \varepsilon_{\mathrm{a}}(\lambda) \cdot \left(I_{\mathrm{p}}(\lambda, T) - I(\lambda)\right) \,. \tag{4.39}$$

With the optical density [see definition in (4.21)]:

$$dD = \varepsilon_{\mathrm{a}}(\lambda)\; ds \;\text{and}\; ds = \frac{1}{\varepsilon_{\mathrm{a}}(\lambda)} dD \,.$$

The above equation becomes:

$$\frac{dI(\lambda)}{dD} = I_{\mathrm{p}}(\lambda, T) - I(\lambda) \,. \tag{4.40}$$

This latter equation is also known as Schwarzschild equation (van de Hulst, 1980).

4.4 Light Attenuation in the Atmosphere

When considering the question of the attenuation of radiation in the atmosphere, two (extreme) cases can be distinguished, as illustrated in Fig. 4.3: wide beams (WB, as e.g. the illumination of the Earth's atmosphere by the Sun) and narrow beam (NB, as e.g. the light beam emitted by a search-light type DOAS light source).

4.4.1 Wide Beams in the Atmosphere, the Two-Stream Model

In this section, we consider the transport of solar radiation in the atmosphere. Compared to the general form of (4.40), the problem is simplified by assuming a flat, horizontally homogeneous, infinite atmosphere. Therefore, only the vertical component of the radiance has to be considered, as sketched in Fig. 4.4.

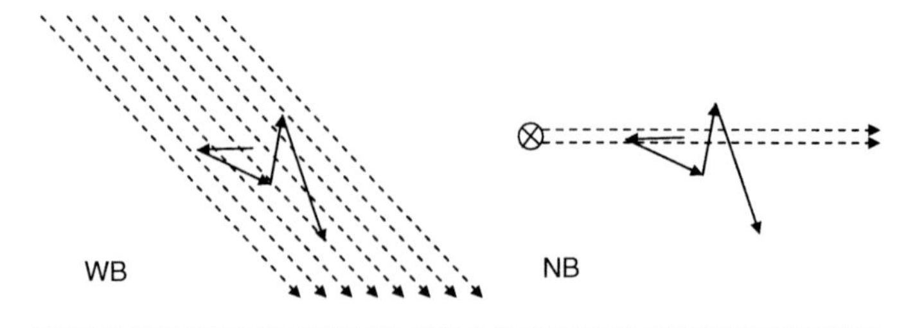

Fig. 4.3. Difference between WB (**left**) as e.g. the illumination of the Earth's atmosphere by the Sun, and NB (**right**) as e.g. the light beam emitted by a search-light type DOAS light source. In the NB case, the probability of a photon being scattered back into the beam after being scattered out of the beam is negligible. In the case of a WB, the lateral (i.e. perpendicular to the propagation of the incident radiation flux) radiation flux can be neglected. Therefore, the scattering has only the effect of reflecting some of the incoming light, as can be calculated by, e.g., a two-stream model

The attenuation dI of the radiance I in a layer of thickness dz is given by (4.27):

$$dI = -I \cdot dD = -I \cdot \varepsilon_a \, dz \; . \tag{4.41}$$

This is, however, only true for photons traversing the layer in vertical direction; in general, (angle of incidence $\vartheta \neq 0°$) the transmission factor T_z of the layer becomes:

$$T_z = \frac{I(\lambda, z_2)}{I(\lambda, z_1)} = e^{-D/\cos\vartheta} \; . \tag{4.42}$$

Since the thermal radiation impinges onto the layer from all directions (Φ denotes the azimuth angle), the average transmission factor is given by:

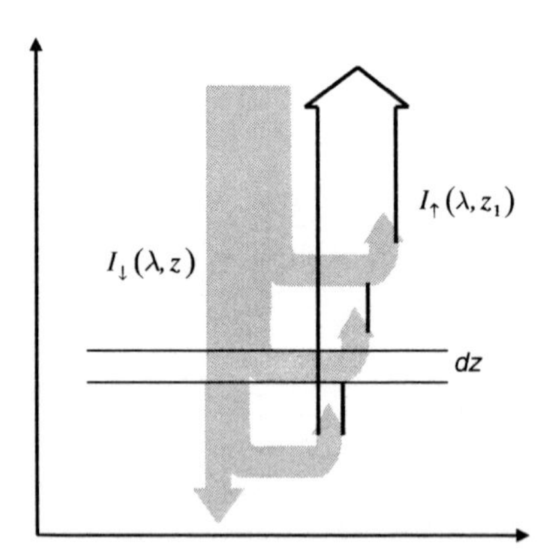

Fig. 4.4. The two-stream model: In a horizontally homogeneous atmosphere, radiation transport can be represented by just two 'streams of radiation', the downwelling flux and the upwelling flux

$$T_{\mathrm{z}}(\lambda, z_1, z_2) = \frac{1}{\pi} \int_0^{2\pi} \int_0^{\pi/2} \exp\left(\frac{\tau(z_1, z_2)}{\cos\vartheta}\right) \cos\vartheta \cdot \sin\vartheta \, d\vartheta \, d\Phi \,. \tag{4.43}$$

Only the layers above the reference height z contribute (according to their temperature and absorbtivity or emissivity) to the downwelling radiance $I_{\downarrow}(\lambda, z)$:

$$I_{\mathrm{p}}(\lambda, T(z')) \cdot \varepsilon_{\mathrm{a}}(\lambda, z') \, dz \,. \tag{4.44}$$

The contribution from the height z' must be weighted by the transmission factor T_z $(\lambda,\ z,\ z')$:

$$dI_{\downarrow}(\lambda) = I_{\mathrm{p}}(\lambda, T(z')) \cdot \varepsilon_{\mathrm{a}}(\lambda, z') \cdot T_{\mathrm{z}}(z, z') \cdot dz' \,. \tag{4.45}$$

Thus, the total downwelling radiance is given by the integral:

$$I_{\downarrow}(\lambda, z) = \int_z^{\infty} I_{\mathrm{p}}(\lambda, T(z')) \cdot \varepsilon_{\mathrm{a}}(\lambda, z') \cdot T_{\mathrm{z}}(z, z') \cdot dz' \,. \tag{4.46}$$

In an analogous manner we obtain the upwelling radiance $F_{\uparrow}(\lambda, z)$:

$$\begin{aligned} I_{\uparrow}(\lambda, z) &= I_{\mathrm{p}}(\lambda, T_{\mathrm{ground}}) \cdot \varepsilon_{\mathrm{ground}} \cdot T_{\mathrm{z}}(0, z) \\ &\quad + \int_0^z I_{\mathrm{p}}(\lambda, T(z')) \cdot \varepsilon_{\mathrm{a}}(\lambda, z') \cdot T_{\mathrm{z}}(z, z') \cdot dz \,, \end{aligned} \tag{4.47}$$

where T_{ground} and $\varepsilon_{\mathrm{ground}}$ denote temperature and emissivity of the ground surface, respectively. The net radiance is then composed of both components (up- and downwelling radiances):

$$I_{\mathrm{n}}(\lambda, z) = I_{\uparrow}(\lambda, z) - I_{\downarrow}(\lambda, z) \,. \tag{4.48}$$

4.4.2 Narrow Beams in the Atmosphere

In the case of NB the probability of a photon being scattered back into the beam after being scattered out of the beam is generally negligible (see Fig. 4.3); therefore, extinction can be treated like absorption, which is given by Lambert–Beer's law, (4.31) with the optical density:

$$D = \int (\varepsilon_{\mathrm{a}}(\lambda) + \varepsilon_{\mathrm{s}}(\lambda)) \, ds \,. \tag{4.49}$$

4.5 The Effect of Atmospheric Refraction (El-Mirage Effects)

Due to the atmospheric pressure gradient, the air density, and thus the index of refraction, decreases with altitude (see Chap. 2). This effect is somewhat counteracted with the decrease in temperature with altitude. For example, in the lowest 100 m of the atmosphere, the pressure gradient alone would lead to a reduction in air density of about 1.25% (assuming a scale height of 8000 m), while the dry adiabatic temperature gradient alone would enhance the air density by about 0.34%, thus leading to a net decrease in air density of about 0.9%.

The corresponding change in the index of refraction of air at 550 nm would be from $n_0(z=0) = 1.0002930$ to $n(z=100) = 1.0002904$. Since the optical path length is $L_{\text{opt}}(n) = L_0/n$, this leads to a radius of curvature, where L_0 denotes the length of any horizontal path and ΔL the optical path difference at the two altitudes $\Delta L_{\text{opt}} = L(z=0, n_0) - L(z=100, \text{n})$:

$$R = \frac{\Delta z \cdot L}{\Delta L_{\text{opt}}} \approx \frac{10^2 \times 10^4}{0.03} \approx 3.3 \times 10^7 \text{ m} . \tag{4.50}$$

This radius of curvature is about five times the Earth's radius. In other words, the deviation of an actual light path from the straight line on a 10-km path is about 1.5 m downwards, while a 'straight' line always following the curvature of earth would deviate by 7.8 m.

For normal operation of active DOAS instruments, this poses no difficulties. However, problems can arise if the deviation from a straight line changes due to anomalous profiles of index of refraction in the atmosphere. This can happen due to strong heating by solar irradiation near the ground, which causes the well-known 'El-Mirage' effect. On the other hand, strong nighttime inversions in the lower atmosphere can offset or even reverse the adiabatic temperature gradient (for details, see Platt et al., 2007). In summary, the apparent position of the light source in an active, bistatic DOAS system can change by several metres in the vertical, depending on the local temperature gradient. Since the image of the light source is usually roughly the size of the entrance aperture (see Chap. 7), realignment of the receiving telescope pointing direction will be needed under conditions of strong and changing temperature gradients. Unattended operation under these conditions will require automatic aligning systems, even if the mechanical mounting of the optics is perfectly stable. For a system involving retro-reflectors, the effect on the returned light should be less severe, as outlined in Chap. 7 (Sect. 7.4.9).

4.6 The Effect of Atmospheric Turbulence

Atmospheric turbulence can influence the propagation of light beams in the atmosphere (e.g. Wolf, 1979). Turbulence can be caused either by mechanical effects, i.e. wind, or by thermal convection. Typical relative density fluctuations Δ N/N are about $3 \times 10^{-3}\,\text{K}^{-1}$, while velocity fluctuations of $1\,\text{m}\,\text{s}^{-1}$

would correspond to relative density fluctuations on the order of 10^{-5}. In either case, turbulent motion can be characterised by the velocity structure function originally introduced by Kolmogorov (1941), which is defined as (see Andrews, 2004):

$$D_{\mathrm{v}}(\Delta x) = \left\langle (v\,(x) - v\,(x + \Delta x))^2 \right\rangle \approx C_{\mathrm{v}}^2 \cdot \Delta x^{2/3} \;, \tag{4.51}$$

where $v(x)$ denotes the velocity of the flow at point x, and ${C_{\mathrm{v}}}^2$ the velocity structure constant. The angle brackets indicate the ensemble average over many pairs of points in space separated by the distance Δx. Similarly, a temperature structure function D_{T} can be defined:

$$D_{\mathrm{T}}(\Delta x) = \left\langle (T\,(x) - T\,(x + \Delta x))^2 \right\rangle \approx C_{\mathrm{T}}^2 \cdot \Delta x^{2/3} \;. \tag{4.52}$$

Since n − 1 (denoting the refractive index of air) varies inversely proportional to the air density, a refractive index structure function D_{n} can be derived from D_{T}:

$$D_{\mathrm{n}}(\Delta x) = \left\langle (n\,(x) - n\,(x + \Delta x))^2 \right\rangle \approx C_{\mathrm{n}}^2 \cdot \Delta x^{2/3} \;, \tag{4.53}$$

where $n(x)$ denotes the index of refraction at point x in space. The refractive-index structure constant ${C_{\mathrm{n}}}^2$ ranges from $10^{-17}\,\mathrm{m}^{-2/3}$ (weak turbulence) to $10^{-13}\,\mathrm{m}^{-2/3}$ (strong turbulence). For two points in the atmosphere separated by e.g. $\Delta x = 1\,\mathrm{m}$, this translates into average differences in the refractive index $\Delta n = n(x) - n(x + \Delta x)$ around $\Delta n \approx 3 \times 10^{-9}$–$3 \times 10^{-7}$. While these differences are very small, there are a large number of these turbulence elements diffracting the beam, as sketched in Fig. 4.5. This leads to 'beam wander', i.e. random deflections of a beam having propagated a distance L through a turbulent atmosphere. The instantaneous beam radius r_{i} is seen when observing for a short time $t << t_0$, where the time constant t_0 is on the order of:

$$t_0 = \frac{r_i}{v} \;.$$

The square of the average deflection of the beam centre is given by the following empirical formula (Andrews, 2004):

$$\left\langle r_{\mathrm{c}}^2 \right\rangle = 2.87 \cdot \frac{C_{\mathrm{n}}^2 \cdot L^3}{r_{\mathrm{i}}^{1/3}} \quad \text{or} \quad r_{\mathrm{c}} \approx \sqrt{\left\langle r_{\mathrm{c}}^2 \right\rangle} = \sqrt{2.87 \cdot C_{\mathrm{n}}^2} \cdot \frac{L^{3/2}}{r_{\mathrm{i}}^{1/6}} \;. \tag{4.54}$$

This would lead to average deflections of a $r_{\mathrm{i}} = 10\,\mathrm{m}$ beam over a 10-km path of 3.6 mm to 0.36 m for weak or strong turbulence, respectively. While this is very small, it should be noted that r_{c} does not vary much with the beam diameter [see (4.57)]; thus, the relative wander would be larger for smaller beams, but even more importantly the coaxial DOAS optics (see Sect. 7.9.3) relies on turbulence to scatter light from the transmitted beam into the field of view of the receiving telescope (besides the lateral beam offset by the retro reflector).

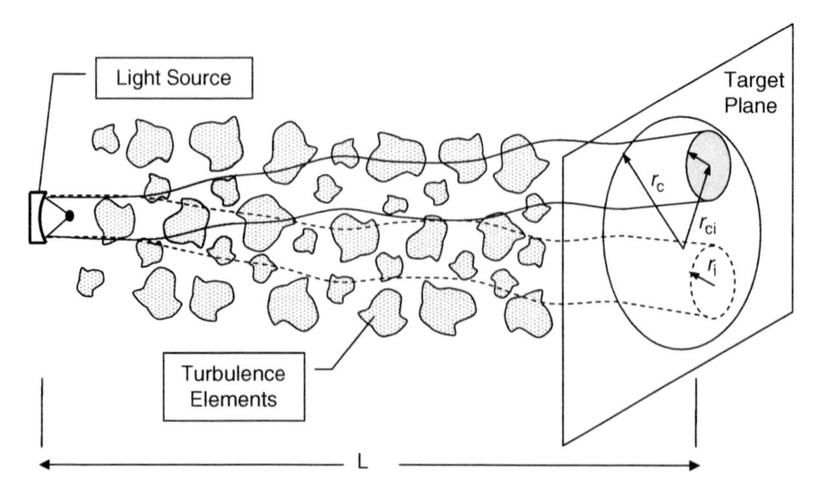

Fig. 4.5. Sketch of the beam wandering process caused by turbulence. At any given time, the 'instantaneous beam' (with an instantaneous radius r_i) arrives at a different position having the distance r_ci from the centre of a circle with radius r_c encompassing the long-term (i.e. for times $t >> t_0$) average positions

4.7 Practical Considerations About Radiation in the Atmosphere

There are many effects of radiation transport that are visible in the atmosphere. For instance, the observation that the sky is blue and the clouds are white is a direct consequence of the different wavelength dependence of Rayleigh scattering compared to Mie scattering. Another important consequence of light scattering and extinction in the atmosphere is the reduction of visibility by haze, smog, dust, or fog. It is an everyday experience that distant objects cannot be seen under conditions of increased scattering.

A quantitative estimate can be made by considering the view of a black object (e.g. a building or a mountain) of area A_0 at a distance L in a homogeneous atmosphere with the scattering coefficient ε. The contrast, i.e. the difference in the brightness, between the object and the background is then given by the following calculation.

When looking at a black object (area A_0), the scattered radiation intensity I_R received by the observer (see Fig. 4.6) is given by:

$$dI_\mathrm{R}(x) = dI_\mathrm{S}(x) \cdot \frac{A_\mathrm{E}}{x^2} \cdot e^{-\varepsilon x} \qquad (A_\mathrm{E} = \text{area of the eye})\,.$$

With the intensity I_S scattered in a volume element with area $A(x)$ and thickness dx at distance x from the observer:

$$dI_\mathrm{S}(x) = \alpha \cdot A(x)\, dx \cdot \varepsilon = \varepsilon\, A_0 \left(\frac{x}{L}\right)^2 dx \quad (\text{where } \alpha \text{ is a constant})\,, \quad (4.55)$$

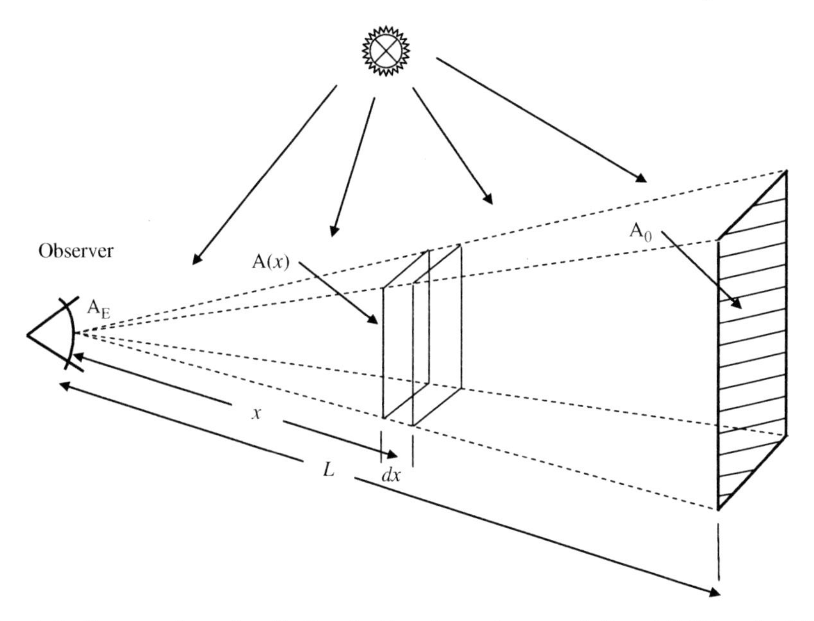

Fig. 4.6. Propagation of radiation in the atmosphere and the visibility of a black surface A_0

we obtain:

$$dI_{\mathrm{R}}(x) = \alpha\varepsilon\, A_0 \left(\frac{x}{L}\right)^2 dx \cdot \frac{A_{\mathrm{E}}}{x^2} \cdot e^{-\varepsilon\, x} = \alpha\varepsilon \cdot \frac{A_0 A_{\mathrm{E}}}{L^2} \cdot e^{-\varepsilon x} dx\,. \tag{4.56}$$

Integration yields the flux received by the eye:

$$I_R = \int_0^L dI_R(x) = \alpha \cdot \frac{A_0 A_E}{L^3} \cdot \left(1 - e^{-\varepsilon\, L}\right)\,. \tag{4.57}$$

For $L \to \infty$, the received intensity I_{R} must be equal to the background intensity I_0, so we obtain:

$$I_{\mathrm{R}} = I_0 \cdot \left(1 - e^{-\varepsilon\, L}\right) \text{ with } I_0 = \alpha \cdot \frac{A_0 A_{\mathrm{E}}}{L^3}\,. \tag{4.58}$$

The relative difference between the radiation intensity received from the black object I_{R} compared to the flux I_0 from its surrounding, the contrast C, is given by:

$$C = \frac{I_{\mathrm{R}} - I_0}{I_0} = \frac{I_0 \cdot \left(1 - e^{-\varepsilon\, L}\right) - I_0}{I_0} = e^{-\varepsilon L}\,. \tag{4.59}$$

Typically, objects are still visible at a contrast between 0.01 and 0.02; thus, the visibility range L_{V} becomes:

$$C = e^{-\varepsilon L_V} \Rightarrow L_V = -\frac{\ln C}{\varepsilon} \approx (3.9 \ldots 4.6) \frac{1}{\varepsilon} = (3.9 \ldots 4.6) \cdot L_E \,, \qquad (4.60)$$

where $L_E = 1/\varepsilon$ and $\varepsilon = \varepsilon_R + \varepsilon_M$, i.e. the sum of the extinction coefficients due to Rayleigh and Mie scattering, as described in Sect. 4.2, where some typical values for L_V are listed in Table 4.6. Of course this is a quite idealised calculation. In reality, the minimum discernible contrast might be lower or higher than assumed. Nevertheless, the estimate of L_E from the visibility of several objects at different distances is quite robust, since only the logarithm of C enters in the calculation (4.63).

5

Measurement Techniques for Atmospheric Trace Gas Concentrations and Other Parameters

Our knowledge about the physical and chemical processes in the atmosphere relies to a large extent on measurements of relevant parameters like temperature or ozone mixing ratio throughout the atmosphere. For instance, the 'meteorological parameters' such as temperature, humidity, pressure, radiation intensities, or wind speed and direction have routinely been measured in the atmosphere. In this book, we focus on measurements of trace gas concentrations and quantities related to chemical processes in the atmosphere, including the intensity of the radiation field. These are the experimental prerequisites for understanding the physico-chemical processes in the earth's atmosphere. In addition, measurements are made in model atmospheres, e.g. smog chambers or environmental chambers. As the name 'trace constituents' suggests, the species to be investigated here are present at minute quantities only (see Chap. 2). Therefore, the determination of trace gas concentrations in the atmosphere constitutes a challenge for the analytical techniques employed. In summary, measurements of atmospheric parameters are done for a variety of purposes:

- Weather forecast and climatological studies
- Meteorological research
- Climate studies
- Pollution control
- Monitoring of the state of atmosphere
- Atmospheric chemistry research

In recent years, the term 'chemical weather' was coined to encompass information on the trace gas composition of, say, air pollutants of the atmosphere.

U. Platt and J. Stutz, *Measurement Techniques for Atmospheric Trace Gas Concentrations and Other Parameters*. In: U. Platt and J. Stutz, Differential Optical Absorption Spectroscopy, Physics of Earth and Space Environments, pp. 113–134 (2008)
DOI 10.1007/978-3-540-75776-4_5

5.1 History of Measurement Techniques

After the discovery of a strong, characteristic smell near electric discharges by van Marum in 1785 and the identification and naming of ozone by Christian Friedrich Schönbein in 1839, this species was one of the first atmospheric trace gases to be investigated. An earlier technique relied on the use of 'Schönbein Paper' (Schönbein, 1840), a paper impregnated with iodide (e.g. KI) and starch. Upon exposure to the ambient air and hence to the ozone dispersed in it, the following reactions produced a blue colour:

$$2\,H^+ + 2\,I + O_3 \rightarrow I_2 + O_2 + H_2O \qquad \text{(R5.1)}$$

$$I_2 + \text{starch} \rightarrow \text{blue colour} \qquad \text{(R5.2)}$$

Together with a suitable colour scale, semi-quantitative measurements could be made. Later, it turned out that the amount of blue colour strongly depends on relative humidity and wind speed, and so these relatively large data set from the 19th century could not be used for quantitative studies.

These problems were avoided by another technique, which was also introduced during the 19th century. Here, air is bubbled through an arsenite-containing solution, which is oxidised to arsenate (Volz and Kley, 1988):

$$O_3 + AsO_3^{3-} \xrightarrow{I^-/I_2} O_2 + AsO_4^{3-}. \qquad \text{(R5.3)}$$

Measurements near Paris during 19th century (1876–1910) [the so-called Montsouris series (Albert-Levy, 1886)] allowed Volz and Kley (1988) to reconstruct the much lower pre-industrial ozone content of the troposphere.

Besides the 'wet chemical' techniques, optical absorption spectroscopy was very closely linked to early research on atmospheric composition. For instance, Alfred Cornu in 1878 (Cornu, 1879) and Sir Walter Noel Hartley (Hartley, 1880) showed that a sharp drop in solar intensity below 300 nm was due to atmospheric ozone. This was later used by several researchers to determine the atmospheric O_3 column density (Fabry, 1913; Dobson and Harrison, 1926 (see also Chap. 6); Roscoe and Clemitshaw, 1997).

5.2 The Role of Measurements in Atmospheric Chemistry

As mentioned earlier, measurements of atmospheric parameters can serve the goal of coming to a better understanding of our atmospheric environment; however, this can be done in quite a number of different ways. They can be characterised by the time scale of measurements, i.e. there may be specialised 'measurement campaigns' aimed at investigating certain phenomena or 'long-term studies' to follow the evolution of chemical changes. In addition, we might look at the spatial scale of investigation, there may be interest in local, regional or global scale phenomena. Apart from measurements made primarily for scientific purposes, a large number of measurements of atmospheric trace gases are made every day to document the pollution level of our atmosphere and to alert the public of extreme pollution events. In trying to group the measurement efforts, we could use the following criteria:

5.2.1 Long-term Observations

There is great interest in changes in the composition of our atmospheric environment. Some well-known examples include the rise of CO_2 mixing ratio directly observed since the 1950s (e.g. Keeling et al., 1970). Since then, many other trace gases were followed, e.g. chlorofluorocarbon species (CFMs), methane, or the thickness of the stratospheric ozone layer (i.e. the total atmospheric O_3 column density).

Therefore, long-term observations are frequently aimed at monitoring gradual changes in the trace gas composition of atmosphere. They include observations of:

- trends in stratospheric ozone
- change of stratospheric chemistry, e.g. as realised in the Network for the Detection of Stratospheric Change, (NDSC)
- evolution of species supplying halogens to the stratosphere (e.g. CFC and HCFC species)
- trend of the tropospheric ozone mixing ratio (Global Atmospheric Watch, GAW)
- trends of greenhouse gases (CO_2, CH_4, or N_2O)
- trends of gases indicating the atmospheric oxidation capacity (i.e. the ability of the atmosphere to remove trace gases). For instance, ozone, CH_3CCl_3 and ^{14}CO are monitored for this purpose.

In this context, the 'operator dilemma' should be noted. The measurement of a given set of species with the same technique over an extended period of time is frequently not considered a scientific challenge; on the other hand, the success of the data series hinges on the very careful conduction of measurements. Here, the psychological side of the project may be as important as the technological aspects.

5.2.2 Regional and Episodic Studies

Regional and episodic studies seek to investigate the causes, extent, and consequences of air pollution. While routine monitoring is an issue, many fundamental questions can be investigated only by observations made on a regional scale. Typical measurement tasks in this context are:

- Monitoring of air pollutants (like O_3, SO_2, NO, NO_2, hydrocarbons)
- Investigation of urban plume evolution (e.g. with respect to O_3 formation downwind of source regions)
- Study of continental plumes
- Observation of the Antarctic Stratospheric Ozone Hole
- Investigation of polar boundary-layer ozone loss events (the 'tropospheric ozone hole') (Platt and Lehrer, 1995)

5.2.3 Investigation of Fast in-situ (Photo)Chemistry

Studies in 'Smog'-Chambers (frequently called 'Reaction Chambers' or 'Photo-Reactors') allow the suppression of transport processes; thus, the effect of chemistry alone can be investigated. In fact, the phenomenon of tropospheric ozone formation was observed in smog-chambers long before the chemical mechanism (see Chap. 2) was discovered. The main disadvantage of smog chambers is the presence of surfaces; thus, care should be taken to avoid artefacts that may arise from chemical processes at the chamber walls. In order to minimise these problems, very large smog chambers with volumes exceeding $100\,m^3$ were recently built; these facilities offer surface/volume ratios below $1\,m^{-1}$. However, investigation of fast (time scale of seconds) in-situ chemical and photochemical processes (see Sect. 2.5) in the open atmosphere allows us to largely neglect the effect of transport, since transport takes place only at longer time scales. Thus, it is possible to study the chemical processes directly in the atmosphere. In particular, this is true for free radical (OH/HO_2) photochemistry, where the lifetime of the reactive species is on the order of seconds.

5.3 Requirements for Measurement Techniques

Useful measurement techniques for atmospheric trace species should fulfill two main requirements. First, they must be sufficiently sensitive to detect the species under consideration at their ambient concentration levels. This can be a very demanding criterion; for instance, species present at mixing ratios ranging from as low as 0.1 ppt (mixing ratio of 10^{-13}, equivalent to about 2×10^6 molecules/cm^3) to several parts per billion (ppb) (1 ppb corresponds to a mixing ratio of 10^{-9}) can have a significant influence on the chemical processes in the atmosphere (Logan et al., 1981; Perner et al., 1987a). Thus, detection limits from below 0.1 ppt up to the ppb range are required, depending on the application.

Second, it is equally important for measurement techniques to be specific (or selective), which means that the result of the measurement of a particular species must be neither positively nor negatively influenced by any other trace species simultaneously present in the probed volume of air. Given the large number of different molecules present at ppt and ppb levels, even in clean air (see Chap. 2), this is also not a trivial condition.

In practice, there are further desirable properties of a measurement technique, including simplicity of design and use of instruments. In addition, the capability of real-time operation (as opposed to taking samples for later analysis) and the possibility of unattended operations are issues. Further considerations are weight, portability, and dependence of the measurements on ambient conditions. To date, no single measurement technique can – even nearly – fulfill all the diverse requirements for trace gas measurements in the atmosphere. Therefore, many different techniques have been developed (e.g. Clemitshaw, 2004).

5.4 Grouping Measurement Techniques in Categories

An important criterion of the available techniques is, therefore, the degree of specialisation; for instance, there are instruments for the measurement of a single species or parameter (like the thermometer that can measure only the temperature). Simply speaking, we distinguish between specialised techniques ('box per species techniques') and universal techniques capable of determining many parameters with a single instrument.

Another fundamental property of instruments is their ability to make **in-situ** or **remote sensing** measurements. While in-situ measurements come close to the ideal to determine trace gas concentrations in a 'spot' in space that is usually very close to the instrument, remote sensing techniques allow measurements from a large distance, perhaps as far as from a satellite instrument in the earth's orbit. On the other hand, remote sensing instruments usually average the trace gas concentrations over a large volume of air. Remote sensing techniques always rely on the sensing of electro-magnetic radiation, i.e. they are spectroscopic methods.

To date, a large variety of measurement techniques for atmospheric trace gases (and other atmospheric parameters) have been developed. A few examples of instruments belonging to either category include:

- Gas chromatography (GC, universal technique, in-situ)
- Optical spectroscopy (universal technique, in-situ, remote sensing)
- Mass spectrometry (MS, universal technique, in-situ)
- Chemiluminescence (e.g. for the detection of NO or O_3)
- Chemical amplifiers for the detection of peroxy radicals (e.g. Cantrell et al., 1984, 1993; Clemitshaw et al., 1997)
- Photoacoustic detection (e.g. Sigrist, 1994b)
- Electrochemical techniques
- Matrix isolation – electron spin resonance (MI-ESR) (e.g. Mihelcic et al., 1985)
- Derivatisation – HPLC or Hantzsch reaction (e.g. for determination of CH_2O) (Fung and Grosjean, 1981; Nash, 1953)

- 'Bubbler' combined with wet chemistry, colorimetry (e.g. West and Gaeke, 1956; Saltzman, 1954), or ion-chromatography (IC)

In this context, spectroscopic techniques are a powerful variety. These techniques are highly sensitive, very specific, universally usable, give absolute results, and have the potential for remote sensing. It is, therefore, not surprising that spectroscopic techniques assume a unique role among many other

Table 5.1. Overview of species of relevance to atmospheric chemistry research and measurement techniques

Species	UV/vis	FT-IR	TDLS (IR)	GC	MS (CIMS)	Fluoresc. Chemolum.	Other
NO	O	O				+	
NO_2	+	O	+			+	MI-ESR[a]
NO_3	+						MI-ESR, LIF
HNO_2	+						Denuder
HNO_3		O	O				Denuder
OH	+				O		LIF
HO_2/RO_2			?		+		LIF, Ch. A[b]
H_2O_2		O	+			+	
O_3	+	O	O			O	Electroch.[c]
HCHO	+	O	+				Derivat.[d]
RCHO							Derivat.[d]
Alkanes				+			
Olefines				+	O		
Aromatic	+			+	O		
CO				+		+	
DMS				+			
SO_2	+					+	
N_2O			+	+			
CFC's		+		+			
HX[e]		+					Wet. Chem.
XO[e]	+				O	+	
HOX[e]			?				

Symbols denote: well measurable(+), measurable(O), not measurable (empty field).

Abbreviations: UV/vis, UV/visible absorption spectroscopy; FT-IR, Fourier-Transform IR Spectroscopy; TDLS, tunable diode laser spectroscopy; GC, gas chromatography; MS (CIMS), mass spectrometry (Chemical Ionisation MS).

[a]Matrix isolation – electron-spin-resonance.
[b]Chemical amplifier.
[c]Electrochemical cell.
[d]Derivatisation + HPLC.
[e]X = halogen atom (F, Cl, Br, I).

methods that are in use today. In this chapter, we will, therefore, have a closer look at spectroscopic techniques.

Today, atmospheric chemistry has a comprehensive arsenal of measurement techniques at its disposal, Table 5.1 gives an overview of popular trace gas measurement techniques for a series of key species relevant for studies of atmospheric chemistry. Among a large number of specialised techniques (like the gas-phase chemoluminescence detection of NO), universal techniques are of great interest due to their relative simplicity – a single (though perhaps relatively complicated by itself) instrument can measure a large range of species – these techniques have gained interest.

5.4.1 In-situ Versus Remote Sensing Techniques

Remote sensing techniques allow detection of the properties of an object from a distance. Using these techniques, the trace gas composition in atmosphere can be measured at a point that is remote from the probing instrument. Examples are LIDAR instruments (see Sects. 5.7.3 to 5.7.5) and observations of trace gas distributions from space. In fact, most DOAS applications constitute a remote detection of trace gases. In contrast to that, in-situ instrumentation measures trace gas concentrations (or other parameters) at a particular location – ideally at a point in space. There are a series of applications where localised measurements are very desirable, e.g. the determination of strong spatial gradients. Since many trace gases have a strong vertical gradient close to the ground, the observation of these gradients requires measurements that are localised at least in one (vertical) dimension. It should be noted, however, that in-situ measurements frequently require relatively long integration times, i.e. they average the concentration over a period (t_m) of time. As a consequence, they will average over the distance d_m given by:

$$d_m = t_m \cdot v_w,$$

where v_w denotes the wind speed.

Thus, an integration time of 5 min at $2\,\mathrm{m\,s^{-1}}$ wind speed already translates to a spatial average of $d_m = 600\,\mathrm{m}$.

5.5 Experimental Evidence for the Presence of Radicals in the Atmosphere

As an example of the combination of measurements by different technologies to solve a difficult problem in atmospheric chemistry, we briefly discuss both direct and indirect evidence for the presence of free radicals in the atmosphere.

Direct detection of free radicals has proven notoriously difficult because of the high reactivity and low concentration of these species in the atmosphere. For instance, although OH-initiated reaction chains are central to our understanding of atmospheric chemistry (see Sect. 2.4), direct experimental verification of this theory took a long time. From the suggestion of the central role of OH in the degradation of trace gases and ozone formation (see Chap. 2) in the early 1970s, it took more than a decade before the first reliable direct observations of atmospheric OH became available (see Table 5.2c). Only during the last decade have tropospheric OH measurements of very high quality been reported (e.g. Dorn et al., 1996; Hofzumahaus et al., 1996; Brauers et al., 1996; Platt et al., 2002; Bloss et al., 2003; Holland et al., 2003; see also Table 5.2). Although these data match the model predictions, they do not yet constitute a data set that covers all the conditions of the atmosphere. On the other hand, there has been a substantial amount of indirect evidence for the

Table 5.2. Evidence for the presence of OH radicals in the troposphere

Evidence	Literature (not intended to be complete)
Indirect	
Photochemical ozone formation	e.g. Crutzen, 1974
Pattern of VOC – degradation	e.g. Calvert, 1976
NO_X – dependence of peroxide formation	Tremmel et al., 1993
Global budgets of CO, ^{14}CO, CH_3CCl_3	Volz et al., 1981; Brenninkmeijer et al., 1992; Prinn et al., 1987
Semi-direct	
Detection by Chemical Ionisation Mass-Spectrometry	Eisele and Tanner, 1991; Eisele et al., 1994, 1996; Berresheim et al., 2000
Detection by in-situ oxidation of ^{14}CO	Campbell et al., 1979; Felton et al., 1990
Direct	
Long-Path Differential Optical Absorption Spectroscopy (DOAS) in the UV	Perner et al., 1976, 1987b; Platt et al., 1988; Mount and Eisele, 1992; Mount, 1992; Dorn et al., 1996; Brandenburger et al., 1998
Laser-Induced Fluorescence (LIF)	Hard et al., 1992, 1995; Holland et al., 1995, 1998, 2003; Hofzumahaus et al., 1996, 1998; Brauers et al., 1996; Dorn et al., 1996; Brune et al., 1995, 1998; Creasey et al., 1997; Hausmann et al., 1997; Carslaw et al., 1999; George et al., 1999; Kanaya et al., 2000, 2001; Heard and Pilling 2003; Hard et al., 2002; Bloss et al., 2003; Holland et al., 2003

presence of OH radicals in the atmosphere. In addition, a number of techniques were developed that, while not identifying the OH molecule directly, gave very good and quantitative measure for OH abundance [like CO, ^{14}CO (Volz et al., 1981; Brenninkmeijer et al., 1992), CH_3Cl (e.g. Prinn et al. 1987)]. In the following section, these methods shall be referred to as semi-direct. Table 5.2 summarises the available evidence for the presence of OH radicals in the troposphere. Today it can be concluded that, although direct, unequivocal identification of these species in the atmosphere can only be performed in a few cases; there is an enormous amount of indirect evidence of their presence. Thus, the central importance of HO_X for atmospheric chemistry can be assumed to be proven beyond reasonable doubt.

In the case of peroxy radicals, the situation is less satisfactory. To date, there are very few direct measurements of HO_2 and some other peroxy radicals by matrix isolation – electron spin resonance detection (see Table 5.3). Infrared spectroscopic detection of HO_2 has not yet been achieved in the troposphere. On the other hand, there are techniques for indirect

Table 5.3. Evidence for the presence of HO_2/RO_2 radicals in the troposphere

Evidence	Literature (not intended to be complete)
Indirect	
Ozone formation	Crutzen, 1974
Pattern of H_2O_2 (org. peroxide) formation	Tremmel et al., 1993
Change in the Leighton Ratio ('Missing oxidant')	Parrish et al., 1986
Semi-direct	
OH–LIF after titration by NO	Hard et al., 1995; Brune et al., 1995, 1998; George et al., 1999; Creasey et al., 1997; Kanaya et al., 2000; Hofzumahaus et al., 1996; Holland et al., 2003; Bloss et al., 2003
Chemical Amplifier (ROX-Box) Measurements	Stedman and Cantrell, 1981; Cantrell et al., 1984, 1993; Hastie et al., 1991; Monks et al., 1996; Clemitshaw et al., 1997; Carslaw et al., 1997; Perner et al., 1999; Mihelcic et al., 2003
Direct	
Matrix-Isolation	Mihelcic et al., 1985, 1993, 2003
Electron-Spin-Resonance Detection	
Tuneable Diode-Laser Spectroscopy (TDLS)	Werle et al., 1991

detection of HO_2 available that rely on the conversion to OH via reaction with NO (Table 5.3). Hydroxyl radicals can then be detected by the well-developed LIF technique. Another widely used 'semi direct' technique employs a chemical amplifier (ROX-Box) to convert NO added to the sampled airflow into NO_2. Although this basically simple technique appears to have severe limitations, it has been demonstrated that it can be used to investigate the tropospheric HO_X system, especially in clean air (e.g. Cantrell et al. 1984; Monks et al., 1996).

The first direct detection of NO_3 radicals (by DOAS) in the troposphere dates back to 1979 (see Table 5.4), but the idea of considerable influence of NO_3 on tropospheric chemical cycles only slowly gained acceptance. Since there appears to be no simple way to derive the global (or regional) average NO_3 concentration, this quantity has to be determined from long-term measurement series, which have recently begun (e.g. Heintz et al., 1996; Plane and Nien 1991]. Again, there is considerable indirect evidence for the importance of NO_3 radical reactions in the troposphere; these include (1) the lack of seasonal dependence of the NO_Y deposition (e.g. Calvert et al., 1985), suggesting

Table 5.4. Evidence for the presence of NO_3 radicals in the troposphere

Evidence	Literature (not intended to be complete)
Indirect	
Seasonal variation of the NO_Y deposition	Calvert et al., 1985
Night-time observation of RO_2 formation	Platt et al., 1990; Mihelcic et al., 1985; Gölz et al., 1997
Night-time degradation pattern of VOCs	Penkett et al., 1993
Semi-direct	
Conversion to NO + chemoluminescence	-
Direct	
Differential Long-Path Absorption Spectroscopy (LP-DOAS)	Noxon, et al., 1980; Noxon, 1983; Platt et al., 1979, 1981, 1984; Plane and Nien, 1991; Heintz et al., 1996; Aliwell and Jones, 1998; Allan et al., 1999, 2000; Martinez et al., 2000; Geyer et al., 2001, 2002, 2003
Cavity Ringdown Spectroscopy	King et al., 2000; Brown et al., 2001–2003
Matrix-Isolation Electron-Spin-Resonance Detection	Mihelcic et al., 1985, 1993, 2003
Laser-Induced Fluorescence (LIF)	Wood et al., 2003; Matsumoto et al., 2005

non-photochemical NO_X–NO_Y conversion via NO_3–N_2O_5 during winter; (2) observation of night-time radical formation (e.g. Mihelcic et al., 1993; Gölz et al., 2001); (3) degradation patterns of hydrocarbons (Penkett et al., 1993), details are discussed in Sect. 2.6.

Recently, a substantial amount of direct and indirect evidence has been accumulated for the presence of halogen atoms and halogen oxide radicals in certain parts of the troposphere (e.g. Wayne et al., 1991; Platt and Hönninger, 2003; see also Table 5.5). For instance, the degradation pattern of hydrocarbons, in particular in the Arctic as well as at remote mid-latitude areas, points to a noticeable role of Cl and Br atoms in the oxidation capacity of the troposphere. In addition, the events of ozone loss on extended areas in the polar

Table 5.5. Evidence for the presence of halogen radicals in the troposphere

Evidence	Literature (not intended to be complete)
Indirect	
'Hydrocarbon clock' observations (Cl, Br-atoms),	Solberg et al., 1996; Jobson et al., 1994; Rudolph et al., 1999; Singh et al., 1996; Wingenter et al., 1996; Ramacher et al., 1997, 1999; Röckmann et al., 1999; Platt et al., 2004
Sudden loss of ozone in the (polar) boundary layer	e.g. Barrie et al., 1988; Platt and Lehrer, 1996
Semi-direct	
Atomic fluorescence after titration by NO (BrO, ClO)	Toohey et al., 1996
Chemical Amplifier (ROX-Box) measurements	Perner et al., 1999
Direct	
Differential Long-Path Absorption Spectroscopy (LP-DOAS), detection of BrO, ClO, IO, OIO	Hausmann and Platt, 1994; Unold, 1995; Tuckermann et al., 1997; Platt and Lehrer, 1996; Kreher et al., 1997; Hegels et al., 1998; Wagner and Platt, 1998; Richter et al., 1998; Martínez et al., 1999; Alicke et al., 1999; Allan et al., 2000, 2001; Wittrock et al., 2000a; Frieß et al., 2001; Wagner et al., 2001; Matveev et al., 2001; Leser et al., 2002; Stutz et al., 2002; Hönninger and Platt, 2002, v. Roozendael et al., 2002; Pundt et al., 2002; Bobrowski et al., 2003; Hollwedel et al., 2004; Hönninger et al., 2004b,c; Frieß et al. 2004a; Saiz-Lopez, 2004a,b; Bobrowski, 2005; Zingler and Platt, 2005

boundary layer are best explained by BrO- and ClO-catalysed ozone destruction (e.g. Barrie and Platt, 1997), as discussed in Sect. 2.9.

In summary, we have a considerable body of indirect as well as direct evidence for the presence of several types of free radicals in the atmosphere.

5.6 Spectroscopic Techniques

To date, no single measurement technique can – even nearly – fulfill all the requirements; therefore, in a particular application the selection of a technique will be based on the specific requirements: What species are to be measured? Is simultaneous determination of several species necessary? What is the required accuracy, time resolution, and spatial resolution? Other aspects to be considered are logistic requirements like power consumption, mounting of light sources or retro-reflectors (see below), or accommodation of the instrument on mobile platforms.

Spectroscopic techniques can be broadly divided, on one hand, into methods relying on absorption of radiation directed to the sample from some source, or, on the other hand, into spectroscopic analysis of radiation emitted by the sample itself. Further divisions can be made according to the wavelength range used and thus the kind of internal state of the sample molecules the radiation interacts with. Figure 5.1 illustrates these different categories by giving a 'family tree' of spectroscopic techniques. In this section we give a brief overview of some spectroscopic techniques used to analyse atmospheric composition.

5.6.1 Microwave Spectroscopy

Spectroscopy in the microwave and sub-mm wavelength range can, in principle, be used in active and passive configurations. To our knowledge, no detection of atmospheric gases with active microwave instruments (i.e. with instruments employing their own radiation sources) has been made. However, the measurements of atmospheric parameters (clouds, precipitation, and turbulence) with RADAR techniques are common.

Passive measurements of stratospheric species by microwave and sub-mm wave emission are state of the art (e.g. Janssen, 1993). This technique registers the thermal emission due to rotational transitions of the atmospheric molecules. For instance, the ClO-molecule radiates at $\nu_o = 649.448\,\mathrm{GHz}$, corresponding to 0.46 mm wavelength ($18_{1/2} \rightarrow 17_{1/2}$ transition of ^{35}ClO), but HOCl is also detectable by this technique. The line width is dominated by collisional (pressure) broadening, which by far dominates over Doppler-broadening ($\Delta\nu_D/\nu_0 \approx v_{\mathrm{molec}}/c \approx 10^{-6}$), as described in Sect. 3.6. While this strong variation in line width with pressure, and thus altitude of the absorbing molecule, allows the retrieval of vertical profiles from a thorough analysis of the recorded

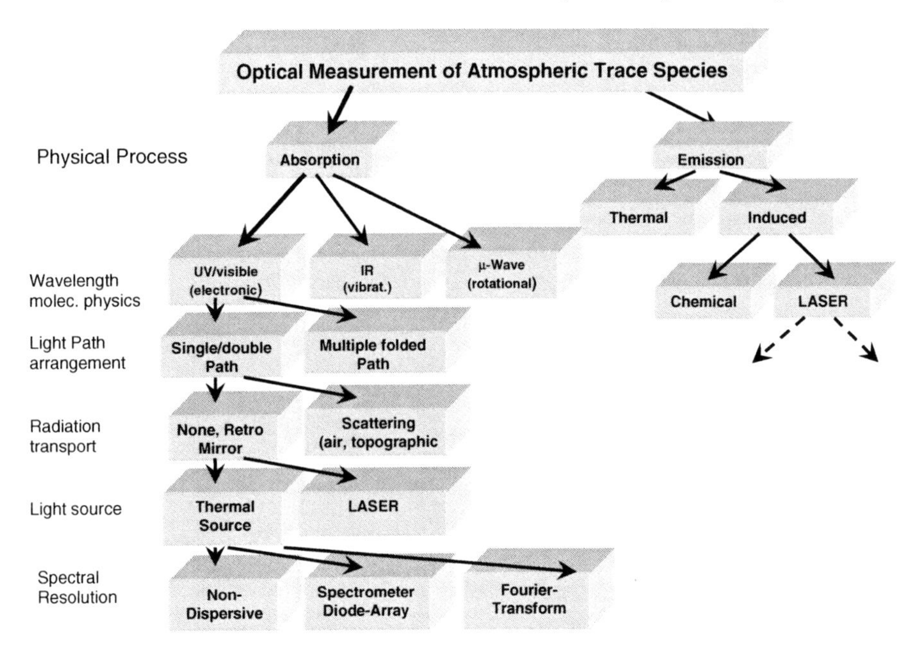

Fig. 5.1. 'Family tree' of spectroscopic measurement techniques for atmospheric trace species. Absorption spectroscopy can be categorised by the type of light source – natural (e.g. sunlight) or artificial (e.g. incandescent-, arc lamps, or lasers), the arrangement of the absorption path, the method of obtaining spectral information (non-dispersive, dispersive techniques, type of spectrometer, etc.), or other characteristics of the particular technique

line shape, it also limits the detection sensitivity at an atmospheric pressure exceeding a few mbars. Thus, for ClO, the best sensitivity is reached in the upper stratosphere, while detection at lower stratospheric or even tropospheric altitudes (i.e. below about 15–25 km) is usually difficult (e.g. Janssen, 1993; Klein et al., 2002; Livesey et al., 2003). The microwave and sub-mm wavelength range is presently not used for measurements in the troposphere; due to the relatively large pressure broadening of the lines, it would probably require measurements at reduced pressure.

5.6.2 IR Spectroscopy

IR spectroscopy has been in use for several decades, and was initially developed for the detection of atmospheric CO_2 by non-dispersive instruments ['Ultra-Rot AbSorption' (German for 'ultra-red absorption'), URAS]. More modern instruments are based on Fourier transform (FT) techniques to measure HNO_3, CH_2O, HCOOH, H_2O_2, and many other species on km path length multiple-reflection cells (e.g. Pitts et al., 1977; Tuazon et al., 1980). The sensitivity is in the low ppb range; thus, those instruments appear to be best suited for studies of polluted air. The technique has been applied in three modes of operation:

1. Active operation, where an artificial broadband light source, usually a thermal radiator, is used (Pitts et al., 1977; Tuazon et al., 1980; Galle et al., 1994).
2. Active operation, where a tuneable diode laser serves as light source. During the last two decades, tunable diode laser spectrometers (TDLS) have been developed as field-usable instruments, which were successfully used to measure HNO_3, NO, NO_2, CH_2O, H_2O_2 and other species at sub-ppb levels (e.g. Harris et al., 1989; Schiff et al., 1990; see also Sect. 5.7.1).
3. Passive operation using thermal emission from the trace gases under consideration. A particular example is the Michelson Interferometer for Passive Atmospheric Sounding (MIPAS) (Fischer, 1993; Fischer and Oelhaff, 1996; Notholt et al., 1997).

5.6.3 UV/Visible Absorption Spectroscopy

At ambient temperatures, thermal emission of UV/visible radiation is completely negligible; nevertheless, active and passive arrangements can be used, the latter technique relying on radiation from naturally hot objects, e.g. the sun.

The most common optical arrangement is shown in Fig. 6.1; the light source and the receiving system are typically separated by several kilometres. UV/visible absorption spectroscopy can be used in both 'active' and 'passive' modes; in the active mode, an artificial light source (see Chap. 6) provides the radiation, while in the passive mode natural light sources, usually the sun or stars, act as source of radiation. The strength of these techniques lies in good specificity, the potential for real-time measurements, and the absence of wall losses. In particular, the first property makes spectroscopic techniques more suitable for the detection of unstable species like OH radicals or nitrate radicals (see Chap. 10). Limitations in the systems using a separated light source and receiving system are due to logistic requirements (the need for electric power at two sites several kilometres apart, but in sight of each other) in the case of unfolded path arrangements; in addition, conditions of poor atmospheric visibility (see Sect. 4.7) can make measurements with this technique difficult.

LIDAR techniques, on the other hand, combine the absence of wall losses and the good specificity with somewhat smaller logistic requirements and the capability to make range-resolved measurements. In contrast to LIDAR, most of the above systems can only make point or path-averaged measurements. Unfortunately, this advantage of LIDAR is usually obtained at the expense of reduced sensitivity (see Sect. 5.7.3).

5.7 Selection Criteria for Spectroscopic Techniques

The important technical criteria of spectroscopic instruments are the wavelength region used (see Fig. 5.1), the physical principle (i.e. absorption or emission spectroscopy), the arrangement of the light path (path in the open atmosphere or enclosed – frequently folded – path), or the type of light source used. The following techniques are presently employed to measure atmospheric trace gases:

- Tunable Diode Laser Spectroscopy (TDLS)
- Photo Acoustic Spectroscopy (PAS)
- Light Detection And Ranging (LIDAR)
- Differential Absorption LIDAR (DIAL)
- Laser-Induced Fluorescence (LIF)
- Cavity-Ringdown Spectroscopy (CRD)
- Mask Correlation Spectroscopy
- Differential Optical Absorption Spectroscopy (DOAS)

While, in principle, most techniques (i.e. TDLS, LIDAR, DIAL, LIF, and DOAS) allow light paths (or sensitive volumina, in the case of LIF) in the open atmosphere, some techniques lend themselves more to enclosed light paths.

5.7.1 Tuneable Diode Laser Spectroscopy (TDLS)

The idea of tunable diode laser spectroscopy is to use a narrow-band, frequency-variable laser source to scan a suitable trace gas absorption structure. This type of laser can be frequency tuned e.g. by varying the diode current. In practice, a sawtooth-like waveform of the laser current results essentially in a linear scan across a certain spectral interval, typically a small fraction of a wave number. Thus, in principle, no further wavelength selective elements are required. Early applications relied on lead–salt (Harris et al., 1989; Schiff et al., 1990; Sigrist, 1994) diode lasers operating in the mid-infrared (about 3–30 μm), thus covering the fundamental vibrations of many molecules of atmospheric interest. The disadvantages of lead–salt diode lasers are their limited commercial availability and the requirement for cooling to temperatures around (and frequently below) the boiling point of liquid nitrogen. While diode lasers are very compact and low power-consuming devices, the need for liquid nitrogen cooling or closed cycle He-coolers and vacuum thermal insulation, in practice, offsets these advantages. In recent years, further development of tuneable diode lasers yielded devices operating around room temperature in the near IR (about 1.6–2 μm); here communication laser diodes operating at room temperatures are available, which allow very compact and lightweight instruments (e.g. Durry et al., 1999; Gurlit et al., 2005). On the other hand, in the near IR, typically weaker overtone-bands are available. In addition, recently, quantum cascade lasers became available, which can generate narrow band emission even in the mid-IR (e.g. Kosterev and Tittel, 2002; Jimenez et al., 2004).

During the last two decades, TDLS became field-usable instruments, and were successfully used to measure HNO_3, NO, NO_2, CH_2O, and H_2O_2 at sub-ppb levels (Harris et al., 1989; Schiff et al., 1990). In the usual arrangement coupled to a multiple reflection cell, the strength of TDLS lies in the mobility of the instrument, allowing measurements on board ships and airplanes, combined with high sensitivity. The limitations are due to the necessity to operate under low pressure (in many applications), thus introducing possible losses at the walls of the closed measurement cells. Moreover, at present, diode-laser technology is still quite complex. Therefore, the development of laser diodes is in demand, and diodes are mainly developed for applications where a mass market is seen (CD players, fibre optical communication, etc.). While the developments in the latter areas are quite impressive, diodes optimised for spectroscopic purposes, especially in the visible and ultraviolet spectral ranges, are still difficult to obtain.

5.7.2 Photo Acoustic Spectroscopy (PAS)

Photo Acoustic Spectroscopy is based on the detection of the pressure change occurring as a consequence of the temperature rise due to the energy absorbed by trace gas molecules illuminated by radiation at suitable frequency. Obviously, the absorption of radiation by atmospheric trace gases is very minute, thus giving rise to only small temperature and pressure changes. On the other hand, extremely sensitive devices for detecting periodic atmospheric pressure changes are available in the form of microphones for the detection of sound waves (e.g. Sigrist, 1994).

Thus, a typical PAS instrument consists of an intensity-modulated light source – usually a laser – illuminating the interior of a cell (e.g. a tube of a few millimetres of inner diameter and a fraction of a meter length) equipped with a microphone 'listening' to the periodic pressure changes occurring at the frequency of the modulation. The weak signal is sometimes enhanced by tuning the modulation frequency to an acoustic resonance of the absorption cell, leading to a signal increase by the 'quality factor' Q of the acoustic resonator (Sigrist, 1994).

Note that photoacoustic detectors directly measure the difference $I_0 - I$ (approximately proportional to $D \cdot I_0$ and thus the trace gas concentration) rather than determining I and I_0 separately.

5.7.3 Light Detection And Ranging (LIDAR)

The term LIDAR was coined to resemble the well-known RADAR (RAdiowave Detection And Ranging) (e.g. Rothe et al., 1974; Hinkley, 1976; Svanberg, 1992; Sigrist, 1994). In principle, short pulses of a strong, collimated light source, typically a pulsed laser, are emitted into the atmosphere. By analysing the temporal evolution of the intensity backscattered from the atmosphere, the spatial distribution of scattering and extinction along the direction of the emitted (and received) radiation can be deduced, as illustrated in Fig. 5.2.

The distance from which the radiation intensity (under consideration) is scattered back is given by:

$$R = \frac{c \cdot t}{2} ,$$

where c is the speed of light, and t is time (after emission of the laser pulse) at which the signal reaches the detector.

The general LIDAR equation is:

$$E(\lambda, R, \Delta R) = K \cdot E_0 \cdot c_S(R)\, \sigma_{SR} \frac{\Delta R}{R^2} \cdot \\ \times \exp\left(-2 \int_0^R \left[\sigma_A(r) \cdot c_A(r) + \sigma_S(r) \cdot c_S(r)\right] dr\right) , \quad (5.1)$$

with the following meanings to the terms:

$E(\lambda, R, \Delta R)$ denotes radiation energy received from a volume of air between R, $R + \Delta R$;
ΔR is the distance interval to be averaged over (corresponds to $\Delta t = \frac{2\Delta R}{c}$);
K is a constant of the system (describes the influence and size of the receiving system);
E_0 is the radiation energy emitted by the laser;
$c_s(R)$is the concentration of backscattering centres (molecules, aerosol particles);
σ_{SR} is the backscatter cross-section;
σ_A is the absorption cross-section (aerosol or gas);
$c_A(R)$ is the concentration of the absorbers (i.e. gas molecules);
σ_S is the total scattering cross-section (note that $\sigma_S \neq \sigma_{RS}$!).

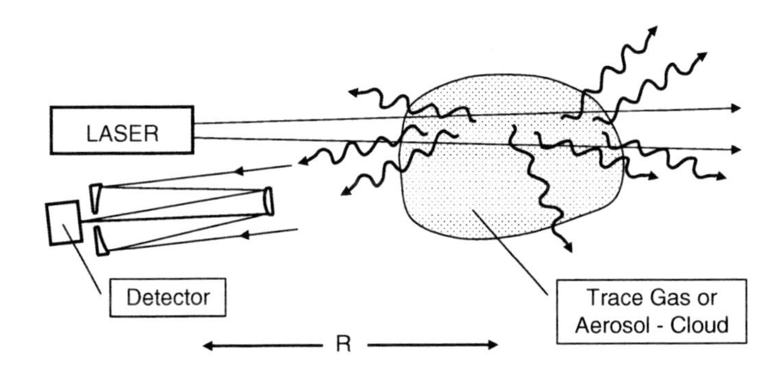

Fig. 5.2. The principle of the LIDAR technique

It is worth noting that the returned LIDAR signal is proportional to R^{-2}, while the returned RADAR signal varies with R^{-4}. This is due to the fact that the scattering volume increases with R^2, while in the case of RADAR the object (e.g. an airplane) is of constant size, and thus both the transmitted signal arriving at the target and the returned signal reaching the receiver decrease proportional to R^{-2}.

The LIDAR techniques combine the absence of wall losses with the unique capability to make range-resolved measurements along the line of sight. In contrast to that, the systems discussed earlier can only make point measurements or path-averaged measurements. In the form described up to this point, LIDAR systems have been successfully used to map aerosol distributions. There are, however, two central problems with LIDAR systems:

1. The signal [see (5.1)] depends on backscattering as well as on total scattering (or extinction) properties of the atmosphere, which are (in the case of Mie scattering) difficult to separate (the ratio σ_S/σ_{SR} is usually unknown and must be estimated from a-priori information unless the Raman–LIDAR technology is used).
2. The backscattered signal is usually very weak, and thus even when high-energy lasers are used, only a relatively small number of photons is received. Therefore, the above advantages are usually obtained at the expense of high sensitivity.

5.7.4 Differential Absorption LIDAR (DIAL)

In order to measure the distribution of trace gases, at least two different wavelengths are required (in contrast to the 'ordinary' (= aerosol-)LIDAR) (Svanberg, 1992; Sigrist, 1994). The DIAL equation is obtained by dividing two general LIDAR equations (5.1) for two different wavelengths λ_1 and λ_2, respectively:

$$\frac{E(\lambda_2, R)}{E(\lambda_1, R)} = \exp\left(-2\left(\sigma_A(\lambda_2) - \sigma_A(\lambda_1)\right) \cdot \int_0^R c_A(r)\, dr\right) . \qquad (5.2)$$

Thereby it is assumed that σ_{SR} and σ_S are the same for λ_1 and λ_2, respectively. This assumption appears justified as long as $\Delta\lambda = \lambda_2 - \lambda_1$ is sufficiently small (a few nanometres). Obviously, the two wavelengths are chosen so that (while their difference is as small as possible) the difference in the absorption cross-sections $\Delta\sigma = \sigma_A(\lambda_2) - \sigma_A(\lambda_1)$ is as large as possible, i.e. one wavelength is at the centre of an absorption line, the other besides the line.

5.7.5 White Light LIDAR

Recently, suggestions and first experiments of two varieties of the LIDAR technique were made using broadband 'white light' sources in conjunction with wavelength dispersing detectors making use of the DOAS principle, see Sect. 5.7.9. If these techniques prove viable, they could combine the advantages of LIDAR and DOAS. Presently, there are two approaches:

1. A LIDAR principle that combines the differential UV-visible absorption spectroscopy and the classical LIDAR technique. The novel element of the system is the use of an imaging spectrometer in conjunction with a two-dimensional CCD detector array to simultaneously spectrally and temporally resolve backscattered radiation. The 'white' light (i.e. broadband radiation) required for this LIDAR approach is provided either by a flash-lamp or a broadband dye laser (of 10–20 nm full width at half maximum tuneable across the UV-visible spectral region). Thus, simultaneous range-resolved measurement of multiple molecular species is possible by the DOAS technique. A first application demonstrated absorption studies in the spectral regions where NO_3 and H_2O absorb using both elastic and inelastic (Raman) backscattered radiation (Povey et al., 1998; South et al., 1998). In addition, the technique has applicability for a wide range of molecules, including O_3, NO_2, and other spectrally structured absorbers, and for atmospheric temperature sounding, which may be derived from either rotational Raman return or temperature-dependent absorptions such as those of O_2.
2. In this approach, ultra-short (femtosecond) laser pulses produce a 'beam' of white light in the atmosphere, which can then be observed with a separate wavelength-dispersing detector system. Atmospheric constituents will leave their spectroscopic fingerprint on the distance between the 'beam' and the detector, which can then be detected by DOAS techniques (e.g. Rairoux et al., 2000; Kasparian et al., 2003).

5.7.6 Laser-Induced Fluorescence (LIF)

Laser-Induced Fluorescence is a very sensitive and specific technique for the detection of atmospheric trace gases. It relies on the excitation of trace gas molecules by absorption of (laser) radiation, the frequency of which corresponds to a suitable transition from the molecule's ground state to an (electronic) excited state, see Figure 3.6. The number of fluorescence photons detected is then proportional to the atmospheric trace gas concentration.

A central problem with LIF is the separation of Rayleigh scattered exciting radiation from the fluorescence signal. While Rayleigh scattering cross-sections of air molecules (i.e. O_2 and N_2) are typically six to eight orders of magnitude smaller than trace gas absorption cross-sections, typical trace gas mixing ratios are 10^{-12} to 10^{-9}. Combined with the fact that fluorescence efficiencies can be orders of magnitude below unity, scattered excitation radiation must be suppressed by many orders of magnitude. Several techniques have been developed to achieve this goal, which include:

1. Excitation at higher (vibrational) levels above the excited state in which fluorescence can be observed. The detected wavelengths (red) are thus shifted from the excitation wavelength, and can be separated by suitable filters (wavelength filtering).
2. Use of short laser pulses for excitation. The fluorescence signal can be recorded after the excitation signal has decayed and thus separated. This technique frequently requires low pressure in order to avoid too rapid a decay of the fluorescence signal by collision quenching and hence its denomination 'temporal' or 'baric' filtering (Hard et al., 1979).

Other potential problems in the technique are photochemical formation of species to be detected by excitation radiation [e.g. OH by O_3 photolysis (Ortgies et al., 1980; Davis et al., 1981; Shirinzadeh et al., 1987)], or saturation of the transition under consideration. In addition, absolute calibration from 'first principles' has been found to be extremely difficult (e.g. Hofzumahaus et al., 1996, 1998; Holland et al., 2003), and so experimental calibration is necessary (e.g. Hofzumahaus et al., 1996, 1997, 1998; Schultz et al., 1995).

5.7.7 Cavity-Ringdown (CRDS) and Cavity Enhanced Spectroscopy (CEAS)

Cavity-Ringdown and Cavity Enhanced Spectroscopy are relatively new techniques that make use of the effect of atmospheric absorbers on the quality of a usually passive optical resonator (a 'cavity'), see also Sect. 7.8.3 (e.g. Paldus and Zare, 1999; Brown, 2003). After initially exciting the resonator with a laser and then switching off the laser, the decay (or 'ringdown') of the resonator is observed, which is influenced by the extinction of light due to trace gases present in the resonator cavity. Alternatively, if excited with continuous radiation, the resonator will exhibit certain attenuation, which can be monitored. The energy loss per 'round trip' of the radiation in an optical resonator with no absorber is given by the reflectivity of the two mirrors, see Sect. 7.8.3. In the case of very weak absorbers, the quality, Q, of the cavity is only determined by mirror reflectivity, R_M, which can reach 0.9999; thus, effective light path lengths on the order of many kilometres can be reached under favourable conditions with resonator lengths below 1 m. On the other hand, in some respects, the limitations of CRDS are similar to the limitations of DOAS in that the effective light path will be limited by visibility. CRDS has been developed into a useful tool for laboratory measurements; several measurements in the atmosphere, mostly of NO_3 radicals, were reported (e.g. King et al., 2000; Ball and Jones, 2003; Brown et al., 2001–2003; Simpson 2003; Ball et al., 2004).

5.7.8 Mask Correlation Spectroscopy (COSPEC)

Mass Correlation Spectroscopy was originally conceived as a tool for oil exploration by the detection of atmospheric iodine (I_2) vapour, which was assumed to be frequently associated with oil deposits. The fundamental idea of COSPEC is to analyse the absorption features (i.e. absorption bands) of atmospheric constituents imprinted on scattered sunlight. In modern nomenclature, it is a 'passive' instrument (see Chap. 6). The instrument allowed some degree of remote sensing of atmospheric trace gases. While the original idea of I_2-vapour detection did not prove practical, other applications were found quickly, which included the observation of tropospheric column densities of SO_2 and NO_2 in industrial emission (e.g. Giovanelli et al., 1979; Beilke et al., 1981; Redemann Fischer, 1985) or volcanic plumes (Hoff, 1992).

The COSPEC instrument was originally designed in 1960s (Barringer et al., 1970; Davies, 1970; Davies et al., 1975); it uses an innovative opto-mechanical correlator (e.g. Davies, 1970; Millan and Hoff, 1977; Millan, 1972, 1978, 1980) to identify and quantify the spectra of atmospheric trace gases. In this way, a remarkably compact and reliable instrument could be built, since spectral recognition of the molecules to be measured was done by special opto-mechanics (Fig. 5.3) with relatively simple electronics and without a computer. Since neither fast-scanning, multi-channel detectors nor powerful computers were available at that time, the mask correlation technique was an attractive way to measure total columns.

By making the instrument mobile and horizontally traversing emission plumes, a total integrated concentration cross-section of the plumes can be obtained (Giovanelli et al., 1979). After multiplication with the concentration weighted wind component perpendicular to the cross-section, the total emission from the source (e.g. in $kg\,s^{-1}$) can be deduced.

For around 30 years, the COSPEC instrument has been the principal tool for remote surveillance of volcanic plumes. Although it was not originally designed for volcanological research, it had a major impact on the discipline, not only in volcano monitoring and eruption forecasting but also in estimating the global contribution of volcanic volatiles to the atmosphere (e.g. Weibring et al., 1998). By allowing the measurements of SO_2 and NO_2 flux in airborne plumes, it has fulfilled a vital role in the management of many volcanic crises, notably at Pinatubo in 1991 (Hoff, 1992), and recently at Soufrière Hills Volcano. The COSPEC instrument was a major technological innovation at its time, but is now outdated in several important respects. Problematic aspects of its concept are interferences from other gases, non-linearity, solar Fraunhofer lines, polarisation effects, and multiple scattering in aerosol and clouds. In particular, each compound to be measured required its individual mask, and the system is sensitive to wavelength shifts due to mechanical distortions and temperature variations.

5.7.9 Differential Optical Absorption Spectroscopy (DOAS)

As will be described in detail in the following chapters, the central idea of DOAS is to make use of structured absorption of many trace gases of atmospheric interest while ignoring the rather 'smooth' extinction features due to

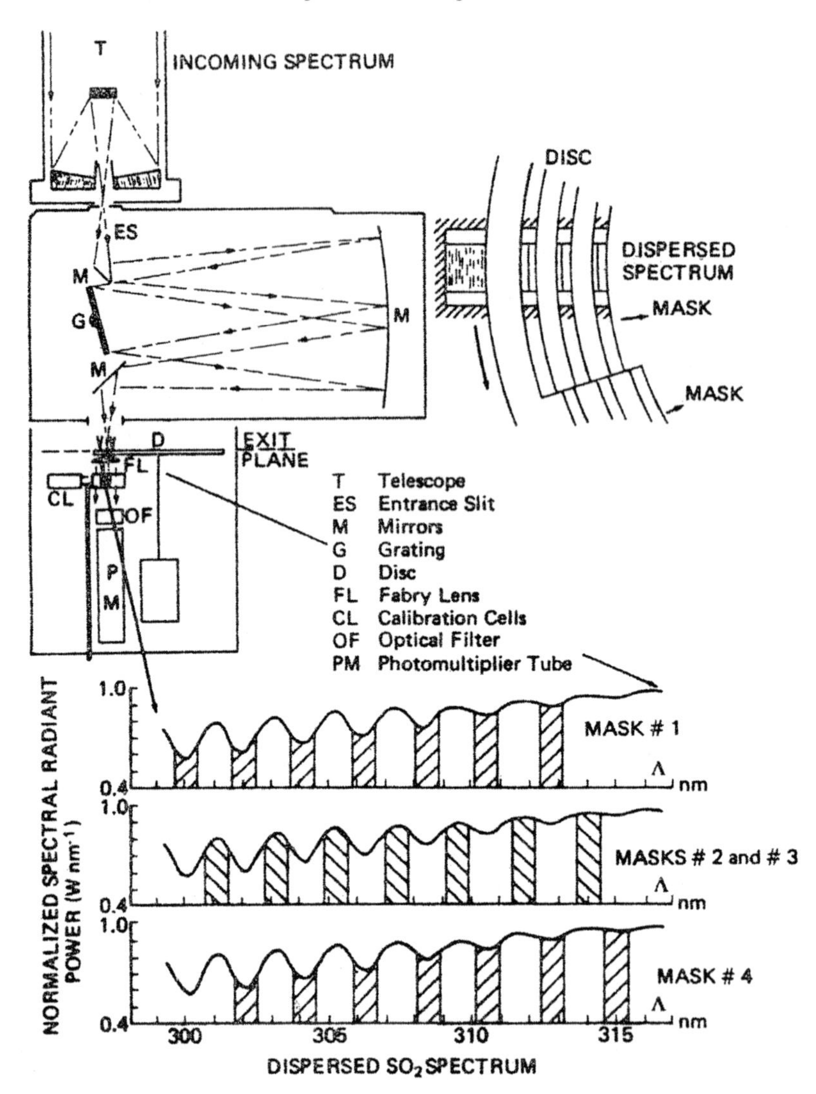

Fig. 5.3. The principle of a COSPEC instrument (from Barringer et al., 1970)

aerosol or instrumental effects. The DOAS principle and its technical realisation builds on some of the earlier developments, superseding them in many respects and supplementing other spectroscopic techniques (e.g. Platt and Perner, 1980; Platt, 1994, 2000; Plane and Smith, 1995; Roscoe and Clemitshaw, 1997; Russwurm, 1999; Finlayson-Pitts and Pitts, 2000; Clemitshaw, 2004).

6

Differential Absorption Spectroscopy

Absorption spectroscopy is a well-established tool for the analysis of the chemical composition of gases. As such, it has played a prominent role in the discovery of the physical and chemical properties of the earth's atmosphere.

6.1 The History of Absorption Spectroscopy

Spectroscopic studies of the earth's atmosphere date back more than 100 years. Some milestones in the investigation of atmospheric composition through spectroscopy include:

1879 – Marie Alfred **Cornu** concludes from the change of the edge of the intensity decay in the UV that a trace species in the earth's atmosphere must be causing the UV-absorption (Cornu, 1879).

1880 – Sir Walter Noel **Hartley** discovers the absorption of UV-radiation (below 300 nm) by ozone. This led to the name Hartley-bands for ozone absorption below 300 nm (Hartley, 1880, 1881).

1880 – M. J. **Chappuis** discovers the absorption of visible light by ozone, which is today called the Chappuis-band. Chappuis also speculates that light absorption by ozone is the reason for the blue colour of the sky (Chappuis, 1880).

1890 – Sir William **Huggins** discovers a new group of lines in the spectrum of Sirius, which are later explained by Fowler and Strutt as absorption of terrestrial ozone. The long wavelength UV-bands of ozone are, therefore, called Huggins-bands today.

1904 – Discovery of the infrared absorption of ozone near 4.8, 5.8, and 9.1–10 µm by Knut Johan **Ångström**.

1913 – Balloon measurements of the UV absorption of ozone up to 10 km altitude by Albert **Wigand** showed essentially no change with altitude.

1918 – John William **Strutt** (better known as Lord **Rayleigh**) concludes that atmospheric ozone must reside in a layer above 10 km altitude above the surface.

U. Platt and J. Stutz, *Differential Absorption Spectroscopy*. In: U. Platt and J. Stutz, Differential Optical Absorption Spectroscopy, Physics of Earth and Space Environments, pp. 135–174 (2008)
DOI 10.1007/978-3-540-75776-4_6

1920 – First ozone column measurements were made by Charles **Fabry** and Henri **Buisson**, who determine a column of about 3 mm (at atmospheric pressure), with large variations.

1925 – First application of a dedicated ozone spectrometer by Gordon Miller Bourne **Dobson** (Dobson and Harrison, 1926).

1926 – Paul **Götz** confirms the theory of an ozone layer by observing the so-called 'Umkehr' effect, and determines its altitude to be about 25 km.

1934 – Direct observation of the ozone layer by UV-spectroscopy by Erich **Regener** (TH Stuttgart).

1948 – Marcel **Migeotte** (Ohio State University) discovers methane and carbon monoxide in the earth atmosphere by near-infrared absorption spectroscopy (Migeotte, 1948, 1949).

1950 – Discovery of the emission bands of the hydroxyl radical (OH) in the nightglow (the Meinel bands of the OH-radical). As a consequence, HO_x-chemistry is viewed in connection with ozone chemistry by David R. **Bates** and Marcel **Nicolet** (1950), and Bates and Witherspoon (1952).

1975 – First detection of OH in the atmosphere by Dieter **Perner** and colleagues using differential optical absorption spectroscopy (Perner et al., 1976).

This list illustrates the role that spectroscopy has played in the measurement of reactive trace gases in the atmosphere, most notably ozone. The reader may notice that the identification and quantification of gases was primarily accomplished by the analysis of atmospheric absorptions. This is still the case in most current applications of atmospheric spectroscopy. The use of emission bands is restricted to the thermal infrared wavelength region (see Chap. 5) or to the excited gas molecules in the upper atmosphere, which emit light at higher energies, i.e. shorter wavelength. Both applications are in use today, but are not the topic of discussion in this book.

The initial use of spectroscopy in the atmosphere concentrated on the identification of various gases. Soon, however, this method was put in use to quantify the concentrations (or column densities) of these species. In particular, the contributions of Dobson, who constructed the first instrument for the regular measurement of atmospheric ozone, should be singled out (Dobson and Harrison, 1926).

This chapter focuses on a modern method to quantitatively measure a large variety of trace gases in the atmosphere. DOAS is now one of the most commonly used spectroscopic methods to measure trace gases in the open atmosphere. At the beginning, we give a general introduction to absorption spectroscopy and DOAS. This is followed by an overview of different experimental approaches of DOAS and a discussion of the precision and accuracy of this method. The last section of this chapter is dedicated to a rigorous mathematical description of the various DOAS applications.

6.2 Classical Absorption Spectroscopy

The basis of the early spectroscopic measurements, and many present quantitative trace gas analytical methods in the atmosphere and the laboratory, is Lambert–Beer's law, often also referred to as Bouguer–Lambert law. The law was presented in various forms by Pierre Bouguer in 1729, Johann Heinrich Lambert in 1760, and August Beer in 1852. Bouguer first described that, 'In a medium of uniform transparency the light remaining in a collimated beam is an exponential function of the length of the path in the medium'. However, there was some confusion in the naming of this law, which may be either the name of individual discoverer or combinations of their names. In this book, we have referred to it as Lambert–Beer's law.

A variety of spectroscopic techniques make use of the absorption of electromagnetic radiation by matter (Fig. 6.1). In a formulation suitable for the analysis of gaseous (or liquid) absorbers, Lambert–Beer's law can be written as:

$$\mathrm{I}(\lambda) = \mathrm{I}_0(\lambda) \cdot \exp\left(-\sigma(\lambda) \cdot \mathrm{c} \cdot \mathrm{L}\right) \ . \tag{6.1}$$

Here, $I_0(\lambda)$ denotes the initial intensity of a light beam emitted by a suitable source of radiation, while $I(\lambda)$ is the radiation intensity of the beam after passing through a layer of thickness L, where the absorber is present at a uniform concentration of c. The quantity $\sigma(\lambda)$ denotes the absorption cross-section at wavelength λ. The absorption cross-section as a function of wavelength is a characteristic property of any species. The determination of the light path length, L, is usually trivial for active DOAS applications (see Chap. 4). Once those quantities are known, the average trace gas concentration, c, can be calculated from the measured ratio $I_0(\lambda)/I(\lambda)$:

$$\mathrm{c} = \frac{\ln\left(\frac{\mathrm{I}_0(\lambda)}{\mathrm{I}(\lambda)}\right)}{\sigma(\lambda) \cdot \mathrm{L}} = \frac{D}{\sigma(\lambda) \cdot \mathrm{L}} \ . \tag{6.2}$$

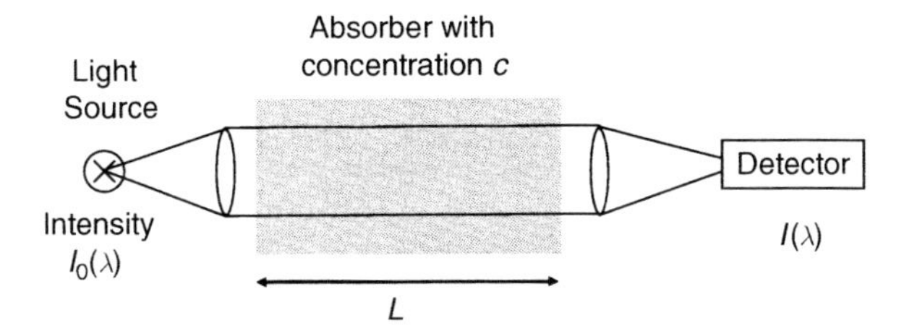

Fig. 6.1. The basic principle of absorption spectroscopic trace gas detection. A beam of light passes through a volume of length L containing the absorber with concentration c. At the end of the light path the intensity is measured by a suitable detector

The expression

$$D = \ln\left(\frac{I_0(\lambda)}{I(\lambda)}\right) , \tag{6.3}$$

is called the optical density of a layer of a given absorber. Note that, in the literature, the decadal as well as the natural logarithm is used in the definition of optical density. In this book, we will exclusively use the natural logarithm.

Equation (6.2) is the basis of most absorption spectroscopic applications in the laboratory, where the intensities $I(\lambda)$ and $I_0(\lambda)$ are determined by measurements with and without the absorber in the light beam.

However, the application of Lambert–Beer's law is more challenging in the open atmosphere. Here, the true intensity $I_0(\lambda)$, as it would be received from the light source in the absence of any atmospheric absorber, is difficult to determine. It would involve removing the air, or more precisely the absorbing gas, from the atmosphere. While this may seem to present a dilemma rendering atmospheric absorption spectroscopy useless in this case, the solution lies in measuring the so-called 'differential' absorption, i.e. the difference between the absorptions at two different wavelengths. This principle was used by Dobson in the 1930s to determine the total column of atmospheric ozone. In an ingenious experimental setup, the Dobson spectrometer compares the intensity of direct solar light of two wavelengths – λ_1, λ_2 – with different ozone absorption cross-section, $\sigma_1 = \sigma(\lambda_1)$, $\sigma_2 = \sigma(\lambda_2)$ (Dobson and Harrison, 1926).

6.3 The DOAS Principle

A schematic setup of an experiment to measure trace gas absorptions in the open atmosphere is shown in Fig. 6.2. Similar to Fig. 6.1, light emitted by a suitable spectral broadband source with an intensity $I_0(\lambda)$ passes through a volume with absorbers (here the open atmosphere), and is collected at the end of the light path. As the light travels through the atmosphere, its intensity is reduced through the absorption of a specific trace gas. However, it also undergoes extinction due to absorption by other trace gases, and scattering by air molecules and aerosol particles. The transmissivity of the instrument (mirrors, grating, retro-reflectors, etc.) will also decrease the light intensity, as will the light beam widening by turbulence. By expanding Lambert–Beer's law, one can consider the various factors that influence the light intensity by an equation that includes the absorption of various trace gases with concentration c_j and absorption cross-sections $\sigma_j(\lambda)$, Rayleigh and Mie extinction, $\varepsilon_R(\lambda)$ and $\varepsilon_M(\lambda)$ (described by $\varepsilon_R(\lambda) \approx \sigma_{R0}(\lambda) \cdot \lambda^{-4} \cdot c_{\mathrm{AIR}}$ and $\varepsilon_M(\lambda) = \sigma_{M0} \cdot \lambda^{-n} \cdot N_A$, respectively; see Chap. 4), and instrumental effects and turbulence, summarised in $A(\lambda)$:

$$I(\lambda) = I_0(\lambda) \cdot \exp\left[-L \cdot \left(\sum (\sigma_j(\lambda) \cdot c_j) + \varepsilon_R(\lambda) + \varepsilon_M(\lambda)\right)\right] \cdot A(\lambda) . \tag{6.4}$$

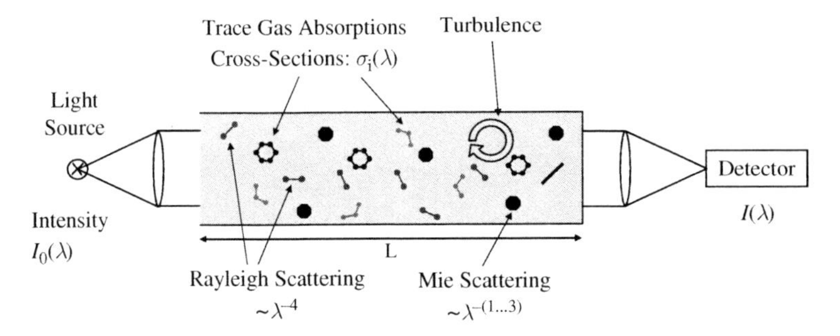

Fig. 6.2. Sketch of an experiment to measure trace gas absorptions in the open atmosphere

To determine the concentration of a particular trace gas, it would, in principle, be necessary to quantify all other factors influencing the intensity. In the laboratory, this can be achieved by removing the absorber from the light path. In the atmosphere, however, where this is impossible, the multiple factors influencing the intensity pose a dilemma.

Differential optical absorption spectroscopy overcomes this challenge by using the fact that aerosol extinction processes, the effect of turbulence, and many trace gas absorptions show very broad or even smooth spectral characteristics. Certain trace gases, however, exhibit narrowband absorption structures. The foundation of DOAS is thus to separate broad- and narrowband spectral structures in an absorption spectrum in order to isolate these narrow trace gas absorptions (Fig. 6.3). The broad spectrum is then used as a new intensity spectrum $I_0'(\lambda)$, and Lambert–Beer's law can again be applied to the narrowband trace gas absorptions.

Figure 6.3 illustrates the separation of the narrow- and broadband structures for one absorption band, both for the absorption cross-section and the intensity:

$$\sigma_j(\lambda) = \sigma_{j0}(\lambda) + \sigma_j'(\lambda) \tag{6.5}$$

σ_{j0} in (6.5) varies 'slowly' with the wavelength λ, for instance describing a general 'slope', such as that caused by Rayleigh and Mie scattering, while $\sigma_j'(\lambda)$ shows rapid variations with λ, for instance due to an absorption band (see Fig. 6.3). The meaning of 'rapid' and 'slow' variation of the absorption cross-section as a function of wavelength is, of course, a question of the observed wavelength interval and the width of the absorption bands to be detected. Inserting (6.5) into (6.4), we obtain:

$$\begin{aligned} I(\lambda) = I_0(\lambda) \cdot \exp\left[-L \cdot \left(\sum_j (\sigma_j'(\lambda) \cdot c_j)\right)\right] \cdot \\ \exp\left[-L \cdot \left(\sum_j (\sigma_{j0}(\lambda) \cdot c_j) + \varepsilon_R(\lambda) + \varepsilon_M(\lambda)\right)\right] \cdot A(\lambda), \end{aligned} \tag{6.6}$$

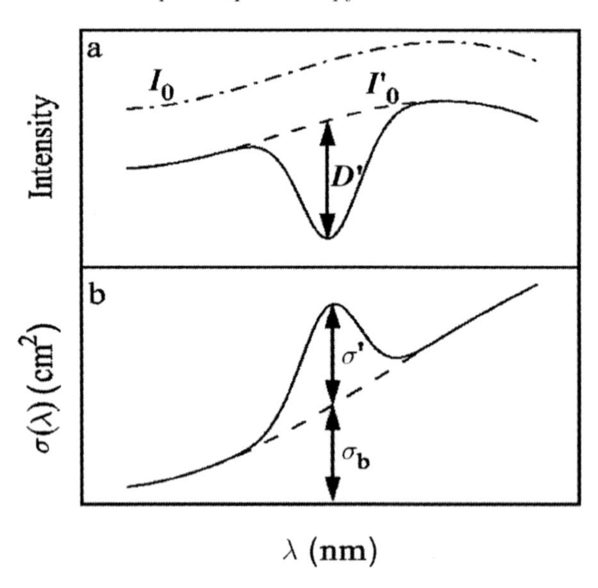

Fig. 6.3. Principle of DOAS: I_0 and σ are separated by an adequate filtering procedure into a narrow (D', and σ') and broad band part (I'_0 and σ_b)

where the first exponential function describes the effect of the structured 'differential' absorption of a trace species, while the second exponential constitutes the slowly varying absorptions as well as the influence of Rayleigh and Mie scattering. The attenuation factor $A(\lambda)$ describes the broad wavelength-dependent transmission of the optical system used and turbulence. Thus, we can define a quantity I'_0 as the intensity in the absence of differential absorption:

$$I'_0(\lambda) = I_0(\lambda) \cdot \exp\left[-L \cdot \left(\sum_j (\sigma_{j0}(\lambda) \cdot c_j) + \varepsilon_R(\lambda) + \varepsilon_M(\lambda)\right)\right] \cdot A(\lambda) \ . \quad (6.7)$$

The corresponding differential absorption cross-section $\sigma'_j(\lambda)$ is then substituted for $\sigma_j(\lambda)$ in (6.1) and (6.2). $\sigma'_j(\lambda)$ is determined in the laboratory (i.e. taken from literature data), just like $\sigma_j(\lambda)$. Likewise, a differential optical density, D', can be defined in analogy to (6.3) as the logarithm of the quotient of the intensities I'_0 and I_0 (as defined in (6.7) and (6.6), respectively):

$$D' = \ln \frac{I'_0(\lambda)}{I(\lambda)} = L \cdot \sum_j \sigma'_j(\lambda) \cdot c_j \ . \quad (6.8)$$

Atmospheric trace gas concentrations can then be calculated according to (6.2), with differential quantities D' and $\sigma'(\lambda)$ substituted for D and $\sigma(\lambda)$, respectively. A separation of the different absorptions in the sum of (6.8) is possible because the structures of the trace gases are unique, like a fingerprint (see Sect. 6.5).

Both the separation of broad and narrow spectral structures and the separation of the various absorbers in (6.8) require the measurement of the radiation intensity at multiple wavelengths. In fact, DOAS measurements usually observe the intensity at 500–2000 individual wavelengths to accurately determine the concentrations of the various absorbing trace gases. The use of multiple wavelengths is an expansion of the principles used, for example, by Dobson, which were based on two or four wavelengths.

The use of differential absorptions over an extended wavelength range has a number of major advantages. Because the transmission of optical instruments typically shows broad spectral characteristics, no calibration of the optical properties or their change with time is necessary. This often makes the instrumentation much simpler and less expensive. The use of a multitude of wavelengths allows the unique identification of trace gas absorptions. A further major advantage of this approach is the opportunity to observe and quantify extremely weak absorptions corresponding to optical densities around $D' = 10^{-4}$. In particular, the ability to use very long light paths in the atmosphere, in active DOAS applications sometimes up to 10–20 km long (passive DOAS applications can reach 1000 km), increases the sensitivity of DOAS and, at the same time, provides spatially averaged values.

Before giving a more rigorous mathematical description of the DOAS method, the basic experimental setups, the trace gases that are commonly measured, and the typical detection limits of DOAS will be reviewed in the following sections.

6.4 Experimental Setups of DOAS Measurements

The DOAS principle as outlined earlier can be applied in a wide variety of light path arrangements and observation modes (Fig. 6.4). To provide a general overview of different setups, we introduce a classification system that will later be used in the description of the different analysis methods (see Sect. 6.7 and Chap. 8) and the technical details (Chap. 7).

According to their light sources, we distinguish between **active** and **passive** DOAS. In short, active DOAS uses artificial light, while passive DOAS relies on natural light sources, i.e. solar, lunar, or stars. An overview of the most common experimental setups illustrates the breadth of DOAS applications that are in use today (Fig. 6.4).

6.4.1 Active DOAS

Active DOAS applications have one thing in common – they rely on an artificial light source coupled to an optical setup that is used to send and receive

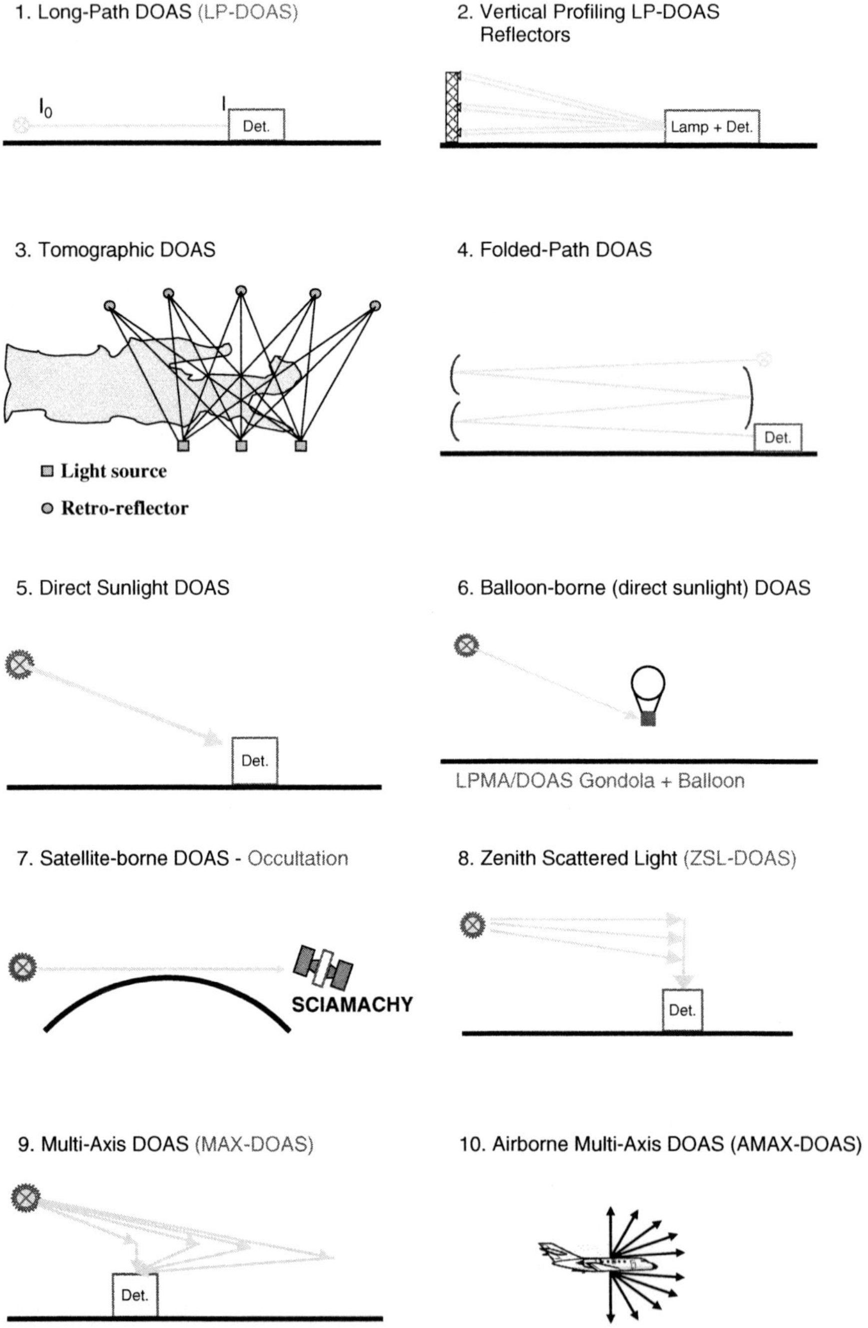

Fig. 6.4. The DOAS principle can be applied in a wide variety of light path arrangements and observation modes using artificial (1–4) as well as natural direct (5–7) or scattered (8–14) light sources. Measurements can be done from the ground, balloons, aircrafts, and from space

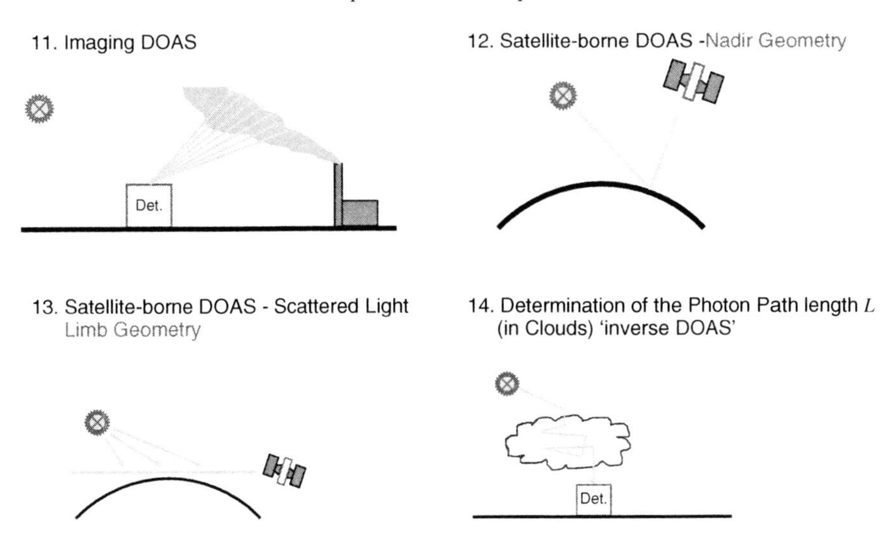

Fig. 6.4. Continued

light in the atmosphere. Spectroscopic detection is achieved by a spectrometer at the end of the light path. In general, active DOAS is very similar to classical absorption spectroscopy, as employed in laboratory spectral photometers. However, the low trace gas concentrations in the atmosphere require very long light paths (up to tens of kilometres in length, see above), making the implementation of these instruments challenging (see Chap. 7 for details). Active DOAS applications are typically employed to study tropospheric composition and chemistry, with light paths that are often parallel to the ground. In addition, active DOAS systems are also used in smog and aerosol chamber experiments.

The earliest applications of active DOAS, i.e. the measurement of OH radicals (Perner et al., 1976), used a laser as the light source along one single path (Fig. 6.4, Plate 1). This **long-path DOAS** setup is today most commonly used with broadband light sources, such as xenon-arc lamps, to measure trace gases such as O_3, NO_2, SO_2, etc. (e.g. Stutz and Platt, 1997a,b). Expansion of this method involves folding the light beam once by using retro-reflectors on one end of the light path (Axelsson et al., 1990). This setup simplifies the field deployment of long-path DOAS instruments. In addition, applications that use multiple retro-reflector setups to probe on different air masses are possible. Figure 6.4, Plate 2, shows the setup that is used to perform vertical profiling in the boundary layer with one DOAS system. An expansion that is currently under development is the use of multiple crossing light paths to perform tomographic measurements (Fig. 6.4, Plate 3).

In applications where detection in smaller air volumes with high sensitivity is required, **folded-path DOAS** is often used (Fig. 6.4, Plate 4) (e.g. Ritz et al., 1992). Because the light can pass the multiple reflection cells in these systems up to 144 times, long light paths can be achieved in small air volumes. These systems are the most common DOAS setups in laboratory applications, where interference by aerosols makes the use of classical absorption spectroscopy impossible (e.g. smog and aerosol chambers). Folded-path DOAS has also been used for the same applications as long-path DOAS (e.g. Alicke et al., 2003; Kurtenbach et al., 2002). In particular, the use of laser to measure OH has been successful (see Chap. 10).

Active DOAS measurements have contributed to the discovery and quantification of a number of important atmospheric trace species, most notably the radicals OH and NO_3 (Perner et al., 1976; Platt et al., 1979). The elegance of active DOAS is that the expanded Lambert–Beer's law (6.4) can be directly applied to the calculation of trace gas concentrations based only on the absorption cross-section, without the need for calibration of the instrument in the field. This gives active DOAS high accuracy and, with the long light paths, excellent sensitivity.

6.4.2 Passive DOAS

Passive DOAS utilises light from natural sources. The two most important sources are the sun and the moon. However, the use of light from other stars has also been reported. While the measurement of light directly from moon and stars is possible, sunlight offers two alternatives: direct sunlight and sunlight scattered in the atmosphere by air molecules and particles. We will further subdivide passive DOAS applications into **direct** and **scattered light** measurements (see also Chap. 11).

Direct measurements use the sun, moon, or stars as light sources, and thus share the advantage of active DOAS of directly applying Lambert–Beer's law. However, since the light crosses the entire vertical extent of the atmosphere, a direct conversion of absorptions to concentrations is not possible. Instead, the column density, i.e. the concentration integrated along the path, is the direct result of these measurements. Only by using geometric and radiative transfer calculations can these measurements be converted into vertically integrated column densities (VCD) or vertical concentration profiles. The most common example for VCDs is the total ozone column, which is measured in Dobson units. Figure 6.4 gives several examples for direct passive DOAS setups. Besides direct measurements of sun, moon, and star light from the ground (Fig. 6.4, Plate 5), balloon-borne solar measurements have been very successful (Fig. 6.4, Plate 6). The measurements during the ascent provide vertical profiles of various trace gases. With the recent deployment of space-borne DOAS instruments, i.e. SCIAMACHY, occultation measurements (Fig. 6.4, Plate 7) have also become possible.

Scattered sunlight measurements are more universally used in passive DOAS since they offer the largest variety of applications. The measurement of scattered light from the zenith (Fig. 6.4, Plate 8) was one the earliest applications of passive DOAS (see Sect. 11.2), and has contributed considerably to our understanding of stratospheric chemistry (e.g. Mount et al., 1987; Solomon et al., 1987, 1988, 1989). In addition, zenith scattered light has also been used to study the radiative transport in clouds (Fig. 6.4, Plate 14), which is an important topic in climate research (Pfeilsticker et al., 1998b, 1999). A more recent development of passive scattered DOAS is the use of multiple viewing geometries (Fig. 6.4, Plate 9). This multi-axis DOAS (MAX-DOAS) uses the fact that, at low viewing elevations, the length of the light path in the lower troposphere is considerably elongated (e.g. Hönninger et al., 2004). It is thus possible to probe the lower troposphere sensitively. In addition, vertical profiles can be derived if enough elevation angles are measured. MAX-DOAS can also be employed from airborne platforms, allowing the measurements below and above the flight altitude (Fig. 6.4, Plate 10), as well as determination of vertical concentration profiles. An expansion of MAX-DOAS, which is currently under development, is Imaging DOAS (Fig. 6.4, Plate 11), where a large number of viewing elevations are measured simultaneously to visualise pollution plumes.

Over the past decade, DOAS has also been used for satellite-borne measurements (Fig. 6.4, Plates 12, 13), which use sunlight scattered either by the atmosphere, the ground, or both (see Sect. 11.5). Two viewing geometries of these measurements are possible (for details and examples, see Chap. 11). In the nadir geometry, the DOAS system looks down towards the earth's surface. Instruments such as GOME provide global concentration fields of trace gases, such as O_3, NO_2, and HCHO. The SCIAMACHY instrument also employs measurements in limb geometry, which allow the determination of vertical trace gas profiles with high resolution (Fig. 6.4, Plate 13).

The advantage of passive DOAS applications is the relatively simple experimental setup. For example, scattered light measurements require only small telescopes. In addition, no artificial light source is needed. However, a number of additional challenges have to be addressed in passive DOAS applications. Because solar and lunar light is spectrally highly structured, special care needs to be taken. To detect very small trace gas absorptions, the strong Fraunhofer bands must be accurately measured. In addition, the fact that the light source structure contains narrow and deep absorptions also makes the application of the DOAS technique, which was outlined in Sect. 6.3, more difficult. This will be discussed in more detail below. The largest challenge in using passive DOAS is the conversion of the observed column densities to vertical column densities, concentrations, and vertical profiles. This is, in particular, the case for scattered light setups, where the length of the light path is difficult to determine. The interpretation of these measurements, therefore, must be based on detailed radiative transfer calculations (see Chap. 9).

6.5 Trace Gases Measured by DOAS

The separation of broad and narrow spectral structures (Sect. 6.3), while making absorption spectroscopy usable in the atmosphere, restricts DOAS measurements to trace gases that have narrow band absorption structures with widths narrower than ~10 nm. In theory, all gases that have these narrow absorption bands in the UV, visible, or near IR can be measured. However, the concentrations of these compounds in the atmosphere, and the detection limits of today's DOAS instruments, restrict the number of trace gases that can be detected. As DOAS instruments improve in the future, this list will most likely grow.

Figures 6.5 and 6.6 show the absorption cross-sections of a number of trace gases that are regularly measured by DOAS. A number of features about these cross-sections should be pointed out here. First and most importantly, each trace gas spectrum has a unique shape. Most of the trace gases only absorb in certain wavelength intervals. However, many spectral regions can contain a large number of simultaneous absorbers. For example, between 300 and 400 nm, the following trace gases will show absorption features if they are present at high enough concentrations: O_3, SO_2, NO_2, HONO, HCHO, and BrO. Because of their unique spectral structure, a separation of the absorptions is possible. From (6.8), it is clear that spectral regions with higher σ' will show the largest optical densities. These spectral intervals are thus preferred for DOAS measurements since the sensitivity improves in these wavelength regions. In principle, each trace gas has an optimal wavelength interval. In practice, however, one has to often compromise in the choice of the wavelength interval to measure more than one trace gas simultaneously. Because expanding the wavelength window reduces the spectral resolution of typical grating spectrometers, the sensitivity is also reduced.

The choice of trace gases thus depends on the specific application. In Fig. 6.7, we have attempted to aid in the choice of the best wavelength region by visualising the detection limits of an extended set of trace gases for long-path applications in the troposphere.

It should be added that a number of other trace gases, besides those shown in Figs. 6.5 and 6.6, can be measured. Table 6.1 gives an overview of the various trace gases measured by DOAS, including stratospheric trace gases. At shorter wavelengths, the usable spectral range of DOAS is limited by rapidly increasing Rayleigh scattering and O_2 absorption (Volkamer et al., 1998). Those effects limit the maximum light path length to ~200 m in the wavelength range from 200–230 nm, where, for instance, the sole usable absorption features of species such as NO (Tajime et al., 1978) and NH_3 are located (see Fig. 6.7 and Table 6.1).

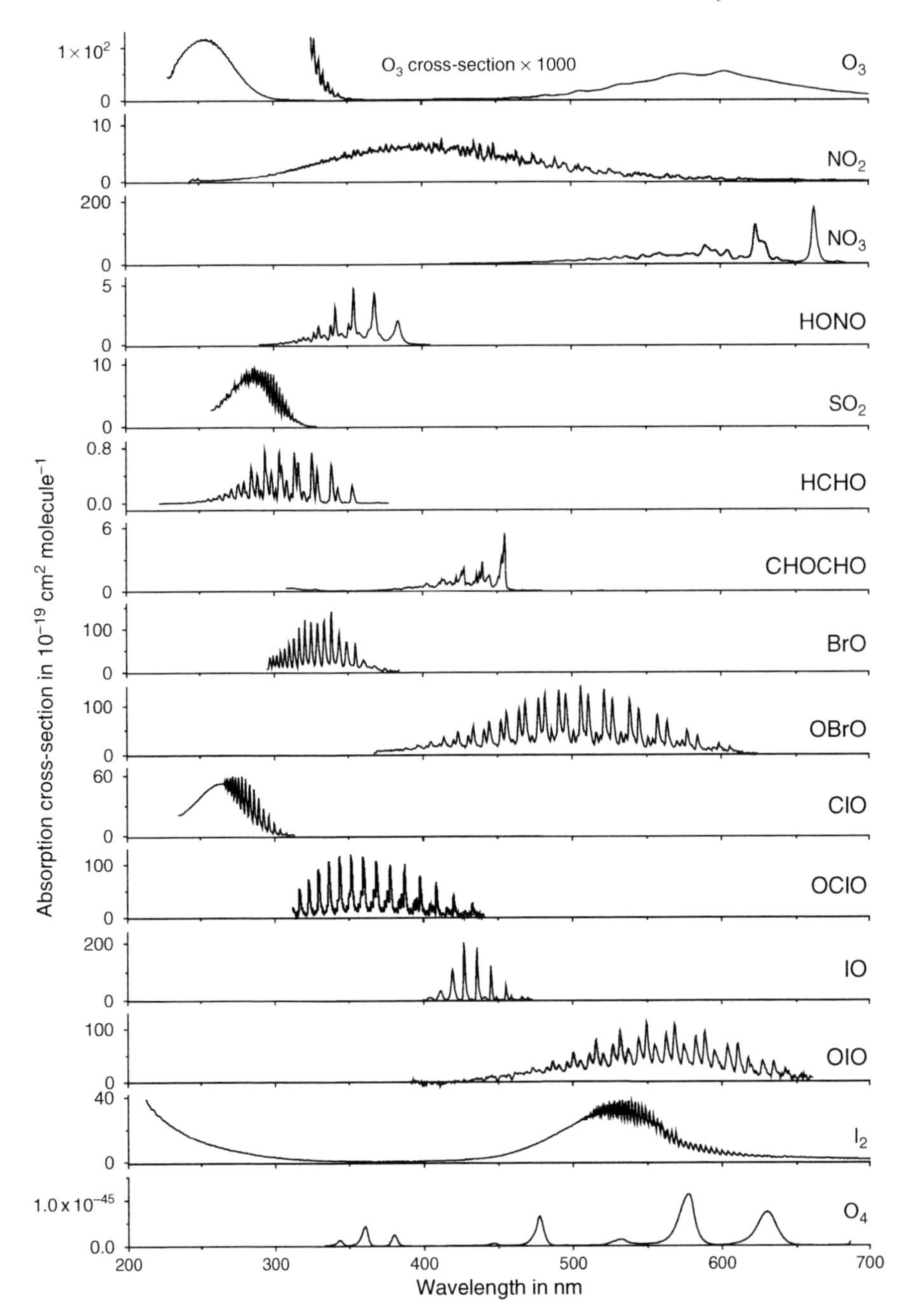

Fig. 6.5. Details of the absorption cross-section features of a number of species of atmospheric interest as a function of wavelength (in nm). Note the 'fingerprint' nature of the different spectra

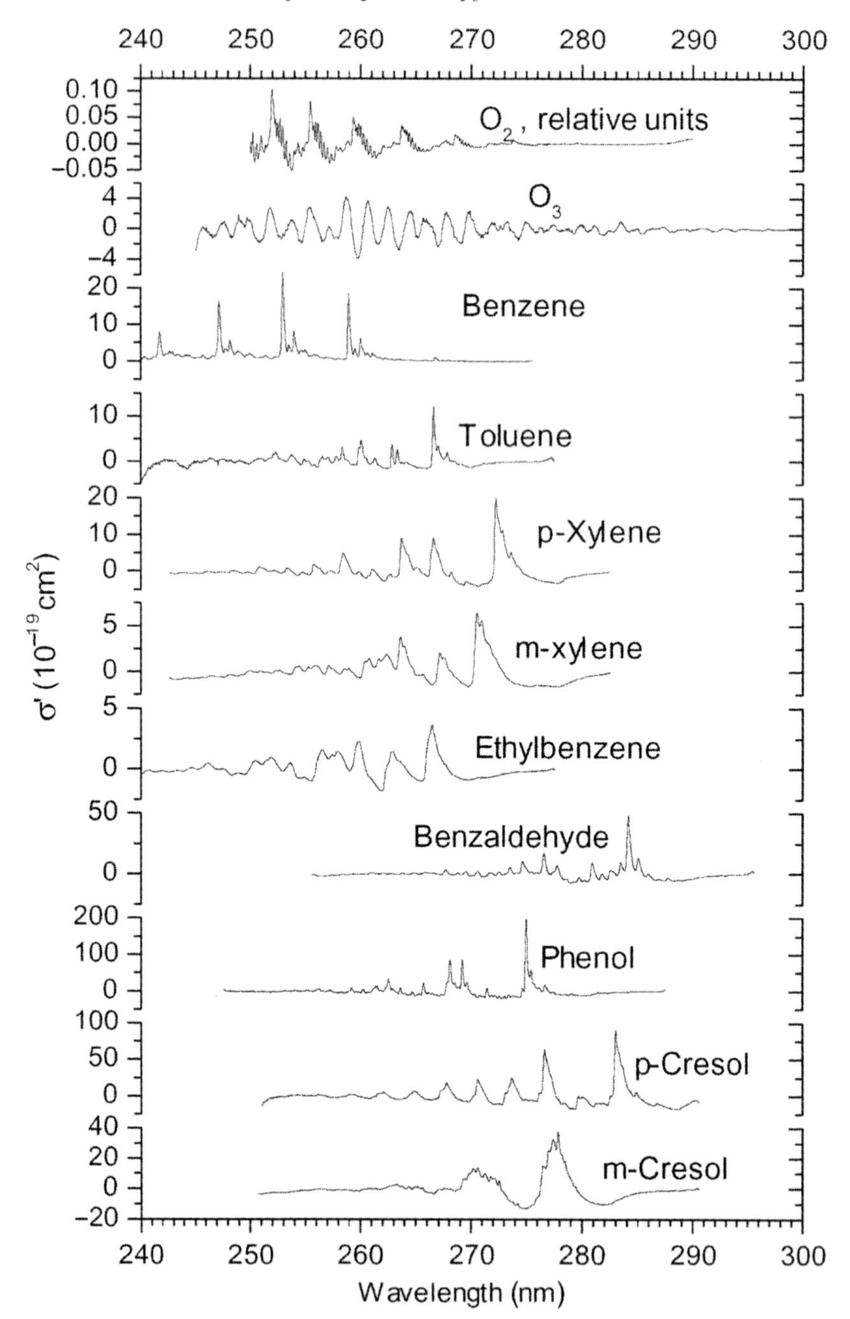

Fig. 6.6. Details of the differential absorption cross-section features of a number of monocyclic aromatic species, O_2, and O_3 as a function of wavelength (in nm). Note the 'fingerprint' nature of the different spectra

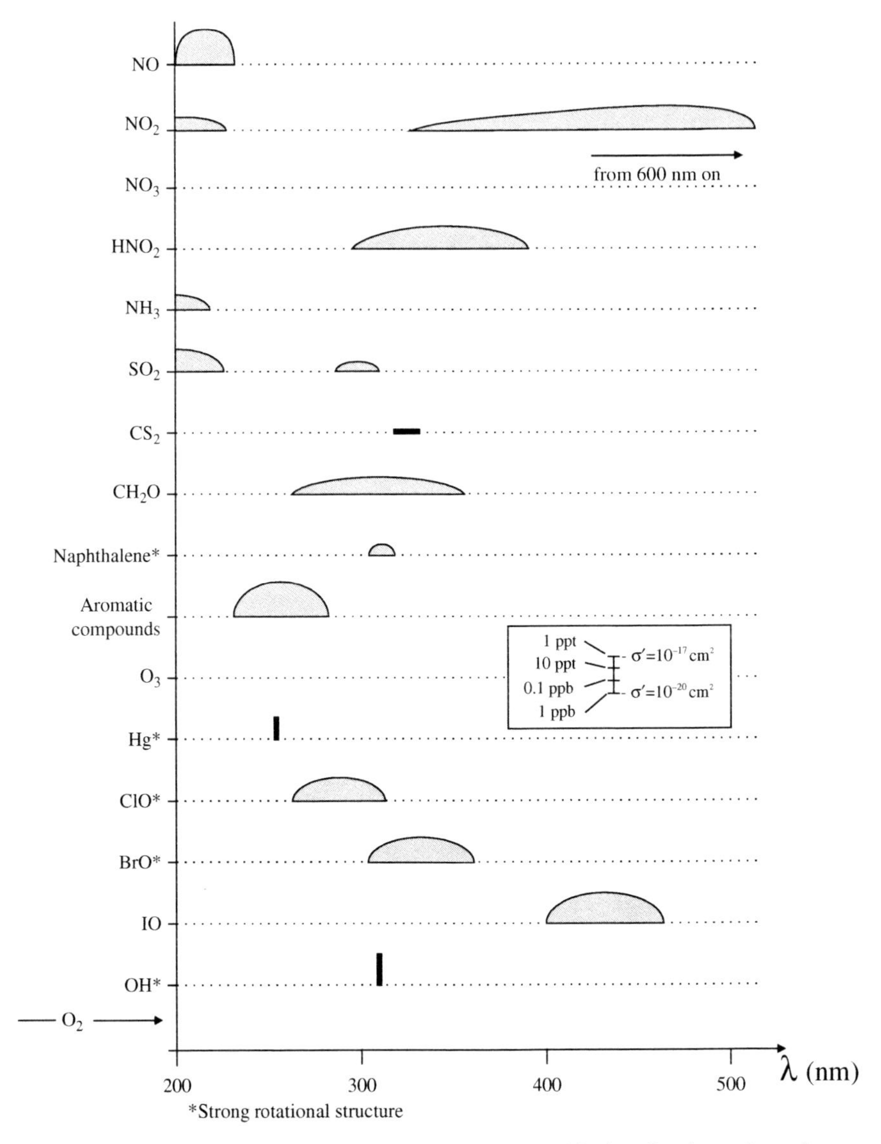

Fig. 6.7. Overview of UV-visible spectral ranges usable for the detection of atmospheric trace gases. Vertical scale: Log of absorption cross-section of the molecule (10^{-20} to 10^{-17} cm^2/molec.), as well as an approximate detection limit at 10 km light path length (1 ppt to 1 ppb, see insert). Molecules exhibiting strong rotational structure at atmospheric pressure are noted by an *asterisk* (*)

Table 6.1. Substances detectable by active UV/visible absorption spectroscopy. Detection limits were calculated for a minimum detectable optical density of 5×10^{-4}

Species	Wavelength interval (nm)	Approximate differential absorption cross-section 10^{-19} cm^2/molec.	Column density detection limit 10^{15} molec./cm	Detection limit (5 km light path) ppt
SO_2	200–230	65	0.077	62[a]
	290–310	5.7	0.88	70
CS_2	320–340	0.4	13	1000
NO	200–230	24	0.21	167[a]
NO_2	330–500	2.5	2.0	160
NO_3	600–670	200	0.025	2
NH_3	200–230	180	0.028	22[a]
HNO_2	330–380	5.1	0.98	78
O_3	300–330	0.1	50	4000
H_2O	above 430	$3 \cdot 10^{-4}$	$1.67 \cdot 10^4$	$1.33 \cdot 10^4$
CH_2O	300–360	0.48	10	830
ClO	260–300	35	0.14	11

OClO	300–440	107	0.047	3.5
BrO	300–360	104	0.048	4
OBrO	400–600	113	0.044	3.5
I_2	500–630	18	0.28	22
IO	400–470	170	0.029	2
OIO	480–600	110	0.045	3.6
Benzene	240–270	21.9	0.23	180[a]
Toluene	250–280	12.8	0.39	310[a]
Xylene (o/m/p)	250–280	2.1/6.6/20.3	2.4/0.76/0.25	2000/650/210[a]
Phenol	260–290	198	0.0025	20[a]
Cresol (o/m/p)	250–280	20.1/31.8/87.2	0.25/0.16/0.06	200/135/50[a]
Benzaldehyde	280–290	44	0.11	90[a]
Glyoxal	400–480	10	0.5	40

[a]500-m light path

6.6 Precision and Accuracy of DOAS

Selectivity, precision, and accuracy are important aspects of an analytical method. We thus give a brief overview of the different sources of random and systematic errors that must be considered in DOAS. A detailed mathematical treatment of these errors will be given in Chap. 8.

Selectivity describes the ability of an analytical method to clearly distinguish different trace gases in a measurement. In wet-chemical methods, for example, it is often challenging to identify species with similar chemical properties. Spectroscopic methods, on the other hand, provide excellent selectivity, since the absorption structures of different trace gases are a unique property of each compound. This property has been the basis of spectroscopic studies in the laboratory for many years and transfers directly to DOAS. As illustrated in Fig. 6.5, each differential absorption cross-section is unique. The numerical separation procedures described in Chap. 8 are able to separate 10 or more overlaying absorptions without creating cross-sensitivities. A condition for the successful differentiation of the trace gases is, however, that the absorption cross-section of a certain gas does not accidentally contain absorption structures of other trace gases due to, say, impurities in the sample when recording it. Such absorptions could cause interferences in the numerical analysis procedure. Careful analysis of the cross-sections can prevent this problem, and with today's high-quality laboratory techniques, cross-sections are typically pure. It should be noted here that DOAS is also able to distinguish different isomers of a species, for example, of aromatic hydrocarbons, such as the various xylenes.

The precision of DOAS measurements is mostly determined by the quality of the instrument and atmospheric conditions. Different sources of errors that influence the precision *must* be considered for the various DOAS applications (see Table 6.2). A common and often the dominating source of error in all applications is random noise in the spectra. The most important noise source originates from photon statistics, and is thus unavoidable. In general, the precision increases as more photons are collected. In addition, the noise of the detector can play a role when light levels are low. With today's technology, however, detector noise is often a minor problem. In many DOAS applications, unexplained random spectral structures are encountered. These structures have a different origin than pure photon or detector noise. In most cases, their spectra show variations that act simultaneously on a number of neighbouring detector channels. For active DOAS applications, the random spectral interferences (also known as 'optical noise') are often the primary limitation of the precision. The influence in passive DOAS measurements is less severe. A detailed discussion of the treatment of these structures is given in Chap. 8. Finally, it should be noted here that both noise and unexplained spectral structures can be determined for each individual absorption spectrum during the analysis routine. Variations in the environmental conditions, such as changes in atmospheric transmission, will automatically be reflected in these uncertainties.

Table 6.2. Errors that influence the precision and accuracy of DOAS

Type of instrument	Precision	Accuracy
Active DOAS	• Noise • Unexplained spectral structures • Insignificant: path length	• Noise • Unexplained spectral structures • Insignificant: path length • Accuracy of absorption cross-section
Passive direct light DOAS	• Noise • Removal of Fraunhofer bands • Temperature-dependent absorption cross-section • Unexplained spectral structures	• Noise • Removal of Fraunhofer bands • Temperature-dependent absorption cross-section • Unexplained spectral structures • Accuracy of absorption cross-section
Passive scattered light DOAS	• Noise • Removal of Fraunhofer bands • Temperature-dependent absorption cross-section • Unexplained spectral structures • Path length/radiative transfer	• Noise • Removal of Fraunhofer bands • Temperature-dependent absorption cross-section • Unexplained spectral structures • Path length/radiative transfer • Accuracy of absorption cross-section

The precision of DOAS can also be influenced by uncertainties in the determination of the absorption path length. In the case of active and passive direct DOAS, path lengths can be determined with high accuracy and thus contribute little to the errors in these applications. In contrast, absorption path length determination in passive scattered light DOAS applications is a major challenge. Only through detailed radiative transfer calculations is it possible to convert observed column densities to vertical column densities or concentrations. While the radiative transfer models used in this conversion are well tested, they rely on correct initialisation. Information on the vertical distribution of the trace gases and aerosol, as well as the optical properties of aerosol particles, is required. Since this information is often not available with sufficient detail, the results of the radiative transfer calculations are inherently uncertain. While we treat this uncertainty as a random effect due to its dependence on random temporal changes in the atmosphere, one can argue that this uncertainty may contribute more to systematic uncertainties that influence the accuracy. DOAS offers the opportunity to validate the radiative transfer and determine the uncertainty of this calculation by analysing trace gas absorptions from gases with known concentrations, such as O_2 and O_4. More information on the radiative transfer calculations, input parameters, and results is given in Chap. 9.

Two other uncertainties are specific for passive DOAS applications. In all passive DOAS applications, the spectral structure of sunlight, the Fraunhofer bands, needs to be accurately removed. Because the optical densities of these bands are, in most cases, much larger than those of the trace gas absorptions, the uncertainty in the removal or modelling of the Fraunhofer structure directly influences the uncertainty of the trace gas measurements. Moreover, due to Raman scattering (see Sect. 4.2.3) in the atmosphere, the Fraunhofer lines are distorted to varying degrees (the 'Ring Effect', see Sect. 9.1.6). This also needs to be compensated for. Another error in passive DOAS applications stems from the temperature dependence of the various absorption cross-sections. Even if this dependence is known, the often unknown temperature structure of the atmosphere, the location of the trace gas, and, in case of scattered light applications, the uncertain light path leads to problems in the interpretation of the data.

The determination of the accuracy of any atmospheric measurement is challenging since, in principle, it would require the knowledge of true concentration of the respective trace gas. However, in many cases, it is possible to estimate the various sources of errors, such as sampling artifacts in in-situ methods, from controlled experiments. Fortunately, this is not necessary for DOAS. Sampling artifacts and losses, as well as chemical transformations occurring after sampling that are often encountered by other methods, do not have to be considered in DOAS, since the measurements are made in the open atmosphere without disturbing or influencing the trace gases. Noise, unexplained spectral structures, Fraunhofer band removal, and temperature dependent cross-section influence accuracy in the same way they influence precision. However, the radiative transfer methods used to interpret passive scattered measurements may introduce additional non-random errors that can influence the accuracy, but not the precision, of a measurement. The accuracy of the cross-sections, which is often the largest factor influencing the accuracy of DOAS, is typically in the range of 1–10% (see Appendix B). These numbers will most likely improve in the future, as more laboratory measurements become available.

One of the most important properties of analytical techniques is the detection limit, which gives the smallest possible trace gas amount that can be detected. It is difficult to give this number for all possible DOAS applications, in particular with different path lengths between the setups shown in Fig. 6.4. To provide an overview of the detection limits of DOAS, we have thus listed two different values in Table 6.1. The first is the lowest column density in units of molecules/cm^2 that can be measured, assuming a minimum detectable optical density of 5×10^{-4}. To determine the detection limit for a specific application, this number must be divided by the path length in units of centimetres. The last column in Table 6.1 shows the mixing ratios that can be reached for an active DOAS system with a path length of $L = 5$ km for trace gases absorbing above 300 nm, or $L = 500$ m for trace gases absorbing below 300 nm. Figure 6.7 offers a more graphical view on the same data, and also facilitates the choice of the wavelength with the lowest detection limits.

In summary, DOAS is a highly versatile, selective, and accurate technique. The accuracy is primarily determined by the known uncertainties of the absorption cross-sections used. One of the main advantages of DOAS is the ability to determine the precision of a single measurement based on the analysis of the absorption spectrum. In principle, a DOAS measurement is a "spectral photograph" of the atmospheric composition that allows the identification of its components and the determination of the uncertainty of this measurement.

6.7 Mathematical Description of the DOAS Approach

The following section provides a mathematical description of DOAS, including the influence of the actual measurement on the shape of the absorption structures. This description is the basis of most other topics discussed in this book. While the basic principles in Sect. 6.3 still apply, their application to the analysis of DOAS measurements is limited due to the omission of some important aspects of the DOAS process. An expanded discussion of the principle of DOAS illustrates the approach that needs to be taken to overcome some of the challenges faced in today's wide variety of DOAS applications.

6.7.1 Fundamentals of the DOAS Approach

We begin our description by considering an idealised experimental setup of a DOAS instrument shown in Fig. 6.8, which serves as a model to describe the different applications shown in Fig. 6.4.

Light of intensity $I_0(\lambda)$ emitted by a suitable source passes through the open atmosphere and is collected by a telescope. As it passes through the atmosphere, the light undergoes extinction due to absorption by different trace gases, and scattering by air molecules and aerosol particles (see Chap. 4). The intensity $I(\lambda, L)$ at the end of the light path is given by (6.9), using Lambert–Beer's law. The absorption of a trace species j is characterised by its absorption cross-section $\sigma_j(\lambda, p, T)$, which depends on the wavelength λ, pressure p, and temperature T, and by its number concentration $c_j(l)$ at the position l along the light path. The Rayleigh extinction and Mie extinction by aerosols is described by $\varepsilon_R(\lambda, l)$ and $\varepsilon_M(\lambda, l)$. $N(\lambda)$ is the photon noise, which depends on $I(\lambda, L)$. For simplicity, we have omitted the influence of the instrument spectral characteristics and atmospheric turbulence, $A(\lambda)$, described in Sect. 6.3. The spectrum at the entrance of the spectrograph (Fig. 6.8a) arises from light that passed the atmosphere with several absorbers over length L.

$$\mathrm{I}(\lambda, \mathrm{L}) = \mathrm{I}_0(\lambda, \mathrm{L}) \cdot \exp\left[-\left(\int_0^L \sum_j \left(\sigma_j(\lambda, p, T) \cdot c_j(l)\right) + \varepsilon_R(\lambda, l) + \varepsilon_M(\lambda, l)\, dl\right)\right] + N(\lambda)\,. \tag{6.9}$$

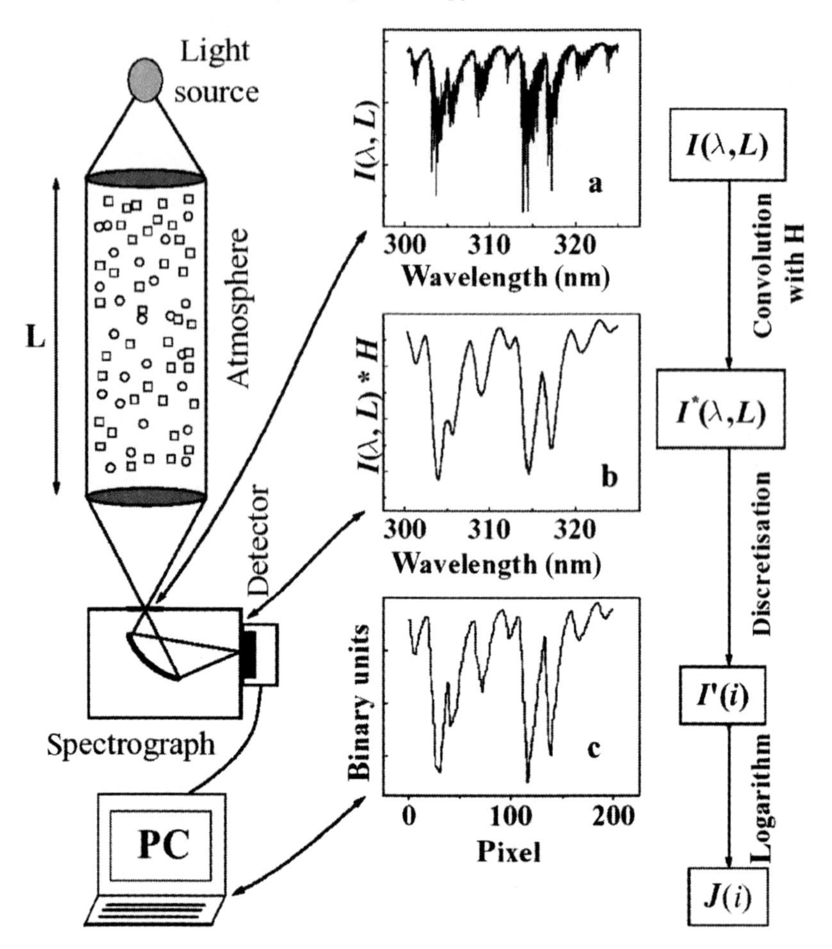

Fig. 6.8. Schematic view of a DOAS instrument used to measure trace gas concentrations. Collimated light undergoes absorption processes on its way through the atmosphere. In (**a**), an example of this light entering the spectrograph is given, where HCHO is assumed to be the only absorber and the lights source has smooth spectral characteristics. This absorption spectrum shows the vibrational–rotational structure of the absorption bands. (**b**) The same spectrum convoluted by the spectrograph instrumental function reaches the detector. In the detector, the wavelength is mapped to discrete pixels. This spectrum (**c**) is then stored in the computer and can be analysed numerically (from Stutz and Platt, 1996)

Special consideration must be given to passive scattered light DOAS. The light observed in these applications follows complicated paths in the atmosphere, which are determined by the solar position in the sky, the viewing direction of the telescope, and most importantly the spatial distribution of air molecules and aerosol particles. The measured intensity is the sum over intensities of different light beams. Each of these beams has travelled on an individual path with a specific length through the atmosphere. One can introduce a simplified mathematical formulation of this phenomenon by describing the measured total intensity as an integral over a light path distribution function and a path length dependent $I_0(\lambda, L)$. It is then possible to show the validity of an approximation for cases with weak absorptions, which describes the measured intensity with (6.9). In this approximation, $I_0(\lambda, L)$ represents the total intensity, and L represents the intensity-weighted average path length. In particular, the average path length will be discussed in the context of radiative transfer calculations in Chap. 9. A number of approaches to solve the DOAS problem for scattered light measurements will be described in Chaps. 9 and 11.

In most DOAS instruments, light of intensity $I(\lambda, L)$ is focused on the entrance of a grating spectrograph, with a detector recording the spectrum. Due to the limited resolution of the spectroscopic instruments, the shape of spectrum $I(\lambda, L)$ changes. The mathematical description of this process is a convolution of $I(\lambda, L)$, with the instrument function H of the spectrograph: $I^*(\lambda, L) = I(\lambda, L)*H$. Figure 6.8b shows the spectrum I after convolution with a typical instrument function H.

$$I^*(\lambda, L) = I(\lambda, L)*H = \int I(\lambda - \lambda', L) \cdot H(\lambda')d\lambda' \,. \tag{6.10}$$

During the recording by a detector, the wavelength range is mapped to n discrete pixels/channels, numbered by i, each integrating the light in a wavelength interval from $\lambda(i)$ to $\lambda(i+1)$. This interval is given by the wavelength-pixel-mapping Γ_I of the instrument. In the case of a linear dispersion, $\Gamma_I : \lambda(i) = \gamma_0 + \gamma_1 \times i$, the spectral width of a pixel is constant $\Delta\lambda(i) = \lambda(i+1) - \lambda(i) = \gamma_1$. The signal $I'(i)$ seen by a pixel i (omitting the response of individual pixels) is given by:

$$I'(i) = \int_{\lambda(i)}^{\lambda(i+1)} I^*\left(\lambda'\right) d\lambda' \,. \tag{6.11}$$

In general, the wavelength-pixel-mapping Γ_I of the instrument can be approximated by a polynomial:

$$\Gamma_I : \quad \lambda(i) = \sum_{k=0}^{q} \gamma_k \cdot i^k \,. \tag{6.12}$$

The parameter vector (γ_k) determines the mapping of pixel i to the wavelength $\lambda(i)$. A change in parameter γ_0 describes a spectral shift of the spectrum. Changing γ_1 squeezes or stretches the spectrum linearly. Parameters

γ_k of higher k describe a distortion of the wavelength scale of higher order. Changes in the parameter vector (γ_k) can be caused by different measurement conditions of the spectra, as grating spectrometers usually show a temperature drift of 1/10 of a pixel per K. A variation in air pressure, as observed in aircraft measurements, also changes the wavelength alignment due to a change in the index of refraction of air. It is, therefore, necessary to correct these effects in the analysis procedure.

Figure 6.8c shows the discrete spectrum $I'(i)$ as was recorded and stored in a computer. One of the main components of DOAS is the analysis of $I'(i)$ with respect to the different absorbers in the spectrum.

6.7.2 Application of the DOAS Approach in Practical Situations

Equations (6.9) and (6.10) can be combined into one equation:

$$\begin{aligned} I^*(\lambda, L) &= I(\lambda, L) * H \\ &= \int_{-\Delta\lambda}^{\Delta\lambda} I_0(\lambda - \lambda', L) \exp\left(- \int_o^L \sum_j (\sigma_j(\lambda - \lambda', p, T) \cdot c_j(l)) + \varepsilon_R (\lambda - \lambda', l) \right. \\ &\quad \left. + \varepsilon_M (\lambda - \lambda', l) \, dl \right) \cdot H(\lambda') d\lambda' \end{aligned} \tag{6.13}$$

We have omitted the noise term in (6.13), since the following discussion focuses on a mathematical description of the DOAS method that does not consider statistical uncertainties. The noise and its influence on the results of DOAS are the topic of discussion in Chap. 8.

In order to simplify the mathematical treatment of the DOAS approach, we neglect the effect of wavelength discretisation and concentrate on the following discussion of the continuous spectra arriving at the detector, $I^*(\lambda, L)$. The omission of the discretisation step (6.11) has no influence on the general results of our discussion. In particular, when the width of the instrument function H is much larger than the wavelength interval of one pixel, the approximation that the intensity at the centre wavelength of a pixel $I^*(\lambda(i), L)$ is very similar to the integral over the entire pixel i is quite good. If this is not the case, other effects such as aliasing have to be considered. However, since the discussion of these effects does not aid in the understanding of DOAS, we will not consider them in this chapter, and we will assume that the integral over the pixel can indeed be approximated by the intensity of the centre wavelength of the pixel.

Although, from a purely mathematical point of view, the integration of the first integral in (6.13) should extend from $-\infty$ to ∞, in practice it is sufficient to constrain it to a small interval around the instrument function of width $2\Delta\lambda$. Assuming a Gaussian function $H(\lambda') = C \cdot \exp\left[-(\lambda'/\lambda_H)^2\right]$ as an approximation of the instrument function, $\Delta\lambda$ can be chosen as a small multiple of λ_H. For example, choosing $\Delta\lambda = 4 \cdot \lambda_H$ would result in an error

$< 10^{-4}$. Following the general DOAS approach (see Sect. 6.3), we can simplify (6.13) by including all broadband terms in a new $I_0'(\lambda, L)$:

$$I^*(\lambda, L) = \int_{-\Delta\lambda}^{\Delta\lambda} I'_0(\lambda - \lambda', L) \exp\left(- \int_o^L \sum_j \left(\sigma_j'(\lambda - \lambda', p, T) \cdot \rho_j(l) \right) dl \right) \cdot H(\lambda') d\lambda' . \tag{6.14}$$

It should be noted that the separation of spectrally broad and narrow terms leading to (6.14) is not an a-priori defined procedure, and there are many implementations of this separation, such as regressions and Fourier filters (see Chap. 8). Here, we treat this separation in the most general term, without giving a quantitative definition of what we consider as broad or narrow. However, one important aspect that must be mentioned is that any separation is applied to the intensity spectra $I(\lambda)$, and not to the absorption cross-section $\sigma(\lambda)$. This may, at first glance, contradict our original approach in (6.5). However, because we are free in choosing the procedure to separate broad and narrow structures, we can generalise (6.5) to: $\exp(-\sigma(\lambda)) = \exp(-\sigma_0(\lambda)) \cdot \exp(-\sigma'(\lambda))$, where any separation procedure is applied to $\exp(-\sigma(\lambda))$. The choice of separation procedure, which will be discussed in more detail in Chap. 8, will not impact our following discussion.

The DOAS problem can, in principle, be solved by a numerical model of (6.14), where the trace gas concentrations $c_j(l)$ are adjusted to optimise the agreement between $I^*(\lambda, L)$ and the measured spectrum. However, in practice, this is difficult and time consuming, since it requires a non-linear optimisation procedure. A variety of simplifications and adaptations have thus been developed.

We will now present a number of these implementations for different DOAS applications. As an initial simplification, we assume that the differential absorption cross-section $\sigma_j'(\lambda)$ is independent of temperature and pressure. For active DOAS applications, which typically measure a very small altitude interval, this restriction is applicable. For certain passive DOAS applications, this assumption is difficult to sustain and would, in principle, require a rigorous numerical solution of (6.14). There is, however, an approximate solution, which splits $\sigma_j'(\lambda, T, p)$ into a small number of different cross-sections $\sigma_k'(\lambda, T_k, p_k)$ for individual temperature/pressure combinations (note that there is a correlation between pressure and temperature in the atmosphere): $\sigma_j'(\lambda, T, p) = \sum b_k \cdot \sigma_k'(\lambda, T_k, p_k)$. With this approach, $\sigma_j'(\lambda)$ becomes independent of the integration over the path length, and the integration over dl only has to be applied to $c_j(l)$.

Based on our assumption that $\sigma_j'(\lambda)$ is independent of temperature and pressure, or can be split into independent parts, we can now introduce the path-averaged gas concentration:

$$\bar{c}_j = \frac{1}{L} \int_0^L c_j(l) dl . \tag{6.15}$$

Equation (6.14) thus becomes:

$$I^*(\lambda, L) = \int_{-\Delta\lambda}^{\Delta\lambda} I_0'(\lambda - \lambda', L) \cdot \exp\left(-\sum_j \left(\sigma_j'(\lambda - \lambda') \cdot \bar{c}_j\right) \cdot L\right) \cdot H(\lambda')d\lambda' . \tag{6.16}$$

Equation (6.16) is the basis of the following discussion. It is the simplest description of a DOAS measurement containing differential absorption of multiple trace gases, a wavelength-dependent I_0', and the convolution process representing the measurement.

Based on (6.16), we can define a number of parameters that influence the choice of DOAS implementation (see also Table 6.3):

- *The differential optical density of the trace gas absorption.* In general, we will denote differential optical densities $\mathrm{D}' = \mathrm{L} \cdot \sum \sigma_\mathrm{j}'(\lambda) \cdot \bar{\mathrm{c}}_\mathrm{j}$ that are below ~0.1 as 'small'. It will depend on the mathematical treatment of the DOAS implementation if this limit applies to the original or convoluted absorption band. The change in differential optical density due to the convolution depends on the width of the absorption band relative to the instrument function width in the convolution. The change is more pronounced in narrow bands.

Table 6.3. Overview of the different cases for DOAS applications

Case	Diff. OD of absorber	Resolution of spectrometer	Spectrum of source	Approach	Remark
1	Small	High and low	Smooth	Linearise Lambert–Beer's law	Classical case
2	Large	High	Smooth	Lambert–Beer's Law	Classical case
3	Large	Low	Smooth	Nonlinear, modelling of entire equation system	Saturated absorber
4	Small or large	High[a]	Structured	Divide by $I_0(\lambda)$	
5	Small	Low	Structured	High res. Model + Lambert–Beer's Law	
6	Large	Low	Structured	Non-linear, modelling of entire equation system	

[a] $2\Delta\lambda << \lambda_I$ and $2\Delta\lambda << \lambda_B$

- *The spectral resolution or instrument function of the spectrograph–detector in comparison to the spectral width of the absorption bands.* In general, we can distinguish between high-resolution cases, in which the natural width of the absorption band, $\Delta\lambda_B$, is spectrally resolved, and low-resolution cases, in which the spectral width of the bands is not resolved by the measurements. In (6.16), this translates into the condition that $2\Delta\lambda << \Delta\lambda_B$ for spectrally resolved bands and $2\Delta\lambda \geq \Delta\lambda_B$ for unresolved bands. We would caution the reader that a band that looks resolved after a measurement may consist of a number of narrower absorption lines that are not resolved. Thus, the measurement would only show the envelope of this line system.
- *The spectral structure of the light source, $I'_0(\lambda)$.* We will denote 'smooth' as light sources that have structures broader than the resolution of the instrument, for example incandescence or high- pressure Xe lamps have this property. 'Structured' light sources show spectral features that are narrower that the instrument resolution. The most important example in our case is the sun, with its narrow Fraunhofer lines. In (6.16), this leads to the condition that $2\Delta\lambda$ is much smaller than the bandwidth, λ_I, of $I'_0(\lambda)$ for a smooth $I'_0(\lambda)$, and larger for structured $I'_0(\lambda)$.

Six different cases can be identified based on these three parameters (Table 6.3). We will now discuss in more detail how (6.16) has to be applied to determine the desired trace gas column densities for the different situations. The individual cases will be discussed in a general way. However, we will also give the mathematical derivation for the different implementations, wherever possible, for more details see also the 'operator representation of DOAS' (Wenig, 2001; Wenig et al., 2005).

Case 1: Weak Low-or High-resolution Absorbers and a Smooth Light Source

The most basic approach of DOAS applies to measurements using a smooth light source, i.e. active DOAS, where weak absorption features are measured with low or high instrumental resolution.

In this case, Lambert–Beer's law can be linearised by taking the logarithm of the ratio of $I^*(\lambda, L)$ and $I'_0(\lambda, L)$, resulting in the following representation of (6.16):

$$\ln\left(\frac{I^*(\lambda, L)}{I'_0(\lambda, L)}\right) = \sum_j \frac{\bar{c}_j \cdot L}{\alpha_j} \cdot \ln\left[\int_{-\Delta\lambda}^{\Delta\lambda} \exp\left(-\sigma'_j(\lambda - \lambda') \cdot \alpha_j\right) \cdot H(\lambda')d\lambda'\right]. \tag{6.17}$$

Equation (6.17) states that $\ln\left(I^*(\lambda, L)/I'_0(\lambda, L)\right)$ can be described by a sum of spectra of pure trace gas absorptions that are scaled by the average number concentration $\bar{c}_j$ and the path length L (we will discuss the factor α_j later).

However, as is made clear in the rest of this section, this mathematical approach is far more common than one may expect, and some of the other cases discussed later will also rely on it.

Since the linearisation of Lambert–Beer's law is the most common form of DOAS implementation and is the basis of Chap. 8, we will introduce an equation that is loosely based on the discreet form of (6.17). The logarithm of the discreet form of $I^*(\lambda, L)$: $J(i) = \ln(I'(i))$ is described by:

$$J(i) = J_0(i) + \sum_{j=1}^{m} a'_j \cdot S'_j(i) \,. \tag{6.18}$$

$J_0(i)$ is the logarithm of the discretisation of $I'_0(i)$, which was defined in (6.11). The differential absorption structures of the trace gases are described by individual 'reference spectra' $S'_j(i)$, which were also discretised (6.11). a'_j are the scaling factors for the individual discretised reference spectra. We will see that, for different DOAS cases, the reference spectra calculation is different, while (6.18) can still be used. For smooth light sources, the spectra $S'_j(i)$ are determined through the following three equations:

$$S^*_j(\lambda) = \int_{-\Delta\lambda}^{\Delta\lambda} \exp\left(-\sigma'_j(\lambda - \lambda') \cdot \alpha_j\right) \cdot H(\lambda')d\lambda'$$

$$S^*_j(i) = \int_{\lambda(i)}^{\lambda(i+1)} S^*_j(\lambda)d\lambda' .$$

$$S'_j(i) = \frac{1}{\alpha_j} \ln(S^*_j(i)) \tag{6.19}$$

We have introduced a scaling parameter α_j that will play a role for case 3 (see below). This parameter can be interpreted as the product of concentration and path length used in the measurement or simulation of S^*_j. For case 1, where the differential optical densities are small, α_j can be set to unity. In the case of small differential optical density ($\alpha_j = 1$), the first equation in (6.19) is taken from the sum of (6.17), and describes the convolution of the exponential of the negative differential absorption cross-section $\sigma'_j(\lambda)$. The second equation is the discretisation according to (6.11). The last equation is then the logarithm of the discretised spectrum.

However, if the differential optical densities are not small, (6.18) and (6.19) can only be used if values for $\alpha_j \neq 1$ are chosen. The scaling factors

$$a'_j = \frac{\bar{c}_j \cdot L}{\alpha_j} \,, \tag{6.20}$$

in (6.18) are the product of the average number densities and the path length, divided by the reference spectrum scaling factor α_j.

The solution of (6.18) is now a mathematical problem that requires the calculation of the $S'(i)$ through (6.19). For this calculation, the knowledge of instrument function $H(\lambda)$ and the highly resolved absorption cross-section $\sigma'_j(\lambda)$ is required. Both the mathematical approach to solve (6.18) with the goal of retrieving the a'_j factors and the details of the calculation in (6.18) are topics of discussion in Chap. 8. Figure 6.9 shows an example of (6.18) for a real atmospheric measurement. The top spectrum represents $J'(i)$ while the other spectra represent the products of the absorption cross-sections and the scaling factors: $a'_j \cdot S'_j(i)$ (in this case $\alpha_j = 1$).

After describing this 'classical' DOAS approach, we now discuss how this implementation can be derived mathematically from (6.16). We will use two related approximations that are directly valid for small differential optical density of the absorption structures:

$$\ln(x) \approx x - 1 \quad \text{for} \quad 1 - \varepsilon < x < 1 + \varepsilon\,, \tag{6.21}$$

and

$$\exp(x) \approx 1 + x \quad \text{for} \quad -\varepsilon < x < \varepsilon\,. \tag{6.22}$$

These approximations are good for ε close to zero. At $\varepsilon = 0.1$, the error imposed by (6.21) and (6.22) is ~5%.

We will also assume that the spectra are measured using a smooth light source. This condition leads to the approximation that $I'_0(\lambda - \lambda', L) \approx I'_0(\lambda, L)$ in the interval from $\lambda-\Delta\lambda$ to $\lambda+\Delta\lambda$. Finally, we will use the condition that the integral of the instrument function is scaled to unity. This condition can always be met by dividing the integral over $H(\lambda)$ by a constant:

$$\int_{-\Delta\lambda}^{\Delta\lambda} H(\lambda')d\lambda' = 1\,. \tag{6.23}$$

Mathematically, this is only true if $\Delta\lambda = \infty$. In all practical cases, however, the integral can be limited to an interval around H.

We will begin the derivation of (6.18) by taking the logarithm of the ratio of $I^*(\lambda, L)$ and $I'_0(\lambda, L)$ in (6.16):

$$\ln\left(\frac{I^*(\lambda, L)}{I'_0(\lambda, L)}\right) = \ln\left[\int_{-\Delta\lambda}^{\Delta\lambda} \exp\left(-\sum_j \left(\sigma'_j(\lambda - \lambda') \cdot \bar{c}_j\right) \cdot L\right) \cdot H(\lambda')d\lambda'\right]\,. \tag{6.24}$$

In the case of weak absorbers, the integral in (6.24) is close to unity and we can employ (6.21) to approximate the logarithm:

$$\ln\left(\frac{I^*(\lambda, L)}{I'_0(\lambda, L)}\right) \approx \int_{-\Delta\lambda}^{\Delta\lambda} \exp\left(-\sum_j \left(\sigma'_j(\lambda - \lambda') \cdot \bar{c}_j\right) \cdot L\right) \cdot H(\lambda')d\lambda' - 1\,. \tag{6.25}$$

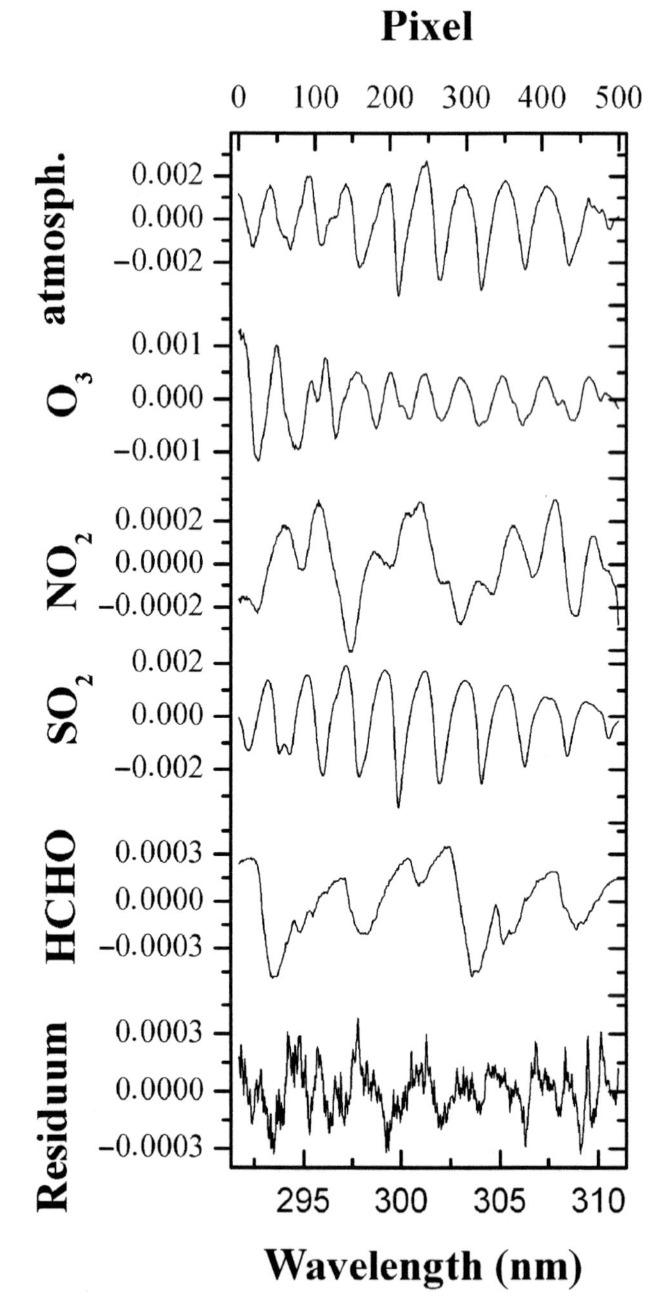

Fig. 6.9. Sample of an atmospheric spectrum (*uppermost trace*) recorded in Heidelberg on August 27, 1994, with overlapping absorptions due to O_3, NO_2, SO_2, and HCHO (traces 2–5 from top, note different scales). The broad spectral structures were described by a fifth order polynomial. The remaining 'residual spectrum' after removal of the absorption structures is shown as the *bottom trace* (from Stutz and Platt, 1996)

This approximation applies to the convoluted absorption structures, which are smaller than the original absorptions before convolution. Using (6.23), this can now be transformed into:

$$\ln\left(\frac{I^*(\lambda,L)}{I'_0(\lambda,L)}\right) \approx \int_{-\Delta\lambda}^{\Delta\lambda} \exp\left(-\sum_j \left(\sigma'_j(\lambda-\lambda')\cdot\bar{c}_j\right)\cdot L\right)\cdot H(\lambda')d\lambda' - \int_{-\Delta\lambda}^{\Delta\lambda} H(\lambda')d\lambda'$$
$$= \int_{-\Delta\lambda}^{\Delta\lambda} \left[\exp\left(-\sum_j \left(\sigma'_j(\lambda-\lambda')\cdot\bar{c}_j\right)\cdot L\right) - 1\right]\cdot H(\lambda')d\lambda' . \tag{6.26}$$

Further applying the approximation for the exponential function (6.22):

$$\ln\left(\frac{I^*(\lambda,L)}{I'_0(\lambda,L)}\right) \approx \int_{-\Delta\lambda}^{\Delta\lambda} \left[-\sum_j \left(\sigma'_j(\lambda-\lambda')\cdot\bar{c}_j\right)\cdot L\right]\cdot H(\lambda')d\lambda' . \tag{6.27}$$

This approximation applies to the absorption structures before convolution, and is thus more restrictive than the approximation used in (6.25). We can now exchange the sum and the integral. In addition, we introduced the factor α_j from (6.20):

$$\ln\left(\frac{I^*(\lambda,L)}{I'_0(\lambda,L)}\right) \approx \sum_j \frac{\bar{c}_j\cdot L}{\alpha_j}\cdot \int_{-\Delta\lambda}^{\Delta\lambda} \left[\sigma'_j(\lambda-\lambda')\cdot\alpha_j\right]\cdot H(\lambda')d\lambda' . \tag{6.28}$$

In the next step of derivation, the approximation for the exponential function (6.22) is applied again (this time, however, in reverse direction):

$$\ln\left(\frac{I^*(\lambda,L)}{I'_0(\lambda,L)}\right) \approx \sum_j \frac{\bar{c}_j\cdot L}{\alpha_j}\cdot \int_{-\Delta\lambda}^{\Delta\lambda} \left[\exp\left(-\sigma'_j(\lambda-\lambda')\cdot\alpha_j\right)-1\right]\cdot H(\lambda')d\lambda' . \tag{6.29}$$

It is important to note here that the errors due to approximations in (6.27) and (6.29) tend to cancel each other. This counterbalance is best if $\alpha_j \approx \bar{c}_j \cdot L$, and actually lifts the restriction to weak absorptions before convolution for the steps from (6.25) to (6.29). The derivation is finalised again using (6.23) and (6.21). The approximation of the logarithm (6.21) is again applied to the convoluted spectral structures.

$$\ln\left(\frac{\mathrm{I}^*(\lambda, L)}{\mathrm{I}'_0(\lambda, L)}\right) \approx \sum_{\mathrm{j}} \frac{\bar{c}_j \cdot L}{\alpha_j} \cdot \int_{-\Delta\lambda}^{\Delta\lambda} \left[\exp\left(-\sigma'_{\mathrm{j}}(\lambda - \lambda') \cdot \alpha_j\right)\right] \cdot H(\lambda')d\lambda' - \int_{-\Delta\lambda}^{\Delta\lambda} H(\lambda')d\lambda'$$

$$\approx \sum_{\mathrm{j}} \frac{\bar{c}_j \cdot L}{\alpha_j} \cdot \ln\left[\int_{-\Delta\lambda}^{\Delta\lambda} \exp\left(-\sigma'_{\mathrm{j}}(\lambda - \lambda') \cdot \alpha_j\right) \cdot H(\lambda')d\lambda'\right]. \qquad (6.30)$$

Equation (6.30) is equivalent to (6.17) and states that in the case of small optical densities after convolution, the DOAS approach can be solved by linearising Lambert–Beer's law. The reference spectra used in this fit have to be calculated as the logarithm of the convolution of the exponential function of high-resolution absorption cross-sections (6.19).

It is important to comment again on the various approximations in the derivation of (6.30). Both approximations – (6.21) and (6.22) – were applied twice, in forward and backward directions. While the approximations were applied to different arguments, there is a high degree of cancellation by the twofold use, which makes the approximation better than one would expect from the single application. In particular, in the case that $\alpha_j = \bar{c}_j \cdot L$, the twofold approximations are quite good. A detailed calculation, which uses a Tailor expansion to the second order, shows that the difference between (6.27) and (6.29) is approximately $\frac{1}{2} \sum_{j \neq k} \sigma_j \sigma_k \bar{c}_j \bar{c}_k L^2$, if $\alpha_j = \bar{c}_j \cdot L$. The derivation shown above is, therefore, more widely applicable than just for case 1, and we will come back to it in case 3. For small absorptions before the convolution, we can set $\alpha_j = 1$, since all the approximations apply in this case without restrictions. The preceding derivation is the basis of most DOAS applications today. Most measurements of O_3, NO_2, HCHO, NO_3, HONO, halogen oxides, etc. in the troposphere have used this approach (see Fig. 6.9).

Case 2: Strong Absorber at High Resolution and a Smooth Light Source

The case of strong absorbers measured with a spectral resolution that is better than the width of the absorption bands is encountered in a number of DOAS applications. The approach described in Case 1 (6.17–6.19) can also be used in this case. However, the mathematical proof is not as obvious, since the approximations of the logarithm and exponential function (6.21 and 6.22) do not apply in the case of strong absorbers. We will, therefore, use an approximation that is based on the fact that the spectral interval covering the instrument function $H(\lambda)$, $2\Delta\lambda$, is smaller than the width of the absorption band. Within this interval, we can approximate the exponential function in (6.16) by a linear function, using the derivative of the exponential function as the slope. In addition, this approximation can also be made by the exponential function of the pure scaled absorption cross-section (6.30). Two new functions, $g(\lambda)$ and $h(\lambda)$, for these exponential functions are introduced:

$$\exp\left(-\sum_{\mathrm{j}}\left(\sigma'_{\mathrm{j}}(\lambda-\lambda')\cdot c_{\mathrm{j}}\cdot L\right)\right)=g(\lambda-\lambda')\approx g(\lambda)+\left.\frac{dg}{d\lambda}\right|_{\lambda}\cdot\lambda'. \qquad (6.31)$$

$$\exp\left(-\sigma'_{\mathrm{j}}(\lambda-\lambda')\cdot\alpha_{\mathrm{j}}\right)=h_j(\lambda-\lambda')\approx h_j(\lambda)+\left.\frac{dh_j}{d\lambda}\right|_{\lambda}\cdot\lambda'. \qquad (6.32)$$

We again consider $I'_0(\lambda, L)$ as being constant within $\lambda \pm \Delta\lambda$. This is certainly true because the resolution of the instrument is very high and, as compared to the interval $2\Delta\lambda$, the change of $I'_0(\lambda, L)$ is small. Applying the approximation in (6.31), the equivalent of (6.24) now becomes:

$$\begin{aligned}\ln\left(\frac{\mathrm{I}^*(\lambda,L)}{\mathrm{I'}_0(\lambda,L)}\right)&=\ln\left(\int_{-\Delta\lambda}^{\Delta\lambda}g(\lambda-\lambda')\cdot H(\lambda')d\lambda'\right)\\&\approx\ln\left(\int_{-\Delta\lambda}^{\Delta\lambda}\left(g(\lambda)+\left.\frac{dg}{d\lambda}\right|_{\lambda}\cdot\lambda'\right)\cdot H(\lambda')d\lambda'\right).\end{aligned} \qquad (6.33)$$

Equation (6.33) is now reorganised by using the fact that some of the terms under the integral are independent of λ':

$$\ln\left(\frac{\mathrm{I}^*(\lambda,L)}{\mathrm{I}'_0(\lambda,L)}\right)\approx\ln\left(g(\lambda)\cdot\int_{-\Delta\lambda}^{\Delta\lambda}H(\lambda')d\lambda'+\left.\frac{dg}{d\lambda}\right|_{\lambda}\cdot\int_{-\Delta\lambda}^{\Delta\lambda}\lambda'\cdot H(\lambda')d\lambda'\right). \qquad (6.34)$$

The derivative of $g(\lambda)$ can now be calculated as:

$$\left.\frac{dg}{d\lambda}\right|_{\lambda}=-g(\lambda)\cdot\frac{d\left(\sum_j\sigma'(\lambda)\cdot c_j\cdot L\right)}{d\lambda}=-g(\lambda)\cdot\sum_j\frac{d\sigma'(\lambda)}{d\lambda}\cdot c_j\cdot L. \qquad (6.35)$$

Equation (6.34) then becomes:

$$\begin{aligned}\ln\left(\frac{\mathrm{I}^*(\lambda,L)}{\mathrm{I}'_0(\lambda,L)}\right)\approx\ln\Bigg(g(\lambda)\cdot\int_{-\Delta\lambda}^{\Delta\lambda}H(\lambda')d\lambda'-g(\lambda)\cdot\sum_j\frac{d\sigma'(\lambda)}{d\lambda}\cdot\bar{c}_j\cdot L\cdot\\\int_{-\Delta\lambda}^{\Delta\lambda}\lambda'\cdot H(\lambda')d\lambda'\Bigg),\end{aligned}$$

and after rearranging:

$$\ln\left(\frac{\mathrm{I}^*(\lambda,L)}{\mathrm{I}_0'(\lambda,L)}\right) \approx \ln\left(g(\lambda)\cdot\int_{-\Delta\lambda}^{\Delta\lambda} H(\lambda')d\lambda'\right)$$
$$+\ln\left(1-\sum_j\left(\left.\frac{d\sigma_j'(\lambda)}{d\lambda}\right|_\lambda\cdot\bar{c}_j\cdot L\cdot\int_{-\Delta\lambda}^{\Delta\lambda}\lambda'\cdot H(\lambda')d\lambda'\right)\right). \tag{6.36}$$

The approximation for the logarithm from (6.21) and the fact that $H(\lambda)$ is normalised to unity can now be used. The application of the approximation is justified because the sum over j in the above equation is small, and thus the argument of the second ln in the equation is near unity.

$$\ln\left(\frac{\mathrm{I}^*(\lambda,L)}{\mathrm{I}_0'(\lambda,L)}\right) \approx \ln\left(g(\lambda)\right)-\sum_j\left(\left.\frac{d\sigma_j'(\lambda)}{d\lambda}\right|_\lambda\cdot\bar{c}_j\cdot L\cdot\int_{-\Delta\lambda}^{\Delta\lambda}\lambda'\cdot H(\lambda')d\lambda'\right). \tag{6.37}$$

After the following transformations,

$$\ln\left(g(\lambda)\right) = -\sum_{\mathrm{j}}\left(\sigma_{\mathrm{j}}'(\lambda)\cdot\bar{c}_{\mathrm{j}}\cdot L\right) = -\sum_{\mathrm{j}}\left(\alpha_j\cdot\sigma_{\mathrm{j}}'(\lambda)\cdot\frac{\bar{c}_{\mathrm{j}}\cdot L}{\alpha_j}\right)$$
$$=\sum_{\mathrm{j}}\left(\frac{\bar{c}_{\mathrm{j}}\cdot L}{\alpha_j}\cdot\ln\left(\exp\left(-\alpha_j\cdot\sigma_{\mathrm{j}}'(\lambda)\right)\right)\right)=\sum_{\mathrm{j}}\left(\frac{\bar{c}_{\mathrm{j}}\cdot L}{\alpha_j}\cdot\ln\left(h_j(\lambda)\right)\right), \tag{6.38}$$

one derives:

$$\ln\left(\frac{\mathrm{I}^*(\lambda,L)}{\mathrm{I}_0'(\lambda,L)}\right) \approx \sum_{\mathrm{j}}\left(\frac{\bar{c}_{\mathrm{j}}\cdot L}{\alpha_j}\cdot\ln\left(h_j(\lambda)\right)\right)$$
$$-\sum_j\left(\frac{\bar{c}_j\cdot L}{\alpha_j}\cdot\left.\frac{d\sigma_j(\lambda)}{d\lambda}\right|_\lambda\cdot\alpha_j\cdot\int_{-\Delta\lambda}^{\Delta\lambda}\lambda'\cdot H(\lambda')d\lambda'\right)$$
$$=\sum_{\mathrm{j}}\frac{\bar{c}_{\mathrm{j}}\cdot L}{\alpha_j}\cdot\left(\ln\left(h_j(\lambda)\right)-\left.\frac{d\sigma_j(\lambda)}{d\lambda}\right|_\lambda\cdot\alpha_j\cdot\int_{-\Delta\lambda}^{\Delta\lambda}\lambda'\cdot H(\lambda')d\lambda'\right). \tag{6.39}$$

This equation can be further transformed by again employing the approximation of the logarithm (6.21) in the reverse direction. This step is followed by a number of transformations based on the properties of the logarithm in (6.23):

$$\ln\left(\frac{I^*(\lambda,L)}{I_0'(\lambda,L)}\right) \approx \sum_j \frac{\bar{c}_j \cdot L}{\alpha_j} \cdot \left(\ln\left(h_j(\lambda)\right) - \int_{-\Delta\lambda}^{\Delta\lambda} \left.\frac{d\sigma_j(\lambda)}{d\lambda}\right|_\lambda \cdot \alpha_j \cdot \lambda' \cdot H(\lambda')d\lambda'\right)$$

$$\approx \sum_j \frac{\bar{c}_j \cdot L}{\alpha_j} \cdot \left(\ln\left(h_j(\lambda)\right) + \ln\left(1 - \int_{-\Delta\lambda}^{\Delta\lambda} \left.\frac{d\sigma_j(\lambda)}{d\lambda}\right|_\lambda \cdot \alpha_j \cdot \lambda' \cdot H(\lambda')d\lambda'\right)\right)$$

$$\approx \sum_j \frac{\bar{c}_j \cdot L}{\alpha_j} \cdot \left(\ln\left(h_j(\lambda) \cdot \int_{-\Delta\lambda}^{\Delta\lambda} H(\lambda')d\lambda' - h_j(\lambda) \cdot \int_{-\Delta\lambda}^{\Delta\lambda} \left.\frac{d\sigma_j(\lambda)}{d\lambda}\right|_\lambda \cdot \alpha_j \cdot \lambda' \cdot H(\lambda')d\lambda'\right)\right)$$

$$\approx \sum_j \frac{\bar{c}_j \cdot L}{\alpha_j} \cdot \ln\left(\int_{-\Delta\lambda}^{\Delta\lambda} h_j(\lambda) \cdot H(\lambda')d\lambda' - \int_{-\Delta\lambda}^{\Delta\lambda} \left(h_j(\lambda) \cdot \left.\frac{d\sigma_j(\lambda)}{d\lambda}\right|_\lambda \cdot \alpha_j \cdot \lambda' \cdot H(\lambda')\right)d\lambda'\right). \tag{6.40}$$

Finally, the derivative of $h_j(\lambda)$:

$$\left.\frac{dh_j(\lambda)}{d\lambda}\right|_\lambda = \exp\left(-\sigma_j'(\lambda) \cdot \alpha_j\right) \cdot -\frac{d\sigma_j'(\lambda)}{d\lambda} \cdot \alpha_j = -h_j(\lambda) \cdot \frac{d\sigma_j'(\lambda)}{d\lambda} \cdot \alpha_j$$

and the definition of $h_j(\lambda)$ (6.32) are introduced:

$$\ln\left(\frac{I^*(\lambda,L)}{I_0'(\lambda,L)}\right) \approx \sum_j \frac{\bar{c}_j \cdot L}{\alpha_j} \cdot \ln\left(\int_{-\Delta\lambda}^{\Delta\lambda} \left(h_j(\lambda) - h_j(\lambda) \cdot \left.\frac{d\sigma_j(\lambda)}{d\lambda}\right|_\lambda \cdot \alpha_j \cdot \lambda'\right) \cdot H(\lambda')d\lambda'\right)$$

$$\approx \sum_j \frac{\bar{c}_j \cdot L}{\alpha_j} \cdot \ln\left(\int_{-\Delta\lambda}^{\Delta\lambda} \left(h_j(\lambda) + \left.\frac{h_j(\lambda)}{d\lambda}\right|_\lambda \cdot \lambda'\right) \cdot H(\lambda')d\lambda'\right)$$

$$\approx \sum_j \frac{\bar{c}_j \cdot L}{\alpha_j} \cdot \ln\left(\int_{-\Delta\lambda}^{\Delta\lambda} h_j(\lambda - \lambda') \cdot H(\lambda')d\lambda'\right)$$

$$\approx \sum_j \frac{\bar{c}_j \cdot L}{\alpha_j \cdot} \ln\left[\int_{-\Delta\lambda}^{\Delta\lambda} \exp\left(-\sigma_j'(\lambda - \lambda') \cdot \alpha_j\right) \cdot H(\lambda')d\lambda'\right] \tag{6.41}$$

This equation is now equivalent to (6.30), showing that the approach of Case 1 is applicable. It should be emphasised that, in this derivation, the approximations were based on the fact that the convolution would change the absorption structure only slightly. In contrast to Case 1, we did not make any assumption about the strength of the absorptions.

Case 3: Strong Absorbers at Low Resolution and a Smooth Light Source

The case that the optical density of an absorber before the convolution in (6.16) exceeds unity is quite common in low-resolution DOAS applications.

Several molecules (e.g. O_2, NO, H_2O, halogen monoxides, some aromatics, and glyoxal) have highly resolved rotational structures in the near UV and visible spectral ranges. It is thus possible that the optical density at the centre of these lines becomes very high. If such an absorption line is measured by a low-resolution instrument, the optical density of the absorption band after convolution is not proportional to the one before convolution.

Figure 6.10 illustrates how the optical density, in this case simply calculated as the logarithm of the ratio of the intensities at the band minimum and its border, for a single narrow absorption line changes after it is convoluted with a instrument function (here assumed to be of Gaussian shape) that is broader than the initial band (see also the bands in the inserts in Figure 6.10). At low values, the optical densities of the convoluted line depend linearly on the optical density of the narrow line. This is the situation we described in Case 1. At increasing values, the two optical densities start to deviate, with the low-resolution OD becoming increasingly smaller than the high-resolution OD. The two optical densities are thus no longer linearly dependent on one another.

The straightforward solution to this problem is to model the entire function (6.16). However, the introduction of the parameter α_j in (6.19) expands the applicability of Case 1 to high optical densities. This is most easily seen for one absorber $j = 1$, when $\alpha_j = \bar{c}_j \cdot L$. In this case, (6.24) and (6.30) become identical. In cases where the value of α_j is close to $\bar{c}_j \cdot L$, (6.18) and (6.19)

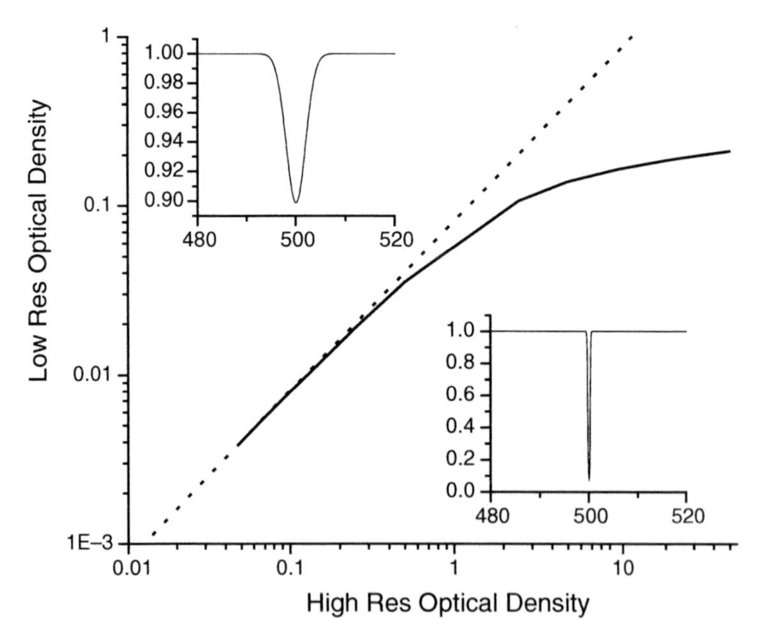

Fig. 6.10. Deviation of the low resolution optical density on the optical density of a narrow absorption band after convolution with a broad instrument resolution

can still be used. However, the initial value of $\alpha_j \approx \bar{c}_j \cdot L$ has to be guessed. Alternatively, a lookup table for reference spectra for different α_j can be used.

While it is clear that, for one strong absorber the linearisation of Lambert–Beer's law can be applied if α_j is guessed correctly, one can show that, in the case of one strong absorber and various weak absorbers, (6.18) and (6.19) still apply. However, α_j for the strong absorber must be guessed correctly. When strong differential absorptions of several absorbers overlay each other, this approach leads to systematic errors in the analysis. This situation is, however, rather uncommon in DOAS applications.

Another effect helps in the application of (6.18) and (6.19) for strong absorbers. For many trace gases, the width of the absorption bands is larger than the width of the instrument function. In this case, the convolution does not change the optical density dramatically, as illustrated in Figure 6.10. For larger optical densities, one operates somewhere between Case 1 and Case 2. By using a good initial guess for α_j, it is then possible to use (6.18) and (6.19), even for fairly strong absorbers. This is, for example, the case for NO_2 and O_3 absorptions, which are dominated by absorption bands that are resolved by typical DOAS instruments.

Because the numerical solution of (6.16) is slow, most DOAS applications rely on (6.18) and (6.19) in their analysis routines. Strong absorbers are taken into account by lookup tables of simulated reference spectra or some other numerical schemes to derive appropriate correction factors (Volkamer et al., 1998; Maurellis et al., 2000; Buchwitz et al., 2000; Frankenberg et al., 2004).

Case 4: Weak Absorbers, High Resolution, and a Structured Light Source

Our fourth case is that of a highly structured light source emission spectrum combined with weak trace gas absorptions and a high spectral resolution. The case of a structured light source is characteristic for the measurement of solar light, either direct or scattered, which shows narrow spectral structures due to Fraunhofer absorption bands in the photosphere of the sun. One can distinguish two situations here: If the spectral width of the instrument function is smaller than the width of the structures of $I_0(\lambda)$, we can apply Case 1. In the case that the width of the instrument function is between the width of $I_0(\lambda)$ and the bandwidth of the absorbers, Case 5 can be applied.

Case 5: Weak Absorbers, Low Resolution, and a Structured Light Source

In the case of weak absorbers, a low spectral resolution, and a structured light source, the solution shown in case 1 has to be adopted. In contrast to Case 1, however, where the convolution of the exponential of the scaled absorption cross-section can be used to describe the absorption spectrum, (6.19) does not directly apply. The following equation must be used to calculate the absorption cross-section instead of (6.19):

$$S_j^*(\lambda)\Big|_{case5} = \frac{\int\limits_{-\Delta\lambda}^{\Delta\lambda} I_0'(\lambda-\lambda',L) \times \exp\left(\sum\limits_j \sigma_j'(\lambda-\lambda') \times \alpha_j\right) \times H(\lambda')d\lambda'}{\int\limits_{-\Delta\lambda}^{\Delta\lambda} I_0'(\lambda-\lambda',L) \times H(\lambda')d\lambda'} . \tag{6.42}$$

In principle, the calculation of (6.42) is very similar to that of (6.19). The only difference is that the convolution now includes $I_0'(\lambda, L)$, and that the result, therefore, has to be normalised to the convolved $I_0'(\lambda, L)$. It is easy to see that, in the case of a smooth light source where $I_0'(\lambda-\lambda', L) \approx I_0'(\lambda, L)$, (6.42) is indeed equivalent to (6.19). Because $I_0'(\lambda, L)$ is used in (6.42), the change in the absorption cross-section is often also referred to as the 'I_0 effect' (Platt et al., 1997; Alliwell et al., 2002; Wagner et al., 2002c).

Before discussing the derivation of (6.42), we will introduce the convolution of $I_0'(\lambda, L)$ with the instrument function, which helps to simplify the equations below:

$$I'^*_0(\lambda, L) = I_0'(\lambda, L) * H = \int I_0'(\lambda-\lambda', L) \cdot H(\lambda')d\lambda' . \tag{6.43}$$

Mathematically, (6.42) can be determined from (6.16). The derivation follows the approach taken in Case 1 [using the approximations from (6.21) and (6.22)], except that the assumption of a smooth light source is dropped, i.e. $I_0'(\lambda-\lambda', L)$ cannot be taken out of the integral. We begin with transforming (6.16) in the same way as in Case 1:

$$\ln\left(\frac{I^*(\lambda,L)}{I'^*_0(\lambda,L)}\right) = \ln\left(\frac{\int\limits_{-\Delta\lambda}^{\Delta\lambda} I_0'(\lambda-\lambda',L) \cdot \exp\left(-\sum\limits_j \left(\sigma_j'(\lambda-\lambda') \cdot c_j\right) \cdot L\right) \cdot H(\lambda')d\lambda'}{I'^*_0(\lambda,L)}\right) . \tag{6.44}$$

As in the above examples, (6.21) can now be applied to approximate the logarithm. The resulting equation can then be further reorganised to extract a factor $\left[I'^*_0(\lambda, L)\right]^{-1}$:

$$\ln\left(\frac{I^*(\lambda,L)}{I'^*_0(\lambda,L)}\right) \approx \left[\int\limits_{-\Delta\lambda}^{\Delta\lambda} I_0'(\lambda-\lambda',L) \cdot \exp\left(-\sum\limits_j \left(\sigma_j'(\lambda-\lambda') \cdot c_j\right) \cdot L\right) \cdot H(\lambda')d\lambda' \cdot \left[I'^*_0(\lambda,L)\right]^{-1} - 1\right]$$

$$= \left[\mathrm{I}_0'^*(\lambda, L) \right]^{-1} \cdot \left[\int_{-\Delta\lambda}^{\Delta\lambda} \mathrm{I}_0'(\lambda - \lambda', L) \cdot \exp\left(-\sum_j \left(\sigma_j'(\lambda - \lambda') \cdot c_j \right) \cdot L \right) \cdot \right.$$

$$\left. \times H(\lambda')d\lambda' - \mathrm{I}'^*_0(\lambda, L) \right]$$

$$= \left[\mathrm{I}_0'^*(\lambda, L) \right]^{-1} \cdot \left[\int_{-\Delta\lambda}^{\Delta\lambda} I_0'(\lambda - \lambda', L) \cdot \left[\exp\left(-\sum_j \left(\sigma_j'(\lambda - \lambda') \cdot c_j \right) \cdot L \right) - 1 \right] \cdot H(\lambda')d\lambda' \right] . \tag{6.45}$$

As in Case 1, the exponential function is then approximated using (6.22):

$$\ln\left(\frac{\mathrm{I}^*(\lambda, L)}{\mathrm{I}_0'^*(\lambda, L)} \right) \approx \left[\mathrm{I}'^*_0(\lambda, L) \right]^{-1} \cdot \left[\int_{-\Delta\lambda}^{\Delta\lambda} \mathrm{I}_0'(\lambda - \lambda', L) \cdot \left(-\sum_j \left(\sigma_j'(\lambda - \lambda') \cdot c_j \right) \cdot L \right) \cdot H(\lambda')d\lambda' \right] . \tag{6.46}$$

This equation is further rearranged, and (6.22) is applied again – this time, however, in reverse order. The application of (6.22) in forward and backward directions lifts the restriction of small absorptions to a certain extent, as already discussed in Case 1.

$$\ln\left(\frac{\mathrm{I}^*(\lambda, L)}{\mathrm{I}_0'^*(\lambda, L)} \right) \approx \left[\mathrm{I}_0'^*(\lambda, L) \right]^{-1} \cdot \sum_j \frac{c_j \cdot L}{\alpha_j} \cdot \left[\int_{-\Delta\lambda}^{\Delta\lambda} \mathrm{I}_0'(\lambda - \lambda', L) \cdot \left[\left(\sigma_j'(\lambda - \lambda') \right) \cdot \alpha_j \right] \cdot H(\lambda')d\lambda' \right]$$

$$\approx \left[\mathrm{I}_0'^*(\lambda, L) \right]^{-1} \cdot \sum_j c_j \cdot L \cdot \left[\int_{-\Delta\lambda}^{\Delta\lambda} \mathrm{I}_0'(\lambda - \lambda', L) \cdot \left[\exp\left(\sigma_j'(\lambda - \lambda') \cdot \alpha_j \right) - 1 \right] \cdot H(\lambda')d\lambda' \right] \tag{6.47}$$

In the next step, this equation is again rearranged (remembering the normalisation of $H(\lambda)$ to unity):

$$\ln\left(\frac{I^*(\lambda, L)}{I'^*_0(\lambda, L)}\right) \approx \left[I'^*_0(\lambda, L)\right]^{-1} \cdot \sum_j c_j \cdot L \cdot$$

$$\left[\int_{-\Delta\lambda}^{\Delta\lambda} I'_0(\lambda - \lambda', L) \cdot \left[\exp\left(\sigma'_j(\lambda - \lambda')\right) \cdot \alpha_j\right] \cdot H(\lambda')d\lambda' - 1\right]$$

$$\approx \sum_j c_j \cdot L \cdot \left[\left[I'^*_0(\lambda, L)\right]^{-1} \cdot \int_{-\Delta\lambda}^{\Delta\lambda} I'_0(\lambda - \lambda', L) \cdot \right.$$

$$\left.\left[\exp\left(\sigma'_j(\lambda - \lambda')\right) \cdot \alpha_j\right] \cdot H(\lambda')d\lambda' - 1\right] \tag{6.48}$$

Finally, (6.21) is applied to transform the last term in the sum to a logarithm:

$$\ln\left(\frac{I^*(\lambda, L)}{I'^*_0(\lambda, L)}\right) \approx \sum_j c_j \cdot L \cdot \ln\left[\frac{\int_{-\Delta\lambda}^{\Delta\lambda} I'_0(\lambda - \lambda', L) \cdot \left[\exp\left(\sigma'_j(\lambda - \lambda')\right)\right] \cdot H(\lambda')d\lambda'}{\int_{-\Delta\lambda}^{\Delta\lambda} I'_0(\lambda - \lambda', L) \cdot H(\lambda')d\lambda'}\right]. \tag{6.49}$$

It is clear from (649) that $I^*(\lambda, \mathrm{L})$ can be analysed by a classical DOAS approach (Case 1) if the differential absorption cross-sections are calculated by (6.42). The discussion about the restrictions imposed by the multiple applications of the approximations of the logarithm and the exponential functions are applied here, in the same manner as in Case 1 (see above).

Case 6: Strong Absorbers, Low Resolution, and a Structured Light Source

This case applies to similar trace gases as those described in Case 3, however measured with passive DOAS instruments. As in Case 3, mathematically correct solution can be achieved in many cases by using (6.18) and (6.42), if α_j are chosen close to $\bar{c}_j \cdot L$. A lookup table approach, as well as a number of other numerical approximations, can also be used (e.g. Maurellis et al., 2000).

The discussion above illustrates that, in most cases, the DOAS problem can be solved by a linearisation of Lambert–Beer's law, using (6.19) in the case of active DOAS applications and (6.42) in the case of passive DOAS applications. The non-linearity that is imposed by the convolution of strong absorbers can, in many cases, be overcome if the factor α_j is close to the product $\bar{c}_j \cdot L$. The approach shown in (6.18), (6.19), and (6.42) is thus the most important implementation of DOAS. Its numerical solution will be discussed in detail in Chap. 8.

7

The Design of DOAS Instruments

In recent years, a large variety of designs of active and passive DOAS systems (see Chaps. 6, 10, and 11) as well as of its optical, mechanical, electronic, and software subsystems were developed in response to the requirements of atmospheric research and monitoring. Many of the characteristics, design considerations, and trade-offs are common to several or all varieties of DOAS systems. In this chapter, we will describe the options and design criteria for the DOAS instrument, as well as for the required subsystems. In particular, we describe the various decisions that have to be made in design and operation procedures of a DOAS system, once the basic arrangement has been selected.

7.1 Design Considerations of DOAS Instruments

Obviously, the design of a particular DOAS system must be based on many requirements such as high sensitivity, spectral properties (resolution and spectral range), and practical aspects (cost, weight, and power consumption). In particular, the instrument must have the capability to detect very weak absorption features, as optical densities (see Chap. 6) around 10^{-3} are a typical signal. From the standpoint of intended application, we can identify the following requirements:

- Scientific (in situ) studies of atmospheric chemistry are frequently performed with ground-based active DOAS systems, although passive, in particular, multi-axis DOAS (MAX-DOAS) systems are gaining importance. Many scientific studies require high sensitivity (low detection limit), specific detection, and a large degree of flexibility (different light path lengths, wide range of species to be measured and thus a large accessible spectral range). Automated operation and simplicity of design is usually not a priority.
- Air pollution monitoring, and in particular long-term observations are, at present, usually performed with active DOAS systems.

U. Platt and J. Stutz, *The Design of DOAS Instruments*. In: U. Platt and J. Stutz, Differential Optical Absorption Spectroscopy, Physics of Earth and Space Environments, pp. 175–285 (2008)
DOI 10.1007/978-3-540-75776-4_7

Here, automated design, simple operation, and low cost are requirements. To some extent, this can be realised at the expense of outmost sensitivity and flexibility of the instrument.

- Monitoring of the stratospheric composition (e.g. as an 'early warning' system) for future changes, for instance by the 'network for detection of stratospheric change' (NDSC), is mostly performed by passive spectroscopy.
- Long-term (ground-based) studies of atmospheric gases require an automated design which needs little supervision. Thus, for the monitoring of stratospheric gases zenith scattered light (ZSL) (i.e. passive) DOAS is the method of choice. For tropospheric species active as well as passive instruments (e.g. MAX-DOAS) can be considered.
- Global and long-term observation of stratospheric and tropospheric trace gas distributions is achieved by satellite measurements, where the DOAS technique has been proven to lead to considerable progress, in particular with respect to the quantitative measurement of tropospheric species (e.g. Borrell et al., 2003). Naturally, unattended operation is a prerequisite for satellite instruments. But at the large data rates, rapid spectral evaluation can also be a challenge.

From a technical standpoint, the central aspects of any DOAS system can be summarised in the following groups.

1. Active or passive operation: Active systems can work in a wider spectral range (i.e. below 300 nm) and also in the absence of sunlight at night, but are relatively complex, both in technical design and logistic requirements. Passive instruments tend to require much simpler hardware (no light source, smaller telescope, etc.), but are usually restricted to operation during daylight hours.
2. The light throughput, i.e. the question how the amount of light reaching the detector can be maximised under the given boundary conditions: This is an important requirement due to the frequently low optical densities of trace gas absorption structures to be measured (see above). This, in turn, demands recording of the optical spectra at low noise levels, which are usually determined by photon noise. For active systems, the light throughput is determined by both light source and receiving optics. In passive systems only the latter can be optimised.
3. The spectral range observed, which determines the number of species to be observed: While a large spectral range is certainly of advantage, its selection will usually be a compromise because increasing the spectral range may affect spectral resolution. It may also be difficult to obtain a uniform signal throughout a large spectral range.
4. The spectral resolution, which should be sufficient to resolve the spectral structures to be observed: Again a compromise will have to be made. A better spectral resolution can improve the detection limit and the ability of the instrument to de-convolute overlapping bands of different species

(see Chap. 8); on the other hand, improvement of the resolution comes at the expense of light throughput and size (and therefore cost) of the instrumentation.
5. The spectral stability of the instrument: Since DOAS relies on the detection of weak absorptions, the instruments must have the capability to eliminate the effect of overlaying, stronger structures (due to Fraunhofer spectra, lamp emission spectra, or other atmospheric absorbers with strong, overlapping spectra) in the same wavelength range. If spectral shifts occur between the recording of the measurement spectrum and overlapping structures, e.g. the Fraunhofer reference, the performance in this area is degraded.

7.2 Key Components of DOAS Systems

Active instruments based on UV-visible absorption spectroscopy have been designed in a large variety of ways. For instance, broadband lasers were used as a light source (e.g. Rothe et al., 1974; Perner et al., 1976; Amerding et al., 1991; Dorn et al., 1993, 1996, 1998; Brauers et al., 1996, 1999; Brandenburger et al., 1998). However, the majority of the designs rely on arc lamps as light sources (e.g. Bonafe et al., 1976; Kuznetzov and Nigmatullina, 1977; Noxon et al., 1978, 1980; Platt, 1977; Platt et al., 1979; Platt and Perner, 1980, 1983; Johnston and McKenzie, 1984; Dorn and Platt 1986; Axelsson et al. 1990a,b, 1995; Plane and Smith 1995; Harder et al. 1997; Flentje et al., 1997; Allan et al., 1999, 2000; Geyer et al., 1999, 2001a,b,c; Hebestreit et al., 1999; Veitel et al., 2002; Yu et al., 2004; Saiz-Lopez et al., 2004a,b; Zingler and Platt, 2005). More details and examples are given in Chap. 10.

Passive instruments deriving trace gas column densities from the analysis of scattered or direct sunlight or direct moonlight have also been in wide use (e.g. Noxon, 1975; Bonafe et al., 1976; Kuznetzov and Nigmatullina, 1977; Millan, 1978, 1980; Noxon et al., 1978, 1980; Wagner, 1990; Burrows et al., 1999; Ferlemann et al., 1998, 2000; Pfeilsticker et al., 1999a,b; von Friedeburg et al., 2002, 2005; Wagner et al., 2002a,b,c; Galle et al., 2003; Bobrowski et al., 2003; Heckel, 2003; Heckel et al., 2005; Leser et al., 2003; Hönninger et al., 2004a,b,c; Wittrock et al. 2004; Heue et al. 2005; Lohberger et al. 2004; Wang et al. 2005). More details and examples are given in Chap. 11.

Depending on the particular type of application and desired measurement, a DOAS system consists of the following key components:

- Receiving optics (telescope)
- Spectrometer and connecting optics
- Detector, analogue-to-digital converter (ADC) and electronics
- Computer and auxiliary electronics
- Software for evaluation

- Database for absorption cross-sections of species (atoms or molecules) to be measured

In the case of artificial light source (active) DOAS, additional components are required:

- Artificial light source (lamp)
- Optics to arrange the light path (transmitting telescope, retro-reflector, or multi-reflection optics)

The components can be arranged in many different ways. For instance, a search-light type arrangement consisting of an artificial light source coupled to a transmitting telescope can project a light beam across a distance in the open atmosphere. The other end of the light path would be defined by a receiving telescope collecting (a usually small) part of the transmitted beam and transferring the light to a spectrometer coupled to a suitable detecting system. Details of the various possibilities are given in Sect. 7.9.

7.3 Light Sources for Active DOAS

To be suitable for DOAS applications, light sources must fulfill several important requirements:

1. Minimal spectral intensity variation [i.e. $I_0(\lambda) = \text{constant}$] is required, in particular at the scale of molecular vibrational bandwidths (i.e. around ≈1–10 nm see Chap. 6). In other words, the light sources should ideally emit 'white' light with a 'smooth' spectral characteristic.
2. As described earlier, the detection limit is improved if a large number of photons are recorded; thus, high intensity of the light source is required.
3. In active DOAS systems, the light has to be projected over a long path; therefore, a good beam collimation is required. For a thermal light source, this is equivalent to a request for high luminous intensity (high emitted power per unit area of the radiator and wavelength interval).
4. In particular, for long-term applications (e.g. in monitoring applications), long lifetime and low cost of light sources are also important.

Clearly, there is no ideal light source; thus, DOAS measurements are made using a variety of artificial light source types.

7.3.1 Characteristics of Artificial Light Sources

The advantage of artificial light sources is that they are always available, while sunlight, moonlight, or starlight can only be used under certain circumstances (day, full-moon, no cloud cover, etc.) and during certain periods of time. In addition, the useable spectral range of artificial light sources can extend further into the UV, as well as (in comparison to scattered light applications) into the IR. For instance, light from extra-terrestrial sources that reaches the earth's surface always has to penetrate the ozone layer (see Chap. 2); thus, there will be virtually no intensity below 280 nm. Measurements with UV radiation at wavelengths as short as 200 nm can be made in the troposphere with suitable artificial light sources. On the other hand, the intensity of Rayleigh scattered sunlight for example as detected by ZSL–DOAS, becomes very small at near-IR wavelengths; thus, artificial light sources (or direct sunlight) can offer advantages in these situations. A summary of available light sources which have been used for DOAS applications and some of their properties is given in Table 7.1.

A basic distinction can be made between thermal light sources, such as incandescent lamps or arc lamps, and non-thermal light sources, such as light emitting diodes (LEDs), atomic emission lamps, fluorescent lamps, and laser light sources. Criteria for the selection of light sources are spectral brightness (W/unit of radiating area and wavelength interval) and spectral interval.

The radiation energy density $\rho(\nu, T)$ (in J/m^3) as a function of frequency ν and temperature T in a cavity (Hohlraum) is given by:

$$\rho\,(\nu,\mathrm{T}) = \frac{8\pi\,\mathrm{h}\nu^3}{\mathrm{c}^3}\cdot\frac{\mathrm{d}\nu}{\mathrm{e}^{\mathrm{h}\nu/\mathrm{kT}}-1}, \tag{7.1}$$

where h, k, and c denote Planck's and Boltzmann's constants and the speed of light, respectively. By substituting $\nu = \frac{c}{\lambda}$ and $d\nu = \frac{d\nu}{d\lambda}d\lambda = -\frac{c}{\lambda^2}d\lambda$, (7.1) can be converted into the corresponding wavelength-dependent expression:

$$\rho\,(\lambda,\mathrm{T}) = \frac{8\pi\,\mathrm{hc}}{\lambda^5}\cdot\frac{\mathrm{d}\lambda}{\mathrm{e}^{\mathrm{hc}/\lambda\mathrm{kT}}-1}, \tag{7.2}$$

Table 7.1. Comparison of different light sources used for active DOAS applications

Type of lamp	Brightness	Typical input power W	Spectral range nm	Typical lifetime hours
High pressure Xe-Arc	High	75 to several 1000	<200–3000	200–2000
D-Arc	Medium	25	180–300	
Incandescent lamps	Low	10–500	300–3000	50–1000 depending on voltage
(Broadband) laser	Very high		<0.3 nm in a 200–3000 nm, interval	Dependent on type
Light emitting diodes (LED)	Low-medium	0.05–4.0	350[a] to several µm, certain wavelengths	>10000–100000

[a]Rapid technological advances towards lower wavelengths LEDs are being made.

Fig. 7.1 shows some examples of $\rho(\lambda)$ for cavities at different temperatures. For thermal emitters, the former quantity is essentially given by the Planck function, and thus the temperature T and the emissivity ε_L of the radiating area. The emissivity is unity for a perfect 'black body' i.e. an object that absorbs all incident light. Kirchhoff's law states:

$$\alpha_L(\lambda) = \frac{\text{Radiation absorbed by the object}}{\text{Radiation incident on the object}} = \frac{\text{Radiation emitted by the object}}{\text{Radiation emitted by a black body}} = \varepsilon_L(\lambda) . \quad (7.3)$$

Typical temperatures and emissivities are: $T \approx 3000\,\text{K}$, $\varepsilon_L = \varepsilon_{\text{tung}} \approx 0.35$ for the tungsten filament of incandescent lamps and $T \approx 6000$–$10000\,\text{K}$, $\varepsilon_{\text{Xe}} \approx 0.03 \ldots 0.4$ for high pressure Xe arc lamps. Radiation energy density and radiation flux (in W/m^2) from a surface are related by:

$$F_p(\lambda, T) = \frac{\varepsilon_L \cdot c}{4\pi} \cdot \rho(\lambda, T) . \quad (7.4)$$

This leads to the radiation flux from a radiating surface with the emissivity ε (from a lamp filament or arc) in units of W/m^2, per steradian and wavelength interval $d\lambda$:

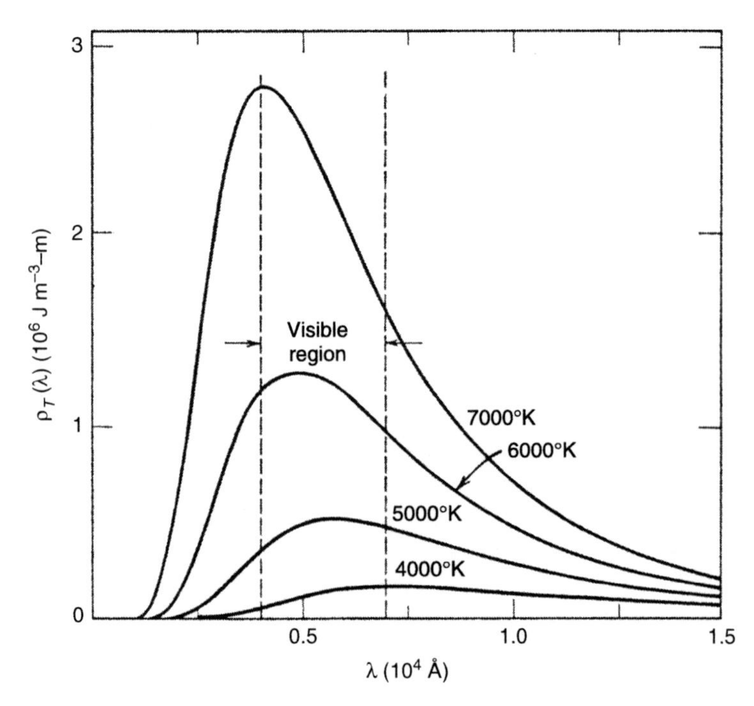

Fig. 7.1. Spectral intensity distributions of black bodies at 4000 K–7000 K (from Eisberg-Resnik, 1985)

$$F_p\,(\lambda, T) = \varepsilon_L \cdot \frac{2\,hc^2}{\lambda^5} \cdot \frac{d\lambda}{e^{hc/\lambda kT} - 1}, \tag{7.5}$$

or expressed as a function of ν in the frequency interval $d\nu$:

$$F_p\,(\nu, T) = \varepsilon_L \cdot \frac{2\,h\nu^3}{c^2} \cdot \frac{d\nu}{e^{h\nu/kT} - 1}. \tag{7.6}$$

Incandescent lamps are inexpensive, easy, and relatively safe to operate. The spectrum of incandescent lamps is very similar to that of a black-body radiator at 2800–3000 K, with a maximum in the near IR (1.0–2 μm, see Fig. 7.2). There is little emission in the UV below 400 nm. In practice, quartz–iodine lamps ('halogen lamps') are used because they reach somewhat higher filament temperatures and last longer than traditional incandescent lamps. However, the main advantage of quartz–iodine lamps for spectroscopy in the UV lies in the fact that their envelope is made of UV-transmitting quartz (as the name suggests). There is no problems with using quartz–iodine lamps of the type widely used for home illumination purposes, (if UV output is desired, care must taken that the lamps are not of the 'UV-protected' type). The output can be greatly increased by operating the lamps at 20–50% higher voltages than nominal, of course at the expense of a shortened lifetime (which normally

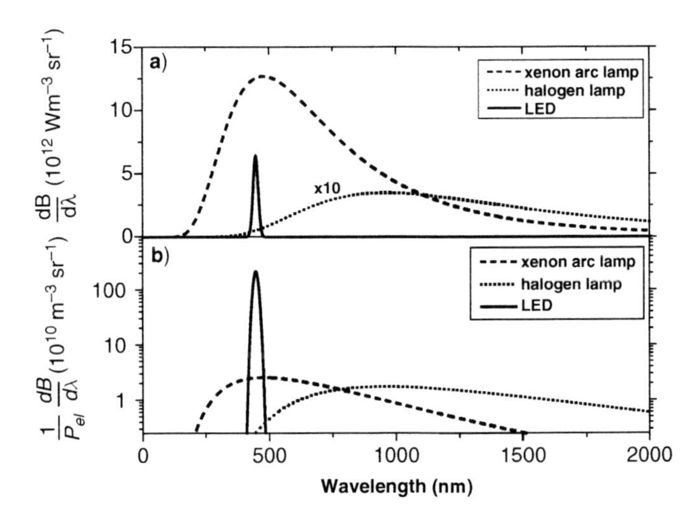

Fig. 7.2. Approximate intensity distributions of light sources. The **top panel** shows the spectral intensity distributions of three light sources: (1) An incandescent lamp (represented by a Planck's spectrum at $T = 3000\,\mathrm{K}$ and an emissivity of $\varepsilon_{\mathrm{tung}} = 0.35$). For clarity the intensity is multiplied by 10, (2) a Xenon-arc lamp (Osram XBO 450 W/2, represented by a Planck's spectrum at $T = 6000\,\mathrm{K}$ and $\varepsilon_{\mathrm{Xe}} = 0.4$), and (3) a LED (Luxeon LXHL-LR3C royal blue LED operated at 2.6 W electrical input power). **Bottom panel**: Same figures as above, intensity on a log scale and normalised to the electrical input power of the light sources (adapted from Kern et al., 2006)

is in the 50–1000 h range). In some cases, absorption lines due to I_2 were observed in quartz–iodine lamps (Senzig, 1995).

Since brightness is more important than total light flux, lamps of higher power than 20–50 W will be of no advantage. Higher-power lamps have larger radiating areas (i.e. filament sizes), but the emitted radiation per unit area stays the same.

Arc Lamps (Xenon and Deuterium): Due to their arc temperature ($T \approx 6000$–10000 K, $\varepsilon \approx 0.1$–0.4) being much higher than that of any filament, xenon-arc lamps (Fig. 7.3) have – compared to incandescent lamps – a much

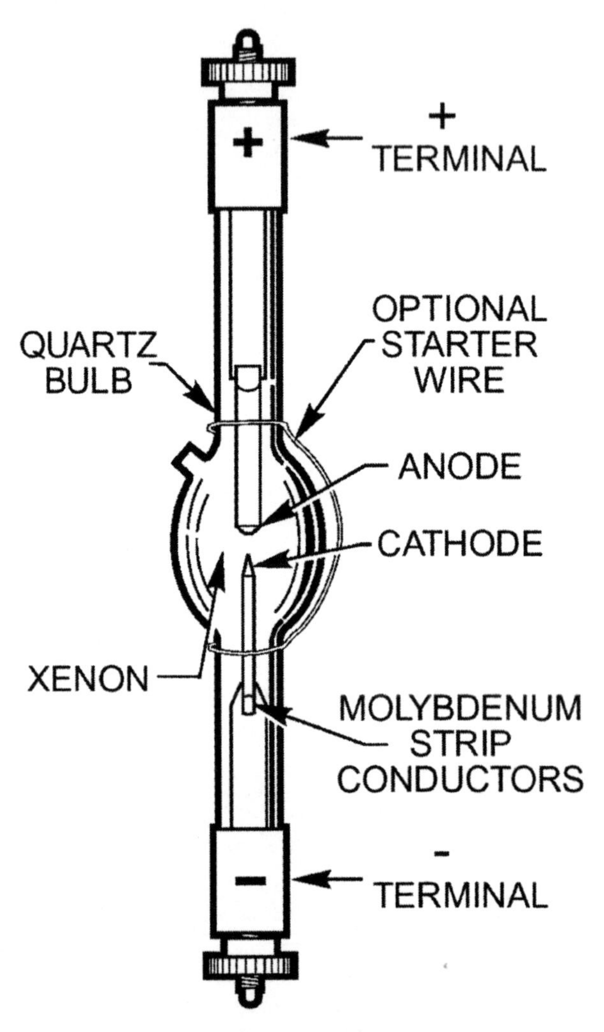

Fig. 7.3. Mechanical outline of a standard high-pressure Xe-arc lamp. The bulb of the lamp is made of quartz, the electrode terminals of metal. A very small arc (around a millimetre long, see text) develops at the centre of the 'bulb' (see Fig. 7.5) (figure courtesy of Hamamatsu Photonics)

larger brightness. They emit an essentially smooth spectrum in the spectral range of interest, with the exception of several weak lines and a group of emission lines between 400 and 450 nm (see Fig. 7.4; Larche, 1953). Xe-arc lamps are available with power ratings from about 75 W to 25 kW and more. As in the case of incandescent lamps, the brightness of the arc is (for a given type of design) largely independent of the absolute power of the lamp.

For special purposes other types of arc lamps are employed. For instance, good UV intensity is obtained from **Deuterium lamps**.

On the other hand, arc lamps, in particular Xe-arc lamps, are relatively difficult to operate. While running at about 18 V DC, they require high voltage (about 25 kV) for ignition, and about 60 V during the initial seconds of warm-up. As in any discharge lamp, there is a need to stabilise the lamp current. The high-voltage burst during ignition, typically provided by a Tesla transformer in the 'lamp igniter' circuit, leads to emission of a strong pulse of electromagnetic radiation, which can be dangerous for other electronic parts of the DOAS systems, such as nearby personal computers. In addition, the high internal pressure (30 atm and more in the cold lamp and two to three

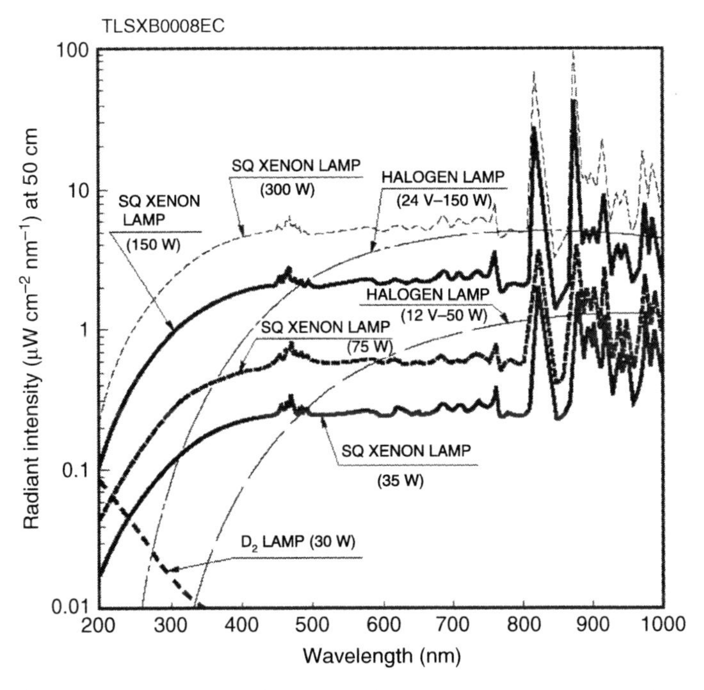

Fig. 7.4. Spectral intensity distribution (brightness) of high pressure Xe-arc lamps (Hamamatsu) in comparison to Deuterium and incandescent lamps. Note that the radiant intensity at 0.5 m distance is plotted. The luminous intensity of high-pressure short-arc Xe lamps is far superior to incandescent lamps throughout the visible and near IR spectrum (figure courtesy of Hamamatsu Photonics)

times as much during operation) of the lamps poses an explosion hazard. Therefore, precautions must be taken during operation, as well as when changing a burned out bulb. Typically, Xe-arc lamps are mounted on a lamp housing that provides electromagnetic shielding against the electromagnetic pulse generated during lamp ignition, as well as protection of the operators against UV radiation and consequences of possible lamp explosions.

Warning: Most arc lamps (including some atomic line calibration lamps) are powerful sources of UV light, with wavelengths in the UVA to UVC range. This radiation represents a severe hazard to eyes and can cause sunburn and ultimately skin cancer. Thus appropriate protection (goggles, radiation tight clothing, sun lotion) must always be used when operating these light sources.

Xe arc lamps are operated at very high internal pressure, thus there is a severe hazard of explosion (of the operating and non-operating lamps. Appropriate protection (safety goggles, protective clothing, etc.) must always be used when handling these light sources.

Xe-arc lamps come in several varieties (see, for example, Table 7.2):

- Standard lamps (e.g. Osram XBO 450 or Hamamatsu SQ – lamps, see Fig. 7.3) emit a Planck spectrum corresponding to a black body of about 6000–8000 K superimposed with some emission lines (see Fig. 7.4), which, due to their width (on the order of 1 nm), can interfere with the DOAS data evaluation. Typical bulb shapes are cylindrical with electrode connections at both ends. Electrode gaps are on the order of millimetres (for instance 4.5 mm in the case of a 450-W lamp). Cross-sections of the intensity distribution across the arc are given in Figs. 7.5 and 7.6. Normally, the lamps can only operate in vertical orientation with the anode at the top. Typical lifetimes are on the order of 1000–2000 h.
- High-intensity lamps (e.g. Osram XBO 500, Narva XBO 301) typically have a much smaller electrode gap than standard Xe-arc lamps, for

Table 7.2. Characteristics of several high pressure Xenon short-arc lamps (from Hermes, 1999)

Lamp manufacturer and type	Size of Arc (50% point) horiz. × vertical µm	Figure	Size of arc (data sheet) horiz. × vertical mm	Intensity in 200 µm circle around brightest spot
Osram XBO 450	770 × 530	7.6	0.9 × 2.7	5.910
Hanovia 959C 1980	380 × 220	7.7	0.3 × 0.3	87.500
Narva XBO 301	300 × 180	7.8	0.2 × 0.3	131.450

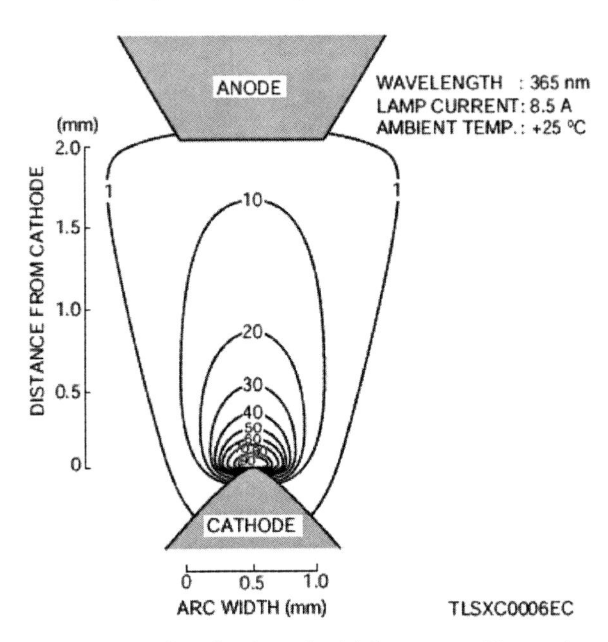

Fig. 7.5. Spatial intensity distribution of a high-pressure Xe-arc lamp. Lines indicate locations of equal luminance (in $10^3\,\mathrm{cd\,cm^{-2}}$) (figure courtesy of Hamamatsu Photonics)

instance <1 mm for a 500-W lamp. They also emit a Planck spectrum corresponding to a black body of about 8,000–10,000 K. More importantly, compared to standard lamps, the radiation is coming from a much smaller area. This can be seen from cross-sections of the intensity distribution across the arc in Figs. 7.7 and 7.8. In addition, the intensity of the emission lines superimposed on the continuum is typically much smaller, and their width is larger; thus, their interference with the DOAS data evaluation is greatly reduced. On the other hand, the lifetime of high intensity lamps is generally much shorter (on the order of 200 h) compared to standard lamps.

- In ozone-free lamps the bulb of the lamp is made of a specially doped type of quartz that blocks radiation below 242 nm (the threshold wavelength for O_2 photolysis). Depending on the type of lamp radiation, output may be degraded at wavelengths as long as 300 nm. Ozone-free lamps are usually more readily available (since they are now the standard in movie projectors). Absence of ozone pollution is also an advantage in the operation of DOAS instruments. On the other hand, care must be taken that the lamp output at short wavelengths is not compromised.

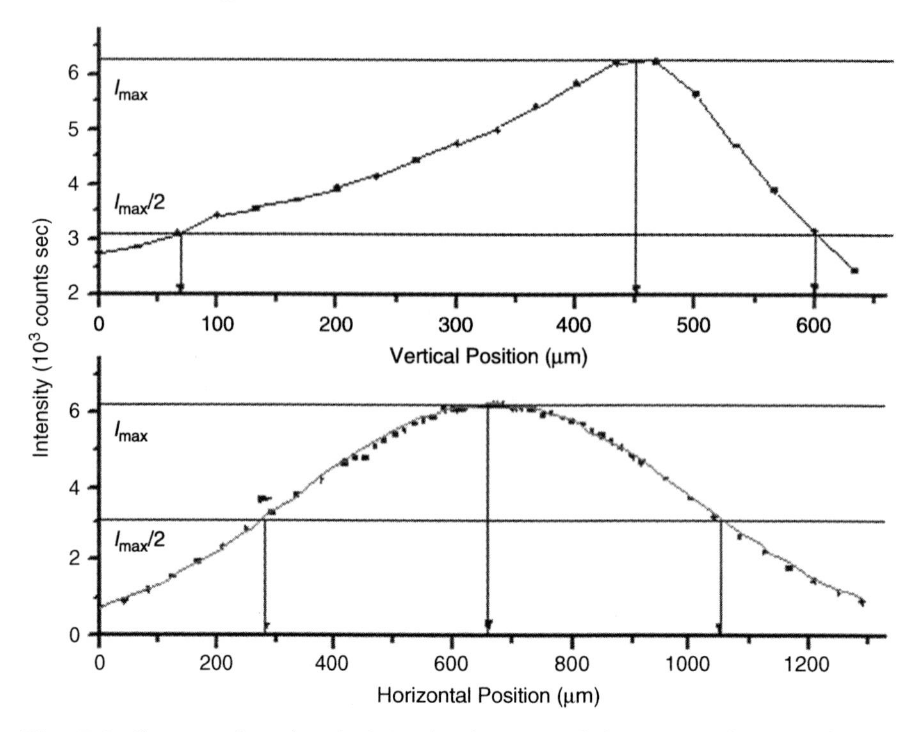

Fig. 7.6. Cross-sections (vertical: in the direction of the current, horizontal: perpendicular to current flow at the plane of highest brightness) of the intensity distribution of a standard 450 W high-pressure Xe-arc lamp (Osram XBO 450; from Hermes, 1999)

- Ozone-producing lamps are sometimes offered with different qualities (with respect to radiation absorbing impurities) of quartz (e.g. Suprasil and Infrasil).

Arc Lamp Stability: Short-term stability is measured over seconds, while long-term stability is measured over minutes, hours, or even days. Short-term stability is affected by arc 'wander', 'flare', and 'flutter'. Arc wander is the movement of the attachment point of the arc on the cathode surface. Typically, the arc circles around the conical cathode tip (see Fig. 7.5) and takes several seconds to complete a full circle. If arc flares occur, there are momentary changes in brightness as the arc moves to an area on the cathode with a better emissive quality than the previous attachment point. Arc flutter is the rapid side-to-side displacement of the arc column as it is buffeted by convection currents in the xenon gas, which are caused by the gas being heated by the arc and cooled by the envelope walls (Fischer, 1987). Arc wander and flare can sometimes be reduced by a slight decrease in the operating current.

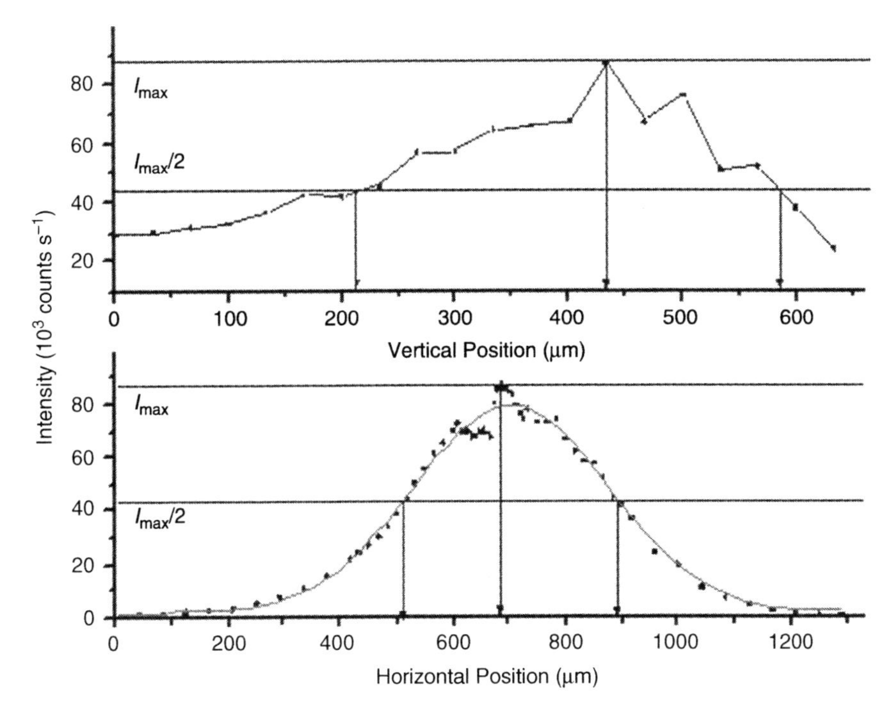

Fig. 7.7. Cross-sections (vertical: in the direction of the current, horizontal: perpendicular to current flow at the plane of highest brightness) of the intensity distribution of a 'high brightness' 500-W high-pressure Xe-arc lamp (Hanovia 959C 1980; from Hermes, 1999)

Xe-arc Lamp Lifetime: Lamp life varies dependent on the type. Specified lifetimes usually range from 200 to 2000 h. The useful life of compact arc lamps is determined by several factors: (1) The decrease of luminous flux due to deposit of evaporated electrode material at the inner wall of the envelope (bulb). The lamp bulb visibly 'blackens', (2) The increase of arc instability, and/or (3) Failure of the lamp to ignite or burn within specified parameters.

Frequent ignition accelerates electrode wear and blackening of the envelope. Usually, the average quoted lamp life is based on approximately 20 min of operation for each ignition. The end of the lamp life is the point at which the UV output has decreased by approximately 25%, the arc instability has increased beyond 10%, or the lamp has ceased to operate. Lamps should be replaced when the average lamp life has been exceeded by 25%.

As the lamp ages, the operating voltage will increase. Lamp current should be decreased to maintain output until the minimum operating current is reached. At this time the lamp should be replaced.

LEDs are potentially useful light sources since they generally emit a smooth spectrum with an emission bandwidth of 10–60 nm. There is a steady improvement of LED brightness and an extension of the spectral range (of

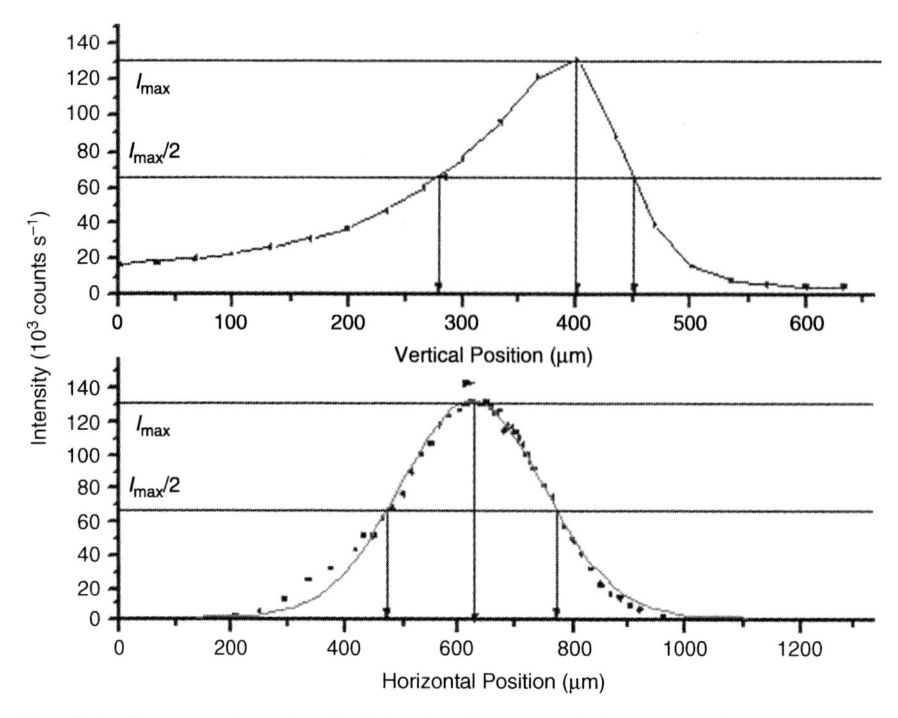

Fig. 7.8. Cross-sections (vertical: in the direction of the current, horizontal: perpendicular to current flow at the plane of highest brightness) of the intensity distribution of a 'high brightness' 300-W high-pressure Xe-arc lamp (Narva XBO 301; from Hermes, 1999)

room-temperature devices) from the original near-IR towards shorter wavelengths (e.g. recently marked by the introduction of GaN devices) (Hermes, 1999; Kern, 2004; Kern et al., 2006).

Presently, LED's are commercially available from the near-UV (about 350 nm) throughout the visible spectral range (starting with blue at about 450 nm) and into the near-IR. Spectral intensity distributions are given in Fig. 7.9. While the emitted total power is orders of magnitude smaller than that of thermal light sources and most laser types, it has to be kept in mind that emission only occurs in a relatively narrow spectral band and, more importantly, from a very small area on the order of 100 μm diameter (Kern, 2004; Kern et al., 2006). In fact, the energy emitted per unit area from many modern LEDs can be comparable to at least that of incandescent lamps. Compared to the electrical input power, LEDs probably give the best spectral intensity (Fig. 7.2) of any light source, with the exception of lasers.

Lasers (broadband) give a far lower beam divergence and higher spectral intensity than thermal light sources, but are usually much more complex devices (although this may change with the availability of room-temperature diode lasers for the visible and UV spectral regions). However, the spectral

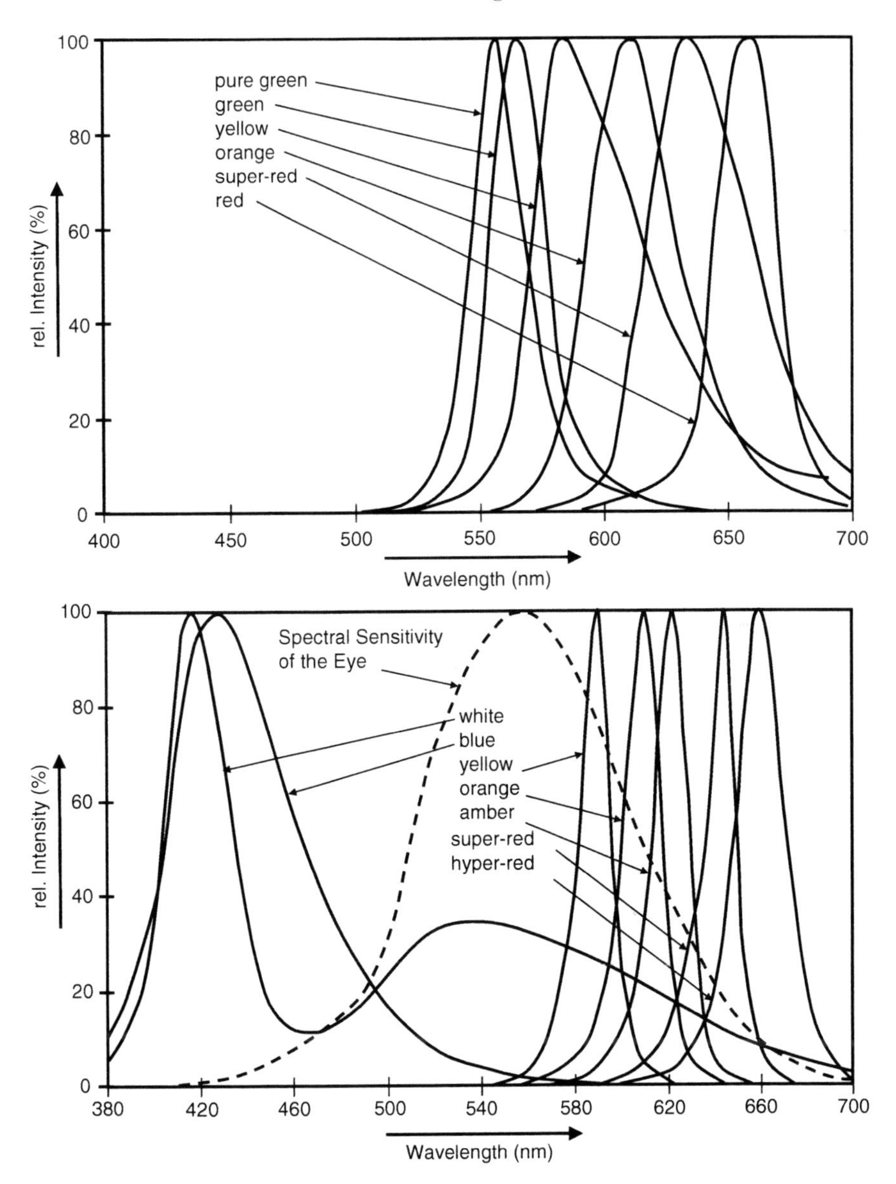

Fig. 7.9. Spectral intensity distribution of a series of LEDs. **Upper panel**: Standard LEDs. **Lower panel**: High brightness GaN LEDs. The dashed line indicates the eye sensitivity

emission bandwidths and tuning ranges of lasers are usually quite narrow. Since the standard DOAS operation requires 'white light' sources, it is usually difficult to observe molecular absorption bands with laser light sources. Therefore, lasers have rarely been used as DOAS light sources. A notable exception is the measurement of OH radicals (together with SO_2, CH_2O, and naphthalene, see Chap. 6), where several rotational lines (of 1–2-pm spectral width) are observed in a spectral interval of about 0.2–0.3 nm, which can be covered by a 'broadband' laser. Presently, developments are under way to improve the spectral emission ranges of dye lasers (Dorn et al., 1996; South et al., 1998) for this, or similar, applications.

7.3.2 Natural Light Sources

Natural light sources include light sources outside the atmosphere i.e. sun, or moon, or even starlight. These light sources provide a light path through the entire atmosphere. The instrument becomes simpler, since no provisions for operating a light source have to be made (see Chap. 11).

The largest problem with natural light sources is the fact that their structured spectrum contains many Fraunhofer lines (see Chap. 9). In addition, the brightness of natural light sources is less than that of good artificial light sources. Other limitations come from the fact that sun, moon, or starlight can only be used under certain circumstances (day, full moon, no cloud cover, etc.) and during certain periods of time (see Sect. 7.3.1).

Direct sunlight allows the determination of the total column density of atmospheric trace gases. The average trace gas concentration measured along the line of sight from the instrument to the sun can be converted to the vertical column density (see Chap.9). Due to the high brightness of the solar disk, the solar intensity is very high. Sun-following optics with moving parts are required to direct the sunlight into the instrument. However, measurements are only possible if the direct sunlight is not blocked by clouds.

A complication with sunlight in comparison with most artificial light sources is the presence of strong absorption features due to the solar atmosphere, the so-called Fraunhofer lines (see Fig. 7.10). Fraunhofer lines are quite strongly structured, and the spectral width of the structures is narrower than the resolution of low-resolution DOAS instruments. The structures are more prominent in the UV, and have large optical densities.

Further complications arise from the fact that the optical densities of the Fraunhofer lines are not constant across the solar disk. This effect is known as 'centre-limb darkening' (Bösch, 2002; Bösch et al., 2003). Therefore, care has to be taken to always observe the full solar disk This can constitute a problem in occultation studies where radiation originating from the (lower) rim of the solar disk is passing through denser parts of the atmosphere, and thus will be more strongly attenuated (Bösch, 2002; Bösch et al., 2003).

Since the contribution of scattered light in direct sunlight is very small, there is no noticeable 'ring effect' (see Chap. 9) in the spectra.

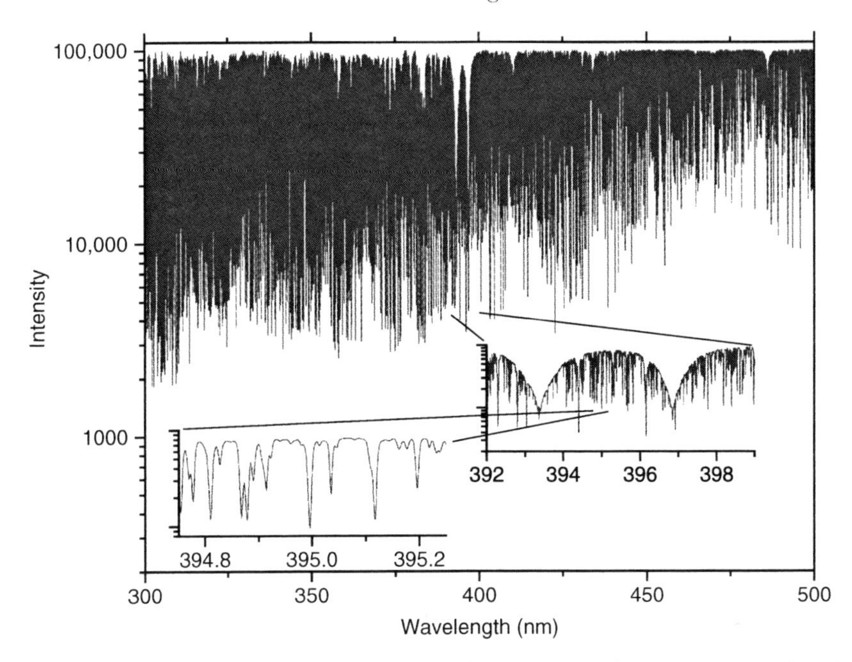

Fig. 7.10. Section of a high-resolution solar spectrum between 300 and 500 nm (from Kurucz et al., 1984)

Direct moonlight observations are subject to essentially the same advantages and restrictions as direct sunlight. However, compared to direct sunlight, the intensity of moonlight is by about five orders of magnitude lower. Of course moon-following optics are required. In addition, moonlight can only be used around the time of full moon. If the moon is not full, brightness (in addition to the smaller visible area) rapidly diminishes because of the large illumination and observation angles under which the lunar surface is seen. On the other hand, moonlight measurements offer the advantage of night-time measurements that allow the study of the abundance of photo-labile species, which cannot be observed with direct or scattered sunlight. Since moonlight is scattered sunlight, the spectral features with respect to Fraunhofer lines are the same as in direct sunlight. Since the moon reflects light from all parts of the solar disk, the centre-limb darkening effect (Bösch et al., 2003) should be absent.

Sunlight scattered in the atmosphere is weaker in intensity than direct sunlight by about four orders of magnitude. Nevertheless, DOAS measurements using scattered sunlight have several advantages. No sun-following optics and hence no moving parts are required, which makes unattended operation very simple. In addition, measurements are also possible in the presence of clouds. Disadvantages include much more complicated calculation of vertical trace gas column densities from the primary 'slant-' (or in this case rather 'apparent') column densities, particularly if Mie scattering (due to haze or clouds) is involved. These conversions typically involve detailed radiation transport calculations. In addition, the observed spectra are distorted due to (rotational) Raman scattering, leading to a 'filling in' effect of solar Fraunhofer lines at large SZA, the 'ring effect' (Grainger and Ring, 1962), see Sect. (9.6).

Since the atmosphere scatters light very uniformly from all parts of the solar disk in the same way, the centre-limb darkening effect should not be large.

In principle, scattered light from each part of the sky can be used. However, certain arrangements have proven to be particularly popular (see Sects. 7.10.2 and 7.10.3), i.e. ZSL-DOAS and off-axis DOAS. The term 'off axis' refers to a viewing direction 'off' the zenith direction (e.g. Smith and Solomon, 1990; Sanders et al., 1993). The term MAX-DOAS was also recently introduced (e.g. Hönninger and Platt, 2002; Wittrock et al., 2004; Hönninger et al., 2004a). More details are given in Chaps. 9 and 11.

7.3.3 Calibration Light Sources

Wavelength calibration of DOAS spectrometers and determination of their instrument function (see Sect. 7.5) are usually done by recording light with known, usually narrowband, spectral features. This can either be accomplished by subjecting white light to structured and spectrally stable absorption by (sufficiently inert) trace gases in absorption cells (cuvettes), or by suitable glasses, e.g. holmium glass. Alternatively, light sources emitting spectrally structured radiation, such as lasers, atomic emission lamps, or the Fraunhofer structure of sunlight, can be used.

Lasers, in particular gas lasers and some solid-state lasers, can have well-defined emission wavelengths; thus, they are ideally suited for precise wavelength calibration. In addition, the narrow spectral bandwidth ($<10^{-3}$ nm in the visible), is a good approximation of a single wavelength, the response to which directly gives the instrument function of the spectrometer–detector combination. On the other hand, many types of lasers are still expensive and bulky. Some small and inexpensive laser types, such as semiconductor lasers, show a large variation of the emitted wavelength with temperature; thus, they can only be used for determining the instrument function, not the wavelength calibration. For the latter purpose, He–Ne lasers or some solid-state lasers (Nd:Yag) are useful, since their wavelength is determined by atomic transitions.

Atomic emission lamps are a very inexpensive tool for both wavelength calibration and instrument function determination of DOAS spectrometers. They consist of a small glass or quartz envelope filled with a small amount of the particular element. Two electrodes allow flow of current in a low-pressure discharge. In the case of noble gases (He, Ne, Ar, Kr and Xe), a few millibars of the gas are the only component. For lamps showing metal lines (e.g. Hg,

Na, and Cd), Ne is also usually added in order to sustain an initial discharge current to heat the lamp until the metal exhibits sufficient vapour pressure to show emission lines. Depending on the active filling, the lamps emit many discrete emission lines of well-known (usually to a few pm or better) spectral positions (see Table 7.3). Line widths are typically on the order of 1 pm, but depend somewhat on operating conditions (they tend to be wider if the lamp is operated at higher temperature). Some care has to be taken to avoid using lines that are actually multiplets, and which might not be resolved at the spectral resolution of typical DOAS instruments (typically 0.1 nm). It should also be noted that the relative intensities of the various lines can vary considerably from lamp to lamp and also with the operating conditions of a particular lamp. Atomic emission lamps are available in many different configurations and fillings from industry. A simple variety of Ne-discharge lamps is used to illuminate operating indicators in electric devices. Most widely used are mercury, neon, and cadmium lamps. Spectral positions of selected lines of cadmium, mercury, hydrogen, neon, and zinc are given in Appendix A. Some mercury

Table 7.3. Recommended wavelengths (air) and wave numbers (vacuum) for selected Hg spectral lines emitted by pencil-type lamps (adapted from Sansonetti et al., 1996)

Line intensity[a]	Wavelength[b] (nm) in air	Wave number (cm^{-1}) in vacuum
3000000	253.6521	39412.236
160	289.3601	34548.888
2600	296.7283	33691.025
280	302.1504	33086.464
2800	312.5674	31983.828
1900	313.1555	31923.765
2800	313.1844	31920.819
160	334.1484	29918.220
5300	365.0158	27388.271
970	365.4842	27353.171
110	366.2887	27293.096
650	366.3284	27290.138
4400	404.6565	24705.339
270	407.7837	24515.883
34	434.7506[b]	22995.229
10000	435.8335	22938.095
10000	546.0750	18307.415
1100	576.9610	17327.389
1200	579.0670	17264.372

[a]Intensities are relative values based on irradiance values from Reader et al. (1996) with the intensity of 436 nm set arbitrarily to 10,000

[b]The wavelength uncertainty is 0.0001 nm, with the exception of that of the 434.7506-nm line

line positions, together with approximate intensities (Sansonetti et al., 1996), are summarised in Table 7.3. More line positions may be found in the literature, e.g. in the CD-ROM and internet database by Kurucz and Bell (1995).

7.4 Optical Elements for DOAS Systems

Besides a light source, telescope, spectrometer, and detector, a DOAS system also requires further optical components to efficiently transfer the radiation energy from one component to the next. Most of the above listed building blocks consist of elementary optical elements such as plane or curved mirrors, lenses, prisms. In this section, we give a brief overview of these elements, their principles of operation, and some properties. A full treatment of the subject is clearly beyond the scope of this book, and the reader is referred to textbooks of optics such as Smith and King (2000), Hecht (2002), and Bergman (1966).

7.4.1 Some Principles of Optics

Electromagnetic radiation, propagating as electromagnetic waves, may conveniently be represented as rays, which are geometric lines giving the direction along which the electromagnetic energy flows. Optical elements change the direction of these rays. Alternatively, electromagnetic waves may be characterised by their wave fronts, which are perpendicular to the direction of the rays. The action of optical elements may be regarded as changing the shape of the wave fronts. Visualisation, design, and calculation of optical systems (i.e. systems consisting of optical elements) are conveniently done by 'ray tracing'.

In the approximation of optical elements being large compared to the wavelength of the radiation, we can use a 'geometrical optics' approach. Radiation propagates along *straight lines* until *reflected* by mirrors or *refracted* by dense, but transparent material. Accordingly, important optical elements are plane and curved surface mirrors, prisms, and lenses.

Radiation incident on a surface is either absorbed, reflected, or transmitted. Neglecting absorption and transmission, the reflection from a flat surface is said to be either specular from a very smooth mirror-like surface, or diffuse from a matte-finish surface. Of course the terms specular or diffuse denote extreme cases, which can only be approximated in real reflectors, as shown in Fig. 7.11.

In the case of specular reflection (Fig. 7.11, left diagram), the light is reflected at the surface like an elastic sphere bouncing back from a hard surface. Incoming and reflected rays are in a plane with angle of incidence α between the normal of the surface and the incident ray. α is equal to the exit angle α' between the normal of the surface and the exiting ray.

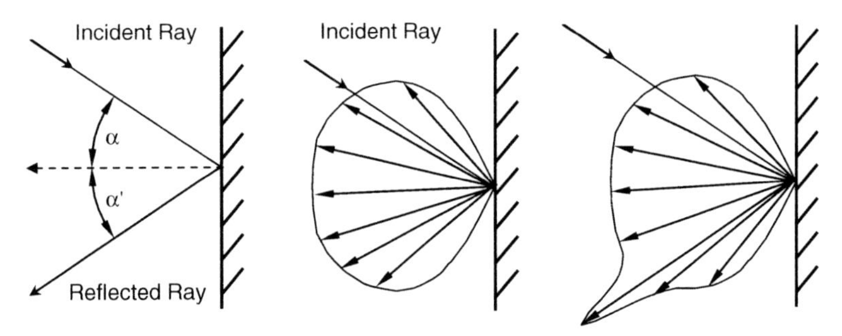

Fig. 7.11. Purely specular (**left**) and purely diffuse (**centre**) reflection from a flat surface are idealised cases. In reality, these cases can only be approximated and usually a combination is observed (**right**)

Radiation incident on transparent surfaces is only partially reflected. Another part of the radiation intensity enters the medium and its direction of propagation is refracted. The direction of the refracted beam (see Fig. 7.12) is given by Snell's law (Snellius law):

$$\frac{\sin\alpha}{\sin\beta} = \frac{n_2}{n_1} \quad \text{or} \quad \sin\alpha = \sin\beta \cdot \frac{n_2}{n_1}. \tag{7.7}$$

The degree of refraction is determined by the indices of refraction (refractive indices) n_1 and n_2 of the two materials. A ray approaching the surface from within the material with the higher index of refraction (n_2 in our example) will exit at $\alpha \geq 90°$, and therefore not exit the material at all if:

$$\sin\alpha = \sin\beta \cdot \frac{n_2}{n_1} \geq 1 \quad or \quad \sin\beta \geq \frac{n_1}{n_2}. \tag{7.8}$$

The ray will undergo **total internal reflection**. Total internal reflection is essentially loss free, if the reflecting surface (boundary between materials with

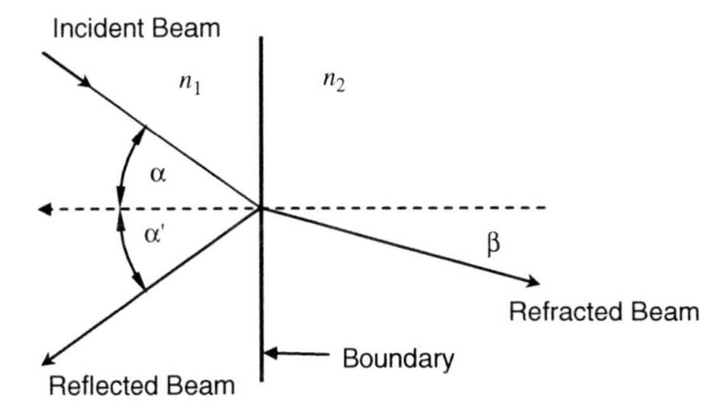

Fig. 7.12. Refraction and (specular) reflection from transparent surfaces. In this example, $n_1 < n_2$; for instance, the ray may enter from air (**left**) through the boundary between the media into glass (**right**)

different index of refraction) is clean. It is used in retro-reflector prisms (see Sect. 7.4.9)

Reflected Intensity: The reflectance $R(\lambda) = I_{reflected}/I_{incident}$ is different for the two polarisation directions; parallel to the plane of incidence and perpendicular to the plane of incidence. It is given by the Fresnel equations:

$$R(\alpha,\beta,\lambda)_{perpendicular} = \left(\frac{n_1\cos\alpha - n_2\cos\beta}{n_1\cos\alpha + n_2\cos\beta}\right)^2 = \left(\frac{\sin(\alpha-\beta)}{\sin(\alpha+\beta)}\right)^2 \quad (7.9a)$$

$$R(\alpha,\beta,\lambda)_{parallel} = \left(\frac{n_2\cos\alpha - n_1\cos\beta}{n_1\cos\beta + n_2\cos\alpha}\right)^2 = \left(\frac{\tan(\alpha-\beta)}{\tan(\alpha+\beta)}\right)^2, \quad (7.9b)$$

where the wavelength dependence is derived from the wavelength dependence of the index of refraction $n(\lambda)$. The last term in (7.9a, b) is derived by introducing Snell's law (7.7) into the preceding term. The total reflectance (for unpolarised light) is given by:

$$R(\alpha,\beta,\lambda) = \frac{1}{2}(R_{parallel} + R_{perpendicular}) = \frac{1}{2}\left(\frac{\sin^2(\alpha-\beta)}{\sin^2(\alpha+\beta)} + \frac{\tan^2(\alpha-\beta)}{\tan^2(\alpha+\beta)}\right), \quad (7.9c)$$

Note that the dependence on the indices of refraction in (7.9a – c) is implicit, since the equations depend on both angles α and β. The dependence on β may be eliminated by introducing $\cos\beta = \sqrt{1-\sin^2\beta} = \sqrt{1-\frac{n_1^2}{n_2^2}\cdot\sin^2\alpha}$ into (7.9a – c).

The transmittance $T(\lambda) = I_{transmitted}/I_{incident} = 1 - R(\lambda)$ is given by:

$$T(\lambda)_{perpendicular} = \left(\frac{2n_1\cos\alpha}{n_1\cos\alpha + n_2\cos\beta}\right)^2 = \left(\frac{2\sin\beta\cdot\cos\alpha}{\sin(\alpha+\beta)}\right)^2, \quad (7.9d)$$

$$T(\lambda)_{parallel} = \left(\frac{2n_1\cos\alpha}{n_1\cos\beta + n_2\cos\alpha}\right)^2 = \left(\frac{2\sin\beta\cdot\cos\alpha}{\sin(\alpha+\beta)\cdot\cos(\alpha-\beta)}\right)^2, \quad (7.9e)$$

At normal incidence, i.e. $\alpha = 0$ (7.9a – c) coincide and simplify to:

$$R(n_1,n_2) = \left(\frac{n_1-n_2}{n_1+n_2}\right)^2, \quad (7.9f)$$

For typical indices of refraction of $n_1 = 1.0$, $n_2 = 1.5$ (e.g. refraction at an air–glass interface), and normal incidence ($\alpha = 0$), a reflectance of $R = 0.04$ is obtained. Fig. 7.13 shows an example of the variation of $R_{parallel}(\alpha)$, $R_{perpendicular}(\alpha)$, and $R(\alpha)$ with α for the above indices of refraction. At small angles of incidence (up to about 20°), the reflectance stays near 4%, sharply increasing to unity as α approaches 90°. Note that $R_{parallel}$ has a minimum with $R_{parallel} = 0$ at some angle of incidence α_B. This angle, where polarised light is not reflected and hence $R_{parallel}(\alpha_B) = 1$, is called the Brewster angle.

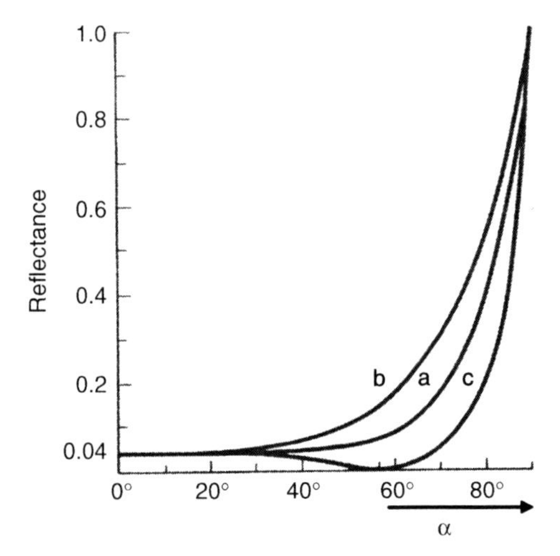

Fig. 7.13. Intensity reflected from a flat, dielectric surface. Curve (**a**) shows total reflectance $R(\alpha)$ for the unpolarised incident light. Curves (**b**) and (**c**) show the reflectances for polarised light with perpendicular and parallel polarisation, respectively. Note the zero reflectance, and thus 100.00% transmission, for parallel-polarised light at the Brewster angle

The index of refraction of (transparent) materials varies as a function of wavelength. Some examples of various glass and quartz types are shown in Fig. 7.14.

The index of refraction of gases is a function of their density, which in turn are functions of temperature T and pressure p.

$$n\,(p,T) = \frac{p}{p_0} \cdot \frac{T_0}{T} \cdot (n_0 - 1) + 1. \tag{7.10}$$

For dry air, $n_0 \approx 1.000293$ denotes the index of refraction at 589 nm, standard pressure and temperature, $p_0 = 1\,\text{atm}$ (101325 Pa) and $T_0 = 273.15\,\text{K}$ (0°C), respectively.

Reflection on Metals: Metals can be treated as materials having a given index of refraction, but also an attenuation coefficient due to the electric conductivity of metals. In the case of a zero attenuation coefficient and a very high index of refraction, Snell's law (7.7) would be applicable and the reflectivity would become very high. The sketch of a typical metal reflectivity as a function of polarisation (parallel or perpendicular to the plane of incidence) in Fig. 7.15 shows that the reflectance of metals indeed behaves qualitatively similarly to the reflectance of dielectric materials.

Fig. 7.16 shows reflectance curves R as a function of wavelength for the popular mirror coatings silver, gold, copper, and aluminium. In the red and infrared spectral regions, gold has superior reflectance (R ≈ 0.99, hence its reddish colour). Silver is almost as good a reflector (R ≈ 0.98) in the visible

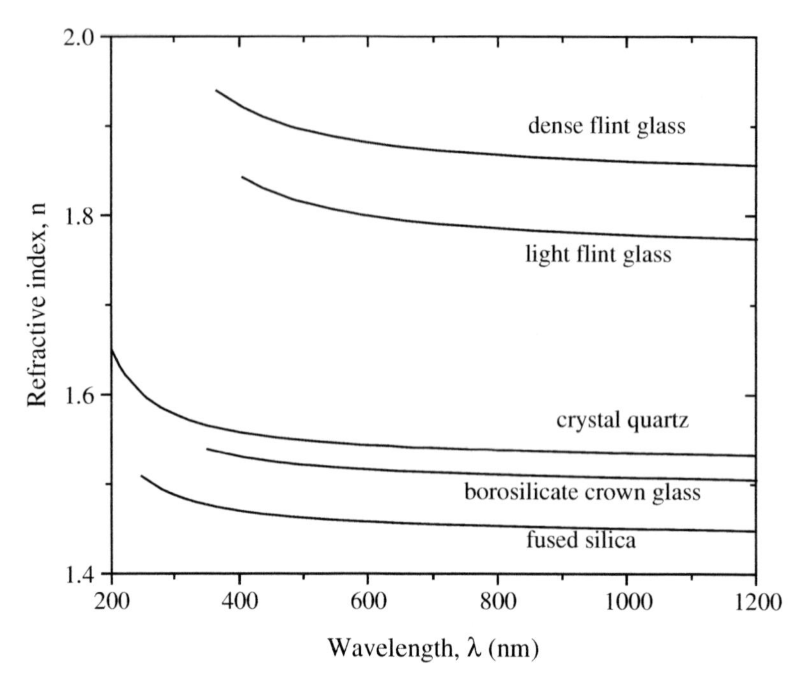

Fig. 7.14. Index of refraction as a function of wavelength for various glasses, quartz, and acrylic plastic (data courtesy of Schott)

and near-IR spectral ranges. Aluminium is much less reflective than the noble metals, but superior in the UV. Note that optical mirrors are usually first surface mirrors, i.e. the reflective coating is on the surface oriented towards the incident radiation. Therefore, the metal surfaces, e.g. silver and to an even larger extend aluminium mirrors, are susceptible to corrosion and normally require protective layers, usually a thin overcoat of SiO or MgO.

7.4.2 Mirrors

The reflection by plane mirror surfaces is shown in Fig. 7.11, with the angle of incidence equal to the angle of reflection. Curved mirror surfaces, however, can have imaging properties. Here, we discuss the imaging properties of a spherical, concave mirror shown in Fig. 7.17. A sufficiently small surface element of any curved mirror can be considered plane.

A ray entering parallel to the mirror axis (solid line in Fig. 7.17) at distance d will hit a mirror with a curvature of radius r, at point A. It is incident at an angle α with the normal to the mirror surface at A (dashed line AM in Fig. 7.17). The ray is reflected at an angle of α and intersects the mirror axis at point F. The reflected ray forms an angle of $\alpha' = \alpha$ with the normal intersecting the mirror axis at point M. Thus, AMF is an isosceles triangle with $a = f$ and:

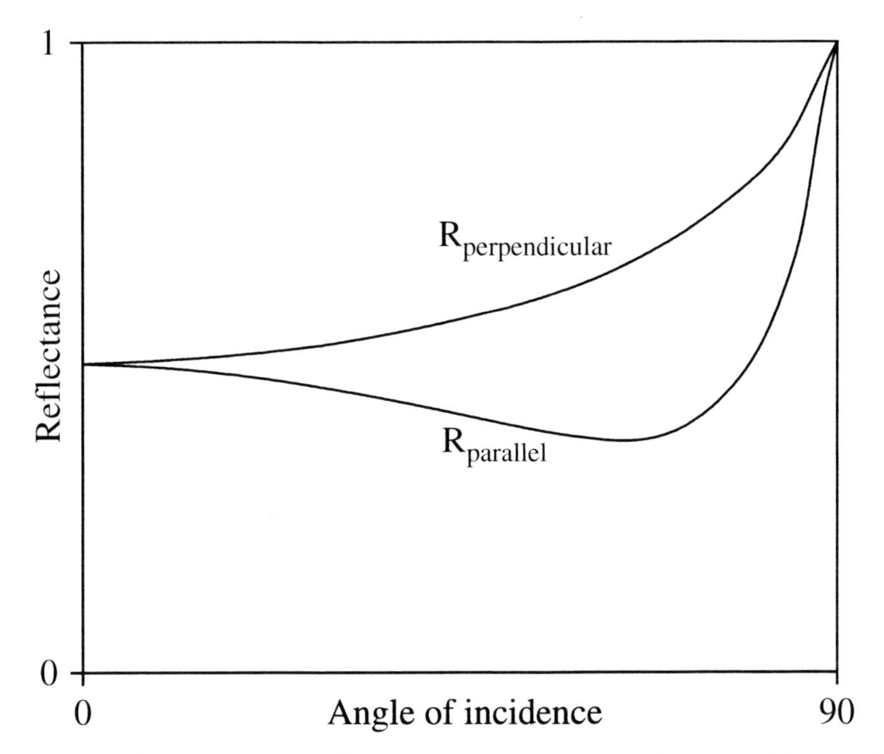

Fig. 7.15. Typical reflectance ($R_{perpendicular} = R_\perp$) for linearly polarised beams of radiation incident on a metal as a function of angle of incidence. $R_{parallel}$ drops to a minimum at the so-called 'principal angle of incidence', marked by a dot

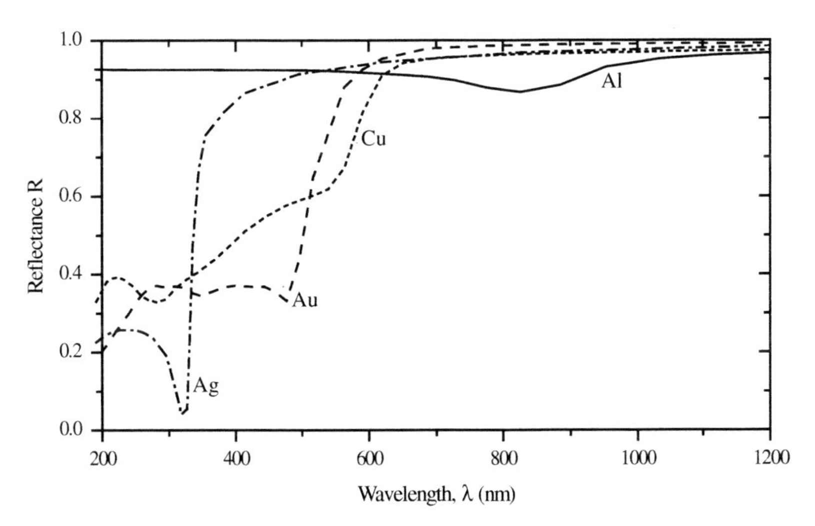

Fig. 7.16. Reflectances of silver, gold, copper, and aluminium as a function of wavelength (CRC, 2008)

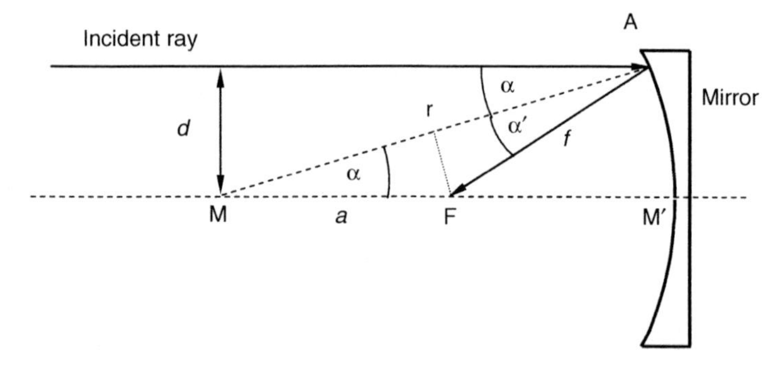

Fig. 7.17. Imaging by a (concave) spherical mirror, with radius of curvature r. Line AM denotes the normal to spherical surface at point A. The angle of incidence is α. In the paraxial approximation ($\cos\alpha \approx 1$), all rays incident parallel to the optical axis (MM′) of the mirror (at distance d) will be reflected such that they cross the focal point F at the focal length (AF = FM′ = f) of $f = r/2$

$$\cos\alpha = \frac{r/2}{a} \quad a = f = \frac{r}{2\cos\alpha} \quad \text{and also } \sin\alpha = \frac{d}{r}. \tag{7.11}$$

For rays entering very close to the optical axis ('paraxial' rays), α will become very small, and thus $\cos\alpha \approx 1$. In this 'paraxial approximation', a spherical mirror has the focal length $f = r/2$.

As a more general case, we consider a concave mirror, as shown in Fig. 7.18. All rays incident at an arbitrary distance d from the optical axis must travel the same distance from points B on plane BB′ via A to F. Thus, the distances BA + AF must be equal to BC (where C is on plane CC′) for all values of d. A curve containing all points with the same distance between a line (i.e. CC′)

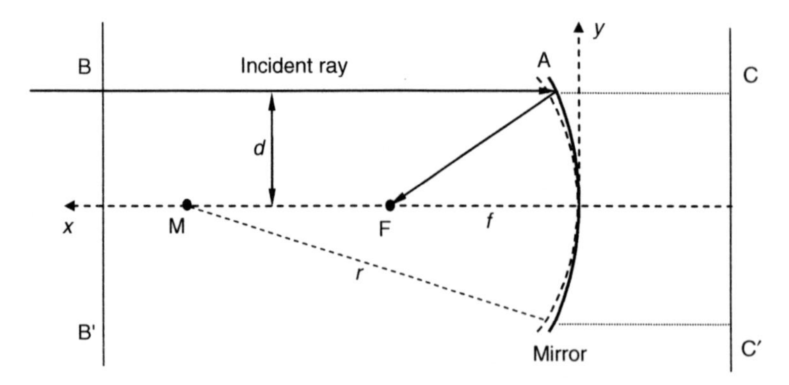

Fig. 7.18. Imaging by a parabolic mirror. For any ray incident parallel to the optical axis the distance BA + AF must be equal to BC i.e. constant. A parabolic mirror can be aproximated by a spherical surface as shown by the dashed line

and a point (F in this case) is a **parabola**. The points of the parabola are related by the formula:

$$x = \frac{1}{4f} \cdot y^2; \quad y = \sqrt{4f\,x} = 2\sqrt{f\,x}. \tag{7.12}$$

A parabola can be approximated by a sphere given by $y^2 + (x - r)^2 = r^2$ (centre of the sphere M is shifted from the origin by r) or $x^2 - 2rx + y^2 = 0$. Solving for x gives:

$$x = r \pm \sqrt{(r^2 - y^2)}. \tag{7.13}$$

Expansion in a binominal series gives:

$$x = \frac{y^2}{2r} + \frac{y^4}{2^2 2! r^3} + \frac{3y^6}{2^3 3! r^5} + \ldots \tag{7.14}$$

The first term of the expansion corresponds to the equation of a parabola (7.12) with $f = r/2$; the remaining terms describe the deviation between a sphere and a parabola:

$$\Delta x = \frac{y^4}{8r^3} + \frac{3y^6}{48r^5} + \ldots \tag{7.15}$$

For small values of y (i.e. paraxial rays), the deviation will be relatively small and spherical, and parabolic curves will be indistinguishable.

A reflecting parabolic (concave) mirror with the focal length f images any object (size: G) further away from the surface than $2f$ in reduced size (B) at a distance between f and $2f$ (see Fig. 7.19). The relationship between focal length f, distance of object and image from the vertex of the parabola g and b, respectively is given by:

$$\frac{1}{f} = \frac{1}{g} + \frac{1}{b}. \tag{7.16}$$

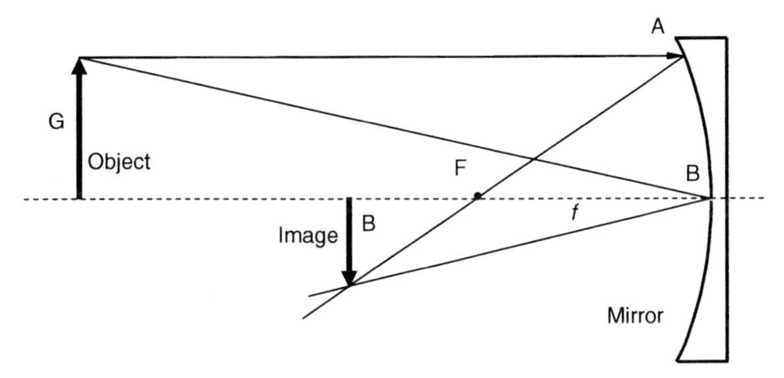

Fig. 7.19. Imaging by a parabolic mirror. An object (G) further away from the surface than $2f$ will be imaged in reduced size (B) at a distance between f and $2f$

7.4.3 Prisms

Prisms change the direction of rays by refraction. There are two refracting surfaces (see Fig. 7.20, apex angle γ). A ray entering at an angle of incidence, i.e. the angle between the direction of the ray and the normal on the surface, β_1 at the left surface is refracted and continues at an angle β_2 inside the prism, until exiting at angle β_3 at the right surface, thereby being refracted again outside of the prism at an angle β_4. The different angles are connected by Snell's law of refraction (7.7):

$$n_1 \sin \beta_1 = n_2 \sin \beta_2 \text{ and } n_2 \sin \beta_3 = n_1 \sin \beta_4,$$

where n_1 and n_2 denote the indices of refraction outside and inside the prism, respectively. The total deviation β between the incoming and outgoing beam is $\beta = \beta_1 - \beta_2 - \beta_3 + \beta_4$. Since in the triangle OAB, $\gamma = \beta_2 + \beta_3$, we obtain $\beta = \beta_1 + \beta_4 - \gamma$.

For small angles γ we have:

$$\beta = (n - 1) \cdot \gamma$$

Prisms can be used to separate radiation of different wavelengths in spectrometers. Putting polychromatic light through a prism will result in the rays being refracted in different directions with the total angular deviation:

$$\frac{d\beta}{d\lambda} = \gamma \cdot \frac{dn}{d\lambda}, \tag{7.17}$$

where $dn/d\lambda$ denotes the change of the refractive index with wavelength, the dispersion (see Fig. 7.14), which for most glass types is on the order of $10^{-4}\,\text{nm}^{-1}$.

7.4.4 Lenses

Treating a lens as a series of thin prisms leads to an expression relating the radii of curvature of both lens surfaces to the focal length of the lens, the 'lensmaker's formula':

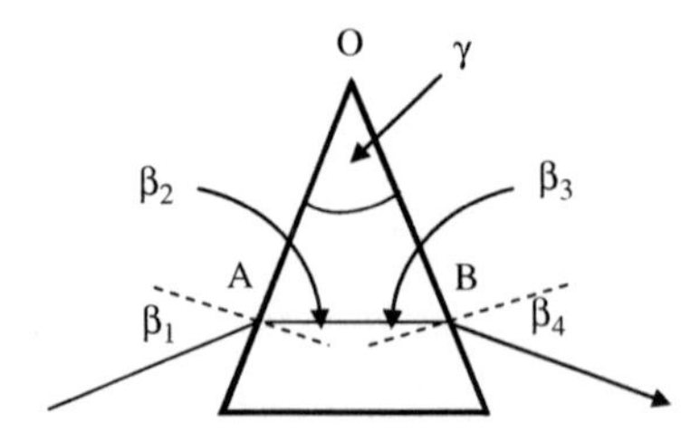

Fig. 7.20. Sketch of ray propagation through a prism

$$\frac{1}{f} = (n-1) \cdot \left(\frac{1}{r_1} - \frac{1}{r_2}\right). \tag{7.18}$$

Note that the difference in the indices of refraction n of the lens material and the material around the lens (here set to unity) enters the equation. For lenses, the Cartesian **sign convention** of r_1 and r_2 is such that the radii of surfaces, for which the centre of curvature is on the opposite side of the light source, are defined as positive. If the centre of curvature is on the same side as the light source, r is defined as negative. For instance, for the lens in Fig 7.21, r_1 would be defined as positive, while r_2 would be negative. Therefore, in the case of a biconvex lens, the focal length would be:

$$\frac{1}{f} = (n-1) \cdot \left(\left|\frac{1}{r_1}\right| + \left|\frac{1}{r_2}\right|\right).$$

Thus, (7.18) also can also be applied to biconcave or concave–convex lenses. An important property of lenses is the relatively strong dependence of the focal length on the index of refraction, with $f \propto 1/(n-1)$, and thus on the wavelength of the radiation, since n varies considerably with the wavelength, in particular in the UV (see Fig. 7.14).

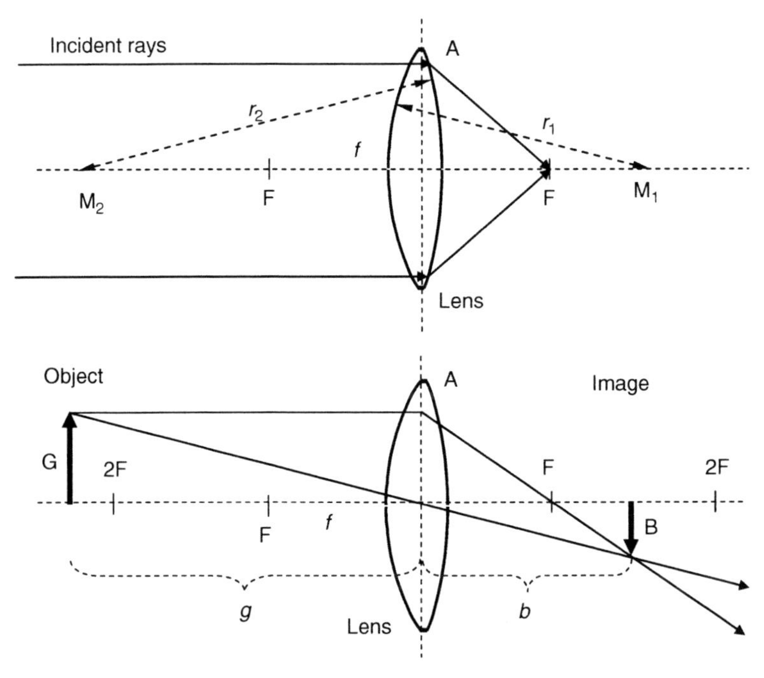

Fig. 7.21. Imaging by a thin biconvex lens. **Upper panel**: relationship between radii of curvature and 'focal length (see 'lensmaker's formula'). **Lower panel**: Imaging of object with extension G into image with extension B. The vertical dashed lines indicate the 'principal plane' of the lens i.e. the position of an idealised, infinitely thin lens

A convex lens can produce an image of an object. The Gaussian lens formula relates the distance g between the object and the lens to the distance b between lens and image:

$$\frac{1}{f} = \frac{1}{g} + \frac{1}{b}, \tag{7.19}$$

where f denotes the focal length of the lens. An example is outlined in Fig. 7.21. Note that the formula is the same as (7.16), but object and image are on opposite sides of the optical element (lens). The size B of the image can be calculated from the relationship:

$$\frac{\mathrm{B}}{\mathrm{G}} = \frac{b}{g} = \mathrm{M}, \tag{7.20}$$

with the magnification M. The angular extents of object and image (i.e. the angle under which the object or image are seen from the position of the lens) are given by:

$$\tan\beta = \frac{\mathrm{B}}{b} \quad \text{and} \quad \tan\gamma = \frac{\mathrm{G}}{g}. \tag{7.21}$$

7.4.5 Apertures, Entendue, Lagrange Invariant

The amount of radiation power gathered by an optical system (e.g. consisting of mirrors or lenses) will be directly proportional to its entrance area. A large 'clear entrance' area will usually gather more light than a smaller one. Of course, this will only be true if all of the entrance area is illuminated by the source. The size (linear extent) of the image produced by an optical system will vary proportional to its focal length f. Since a given amount of light enters through the entrance area, the irradiance in the focal plane of an optical system of given entrance area, will be inversely proportional to the image area, and hence f^2.

Most optical elements are circular in shape, and thus their area is proportional to the square of their diameter $D = 2R$ (R: radius of the optics). The ability to produce a bright image is therefore proportional to $(D/f)^2$. A customary way to specify this ability is thus the relative aperture D/f or its inverse, the focal ratio or **f-number** f/D. Frequently, the f-number is written as $f/\#$. Optical systems with a smaller f-number will therefore produce a brighter image. In fact, comparing a system with a given $(f/\#)$ to one having $(f/\#)/2$ will result in four times the brightness of the image in the system with $(f/\#)/2$. Another way to specify the ability of an optical system is its **numerical aperture** (NA), which is defined as R/f. Both quantities are related by:

$$F = {}^{f}\!/_{\#} = \frac{1}{2\,(\mathrm{NA})}. \tag{7.22}$$

The **entendue**, which is defined as the product of the entrance area A times the aperture solid angle Ω: $E = A \cdot \Omega$, is a constant of optical systems. In other words, the entendue of the light entering an optical system is as large as the entendue of the light leaving it. For example, reduction of the size B (and thus the area in proportion to B^2) of an image e.g. by a lens, will increase the aperture solid angle, thus keeping the entendue constant.

Object and image, respectively, see the lens in Fig. 7.21 under the angles α_G and α_B. Using the lens radius R_L we derive:

$$\tan \alpha_G = \frac{R_L}{g} \quad \text{and} \quad \tan \alpha_B = \frac{R_L}{b},$$

Considering $g \cdot \tan \alpha_G = R_L = b \cdot \tan \alpha_B$ and (7.20) we get:

$$G \cdot \tan \alpha_G = B \cdot \tan \alpha_B = \mathrm{L}, \tag{7.23}$$

where L denotes the **Lagrange invariant**. L is another conserved quantity.

7.4.6 Diffraction at Apertures

Radiation passing through apertures of finite size or grazing edges will undergo diffraction, meaning that part of the incident radiation deviates from its original direction.

In the case of a parallel beam passing through a **circular aperture** (i.e. a circular hole in a piece of non-transparent matter oriented perpendicular to the propagation direction of the radiation, or a circular mirror) with radius R, there will be concentric rings of diffracted radiation around the centre beam. The angle ϑ between the diffracted radiation and the original beam direction and its intensity $\mathrm{I}(\vartheta)$ is given by:

$$I(\vartheta) = I_0 \left(\frac{2\mathrm{J}_1(x)}{x} \right)^2 \quad \textit{with } x = \frac{2\pi R}{\lambda} \cdot \sin \vartheta. \tag{7.24}$$

$\mathrm{J}_1(x)$ denotes the Bessel function of first degree, λ the wavelength, and I_0 the original intensity. The diffraction angles and intensities of the first few minima and maxima are given in Table 7.4

Table 7.4. First maxima and minima of radiation diffracted at a circular aperture

Maximum/minimum no.	I/I_0	$\frac{\sin \vartheta}{\lambda/R} = \sin \vartheta \cdot \frac{R}{\lambda}$
Min. 1	0	0.61
Max. 1	0.0175	0.815
Min. 2	0	1.08
Max. 2	0.00415	1.32
Max. 3	0.00160	1.85

For a very long rectangular aperture with width d (i.e. a slit with a length $>> d$), the diffracted intensity is given by:

$$I(\vartheta) = I_0 \frac{\sin^2 x}{x^2} \text{ with } x = \frac{\pi d}{\lambda} \cdot \sin \vartheta. \tag{7.25}$$

Thus, the first minimum is at $x = \pi$ with $\sin \vartheta = \lambda / d$.

7.4.7 Quartz-fibres, Mode Mixers, and Cross-section Shaping

Quartz (or glass) fibres rely on total internal reflection (see Sect. 7.4.1) to conduct radiation. In practice, the total reflection does not occur between the quartz (or glass) and the surrounding air, but rather between the (relatively) high index of refraction (n_f) core of a fibre clad with a outer layer of material with a lower index of refraction (n_c) (see Fig. 7.22).

A maximum angle of incidence $\alpha_\mathrm{i} = \alpha_\mathrm{max}$ exists below which the light entering the fibre is actually transmitted by total internal reflection. The external α_i corresponds to an internal α_t due to refraction at the boundary between the ambient medium (e.g. air, index of refraction $n_0 \simeq 1$) and the core of the fibre (index of refraction n_f). A detailed calculation yields:

$$\sin \alpha_\mathrm{max} = \sqrt{n_f^2 - n_c^2}.$$

The NA and f-number of a glass fibre are given by:

$$\mathrm{NA} = \sqrt{n_f^2 - n_c^2} \qquad f/_{\#} = \frac{1}{2\sqrt{n_f^2 - n_c^2}}. \tag{7.26}$$

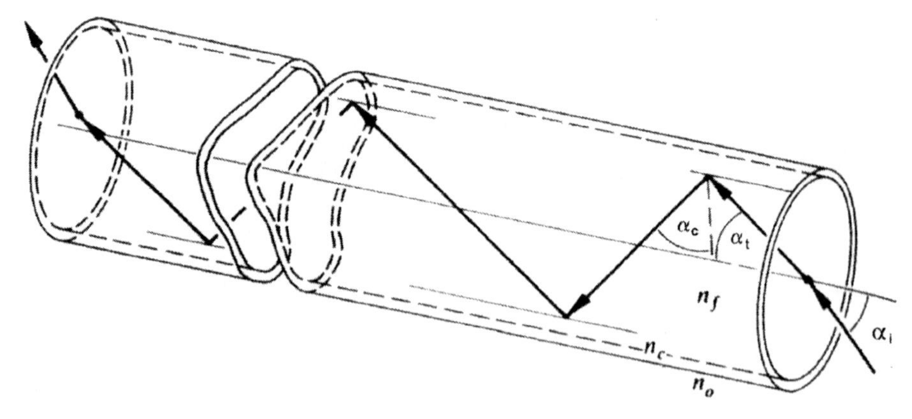

Fig. 7.22. Propagation of radiation in an optical fibre

Fibres are conveniently employed to connect optical elements within DOAS systems, for example the receiving telescope to the spectrometer. This arrangement gives freedom in the placement of the components; for instance, the telescope can be placed outside, while the spectrometer is sheltered. The spectrometer does not need to be physically attached to the telescope (see also Sect. 7.9.3).

Optical fibres can in fact be an important optical component, improving the overall sensitivity of the system considerably. This is due to several properties of optical fibres:

1. Non-uniform illumination of the field of view of the spectrometer can cause residual structures in the spectra, which may limit the minimum detectable optical density to $2-10 \times 10^{-3}$ (Stutz and Platt, 1997a). Introducing 'mode mixers' into the quartz fibre connecting the receiving telescope to the spectrometer can greatly reduce this effect. Figure 7.23 shows a sketch of a mode mixer arrangement.
2. Most fibres scramble the polarisation state of the transmitted radiation, thus effectively removing the polarisation sensitivity of DOAS instruments. Polarisation sensitivity can be a problem in DOAS instruments; in particular in passive, scattered light applications, where the incoming radiation is, in general, partially polarised, see Chap. 4.
3. Fibre bundles can be used to change the cross-section of a light bundle. This property can be used to match the approximately circular (see Sect. 7.3.1) shape of e.g. a Xe-arc light source to the rectangular aperture of the spectrometer entrance slit. Figure 7.24 shows an example in which a bundle of 19 fibres is arranged in an approximately circular shape at the entrance and a column at the exit.

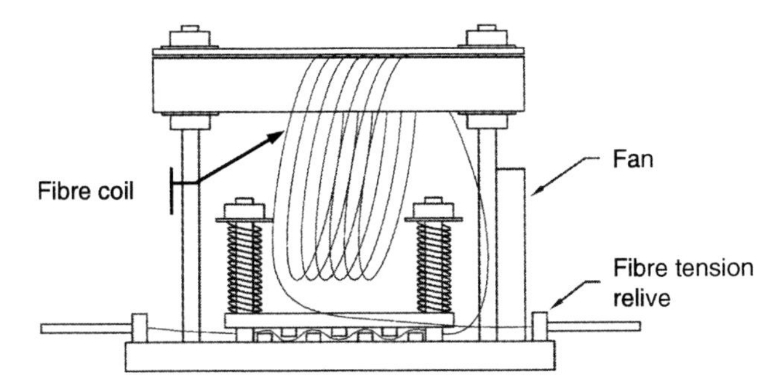

Fig. 7.23. Sketch of a fibre-mode mixer after. The device consists of two elements: (1) A fibre coil (top part of the figure), which is slightly 'shaken' by an air stream generated by the fan and (2) a section of the fibre that is put under mechanical stress by a spring loaded plate introducing several sharp bends in the fibre (bottom part of the figure) (adapted from Stutz and Platt, 1997a)

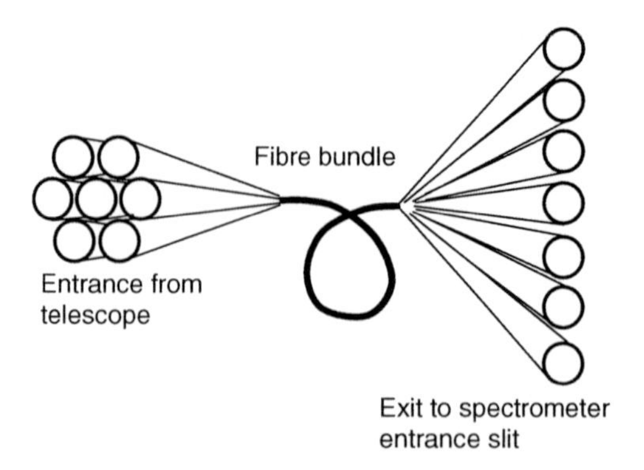

Fig. 7.24. Cross-section change from a circular entry to a column at the exit by an optical fibre bundle

7.4.8 Filters

In DOAS systems spectral filters are usually required for several purposes. These include blocking unwanted spectral orders (see Sect. 7.5.1) or reducing the amount of stray light in the spectrometer (see Sect. 7.5.4) by limiting the spectral bandwidth of the radiation entering the instrument. In addition, it is desireable to eliminate the visible part of the light from the light source in order to avoid blinding humans and make the radiation emitted by the DOAS instrument less visible.

Filters can either be 'colour glass' type (e.g. Schott UG5), thin film partially reflecting filters, or thin-film interference filters. Figure 7.25 shows the transmission curve for a several UG-type colour glass filters used to block visible light and thus reduce spectrometer stray light.

Under certain circumstances the radiation intensity may be too large for a given detector, for example when observing direct sun signals by an instrument optimised for scattered light operation. In these cases, optical attenuation is required. While this can be achieved by several measures, such as defocusing or reducing the aperture, grey filters offer a simple and clean way to reduce the signal level essentially without changing other properties of the optical system. Grey filters are usually characterised by an attenuation factor $A = \mathrm{I}/\mathrm{I}_0$ (I_0 and I denoting incident and transmitted intensities, respectively), frequently expressed as optical density base 10 [$\log^{10}(A)$ or $\lg A$]. The attenuation A should ideally be independent of the wavelength. There are three popular types of optical attenuators, grey glass filters, which rely on the optical absorption by suitable glass types, thin-film partial reflective surfaces, and mesh-type devices. The latter attenuator types are useable from the UV to near-IR, while grey glass filters are usually limited to the visible spectral range. A potential difficulty (in particular for DOAS applications)

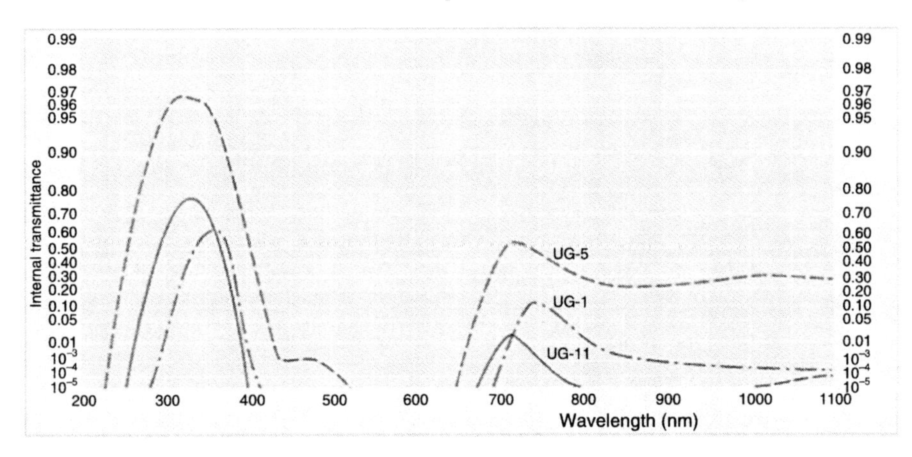

Fig. 7.25. Sample transmission curve of colour glass filters (UG-1, UG-5, UG-11 from Schott, Mainz, 2 mm thickness). These filters are used to reduce spectrometer stray light in DOAS applications. Note that the second transmission maximum in the red/near-IR region is where the sensitivity of silicon detectors usually reaches its maximum (figure courtesy of Schott)

of reflective attenuators is the Fabry–Perot etalon effect, which is caused by interferences in the thin optical reflective layer.

7.4.9 Retro-reflectors

The purpose of retro-reflectors is to return a beam of light back to the sending-receiving telescope of the DOAS instrument. While, in principle, plane mirrors could be used [see e.g. Perner et al. (1976) and Hübler et al. (1984)], there are several types of optical arrangements that return the incident light exactly (although with some lateral offset) into the direction of incidence, see Fig. 7.26 (Eckhardt, 1971; Sugimoto and Minato, 1994). The most popular design is the 'corner-cube' retro-reflector, also called 'trihedral' retro-reflector or triple retro-reflector [see e.g. Eckhardt (1971)]. It consists of three plane, reflecting surfaces arranged at right angles (Fig. 7.26). Thus, the reflecting surfaces can be viewed as being arranged like the three surfaces that meet in the corner of a cube. The radiation would incite from the direction of the cube's main diagonal. In a typical active DOAS instrument, a light beam from a search light-type light source would be directed towards one or several retro-reflectors mounted at a distance equal to half the total desired light path (see Fig. 6.2). The necessary angular precision of the retroreflectors, measured as the angle between incoming and outgoing beams, is on the order of a few seconds of arc (about 10^{-5} radians), which usually requires specially made high-precision devices. An additional property of retro-reflectors is reduction of the effect of atmospheric turbulence on the light beam. As illustrated in

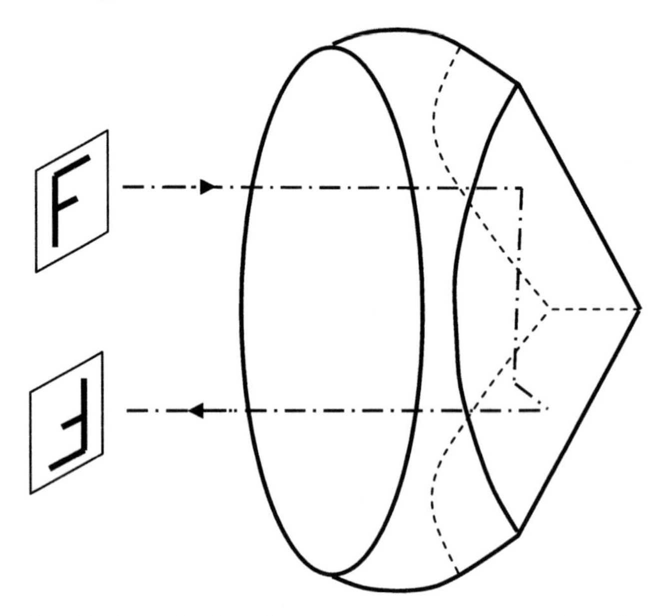

Fig. 7.26. Reflections in a corner cube retro-reflector prism

Fig. 7.27, this is due to the capacity of corner cube retro-reflectors to return the incident light exactly (although with some lateral offset) into the direction of incidence.

There are two varieties of the 'corner cube' retro-reflector design in use: Hollow corner cubes made of three flat (first surface) mirrors mounted at very precise 90° angles. In order to protect the mirror surfaces, which are difficult to recoat without taking the assembly apart, quartz front panes are sometimes added. These front panes represent four additional surfaces to the

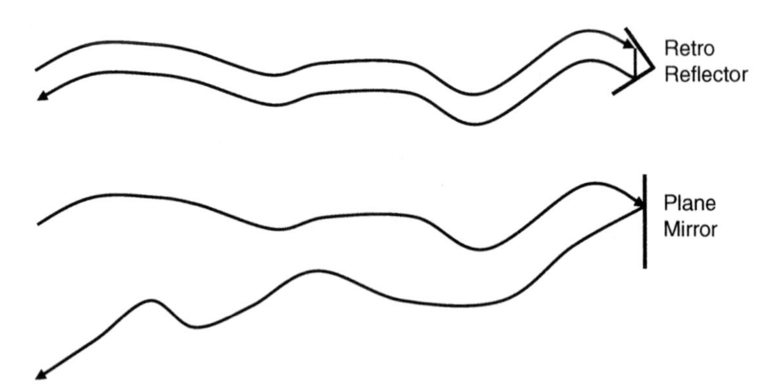

Fig. 7.27. Compensation of turbulent beam dispersion by a (corner cube) retro-reflector arrangement (**upper panel**), in comparison to reflection by a plane mirror (**lower panel**).

light returned by the retro-reflector leading to additional light loss due to Fresnel reflection [see (7.9) and Table 7.5].

In contrast, prism-type retro-reflectors need no additional protection of front surfaces. The reflection at the rear surfaces of prism-type retro-reflector elements can either be due to total internal reflection or by a coating applied to the outside of the flat corner cube surfaces of the prism. Both approaches have advantages and drawbacks. In the case of internal reflection, the efficiency of returning light is the highest of all types described here, the only loss being the reflection off the (usually circular) surface of the prism. On the other hand, the useable range of acceptance angles is smallest because the limiting angle of total reflection is exceeded at higher incidence angles (Rityn, 1967). In contrast, the coated corner cube prism has relatively large reflectance losses, in particular in the UV wavelength range. Typical reflectivities are on the order of $R \approx 0.8$–0.9 (see Fig. 7.16); for three reflections, this amounts to an overall efficiency of 40–50% including the 4% losses ($R_S \approx 0.96$) occurring twice at the surface of the prism, see Table 7.5. On the positive side, the coated corner cube prism features the largest acceptance angle range (much larger than in the internal reflectance case, but also larger than the hollow cube corner), since all incident rays are diffracted towards the symmetry axis, i.e. become lower inside the prism (Rityn, 1967).

In order to save weight, an array of small retro-reflector elements is typically used, rather than a single large unit. Typically, reflector prism sizes

Table 7.5. Characteristics of different corner cube retro-reflector designs

Type of retro-reflector	Reflectance	Acceptance angle range Degrees	Comment
Hollow cube corner	R^3 typ. 50 %	$\approx$ 32–56	Least expensive type for a given aperture. Vulnerable to dirt
Hollow cube corner with protective pane	${R_S}^4 \cdot R^3$ typ. 40 %	$\approx$ 32–56	Protected, but more expensive, while being less efficient
Corner cube prism, coated	$R_S^2 \cdot R^3$ typ. 45 %	60–90	Type of choice for wide angle applications
Corner cube prism, total internal reflection	R_S^2 typ. 92 %	20–60	Good compromise between protection and economy, most widely used type

R = reflectance of one of the corner cube surfaces, R_S = reflectance of a surface perpendicular to the light beam, e.g. of a protective (quartz) cover pane

are on the order of 60 mm diameter. Special design variants have also been explored (Minato et al., 1992; Minato and Sugimoto, 1998).

There are several limitations to watch, however: For most applications the reflectors must be of very high precision. For instance, if a deviation of the returning beam of a typical retro-reflector diameter (60 mm, see above) over a distance of 5 km is desired, this corresponds to an angle between incoming and outgoing light beams of about $\delta = 0.06\,\mathrm{m}/5000\,\mathrm{m} \approx 1.2 \times 10^{-5}$ radian ($\approx 7 \times 10^{-4}$ degrees or about 2 s of arc).

Another limitation is due to diffraction at the retro-reflector aperture. The angle of diffraction for the first maximum of a circular aperture, see Table 7.4, is $\alpha \approx \sin\alpha \approx 0.815 \cdot \lambda/R$. This results in $\alpha \approx 1.25 \times 10^{-6}$ radians at a wavelength of $\lambda = 500\,\mathrm{nm}$, and a diameter of a retro-reflector element of $D = 2R = 63\,\mathrm{mm}$ (2.5 inches). Thus, for the customary size of retro-reflector elements, diffraction limitation is already close to the tolerable limits for long light paths.

7.5 Spectrometers/Interferometers for DOAS Systems

A central and critical component of all DOAS systems is a device which serves to separate the individual wavelength intervals, so that the intensity in these intervals can be measured. There are two fundamentally different approaches in use, spectrometers and interferometers.

Spectrometers separate radiation of different wavelengths into different spatial directions, where the spectrum can be measured by a spatially resolving detector (see Sect. 7.6). Spectrometers are based on two basic physical principles:

(a) Refraction by prisms (see Sect. 7.4.3), which rely on the dispersion in transparent optical materials such as quartz, glass, or plastic. Advantages of prism spectrometers are the absence of overlapping orders (as in the case of grating spectrometers). In addition, their wavelength separation (linear dispersion) is larger at shorter wavelengths. Disadvantages are limited resolution ($\lambda/\Delta\lambda <$ a few 10,000 for practical purposes) and a non-linear wavelength scale. Today, prism spectrometers are rarely used, although there may be many advantages in DOAS applications.

(b) Diffraction by gratings (see Sect. 7.5.1), which make use of spatial interference effects. Advantages of gratings are potentially very high resolution ($\lambda/\Delta\lambda$ in excess of 100,000 is possible), approximately linear dispersion, and the ability to build very simple instruments with concave gratings (see below). One disadvantage is the overlapping orders, i.e. a given point in the focal plane simultaneously receives light with wavelengths of $N \cdot \lambda$ (where $N = 1, 2, 3, \ldots$ denotes a positive integer).

Interferometers make use of interferometric principles to effectively produce (and record) the Fourier transform (FT) of the spectral intensity distribution of the received radiation. Although, in principle, spatial as well as temporal interferograms can be produced, in practice only the latter principle is in widespread use, see Sect. 7.5.3.

7.5.1 Diffraction Gratings

A grating can be viewed as a series of equally spaced slits (distance between slits = d), which transmit or reflect light, while the area in between does not. The angles β at which the rays, incident under angle α, from the slits interfere constructively are given by:

$$d\,(\sin\alpha + \sin\beta) = N\lambda. \tag{7.27}$$

The left side of (7.27) gives the optical phase difference (see Fig. 7.28) which, for constructive interference, must be equal to an integer (N) multiple of the wavelength λ. The integer N is called the **order** of the spectrum. In practice, reflective gratings are used instead of transmission gratings, as illustrated in

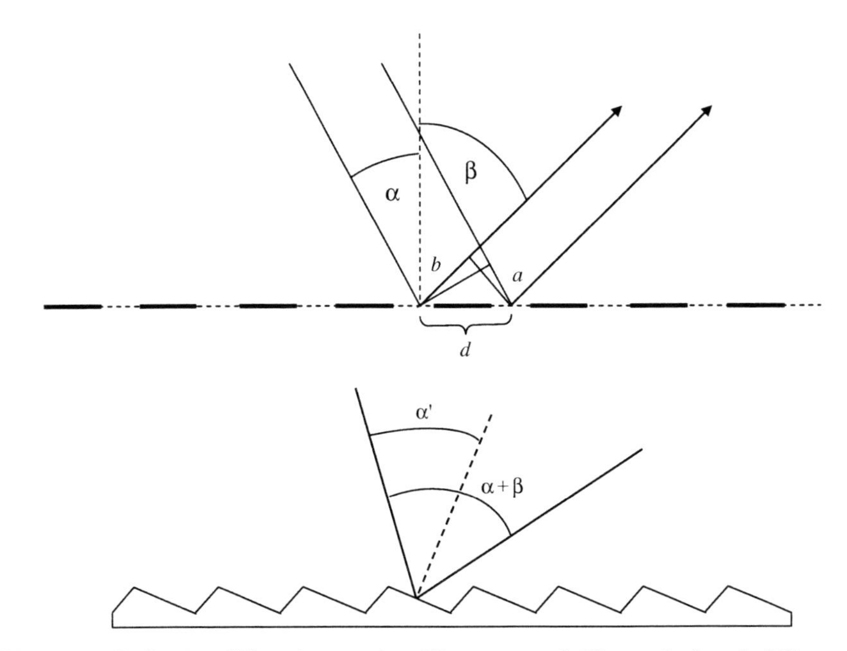

Fig. 7.28. Reflective diffraction grating. **Upper panel**: The optical path differences for the incident ray are $a = d \cdot \sin\alpha$, for the exiting ray $b = d \cdot \sin\beta$. The total path difference is $a+b$. **Lower panel**: The blaze angle of a grating is defined as the angle $\alpha+\beta$ where the refracted beam has the same direction as a specular reflected beam would have

Fig. 7.28. Orders with $N\lambda/d < 2$ can be observed, i.e. in first order d should be larger than $\lambda/2$.

The **angular dispersive power** is defined as the rate of change $\partial\beta/\partial\lambda$ of the angle of emergence β versus wavelength. For a given angle of incidence α and a given order N, we obtain:

$$\frac{\partial\beta}{\partial\lambda} = \frac{1}{d} \cdot \frac{N}{\cos\beta} = g \cdot \frac{N}{\cos\beta}. \tag{7.28}$$

The number of grooves per unit length (e.g. grooves per millimetre) $1/d$ is also known as **grating constant** g. Thus, $\partial\beta/\partial\lambda$ is independent of the angle of incidence, but increases with order and grating constant. It also becomes larger at large angles of emergence, where $\cos\beta$ is small. For instance, for an angle of emergence $\beta = 60°$ ($\cos\beta = 0.5$), $g = 1200$ grooves per millimetre ($d = 0.8333\,\mu\text{m}$ or $1/d = 1.2 \times 10^6\,\text{m}^{-1}$), and for first order one obtains $\partial\beta/\partial\lambda = 2.4 \times 10^6$ radians m^{-1} or 2.4×10^{-3} radians nm^{-1}. Thus, two parallel bundles of radiation with wavelength 1 nm apart will emerge as two bundles propagating at an angle of 2.4 mradians.

Ruled Versus Holographic Gratings: Depending on the manufacturing process, there are two popular types of gratings: ruled gratings and holographic gratings. The former are produced by mechanically drawing grooves into a suitable substrate. Since this is a lengthy process, the gratings commonly used in spectrometers are replicated from a ruled 'master' grating (sometimes in several generations). In contrast, holographic gratings (interference gratings) are produced by recording the interference pattern of two (laser) light beams, e.g. Lerner and Thevenon (1988). Since all grooves are formed simultaneously, holographic gratings are virtually free of random and periodic deviations in line positions, and thus they have no 'ghosts' (see below) and low levels of stray light. In addition, holographic techniques allow more freedom in producing gratings with uneven (continuously varying) line spacing. On the other hand, ruled gratings are more easily blazed, see below.

Gratings can be produced such that the reflecting surfaces are inclined with respect to the plane of the grating, as shown in the lower panel of Fig. 7.28. These gratings are known as 'blazed' gratings. The **blaze angle** of a grating is defined as the angle $\alpha + \beta$ where the refracted beam has the same direction as a beam, which is specular reflected from the sawtooth-shaped grooves of the grating (see Fig. 7.28, lower panel). For conditions where the angle between incident and refracted beam equals the blaze angle, the highest grating efficiency is achieved.

In principle, a grating distributes the incident radiation energy into an infinite number of orders. In practice, most of the energy is diffracted in a few or only one order (see blaze angle). The fraction of the incident energy reaching the strongest order is called the **efficiency** of the grating. Generally, the grating efficiency is a function of wavelength and the polarisation of the light. Typical maximum grating efficiencies are on the order of 20–80%.

For a particular application, it must be ensured that the grating efficiency is sufficient in the desired spectral range. Note that a low grating efficiency not only reduces the light throughput of the spectrometer, but may also considerably increase its stray light level, since low efficiency usually means that the missing radiation is diffracted in other orders.

Another point to watch, in particular in scattered light applications, where the incident radiation is usually partially polarised, is the polarisation dependence of the grating efficiency.

Grating Anomalies (Wood's Anomaly): Real gratings can exhibit a number of prominent deviations from the ideal behaviour described earlier. In particular, individual grooves (or groups of grooves) can be misplaced randomly or periodically. The latter imperfection gives rise to **grating ghosts**. An example for grating ghosts is the Rowland ghost caused by periodic misplacement of lines in the ruled grating. If we assume that each nth line is displaced by an amount $\Delta d << d$ from its nominal distance d to its neighbouring line, then the resulting grating can be viewed as an overlay of three gratings with slightly different line spacing $d - \Delta d$, d, and $d + \Delta d$. Such a grating will consequently produce two additional spectra, or 'ghosts', i.e. each spectral line will have two satellites.

An anomaly of a different kind is the **Wood anomaly** (discovered by R. W. Wood, 1902). Many gratings show sharp changes of the diffracted intensity within narrow spectral regions. Typical reflection gratings produce a series of orders of diffracted light. At some critical wavelength, the diffracted light falls back into the plane of the grating. The light which would be sent into the forbidden region is redistributed back into the allowed orders, and appears as an addition to the spectral response, with a sharp increase at the critical wavelength and a steep decline to the red. This effect acts as an enhancement of the grating efficiency, as if the light from two orders are combined. A Wood's anomaly is almost entirely polarised perpendicular to the grating grooves.

7.5.2 Spectrometers

The purpose of a **spectrometer** is to separate and re-image radiation of different wavelengths in order to record spectral signatures. In principle, this can be done in several ways. For instance, by a scanning **monochromator**, which selects and transmits a narrow spectral interval of the incoming radiation, in connection with a suitable detector. A monochromator can be thought of as an (usually adjustable) optical band-pass filter. Alternatively, a **spectrograph**, which images a range of wavelengths at the exit focal plane for detection by multi-channel detector or photographic film, can be used. Since spectrometer is the more general term, we shall use this term in the following.

Traditionally, the spectrometer of a DOAS instrument produces a 'spectrum' i.e. spatially separates the radiation into wavelength intervals, so that the intensity in the different wavelength intervals can be recorded by a suitable detector capable of resolving spatial differences in radiation intensity. The latter task can be performed by detectors using moving parts to convert spatial separation in temporal separation (e.g. a moving exit slit), or – in recent designs – by a collection of many sensors (detector arrays) that simultaneously record spatial variations in intensities.

It should be noted, however, that generation of classical spectra is not really required for DOAS applications, rather a discernible signature must be generated that allows distinction between relevant atmospheric absorbers and recording of their presence in the light path with sufficient sensitivity.

A step in this direction was made by early COSPEC spectrometers (e.g. Millan, 1972), which used multiple entrance slits paced to superimpose periodic trace gas (e.g. SO_2) spectra in phase.

As mentioned earlier, the 'classic' spectrometer design rests on the two optical principles of prism spectrometers and grating spectrometers, which are used in several design variants.

The Prism Spectrometer: In the early days of optical spectroscopy (e.g. Dobson, 1926), prism spectrometers (see Fig. 7.29) were the instrument of choice. The angular dispersion is given by (7.17). In recent decades, the grating spectrometer has become the dominating spectrometer design principle. This is due to its inherent advantages, such as the almost linear dispersion curve [see (7.28)], and the higher resolution at a given size of the instrument combined with great advances in manufacturing of diffraction gratings. Nevertheless, it should be said that the prism spectrometer has some merits which may make it worthwhile to consider its future use in DOAS systems: The non-linearity of the dispersion can be corrected with modern software, in fact higher resolution would usually be required in the UV than in the visible spectral range. Also the absence of higher orders could be an advantage.

Plane Grating Spectrometers: The Czerny–Turner-type spectrometer (see Fig. 7.30a) and variants have been traditionally selected for DOAS instruments (e.g. Platt et al., 1979; Platt and Perner, 1983). Besides the grating, it requires at least two additional reflecting surfaces (collimating and camera

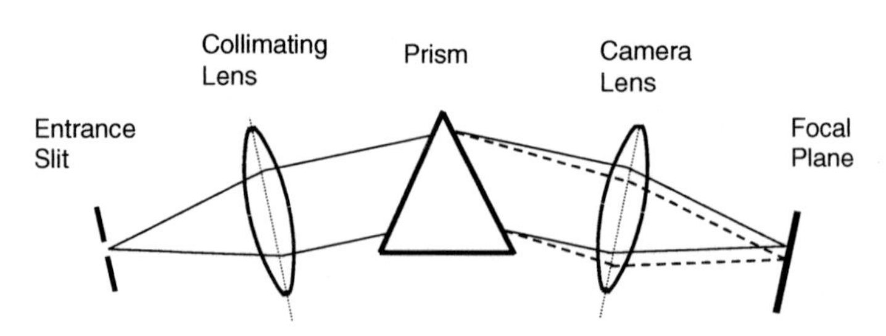

Fig. 7.29. Optical outline of a typical prism spectrometer

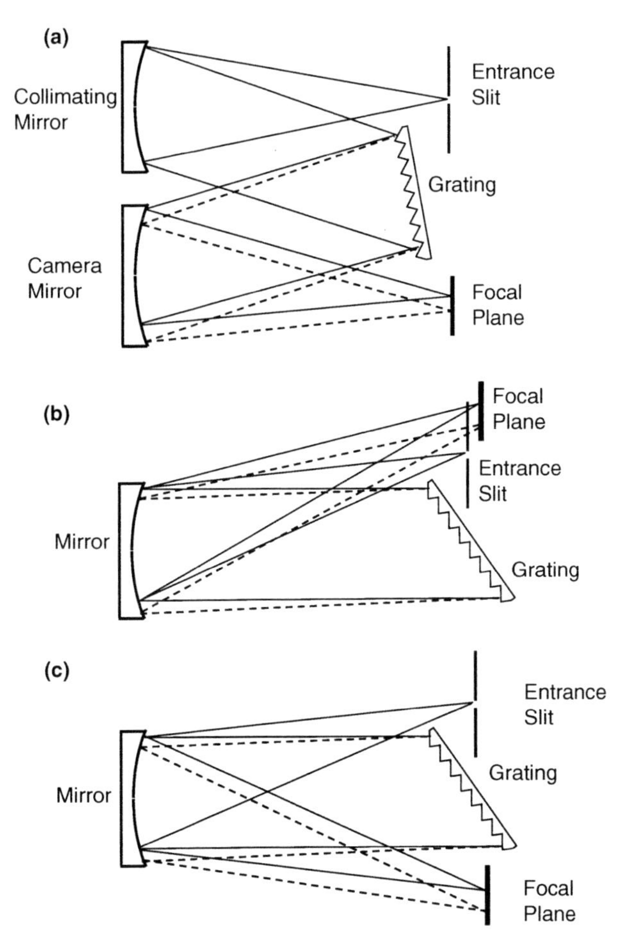

Fig. 7.30. Optical outlines of different variants of grating spectrometers. (**a**) Czerny–Turner; (**b**)Littrow; (**c**) Fastie–Ebert

mirror). The additional reflection losses at these surfaces are usually offset by high-efficiency (60–80%) plane blazed gratings. Furthermore, the wavelength range is readily selected by rotating the grating.

Since most of the incident light is directed into the desired diffraction order, stray light caused by reflection of unwanted orders off the spectrometer housing is minimised. The linear dispersion $\partial s/\partial\lambda$ of a grating spectrometer is given by $\partial\beta/\partial\lambda$ (7.28) multiplied by the focal length of the camera mirror:

$$\frac{\partial s}{\partial \lambda} = f\frac{\partial \beta}{\partial \lambda} = \frac{f}{d} \cdot \frac{N}{\cos\beta} = fg \cdot \frac{N}{\cos\beta}. \tag{7.29}$$

In special applications, for example if only a small aperture is required, Fastie–Ebert-type (Fig. 7.30c) or Littrow-type spectrometers (Fig. 7.30b) are also often used. While the design of these spectrometer types appears to be simpler, since a single mirror doubles as collimating mirror and camera mirror, the disadvantages include larger size (in the case of Fastie–Ebert) and the requirement to work at a wavelength where the angle of incidence and the angle of refraction of the grating are equal (Littrow-type spectrometer).

Concave (Holographic) Grating Spectrometers: Modern holographic grating technology (see Sect. 7.5.1) allows the design of spectrometers only consisting of entrance slit, concave grating, and detector (i.e. optomechanical scanning device (OSD) or diode array, see Fig. 7.31 and Sect. 7.6). In contrast to the traditional Rowland mounting, that also uses a concave grating as the only imaging element in the spectrometer, modern concave holographic gratings can produce a flat spectrum over a certain range (hence the name flat-field spectrometers).

The partly low efficiency of most holographic gratings (20–40%, see Sect. 7.5.1) is compensated by the reduced losses due to the low number of optical surfaces. While the stray light produced by holographic gratings is lower compared to ruled gratings, the higher intensity of unwanted orders may increase the stray light level in the spectrometer. On the other hand, the unused orders are also focused and can thus be removed relatively easily by light traps. A disadvantage of most flat field gratings is their limitation to a relatively small spectral range that cannot normally be changed.

Imaging Spectrometers: The term 'imaging spectrometer' refers to spectrometers that can image a position along the entrance slit onto a corresponding position in the focal plane. The imaging direction is perpendicular to the dispersion direction. Thus, usually two-dimensional detectors (typically 2D CCD detectors) in the focal plane record multiple spectra and each point along the entrance slit corresponds to an individual spectrum in the

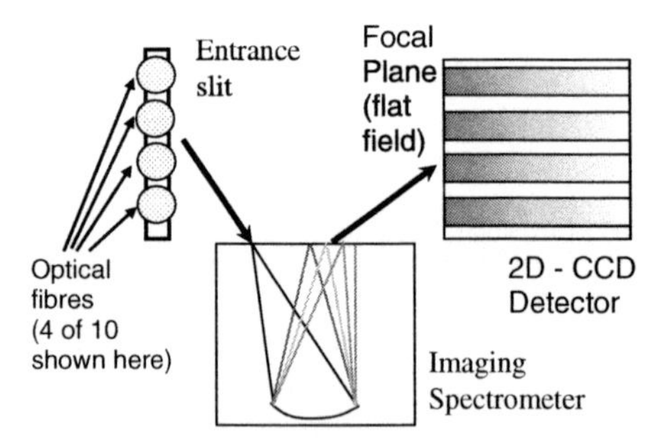

Fig. 7.31. Sketch of an imaging spectrometer. Here, a concave grating spectrometer is shown. But, in principle, any type of spectrometer can be used. In this example, $n_y = 4$ optical fibres could connect to four individual telescopes looking at different viewing angles. In reality, frequently larger numbers of fibres (e.g. 10; Heismann, 1996) are present (figure courtesy of T. Wagner)

focal plane. As illustrated in Fig. 7.31, there are n_x pixels in the 'dispersion direction' determining the spectral sampling, similar to a one-dimensional diode array (or CCD detector). In the 'slit direction', n_y rows of n_x pixels each provides spatial resolution. Theoretically, the imaging spectrometer behaves like n_y individual spectrometers, each with a spectral sampling of n_x pixels (the spectral resolution of course will depend on dispersion and instrument function). In practice, the imaging qualities of most spectrometers perpendicular to the dispersion direction are limited to many pixels in y direction.

To date, imaging spectrometers are used in several applications:

- MAX-DOAS (see also Sect. 7.10.3). In this application, a column of fibre exits illuminates the entrance slit. Each fibre (or set of fibres) is coupled to a telescope observing the sky under an individual angle.
- One-dimensional imaging in order to resolve e.g. vertical trace gas profiles.
- Two-dimensional spectral imaging. Since only one dimension is covered by the slit, other scanning mechanisms (e.g. mechanical; Lohberger et al., 2004) must be employed by the other dimension. An example is a satellite or aircraft, the movement of which provides the second scanning dimension (e.g. Laan et al., 2000; Levelt et al., 2000).

Applications such as these can be seen as generating 'super colour images', where each pixel has not just three colours, but encompasses an entire spectrum with several hundred (or more) spectral elements.

7.5.3 Interferometers (FT Spectrometry)

Interferometers make use of the wave properties of electromagnetic radiation in order to effectively produce (and record) the Fourier Transform (FT) of the spectral intensity distribution in the received radiation. Therefore, interferometers are frequently called FT spectrometers. Although, in principle spatial as well as temporal interferograms can be produced in practice essentially all instruments are based on the latter principle (see Fig. 7.32).

FT spectrometers have been in use for IR absorption spectroscopy for a long time (FT-IR) (e.g. Tuazon et al., 1980; Galle et al., 1994; Notholt et al., 1997; Hanst and Hanst, 1994). However, technological advances have only recently made their application in the UV/visible spectral range (now called FT-UV), and thus for DOAS, possible (Colin et al., 1991; Vandaele et al., 1992).

7.5.4 Characteristics of Spectrometers

As shown earlier, there are a large variety of spectrometer designs with widely varying characteristics, such as:

Spectral dispersion – The spatial separation of radiation of different wavelengths in the focal plane of the instrument is referred to as spectral

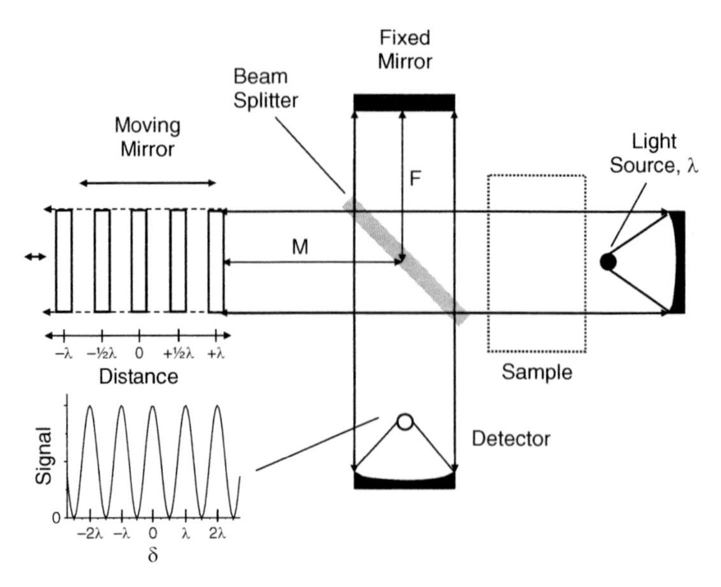

Fig. 7.32. Optical outline of a FT spectrometer

dispersion. This dispersion is given by $ds/d\lambda$ measured in, e.g., mm/nm, where s denotes the linear distance in the focal plane and λ the wavelength, see (7.29).

Spectral resolution – The capability of the instrument to separately measure the radiation intensity at different wavelengths is characterized by the spectral resolution, $\Delta\lambda_R$. There are several definitions of the spectral resolution, usually based on the capability to separate the intensity of two monochromatic light sources (atomic emission lines) simultaneously observed by the instrument. A more general and meaningful term is the instrument function [or slit function $H(\lambda\text{–}\lambda_0)$ at the wavelength λ_0], which is defined as the instrument's response to a delta function $\delta(\lambda\text{–}\lambda_0)$ as input intensity distribution. The **spectral resolution** $\Delta\lambda_R$ at the wavelength λ_0 can be seen as e.g. the half-width of $H(\lambda\text{–}\lambda_0)$.

Flatness – Flatness of the focal plane determines how the spectral characteristics change in the focal plane. This determines how well a 1D or 2D detector of a given size can be accommodated. Typically the focal plane is reasonably flat in the dispersion direction (i.e. over ~25 mm or more). Imaging spectrometers are necessary to achieve a flat field perpendicular to the dispersion direction.

Thermal Stability – Experience and theoretical calculations show that even slight variations in the position of a recorded atmospheric spectrum with respect to a reference spectrum can lead to relatively large errors in the spectral fit (see Chap. 8). Changes in dispersion have the same effect, since they lead to spectral shift of one portion of the spectrum compared to another.

In practice, shifts or dispersion changes as small as 0.01, spectral pixels can lead to noticeable degradation of detection limits and accuracy of DOAS instruments. Although shift and change in dispersion can be compensated for by evaluation software (see Chap. 8), occurrence of these effects will always degrade the result. It is, therefore, of great advantage to minimise spectral shift and change in dispersion. In practice, these effects are largely caused by a change in the temperature of the spectrometer. Thus, either the susceptibility of spectrometers to temperature changes should be minimised (e.g. by use of material with low thermal expansion) and/or the temperature changes must be kept at a minimum. However, air density changes due to changes in temperature also lead to a small drift as is explained below.

Although efforts are being made to design spectrometers, with low thermal effects (e.g. made from Zerodur), the most widely used approach is to thermostate the entire spectrometer subsystem. This is usually accomplished by insulating the instrument and heating it to a temperature above the highest expected ambient temperature. Typical temperature stabilities reached in field experiments are on the order of ± 0.1–1 K. Typical wavelength shifts are 0.05 nm/degree. Thermostating the detector is usually reached as a by-product of the necessary cooling. In some designs, both functions were combined by cooling and thermostating at a low temperature the entire spectrometer–detector assembly (Hönninger, 2002; Bobrowski et al., 2003).

Pressure (Air Density) Effects – Changes in ambient atmospheric pressure (and temperature) will usually also lead to a wavelength shift since the index of refraction of air n_{air} (p, T), the speed of light, and thus the wavelength of the radiation changes, see (7.10).

$$\lambda = \frac{c}{\nu} = \frac{c_0}{\nu} \cdot \frac{1}{n_{air}}. \tag{7.30}$$

In a grating spectrometer, this will alter the dispersion of the instrument.

Stray Light – Stray light requires careful consideration when selecting or designing a spectrometer. Sources of stray light include: Light scattered from optical elements (grating and mirrors), reflection of unused diffraction orders off the spectrometer walls, reflection of unused portions of the spectrum from walls near the focal plane, and reflections from the detector surface (e.g. Pierson and Goldstein, 1989). A particular form of stray light is due to 'grating ghosts', i.e. portions of radiation diffracted in unwanted directions (see Sect. 7.5.1). An important source of stray light is due to incorrect illumination of the spectrometer. If the F-number of the illumination exceeds that of the spectrometer, radiation will overfill the collimating mirror and hit interior walls of the spectrometer, where it can be reflected to the detector.

Note that stray light tends to be comparatively high in spectrometers filtering a relatively broad wavelength interval from a continuous spectrum, as in DOAS applications. To illustrate this point, we consider a typical stray light level of 10^{-5} for a Czerny–Turner spectrometer (Pierson and Goldstein, 1989). This figure gives the fractional light intensity found anywhere in the spectrum when a single (laser) line of very narrow spectral width enters the spectrometer. The actual width of the line seen in the focal plane will then be equal to the spectral resolution of the instrument. Continuous light entering the spectrometer can be thought of being composed of a series of lines spaced at center-to-center distances equal to their width (i.e. the spectral resolution of the instrument). Since the total spectral range of light entering the spectrometer (for instance from 300 to 600 nm) is on the order of 1000 times larger than the spectral resolution (of typically 0.3 nm for a low resolution DOAS instrument) the level of stray light (per wavelength interval) is roughly three orders of magnitude larger than would be expected from the single line definition usually considered. Thus, in DOAS applications, stray light levels I_{SL}/I can be expected to be closer to 10^{-2} than to 10^{-5}. The effect of stray light is to reduce the recorded optical density D, as shown in Sect. 8.7.2.

Reentrant Light: In many spectrometers (irrespective of type) where a linear or matrix solid-state detector array is used, there might be reflection of radiation from the detector surface back to the grating. The solution is to either tilt the array up to the point that the resolution begins to degrade or, if the system is being designed for the first time, to work out of plane.

In some Czerny–Turner spectrometers, a diffracted wavelength other than that at which the instrument is set may hit the collimating mirror (see Fig. 7.30) and be reflected back to the grating, where it may be rediffracted and directed to the detector. If this problem is serious, a good solution is to place a mask perpendicular to the grooves across the centre of the grating. The mask should be the same height as the entrance slit (and detector array). If the wavelength is known, it is possible to calculate the point on the grating where the radiation of the reflected wavelength hits. In this case, the only masking necessary is at that point.

Light Throughput of Spectrometers – An important characteristic of any spectrometer is its light throughput, L_t. It is defined as the product of acceptance solid angle Ω, effective entrance aperture A (i.e. the area of the entrance slit), and the light transmission T of the instrument:

$$L_t = \Omega \cdot A \cdot T = E \cdot T \, (E = \text{Entendue, see Sect. 7.4.5})$$

In principle, L_t is affected by many parameters including:

- The aperture ratio ($f/\#$) of the instrument
- The entrance slit dimensions
- The light losses inside the instrument.

It is interesting to note that the light throughput is approximately inversely proportional to the square of the spectral resolution, which scales with the width, and thus the areas A of both the entrance slit and the detector pixels. Thus, $L_t \propto 1/\Delta\lambda_R$.

7.6 Detectors for UV/Visual Spectrometers

The basic function of detectors for UV/vis spectrometers is to convert spatial radiation intensity differences $I(\lambda)$ into a signal $I_D(\lambda)$ that can be transferred to a computer for further evaluation.

$$I_D(\lambda) = \alpha \cdot I(\lambda), \tag{7.31}$$

where α denotes the detector sensitivity. Detectors must rapidly capture a spectrum (quasi-) simultaneously in order to minimise additional noise due to atmospheric turbulence. Good quantum efficiency is also desirable. However, the most significant requirement is the capability to reliably detect relative intensity differences within the spectrum on the order of 10^{-4}. This latter requirement implies low noise and excellent linearity (see below), as well as sufficient sampling of the information. Presently, several different detector designs are in use for DOAS spectrometers. OSDs, which record the spectra sequentially, make use of only a small fraction of the light, and thus can be regarded as outdated for scientific applications. However OSDs are still found in several commercial DOAS instrument designs. Recent scientific instruments employ solid-state detector arrays known as photo diode array (PDA) detectors, CCD array detectors, and complementary metal oxide semiconductor (CMOS) array detectors.

7.6.1 Geometrical Focal Plane Sampling Requirements

The requirements for DOAS sensors can be described by a number of properties:

Geometry – For spectroscopic purposes, pixels with an aspect ratio that is ratio narrow in the direction of the scan (i.e. dispersion direction of the spectrometer) and wide in the direction perpendicular to the scan are required. Typically, a pixel width of a fraction (say 1/6th, see below) of the width of the spatial instrument function is desirable. The spatial instrument function is the resulting spatial intensity distribution in the focal plane of a spectrometer as a response to a delta peak of monochromatic light at the entrance of the spectrometer.

Taking an entrance slit width of 50 µm corresponding to, for example, a width of the spatial instrument function of 75 µm, a detector pixel width on the order of $w_D = 15$ µm would be required.

On the other hand, the 'height' (extension perpendicular to scan direction or dispersion direction) of each pixel should be at least as large as the height

h_S of the illuminated area in the focal plane. For a typical case of a one-to-one imaging spectrometer, h_S is at least as large as the illuminated height of the entrance slit. For active DOAS systems employing an optical fibre (or fibre bundle) to transfer the light from the receiving telescope to the spectrometer h_S (and therefore the detector height h_D), is around 100 μm to 500 μm (see Sect. 7.4.7). For passive systems observing the scattered sky light, the imaged object (i.e. the sky) is very large and the light throughput is limited by the entrance area of the spectrometer (and its aperture ratio, which we shall not consider here). Since the slit width is determined by the required spectral resolution, the only free parameter is the slit height, which should be maximised. Thus, large pixel heights h_D of the order of several millimetres are of advantage.

Readout Time – In active DOAS applications with large optics (see Sect. 7.9.3) a signal level in the spectrometer focal plane of 10^9 photoelectrons per second and 25 μm resolution element can be reached. For passive DOAS systems observing scattered sunlight (zenith or off-axis geometry), typical signal levels are about two orders of magnitude longer. The detector readout has to be fast enough to keep up with the signal accumulation. In fact it would be desirable that the actual readout time is short compared to the exposure time required to fill the detector pixels near full-well capacity. The full-well capacitance of a typical (one-dimensional) CCD detector is of the order of 10^5 electrons (see Table 7.6), this means that in active DOAS applications, up to 10^4 complete readout processes per second may be required. For a 2000-element linear CCD sensor, this figure would correspond to about 20 MHz pixel frequency. This might be by a factor of 4 better for the CCD's listed in Table 7.6, because their pixel width – and thus signal intensity – is a factor of 2 lower than assumed (13 or 14 μm instead of 25 μm as assumed earlier), but still some CCDs fall short by a factor of 5 from the above requirement. In particular for 2D-CCD arrays, which have comparable full well capacities, the readout time can be a limitation.

Photodiode arrays have a much larger full well capacity of $6\times10^7-1.2\times10^8$ electrons per pixel and the read out times can be considerably lower. However, due to their electronic layout, the readout frequency is typically limited to ~100 kHz.

Co-adding Readouts and Exposure Time – In order to achieve the desired signal-to-noise ratios (e.g. $R = 10^4$), a sufficiently large number N of photoelectrons needs to be recorded so that $N > R^2$ and $N > B(B =$ combined effects of all fixed noise sources, e.g. detector readout noise and preamplifier noise), see (7.50). Frequently, the desired number of photoelectrons cannot be accumulated in a single scan, and thus several scans have to be co-added (see Sects. 7.6.2 and 7.6.3). For a good N/B ratio of a single scan, a signal level not too much below the maximum detector signal is desirable. On the other hand, 'overexposure' e.g. a signal larger than the maximum allowable detector signal (e.g. due to pixel saturation or ADC overload) must be avoided. Various schemes for 'exposure control' have been developed, mostly

Table 7.6. The characteristics of PDAs versus CCD and CMOS sensors

Type	PDA	CCD	CMOS
Characteristic Geometry	1-D only	Available in 2-D (usually) and 1-D	Available in 2-D (usually) and 1-D
Readout Noise (electrons)	High readout noise (several 10^3 electrons)	Low readout noise ($\approx$ 1–20 electrons)	Low readout noise ($\approx$ 10 electrons)
Typical pixel capacity (electrons)	10^8	10^5	10^5
Typical noise due to photoelectron statistics, relative S/N	10^4 10^{-4}	300 3×10^{-3}	300 3×10^{-3}
Dominant noise source	Photoelectron statistics, if full well capacity can be used	Usually photoelectron statistics	Usually photoelectron statistics
Dark current	100–1000 s to saturate pixel	100–1000 s to saturate pixel	100–1000 s to saturate pixel
Signal-dependent dark current	Yes	Yes	Unknown
Typical readout time	20 ms	10 ms to 1 s (2-D)	10 ms to 1 s (2-D)
Spectral range	NIR (Silicon limit) to UV (200 nm)	NIR (Silicon limit) to UV, many types	
Quantum efficiency	Good	Excellent (back-thinned arrays)	Good
On-chip ADC available	No	No	Yes

based on predicting the future signal on the basis of the already recorded signal. The magnitude of the signal is usually controlled by PMT high voltage (for optomechanical detectors) or exposure time (for solid-state detectors). If radiation signals are fluctuating in intensity, as can occur in many active and passive DOAS applications, overexposure cannot always be avoided. In this case, overexposed spectra must be discarded. Note that each individual spectrum must be checked for overexposure, since it may not be detectable after a few overexposed spectra are 'buried' under a large number of (properly exposed) co-added spectra.

Spectral Sensitivity – Detector must be sufficiently sensitive in the spectral range of interest. For most classical DOAS applications, a spectral range from about 200 to 800 nm is required to cover typical applications (see Chap. 6 and Appendix B). Recent applications, however, extend this range into the near-infrared. Silicon allows measurements from below 200 nm up to about 1100 nm, where the upper wavelength end is set by the band gap of silicon. For measurements at longer wavelengths, detectors based on other semiconductor materials are used. A popular material for wavelengths, up to 1600 nm or even 2400 nm is InGaAs. Single elements and linear arrays based on this material are available.

For PMTs, the long wavelength end of the sensitivity range is given by the requirement that the photon energy must exceed the work function, i.e. work to release an electron from the cathode material. Modern cathode materials (e.g. multi-alkali) give sensitivities to about 600 nm. In contrast, the short wavelength end of the sensitivity range is set by the window material. Glass windows will give about 400 nm, quartz windows $<$200 nm range.

The detector spectral sensitivity is frequently expressed in terms of the quantum efficiency Q:

$$Q = \frac{\text{Number of photoelectrons}}{\text{Number of photons}}. \tag{7.32}$$

Spectral Sampling – In order to obtain a true digital representation of the spectrum projected into the focal plane of the spectrometer, sufficiently fine spectral sampling is required. Obviously, the spatial frequency (e.g. density of detector pixels) should match the spectral resolution of the dispersing instrument. In other words, there must be a sufficient density of pixels compared to the spectral resolution. A rule of thumb is that the spectral width of a detector pixel should be about 10–20% of the FWHM of the instrument function (see Chap. 8). Details are discussed in the following and e.g. by Hofmann et al. (1995), Stutz (1996), Roscoe et al. (1996), and Chance et al. (2005).

Aliasing problems – Aliasing problems are well known from many areas of electronic signal evaluation. Aliasing will arise in cases where the spectral resolution is too high for a given number of detector pixels in the focal plane of the spectrometer (or if there are too few pixels in the focal plane sensor in comparison with the spectral resolution of the spectrometer). Fig. 7.33 illustrates this problem for the case of discrete absorption lines. In this example, the position of an emission line (smooth Gaussian in A) is shifted by the half-width of a pixel (B). As can be seen, there can be a drastic change in line shape of the digitised representation, shown as a histogram-like curve in panels A and B, which actually describe the same line shape.

It is tempting to make 'best possible use' of a given number of pixels and thus to specify only the minimal twofold oversampling (Nyquist criterion). However, practical experience and numerical modelling show there must be

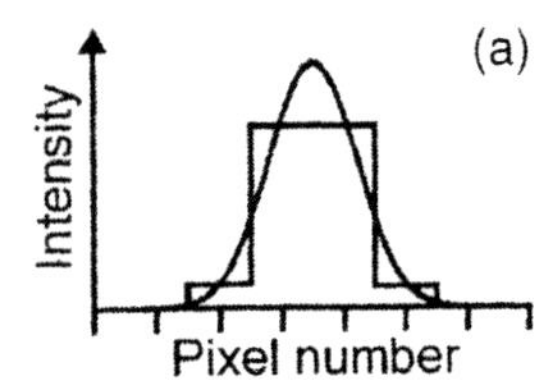

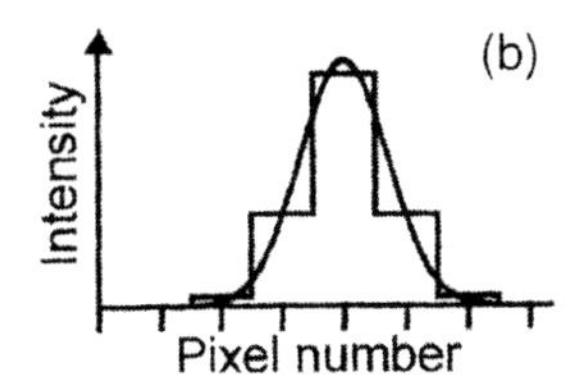

Fig. 7.33. Aliasing problems occur if there are too few pixels in the focal plane sensor in comparison with the spectral resolution of the spectrometer. If in this example, the position of an emission line (**a**) is shifted by a half width of a pixel (**b**), there can be a drastic change in line shape of the digitised representation. The histogram-like curve in panels A and B actually describe the same line shape

a sufficiently large number of signal samples $I(i)$ (or detector pixels) across one 'resolution interval'. Sample calculations demonstrate that on the order of 5–10 samples per resolution interval lead to acceptable errors (Hofmann et al., 1995; Stutz, 1996; Roscoe et al., 1996; Platt et al., 1997). In principle, advanced interpolation techniques and other procedures can be used to correct for this problem (e.g. DOAS evaluation of GOME spectra, Chance et al., 2005).

Detector Linearity – Detector linearity is a prerequisite for good sensitivity and measurement precision of DOAS systems. It can be described as:

$$I_{\mathrm{D}}(i) = \alpha \cdot I(i) \quad or \quad \frac{dI_{\mathrm{D}}(i)}{dI(i)} = \alpha = \text{constant}, \tag{7.33}$$

where $I_{\mathrm{D}}(i)$ denotes the signal generated by the ith detector pixel, while $I(i)$ is the radiation intensity reaching this pixel (see Chap. 8). The effect of non-linearity, i.e. the sensitivity α varying as a function of the radiation intensity I, is illustrated by the following consideration. In the limit of small optical densities, the (differential) optical density D of a spectral structure is given by:

$$D = \ln \frac{I_0}{I} \approx \frac{I_0}{I} - 1 = \frac{I_0 - 1}{I} = \frac{\Delta I}{I}. \tag{7.34}$$

In a practical measurement, however, I can only be estimated from the detector signal I_{D}:

$$D_{\mathrm{D}} \approx \frac{\Delta I_{\mathrm{D}}}{I_{\mathrm{D}}}.$$

Assuming for the sake of simplicity, two different sensitivities α_1 and α_2 for low and high signals, respectively (see Fig. 7.34), we obtain:

$$D_{\mathrm{D}} \approx \frac{\Delta S_{\mathrm{D}}}{S_{\mathrm{D}}} = \frac{\alpha_2 \Delta I}{\alpha_1 I} = \frac{\alpha_2}{\alpha_1} \cdot D. \tag{7.35}$$

Thus, the measured optical density (and consequently trace gas level) would deviate from the true value by the factor α_2/α_1. This could be interpreted as

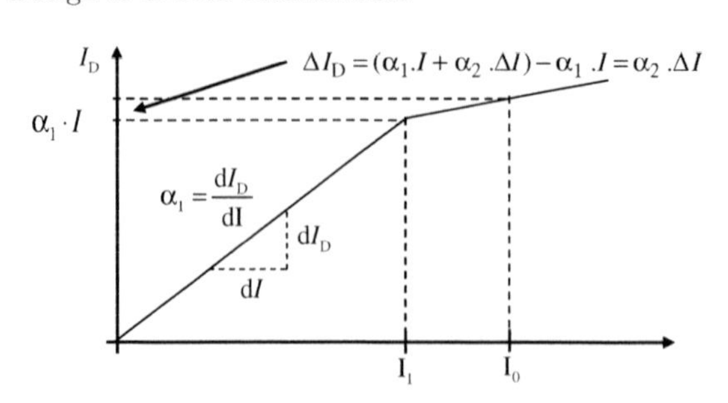

Fig. 7.34. Insufficient linearity of a detector can cause incorrect determinations of the optical density

the following requirement: If a 1% precision of the trace gas level is desired, the sensitivity $\alpha(I)$ cannot deviate more than 1% from its average value at any I. While this may be sufficient if there is only one dominating absorber present, there might be problems if weak spectral structures are to be detected in the presence of strong ones. For instance, in passive DOAS systems most absorbers are weak compared to the Fraunhofer bands. Here, distortion of the strong spectral structure due to non-linearity of the detector can introduce large errors in the determination of weak spectral absorption structures. Consequently, a much better detector linearity than estimated above might be required.

7.6.2 Optomechanical Scanning Devices and Photomultiplier Tube

Optomechanical scanning devices scan a section of the dispersed spectrum, typically several nanometres to 100 nm, in the focal plane of the spectrograph by a moving exit slit, as shown in Fig. 7.35. In the most common design, radial slits are etched in a thin metal disk ('slotted disk') rotating in the focal plane (see Fig. 7.36). At a given time, one particular slit serves as an exit slit. The start of a scan is determined by an infrared light barrier. The light passing through this slit is received by a photomultiplier tube (PMT), the output signal of which is digitised by a high-speed ADC and recorded by a computer. The signal is adapted to the ADC range by variable preamplifier gain and/or changing the PMT high voltage. During one scan (i.e., one sweep of an exit slit over the spectral interval of interest), several hundred digitised signal samples are taken. Consecutive scans are performed at a rate of typically 100 scans per second, and are digitised into 500–1000 'channels', which are signal averaged by the software (thus individual samples of the signal are recorded at 50–100 kHz frequency), thereby meeting the requirement of detection of very small signal differences.

Alternatively, vibrating exit slits can be used. However, disadvantages include less uniform scanning speed and possible influences of the vibration on other parts of the optics.

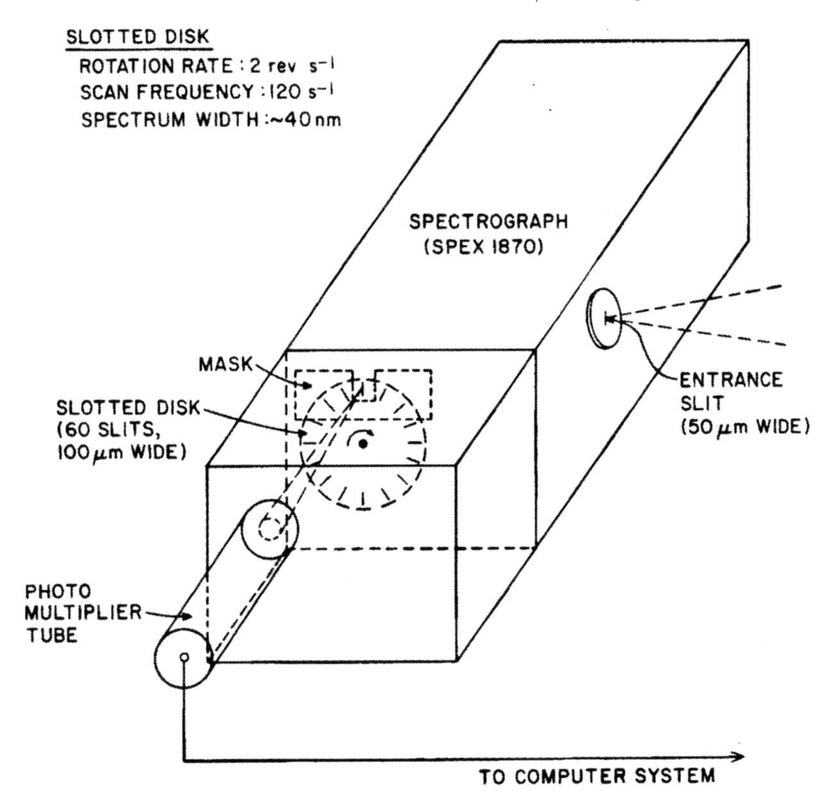

Fig. 7.35. Sketch of an OSD, the 'slotted disk' design. A series of (in this case 60) radial slits are etched in a thin metal disk (*'slotted disk'*) rotating in the focal plane.

Since a single scan takes ≈10 ms, the effect of atmospheric scintillations is negligible, because the frequency spectrum of atmospheric turbulence close to the ground peaks around 0.1–1 Hz and contains very little energy at frequencies above 10 Hz. In addition, typical spectra obtained during several minutes of integration time represent an average over 10,000–40,000 individual scans. Thus, effects of noise and temporal signal variations are very effectively suppressed. In fact, even momentarily blocking the light beam entirely (e.g., due to a vehicle driving through the beam) has no noticeable effect on the spectrum.

The main disadvantage of the optomechanical scanning of absorption spectra is the large multiplex loss, since only the section of the spectrum passing through the exit slit reaches the detector. Only on the order of 1% of the light is utilised, making relatively long integration times necessary.

The conversion of the radiation intensity to a detector signal is performed by the PMT. The design of the voltage divider chain supplying the voltages for the dynodes of the PMT is important. Since DOAS detectors are 'high anode current' applications (anode currents can reach $I_A = 100\,\mu A$), relatively high

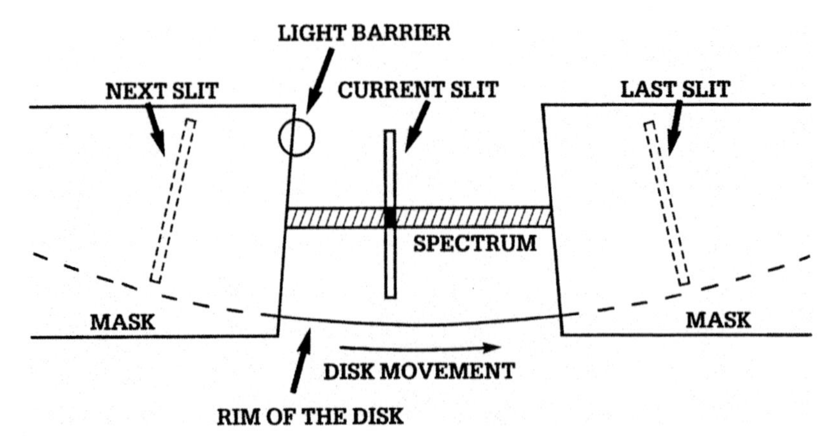

Fig. 7.36. OSD, the'slotted disk'. A mask allows only one slit to serve as an exit slit at any given time. The start of a scan is signalled by an infrared light barrier. The light passing through this slit is received by a PMT. (from Platt and Perner, 1983)

currents in the voltage divider (at least $10 \cdot I_A$) are required in order to not compromise detector linearity (see Sect. 7.6.1). Alternatively 'active voltage dividers' using transistors can be employed. Another important aspect is the response of the PMT electronics to transients. Commercial voltage dividers frequently employ capacitors between dynodes to improve response for light pulses; this is not needed or desirable for DOAS applications. However, PMT detectors are inherently fast response devices, so a low-pass filter is usually required before or within the preamplifier, supplying a time constant roughly equivalent to the time to record one detector pixel (typically $\tau \approx 20\,\mu s$). Finally, PMT photocathodes must be chosen according to the desired long wavelength end of the response. In general, cathodes with higher sensitivity at longer wavelength have higher dark currents. Available cathodes (e.g. multialkali, S20) limit the spectral range to the long wavelength end of the visible range.

7.6.3 Solid-state Array Detectors and Characteristics

Solid-state detector arrays consist of a large number, typically several hundred or more, of individual photodetectors (i.e. photodiodes) arranged in a linear row or in a two-dimensional, matrix-like fashion. They are, in principle, far superior to OSD in recording absorption spectra, since radiation of all wavelengths is recorded simultaneously rather than sequentially. This so-called 'multiplex advantage' amounts to about two orders of magnitude more photons recorded in typical systems. Accordingly, the measurement time can be shortened by the same factor or, alternatively the photon noise be reduced by about one order of magnitude. Additional advantages include the absence of moving parts, and nearly simultaneous sampling of all wavelength intervals. In addition, no high voltage is required and the quantum efficiency (at least in the red and near-infrared spectral regions) is better than that of most PMT cathodes (see e.g. Yates and Kuwana, 1976).

There are three basic designs of solid-state multi-element photodetectors, which are sketched in Fig. 7.37: PDA, CCD, and CMOS devices. Either design principle allows construction of linear or two-dimensional arrangements of hundreds to many thousands of individual photosensors, which can be read out sequentially.

In the case of **PDAs**, individual photodiodes are arranged on a silicon chip, and are read out by sequentially connecting them (via a field effect transistor) one by one to a common readout line, that conducts the accumulated charge to a readout amplifier. There may be more than one readout line (e.g. one for all even numbered, one for all odd-numbered diodes), and thus the readout time is reduced accordingly. PDAs are usually only available as one-dimensional (linear) arrays.

In contrast, **CCDs** have no readout line. Rather, the charge of each sensor is sequentially transferred through all pixels until it reaches a readout amplifier connected to the very last pixel in the row. Advantages of CCDs over PDAs include a simpler design, since no readout transistors and associated driving circuits are required, combined with much smaller output capacitance. The output capacitance is equivalent to one pixel, while the capacitance of a typical PDA readout line is equivalent to dozens or hundreds of pixels. As a consequence, the readout noise of a typical CCD is about two orders of magnitude lower than that of a PDA (see Table 7.6). On the other hand, the charge transfer efficiency of CCDs is limited to values of typically 99.999%, leading to some cross-talk between pixels, which is absent in PDAs. Usually, CCDs are produced as linear (1D) or area (2D) sensors. In the latter case, they provide additional information, which can be utilised e.g. to simultaneously record several spectra. However, the capacitance of the individual elements is much larger in PDAs, as outlined in Table 7.6. Therefore, the total noise can be limited by photoelectron shot noise, despite the higher readout noise of PDAs.

CMOS sensors are similar to PDAs, in that they are read out by sequentially connecting the individual pixels (via a field-effect transistors) one by one to a common readout line, which conducts the accumulated signal to a readout terminal. In contrast to PDAs, amplifiers (and sometimes reset gates) are frequently integrated directly at each pixel. Another advantage is that the CMOS process allows integration of other devices such as amplifiers and ADC directly on the sensor chip. There is also typically random access to each pixel. If amplifiers are attached to each pixel non-destructive readout is possible, allowing simple 'exposure control'.

A more subtle point concerns possible differences in exposure times for the different pixels. In PDAs, all pixels have the same length of exposure, but the time intervals do not quite overlap since exposure takes place from

(a) 1D-PDA or CMOS

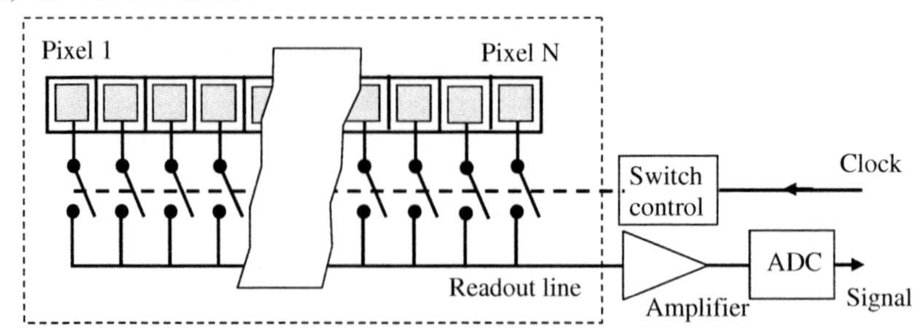

(b) 1D-CCD

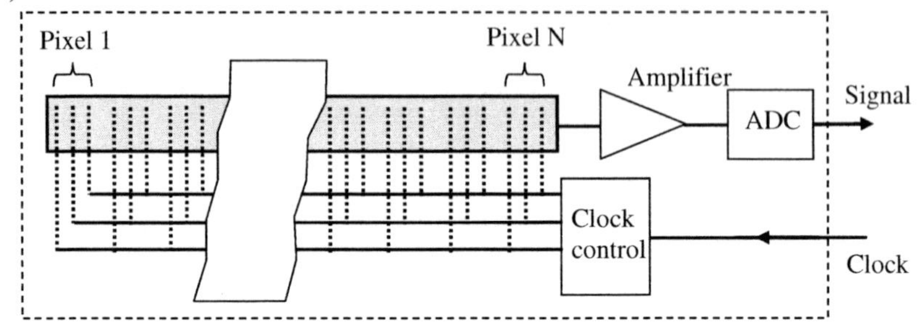

(c) 2D-CCD

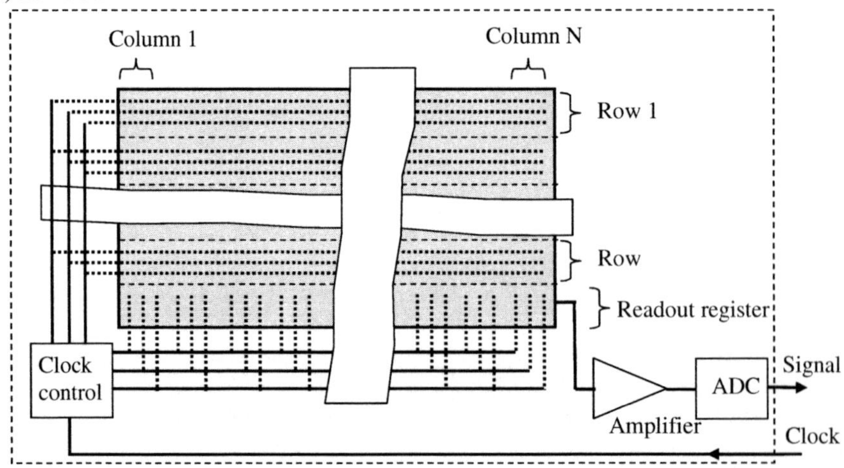

Fig. 7.37. Comparison of linear and two-dimensional solid-state photodetector array design principles. Shown are (**a**) layout of PDA or CMOS, (**b**) One-dimensional CCD, (**c**) Two-dimensional CCD devices

readout to readout. Readout is thus earlier for the first diode than the last, and thus the exposure time interval for the last diode is shifted in time (later) by the readout time. This may be of consequence at relatively short exposure times (compared to the readout time) and varying light levels. On the other hand, linear CCD detectors usually have a readout gate that transfers the charge of all pixels simultaneously into the readout CCD register, which is not exposed to light. Therefore, exposure time for all pixels covers exactly the same interval. In particular, in 2D sensors exposure to radiation during readout can be a problem. These problems can be overcome by using a shutter to darken the detector during readout.

Although there are clear advantages of solid-state array detectors, these devices pose some problems when used to record weak absorption features in DOAS instruments. First, the signal dependence of the dark current of the photodiodes has to be considered (see below). Another problem is the presence of **Fabry–Perot etalon structures** in the recorded spectra, which are caused by the protective overcoat of many commercially available detector arrays and also by possible vapour deposits on the array (Mount et al., 1992; Stutz and Platt, 1992, 1996), see Fig. 7.38 for an example.

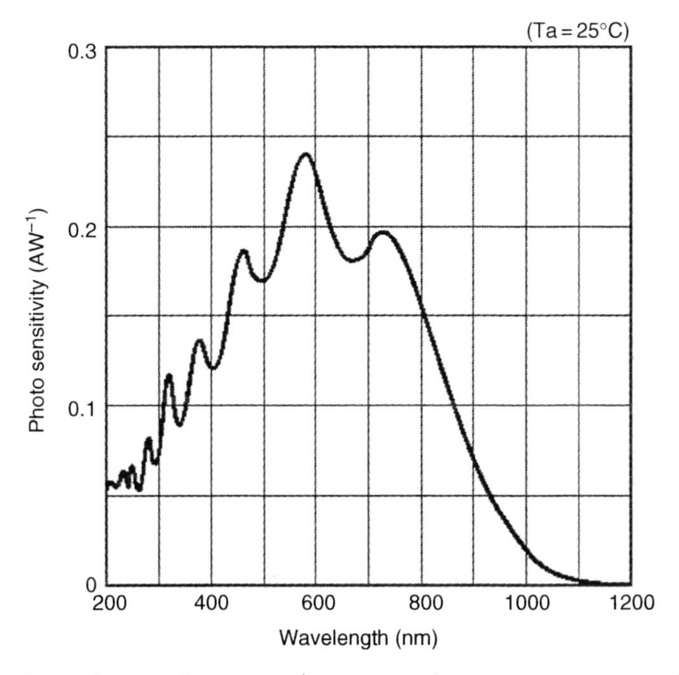

Fig. 7.38. Typical spectral response (in ampere photocurrent per watt of incident radiation) of a PDA detector (Hamamatsu Model S3904-1024). Note strong modulation of response due to the Fabry–Perot etalon effect of the overcoat of the detector chip (from Hamamatsu data sheet NMOS Linear Image Sensors S3900, 3901-1024Q, 3904-2048Q, figure courtesy of Hamamatsu Pholonics)

Particular problems are due to changes in the thickness of the layer of vapour deposit causing continuous changes in the 'etalon' modulation of the array sensitivity. In any case, diode sensitivity (and dark current) varies $>\pm 1\%$ between individual diodes. This usually constant fixed pattern must be compensated for.

Presently, solid-state array detectors have clear advantages, in particular in applications where only low-light levels are available (as in ZSL observation of stratospheric species). For the observation of low optical densities ($D \approx 10^{-3}$–10^{-4}) OSDs appear to give comparable results, however, at inferior light utilisation and thus much longer exposure times. Solid-state array detectors can be described by a number of characteristic features:

Spectral Sensitivity (Quantum Efficiency) – The most popular semiconductor material for solid-state detectors is silicon. Its spectral response extends from <200 nm to about 1100 nm, see Fig. 7.38. The long wavelength (or lower photon energy) end of the range is set by the band gap of silicon, while there is no hard limitation at the short wavelength end. In practice, thin active Si layers are required to obtain good UV sensitivity. While UV-optimised silicon devices (in particular back-illuminated CCDs, see below) can have >20% quantum efficiency below 300 nm, many detectors show little sensitivity below 400 nm. A simple way to obtain UV sensitivity consists of applying an UV-fluorescent coating (such as Metachrome II or Lumogene) on the surface of the device that converts the UV to visible (largely green) light at a wavelength near the maximum of the Si response. Disadvantages of this approach include the relatively low UV quantum efficiency of these coatings (around 10%), their limited lifetime, and the reduction of the sensitivity at longer wavelengths. It should be noted that, in some cases, a glass window with little transmission below ≈370 nm can be the limiting factor in the UV sensitivity.

Memory Effect – Illuminating the PDA with light of high intensity generates a signal which can sometimes also be observed (commonly at much lower intensity) in subsequently recorded spectra, although the detector is completely darkened. This effect is called memory effect. It was e.g. studied by Stutz (1996), but its origin is not quite clear. By waiting a few seconds and/or recording some spectra with the shortest possible integration time, it is usually possible to remove this effect (Alicke, 1997; Geyer, 2000).

Dark Current – As the name indicates, the 'dark current' of an optical detector is the current (or more generally speaking its output signal) which is still present if no radiation (within the detectors spectral sensitivity range) is admitted to the device. The dark current can be minimised by detector design and, for a given detector, by reducing its temperature. The dark current decreases exponentially with temperature, and can always be reduced to negligible levels by cooling the array low temperatures (−40°–80°C). The remaining dark signal can, in principle, be determined by observing the detector signal without illumination and then be subtracted from the signal.

Signal-dependent Dark Current – Naive subtraction of the diode array dark signal determined by recording the signal without illumination leads to wrong results. The origin of this observation can be found by inspection of the (simplified) circuit diagram of an individual diode (Fig. 7.39). It shows that the average voltage across the p-n junction is largest (approximately equal to U_o) if the diode is not illuminated. Under illumination (i.e. during an actual measurement), the voltage across the junction necessarily decreases. This apparent paradox is due to the different average voltages across the diode junctions during exposures, when the array is darkened or irradiated. At an exposure close to saturation, the average voltage during the signal integration period will be about $U_o/2$. In consequence, the 'dark signal' obtained by a measurement with no light admitted to the detector will overestimate the actual dark signal by up to a factor of two (Stutz and Platt, 1992). In fact, the effect can be even larger, since the diode capacitance increases with decreasing voltage. Note that this effect occurs with linear diode arrays (using readout switches), as well as with CCD arrays. A detailed explanation of this effect and procedures for its elimination is given by Stutz and Platt (1992).

Fixed Pattern 'Noise': Another potential problem when using solid-state detectors is the high variability of the sensitivity from diode to diode. The diode sensitivity depends on the wavelength of the incoming light and can vary from 1% up to several percent between the diodes of one array (Stutz, 1996). Usually, however, this sensitivity variation is systematic, (the popular term 'fixed pattern noise' is a misnomer), and can be readily corrected. Since DOAS is intended to observe optical densities down to well below 10^{-3}, it is necessary to accurately remove the diode sensitivity structure from a spectrum.

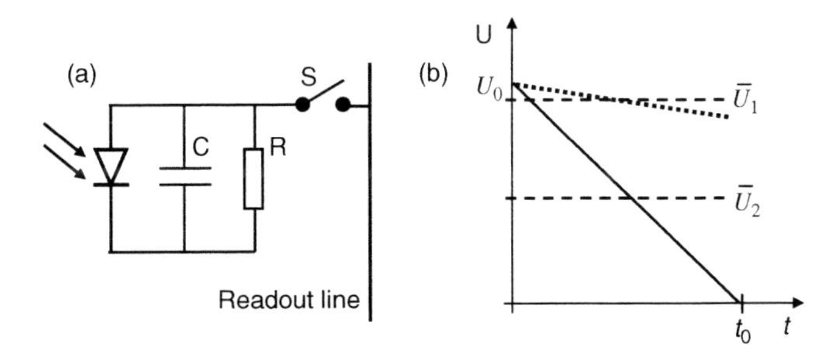

Fig. 7.39. (**a**) Simplified circuit representation of a single pixel in a diode array (or CCD) detector. During readout switch S is closed; (**b**) Temporal evolution of the voltage of a pixel: During acquisition of a dark spectrum the average voltage $\bar{U}_1$ across the capacitance C of the diode is close to the initial voltage U_0, and the dark current $I_{DC} = U/R$ is relatively high. However, in an actual measurement the average voltage $\bar{U}_2$ is always lower (in fact $\bar{U}_2 \approx U_0/2$) if the signal is close to saturation. Therefore, the dark current during a typical measurement will be considerably lower than indicated by a measurement with an un-illuminated detector

An easy way would be the division of each spectrum by a 'white light' lamp reference spectrum scanned just before each spectrum. However, this procedure can lead to additional lamp structures in the spectrum, in particular when Xe high-pressure lamps are used, since Xe emission lines differ in both spectra, due to differences of the illumination of the detector. Note that, in passive DOAS systems, the diode sensitivity structure is usually removed while eliminating the Fraunhofer structure (see Sect. 7.14.3). A common method to eliminate diode sensitivity structures in active long path DOAS spectra is the 'multi-channel scanning technique' (MCST) introduced by Knoll et al. (1990). The basic idea of the MCST is the combination of a multi-channel detection system (e.g. a PDA) with the scanning technique generally used to cover a larger spectral region with a single channel detection system. This technique is described in detail in Sect. 10.5.

7.6.4 PDA Detectors

Figure 7.40 shows the layout and cross-section of a typical silicon PDA detector (S3904-1024 by Hamamatsu) encompassing 1024 diodes, each measuring 0.025 by 2.5 mm, integrated on a single silicon chip. Thus, the total sensitive area measures 25 mm (in dispersion direction) by 2.5 mm. Note that there are two sets of diodes, 'active diodes' and 'dummy diodes'. The latter are read out separately. Taking the difference of 'active' and 'dummy' signals reduces the fixed pattern noise of the device. Data of some popular PDA types are summarised in Table 7.7.

7.6.5 CCD Array Detectors

Another type of solid-state photodetector is the charge coupled device CCD detector. Its name is derived from the readout mechanism. The charge generated by the radiation impinging on the detector is stored in potential wells

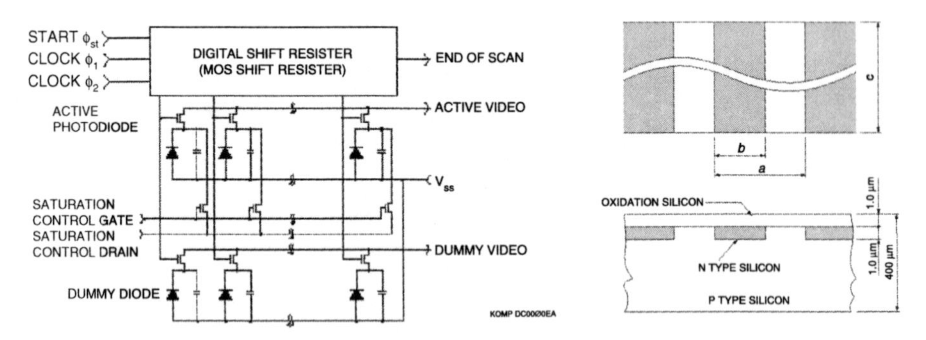

Fig. 7.40. Circuit schematic and layout of a PDA detector (Hamamatsu Model S3904-1024) (from Hamamatsu data sheet NMOS Linear Image Sensors S3900, 3901-1024Q, 3904-2048Q, figure courtesy of Hamamatsu Photonics)

Table 7.7. Characteristics of some linear Photo-Diode Arrays (PDA) for spectroscopic applications

Type (Manufacturer)	Pixel aspect ratio $(h_D \times w_D)$μm	No. of elements N	Sensitive Area mm × mm	Saturation Charge electrons	Max. pixel rate MHz	Dark current pA
S3900-1024, NMOS (Hamamatsu)	2500 × 25	1024	25.6 × 2.5	10^8	2	0.1
S3904-2048, (Hamamatsu)	2500 × 25	2048	51.2 × 2.5	1.6×10^8	2	0.1
S3904-1024Q (Hamamatsu)	5000 × 50	1024	51.2 × 5	6.3×10^8	2	0.4

generated by electrodes on the surface of the semiconductor (silicon). By applying suitable waveforms to the electrodes, the charge can be moved across the semiconductor, usually towards one edge of the device where it is fed to an electrode connected to a readout amplifier. While PDAs are usually one-dimensional (i.e. the individual photodiodes are arranged in a linear row), CCD's are typically two-dimensional. However, one-dimensional devices are also available (e.g. ILX-511 from SONY, see Table 7.8).

Geometry of Linear CCD Detectors: In principle, linear CCDs are in widespread use in the industry, for instance in fax machines, computer scanners, bar code scanners, modern (electronic) copiers, etc. However, in most of these applications two-dimensional originals are scanned, so the CCD line sensor pixels must be (at least approximately) quadratic. Typical pixel dimensions (h_D by w_D) are, therefore, 15 μm by 15 μm or 7 μm by 7 μm (see Table 7.9). For special purposes, some linear CCDs with aspect ratios (h_D by w_D) of about 10 are available (see Table 7.8).

Geometry of Area CCDs: The problem of insufficient pixel height can be overcome by using two-dimensional CCDs and summing up all pixels of a column.

Readout Speed: Readout speed is on the order of 10^6–10^7 pixels per second (1–10 MHz, see Table 7.8). Thus, total readout of a linear CCD takes about 2–0.2 ms. Taking the above requirements for readout speed (about 0.1 ms for a complete readout in active DOAS systems), presently only the fastest systems can match the requirements. Typical devices/systems fall short by about one order of magnitude. Using CCD sensors with *multiple readout taps* (e.g. 8 as in the case of DALSA IT-F6 TDI, see Table 7.8) could provide the required speed, although at the expense of additional electronics (8 ADCs plus associated electronics are required). For two-dimensional CCDs, readout time is an even more severe limitation. For instance, a CCD with 2000 by 100 elements (dispersion direction by vertical) would require 200 ms

Table 7.8. Some CCD sensors for linear spectroscopic applications

Type (Manufacturer)	Pixel aspect ratio ($h_D \times w_D$) µm	No. of elements N	Sensitive area mm × mm	Saturation Charge electrons	Max. pixel rate MHz	Separate readout shift register
ILX 511 (SONY)	200×14	2048	28.67×0.2	1.6×10^5	2	Yes
TCD 1205D (Toshiba)	200×14	2048	28.67×0.2	$\approx 2 \times 10^5$	2	Yes
TCD1304AP (Toshiba)	200×8	3648	29.18×0.2	?	0.6	Yes
IL-C6 (DALSA)	500×13	2048	26.6×0.5	$\approx 2 \times 10^5$	15	Yes
IT-F6 TDI[a] (DALSA)	$13 \times 1248(13 \times 96 \times 13)$	2048	26.6×1.3	4.6×10^5	8×20	Yes
S7030-1007 (Hamamatsu)	24×24	1024×122	24.58×2.93	3×10^5		

[a]UV sensitive to 250 nm.

Table 7.9. Some CMOS sensors for spectroscopic applications

Type (Manufacturer)	Pixel aspect ($h_D \times w_D$) μm	No. of elements $N \times M$	Sensitive area mm × mm	Saturation charge electrons	Dark current electrons/s	Max. pixel rate MHz	Readout noise electrons
Star250[a] (fill factory[b])	25×25	512×512	12.8×12.8		4750	8	74
Star1000[a] (fill factory[b])	15×15	1024×1024	15.4×15.4	120,000	3500	12	39
ACS-1024[a](PVS[c])	13×13	1026×1026	13.21×13.21	300,000	< 3000	30 fps	≈ 100
ELIS-1024[d](PVS[c])	$125 \times 7.8125 \times 64.4$	1024×128	0.125×7.998	800,000–6.4×10^6	12,000	30	1600
LIS-1024[d] (PVS[c])	125×7.8	1024×1	0.125×7.998	8×10^6	?	20	≈ 500
LNL2048R[e] (pixel devices[f])	200×7	2048×1	0.2×14.34	187,500	8000	32	17
S8378-1024Q[d,e] (Hamamatsu)	500×25	1024×1	25.6×0.5	3.9×10^7	2.5×10^5	0.5	≈ 6000

fps = frames per second
[a]On chip 10-bit ADC
[b]Fill factory: http://www.fillfactory.com/index2.htm
[c]Photon vision systems: http://www.photon-vision.com/p-lis.htm
[d]UV sensitive
[e]On chip amplifier(s)
[f]Pixel devices

for a complete readout at 1 MHz. Thus, up to two orders of magnitude are missing. Some help could come from the ability to 'bin' pixels, i.e. combine the charge of several pixels (ideally all) of a column in the analogue readout shift register.

Points to Consider When Using CCDs for DOAS Applications

There is a large and growing variety of CCD detectors available. While in principle linear CCDs are ideally suited for DOAS, a number of crucial parameters have to be evaluated.

1. The pixel width w_D and height h_D of most standard scanner CCDs is appropriate for practically all DOAS applications. For passive DOAS applications higher h_D would be desirable.
2. The UV sensitivity of linear CCDs is frequently not sufficient. Linear CCDs are not available in back illuminated versions. This can partly be circumvented by adding UV fluorescent coating (such as Metachrome II or Lumogene) that converts the UV to visible (largely green) light at a wavelength near the maximum of the Si response. Limitations are due to relatively low quantum efficiency (near 10%) compared to >20% for special UV Si devices or back illuminated (two-dimensional) CCDs.
3. Readout time can be a problem, in particular in active DOAS systems. The fastest linear CCD systems or multiple tap CCDs appear to be just fast enough for active DOAS applications. Two-dimensional CCDs are frequently too slow by about two orders of magnitude for active DOAS applications.
4. The charge transfer efficiency has to be considered. While stated efficiencies >99.999% may sound impressive it has to be kept in mind that the charge generated in e.g. the first pixel is shifted through about 1000 pixels. This may lead to a charge loss of the order of 1%. This charge is distributed across other pixels, thus adding to the cross-talk.

Comparison of the Different Solid-state Detectors: As outlined above, PDA and CCD detectors have their strengths and weaknesses. All allow the detection of light signals at the limit of photoelectron shot noise, if conditions are carefully chosen. The margins for shot noise limited detection are much wider for CCD detectors, provided the ADC resolution is sufficient. Some characteristics are compared in Fig. 7.41. For instance, if there are fewer than about 3000 photoelectrons recorded in a CCD detector with 12-bit ADC, the S/N ratio will be determined by the ADC and no longer by photoelectron statistics (intersection of fat, solid line with dashed line b).

7.6.6 CMOS Detectors

Complementary metal–oxide–semiconductor (CMOS) arrays are a relatively new addition to the family of spectroscopic detectors. Little is currently known

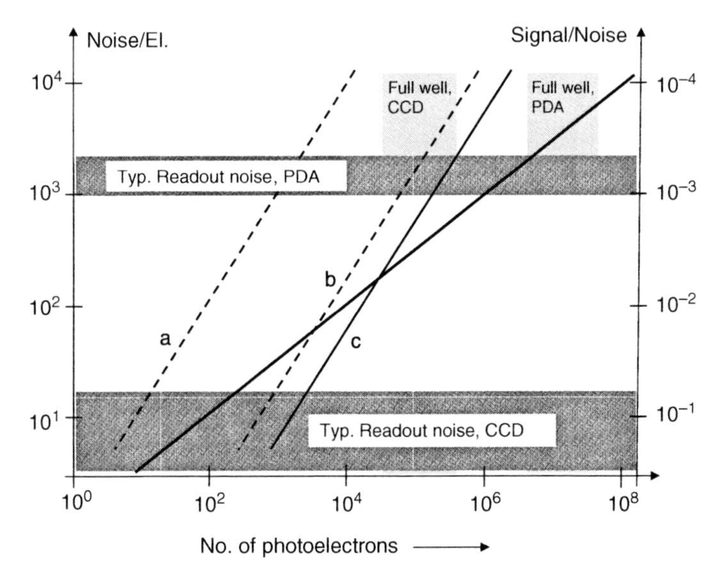

Fig. 7.41. Sketch of signal-to-noise (S/N) ratios as a function of the number of photoelectrons and characteristics of typical CCD and PDA detectors. Fat line: S/N ratio from photoelectron statistics alone; Lines a–c: contribution of the ADC to the S/N ratio of the detector system. (a) 16-bit ADC for a typical CCD detector (6×10^4 photoelectrons full well); (b) 12-bit ADC for a typical CCD detector; (c) 16-bit ADC for a typical PDA (6×10^7 photoelectrons full well)

about their performance in DOAS applications. However, the fact that the detector elements can be integrated on the chip with other electronic components, such as amplifiers and ADC converters, makes them an interesting option. Table 7.9 gives details of currently available CMOS sensors.

7.7 Telescope Designs

In DOAS systems, telescopes are used for two purposes: to collect light from (natural or artificial) light sources and, in active DOAS applications, to transmit a collimated beam of radiation from the light source through the open atmosphere. Basically, the telescopes designs either employ refracting elements, i.e. lenses (refractors) or reflecting elements, i.e. mirrors (reflectors). For DOAS applications, reflectors are often chosen due to absence of chromatic aberrations. Reflectors also need fewer large optical surfaces, which are expensive to manufacture. Traditionally, there are several design choices for reflecting telescopes, which are known as Newtonian, Cassegrain, or Gregor arrangements, as shown in Fig. 7.42. A popular design is the Newtonian, which combines compact and simple optics with easy adjustment. Also, Newtonian telescopes usually form the basis of transmitting–receiving

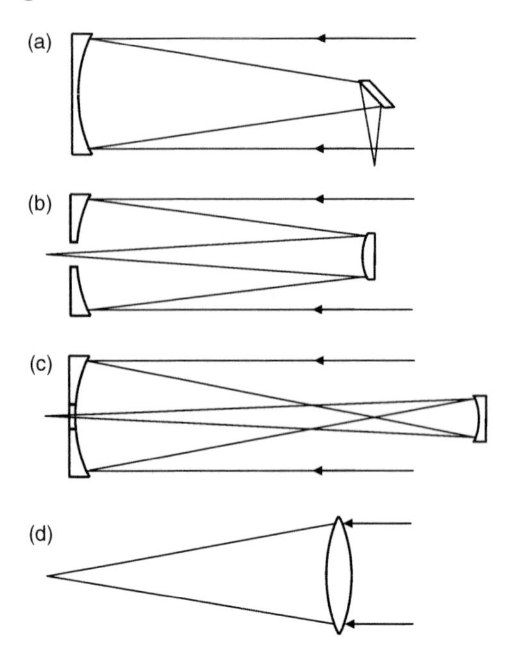

Fig. 7.42. Different telescope designs in use for DOAS applications. Reflectors: (**a**) Newtonian, (**b**) Cassegrain, (**c**) Gregor type telescope, (**d**) Refractor

telescope combination (see Fig. 7.49, Sect. 7.9.3). In contrast, Cassegrain-type telescopes are much more compact, but difficult to manufacture and require strict tolerances in adjustment.

7.8 Optical Multi-pass Systems

In principle, an attractive alternative light path arrangement is to fold the absorption light path in a multiple pass (or multiple reflection) cell (MRC), where the probing beam of radiation traverses the same volume (the 'base path') many times. There are several optical arrangements that allow the multiple folding of light beams. The most widely used are:

- White Cells (White, 1942, 1976)
- Herriot Cells (Herriot et al., 1964; Herriot and Schulte, 1965)
- Passive resonators (Engeln et al., 1998; Ball et al., 2004)

For each design, a number of variants were developed (Horn and Pimentel, 1971; Schulz-DuBois, 1973; Chernin and Barskaya, 1991; Ritz et al., 1992; MacManus et al., 1995; Vitushkin and Vitushkin, 1998; Grassi et al., 2001).

Multi-reflection systems have two main advantages. First, the measurement volume is greatly reduced, thus making the assumption of a homogeneous trace gas distribution much more likely to be fulfilled. Second, the length of the light path can be changed (typically in certain increments of small multiples of the base path, depending on the particular design of the cell). Therefore, it is possible to keep the length of the light path close to its optimum (see Sect. 7.13) for the respective atmospheric conditions.

Disadvantages of multi-reflection cells include the relatively complicated optical design, the requirement for extreme mechanical stability and small reflection losses. Typically, high reflectivity ($R > 0.99$–0.9999) of the cell mirrors over spectral intervals of several 10 nm is required, but difficult to obtain and to maintain in the field (e.g. due to dust settling on the surfaces), in particular for large area mirrors and in the UV. In some applications, the high-radiation intensity inside the folded light path can cause photochemical reactions that may change the concentration of the species to be measured. In addition, the level of stray light originating from the light source is difficult to measure in multi-reflection cells. Another problem can be due to the reaction of the trace gases with the surfaces of optical or mechanical parts of the cell, although 'open path' cells can reduce this effect.

7.8.1 White Multi-reflection Cells

The basic White multi-reflection system (Fig. 7.43; White, 1942) consists of three spherical, concave mirrors of identical radius of curvature (and thus focal length). The distance between the mirror sets is twice the focal length. The front mirror A faces the two side by side back mirrors B and C. Radiation from the light source is focused in point F_0 such that it illuminates back mirror B, which images point F_0 into F_1 on mirror A. Mirror A is oriented such that the radiation originating from F_1 illuminates back mirror C, which in turn images F_1 into F_2. This completes a round trip of the radiation traversing the system four times. F_2 can be thought of acting as a new entrance to the system, with the next round trip producing foci F_3 and F_4. Since F_2 is closer to the axis of the system, F_3 and F_4 will be as well. Ultimately, there will be

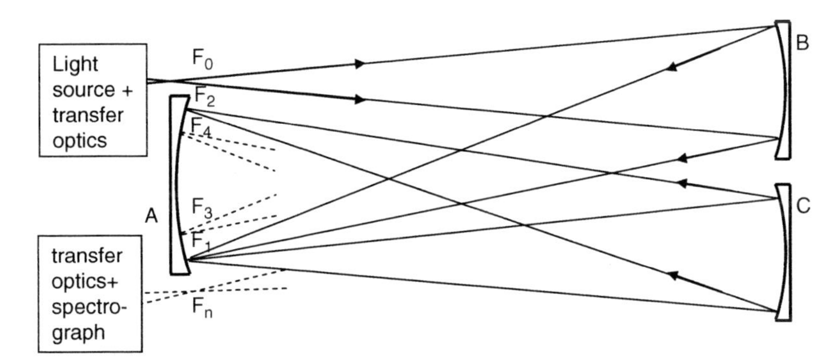

Fig. 7.43. Schematic diagram of a basic White system (adapted from J.U. White, 1942)

two sets of foci F_0, F_2, F_4, ... and F_1, F_3, ... across the surface of 'nesting mirror' A. Usually, mirrors B and C will be tilted slightly such that even and odd numbered reflections of mirror C form two separate rows on mirror A. Finally, one of the images F_N (not shown) falls off the surface of mirror A, thus, leaving the system. Therefore, the number of passes can be adjusted in steps of 4.

In the more advanced White system (White, 1976) in addition to the basic optics, two or three quartz prisms are set up near the front mirror, as shown in Fig. 7.44. In its basic function, the system functions as the 'basic' White system (White, 1942) described earlier. However, the radiation leaving the mirror system after N traverses near the bottom of mirror A is reflected back by prism P1. After making another N traverses, it is returned again by prism P2, giving rise to another $2N$ traverses. Finally, prism P3 reverses the entire set of light paths for a total of $8N$ traverses. Thus, the number of passes can be adjusted in steps of 32 traverses. This scheme not only multiplies the number of passes by 8, but more importantly also considerably increases the stability against mis-alignment of the system. This is due to the fact that each prism acts as retro-reflector, returning the radiation (nearly) exactly back to its original direction in one plane. However, the planes of P1 and P2 are perpendicular to each other, thus compensating misalignments within the system only as long as they do not become too large.

In fact two prisms would be enough to accomplish this effect, as originally described by White (1976). The vertex angle of the prisms is actually slightly smaller than 90° (Ritz et al., 1992). Further optimisations of the design were suggested by Hausmann et al. (1997) and Grassi et al. (2002).

The transmission of radiation through a multi-reflection system is given by the transmission of the transfer optics, the reflectivity R of the mirrors, and the extinction of the air inside the system. The transfer optics consists of

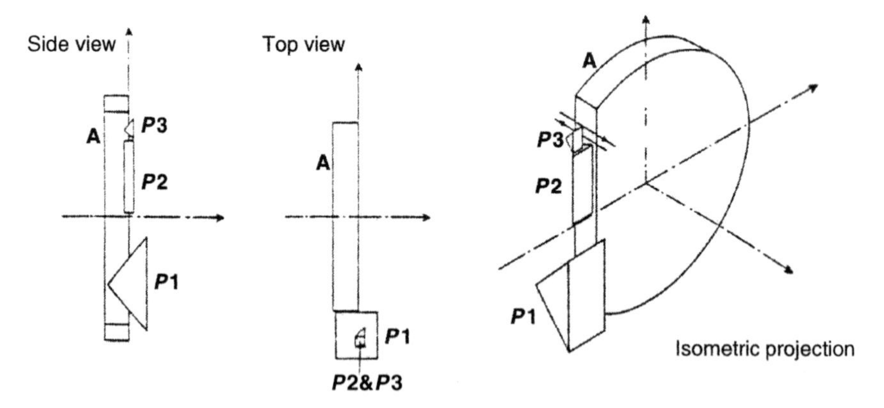

Fig. 7.44. Schematic diagram showing the nesting mirror (**A**) and the prisms P1, P2 and P3 (from Ritz et al., 1992)

two parts. First, optics matching the aperture of the light source to that of the entrance of the multi-reflection cell, and second, optics matching the exit aperture of the multi-reflection cell to the entrance of the DOAS spectrometer. Typically, the desired aperture from a Xe-arc lamp might correspond to an f-number of $F_1 \approx 1$, while a White cell might have $F_2 = 100$ (i.e. $2f$ divided by the diameter of the mirrors B and C). Transfer optics would then image the Xe-arc at a magnification of $F_2/F_1 \approx 100$. At the exit side (assuming the exit $f/\#$ of the white cell to be the same as the entrance $f/\#$), matching the white cell to the entrance of a spectrometer with $F_3 = 5$ would require a reduction in image size by $F_3/F_2 \approx 20$. Given a diameter $d \approx 0.2$ mm of the brightest spot in the arc (see Sect. 7.3.1), the images on nesting mirror A of the White cell (Fig. 7.43) would be 20 mm in diameter; thus, in order to accommodate many reflection mirrors, A has to be quite large. Alternatively, the aperture of the light source (F_1) could be reduced or that of the White cell (F_2) increased.

Overall, the (broadband) transmission of a White multi-reflection cell is given by:

$$T = \frac{I}{I_0} = G \cdot \mathrm{R}_\mathrm{M}^N \cdot e^{-\varepsilon_\mathrm{B}}, \tag{7.36}$$

where G denotes the transmission of the transfer optics, R_M the mirror reflectivity, N the number of reflections, and ε_B the broadband extinction of the air (see Chap. 6).

7.8.2 Herriott Multi-reflection Cells

The optical design of a Herriott-type multi-reflection cell (e.g. Herriott et al., 1964; Herriott and Schulte, 1965; McManus and Kebabian, 1990; McManus et al., 1995; Zimmermann, 2003) is shown in Fig. 7.45. A collimated beam of radiation enters through a hole in (circular) mirror A at a certain distance R_0 from the centre of the mirror. The radiation is then reflected N times between mirrors A and B and leaves through the same hole. The reflections are located

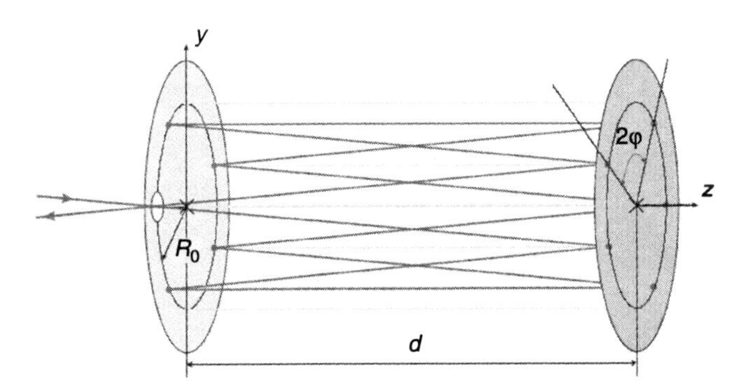

Fig. 7.45. Schematic diagram of a Herriott cell (from Zimmermann, 2003)

on either mirror on circles (or more general on ellipses) with radius R_0. The vertex angle θ in a triangle formed by the centre of mirror A and the locations of two consecutive reflections is given by $\theta = 2\pi M/N$, where M denotes the number of times the reflections circle around the optical axis of the system. The angle θ is determined by the ratio of the distance d between the mirrors and their radius of curvature R:

$$\cos\Theta = 1 - \frac{d}{R}. \tag{7.37}$$

The total absorption path L is given by $L = d \cdot N$. Advantages of Herriott cells are their simple optical setup combined with a good stability of the beam profile due to the re-focussing properties of the optics. In addition, there is some immunity against tilt of the mirrors against each other and lateral translation of the mirrors. A major disadvantage is due to the need for collimated beams. More details on the theory of Herriott cells can be found in Herriott and Schulte (1965), and some details for applications in Zimmermann (2003) and McManus and Kebabian (1990).

7.8.3 Passive Resonators (CEAS, CRDS)

Passive resonators can also provide long optical paths (e.g. Engeln et al., 1998; Brown, 2003; Ball et al., 2004; Fiedler, 2005). This technique has become known as cavity enhanced absorption spectroscopy (CEAS). The basic idea is to introduce white light (intensity I_L) into an optical resonator (see Fig. 7.46) consisting of two mirrors with reflectivity R. Initially, averaged over a wavelength interval larger than the free spectral range, only the fraction $\rho = 1 - R$ will enter the resonator. However, once in the resonator the radiation will be reflected $1/(1 - R)$ times (neglecting other losses) on average. Finally, half the radiation will leave the resonator through mirror 1, the other half through mirror 2, with each fraction having the intensity:

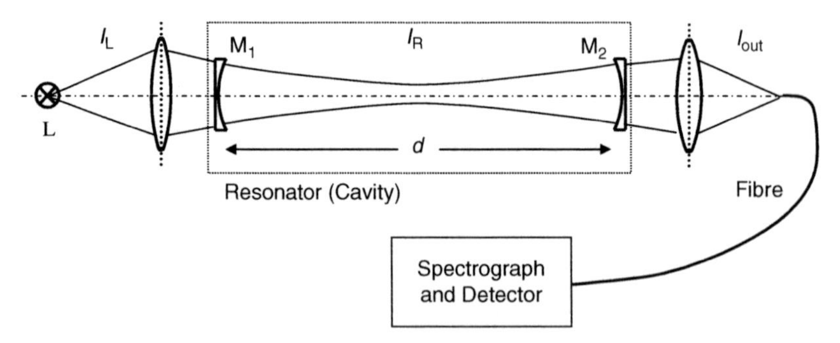

Fig. 7.46. Sketch of a cavity-enhanced absorption-based DOAS system (Ball et al., 2004; Fiedler, 2005). The resonator would be set up in the open atmosphere

$$I_{out} = I_L \times \frac{(1-R)}{2} = I_L \cdot \frac{\rho}{2}. \tag{7.38}$$

If additional (broadband) extinction ε_B is present, the average length of the absorption path will be:

$$\bar{L} = \frac{d}{\rho + \varepsilon_B d}, \tag{7.39}$$

where ε_B denotes all broadband extinctions (due to Rayleigh and Mie scattering, and broadband trace gas absorption). Presently, mirrors with $R > 0.9999$ over a spectral range of several 10 nm can be manufactured, leading to $\rho = 1 - R < 10^{-4}$, and consequently average light paths $\bar{L} > 10^4 \cdot d$ if extinction is neglected. At a base path $d = 1$ m this would correspond to $\bar{L} > 10$ km. The transmitted light would be $I_{out} \approx 5 \times 10^{-5} \cdot I_L$. While this looks small, it is comparable to losses in 'long path' DOAS system using light beams in the open atmosphere, see Sect. 7.11.

Practical problems with CEAS include the need to determine the actual light path, accumulation of dust on the mirrors, and the fact that highest transmission occurs outside the range of high mirror reflectivity.

As can be seen from (7.39), the light path depends not only on the mirror reflectivity, but also on the (broadband) atmospheric extinction, and it can thus vary widely as atmospheric conditions change. Moreover, the reflectivity is likely to change, e.g. due to dust deposition on the mirrors. Typically the length of the light path has to be determined by recording absorption lines of species with known concentration such as O_2, O_4, or water vapour (if the humidity is simultaneously measured), or by cavity ring down spectroscopy (CRDS) (Brown, 2003).

7.9 Active DOAS Systems

Active DOAS systems employ their own light source; therefore, they have to include suitable optics to transmit light through the atmosphere to the spectrometer. The most basic arrangement of a light path in the open atmosphere (see Fig. 7.47) consists of a light source (search light type) defining one end of the light path and a receiving telescope coupled to a spectrometer defining the other. Earlier DOAS instruments (e.g. Perner and Platt, 1979; Platt et al., 1979) were based on this arrangement and they are still in use (in particular in commercial instruments). However, in recent years a series of variants was developed which have clear advantages over the original active DOAS setup.

7.9.1 'Classic' Active Long-path System

The original design of active DOAS systems (see Fig. 7.47) consisted of a receiving telescope (coupled to a dispersing device, e.g. a spectrometer and a telescope) 'looking' into the light beam emitted from a light source, e.g. a

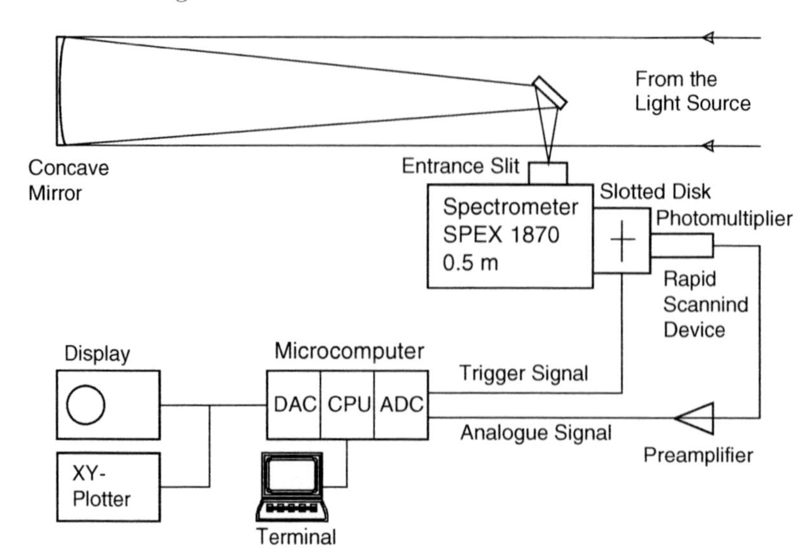

Fig. 7.47. Schematic setup of a 'classic' DOAS system using separate transmitting and receiving telescopes and an optomechanical (rotating 'slotted disk' + photomultiplier) scanning system

search light. This type of setup has been used in many investigations since the 1980s (Perner and Platt, 1980; Platt et al., 1979, 1981; Harris et al., 1982). The instrument averages trace gas absorptions over the extended volume of air. Disadvantages of this arrangement include the requirement to align the optical elements at either end of the light path, and the necessity of power at two location, which frequently complicates the logistic requirements in field campaigns.

In active long-path DOAS systems, a receiving telescope of sufficiently large aperture, usually 150–300 mm, is required to collect the light. The telescope has to be precisely pointed at the light source or retro-reflector array (see Sect. 7.4.9). This can usually only be accomplished by active control of the telescope. Note that even a very rigid mounting of the optical building blocks of the system will usually not solve the problem. Due to an 'El Mirage' effect, the light path will 'bend' due to changing thermal gradients in the atmosphere.

The aperture ratio of the telescope must match that of the spectrometer (or of the connecting Quartz fibre), see Table 7.10.

7.9.2 High-resolution DOAS Spectrometers

Some species, in particular atoms and certain small molecules, exhibit narrow absorption lines (see Table 7.10), which dictate the use of a light source with high spectral intensity. A prominent case is the OH radical, which was

Table 7.10. Data pertinent to high resolution and low resolution DOAS systems for the spectroscopic detection of OH radicals and NO_3 radicals

Radical parameter	OH	NO_3
Wavelength of strongest band or line l/nm	308	662
Line width Γ/nm	0.0017	3.2
Effective abs. cross section σ/cm^2	1.3×10^{-16}	1.8×10^{-17}
Minimum detectable optical density D_0	10^{-4}	10^{-4}
Detection limit C_0 molecules/cm^3 ($L = 5\,km$)	1.5×10^6	1.1×10^7
Interfering molecules	SO_2, CH_2O, CS_2, $C_{10}H_8$	H_2O

reliably detected in the atmosphere for the first time by DOAS (Perner et al., 1976). The required high-spectral energy density can, at present, only be supplied by a laser. Its emission bandwidths can either be broad or narrow compared to the OH lines. The former approach has a fixed laser emission wavelength, but requires combination with a high-resolution spectrometer (Perner et al., 1976, 1987; Hübler et al., 1984; Platt et al., 1987, 1988; Dorn et al., 1988; Hofzumahaus et al., 1991; Mount, 1992). The latter variety of the technique uses a narrowband laser scanning the OH line (Zellner and Hägele, 1985; Amerding et al., 1990, 1992).

As an example, an experiment following the broadband approach is shown in Fig. 7.48. A frequency-doubled dye laser coupled to a beam-expanding telescope serves as a light source. To keep interference at negligible levels, the beam is expanded to about 0.2 m diameter (Perner et al., 1976). The laser emission bandwidth (FWHM) of about 0.15 nm is large compared to the width of the OH line (≈0.0017 nm). The received light is dispersed by a 0.85-m double spectrometer with a resolution comparable to the OH line width (Perner et al., 1987a; Hofzumahaus et al., 1991). Table 7.10 lists some key parameters of a long-path OH spectrometer in comparison to a 'classical' low-resolution DOAS instrument. An important consideration is the amount of self-produced OH, since the intense UV radiation of the laser beam photolyses ozone, yielding O^1D atoms that can react with H_2O to produce OH in the beam.

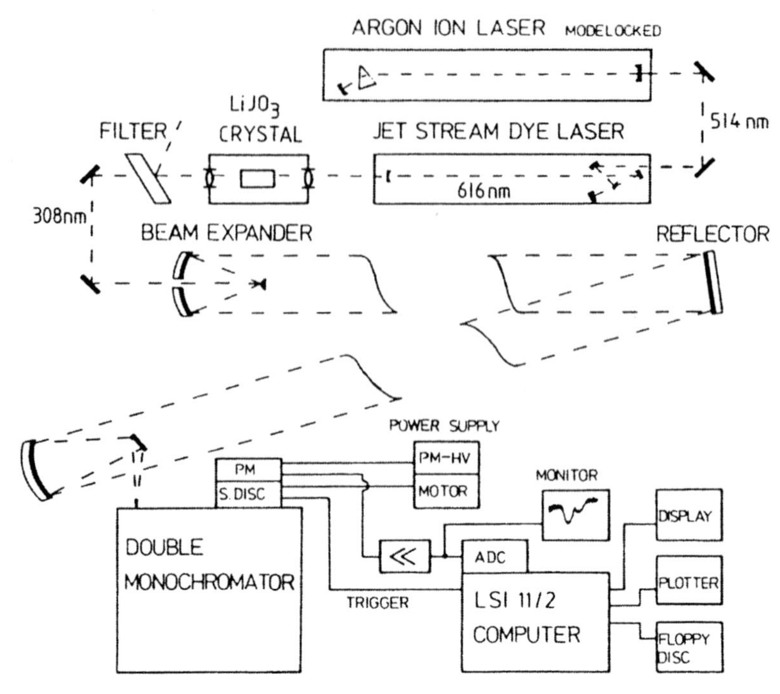

Fig. 7.48. Optical setup of a high-resolution long-path spectrometer. A laser system supplies the high-spectral intensity required due to the high resolution of 0.002 nm (instead of about 0.5 nm for a low-resolution DOAS) (adapted from Platt et al., 1987)

Later designs of this instrument coupled the high-resolution spectrometer to a multi-reflection cell (see Sect. 7.8) to allow more localised measurements and better control of the optical path (Brauers et al., 1996; Dorn et al., 1993, 1996, 1998; Hausmann et al., 1997; Hofzumahaus et al., 1991, 1998; Brandenburger et al., 1998).

7.9.3 Recent Designs of Active Long-path DOAS System

A variation of the basic active DOAS design uses a reflector (usually a corner cube retro-reflector as indicated in Fig. 7.49) to return the light of a source located next to the spectrometer. Thus the total light path length is twice the distance between light source/spectrometer and reflector. Usually, a coaxial merge of transmitting and receiving telescope as shown in Fig. 7.49 is used. This setup was first described by Axelsson et al. (1990). Advantages and limitations are similar to the classic arrangement described in Sect. 7.9.1. However, power and delicate alignment is only necessary at the spectrometer/light source site (Plane and Nien, 1992; Edner et al., 1993; Hausmann

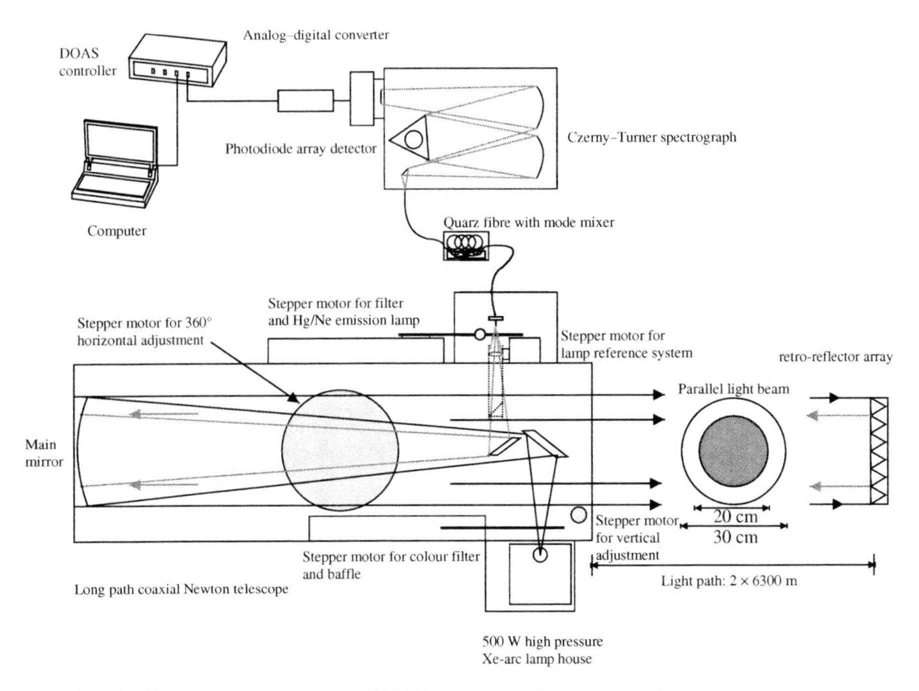

Fig. 7.49. Schematic setup of a DOAS system using a coaxial arrangement of transmitting and receiving telescope in conjunction with a retro-reflector array (Geyer et al., 2001). This type of setup, pioneered by Axelsson et al.(1990), has become the standard for active DOAS systems for research in the recent years (from Geyer et al., 2001)

and Platt, 1994; Heintz et al., 1996; Alicke et al., 2002; Geyer et al., 1999; Allan et al. 1999, 2000). The introduction of retro-reflectors also has the advantage of reducing the effect of atmospheric turbulence on the light beam (see Sect. 7.4.9). This is due to the capability of corner cube retro-reflectors to return the incident light exactly (although with some lateral offset) into the direction of incidence. In addition, the light collected by the receiving section of the telescope is transmitted to the spectrometer via quartz fibre, allowing the introduction of a 'mode mixer' (see Sect. 7.4.7; Stutz and Platt, 1997a), which further reduces the effect of atmospheric turbulence. Because of these advantages, the arrangement shown in Fig. 7.49 has become the standard setup for active long-path DOAS instruments.

Variations of this design use two telescopes for transmitting and receiving the light mounted side by side at close distance. Thus, only a small fraction of the transmitted light is actually collected by the receiving telescope. This drawback can be partially overcome by covering half of the aperture of each retro-reflector element by a wedged window with a very small (fraction of a degree) angle (Martinez et al., 1999). In this way, the light returned by the retro-reflectors is deflected from its initial direction by a small angle. Thus, two beams are reflected, one at either side of the incident beam on the retro-reflector. The receiving telescope is mounted at the position of one of the beams.

7.9.4 DOAS Systems with Optical Multi-pass Systems

The absorption light path may be folded in a multiple reflection cell. The optical arrangements are described in Sect. 7.8. Figure 7.50 shows a typical setup of a DOAS system using a White type open multiple reflection cell. Main elements are the light source, and entrance transfer optics to match the large aperture of the light source to the small aperture of the White multiple reflection system. After traversing the multiple reflection system, secondary transfer optics match the aperture to that of the fibre and spectrometer. Spectrometer, mode-mixer, and detector are of the same design as in the long-path system described in Sect. 7.9.3.

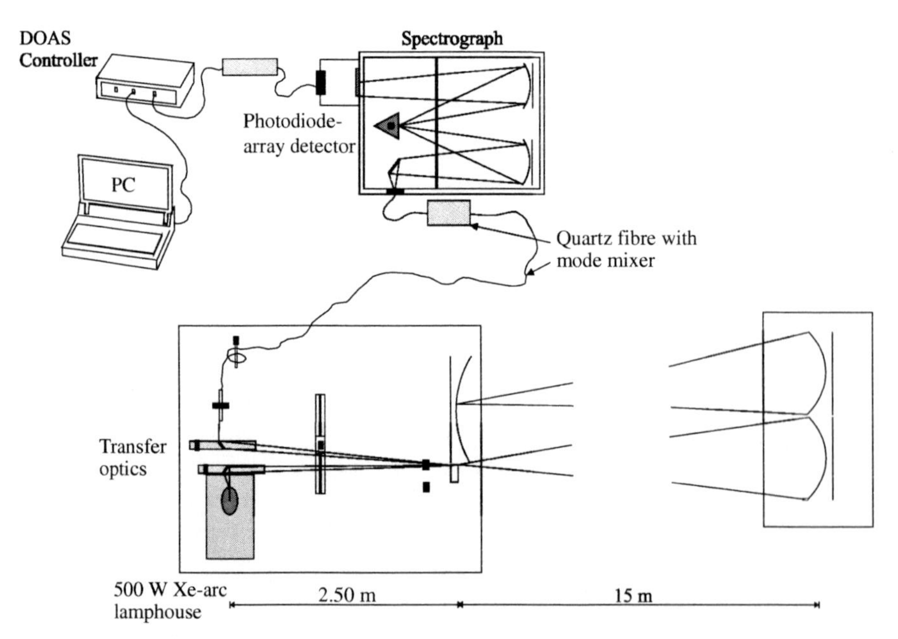

Fig. 7.50. Schematic optical setup of the 'White' multi-pass system with a base path of 15 m, as used during a field campaign in Pabstthum/Germany

7.10 Passive DOAS Systems

Passive DOAS systems measure atmospheric trace gases by using extraterrestrial light sources, e.g. sunlight or moonlight. As described in detail in Chaps. 9 and 11, radiation from these sources traverses the entire atmosphere, and thus the measured 'apparent column density' (ACD) or SCD always contains contributions from all layers of the atmosphere. However, suitable choices of the observation geometry can enhance the contributions of certain layers compared to others. In this way, stratospheric trace gases can be measured, as well as trace gases close to the ground. In fact, to some extent atmospheric trace gas profiles can be derived from passive measurements made with instruments located at the ground. Since a relatively large aperture angle of the telescope can be tolerated (on the order of 0.1–1° instead of 10^{-3}–$10^{-2\circ}$), short focal length are sufficient. Consequently, much smaller telescopes (aperture 25–50 mm) compared to typical active instruments are usually sufficient. Depending on the wavelength and sensitivity requirements, a wide variety of detector designs is used, ranging from non-dispersive semiconductor or photomultiplier detectors, to spectrograph – photodiode array combinations. A number of spectrometer designs are in use, as outlined in Sect. 7.5.

7.10.1 Direct Sun/Moon Setup

Measurements of direct sun or moon light provide information on the trace gas concentration integrated along the light path. The instruments typically consist of a spectrometer-detector combination coupled to a telescope. An optical and mechanical setup is needed to follow the apparent movement of these celestial bodies in the sky. These solar trackers can be quite complex, depending on their pointing accuracy. Care must be taken to reduce the solar light intensity in order to avoid damaging the instrument.

7.10.2 Zenith Scattered Light DOAS

A spectrograph directed to the zenith (Fig. 7.51) observes the optical density of trace gas absorption bands indirectly via the scattering process of the sunlight in different layers of the atmosphere (assuming that the instrument located is north or south of the respective tropic so that it never receives direct sunlight).

DOAS of scattered rather than direct sunlight has been applied by many authors to the determination of vertical column densities (VCDs) of O_3, NO_2 (e.g. Brewer et al., 1973; Noxon, 1975; Pommereau, 1982; Mount et al., 1986; Pommereau and Goutail, 1987; Wahner et al., 1990; Johnston et al., 1992), NO_3 (Sanders et al., 1987; Solomon et al., 1989), OClO (Solomon et al., 1987b; Schiller et al., 1990; Perner et al., 1991; Kreher et al., 1996; Otten et al., 1998),

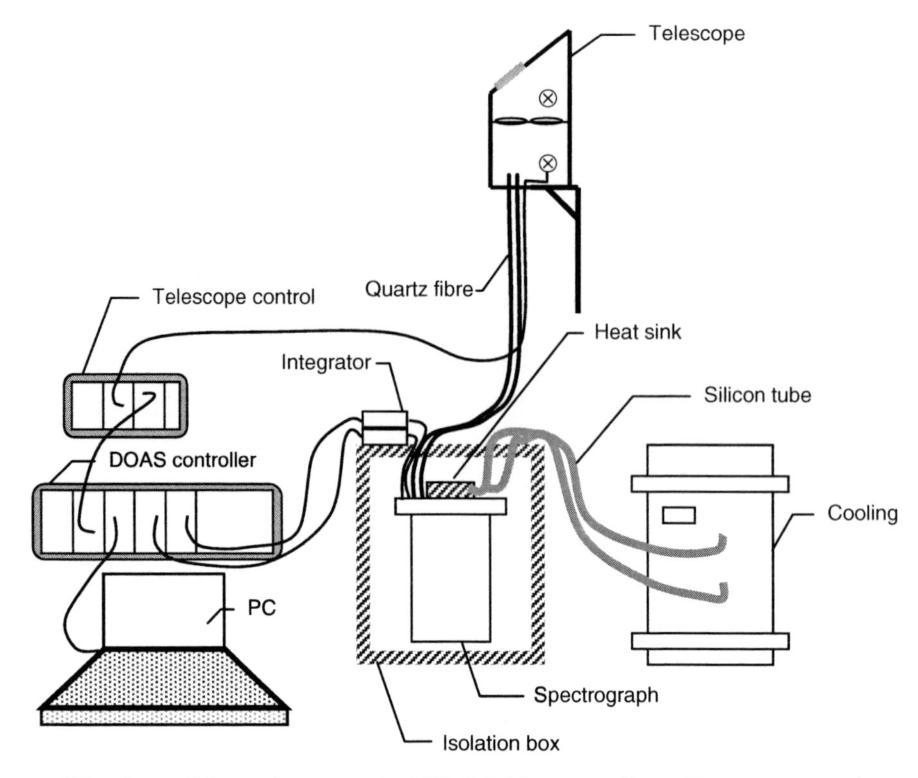

Fig. 7.51. Schematic setup of a ZSL-DOAS system (from Hönninger, 2002)

BrO (Solomon et al., 1989b; Kreher et al., 1997), and $(O_2)_2$ (e.g. Sarkissian et al., 1991; Erle et al., 1995). Examples can be found in Chap. 11.

There are three major advantages of this technique:

1. No adjustment of the telescope direction is required.
2. Stratospheric species can be detected by ground-based instrumentation, even in the presence of (tropospheric) cloud cover.
3. Tropospheric species can be measured, even if they are distributed over the entire troposphere (see off-axis measurements, Sect. 7.10.3).

These advantages come at the expense of relatively difficult calculations of the effective light path length. In addition, several effects of light scattering in the atmosphere have to be carefully compensated for, as outlined in Chap. 9.

Figure 7.51 shows the schematic setup of a ZSL-DOAS system. In this example, two telescopes, each coupled to its own spectrometer, were used to cover different wavelength ranges. Quartz fibres conduct the light from the telescopes to the spectrometers. The quartz fibres also remove the polarisation sensitivity of the instrument.

7.10.3 Off-axis, MAX-DOAS Instruments

A recently developed variant of the ZSL-DOAS technique is off-axis spectroscopy. In off-axis spectroscopy, the instrument is pointed away from the zenith by an 'off axis' angle α. There are several azimuth configurations possible. Frequently an observation direction away from the sun is chosen (e.g. Hönninger et al., 2004a). Advantages of this arrangement are better sensitivity at very large SZAs (>95°) and better sensitivity to trace gases close to the instrument. At an instrument location at the ground, this means better sensitivity to trace species in the boundary layer and lower troposphere. However, as detailed in Chap. 9, the radiation transport calculations for off-axis viewing geometries are more complex than in the case of ZSL-DOAS.

When making a series of off-axis observations at different observation angles $\alpha_1 \ldots \alpha_n$, the growth of the observed trace gas slant column as a function of increasing α gives information about the vertical distribution of the trace gas in the vicinity of the instrument (see Chap. 9). This approach became known as MAX-DOAS (e.g. Heismann, 1996; Hönninger, 2002; Hönninger and Platt, 2002; Leser et al., 2003; Wagner et al., 2004; Heckel et al., 2005) and has since been used in a large and growing number of applications. Some examples are given in Chap. 11.

MAX-DOAS systems are employed for the measurement of atmospheric trace gases by several authors using different technical approaches. Sequential observation at different elevation angles is shown in Fig. 7.52. This approach has the advantage of a relatively simple setup, requiring only one spectrometer and a mechanism for pointing the telescope (or the entire spectrometer–telescope assembly) in different directions. It has therefore been in several studies (Sanders et al., 1993; Arpag et al., 1994; Miller et al., 1997; Kaiser, 1997; Hönninger and Platt, 2002; Leser et al., 2003; Bobrowski et al., 2003; Löwe et al., 2002; Heckel et al., 2005; Bobrowski, 2005). One disadvantage is that the measurements are not simultaneous. Typically, a complete cycle encompassing several observation directions may take several minutes (e.g. Hönninger and Platt, 2002). This can be a problem during periods of rapidly varying radiation transport conditions in the atmosphere, e.g. varying cloud cover, aerosol load, or during sunrise/sunset. In addition, the instrument contains mechanically moving parts, which may be a disadvantage in adverse environments and during long-term operation at remote sites.

The simultaneous observation at different elevation angles solves this problem. However, one disadvantage is the increased instrumental requirement, since a spectrometer and a telescope for each of the observation directions is needed. However, recent development of very compact, low-cost spectrometer/detector combinations, as used in the "Mini-MAX-DOAS" instrument, shown in Fig. 7.53, can be of help here. Alternatively, some simplification may come from using 'two-dimensional' spectrometry, where a single spectrometer is equipped with a two-dimensional (CCD) detector, where one dimension is

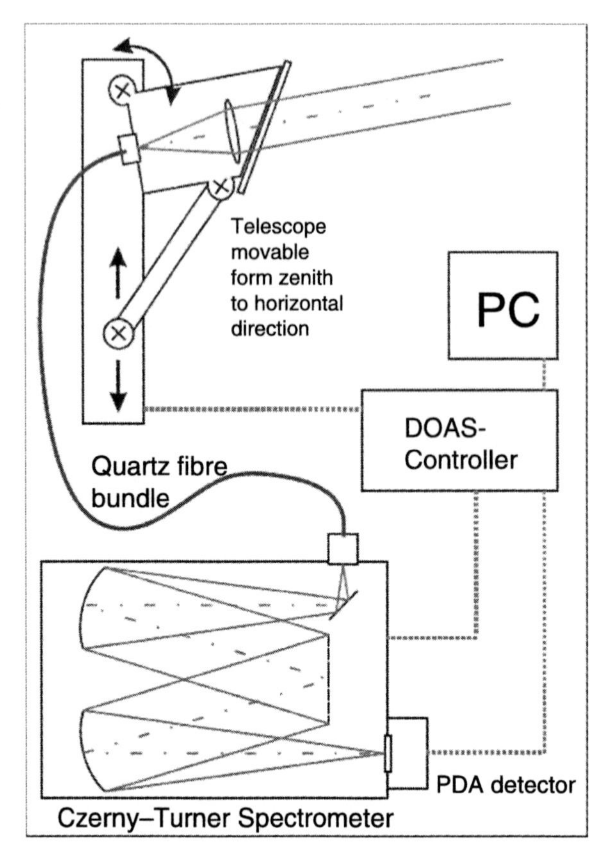

Fig. 7.52. Schematic setup of a MAX-DOAS system. A small telescope coupled to a spectrometer can be pointed to any elevation between horizon and zenith (from Hönninger, 2002)

used for dispersion, and the other for spatial resolution, i.e. different viewing angles, as outlined in Fig. 7.31 (Heismann, 1996; Wagner et al., 2002b; Wang et al., 2003, 2005; Heue et al., 2005).

A more severe problem lies in the fact that the different instruments have to be ratioed against each other in order to eliminate the solar Fraunhofer structure (see Chap. 9). This proves extremely difficult if different instruments with different instrument functions are involved (e.g. Bossmeyer 2002). Although the situation can be somewhat improved by 'cross convolution' (i.e. convoluting each measured spectrum with the instrumental line shape of the respective spectrum it is to be compared with), it appears that, with present instrumentation, the highest sensitivity is reached with sequential observation.

A solution lies in the combination of techniques (1) and (2), i.e. the use of multiple spectrometers (one per observation direction) and moving telescopes (or spectrometer–telescope assemblies). While this approach appears

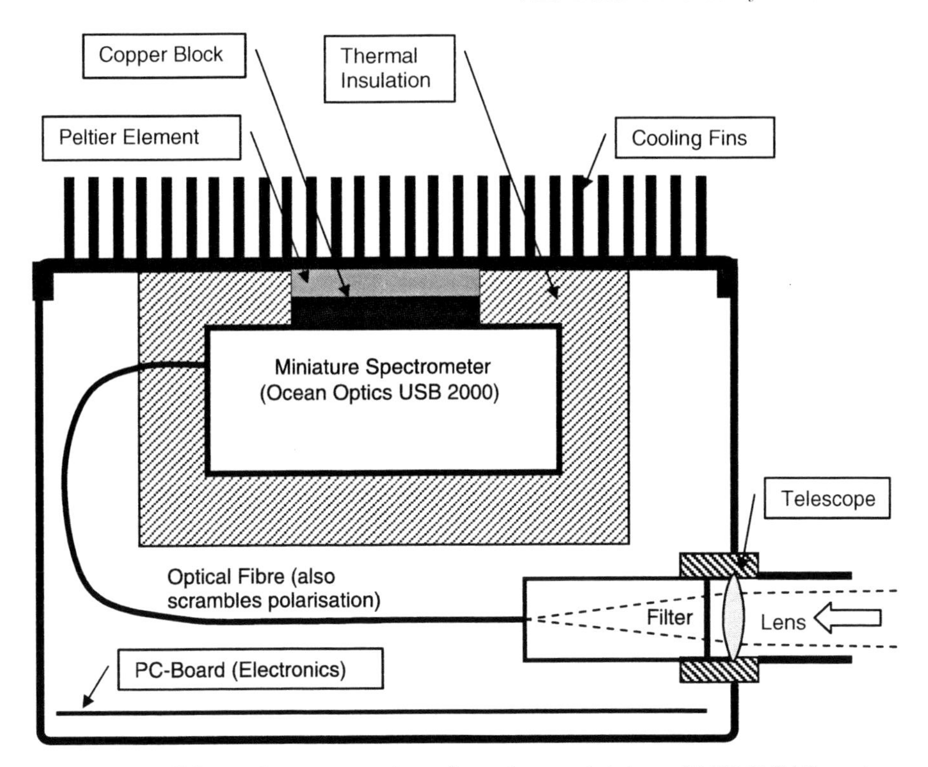

Fig. 7.53. Schematic cross-section through a miniature MAX-DOAS system (Bobrowski, 2005). The entire instrument, including the telescope, can be pointed to any elevation between horizon and zenith

to combine the disadvantages of the two above setups, it also combines their advantages. If each individual instrument sequentially observes, at all elevation angles, and the observation are phased in such a way that at any given time one instrument observes each observation direction, then not only simultaneous observation at each angle is achieved, but also each instrument regularly observes the zenith to record a Fraunhofer spectrum for reference. Instruments based on this approach have been used on the German research vessel Polarstern (see Leser et al., 2003; Bossmeyer, 2002).

7.10.4 Imaging DOAS (I-DOAS) Instruments

Imaging spectroscopy is characterised by acquiring an absorption spectrum on each spatial pixel in a two-dimensional array. The additional, spectral 'dimension' contains information about wavelength-dependent radiation intensities. Thus, the measurement results in a three-dimensional data product with two spatial dimensions and one wavelength dimension, which could be called a 'super colour' or 'hyperspectral' image, in contrast to ordinary colour images containing intensity information of only three wavelength ranges.

There are two commonly used techniques to obtain this information, distinguished by the temporal sequence of acquiring data of the three dimensions. 'Whiskbroom' imaging sensors acquire the spectrum of one spatial pixel at a time. Hence, each of the two spatial dimensions needs to be scanned consecutively. Extending non-imaging sensors with a mechanical (two axis) scanning mechanism leads to this type of instrument. While preserving the major advantage of relatively simple design of the diffractive optics and photon detector elements, the disadvantage of the whiskbroom scanning principle is a linear dependence of total measurement time on the total number of spatial pixels, i.e. the time increases proportional to the square of the desired (linear) image resolution. 'Pushbroom' sensors simultaneously image one spatial direction (vertical in Fig. 7.54) and spectrally disperse this light in order to obtain a spectrum of each pixel of the investigated object. Doing this simultaneously requires a two-dimensional detector. Since scanning is needed only in the one spatial direction (horizontal in Fig. 7.54), the total measurement time is greatly reduced (Lohberger et al., 2004).

Sunlight, which is scattered by atmospheric molecules and particles in the viewing direction of the instrument, serves as the light source. Radiative transfer is characterised by typically small distances between the instrument and the object of investigation, compared to scattering lengths by atmospheric Rayleigh and Mie scattering. Typical extinctions due to Rayleigh and Mie scattering (at e.g. 400 nm) are $4.6 \times 10^{-7}\,\mathrm{cm}^{-1}$ and $4 \times 10^{-7}\,\mathrm{cm}^{-1}$, which result in scattering lengths of 22 km and 25 km, respectively (see Sect. 4.2) In contrast, the typical range of gas plumes being visualised is a few hundred

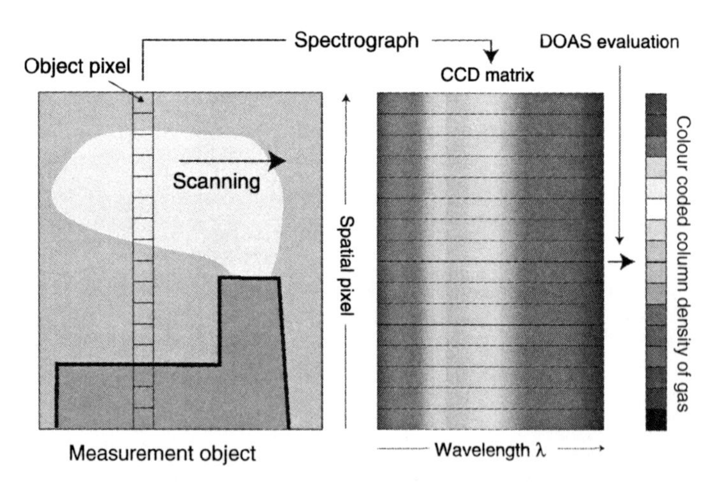

Fig. 7.54. Measurement principle of an I-DOAS instrument. The instrument utilises pushbroom imaging to simultaneously acquire spectra of one spatial direction (i.e. vertical). Spectral information along the second spatial dimension (horizontal) is obtained by scanning the scene using a scanning mirror. The spectra are evaluated using the DOAS technique (from Lohberger et al., 2004)

metres. In many applications, the contribution of photons that are scattered into the instrument without passing the gas cloud are therefore negligible.

7.10.5 Aircraft-based Experiments

DOAS instruments have been used on aircrafts in many applications, primarily using ZSL geometries (Wahner et al., 1989a,b, 1990a,b; Schiller et al., 1990; Pfeilsticker and Platt, 1994; Pfeilsticker et al., 1997a). Advantages of this platform are mobility, and the ability to make measurements at many places in a short period of time and over inaccessible places. In addition, the viewing geometry from an elevated platform provides some altitude information (see Chap. 11). The recently developed variants of the ZSL-DOAS technique, the off-axis and MAX-DOAS techniques are also being used for aircraft measurements (e.g. Wang et al., 2003, 2005; Heue et al., 2004a,b).

As outlined in Fig. 7.55, there are usually a large number of viewing directions. Simultaneous measurements are of great advantage here since, on a rapidly moving platform, the time for an individual measurement is very short if a reasonable spatial resolution is desired. In a recent implementation

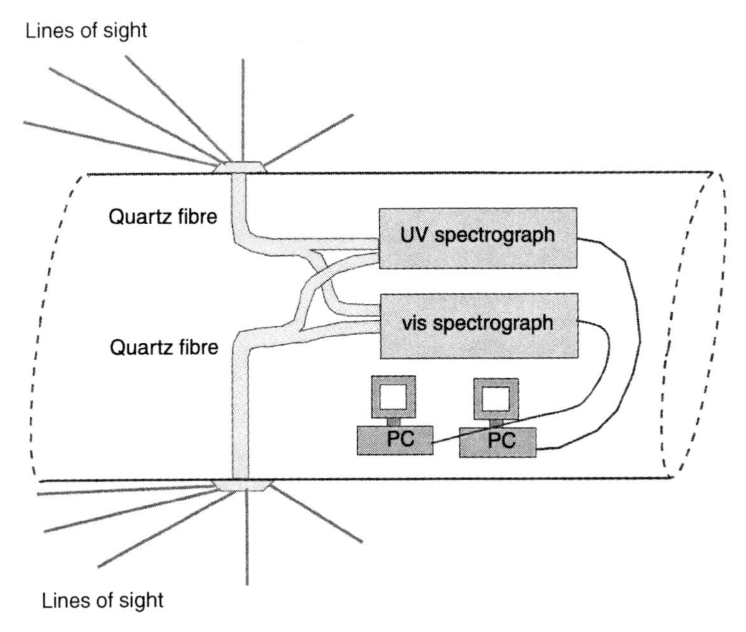

Fig. 7.55. Schematic setup of an airborne MAX-DOAS system (AMAX-DOAS). The scattered sunlight is observed using a set of small telescopes mounted at the upper and lower side of the aircraft body, respectively, and pointing to several elevation angles between zenith and nadir. The light is transmitted via quartz fibre bundles to two spectrographs, where it is analysed. The spectra are then saved on a PC (from Heue et al., 2005)

of the principle (Heue et al., 2005; Bruns et al., 2004), two groups of telescopes were mounted at the top and bottom of the aircraft body. Besides nadir- and zenith-looking telescopes, several telescopes collecting radiation at small angles (e.g. $\pm 2^\circ$) above and below the aircraft were installed. The latter telescopes gave good resolution at atmospheric layers above and below the aircraft, see Chap. 11.

While it is possible to employ a separate spectrometer for each viewing direction, an approach using an imaging spectrometer, as described in Sects. 7.5.2 and 7.10.3, can be of advantage. The fibres from the different telescopes would then be mounted along the entrance slit, as shown in Fig. 7.31. Design goals of aircraft instruments are low weight, resistance to vibration, and unattended operation. Particular problems for aircraft instruments include provisions for continued operation during short-time power interruptions and compensation of the effect of ambient pressure change.

7.10.6 Balloon-borne Instruments

An interesting application of this technique is the determination of vertical trace gas concentration profiles by taking a succession of spectra from a balloon-borne spectrometer. From these spectra, trace gas vertical profiles can be derived in two ways:

1. From a series of spectra recorded at different altitudes during ascent, a series of the column densities $S(z)$ above the balloon altitude z can be derived. After conversion to vertical column densities and differentiation of the concentration profile, $c(z)$ can be derived (e.g. Pommereau and Piquard, 1994a; Pundt et al., 1996; Camy-Peyret et al., 1995, 1999; Ferlemann et al., 1998; Fitzenberger et al., 2000; Bösch, 2002; Bösch et al., 2003).
2. From a balloon floating at high altitude, a series of spectra is taken during sunset (or sunrise). From the derived trace gas column densities $S(\vartheta)$, as a function of SZA and ϑ, the concentration profile $c(z)$ can be derived by an 'onion peeling' approach (Weidner et al., 2005; Dorf et al., 2005).

The required hardware is shown in Figs. 7.56 and 7.57. Figure 7.56 illustrates the mechanical setup of a balloon-borne direct-sunlight DOAS system, which is housed together with an FT-IR instrument on a three-axis stabilised balloon gondola (Camy-Peyret et al., 1995). A moveable mirror directs the sunlight into the spectrometers. Figure 7.57 depicts a cross-section of the pressurised dual spectrometer module.

7.10.7 Satellite Instruments

Given the successful application of differential absorption techniques on the ground and on board aircrafts, satellite-based systems appeared attractive.

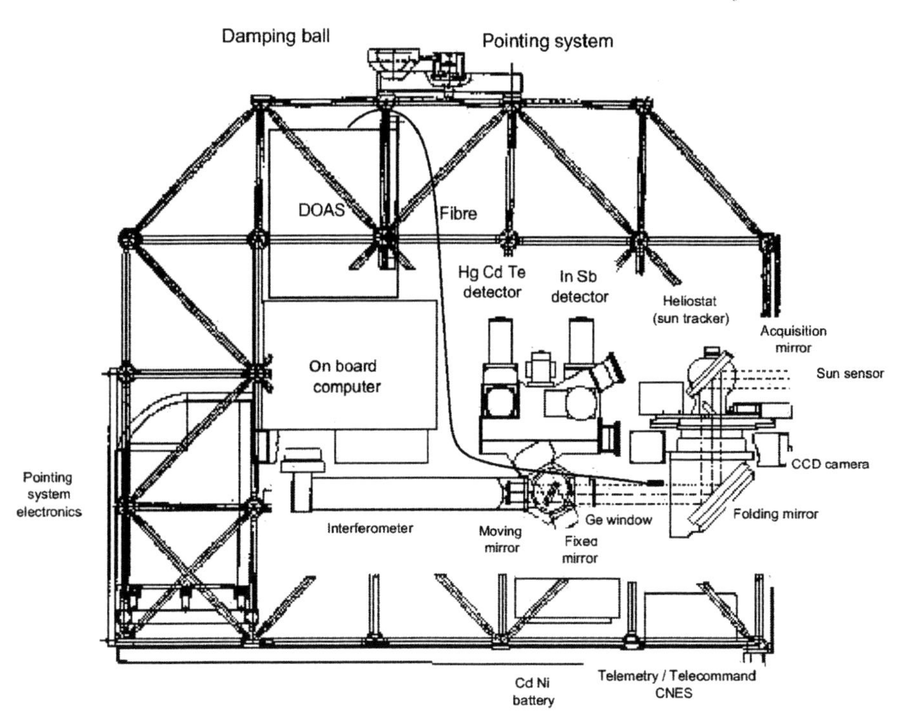

Fig. 7.56. Setup of a balloon-borne direct-sunlight DOAS system. The DOAS instrument is housed together with an FT-IR instrument on a three-axis stabilised balloon gondola (from Camy-Peyret et al., 1995)

Earlier instruments such as the Solar Backscatter UV (SBUV) (Dave and Mateer, 1967) measured the ozone column density by taking the ratios of a few wavelength pairs, essentially bringing the concept of the Dobson spectrometer (Dobson and Harrison, 1926; Dobson, 1968) to space. However, with the launch of the Global Ozone Monitoring Experiment (GOME) on the ESA satellite ERS-2 (GOME Users Manual, 1995) in 1995, DOAS principles have been applied to satellite instruments with great success. Since GOME, several further DOAS-type instruments, in particular the SCanning Imaging Absorption spectroMeter for Atmospheric CHartographY (SCIAMACHY) (e.g. Burrows et al., 1991; Noël et al., 1999) on the ESA ENVIronmental SATellite (ENVISAT), OSIRIS on ODIN (Murtagh et al., 2002), the ozone monitoring instrument (OMI) on the NASA Earth Observing Satellite Aura (EOS-Aura), and others (see Table 7.11) have been launched. Three viewing geometries are utilised for DOAS measurements (see Chap. 11):

1. Nadir view: Looking down from space towards the nadir direction, the sunlight reflected from the earth's surface or atmosphere (called the 'earthshine' in analogy to 'moonshine') can be utilised to obtain information

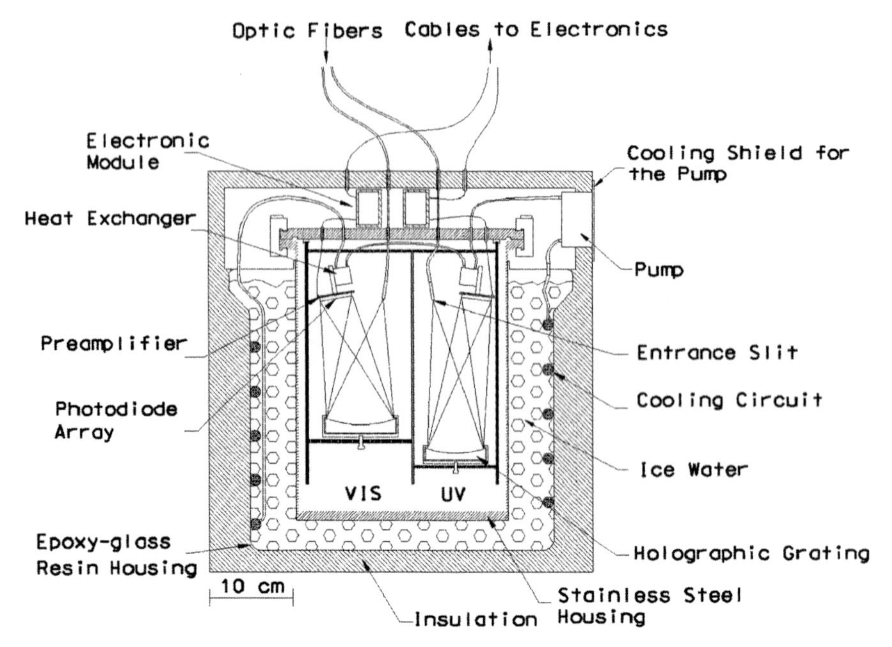

Fig. 7.57. Spectrometer of a balloon-borne direct-sunlight DOAS system (Bösch, 2002). Two spectrometers (one for the visible (VIS) and the other for the ultraviolet (UV) spectral range) receive direct sunlight through a heliostat (see Fig. 7.56) and quartz fibres. The spectrometers are sealed in a stainless steel housing and thermostated to 0°C by a ice-water package inside a Styrofoam insulated housing (from Bösch, 2002)

on atmospheric trace gas column densities. Examples of satellite instruments using this geometry include GOME, ILAS, SCIAMACHY, ODIN, and OMI, as shown in Table 7.11.

2. Limb view: In this geometry, light scattered from the earth's terminator is analysed. By observing the limb at different tangent altitudes, vertical trace gas concentration profiles can be inferred.
3. Occultation: By observing the light of the rising or setting sun, moon, or stars through the atmosphere, vertical trace gas concentration profiles can be deduced.

In addition, there are several types of orbits which are particularly suited for probing the atmospheric composition from satellite:

1. Low earth orbit (LEO), where the satellite circles earth at an altitude around 800 km. Particularly advantageous are sun-synchronous polar orbits, from where the satellite instrument can observe the whole earth each day or within a few days. In the latter case, the time for one orbit can be adjusted such that the instrument will see a specific point on earth at

Table 7.11. DOAS-type satellite instruments and some characteristics

Instrument/platform	Launch - operation until	No. of spec-trome-ters	Spectral ranges nm	No. of pixles	Spectra resolution nm	Viewing geometries
Global ozone monitoring experiment (GOME)/ ERS-2	1995–today	4	237–316	1024	0.22	Nadir
			311–405	1024	0.24	
			405–611	1024	0.29	
			595–793	1024	0.33	
OSIRIS/ODIN	20 February 2001–today	1	280–800	1353 × 286	1.0	Limb
ILAS-I/ADEOS	1 November 1996–30 June 1997	1	753–784	1024	0.1	Solar Occultation
SCIAMACHY/ENVISAT	2002–today	8	240–314	1024	0.24	Nadir, Limb, Solar Occulta-tion, Lunar Occulation
			309–405	1024	0.26	
			394–620	1024	0.44	
			604–805	1024	0.48	
			785–1050	1024	0.54	
			1000–1750	1024[a]	1.48	
			1940–2040	1024[a]	0.22	

(continued)

Table 7.11. continued

Instrument/platform	Launch - operation until	No. of spec-trome-ters	Spectral ranges nm	No. of pixles	Spectra resolution nm	Viewing geometries
			2265–2380	1024[a]	0.26	
Ozone Monitoring Instrument (OMI)/EOS-AURA	15 July 2004–today	1	350–500	780 × 576	0.45–1.0	Nadir
GOME-II/METOP	2006	4	240–315	1024	0.24–0.29	Nadir
			311–403	1024	0.26–0.28	
			401–600	1024	0.44–0.53	
			590–790	1024	0.44–0.53	

[a]InGaAs detectors

essentially the same local time (e.g. 10:30 in the morning in the case of ERS-2)

2. Geostationary orbits where the satellite circles earth in synchronicity with the rotation of earth at an altitude of 36000 km.

An example of the LEO is GOME on board the second European Research Satellite (ERS-2). A schematic outline of the instrument is shown in Fig. 7.58. It consists of a scanning mirror, a set of four spectrometers, and thermal and electronic sub-systems (Burrows et al., 1991, 1999). Light backscattered from earth is collected by the scan mirror, then focused on the entrance slit of the spectrometer by an off-axis parabolic telescope. After being collimated, the light passes through a pre-dispersing prism, which produces an intermediate spectrum within the instrument. The pre-dispersing prism is arranged in such a way that internal reflection results in a polarised beam which is directed towards three broadband detectors monitoring the polarisation state of the light (the polarisation monitoring device, PMD). The PMD observes radiation from the atmosphere in the wavelength ranges 300–400 nm, 400–600 nm, and 600–800 nm. The main beam from the pre-dispersing prism is then directed towards the four individual spectrometer gratings of GOME. Each

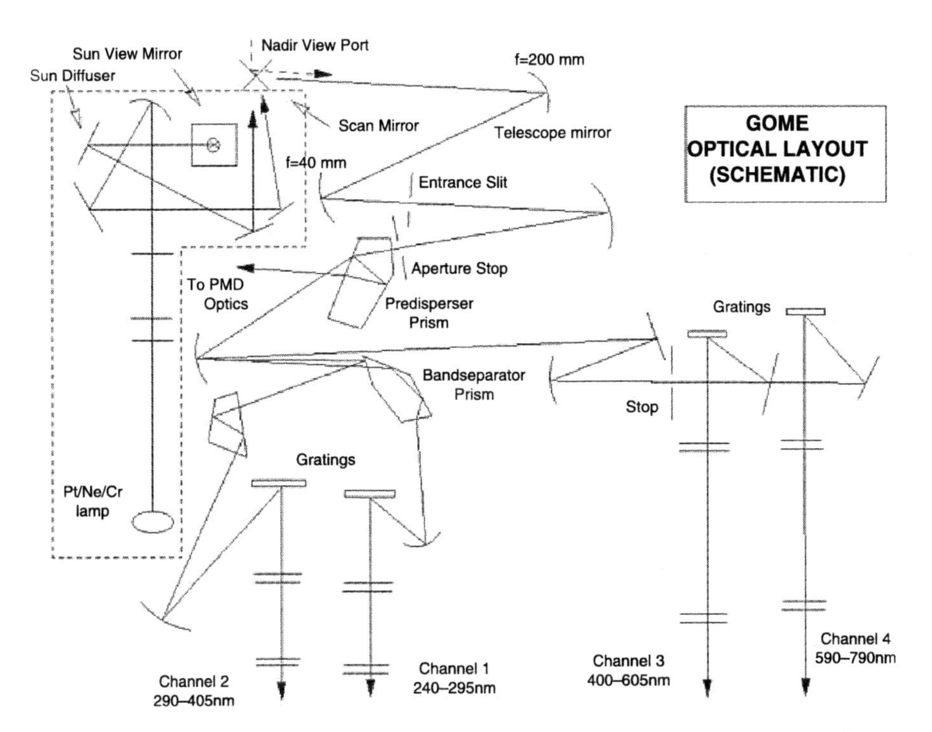

Fig. 7.58. Schematic setup of a satellite DOAS system: GOME on ERS-2 (from Burrows et al., 1999)

spectrometer uses refractive optics and a 1024-element diode array detector. Thus the entire spectral range between 232 and 793 nm can be observed at a spectral resolution of about 0.2 nm and 0.33 nm, below and above 400 nm, respectively.

The GOME instrument is optimised for the collection of the upwelling radiation from earth (nadir view) directly by the scan mirror. The direct, extraterrestrial solar irradiance (the Fraunhofer reference spectrum, FRS) is measured by GOME over a diffuser plate to reduce its intensity prior to being reflected by the scan mirror into the instrument. The spectra and other signals recorded by GOME are transmitted to the ESA ground station at Kiruna. These data are processed for ESA by the GOME data processor at the DLR-PAC (German Atmospheric and Space Research Processing and Archiving Centre). The resulting 'level 1' products, essentially calibrated radiances, and level 2 total ozone column data product are subsequently distributed to the scientific community. ERS-2 was launched on 20 April 1995 into a nearly polar, sun synchronous orbit, having an equator crossing time of 10:30 am in a descending node. The scan strategy of GOME yields complete global coverage at the equator in 3 days. The swath width of GOME is 960 km, divided into three pixels of $320 \times 40\,\mathrm{km}^2$ for the array detector ($20 \times 40\,\mathrm{km}^2$ for the PMDs, which are read out 16 times faster than the arrays; Burrows et al., 1999, and references therein).

Besides ozone, GOME can measure a wide variety of atmospheric trace gases, including bromine monoxide (BrO) (e.g. Hegels et al., 1998; Richter et al., 1998; Wagner and Platt, 1998), chlorine dioxide (OClO) (Burrows et al., 1998; Burrows et al., 1999; Wagner et al., 2001), formaldehyde (HCHO) (Burrows et al., 1999; Chance et al., 2000) sulphur dioxide (SO_2) (Eisinger and Burrows, 1998, and references therein), nitrogen dioxide (NO_2) (Burrows et al., 1998; Leue et al., 2001), and water vapour (H_2O) (Noël et al., 1999; Maurellis et al., 2000b). In addition, ozone and aerosol profiles may also be retrieved (e.g. Hoogen et al., 1999, and references therein).

7.11 Light Utilisation in a Long-path Spectrometer

The spectral noise, and thus detection limit and accuracy of DOAS instruments, usually depends on the amount of light reaching the detector (see Sect. 7.6). Thus, it is important to maximise the fraction of light provided by the light source that reaches the detector. This section describes light throughput of a typical long-path spectrometer. For simplicity, we take the dimension of a single-path DOAS system, as sketched in Fig. 7.59. The results are also valid for folded-path systems (see Sect. 7.9.3).

The fraction x_1 of light emitted from the radiating surface of the light source, e.g. from the filament of an incandescent lamp or arc of a Xe high-pressure lamp (see Sect. 7.3.1), transformed into a light beam, can be approximated as follows:

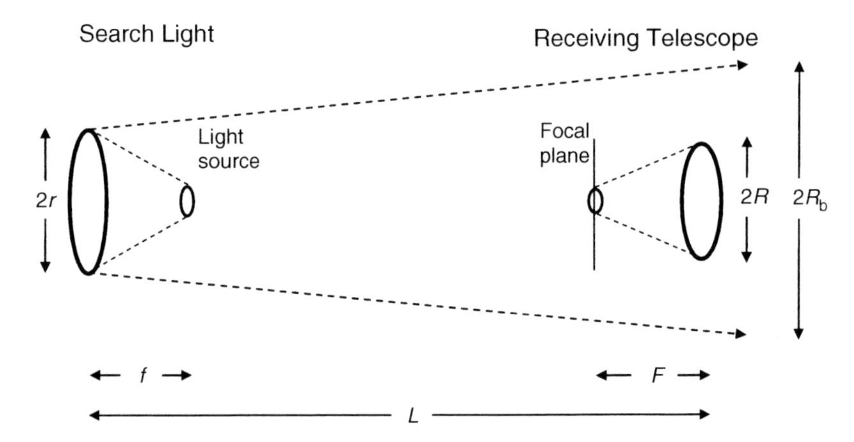

Fig. 7.59. Optical sketch of a single-path active DOAS system. The search light mirror of the transmitting system has radius r and focal length f, the corresponding quantities for the mirror of the receiving telescope are R and F

$$x_1 \approx \frac{r^2\pi}{4\pi f^2} = \frac{1}{4} \cdot \left(\frac{r}{f}\right)^2, \tag{7.40}$$

where r denotes the radius of the light-source (search light) mirror and f its focal length (see Fig. 7.59). Assuming the radiating surface of the lamp to be circular with the radius r_l (and thus the surface $r_l^2\pi$), the divergence of the beam (half angle) is given by:

$$\alpha \approx \frac{r_l}{f}, \tag{7.41}$$

and the radius R_b of the light beam at the receiving telescope located at the distance L from the search light (assuming $R_b >> r$ and setting $\alpha \simeq \tan\alpha$):

$$R_b \approx \alpha \cdot L = \frac{r_l \cdot L}{f}. \tag{7.42}$$

We obtain the fraction x_2 of the emitted light beam actually collected by the telescope:

$$x_2 = \left(\frac{R}{R_b}\right)^2 = \left(\frac{R \cdot f}{r_l \cdot L}\right)^2. \tag{7.43}$$

Combining (7.43) with (7.40) gives the fraction x_e of the total lamp output received by the telescope:

$$x_e = x_1 \cdot x_2 = \frac{1}{4} \cdot \left(\frac{R \cdot r}{r_l \cdot L}\right)^2. \tag{7.44}$$

Note that for a given radius r, x_e is independent of the focal length f of the search light mirror and thus its aperture. On the other hand, x_e is proportional to the product of the areas of the two mirrors and inversely proportional to the radiating area of the lamp, and thus its luminous intensity (radiated energy per unit of arc area). The important quantity is the illumination intensity D_b in the focal plane of the receiving telescope (i.e. the brightness of the focal spot). It is given by:

$$D_b = \frac{P_l \cdot x_e}{(r \cdot F/L)^2 \pi}, \tag{7.45}$$

where P_l denotes the lamp output power and $(r\,F/L)^2\pi = R_f^2\pi$ the area of the focal spot (see below). Substituting x_e from (7.44) yields:

$$D_b = \frac{P_l}{4\pi} \cdot \left(\frac{R}{r_l \cdot F}\right)^2 . \tag{7.46}$$

The illumination intensity of the focal spot neither depends on the length of the light path L (if light extinction in the atmosphere is neglected) nor on the properties of the search light optics. Of course the size of the focal spot (i.e. its radius R_f) will be proportional to the size (r) of the search light mirror, and inversely proportional to the light path length:

$$R_f = \frac{r \cdot F}{L} . \tag{7.47}$$

From (7.46), it appears as if the light throughput of the system could be maximised by choosing the largest possible aperture $A_r = 2R/F$ of the receiving telescope. However, the aperture A_r, must not be larger than the aperture A_s of the spectrometer (if $A_r > A_s$, a portion of the light received by the telescope will simply not be used). Thus, for a given value of A_s, the brightness of the focal spot has a maximum value given by substituting $R = A_s F/2$ into (7.46):

$$D_b = \frac{P_l}{16\pi} \cdot (A_s/r_l)^2 . \tag{7.48}$$

The result is, that for a long-path spectrometer the size of the optical system does not matter, provided the aperture of the receiving system matches that of the spectrometer and the dimensions of the optical components exceed certain 'minimum sizes' given by (7.47).

Depending on the relative sizes of the focus R_f and the entrance slit ($w \cdot h$), three regimes can be distinguished (see Fig. 7.60):

1. w, $h < R_f$: The amount of light entering the spectrometer will be independent of L (except for extinction). This is due to the independence of the illumination intensity from L stated in (7.46).
2. $w < R_f < h$: The amount of light entering the spectrometer will vary nearly in proportion to L^{-1}.
3. $R_f < w$, h: The amount of light entering the spectrometer will vary proportional to L^{-2}

Thus, it can be concluded that the product of the size of the search light mirror and the focal length of the receiving system (the aperture of which

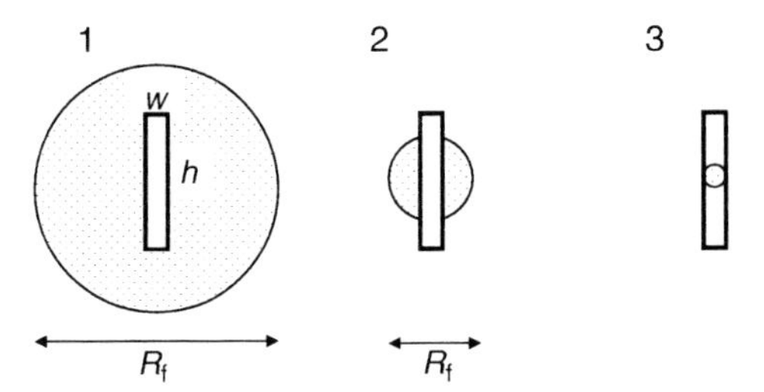

Fig. 7.60. Focusing the image of a retro-reflector or lamp on the spectrometer entrance slit there are three regimes: (1) For w, $h < Rf$, the radiation flux entering the spectrometer will be essentially independent of L (except for extinction); (2) For $w < Rf < h$ the radiation flux is approximately $\propto L^{-1}$; (3) If $Rf < w$, h, the radiation flux is approximately $\propto L^{-2}$

matches the aperture of the spectrometer used) should be chosen as sufficiently large as to allow the system to be in regime 1. Table 7.12 gives examples of the quantities derived earlier. Data are evaluated for typical DOAS systems with radii of search light and receiving telescope mirrors $r = R = 0.15$ m, focal length of the search light mirror, $f = 0.3$ m, and focal length of the telescope mirror $F = 1.5$ m. Further assuming entrance slit dimensions (width w by height h) of 0.05 by 0.2 mm, the light paths for regimes 1–3 are obtained:

Regime 1: $L <$ about 2000 m
Regime 2: $2000 < L <$ about 9000 m
Regime 3: $L >$ about 9000 m

As seen in Table 7.12, the total loss factor from lamp output to the detector is close to 10^{-12} in the case of a optomechanic scanning device with a photomultiplier detector. The total number of photons per second and spectral interval (assuming a division of the spectrum in 400 intervals) reaching the detector can be estimated as follows: A 450-W Xe-arc lamp emits about 2.5 W, corresponding to $dN_0/dt \approx 6 \times 10^{18}$ photons per second, in the form of light in the spectral range 440–460 nm (Canrad-Hanovia). Taking the above loss factor from Table 7.12, about $dN/dt \approx 6 \times 10^{6}$ photoelectrons per second and spectral interval of 0.05 nm width are registered. This figure corresponds to a shot noise limit of about $4 \times 10^{-4}\,\mathrm{s}^{-1/2}$ (1σ).

7.12 Software Controlling DOAS Instruments

The operating and evaluation software of a state-of-the-art DOAS system is a very essential part of the system; it is at least comparable in complexity to the hardware. The tasks of the DOAS operating software include:

(a) Record and store the spectra and associated information (meteorological data such as air pressure and temperature required to convert measured concentrations to mixing ratios).

Table 7.12. Light losses in a long path spectrometer system (10-km light path) using a 'slotted disk'-type optomechanical scanner

Optical element/process		Loss factor (optomechanical scanner)	Loss factor (multi-element solid state detector)
Light source into beam	x_1	0.08	0.08
Geometric losses ($\alpha = 10^{-3}$ radian, $R_b = 17$ m)	x_2	3×10^{-4}	3×10^{-4}
Atmospheric attenuation (Mie and Rayleigh scattering)		0.2	0.2
Entrance slit		1.0	1.0
Spectrograph losses		0.2	0.2
Dispersion (fraction of total intensity received by individual pixel)		5.0×10^{-3}	5.0×10^{-3}
Multiplex losses (400 channels)		2.5×10^{-3}	1.0
Detector quantum efficiency		0.2	0.4
Total loss factor	I_{tot}	1.2×10^{-12}	10^{-9}
Total number N of photoelectrons recorded per pixel and second		$\approx 10^5$	$\approx 10^8$

(b) Cycle through measurement sequences (e.g. take atmospheric spectra, background spectra, and dark current, see Sect. 7.14) and calculate the true atmospheric spectra (see Chap. 8). Make measurements in different wavelength regions, switch to different light paths or observation geometries, etc.
(c) Align optics. For instance, in long-path systems the telescope has to be precisely pointed at the light source or retro-reflector array.
(d) Evaluate spectra and display trace gas concentrations following the principles outlined in Chap. 8. This is the most complex task, and is frequently performed off-line.
(e) Perform auxiliary functions, including calculation of sun and moon positions (e.g. SZA). Support wavelength calibration and instrument function determination procedures.

Table 7.13. Software packages developed for acquiring spectra and DOAS evaluation

Designation	Operating system	Functions	Remark	Reference
MFC	DOS	Acquisition, storage, instrument operation, evaluation		Gomer et al., 1995
WINDOAS	WINDOWS	Evaluation only		Fayt et al., 1992
XDOAS	Linux	Acquisition, storage, instrument operation, evaluation	Adoption of MFC to Linux	Grassi pers. comm., 2003
DOASIS	WINDOWS	Acquisition, storage, instrument operation, evaluation	Formerly 'WINDOAS Heidelberg'	Leue et al., 1999; Kraus, 2005

Several software packets have been developed to perform some or all of the above tasks, as summarised in Table 7.13. In addition, there are several software systems which have been developed for use by individual laboratories and suppliers of commercial DOAS systems.

7.13 Optimising DOAS Instruments

A central question of any DOAS system concerns the light throughput, i.e. the question how the amount of light reaching the detector can be maximised under the given boundary conditions.

The detection limit for a particular substance can be calculated according to (6.9), if the differential absorption cross-section, the minimum detectable optical density D'_0, and the length of the light path are known. In several applications, however, the goals of maximising light throughput and light path length can contradict each other, as is detailed below.

In general, D_0 is determined by: (1) Photoelectron statistics (shot noise), which is a function of light intensity only and (2) by any other noise sources including detector noise, induced 'background absorption structures', or 'optical noise' (see Sect. 7.3). This second class of noise sources, summarised in B, is on the order of $N/S \approx 10^{-4}$ for typical DOAS systems with solid state detectors or OSDs (see Sect. 7.6). Photoelectron shot noise is proportional to $N^{-1/2}$, where N is the total number of photons recorded around the centre of the absorption line during the time interval t_{m} of the measurement. The number of photoelectrons N is related to the power P received by a detector pixel.

$$N = \frac{P \cdot t_{\mathrm{m}} \cdot \eta \cdot \lambda}{hc}, \tag{7.49}$$

where η denotes the detector quantum efficiency (at the given wavelength λ), c the speed of light, and h Planck's constant. Thus, the minimum detectable optical density D_0 will be proportional to:

$$D_0 \propto \sqrt{\frac{1}{N} + B^2}. \tag{7.50}$$

If sufficient light is available (which is rarely the case), D_0 is limited by B (Platt et al., 1979; Platt and Perner, 1983). Using N from Table 7.12, ideally an integration time on the order of minutes for optomechanical detectors and around 1 s for array detectors would be sufficient to make $N^{-1/2} \approx$ B.

While for a given minimum detectable optical density the detection limit improves proportionally to the length of the light path, the actual detection limit, will not always improve with a longer light path. This is because with a longer light path the received light intensity I(λ) tends to be lower, increasing the noise associated with the measurements of I(λ), and thus increasing the minimum detectable optical density. More details are given in Sect. 7.13.1.

7.13.1 Optimum Light Path Length in Active DOAS Systems

An important consideration in the practical application of DOAS is the length (L) of the light path chosen. The value of L has a large influence on the performance of the system:

1. The optical density due to light absorption by trace gases (at a given concentration) is directly proportional to L.
2. The intensity (I) received from the light source generally decreases with the length of the light path (the decrease dI/dL will depend on the details of the setup and L itself, as discussed below). In addition, for folded paths (DOAS systems coupled to multi-reflection cells), I decreases with increasing L.
3. The volume over which the trace gas concentration is averaged (for long path systems) is proportional to L.
4. Long light paths (folded or unfolded) make the DOAS system more vulnerable to extinction by aerosol, haze, or fog.

In the following, we make a few simple considerations which may clarify this relationship and also allow the derivation of detection limits.

For a typical DOAS system, as described in Sect. 7.11, there are no geometrical light losses of path lengths upto ~2 km. However, in addition to geometrical light losses, light attenuation due to atmospheric absorption, as well as to scattering affect the magnitude of the received light signal, is described by:

$$I_{\text{received}} = I_{\text{source}} \exp[-\mathrm{L}(\Sigma(\sigma_{\mathrm{i}}(\lambda)c_{\mathrm{i}}) + \varepsilon_{\mathrm{R}}(\lambda) + \varepsilon_{\mathrm{M}}(\lambda))], \qquad (7.51)$$

or simply:

$$I_{\text{recieved}} = I_{\text{source}} \cdot \exp\left(-\frac{L}{L_0}\right), \qquad (7.52)$$

where the absorption length L_0 reflects the combined effects of broadband atmospheric absorption, Mie and Rayleigh scattering, as given in the second term of (5.7), i.e.:

$$L_0 = \frac{1}{\sum_i (\sigma_{\mathrm{i}0}(\lambda)\, c_{\mathrm{i}}) + \varepsilon_{\mathrm{R}}(\lambda) + \varepsilon_{\mathrm{M}}(\lambda)}. \qquad (7.53)$$

The noise level of a photodetector signal is essentially dependent on the number of photons received and thus is proportional to the square root of the intensity. The signal-to-noise ratio of a DOAS system as a function of the light path length L can be expressed as:

$$\frac{D'}{N} \approx L \cdot G(\lambda) \cdot \exp\left(-\frac{L}{2L_0}\right), \qquad (7.54)$$

where $G(\lambda)$ is constant (regime 1), $G(\lambda) = L^{-1}$ (regime 2), or $G(\mathrm{i}) = L^{-2}$ (regime 3), respectively, as defined in Sect. 7.11 (see also Fig. 7.60). Relationship (7.54) yields optimum D'/N ratios for light path lengths of $L = 2L_0$ (regime 1), $L = L_0$ (regime 2), or a boundary maximum at the shortest light path (regime 3). While there are no lower limits for L_0 in the atmosphere (consider, for example, fog), the upper limits are given by Rayleigh scattering, and the scattering by atmospheric background aerosol, which indicate optimum light path lengths ($2L_0$) in excess of 10 km for wavelengths above 300 nm (see Table 4.6).

Another point to consider is the width G of the absorption line (or band). At a given spectral intensity of the light source $N(\lambda)$, the light intensity around the line centre will be $N(\lambda)d\lambda \approx N(\lambda)\,\Gamma$ (in photons per nanometre and second). Thus D_0 becomes:

$$D_0 \propto \sqrt{\frac{1}{N(\lambda) \cdot \Gamma \cdot t} + B^2}. \qquad (7.55)$$

From (7.55), it becomes clear that the detection of a spectrally narrow absorption line requires a proportionally higher spectral intensity of the light source (see Sect. 7.9.2).

7.13.2 Optimum Spectral Resolution

For a given (grating) spectrometer, the light throughput varies in proportion the square of the spectral resolution, as shown in Sect. 7.5.4 (here expressed as the width Γ_0 of the instrument function in nm). Thus, the light intensity at the output becomes:

$$I \approx I_i \cdot {\Gamma_0}^2. \tag{7.56}$$

The 'signal' in the spectrum is the differential optical density D', which is (for a given species, wavelength, and light path) proportional to the concentration.

In the case of shot noise limitation, the minimum detectable optical density D_0 in the spectrum is inversely proportional to the square root of I, thus:

$$D_0 \sim \frac{1}{\sqrt{I}} \sim \frac{1}{\Gamma_0}. \tag{7.57}$$

In general, the signal as a function of the spectral resolution will vary proportional to the differential absorption cross-section as a function of resolution $\sigma'(\Gamma_o)$. Figure 7.61 (upper panel) shows an example for several bands of NO_2 in the visible spectral region. In principle, $\sigma'(\Gamma_0)$ can be an arbitrary function of Γ_0, which decreases with increasing Γ_0. In the example of Fig. 7.61, $\sigma'(\Gamma_0)$ can be approximated as a linear function:

$$\sigma'(\Gamma_0) \approx \sigma'_0 \, (1 - a\Gamma_0), \tag{7.58}$$

where a is a constant (in the example of Fig. 7.61; $\approx 2.4\,\mathrm{nm}^{-1}$). Thus, in this special case, the signal-to-noise ratio becomes:

$$\frac{D'}{N} = \sigma'_0 \cdot (1 - a\Gamma_0) \cdot \Gamma_0 \propto \Gamma_0 - a \cdot \Gamma_0^2. \tag{7.59}$$

The optimum D'/N is thus obtained at the resolution:

$$\frac{d}{d\Gamma_0}\frac{D'}{N} = 1 - 2a\Gamma_0 = 0 \text{ or } \Gamma_{0,\mathrm{opt}} = \frac{1}{2a} \approx 1.2\,\mathrm{nm} \ (a \approx 2.4\,\mathrm{nm}^{-1}, \text{ see Fig. 7.61}) \tag{7.60}$$

Note that, in general, the instrument function $H(\lambda)$ will not be of Gaussian shape. The spectrum of a given trace species (expressed as optical density), as seen by a particular spectrometer, is obtained by convoluting the true spectrum $D_T(\lambda)$ with the instrument function $H(\lambda)$, as described in Chap. 6.

7.13.3 Optimum Measurement Time

The number of photons recorded in a DOAS spectrum is proportional to the integration time (either of a single detector exposure or as a sum of several co-added spectra). Consequently, the noise in the spectrum (and thus the minimum detectable optical density D_0) will be reduced proportionally to the square root of the integration time, see (7.55). This will not continue

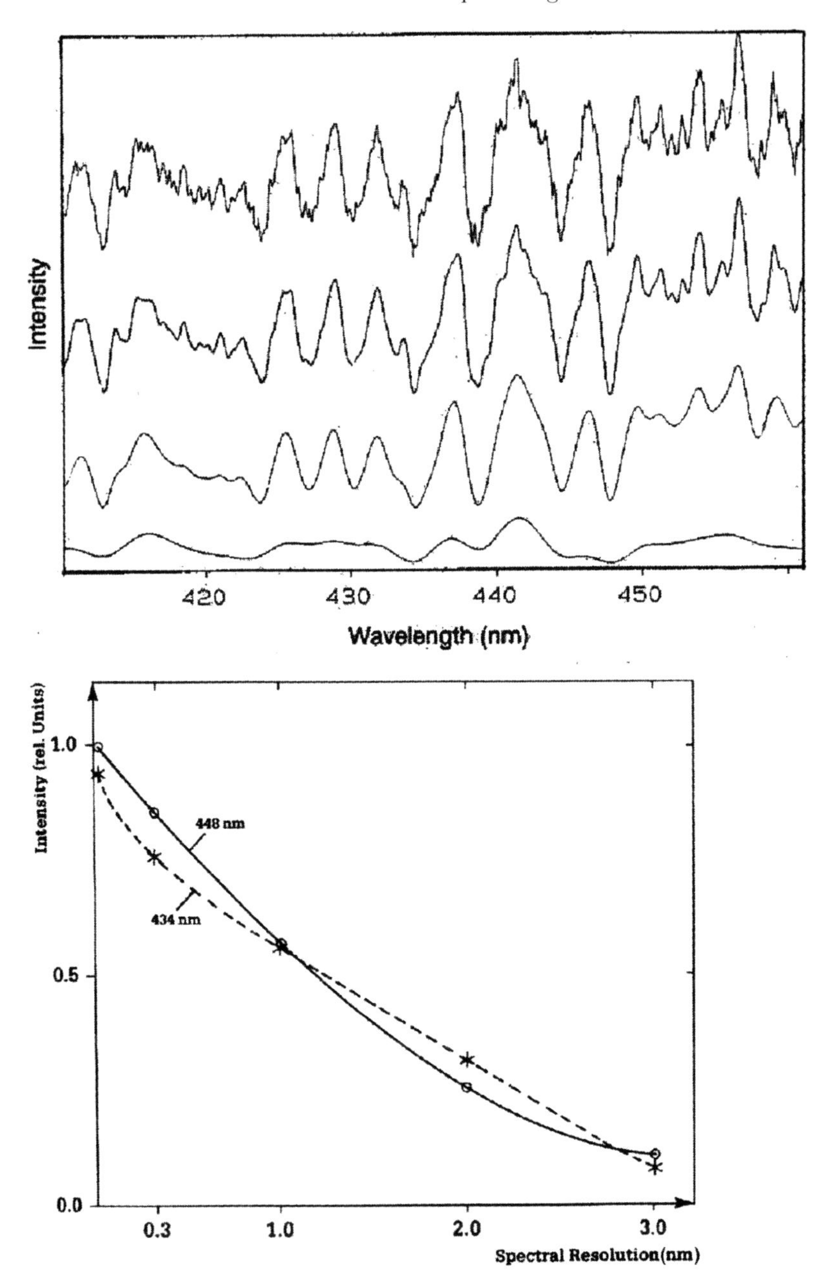

Fig. 7.61. Absorption cross-section of NO_2 as a function of spectral resolution, here expressed as the FWHM Γ_0 of the instrument function $H(\lambda)$. **Upper panel**: Absorption spectrum for different values of the spectral resolution Γ_0, from top to bottom $\Gamma_0 = 0.01, 0.1, 1.0, 3.0\,\mathrm{nm}$; **Lower panel**: Differential optical density for constant NO_2 column density as a function of the spectral resolution Γ_0. There is a nearly linear decay of the optical density as the spectral resolution is reduced, i.e. increasing Γ_0 (adapted from Platt, 1994)

indefinitely, but reach a system dependent minimum noise level B, which is determined by other noise sources in the system. An example for a typical active DOAS system (Tuckermann et al., 1997) using a diode array detector is shown in Fig. 7.62. With increasing signal (i.e. photoelectrons accumulated, given as total binary units), the noise levels are reduced proportionally to the square root of the signal. However, from $\sigma^2 \approx 10^{-8}$–10^{-7} the noise will not be further reduced. This observation shows that there are additional noise sources which are not signal dependent. Under laboratory conditions with a fixed optical setup, noise levels of at least two orders of magnitude lower are reached at sufficiently long integration times. This observation indicates that the noise source dominating long integrations times is not due to electronic noise but is more likely due to optical interference effects (Stutz, 1996; Stutz and Platt, 1997) or processes in the light source. The remaining noise is therefore called 'optical noise'. Obviously, it is of little use to continue a measurement to

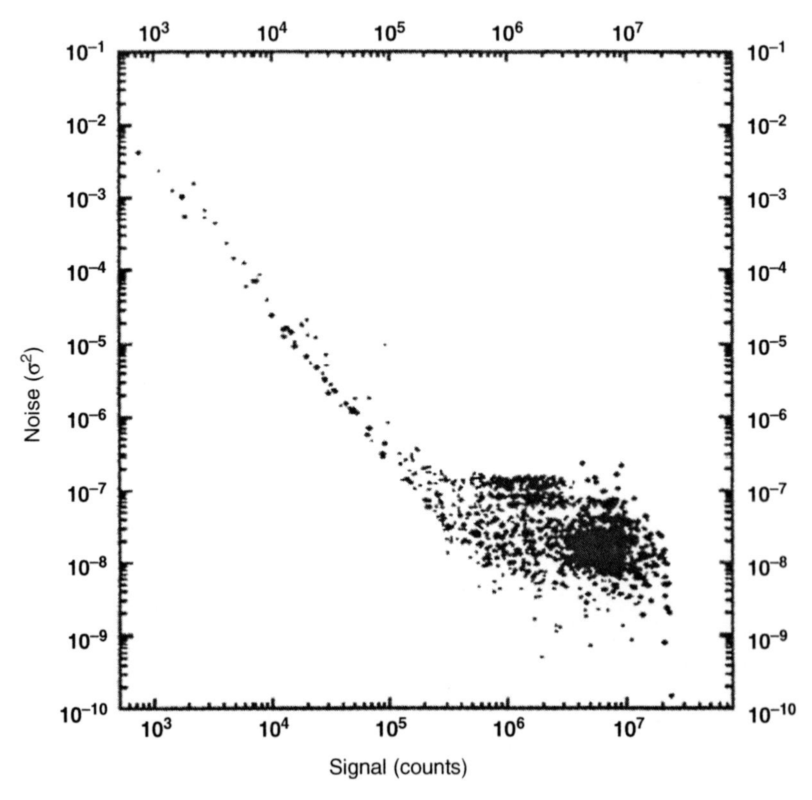

Fig. 7.62. Square of the noise level (σ) as a function of measurement time (log–log scale). Up to a certain limit, there is a linear decrease of σ^2 with the number of photoelectrons recorded (here 1 count signal $\simeq$ 1000 photoelectrons) (adapted from Tuckermann, 1996)

the point where the photoelectron-shot noise becomes much smaller than the combined effect of all other noise sources in the system.

7.14 Measurement Process Control

Like most techniques, DOAS measurements require a number of additional parameters besides the recording and evaluation of trace gas absorption spectra. Several supporting systems are needed for DOAS measurements, particularly if automatic operation is desired. These systems include mechanical actuators for optical alignment of the light path and change of measurement modes, recording of auxiliary parameters required for the interpretation of the measurements (length of the light path, air density, etc.), and monitoring of instrument performance.

In this section, we present a series of 'typical' measurement procedures. Generally, these procedures involve taking measurement spectra and a series of auxiliary spectra (background, dark current, direct light source, calibration, and sometimes scattered sunlight spectra).

7.14.1 Active DOAS Systems – Standard Approach

A complete DOAS measurement using radiation from an artificial light source encompasses taking a series of several different spectra. Here, we give an example for a system using a diode array detector (see e.g. Fig. 7.49), but other arrangements follow procedures similar to those outlined in Fig. 7.63. For a complete DOAS measurement, the following spectra are usually recorded:

1. Atmospheric raw spectrum [ARS, denoted $I_M(i)$]. This is the spectrum of atmospheric absorbers (trace gases) superimposed on a lamp (light source) spectrum. Usually, several individual spectra (N_M) with integration time T_M are co-added.
2. Lamp intensity spectrum [LIS, $I_L(i)$]. This spectrum is recorded by admitting direct light from the lamp to the spectrometer (using some optics to 'bypass' the light path).
3. Dark current (and electronic offset) spectrum [$I_D(i)$]. This spectrum is taken with no light admitted to the detector e.g. by closing a shutter in front of the entrance slit of the spectrometer. By normalising the sum of N_D co-added dark current spectra, we obtain $I_{D1}(i) = I_D(i)/N_D$.
4. A background spectrum [$I_B(i)$] with integration time T_B is sometimes required. Since sunlight scattered into the receiving optics might disturb the measurements, a scattered sunlight or background spectrum is taken by blocking the lamp.

   As a result of steps 1–4, a true atmospheric spectrum I(i) is calculated from the above series of spectra (N_B and N_L denote the numbers of co-added background and LIS, respectively):

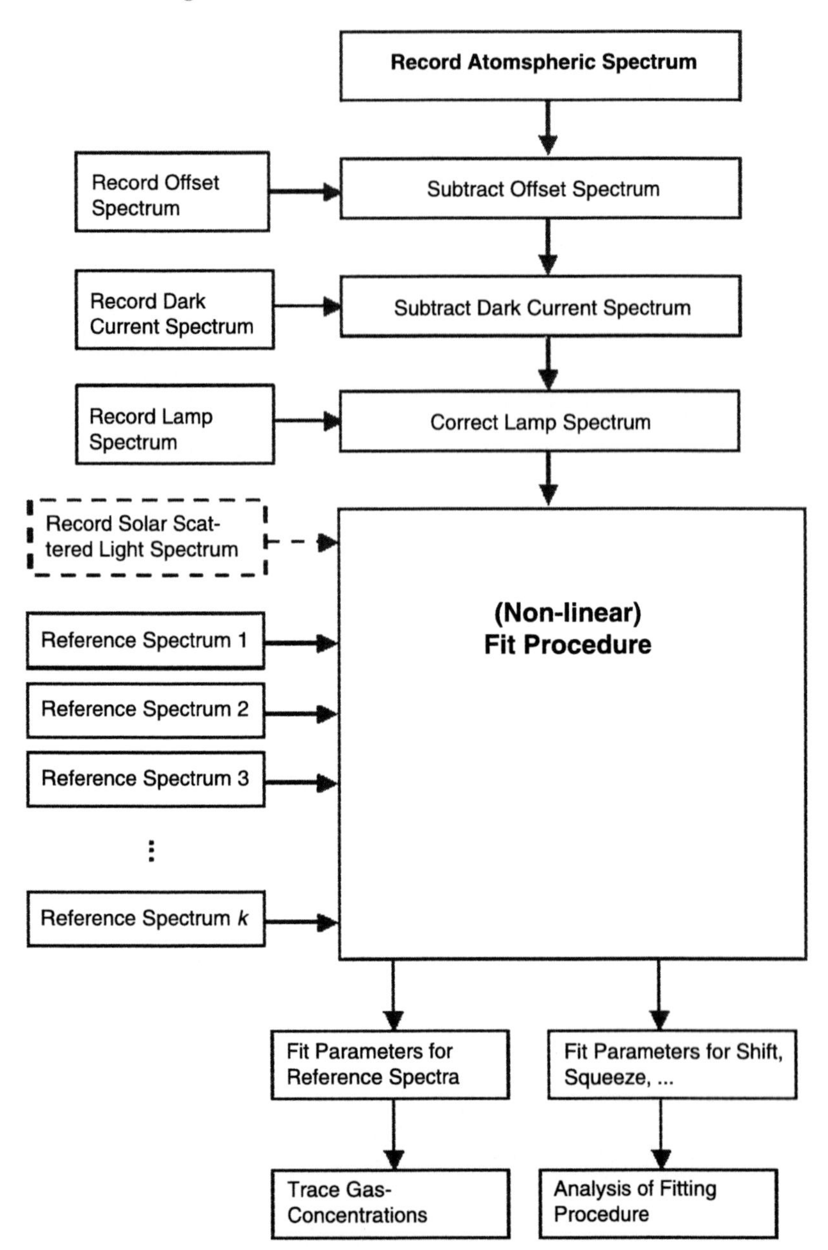

Fig. 7.63. Flowchart of a typical active DOAS measurement and evaluation procedure

$$I(i) = [(I_M - N_M \cdot I_{D1}) - T_M/T_B \cdot (I_B - N_B \cdot I_{D1})] \cdot (I_L - N_L \cdot I_{D1})^{-1}. \quad (7.61)$$

Note that background spectra need not always be taken and compensated for (for instance, at night or at wavelengths below 300 nm where scattered sunlight levels are low). Also, while LIS are always required to compensate for diode-to-diode sensitivity variations (see Sect. 7.6.3), it might not be necessary to take one with each ARS.

In the final steps, the atmospheric spectrum $I(i)$ is evaluated for absorption signatures due to atmospheric gases.

5. A non-linear least squares fit of a series of reference spectra is performed. The reference spectra are usually taken from the literature, and must be adapted to the instrument function of the spectrometer used. This procedure is described in Chap. 8.
6. The resulting fit parameters describing the optical density of the reference spectra are used to calculate the respective trace gas column densities (and their uncertainties).
7. The desired trace gas concentrations are derived by dividing by the length of the optical absorption path. If desired, the concentrations can be converted to mixing ratios.
8. The fit parameters for 'shift' and 'squeeze' of the spectra, which are not directly relevant for deriving trace gas column densities or concentrations, can be used to improve the estimate of the uncertainties and to give general information about the quality of the measurement.

7.14.2 Active DOAS Systems – MCST

A different approach in removing the diode sensitivity variation than taking LIS is the Multichannel Scanning Technique, MCST, introduced by Knoll et al. (1990) (see also Brauers et al., 1995; Stutz, 1996; Alicke, 2000). The basic idea of the MCST is the combination of a multi-channel detection system (PDA) with the scanning technique generally used to cover a larger spectral region with a single-channel detection system. A series of (typically ≈10) ARS $[I_{MM}(i)]$ is taken and stored. The individual spectra are successively shifted in wavelength (for instance by slightly rotating the grating) in such a way that the spectral information is essentially smoothed out when the ARS are co-added (after subracting I_D) to give LIS $[I_{LM}(i)]$. Note that the wavelength difference must be smaller than any spectral structure like absorption bands; otherwise this structure not only remains in the sum spectra, but would also be duplicated near itself. Now, each ARS is divided by the obtained LIS and digitally shifted back to the same wavelength position (for instance of the first ARS in the series) to obtain a true atmospheric spectrum (according to (7.61), with $I_{MM}(i)$ instead of $I_M(i)$ and $I_{LM}(i)$ instead of $I_L(i)$). The advantage of the MCST is that the diode-to-diode sensitivity variations are even more precisely eliminated, since spectra taken to determine these variations are recorded under precisely the same conditions as the ARS. In addition, measurement time is saved, since no separate LIS have to be taken.

7.14.3 Passive DOAS Systems

In passive DOAS systems the primary spectrum consists of absorption features of atmospheric molecules superimposed on a solar (or stellar) spectrum. Raw spectra of zenith scattered skylight (or moonlight or starlight) in the UV/visible are dominated by strong Fraunhofer lines with (in the near UV) optical densities sometimes exceeding unity. Thus, in order to make useful determinations of the weak absorption features of the trace gases in the earth's atmosphere (with optical densities on the order of 10^{-3}), the very accurate removal (with residuals of below 10^{-3} of its initial strength) of the solar spectral structure (LIS) is an indispensable prerequisite. Unlike the case of the artificial light DOAS, the direct solar spectrum is not accessible for ground-based instruments. Therefore, another approach is used:

In order to remove the Fraunhofer structure, spectra taken at different zenith angles of the observed celestial body, ϑ_1 and ϑ_2, are ratioed. The corresponding air mass factors are $A(\vartheta_1)$ and $A(\vartheta_1)$. The observed trace gas slant columns S are $\mathrm{S}(\vartheta_1) = V \cdot A(\vartheta_1)$ and $\mathrm{S}(\vartheta_2) = V \cdot A(\vartheta_2)$, where V is the vertical trace gas column. We obtain for the ratio of the measured optical densities:

$$\frac{D_2}{D_2} = \frac{e^{-\mathrm{S}(\vartheta_2)\cdot\sigma}}{e^{-\mathrm{S}(\vartheta_1)\cdot\sigma}} = e^{-\sigma\cdot[\mathrm{S}(\vartheta_2)-\mathrm{S}(\vartheta_1)]}$$
$$= e^{-\sigma\cdot V\cdot[A(\vartheta_2)-A(\vartheta_1)]}.$$

In other words, the ratio of the optical densities is the optical density based on the difference in the air mass factors (see Chap. 9). For the slant column $\mathrm{S_D} = V\ [A(\vartheta_2) - A(\vartheta_1)]$ the term 'differential slant column' (DSC) is sometimes used, which should not be confused with the use of the term 'differential' in the context of DOAS evaluation of optical spectra.

A complete scattered light DOAS measurement thus consists of taking the following spectra (Again we give an example for a system using a diode array detector):

1. Two ARS (denoted $I_{\mathrm{M1}}(i)$ and $I_{\mathrm{M2}}(i)$) at different (solar) zenith angles. For each spectrum several individual spectra (N_{M}) with integration times T_{M} are usually co-added.
2. Dark current (and electronic offset) spectrum $[I_{\mathrm{D}}(i)]$. By normalising the sum of N_D co-added dark current spectra, we obtain $I_{\mathrm{D1}}(i) = I_{\mathrm{D}}(i)/N_{\mathrm{D}}$.

The procedure is also outlined in Fig. 7.64.

7.14.4 Off-axis Scattered Sunlight DOAS Systems

Off-axis DOAS measurements make use of scattered light spectra recorded at non-zero angles β between zenith and observation direction. If observations

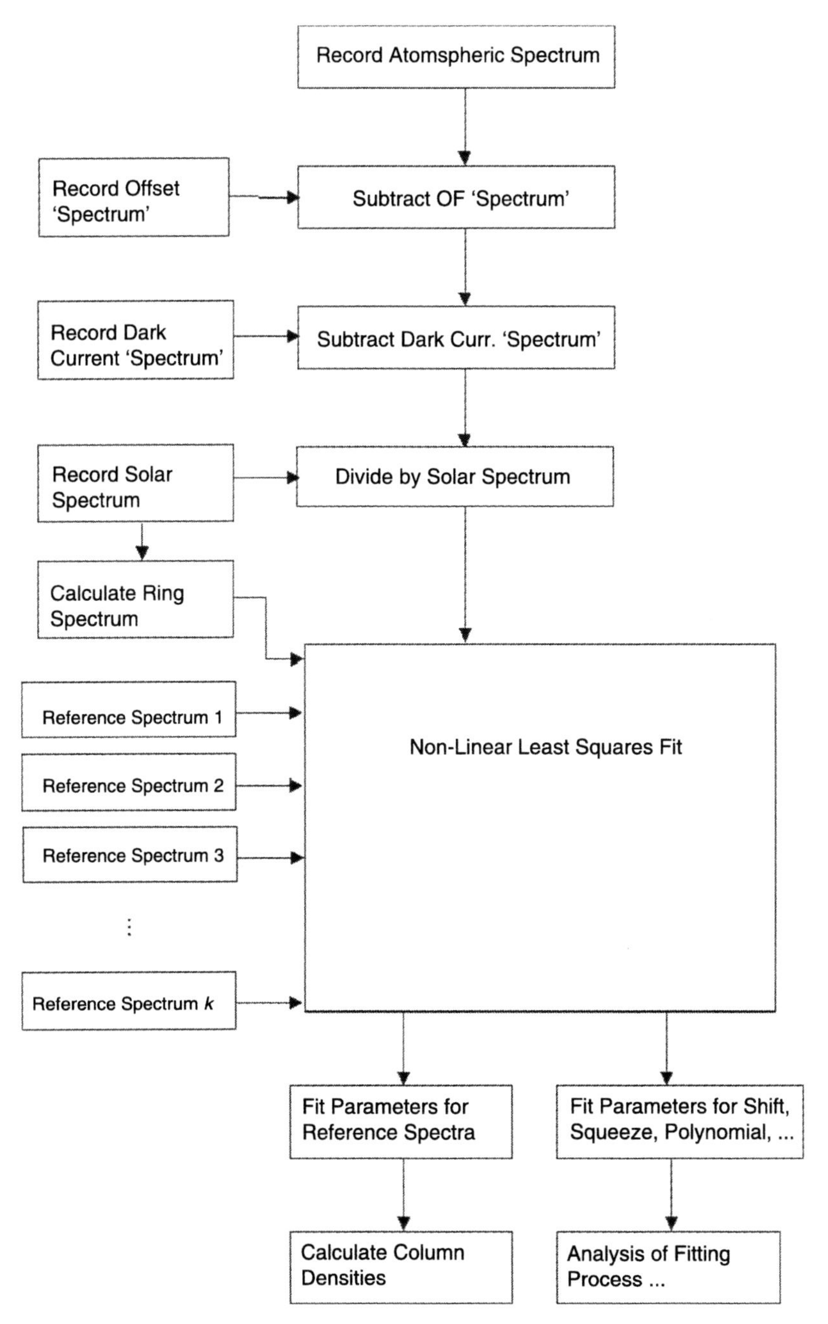

Fig. 7.64. Flowchart of a typical passive DOAS measurement and evaluation procedure.

are made at only one angle β, the difference to ZSL-DOAS lies in a different (usually more complex) air mass factor calculation, and essentially the same procedures for removing the Fraunhofer structure are applied.

On the other hand, observations at several off-axis angles $\beta_1 \ldots \beta_k$ can be made. Now, there are two variable angles β and ϑ. Generally, the absorption due to gases in the upper atmosphere (stratosphere) increases with increasing SZA ϑ, while the absorption due to gases in the lower atmosphere (boundary layer) strongly increases with higher off-axis angles β (see Chap. 9). Similar to the approach using different SZAs ϑ, the Fraunhofer structure can also be removed by ratioing spectra taken at different off-axis angles.

7.15 Mechanical Actuators

The DOAS operating procedures described in Sect. 7.14 require a number of supplemental data, such as offset, dark current and other spectra. These data can be measured manually, but will usually be acquired in an automatic sequence under computer control. In order to switch between the different measurement modes e.g. recording atmospheric spectra, dark current, or observing at different elevation angles, mechanical actuators are needed. In addition, in active DOAS systems (using artificial light sources), alignment of the receiving telescope is often required (due to mechanical instabilities of the optical setup, but also due to diffraction of the light path due to temperature gradients, and thus density and index of refraction, gradients in the atmosphere). These changes in the optical configuration required for the different measurement modes and the optical alignment tasks are usually performed by electrical motors coupled to appropriate mechanics like translation stages. When defined, absolute positions need to be reached. For example, when pointing a telescope to a retro-reflector array. Stepper motor-driven actuators are frequently the devices of choice, although DC-motor–encoder combinations might become important in the future.

For a typical active (artificial light) DOAS system, actuators are usually part of the system:

- 'Short circuit' optics which are moved into the light beam record the lamp emission spectrum.
- A shutter allows the recording of the dark-current spectrum.
- Motorised setting of the spectrometer wavelength allows measurement of spectra in several spectral intervals.
- A filter actuator inserts spectral filters (i.e. to block second order spectra or to reduce stray light) into the optical path.
- Actuators for alignment of the telescope pointing direction (azimuth and elevation) are usually required.
- Alignment of the telescope focus can be motorised in order to accommodate largely different light path lengths.

In summary, about six to seven or more actuators might be needed in a typical artificial-light source DOAS system.

For a typical passive (scattered-sunlight) DOAS system, fewer (about three to four) actuators are required:

- A shutter allows the recording of the dark-current spectrum.
- Motorised setting of the spectrometer wavelength allows measurement of spectra in several spectral intervals.
- A filter actuator inserts spectral filters (i.e. to block second order spectra or to reduce stray light) into the optical path.
- An actuator for aiming of the telescope, usually one but sometimes two axes (azimuth and elevation), may be required for multi-axis measurements.

7.15.1 Stepper Motors

Stepper motors move in precisely known increments. Typical incremental angles (or step sizes) are $1.8°$ (or 200 steps per revolution), but a wide range of different step sizes are available. Special drive electronics ensures that the correct current waveforms are supplied to the motor windings. Modern drive electronics also frequently support fractional steps; i.e. by driving the motor windings by different relative currents the steps can be further divided. Today, many drivers allow a 'half-step' mode, but much smaller 'micro-steps' are also sometimes possible.

Under computer control, such a motor can perform a precisely determined motion pattern. The positions of the motor are repeatable, for instance, different filters mounted on a wheel attached to the motor axis can be selected without further hardware, provided the initial position of the motor is known. This is usually ensured by driving each actuator to a limit switch (and thus to a known mechanical position) during the set-up period of the instrument. In addition, several different telescope pointing directions (i.e. towards several retro-reflectors or light sources) can be selected under software control when azimuth and elevation of the telescope are actuated by stepper motors.

If further mechanical gears are attached to a stepper motor slackness in the mechanics may be present. This is caused by thread clearance in a translation stage. A consequence is that the actuator will reach a different position when approached from different directions of motion. A possible solution is to always make the final move in the same direction, for instance in 'forward direction', which means that moving in the 'backward' direction consists of two phases: first, move backwards by the desired number of steps plus a number of n extra steps, then move 'forward' by n steps.

Disadvantages of stepper motors include their relatively low speed, high cost, and the possibility of 'loosing steps', which is usually not directly detectable by the driving electronics. Furthermore, there might be electrical interference from the rapidly changing currents in the motor windings (the stepper motor is an inherently 'digital device'), for instance with the detector electronics. This problem can usually be avoided by not operating stepper motors during the actual measurement i.e. the recording of optical spectra. However, it should be noted that many commercial stepper motor drivers use a 'switch mode' approach to limit the current to the motor windings. Therefore, although the motor is not moving and the currents to the windings should be 'static', there might be high frequency modulated (usually in the 10 kHz range) currents on the windings. This is difficult to avoid in micro-step mode, but in full or half-step mode, when currents are off or have the same absolute magnitude, a series resistor to the winding might be chosen to limit the current to a point just below the 'desired' current of the motor driver.

In order to minimise the problems of losing steps, a motor should be selected that has sufficiently large torque for the particular application. Additionally, the motor should be operated at a sufficiently low speed.

Stepper motors have a recommended 'starting speed' (=step frequency) and a 'maximum speed'. These numbers are given by the manufacturer for the motor alone and are obviously smaller in any practical application where the motor shaft is connected to additional mechanical elements (like lead screws) which increase the inertia of the motor armature. Also, in practice the motor has to drive a load. A simple approach involves always operating the motor at a same margin below the starting speed. A more sophisticated approach, allowing faster motor operation, starts the motor at its starting speed, followed by a ramp of the step frequency up to the maximum speed. Towards the end of the motion, the frequency is ramped down to the starting speed again. In either case, care has to be taken to avoid motor resonance frequencies. Motor resonance effects can be the cause for errors that are difficult to diagnose later. Motor resonance frequencies are given by the manufacturer, but they apply to the motor alone and will be changed by added load. Note that frequently motor resonance is less pronounced if half-step mode rather than full-step mode is applied.

7.15.2 Other Actuators

Despite their limitations, stepper motors are most commonly used to drive actuators in DOAS systems. Alternatives include DC motors with or without angular encoders. Given a suitable drive electronics encoder, motors can perform almost like stepper motors, with the advantage of not missing steps (since the encoder always allows verification of the motor motion). However, resolution down to a single step is usually not possible.

In some cases, encoders might not be needed; for instance, the position of different filters does not need to be very precise, and thus DC motors with a series of position switches might be sufficient.

For small (linear) displacements piezo actuators might be useful. This type of actuator has seen rapid technological development in recent years.

7.16 Information Needed for Later Analysis

Besides the trace gas column density calculated from the spectra, additional information is required in order to use the data in an orderly way. This information includes obvious data:

- Site of measurement (i.e. geolocation, typically latitude and longitude of the measurement). Obviously, this information will always be required for measurements from mobile platforms (e.g. satellites, aircraft, or ships), but it is good practice to record this information for fixed sites.
- Time and date of measurement.
- Information on type of measurement (active, passive, long-path DOAS, ZSL, multi-axis, etc.).
- Light-path identification (for active systems), length of light path
- Viewing direction (for off-axis DOAS measurements), azimuth and elevation
- Solar (lunar) zenith angle (this information can also be calculated from time, date, and location of the measurement)
- Information on air density. This information is needed to convert the trace gas concentrations into mixing ratios. The air density can be readily determined by recording atmospheric pressure and temperature.
- Technical information, including number of pixels per spectrum, type of light source, type of spectrometer, grating used, type of detector, optical filter, wavelength, etc.

8

Evaluation of DOAS Spectra, Sensitivity, and Detection Limits

One of the central aspects of DOAS is the analysis of atmospheric absorption spectra recorded with instruments described in Chap. 7. As discussed in Chap. 6, the analysis serves different purposes. First and most importantly, trace gas columns (the product of concentration and absorption path length) are derived. This process includes the separation of narrowband from broadband absorption features, and the separation of overlying absorption structures. In many DOAS applications, it is also of advantage to perform an automatic spectral alignment of different absorption cross-sections and absorption spectra. The other routine task of a DOAS analysis is the accurate determination of statistical uncertainties of trace gas columns. In contrast to laboratory experiments, the repetition of measurements under the same experimental conditions in the open atmosphere is impossible. Thus atmospheric observations rarely offer the opportunity to derive errors based on the statistical behaviour of measured concentrations. DOAS overcomes this problem by using the spectroscopic information and the quality of analysis of a single spectrum to derive an error for each measured concentration based on sound physical/statistical principles.

This chapter is dedicated to the introduction and description of the most common DOAS analysis procedures. Various techniques for the modelling of spectra, the determination of associated measurement errors, and as a related topic, the derivation of detection limits for spectroscopic trace gas measurements, have been covered. Most DOAS analysis routines in use today are based on least squares techniques, which will be discussed in the first part of this chapter. The application to the analysis of atmospheric spectra is discussed in the following section. A description of the simulation of absorption spectra required in the analysis is then given. While this chapter is intended to provide an introduction to DOAS analysis, a certain degree of knowledge in mathematical methods is presumed.

U. Platt and J. Stutz, *Evaluation of DOAS Spectra, Sensitivity, and Detection Limits*. In:
U. Platt and J. Stutz, Differential Optical Absorption Spectroscopy, Physics of Earth and Space Environments, pp. 287–328 (2008)
DOI 10.1007/978-3-540-75776-4_8

8.1 Linear Fitting Methods

One of the most common methods to analyse experimental data is the linear least squares fitting technique (Bevington, 1969; Press et al., 1986). The basic idea behind this approach is to minimise the difference between the measured data, $y(x_i)$, where x_i is the independent parameter, and a function, $F(x, a_0, a_1, \ldots)$, which describes the dependence of $y(x_i)$ on the independent parameter x. The model parameters $a_0, a_1, \ldots$ are adapted during the fitting process until the function $F(x, a_0, a_1, \ldots)$ simulates $y(x_i)$ as closely as possible.

To quantify the similarity of $y(x_i)$ and $F(x_i, a_0, a_1, \ldots)$, one introduces χ^2:

$$\chi^2 = \sum_{i=1}^{n} \left(\frac{y(x_i) - F(x_i, a_0, a_1, \ldots)}{\varepsilon(i)} \right)^2 , \tag{8.1}$$

where $\varepsilon(i)$ is the statistical error of the n measured data points $y(x_i)$. The meaning of χ^2 becomes clear if one compares it to the statistical definition of the standard deviation of the difference between $y(i)$ and $F(x_i, a_0, a_1, \ldots)$:

$$\sigma = \sqrt{\frac{1}{n} \sum_{i=1}^{n} (y(x_i) - F(x_i, a_0, a_1, \ldots))^2} . \tag{8.2}$$

One can associate χ^2 with σ via $\chi^2 = \frac{\sigma^2}{\varepsilon^2} n$ if all the errors $\varepsilon = \varepsilon(i)$ are equal. χ^2 is thus a statistical measurement of the difference between measured data and model function. The goal of the fitting routine is to minimise χ^2.

In many cases, the error $\varepsilon(i)$ is omitted from (8.12) by setting $\varepsilon(i) = 1$ for all data points. This is the most commonly used form of linear least square fit, which is called unweighted least squares fit. Both unweighted and weighted [which includes $\varepsilon(i)$] least squares fit will be discussed in the following section. However, it is first necessary to discuss the general assumptions that the least squares method is based on. Later, it will become clear that these assumptions are not always fulfilled in DOAS analysis routines.

1. The model function $F(x, a_0, a_1, \ldots)$ must be linear in the parameters $a_0, a_1, \ldots$. A common example for such a function is a polynomial $F(x) = \sum_{h=0}^{r} a_h \times x^h$.
2. The model function needs to accurately describe the physical characteristics of the problem; otherwise, systematic biases can be introduced in the results of the fit. It is the task of the user to ensure that this assumption is fulfilled, for example, by carefully analysing the residual $R(i) = y(x_i) - F(x_i, a'_0, a'_1, \ldots)$, where $a'_0, a'_1, \ldots$ are the fitting results. $R(i)$ should be statistically distributed around 0 and not show any functional dependence on x. For example, incorrectly using a linear function $F(x, a_0, a_1)$ as a model for a data set that has a quadratic (x^2) component would lead to biases in the results, and $R(i)$ would clearly show a functional dependence.

3. The statistical mean of each error $\varepsilon(i)$ upon repeated measurements has to be 0. In other words, no systematic errors should be imposed on $y(x_i)$. If the mean is not 0, biases may be introduced in the analysis. It is clear that systematic errors are commonly encountered in experimental data. One can, however, soften this requirement by stating that the systematic error of each data point, $y(x_i)$, needs to be much smaller than the respective statistical error, $\varepsilon(i)$. It is crucial that this assumption is tested in a DOAS analysis.
4. The variances of errors $\varepsilon(i)$ need to be known correctly to calculate the error of parameters a_0, a_1,... with the fitting routine. This assumption is often ignored in the application of least squares fit. Unfortunately, as will be discussed later, it is also often violated in spectroscopic measurements.

If all the above assumptions are fulfilled, the least squares fit will give the best estimate for the parameters a_0, a_1,... to minimise χ^2.

8.1.1 Unweighted Linear Least Squares Fit

To simplify the mathematical description of linear least squares method, we will use vector-matrix nomenclature to describe the solution of the fitting problem. The n measurements of $y(x_i)$ will be described by a vector $\vec{y}$, and the m parameters a_0, a_1,...a_{m-1} with a vector $\vec{a}$. The physical content of $F(x_i, a_0, a_1, \ldots, a_{m-1})$ is summarised in a $m \times n$ matrix $\mathbf{X}$. The lines in the matrix denote the functional dependence of each parameter a_j in F, i.e. for $F(x, a_0, a_1) = a_0 + a_1 \times x$, the first line in matrix $\mathbf{X}$ would be $(1, x)$.

The parameter vector $\vec{a}$ that provides the best description for the measured data can then be derived analytically via:

$$\begin{aligned} \vec{a} &= \left[\mathbf{X}^T\mathbf{X}\right]^{-1}\mathbf{X}^T\vec{y} \\ \Theta &= \hat{\sigma}^2\left[\mathbf{X}^T\mathbf{X}\right]^{-1} \\ \hat{\sigma}^2 &= (n-m)^{-1}\left[\vec{y}-\mathbf{X}\vec{a}\right]^T\left[\vec{y}-\mathbf{X}\vec{a}\right] \end{aligned} \tag{8.3}$$

Here, Θ represents the covariance matrix. The square root of diagonal elements of Θ, Θ_{ii}, represents the statistical uncertainty (1σ error) of the parameters a_i:

$$\Delta a_i = \sqrt{\Theta_{ii}}. \tag{8.4}$$

The non-diagonal elements of Θ can be used to determine the correlation between parameters a_i. $\hat{\sigma}$ is the root mean square (rms) of the residuals $R(i)$, and is thus a measure for the average "instrument error" ε, if $n >> m$.

As shown in (8.3), the linear least squares fit is an analytical method. It can be shown that the calculated vector $\vec{a}$ provides the parameters with which $F(x_i, a_0, a_1, \ldots, a_{m-1})$ describes the data set $y(x_i)$ best (Bevington, 1969; Press et al., 1986). The fact that the linear least squares fit is an analytical method explains its popularity in the analysis of many types of data.

8.1.2 Weighted–Correlated Least Squares Fit

The approach described earlier can be expanded by explicitly including the errors $\varepsilon(i)$ for each measurement $y(x_i)$. Mathematically, this can be achieved by introducing the covariance matrix of the errors $(M)_{ij} = E(\varepsilon(i) \times \varepsilon(j))$, where E is the expectance value of $\varepsilon(i) \times \varepsilon(j)$ (note that $E(\varepsilon(i) \times \varepsilon(j))$ is different from $\varepsilon(i) \times \varepsilon(j)$). If $\mathbf{M}$ is known, the following solution for the linear least squares fit can be found.

$$\begin{aligned}\vec{a} &= \left[\mathbf{X}^T\mathbf{M}^{-1}\mathbf{X}\right]^{-1}\mathbf{X}^T\mathbf{M}^{-1}\vec{y} \\ \Theta &= \hat{\sigma}^2\left[\mathbf{X}^T\mathbf{M}^{-1}\mathbf{X}\right]^{-1}. \\ \hat{\sigma}^2 &= (n-(m+r))^{-1}\left[\vec{y}-\mathbf{X}\vec{a}\right]^{\mathrm{T}}\mathbf{M}^{-1}\left[\vec{y}-\mathbf{X}\vec{a}\right]\end{aligned} \tag{8.5}$$

If all the errors are statistically independent from one another, $\mathbf{M}$ is a diagonal matrix. We will see in spectroscopic applications that this is not always the case. If all the errors $\varepsilon(i)$ are equal, $\mathbf{M}$ will be a matrix with a diagonal of 1, and the solution will be equal to the one in (8.3). The meaning of Θ and the determination of Δa_i are the same as described in (8.4).

8.2 Non-linear Fitting Methods

The linear least squares method described earlier offers an elegant way of analysing data. However, the limitation to linearly dependent fit parameters can impose restrictions for certain applications. The problem of spectral alignment cannot be solved by a linear method, and requires non-linear least squares method. It should be noted that it is preferable to use linear methods since they provide an analytical solution to the fitting problem, whereas non-linear routines are iterative and thus often sensitive to the starting conditions. In addition, non-linear methods require long computation times for problems with a large number of parameters.

Here, three different non-linear fitting methods will be presented. In general, one tries to solve a problem where a number of measurements $y(x_i)$ are analysed by fitting a model function, $f(x_i, c_0, c_1, \ldots)$, that is non-linear in the parameters $c_0, c_1, \ldots$. In addition, the approach will rely on minimising χ^2, as defined in (8.11).

8.2.1 Gradient Method

The gradient method is based on finding the minimum of χ^2 by following the steepest descent of χ^2 on the multi-dimensional surface in the $c_0, c_1, \ldots$ space by an iterative process (Press et al., 1986). If one denotes the vector of the parameters $c_0, c_1, \ldots$ during the iteration as $\vec{\gamma}_k$, where k is the number of iterations of the fit, the mathematical equation allowing the calculation of next iteration is as follows:

$$\vec{\gamma}_{k+1} = \vec{\gamma}_k + \text{const.} \times \left[-\nabla \chi^2 \left(\vec{\gamma}_k \right) \right] . \tag{8.6}$$

The equation is based on the calculation of the gradient of χ^2 and a step along the steepest descent, with a width that can be chosen through the constant. For small constants, the iteration will converge very slowly, whereas for large constants there is a high risk that the iteration will step over the minimum of χ^2.

The iteration has to be repeated until a certain convergence condition is met. The choice of this condition depends on the problem and the desired accuracy. One can easily understand that it is difficult to guarantee the convergence of this method at all accuracies. The method also depends on the starting conditions $\vec{\gamma}_0$ for problems that have more than one minimum of χ^2.

8.2.2 Gauß–Newton Method

The Gauß–Newton method is based on the fact that, for a minimum of χ^2, the first derivative of χ^2 in the parameter space has to be zero. By using this fact and applying a Taylor expansion to χ^2, one can derive the following expression for an iteration step in the fitting process (Press et al., 1986):

$$\vec{\gamma}_{k+1} = \vec{\gamma}_k + -\mathbf{D}^{-1} \times \left[-\frac{1}{2} \nabla \chi^2 \left(\vec{\gamma}_k \right) \right] , \tag{8.7}$$

where $\mathbf{D}$ is the Hessian of the our model function $f(x_i, c_0, c_1, \ldots)$, which is approximated by the following expression in the Gauß–Newton method:

$$(\mathbf{D})_{rs} = -\frac{1}{2} \frac{\partial \chi^2}{\partial \gamma_r \partial \gamma_s} \approx \sum_{j=1}^{n} \frac{\partial f(\vec{\gamma}_k)}{\partial \gamma_r} \frac{\partial f(\vec{\gamma}_k)}{\partial \gamma_s} . \tag{8.8}$$

As in the case of the gradient method, a suitable condition for convergence has to be applied to stop the iteration. There is, however, no guarantee that the Gauß–Newton method converges. The method works best close to the minimum of χ^2, where (8.8) is a good approximation.

8.2.3 Levenberg–Marquardt Method

As discussed in the previous sections, the gradient method converges rapidly away from the minimum of χ^2, while the Gauß–Newton method works best close to the minimum. Taking into account the similarity of (8.6) and (8.7), Levenberg (1944) combined the methods in a single algorithm that is more stable and converges faster than the individual methods. By rewriting the vector step in (8.6) as $\boldsymbol{\lambda} \cdot \mathbf{1} \cdot \left[-\nabla \chi^2 (\vec{\gamma}_k) \right]$ with a parameter $\boldsymbol{\lambda}$ that stands for the constant, and by including the 1/2 factor into $\mathbf{D}^{-1}$ in (8.7): $\mathbf{D'}^{-1} = 1/2\,\mathbf{D}^{-1}$, he introduced a matrix $\mathbf{A}$:

$$A = \left(\lambda \cdot \mathbf{1} + \mathbf{D'}^{-1}\right). \tag{8.9}$$

$\mathbf{A}$ can be used to calculate $\vec{\gamma}_{k+1}$ from $\vec{\gamma}_k$ in the iteration:

$$\vec{\gamma}_{k+1} = \vec{\gamma}_k + \mathbf{A}\left[-\nabla\chi^2\left(\vec{\gamma}_k\right)\right]. \tag{8.10}$$

If λ is large, this method favours the gradient method, where λ is the step size. Small λ leads to the use of Gauß–Newton method. For medium-sized λ, a linear combination of the two methods is used.

Marquardt (1963) describes the following recipe for the choice of λ:

1. Choose a starting parameter vector $\vec{\gamma}_0$, and set λ to a moderate value, for example $\lambda = 0.001$.
2. Perform an iteration step and calculate $\vec{\gamma}_{k+1}$ from $\vec{\gamma}_k$ using (8.10). Determine $\chi^2(\vec{\gamma}_{k+1})$ for the new parameter vector.
3. If $\chi^2(\vec{\gamma}_{k+1}) > \chi^2(\vec{\gamma}_k)$, set $\vec{\gamma}_{k+1} = \vec{\gamma}_k$, thus rejecting the new parameter vector, and increase λ by a factor of 10. Return to step 2 to calculate a new parameter vector.
4. If $\chi^2(\vec{\gamma}_{k+1}) \leq \chi^2(\vec{\gamma}_k)$, decrease λ by a factor of 10. Return to step 2 to calculate the next parameter vector.

To stop iteration, the conditions have to be set. For example, one may wish to stop iteration if λ is too large, when the iteration does not seem to converge, or if λ is too small, when the iteration is already very close to the minimum of χ^2. Another possibility is a limit for the change of χ^2, e.g. if χ^2 does not change more than a certain factor, one assumes that the iteration is close to the minimum.

The advantage of the Levenberg–Marquardt method is that it always converges, since the steps that do not reduce χ^2 are rejected while, at the same time, increasing λ. In this case, more weight will be put on the gradient method. The algorithm, on the other hand, will try to make λ smaller and thus give more weight to the Gauß–Newton method, which provides a good approximation close to the minimum of χ^2.

A number of methods have been suggested to calculate the errors of the parameters c_0, c_1,..., which are the result of the fit. Press et al. (1986), for example, suggest a method that assumes that the errors of the original data are normally distributed. To derive the errors, they suggest to recalculate $\mathbf{D'}^{-1}$ (say, by setting $\lambda = 0$ and recalculating $\mathbf{A}$) at the end of iteration. $\mathbf{D'}^{-1}$ can then be interpreted as the covariance matrix of the non-linear fit.

$$\Theta = \mathbf{D'}^{-1} = \frac{1}{2}\mathbf{D}^{-1}. \tag{8.11}$$

The errors can then be calculated in a similar way as described earlier.

$$\Delta c_i = \sqrt{\Theta_{ii}}. \tag{8.12}$$

Cunningham (1993) tested this calculation using Monte Carlo methods and found that it underestimates the error by less than 50%.

8.3 DOAS Analysis Procedure

The DOAS approach was discussed in detail in Chap. 6. It is helpful to recall the most important aspects of the DOAS technique, with respect to the consequences they have for the design of analysis routines.

- The DOAS technique is based on the separation of slow- and fast-varying spectral structures. A DOAS analysis routine has to be able to either high-pass filter the absorption spectra before the analysis or provide the means to do this during the analysis.
- Atmospheric absorption spectra contain overlaying absorption structures that must be separated by the analysis routine. The typical approach is to use absorption reference spectra or cross-sections that are adapted to the instrument resolution.
- In case the reference spectra are not spectrally aligned, which occurs frequently, the analysis routine has to be able to correct this misalignment.
- Because the absorption spectra always contain noise, which may introduce statistical uncertainties, the analysis routine has to be able to calculate errors of the desired trace gas columns.

As discussed in Chap. 6, the DOAS problem can be summarised by (8.13):

$$I(\lambda, L) = I_0(\lambda) \exp\left(\int_o^L \sum_j \left(-\sigma_j(\lambda, p, T) \times \rho_j(l)\right) - \varepsilon_R(\lambda, l) - \varepsilon_M(\lambda, l)\, dl\right) + N(\lambda). \tag{8.13}$$

where $I_0(\lambda)$ and $I(\lambda, L)$ are the light intensity at the beginning and end of the light path, respectively; $\sigma_j(\lambda, p, T)$ is the absorption cross-section of a trace species j, which depends on the wavelength λ, the pressure p, and the temperature T; $\rho_j(l)$ is the number density at position l along the light path; $\varepsilon_R(\lambda, l)$ and $\varepsilon_M(\lambda, l)$ are the Rayleigh-extinction and Mie-extinction, respectively; and $N(\lambda)$ represents the noise.

The measurement changes $I(\lambda, L)$ in two ways. First, the limited resolution of the spectrograph, which can be described by a convolution with the instrument function H, $I^*(\lambda, L) = I(\lambda, L)^* H$, reduces the narrowband spectral structure. In addition, the different optical elements can impose a broad spectral structure onto $I^*(\lambda, L)$. Finally, $I^*(\lambda, L)$ is digitised by the detector $I'(i)$.

In Chap. 6, we discussed various implementations of the DOAS technique. In Case 1, a linearisation of (8.13) after convolution and discretisation is possible. This is the most common example of DOAS implementation today, and we will concentrate on this approach in the following section. In cases with strong trace gas absorptions, the modelling of equation $I'(i)$, including convolution and discretisation, is necessary. This is, in principle, possible using the non-linear least squares fits described earlier. However, in practice this approach is slow and cumbersome. It is thus rarely used, and we will not discuss it further.

Most DOAS analysis software relies on the linearisation of $I'(i)$ by calculating the logarithm $J(i) = \ln(I'(i))$:

$$J(i) = J_0(i) + \sum_{j=1}^{m} a'_j \cdot S'j(i) + B'(i) + R'(i) + A'(i) + N'(i). \qquad (8.14)$$

The narrow absorption structures of trace gases are described by their individual differential absorption structure $S'_j(\lambda) = \ln(\exp(-\sigma'_j(\lambda))^* H)$ calculated from the convolution of the cross-section of trace gas j with the instrument function H (see Chap. 6 for the mathematical rationale to calculate $S'_j(\lambda)$ for different DOAS implementations). The scaling factors $a'_j = c_j \times L$ are then the product of average number densities c_j over path-length L. The broad absorptions of trace gases are represented by $B'(i)$. Variations in the spectral sensitivity of the detector or spectrograph are summarised in $A'(i)$ as a function of pixel number. The extinction by Mie and Rayleigh scattering is represented by $R'(i)$. The noise $N'(i) = \ln(N(\lambda))$ is caused by detector noise and photon statistics. The overlaying absorption structures of several trace gases are represented by the sum in (8.14). In practice, the number of absorbers m can be limited to trace gases with absorption structures sufficiently strong to be detectable with DOAS instruments.

8.3.1 The Linear Model

Most DOAS evaluation procedures are based on a model that describes the DOAS spectra according to (8.14). The logarithm of the discrete measured intensity, $J(i)$, is modelled by a function F(i):

$$F(i) = P_r(i) + \sum_{j=1}^{m} a_j \times S_j(d_{j,0}, d_{j,1}, ..)(i). \qquad (8.15)$$

The absorption structures of trace gases, S_j, are input data to the procedure. The scaling factors a_j are the result of the fit and can be used to calculate the concentration c_j of the respective trace gases with differential absorption cross-section σ'_j by: $c_j = a_j/(\sigma_j \times \mathrm{L})$.

The polynomial, $P_r(i)$, describes the broad spectral structures caused by the characteristics of the lamp $I_0(i)$, the scattering processes $R'(i)$, the spectral sensitivity $A(i)$, and the broad absorptions by the trace gases $B(i)$:

$$P_r(i) = \sum_{h=0}^{r} c_h \cdot (i - i_c)^h. \tag{8.16}$$

The parameter $i_c = \text{int}\,(n/2)$ represents the centre pixel of the spectral region used for evaluation. The polynomial refers to i_c to maximise the influence of non-linear terms. In this approach, the polynomial is responsible for the separation of broad and narrow spectral structures. The degree of filtering depends on the problem. In case of relatively broad absorption features, for example of NO_3, a low-order polynomial has to be used. If a large number of narrow absorption structures were to be analysed, for example of SO_2, a higher order polynomial can be used. An alternative approach to fitting a higher order polynomial is to high-pass filter the absorption spectra and references before using them in (8.15) (see Sect. 8.3.2).

Among the most important aspects of DOAS analysis using the model function (8.15) is the very accurate description of the reference spectra S_j. As discussed earlier, the least squares methods that are most commonly used in the analysis of DOAS spectra (see Sect. 8.1) require that the model function is an accurate description of the physical characteristics of the problem. Reference spectra of poor quality can introduce systematic biases in the analysis.

8.3.2 High- and Low-pass Filtering

The basis of the DOAS method is the separation of narrow and broad spectral structures. While this can be done directly in the analysis routine, this separation is often also performed before the actual fitting process. While a comprehensive survey of available filtering methods is beyond the scope of this book, a number of commonly used filtering approaches will be briefly described below. In addition to high-pass filters, low-pass filters, which reduce the magnitude of noise in the spectra, are also employed. Examples of the various methods described here are shown in Fig. 8.1.

Low-Pass Filters

Low-pass filters are used to reduce the unwanted high frequency components of absorption spectra. These can be caused by noise, pixel-to-pixel sensitivity changes, or other instrumental problems. The following filters are often used in DOAS applications:

Moving average

The moving average approach replaces the content of a channel i by taking the average of the original data point $y(i)$ and the m neighbour channels to the left and right:

$$\hat{y}(i) = \frac{1}{2m+1} \sum_{j=-m}^{m} y(i+j). \tag{8.17}$$

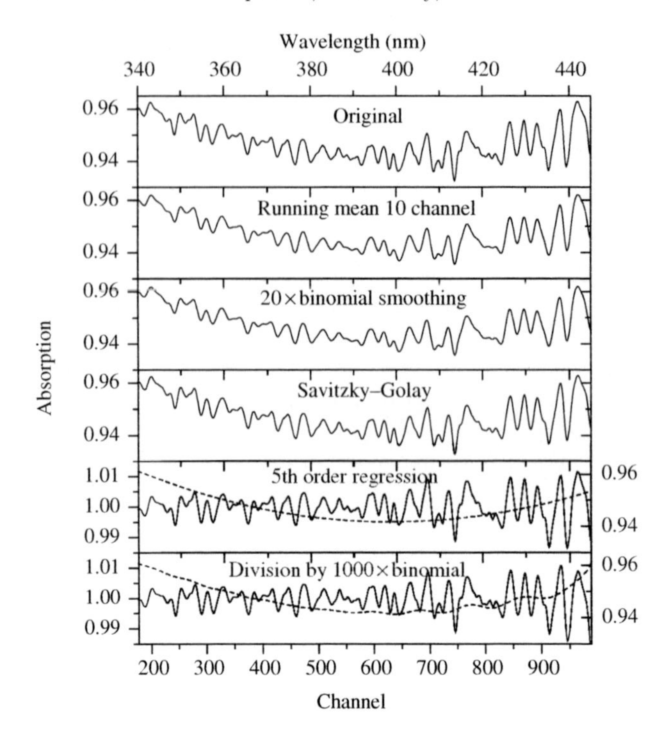

Fig. 8.1. Example of low- and high-pass filtered absorption spectra

The width of the box $2m + 1$ determines the degree of smoothing of the spectrum. A wider box leads to the removal of broader structures. Care must be taken at the edges of the spectrum.

Binomial (triangular) smoothing

In this method, each channel is replaced by a value equal to the weighted average of neighbouring channels. Each channel is given a weighting according to the binomial distribution so that points further from the original channel contribute less. The simplest example of a binomial smoothing approach is a three-point binomial filter with the following weighting coefficients: 0.25, 0.50, 0.25:

$$\hat{y}(i) = \frac{1}{4}y(i-1) + \frac{1}{2}y(i) + \frac{1}{4}y(i+1). \tag{8.18}$$

Equation (8.18) can be applied multiple times, in which case the smoothing will be equivalent to a higher order smoothing.

Polynomial smoothing.

The basic principle of polynomial smoothing method is the fitting of a polynomial of nth degree to the neighbouring channels surrounding a particular channel i in the dataset. The 'smoothed' value of the channel i is then assigned the value of the polynomial at that point. The two parameters influencing the degree of smoothing are the degree of the polynomial and the window over which the polynomial is fitted.

An efficient algorithm for computing the polynomial filter has been described by Savitzky and Golay (1964). It enables rapid calculation of the polynomial coefficients. Another implementation is cubic-spline fits, which invoke additional conditions for the calculation of polynomials, e.g. the first and second derivatives of adjacent curves have to be equal, avoiding discontinuities at the joints.

High-pass Filters

A number of filters are used to remove broad (much larger than typical absorption bandwidth) spectral structures from absorption spectra.

Division by high-pass filtered spectrum

Any low-pass filter can be converted into a high-pass filter by dividing the original spectrum by the low-pass filtered spectrum. This is often used in combination with binomial smoothing or running averages.

Regression

Another common high-pass filter method is division by a polynomial of degree g, which is fitted to the entire spectrum with a linear least squares method. The degree can be chosen according to the expected broadband behaviour. This method is ideal if the dominant broadband contribution originates from Raleigh scattering. Typically, regressions are performed using polynomials with maximum orders of five to nine. The difference to the polynomial low-pass filters described earlier is that, for regression, a large number of channels are used in the fit.

Fourier filters

The most sophisticated means of filtering absorption spectra are methods that are based on Fourier techniques. Here, the spectrum is first transformed into frequency space, for example, by a fast Fourier transform (FFT) calculation. Certain higher or lower frequencies are removed or reduced. After the transformation back into normal space via another FFT, a smoothed spectrum is obtained. There are many different choices for these types of digital filters, and the reader is advised to refer textbooks on this topic.

A word of warning: If filters are used in a DOAS analysis, they must be applied to all spectra – atmospheric absorption and reference spectra – alike. This will ensure that the absorptions in the reference and atmospheric spectra have the same structure. For this reason, it has also become less common to use regressions, since these do not guarantee that the absorptions change equally in both types of spectra.

8.3.3 Wavelength Alignment

While it should be theoretically possible to measure or calculate reference spectra that are well aligned with the atmospheric absorption spectrum and each other, small uncertainties in the wavelength position and dispersion can arise due to thermal changes in the spectrometer, inaccuracy of the grating position, and errors in the wavelength calibration of literature absorption cross-sections. It is, therefore, often necessary to perform a spectral alignment during the analysis. It should be noted that small misalignments do not affect the scaling factor/concentration of the respective trace gas much, but influence other weaker absorbers by introducing unwanted residual structures.

The task of the analysis procedure is to align the reference spectra $S_j(i)$ (wavelength-pixel-mapping Γ_j) to the atmospheric absorption spectrum $J(i)$ (wavelength-pixel-mapping Γ_J). The procedure, therefore, must recalculate the reference spectrum $S_j^*(i)$ with the wavelength-pixel-mapping Γ_J (see Chap. 6 for a discussion on wavelength-pixel-mapping). This can be described as 'shifting and stretching/squeezing' the reference spectrum in wavelength. As Γ_j is a strongly monotonous function, its inverse can be described by a polynomial: Γ_j^{-1}: $x(\lambda) = \sum_{k=0}^{q} \beta_k \times \lambda^k$, where $x(\lambda)$ represents the non-integer 'pixel number' that results from this inverse transformation. $S_j(\lambda)$ can now be calculated from the continuous spectrum $S_j(x)$. This spectrum must be approximated using an appropriate interpolation method, e.g. a cubic spline procedure (Press et al., 1986), on the discrete spectrum $S_j(i)$.

We can now calculate $S_j^*(i)$ with the wavelength-pixel-mapping Γ_J by deriving $S_j(\lambda)$ with Γ_j^{-1} from $S_j(x)$, which is approximated by an interpolation on $S_j(i)$, and then applying Γ_J:

$$S_j(i) \overset{\text{interpolation}}{\longrightarrow} S_j(x) \overset{\Gamma_j^{-1}}{\longrightarrow} S_j(\lambda) \overset{\Gamma_J}{\longrightarrow} S_j^*(i). \tag{8.19}$$

It is possible to combine Γ_j^{-1} and Γ_J into one formula, which links i to x using a polynomial with parameters δ_k:

$$x(i) = x(\lambda(i)) = \sum_{k=0}^{q_s \times q_I} \delta_k \cdot i^k. \tag{8.20}$$

It is advantageous to use a slightly modified version of (8.20), where the spectral alignment parameters $d_{j,k}$, determining the transformation are zero if the wavelength-pixel-mappings of J and S_j are equal:

$$x = i + f_j(i) \quad \text{with} \quad f_j(i) = \sum_{k=0}^{p_j} d_{j,k} \cdot (i - i_c)^k. \tag{8.21}$$

The spectrum $S_j(d_{j,0}, d_{j,1}, \ldots)(i) = S_j^*(i)$ now has the wavelength-pixel-mapping Γ_J, which was calculated with the parameters $d_{j,k}$ using (8.19) and (8.21) and a cubic spline interpolation of $S_j(i)$.

The parameters $d_{j,k}$ are derived by performing a non-linear fit of the model F to the spectrum J with fixed parameters a_j and c_h. If $p_j = 0$, the spectrum S_j is shifted by $d_{j,0}$ pixels. If $p_j = 1$, the spectrum is additionally linearly squeezed or stretched according to parameter $d_{j,1}$. Higher values of p_j represent a squeeze or stretch of higher order. To achieve the best physical description of the spectra, the degree of the squeeze process p_j should be selected for every reference spectrum S_j. Alternatively, one set of parameters $d_{j,k}$ for two or more reference spectra is used if the wavelength calibration is identical for these spectra.

To quantify the importance of a spectral alignment, one may set up a test spectrum J, which is formed from the sum of a single trace gas spectrum $S_g^*(d_{g,0}, d_{g,1}, \ldots)(i) = a_g \times S_g(d_{g,0}, d_{g,1}, \ldots)(i)$, with shift and squeeze represented by the parameters d_{gj} and $(m-1)$ with unaltered reference spectra $S_j^* = a_j \times S_j$:

$$J = 1 + S_1^*(0) + \ldots + S_g^*(d_{g,0}, d_{g,1}, \ldots) + \ldots + S_m^*(0)\,. \tag{8.22}$$

A linear fit based on the model function $F_{\text{sh}} = 1 + a_1^* \times S_1^*(0) + a_2^* \times S_2^*(0) + \ldots + a_m^* \times S_m^*(0)$ is then used to analyse J for different parameters $d_{g,0}, d_{g,1}, \ldots$ in J. In addition, different spectra S_g^* can be shifted. By varying the different parameters and spectra, one can calculate a function:

$$\mathbf{\Phi}_{j,g}(d_{g,0}, d_{g,1}, ..) = 1 - a_j^*, \tag{8.23}$$

which determines the change in the scaling factors a_j^* due to a shift of spectrum S_g^*. For m reference spectra in J, there are m^2 different $\mathbf{\Phi}_{j,g}$. $\mathbf{\Phi}_{j,g}$ illustrates the sensitivity of the fit results to misalignments. Figure 8.3 shows 16 different $\mathbf{\Phi}_{j,g}$ for the spectra in Fig. 8.2 and an alignment by a shift. It can be seen that a shift of one pixel in spectrum S_1, corresponding to approximately 1/20 of the width of the absorption bands, produces a result for spectrum S_3 that differs by 70% from the real value ($\mathbf{\Phi}_{3,1}$; Fig. 4c). This confirms that, without correcting the wavelength calibration, the linear fit procedure can give incorrect results.

8.3.4 Realisation

There are a number of different implementations of a DOAS analysis algorithm. Most of them have one thing in common that they use least squares methods to fit a model function to the absorption spectra. The most popular approach uses linearisation explained in Sect. 8.3.1, which is shown here again as one single-model function:

$$F(i) = \sum_{h=0}^{r} c_h \cdot (i - i_c)^h + \sum_{j=1}^{m} a_j \cdot S_j(d_{j,0}, d_{j,1}, ..)(i). \tag{8.24}$$

The model function is determined by three sets of parameters: (1) $c_0, c_1, \ldots, c_h$ determine an additive polynomial; (2) $a_1, \ldots, a_m$ scale the reference spectra/

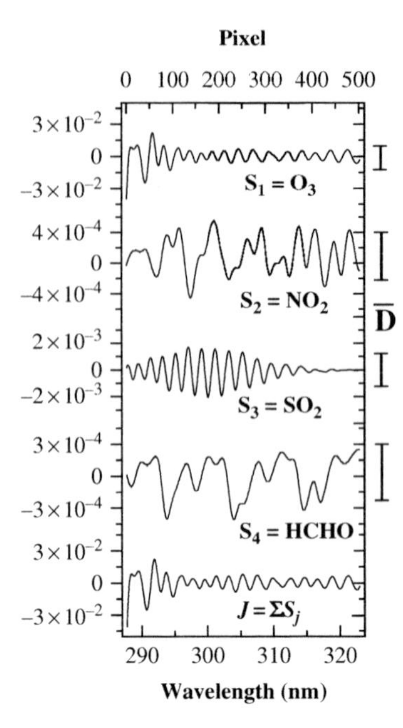

Fig. 8.2. Absorption spectra used for the alignment tests. The spectra were calculated by convoluting high-resolution cross-sections with a Gaussian-shaped instrument function of 1.3 nm half-width. The average optical density $\overline{D}$ of the spectra, 0.0195 for O_3, 0.00066 for NO_2, 0.0023 for SO_2 and 0.00051 for HCHO, is marked with *bars* on the right axis. This corresponds to atmospheric concentrations of 100 ppb for O_3, 2.5 ppb for NO_2, 0.5 ppb for SO_2 and 0.5 ppb for HCHO for a light path of 7.5 km in the troposphere (from Stutz and Platt, 1996)

cross-sections; (3) $d_{j,0}, \ldots$ are the parameters for spectral shift and squeeze for each reference spectrum. The first two sets of parameters are linear in $F(i)$ and can be summarised in a vector $\vec{\beta} = (c_0, c_1, c_2, \ldots, a_0, a_1, a_2, \ldots)$. The third set of parameters is non-linear in $F(i)$ and can only be fitted through non-linear least squares methods. While it is possible to use non-linear procedures to optimise all three sets of parameters, we will discuss here an approach that is based on a combination of linear and non-linear methods, which has the advantage of being faster than a purely non-linear approach.

The fitting process starts with the choice of reasonable starting values for the $d_{j,k}$ parameters. In the first step, the parameter vector $\vec{\beta} = (c_0, c_1, c_2, \ldots, a_0, a_1, a_2, \ldots)$ is determined by a linear least squares method using the measured spectrum $J(i)$ and the coefficient matrix **X**. **X** contains $(r+1)$ polynomial arguments $(i - i_c)^h$ and the m reference spectra S_j, which are shifted according to the parameters $d_{j,k}$:

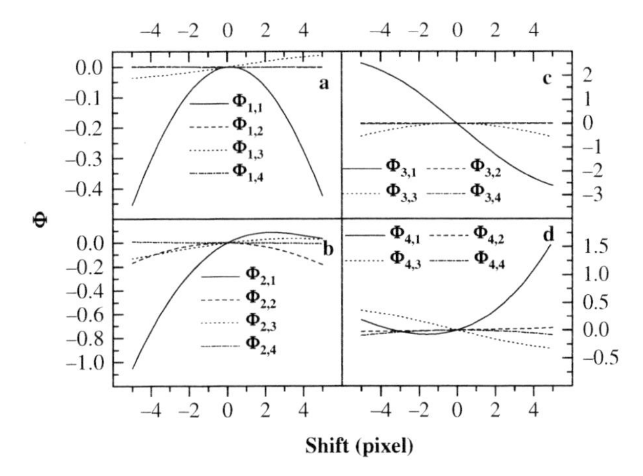

Fig. 8.3. $\Phi_{j,k}$ [see (8.23)] describes the influence on the fit results a_j of the misalignment of spectrum S_k. The $\Phi_{j,k}$ shown in this figure was calculated using the spectra of Fig 8.2. (**a**) shows the influence of the shift of the four different S_j on a_1 of O_3. (**b**), (**c**), and (**d**) show the influence on NO_2, SO_2 and HCHO, respectively. The highest influence is found for the shift of O_3 on SO_2 ($\Phi_{3,1}$ in (c)), where a shift of 1 pixel changes a_3 by 70% (from Stutz and Platt, 1996)

$$
\mathbf{X} = \begin{pmatrix}
1 & (-i_c)^1 & (-i_c)^2 & \cdots & (-i_c)^r & S_1(0) & S_2(0) & \cdots & S_m(0) \\
1 & (1-i_c)^1 & (1-i_c)^2 & \cdots & (1-i_c)^r & S_1(1) & S_2(1) & \cdots & S_m(1) \\
1 & (2-i_c)^1 & (2-i_c)^2 & \cdots & (2-i_c)^r & S_1(2) & S_2(2) & \cdots & S_m(2) \\
\vdots & \vdots & \vdots & \ddots & \vdots & \vdots & \vdots & \ddots & \vdots \\
1 & (n-i_c)^1 & (n-i_c)^2 & \cdots & (n-i_c)^r & S_1(n) & S_2(n) & \cdots & S_m(n)
\end{pmatrix}. \tag{8.25}
$$

The number of columns in $\mathbf{X}$ is given by the number of fitted parameters $(r+1)+m$. The number of lines is equal to the number of pixels $(n+1)$ of the wavelength interval of the analysis. Performing the linear least square procedure gives the best estimate for linear parameters, using the given $d_{j,k}$: $\vec{\beta}^1$.

Next, one iteration of a Levenberg–Marquardt fit is performed to derive the estimate of non-linear parameters $d_{j,k}$. The resulting parameters $d^1_{j,k}$ are used in the successive call of the linear fit, which calculates $\vec{\beta}^2$. This result is then used in the next call of the non-linear fit. The procedure then continues to alternate the two methods, always using the result of the previous method as values for the other fit method (see also Fig. 8.4). This procedure is repeated until one of several stopping conditions for the non-linear fit is fulfilled.

Normally, the fit is aborted when the relative changes of χ^2 in the last step is smaller than a given value (usually 10^{-6}), and thus the fit has converged.

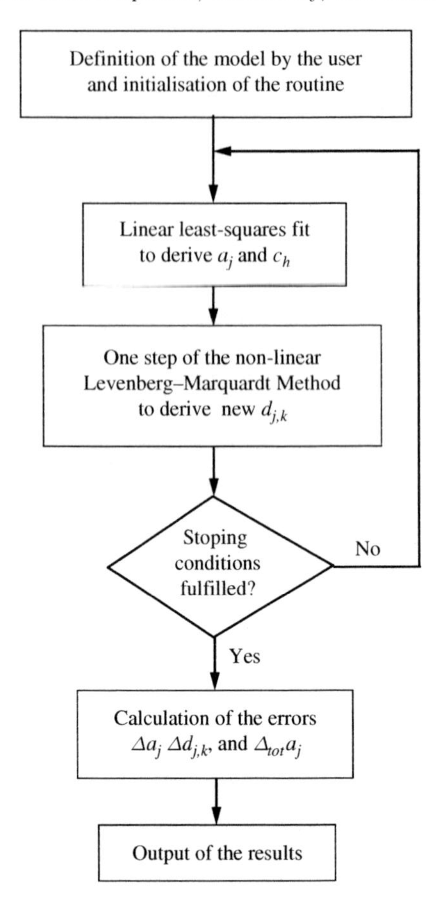

Fig. 8.4. Overview of the linear–non-linear DOAS analysis procedure (from Stutz and Platt, 1996)

The fit also stops if the number of repetitions of the iteration, determined by the user, is exceeded or if the non-linear method becomes unstable, e.g. the λ parameter of the Levenberg–Marquardt method becomes too large or too small.

The analysis procedure is thus a combination of the well-known non-linear Levenberg–Marquardt method (Levenberg, 1944; Marquardt, 1963; Press et al., 1986) determining the coefficients $d_{j,k}$ (see Sect. 8.3.3) and a standard linear least squares fit (Press et al., 1986; Bevington, 1969; Albritton et al., 1976) to derive the coefficients a_j and c_k.

8.3.5 Error Analysis

The calculation of uncertainty of the various fit factors and ultimately the error of concentrations of the respective trace gases are an important part of the analysis routine. The fit method described in the previous section calculates two sets of errors: Δa_j and Δc_h, which are derived by the linear least squares fit using the alignment parameters $d_{j,k}$; and, in the case of an analysis routine with spectral alignment, $\Delta d_{j,k}$, the errors of the alignment parameters, which can be calculated using the approach suggested by Press et al. (1986). Both sets of errors are calculated independently from one another. The discussion in Sect. 8.3.3 showed that the fit factors a_i depend on the alignment parameters $d_{j,k}$ of the spectra. Following the same arguments, the alignment uncertainties $\Delta d_{j,k}$ will increase the uncertainty Δa_j of the linear fit parameters a_i (we ignore the polynomial parameters here since they are not the main result of the analysis, although the same arguments apply). The following section will present a method to estimate this error propagation.

The fact that the residuals of DOAS spectra are often not pure noise, containing structures that are broader and larger than the expected photon noise, needs to be considered in the calculation of errors. The last part of this section will thus be dedicated to the discussion of various methods that have been developed to address this problem. A discussion of the causes of the residual structures is given in Sect. 8.6.

Influence of Alignment Uncertainty

The impact of spectral alignment errors $\Delta d_{j,k}$ on the total errors of the fitting parameters $\Delta_{\text{tot}} a_j$ depends on the shape of different absorption spectra involved. There is no simple equation that connects the alignment uncertainty with the uncertainty Δa_j derived by the linear fit. Therefore, an approximate method has to be found. The fact that the alignment uncertainty of one trace gas spectrum propagates to the uncertainty Δa_j of another gas complicates this approach. To solve this problem, Stutz and Platt (1996) suggest a method that relies on a similar approach to that described to investigate the dependence of a_j on $d_{j,k}$ in Sect. 8.3.3. A test spectrum is calculated based on the original reference spectra that are shifted and scaled based on the results of the fitting procedure, $S_j^*(i) = a_j \times S_j(d_{j,0}, d_{j,1}, \ldots)(i)$. These reference spectra thus represent the best estimate of the trace gas absorptions contained in the original atmospheric spectrum. In addition, spectrum $S_g^*(i)$ is shifted and squeezed as specified by the set of parameter errors $\Delta d_{g,k}$ according to (8.21):

$$x_\pm = i \pm \Delta f_g(i) \quad \text{with} \quad \Delta f_g(i) = \sqrt{\sum_{k=0}^{p_j} \left(\Delta d_{g,k} \times (i - i_c)^k\right)^2}. \tag{8.26}$$

Here, $\Delta f_g(i)$ represents the error of the position of pixel i. The test spectrum is then calculated by:

$$J = 1 + S_1^*(0) + \ldots + S_g^*(\Delta d_{g,0}, \Delta d_{g,1}, \ldots) + \ldots + S_m^*(0)\,. \tag{8.27}$$

A linear fit using the model function $F_{\text{sh}} = 1 + a_1^* \times S_1^*(0) + a_2^* \times S_2^*(0) + \ldots + a_m^* \times S_m^*(0)$ with the references $S_j^*(0)$ is then used to analyse J. In the case of negligible errors $\Delta d_{g,k}$, all the fit parameters a_j^* will be $a_j^* = 1$. If S_g^* in J is shifted and squeezed due to the uncertainty $\Delta d_{g,k}$, the resulting a_j will change. These changes in a_j give an estimate of the propagation of spectral alignment errors of S_i^*, on the fit result for spectra S_j^*. We can use the function $\Phi_{j,g}$ defined above to quantify this dependence. As this error is an absolute value, two spectra $S_i^*(+x)$ and $S_i^*(-x)$ have to be calculated, and two linear fits with the model F_{sh} to the spectrum J have to be performed to derive $\Phi_{j,g}(\pm\Delta f_g)$. The influence of the alignment error of S_g on the error of a_j, $|\Phi_{j,g}|$ can then be calculated by:

$$|\Phi_{j,g}(\Delta f_g)| = \frac{1}{2} \times (|\Phi_{j,g}(+\Delta f_g)| + |\Phi_{j,g}(-\Delta f_g)|) . \tag{8.28}$$

By performing this procedure for all $g = 1, \ldots, m$, one derives a set of $m^2|\Phi_{j,g}|$ values. The set of $|\Phi_{j,g}|$ can now be used to estimate the error $\Delta_{\mathrm{sh}} a_j$ of a_j caused by the uncertainty of the alignment of all spectra included in the fit:

$$\frac{\Delta_{sh} a_j}{a_j} = \sqrt{\sum_{g=1}^{m} (|\Phi_{j,g}(\Delta f_g)|)^2} . \tag{8.29}$$

Assuming that this error and the error $\Delta d_{j,k}$ of the linear fit are independent, the total error of a_j is:

$$\Delta_{\mathrm{tot}} a_j = \sqrt{(\Delta a_j)^2 + (\Delta_{\mathrm{sh}} a_j)^2} . \tag{8.30}$$

This error estimation can be included in the analysis procedure by calculating the $\Phi_{j,g}$ for all reference spectra after the procedure has finished the iteration process to derive $d_{j,k}$.

A Monte Carlo analysis illustrates the behaviour of errors in DOAS analysis (Stutz and Platt, 1996). Noise spectra, $N(\sigma)$, with standard deviations (σ) from 10^{-4} to 3×10^{-2} were added to a spectrum similar to J shown in Fig. 8.5. This test spectrum J was calculated as the sum of references S_j (see Fig. 8.2) that were linearly shifted against each other: $J = 1 + S_1(0.1) + S_2(-1) + S_3(1) + S_4(-0.1) + N(\sigma)$. The individual shifts of the spectra were: O_3, 0.1 pixel; NO_2, −1 pixel; SO_2, 1 pixel; HCHO, −0.1 pixel. To analyse the spectra, a model function allowing individual shifts was used:

$$F = P_4 + a_1 \times S_1(d_{1,0}) + a_2 \times S_2(d_{2,0}) + a_3 \times S_3(d_{3,0}) + a_4 \times S_4(d_{4,0}) .$$

For every level of σ, 800 different artificial noise spectra were calculated. The resulting 800 spectra J were then analysed with the fitting algorithm described earlier. The average results $\overline{a_j}$ and $\overline{d_{j,k}}$ agreed with the values used to calculate J. Figure 8.6a illustrates that $\overline{\Delta d_{j,k}}$ is linearly dependent on noise levels σ, as long as the noise is not too high as compared to the optical density of the reference spectra. It is remarkable that uncertainties in the shift of several pixels to several tens of pixels are found due to noise in the

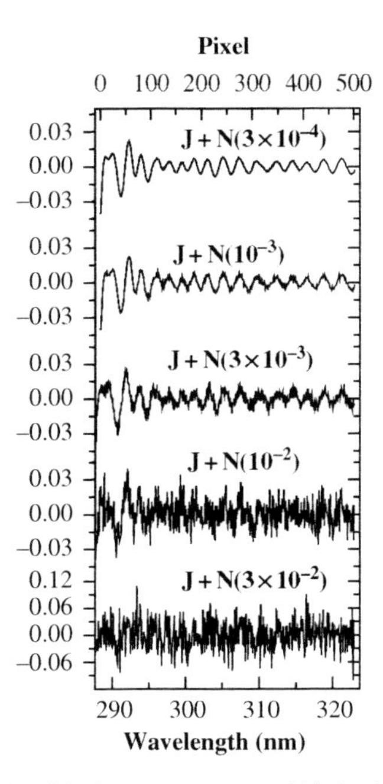

Fig. 8.5. J (Fig. 8.2) with added noise spectra $N(\sigma)$ of different magnitudes from 3×10^{-4} to 3×10^{-2}. The magnitude is defined by the standard deviation σ over the pixel intensities. For high σ, the spectrum J cannot be identified in the noise (from Stutz and Platt, 1996)

spectra. The errors of the shift $\overline{\Delta d_{j,k}}$ agreed well with the true error $\delta(d_{j,k})$, which was calculated from the standard deviation of the fit results $d_{j,k}$ from the true shift.

Figure 8.6b illustrates that the average of the total fit error $\overline{\Delta_{\text{tot}} a_j}$ increases linearly with increasing noise level σ. Figure 8.6c shows how well the error calculation discussed in the previous section describes the statistical fluctuations of the results. The average total errors $\overline{(\Delta a_j)_{\text{tot}}}$ agree with $\delta(a_j)$ within better than 10%, as long as the relative error of a_j is smaller than 0.5. Figure 8.6d shows that the influence of the shift error is, in this case, relatively small as compared to the average error $\overline{\Delta^{\text{lin}} a_j}$ of a linear analysis. The influence of the shift error $\Delta d_{j,k}$ is highest for O_3 and SO_2. A shift uncertainty of a few hundredths of a pixel can, in some cases, increase the total error of a_j by 20–50%, thus increasing the derived trace gas concentration error considerably.

The Effects of Residual Structures

A common problem in the application of the DOAS technique is the occurrence of structures other than noise in the residuum J–F of the fit. These structures may indicate an unknown absorber, or can be caused by the instrument itself, and occur at random in most of the cases. Possible sources will be further discussed below. Here, only random structures will be discussed. Non-random

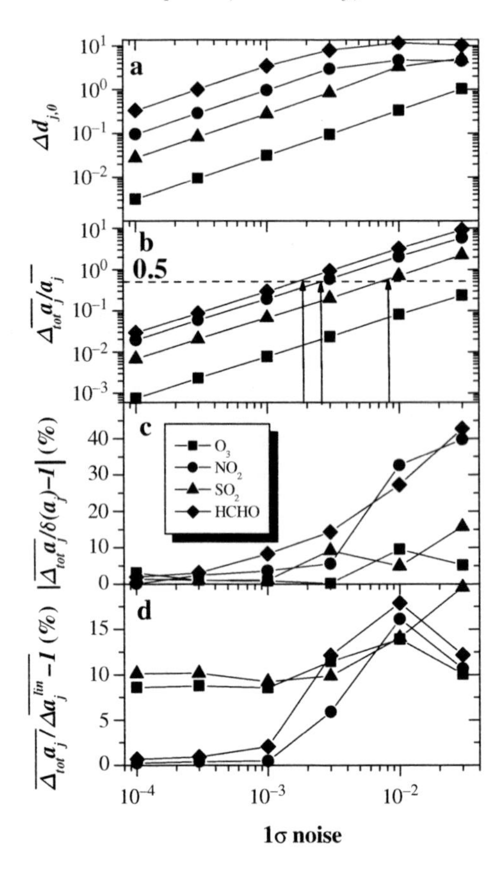

Fig. 8.6. Results of the investigation of the error calculation, including spectral alignment uncertainties (from Stutz and Platt, 1996)

residual structures cause systematic errors in the analysis, which cannot be described by the statistical methods.

In general, the effect of residual structures is to increase the error of the fit parameters a_j and the derived trace gas concentrations. A mathematical reason for this behaviour will be given below. A number of empirical methods have been applied to correct the effect of residuals, most of which are based on the experience of the respective researchers with their instruments. The following lists some of the common methods:

- The root mean square of the residual structures in the fitting window scaled with the absorption cross-section and path length gives an estimate of the detection limit of the instrument. Smith et al. (1997), for example, used this approach during their NO_3 measurements in California.
- A very conservative approach is the use of peak-to-peak size of the residual structure scaling it with cross-section and path length.

- The peak-to-peak size of the residual, divided by a weighting function determined by the number of absorption bands in the fitting window, has also been used.
- The strongest artefact line in the residual closest to the trace gas absorption structure has also been used.

In recent years, other methods based on more statistical arguments have been suggested in literature. Some of those methods will be discussed below.

Figure 8.7 shows examples of residual structures in different DOAS applications. The general features of these spectra, as compared with pure noise spectra as shown in Fig. 8.7, are structures that are several channels wide. One can argue that, if these structures are random, the errors of the channels

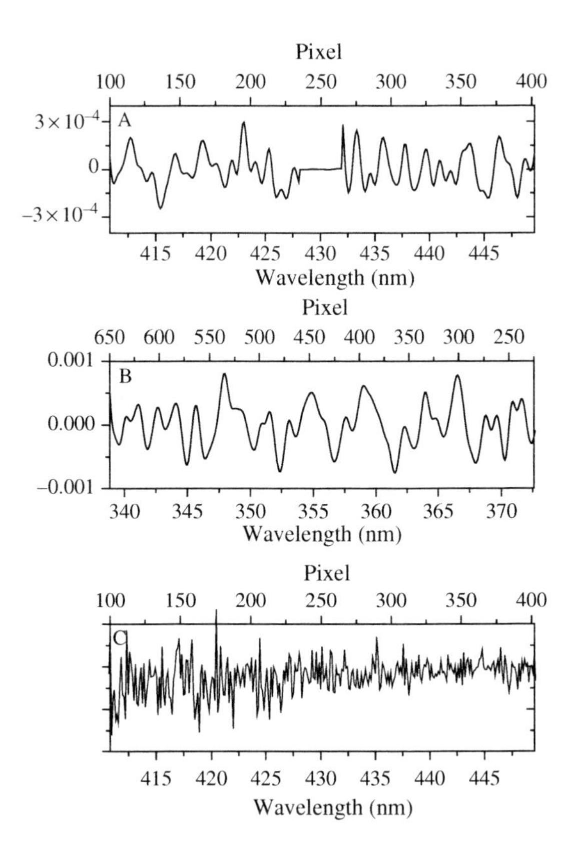

Fig. 8.7. Example of residual structures in different DOAS applications. Plate A is from measurements with a MAX-DOAS system. The flat residual around 430 nm is caused by the exclusion of this interval from the analysis. Plate B shows an example of a LP-DOAS measurement of NO_2 and HONO. Both residuals were filtered to remove random noise and to better illustrate the shape of the residuals. In both cases, the random noise was considerably lower than the residuals. Plate C shows a pure noise spectrum measured using lamp references of a MAX-DOAS system

of such a structure are not independent of each other, i.e. a random event will change all channels of a structure simultaneously (Stutz and Platt, 1996; Hausmann et al., 1999). This argument can also be supported if one recalls that, due to the finite resolution of the spectrometer, a single wavelength is imaged on a number of different pixels determined by the width of the instrument function H. Any random change in the instrument that is strongly wavelength dependent will, therefore, lead to structures that are as wide as, or wider than, the instrument function.

The fact that the errors of individual channels are correlated indicates that an uncorrelated least squares fit is inappropriate in this case. However, all known DOAS analysis programs are based on such an uncorrelated least squares approach.

Stutz and Platt (1996) describe how the correlation between errors may influence the results of a least squares fit. To introduce a correlation between the errors, they used a pure noise spectrum and smoothed it with a moving average. Figure 8.8 compares such a spectrum with a residual from the analysis of an atmospheric absorption spectrum.

The influence of these residuals and the applicability of a correlated least squares fit was tested by Monte Carlo experiments in which several hundred noise spectra were smoothed by a running mean of 9 pixel width. Each of these spectra was added to the test spectrum shown in Fig. 8.2 and analysed with a model function of the sum of the reference spectra, both with an uncorrelated and a correlated linear least square fit without spectral alignment. The changes in the resulting a_j were small, but the calculated errors increased drastically by a factor of 3 (Fig. 8.9) for the uncorrelated least squares fit. The tests also showed that the underestimation of error is independent of the magnitude of the residual.

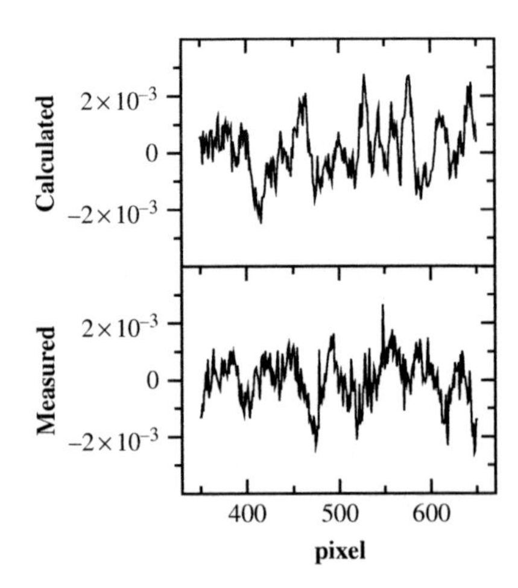

Fig. 8.8. Example of calculated and measured residual structures (from Stutz and Platt, 1996)

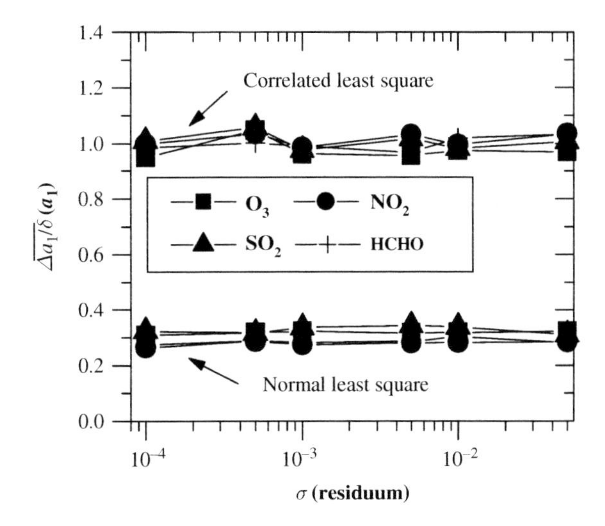

Fig. 8.9. Comparison of the ratio of average fit error and standard deviation of the results of a normal fit and a correlated least squares fit for a spectrum containing residual structures (from Stutz and Platt, 1996)

The correlation of errors was described by the variance–covariance matrix of the errors of the pixel intensity $(\mathbf{M})_{ij} = E\left(\varepsilon_i \times \varepsilon_j\right)$ (8.31), where ε_k is the error of the intensity of pixel k, and E is the expectation value of $\varepsilon_i \times \varepsilon_j$. This matrix can be built for the running mean smoothing, assuming all ε_k to be of equal size 1:

$$M_9 = \frac{1}{81} \times \begin{pmatrix} & & \ddots & & \\ \cdots 1\,2\,3\,4\,5\,6\,7 & 8 & 9\,8\,7 & 6 & 5\,4\,3\,2\,1\,0\,0\,0 \cdots \\ \cdots 0\,1\,2\,3\,4\,5\,6 & 7 & 8\,9\,8 & 7 & 6\,5\,4\,3\,2\,1\,0\,0 \cdots \\ \cdots 0\,0\,1\,2\,3\,4\,5 & 6 & 7\,8\,9 & 8 & 7\,6\,5\,4\,3\,2\,1\,0 \cdots \\ & & & \ddots & \end{pmatrix} . \tag{8.31}$$

M_9 is a symmetric band matrix with 9/81 in the diagonal. The correlated least squares fit was then performed according to (8.5) on pixel 160 to 260 of the test spectra to reduce calculation time.

Figure 8.9 shows the ratio of the average error calculated from both fitting procedures, $\overline{\Delta a_j}$, and the standard deviation of the fit factors, a_i, $\delta(a_j)$. In the case of uncorrelated fit, the errors are underestimated by a factor of ~3. The errors derived with the correlated fit agree, on the other hand, within 5% to the statistical scatter of the fit factors.

This suggests that, if the assumption of correlated channel errors during a DOAS analysis is indeed correct, as assumed by several authors (Stutz and Platt, 1997a; Hausmann et al., 1999), the uncorrelated least squares fit underestimates the error considerably. A correlated fit will give calculated correct errors if the correlation matrix, $\mathbf{M}$, is known.

The most serious limitation in the use of a correlated least squares fit is unknown error correlation in a DOAS residual spectrum. In the case shown above, the correlation of errors was known, and could thus be incorporated into the analysis. For atmospheric spectra, it is likely that the correlation changes with time, and perhaps even from spectrum to spectrum. This would make the determination of $\mathbf{M}$ difficult and time consuming. However, this discussion is important since it puts the treatment of residual structures on a mathematical basis, moving away from purely empirical correction methods. It remains to be explored if correlated least squares will be used in the future.

Other methods that have been suggested are based on other statistical methods or approximations.

Stutz and Platt (1996) suggest, based on their use of the running mean in the study of correlated least squares fit, to simulate residuals using this method, and to use the Monte Carlo experiments to derive correction factors for the errors of an uncorrelated least squares fit. They performed experiments with artificial absorption spectra that contained five identical, non-overlapping Gaussian-shaped absorption lines with half width, τ, ranging from 2.5 to 50 pixels in the different spectra. These spectra were added to noise spectra $N(W)$ that were smoothed by different running means of width $W = 2$ to 64, and scaled to a standard deviation of 1%. For every combination of τ and W, 500 random noise spectra N were calculated, smoothed, and added to $J = S(\tau) + N(W)$. Then the model $F = P_0 + a_1 \times S(\tau)$ was fitted to the spectrum J. The ratio of the standard deviation of the fit parameter $\delta(a_1)$ and the average error $\overline{\Delta a_1}$ calculated by the fit of 500 different residuals is the desired correction factor $C(\tau, W) = \delta(a_1)/\overline{\Delta a_1}$. Figure 8.10a shows $C(\tau, W)$ for different combinations of W and τ. The graph shows that the ratio is lowest for broad structures of the spectrum and the residual. As the correction factor is independent of the magnitude of the residual, this factor can be applied to correct the results and analyse the width of the residuals and the reference spectra. The same test was repeated with the non-linear procedure for $d_{1,0}$ by analysing the error of the shift after shifting $S(W)$ by 1 pixel against its reference and then adding the residuals $N(W)$. Figure 8.10b illustrates a similar behaviour in the linear case.

Hausmann et al. (1999) describe two statistical methods to estimate the true error of a DOAS analysis with residual structures. The first method, called Residual Inspection by Cyclic Displacement (RICD), is based on the assumption that artificial residual spectra can be simulated from true residual spectra by a cyclic shift, i.e. shifting a residual spectrum by k pixel, and appending the k pixel that are shifted out of one side of the spectral window to the other side. To derive the true error of the parameters a_j, they suggest to first analyse the atmospheric absorption spectrum. The residual $R(i) = J(i) - F(i)$ of this fit is then shifted cyclically by k pixel, $R^*(i) = R(i + k)$,

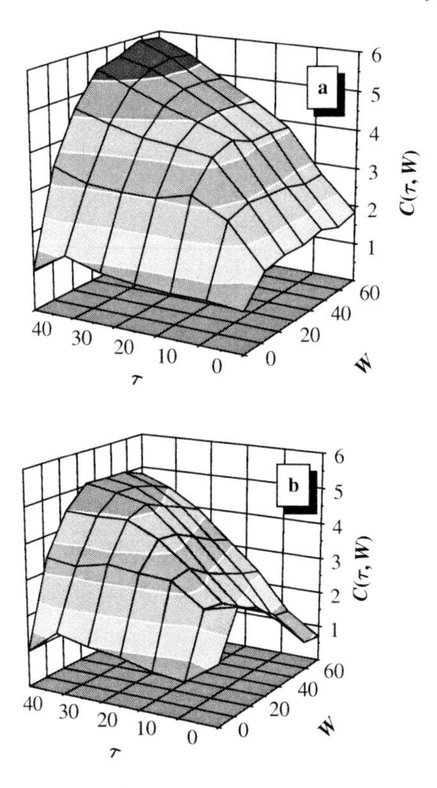

Fig. 8.10. Correction factor $C(\tau, W)$ for the linear and non-linear fit (from Stutz and Platt, 1996)

and all the reference spectra, scaled according to the results of the original fit a_j, are added to this shifted spectrum: $J^*(i) = P(i) + a_1 \times S_1 + a_2 \times S_2 + \ldots + a_m \times S_m + R^*(i)$. Figure 8.11 illustrates the shifting method for an OH radical absorption spectrum. The analysis of $J^*(i)$ with the same procedure gives a new set of fit factors a_j^*. By repeating this procedure with a large number of shift distances k, a set of fit factors a_j^* is derived. By calculating the standard deviation of this data set for each spectrum S_j, one can estimate the error of fit $\Delta_{\mathrm{RCID}} a_j$. It is obvious that, in the case that the residual spectrum $R(i)$ consists of pure noise, this method will yield similar uncertainties to the least squares fit calculated directly through (8.4). Hausmann et al. (1999) found that this method slightly underestimated the true error, $\Delta_{\mathrm{true}} a_j$, of the fitting parameter, which was calculated from the standard deviation a_j calculated for a large number of different unshifted residual spectra. Values in their experiments with pure noise residuals were 5–30% lower than those calculated by the least squares fit. In the presence of residual structures different from noise, a similar bias between true error and error calculated by the RICD

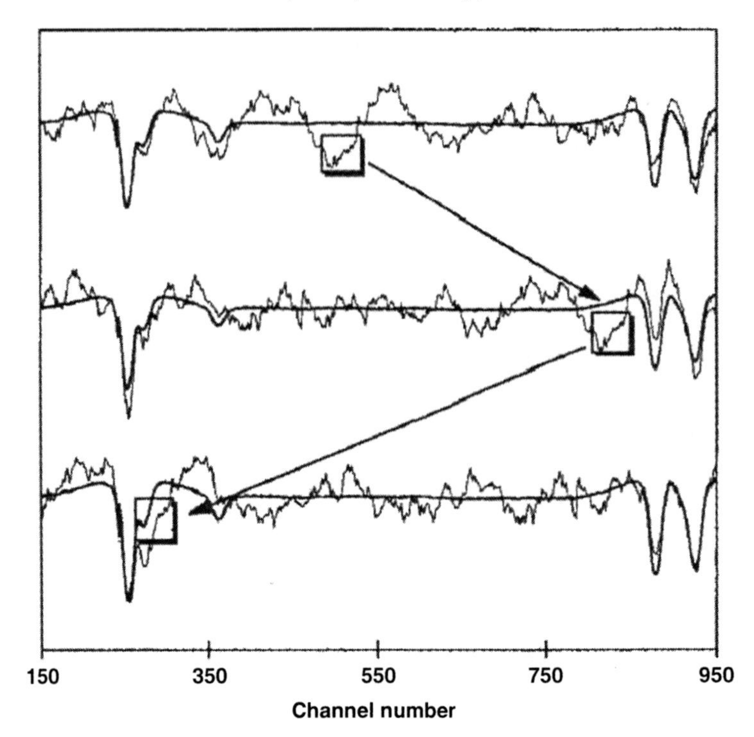

Fig. 8.11. Example of the proposed RICD method (from Hausmann et al., 1999)

method was found (Table 8.1). The true and the RICD errors were, however, considerably larger than the error that was calculated by the least squares fit alone. The factor between true and pure least squares fit was ~4. The precise value was dependent on the individual reference spectrum (Table 8.1). A limitation in this method is the restriction to residuals that have relatively constant peak-to-peak strength across the spectrum.

The other method is based on the bootstrap approach (BS). In short, the bootstrap method, which was introduced by Efron (1979), is a resampling procedure used to estimate the distributions of statistics based on independent observations. An estimate of the sampling distribution is created by randomly drawing a large number of samples from a population, and calculating for each one an associated statistical parameter.

Hausmann et al. (1999) applied the BS method to DOAS by using the absorption spectra and reference spectra as the original data sets. The statistical estimate is derived by a linear least squares analysis described in Sect. 8.1.1. The BS procedure is performed in three steps.

1. A large number (100–1000) of artificial spectra (both absorption and reference spectra) are created. These spectra have the same number of channels as the original spectra. The content is determined by randomly attributing

Table 8.1. Standard deviation of the ratio between calculated error and statistical scattering of the results of the 100 simulated measurement spectra[a]

Estimation methods	Measurement spectra			
	Ratios for the following errors			
	RICD	Bootstrap	BS ⊕ RICD	err(LSQ)
OH	1.43	1.32	0.91	4.27
$C_{10}H_8$	1.30	1.53	0.96	4.86
SO_2	1.20	1.44	0.91	3.91
HCHO	1.36	1.60	1.00	3.98
Deviation from true error	0.75	0.65	1.05	0.25

Table adapted from Hausmann et al. (1999).
[a]In the case of correct error estimation, the ratio between calculated error and statistical scattering of the results is unity.

channels or blocks of channels from the original spectra to channels of the new spectra, as illustrated in Fig. 8.12. This drawing procedure is equally applied (the same pixel of the original is attributed to the artificial spectra) to both atmospheric absorption and reference spectra, yielding new spectra, $J^*(i)$ and $S_j^*(i)$. As illustrated in Fig. 8.12, channels from the original spectrum can occur more than once in the new artificial spectrum, while others may not occur at all.

2. In the second step, the artificial atmospheric absorption spectrum $J^*(i)$ is analysed with a DOAS linear least squares procedure (see Sect. 8.1) that employs the reference spectra with the identical drawing scheme $S_j^*(i)$. This is repeated for each of the 100–1000 different drawings.
3. In the last step, the standard deviation of all the fitting results, i.e. a_j^*, is determined. This is, according to the BS theory, an estimate of the error of the original parameters, a_j.

As illustrated in Fig. 8.12, the block size of resampling can vary from a single channel to as many as hundreds of channels. Hausmann et al. (1999) showed that the results of their calculations are not independent of the block size, and attributed this to the interdependence/correlation between different channels. The dependence on the block size disappears for most spectra when a large block size is used. They recommend a block size of at least 40 channels to account for this effect. At this block size, the error calculated by the BS method, $\Delta_{BS}a_j$, reaches a plateau for trace gases with multiple absorptions in the spectral window, and larger block sizes do not change the results. However, for absorbers with only one or two absorption bands, this plateau is not reached, and the BS method may not be applicable.

The Monte Carlo test performed by Hausmann et al. (1999) (Table 8.1) shows that, as was the case for the RICD method and the correlated least squares fit (Stutz and Platt, 1996), the true error is about a factor of ∼4

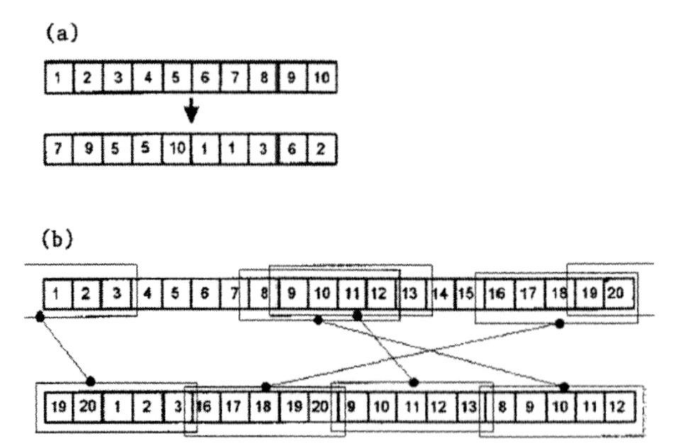

Fig. 8.12. Schematic example of the bootstrap method: (**a**) resampling of single channels, (**b**) block resampling (from Hausmann et al., 1999)

higher than that calculated by the normal uncorrelated least squares fit. The errors calculated by the BS method underestimate the true error by only 30–60%, and is thus a much better estimate of the true error.

The authors argue that, because the RICD and BS methods describe different contributions to the spectral evaluation error caused by residual structures, a combination of the two methods – BS $\oplus$ RCID – is the best estimate. This combination can be calculated by $\Delta_{\text{tot}} a_j = (\Delta_{\text{RCID}} a_j^2 + \Delta_{\text{BS}} a_j^2)^{1/2}$. Since the error of a noise-dominated residual would be overestimated due to the addition of both methods, it is proposed that this error is scaled by a factor of 0.85 to take into account the fact that it is impossible to distinguish between pure noise and residual-dominated spectra.

Validation of Error Analysis Methods

It is challenging to validate the error calculations for a realistic DOAS analysis, i.e. for true atmospheric spectra, due to unknown trace gas amounts in the atmosphere. One older approach is the statistical analysis of fit results for a trace gas that is believed to have much lower concentrations than the detection limit of the DOAS instrument used in its measurement. While this approach can give an approximate check for the correctness of the routine, it is not accurate enough to make precise quantitative statements. The method adopted by Tuckermann et al. (1997) and Alicke et al. (2002) compares the statistical standard deviation of the fit results with the average of errors calculated by the fit for the same data points. Only those data points for which it can be assumed that the concentration of the respective trace gas is much smaller than the detection limit are used. Obviously, this method is limited by the latter assumption which, in many cases, cannot be verified from the DOAS measurements alone.

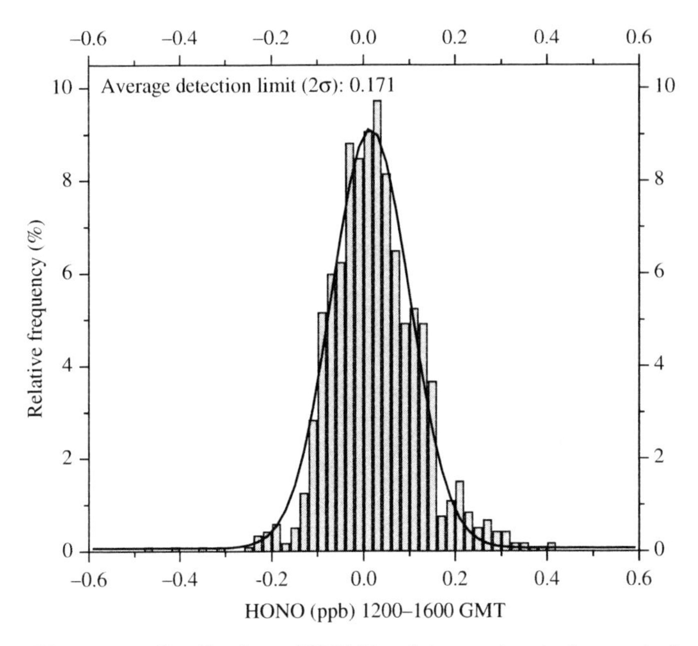

Fig. 8.13. Frequency distribution of HONO mixing ratios during periods of strong actinic flux and expected low HONO concentrations. The relative frequency agrees well with a Gaussian curve (*solid line*, see text) at 15 ppt. The average error of 0.171 ppb agrees well with the width of this distribution (from Alicke et al., 2002)

Figure 8.13 shows the results of such an analysis using HONO measurements in Milan, Italy (Alicke et al., 2002). The histogram of HONO mixing ratios observed by DOAS during times of strong solar radiation scatter resembled a Gaussian distribution, with a maximum of ~0.015 ppb. The width of this distribution was very similar to the average error calculated by the DOAS analysis routine. Based on this analysis, Alicke et al. (2002) concluded that the error calculation based on Stutz and Platt (1996) (see also above) does indeed give good estimates of the true error of DOAS results. The results by Tuckermann et al. (1997) show a similar agreement.

Examples of Analysis Procedures

At the end of this section, we present two examples of the analysis of atmospheric spectra. Figure 8.14 shows an example of the analysis of a spectrum measured using the MAX-DOAS method. The atmospheric spectrum to be analysed, i.e. $J(i)$, is the measurement at 3° elevation displayed in the top trace. The spectra below this trace are the zenith measurements, which serve as Fraunhofer reference, and the scaled references of NO_2, H_2O, and O_4, i.e. $a_j \times S'_j(i)$. Also included in the fit is a spectrum describing the Ring effect

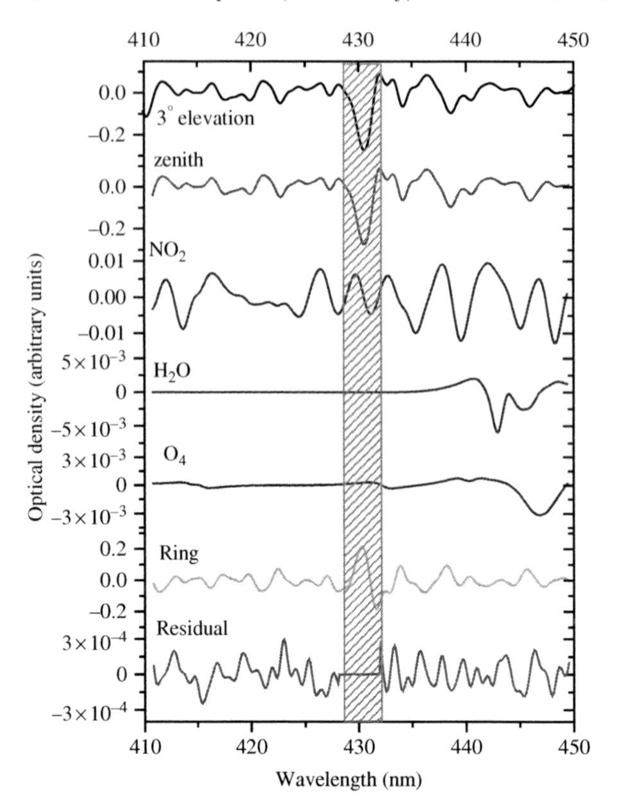

Fig. 8.14. Example of the analysis of a MAX-DOAS spectrum measured in Los Angeles in October, 2003. The *top trace* shows the spectrum measured with a 3° elevation. The *trace below* is the zenith spectrum. The various other traces are the scaled reference spectra of NO_2, H_2O, and the ring-effect spectrum. The *lowest trace* shows the residual structure of this fit. All spectra were high-pass filtered before the analysis

(see Sect. 9.1.6). The bottom trace shows the residual of the fit, which was, in this case, larger than the noise one would expect in the respective spectra.

Figure 8.15 shows the example of measurements of NO_2 and HONO by a long-path instrument in Nashville, TN. All spectra were high-pass filtered before analysis. The middle trace shows a comparison of the scaled HONO reference spectrum and the sum of this spectrum and the residual (bottom trace). This comparison allows the visualisation of the quality of the analysis procedure. If the absorptions cannot be identified in the sum spectrum, the trace gas concentration is below or very close to the detection limit of the instrument. If the absorption bands can be identified, the quality of the fit was good and the measurement of the respective trace gas with a small error is possible.

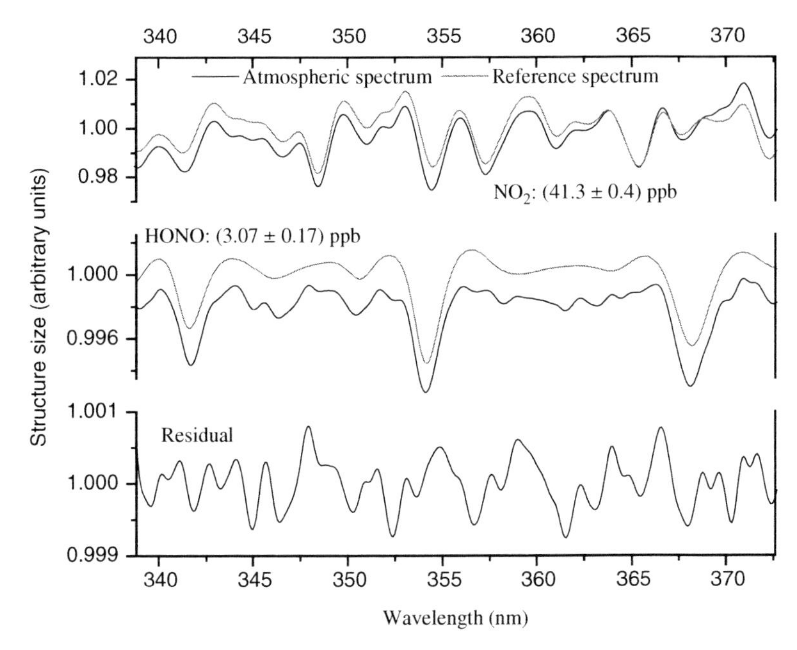

Fig. 8.15. Example of the analysis of a spectrum measured by a long-path DOAS. The two *top traces* show the atmospheric spectrum and the NO_2 reference spectrum after high-pass filtering. The *middle traces* show the HONO reference scaled to the absorption contained in the atmospheric spectrum. Also shown is the sum of this spectrum and the residual, which allows a visual confirmation of the quality of the fit. The *lowest trace* is the residual of the fit. All spectra shown in this figure were filtered to reduce noise in the spectrum

Because of the breadth of DOAS application (see Chap. 6), it is not possible to show examples of all the possible DOAS analysis procedures. Chapters 10 and 11 will, however, show more examples of the implementation of DOAS analysis procedures.

8.4 Determination of Reference Spectra

The most direct, and often most successful, approach to obtaining the correct reference spectra S_j in (8.15) is to measure them with the instrument that is also used to perform atmospheric measurements. For many years, this has been the preferred method for trace gases such as SO_2, NO_2, HCHO, O_3, and aromatics in the troposphere. For SO_2 and NO_2, one can fill the reference cells with sufficient accuracy and long-term stability to transport them into the field. While the use of an SO_2 absorption cell is straightforward, NO_2 cells are more difficult to use due to the temperature-dependent equilibrium of NO_2 with N_2O_4, which leads to temperature-dependent NO_2 concentration in the cell. The measurement of HCHO in the field is often achieved by heating a cell with para-formaldehyde. Despite the higher temperatures in the cell, this approach has been shown to be quite good. The measurement of O_3 in the field is rarely used anymore due to the substantial experimental effort, and the fact that the O_3 absorption cross-section is temperature dependent in the UV and thus difficult to measure in a temporary field setup. Aromatics can be measured by filling the fresh liquid compound into a small cell and relying on the relatively high vapour pressure of these compounds to provide the concentration needed for cell measurement. The advantage of this method is a very accurate determination of the absorption structure, which is particularly important for the different aromatics, which have very similar absorption spectra.

The direct measurement of pure trace gas absorptions has the advantage of providing the best reference spectra, since the convolution with the instrument function (see Chap. 6) is performed through the measurement. The disadvantage of this approach is, however, that for all trace gases, except perhaps SO_2, an absolute calibration of the absorption cross-section is not easily possible. The absorption spectra must be calibrated after the measurement with literature absorption cross-sections. The very accurate description of S_j, however, outweighs the disadvantage of this calibration.

8.4.1 Theoretical Basis of Reference Spectra Simulation: Convolution

For trace gases that are too reactive to be measured in the field in their pure form and to calibrate the measured reference spectra, one has to rely on the simulation of the absorption spectra for the respective instrument. The first step in this simulation is the determination of the instrument function of the spectrometer, H. For most low-resolution DOAS applications, atomic emission lines from low-pressure lamps provide an effective means to determine H. Because the natural line width of the emission lines is much smaller (~10 pm) than the typical resolution of DOAS spectrometers (~0.1 – 1 nm), the emission line measured with the respective spectrometer–detector system is a very good approximation of H (see Fig. 8.16).

In addition to the emission lines, the accurate determination of the wavelength to pixel mapping is also required in the simulation. This can be determined by using a number of well-known positions of atomic emission lines, and/or by using known absorption structures of trace gases or the solar Fraunhofer lines. Most high-quality spectrometers display a nearly linear dependence of channel numbers and wavelengths with a very small non-linear component. Less expensive spectrometers, however, can have more complex dependencies. However, in most cases, a polynomial of degree 3 describes the wavelength-pixel mapping sufficiently.

As we discussed earlier, the measurement of an absorption spectrum changes its shape by reducing its spectral resolution. Mathematically, this

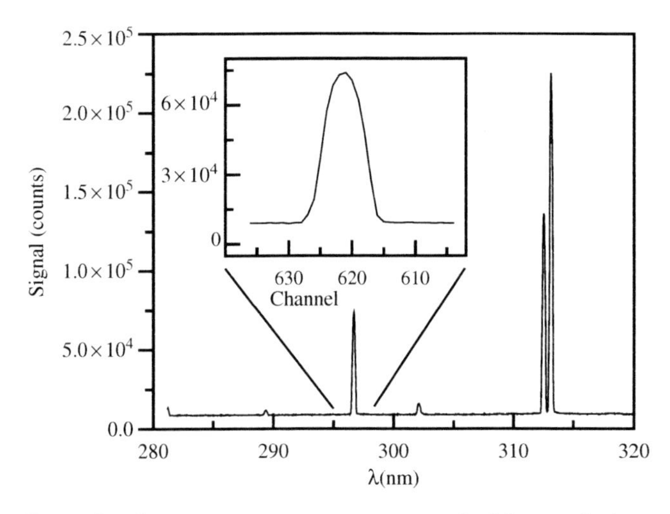

Fig. 8.16. Example of a mercury spectrum measured with a typical spectrometer–detector combination used for DOAS. The insert shows the shape of the mercury atomic emission line (see Appendix A) that is used for a convolution

process can be expressed by a convolution of the absorption spectrum $I(\lambda)$ with the instrument function $H_{\mathrm{sp}}(\lambda)$ of the spectrometer via:

$$\begin{aligned} I^*(\lambda) &= \int I(\lambda-\lambda') \cdot H_{sp}(\lambda')\, d\lambda' \\ &= I(\lambda')^* H_{sp}(\lambda') . \end{aligned} \tag{8.32}$$

In addition, most detectors show certain sensitivity overlap between adjacent channels. For example, in the case of a photodiode array, ~6.5% of the signal of a pixel stems from each neighbour. This can again be represented by a convolution of, in this case, $I^*(\lambda)$ with the detector response function, which will be called $w(i, \lambda)$, where i is the channel number and λ_i the mid-wavelength of the respective channel.

$$I'(i) = \int I^*(\lambda_i - \lambda') \cdot w(i, \lambda')\, d\lambda' . \tag{8.33}$$

In reality, it is often impossible to distinguish the convolution by the spectrograph and the detector. By combining (8.32) and (8.33), it can be seen that both processes can be combined to a single instrument function $H = H_{\mathrm{sp}}{}^* w(i, \lambda)$. During the measurement of an atomic emission line, the combined instrument function, H, is measured. The mathematical idea behind this measurement is that an atomic emission line is narrow as compared to the resolution imposed by H. Thus, it can be approximated by a δ-function:

$$I^*_{AEL}(\lambda) = \int \delta(\lambda-\lambda') \cdot H(\lambda')\, d\lambda' = H(\lambda) . \tag{8.34}$$

While in many spectrometers one can assume that H is independent of the wavelength, we have encountered a number of instruments where H is a function of wavelength $H(\lambda)$. This dependence can lead to different emission line widths on both ends of the recorded spectrum.

8.4.2 Practical Implementation of Reference Spectra Simulation

To simulate the absorption spectra, and thus obtain reference spectra for (8.15) for a specific spectrometer–detector combination, the following steps have to be performed:

1. Choose an accurate absorption cross-section $\sigma(\lambda)$ for the respective trace gas. Ideally, the spectral resolution of this cross-section should be much better than the spectral resolution of the spectrograph–detector system. While finding a highly resolved absorption cross-section is sometimes challenging, it has been shown that a resolution at least 10–100 times better is necessary to avoid problems with this approach (see below). Because the cross-section is also digitised with a step interval of $\Delta\lambda_\sigma$, we will call it $\sigma(j)$.
2. Determine the wavelength to pixel mapping, for example by using an atomic emission line spectrum. If the dispersion is linear, the distance between two pixels will be $\Delta\lambda_{AEL}$.
3. Choose the atomic emission lines that can be used for convolution. Only single lines with a good signal-to-noise ratio should be used. If possible, choose several lines in different positions in the spectrum to take into account the possible changes of H with the wavelength.
4. Because the convolution applies to the intensity, an absorption spectrum has to be calculated next:

$$I_{ref}(j) = I_0(j) \exp\left(-A \cdot \sigma(j)\right) . \tag{8.35}$$

   We can distinguish two cases for this calculation. In the case of smooth light sources, such as those provided by artificial lamps, $I_0(j)$ can be set as constant, e.g. $I_0(j) = 1$. If the light source has strong narrow-band spectral features, e.g. the sun, it is often better to use the known $I_0(j)$ of the light source. We will come back to this point later. The parameter A represents the trace gas column (concentration × absorption path length). In general, A should be chosen to be similar to the expected column. If the absorber is weak, the precise value of A is not very important. If the absorber is strong or there is a risk that some of the high-resolution features of the spectrum are in or close to saturation, one must choose A as close to the real column as possible to simulate saturation effects accurately. This is, for example, often the case for H_2O and O_2.
5. In the next step, the convolution is performed. To make this step simple, one needs to bring I_{ref} and the mercury lines chosen for the convolution to the same wavelength grid. We have successfully used polynomial

extrapolation routines to expand the mercury lines to the $\Delta\lambda_\sigma$ grid. After this expansion, the convolution is performed using a suitably chosen wavelength/channel window from j_{low} to j_{up}:

$$I^*_{ref}(j) = \sum_{j_{low}}^{j_{up}} I_{ref}(j - j')H(j'). \tag{8.36}$$

6. Finally, the convoluted spectrum $I_{\text{ref}}{}^*(j)$ with a wavelength grid of $\Delta\lambda_\sigma$ needs to transferred to the grid determined by the wavelength pixel mapping. The content of one pixel can be calculated by integrating $I_{\text{ref}}{}^*(j)$ from the lower to the upper wavelength border of the pixel:

$$I_{ref}{}^{**}(j) = \int_{\lambda(j)-\frac{\Delta\lambda}{2}}^{\lambda(j)+\frac{\Delta\lambda}{2}} I_{ref}{}^*(\lambda_k)d\lambda. \tag{8.37}$$

7. To calculate the spectrum in the form that can be used in the model function of the fit (8.15), the logarithm of the ratio of $I_{\text{ref}}{}^{**}(j)$ and $I_0{}^{**}(j)$ has to be calculated. If I_0 is wavelength dependent, the same convolution as in (8.35–8.37) with $A = 0$ needs to be performed before the division.

$$S_{ref}(j) = -\ln\left({I_{ref}{}^{**}(j)}/{I_0{}^{**}(j)}\right). \tag{8.38}$$

The reference spectrum $S_{\text{ref}}(j)$ calculated in (8.38) should be an accurate representation of the respective trace gas absorption, if all the input data were correct.

Figure 8.17 shows a comparison of the measured and calculated spectrum of SO_2. The calculation was performed with the absorption cross-section of SO_2 and the mercury line shown in Fig. 8.16. The correlation coefficient between the two spectra is 0.9999, showing the excellent agreement that can be achieved with this method.

8.4.3 Optimum Resolution of Literature Reference Spectra

Our method relies on the assumption that the spectral resolution of the literature absorption cross-section can be ignored. It is worthwhile to consider when this assumption is fulfilled. A simple calculation based on Gaussian-shaped instrument functions reveals that, to determine an optical density better than $\varepsilon\%$, the spectral resolution (here expressed κ_{AXS} and κ_H as the half-width of the Gaussian instrument function for the absorption cross-section and the instrument) has to fulfill the following condition:

$$\kappa_{AXS} < \frac{\varepsilon}{100\%} \cdot \kappa_H. \tag{8.39}$$

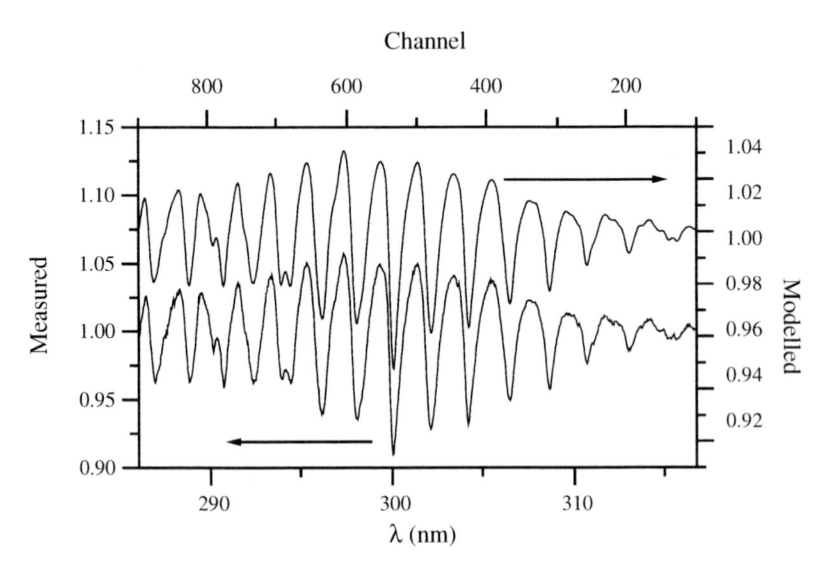

Fig. 8.17. Comparison of a measured and modelled SO_2 absorption spectrum. The mercury line shown in Fig. 8.16 and the absorption cross-section by Vandaele et al. (1994) were used in the simulation. The correlation between the two spectra is 0.9999

To describe the half-width of an absorption band better than $\varepsilon\%$, the following condition has to be met:

$$\kappa_{AXS} < \sqrt{\frac{\varepsilon}{100\%}} \cdot \kappa_H . \tag{8.40}$$

These two equations illustrate that the spectral resolution of the literature absorption cross-section used from the convolution process should be at least 10 to 100 times better than the instrument resolution. Fortunately, the data set of high-resolution absorption cross-sections has been increasing considerably over recent years.

Finally, we will discuss what approaches are available if a high-resolution absorption cross-section is not available. If both the instrument function of the absorption cross-section and of the instrument are available, it is possible to convolute the atmospheric absorption spectrum with the instrument function of the absorption cross-section, and to convolute the cross-section with the instrument function of the spectrograph used in the atmospheric measurement following the same strategy as outlined earlier. Both the absorption cross-section and the absorption spectrum will be based on the product of the two instrument functions. Any high-resolution absorption cross-section that is included in the analysis will have to be convoluted with both instrument functions to match the resolution. This also illustrates that it is essential to report absorption cross-section with the instrument function that was used for their measurement.

8.5 Detection Limits

It is now interesting to derive the detection limit for the spectra S_j at a given noise level σ. We will define the detection limit by the trace gas concentration, which can be determined with a relative error of $\Delta a_j / a_j = 0.5$. For example, Figure 8.6b can be used to derive the detection limit of the trace gases for a given noise level σ. Unfortunately, this method is very cumbersome, as the Monte Carlo experiments are very time-consuming.

A faster but less exact method is the analysis of the pure linear problem, neglecting the uncertainties of the wavelength-pixel-mapping. This estimation of the detection limits by calculating $\Delta a_j = \sqrt{\Theta_{jj}} = \sigma \times \sqrt{\left((\mathbf{X}^T\mathbf{X})^{-1}\right)_{jj}}$ (8.4) with known σ still requires the calculation of the covariance matrix Θ. Assuming that the absolute value of the linear correlation coefficient between the references is smaller than 0.1–0.2, (8.4) can be simplified further. After this simplification, an expression for the smallest $\overline{D}$ detectable can be derived. For a given noise level σ and a number of pixels n, the detection limit $\overline{D}_{\text{limit}}$ can be approximated by:

$$
\begin{aligned}
\overline{D}_{\text{limit}} &\approx \sigma \times \frac{6}{\sqrt{n-1}} \qquad \text{(a)} \\
\sigma_{\text{limit}} &\approx \frac{\sqrt{n-1}}{6} \times \overline{D} \qquad \text{(b)}\,.
\end{aligned} \tag{8.41}
$$

This expression was validated by Stutz and Platt (1996) with their results of the Monte Carlo experiments with a number of pixels $n = 500$. In Fig. 8.6b, the noise levels σ_{limit} given by (8.41b) for the $\overline{D}$ of NO_2, SO_2, and HCHO, respectively, are marked by arrows. The arrows point to the curve of relative error $\Delta a_j / a_j$ of the respective trace gas. The estimated detection limits are within 10–20% of the value derived in the Monte Carlo tests. Admittedly, this is only a rough estimation, but it can be performed on every spectrum without complicated calculations.

From the data in Fig. 8.9b, it can be seen that it is possible to analyse spectra with (1σ) levels of pure uncorrelated noise higher than the typical optical densities of the absorption bands. At the detection limit, where the 1σ error of a_j is $\Delta a_j \approx 0.5 \times a_j$, the noise level can be about a factor 3.5 larger than the average optical density $\overline{D}$ for a wavelength region with 500 pixels. Therefore, the analysis of spectra where absorption structures cannot be identified with the eyes is possible. The detection limits are, in this case, smaller than the intuitively expected values.

In the presence of residual structures, a correction factor of 3–4 (Stutz and Platt, 1996; Hausmann et al., 1999) has to be included in (8.41). As a consequence, the detection limits increase by the same factor, and the smallest detectable absorption structure will be of similar size than the residual.

8.6 Residual Spectra

Residual structures influence the spectral analysis and the general performance of atmospheric DOAS measurements. In particular, the error determination is challenging if structures other than pure noise are present. While a number of approximation methods to determine the errors of the derived trace gas levels have been developed, an accurate statistical description of the effect of residual structures is still not available.

A more serious problem than the effect on the analysis is the obstacle that residual structures pose to further improvement of the detection limits of DOAS. Most instruments, in particular the active DOAS systems, show residual structures that are much larger than the noise levels one would expect from photon statistics. The residual thus determines the smallest optical density that can be analysed. Reducing or removing residual structures is thus an important undertaking.

Solutions to numerically solve the residual challenge have thus far not been very successful. Often, the reduction of the residuals, for example, by including an average residual spectrum or some other estimate of the residual reduces the random error in the analysis, however, at the price of increasing the systematic uncertainty of the measurement. The solution of the residual problem is thus most likely the study of the sources of these residuals and the improvement of DOAS instruments.

While we do not know, in many cases, what causes the residual structures in the DOAS analysis, we will discuss here a number of obvious candidates that could be responsible. We will begin with listing some trivial examples that are often encountered in DOAS applications. These cases can be identified through a careful analysis of the residual. It should be mentioned here that the residual spectrum often offers the opportunity to study the performance of an analysis. For expert DOAS users, the shape of the residual as compared to the reference spectra can give indications of possible mistakes in the analysis.

The following examples of residual structures are often easily fixed:

- The omission of an absorber in the analysis will lead to a residual that is often similar to the spectral structure of the absorber. However, the residual will not be identical to the absorber, since the analysis routine will try to reduce the χ^2 by finding linear combinations of the reference spectra to describe the missing absorber.
- Different resolutions of the atmospheric spectrum and the reference spectra will lead to residuals that often look like the second derivative of the spectra. The same applies if different low-pass filters are used for different spectra.
- Spectral misalignments between the spectra in an analysis will lead to residuals that are similar to the first derivative of one or more of the spectra.

- The presence of strong absorbers that are included in a classical DOAS analysis (see Chap. 6) can also lead to specific spectral structures.

While these examples are fairly obvious, the random spectral structures often found in DOAS applications are more elusive. In the following paragraphs, we will speculate about the causes of these residuals.

- One of the causes for residual structures that has been discussed in literature (Stutz and Platt, 1997) is the direction sensitivity of solid-state detectors. Because in many DOAS applications the illumination of the spectrograph detector system varies with time, the response of different pixels of the detector will also change. This change in response will cause spectral structures that cannot be easily removed during the analysis. Optical setups and measurement methods have been developed to reduce this effect (see Chap. 10). It is, however, unclear whether the detector sensitivity can be entirely removed by these methods.
- Any inaccuracy in the reference spectra can also lead to residual structures. With further improvement of laboratory measurements of absorption cross-sections, it will become clear in future if this indeed can explain some of the residuals.
- Another possible source of residual structures could be deficiencies in the spectrographs that are commonly used in DOAS applications. Stray light in the spectrometer, for example, could cause an uneven additional signal on top of the original spectrum. The magnitude of the stray light signal would only have to be of the size of the residual, i.e. in the range of 0.1% of the original light intensity. Stray light of this magnitude is commonly encountered in grating spectrometers.

These examples illustrate that very small deficiencies in the optical instrumentation or input data for the analysis can lead to residual structures of the size found in DOAS. We would like to remind the reader that, although the residuals are a serious problem for DOAS, they are indeed a very small signal, often equivalent to optical densities below 0.1%.

Few attempts have been made to systematically study the source of the residuals based on their specific structure. A notable exception is the work by Ferleman (1998) who used principal component analysis of the residuals of a large set of passive DOAS measurements. His results illustrate that statistical methods could potentially provide important information on the nature of residuals.

8.7 Systematic Errors in the Analysis

Here, we will discuss systematic uncertainties that can occur in the DOAS analysis. While there are many ways how systematic uncertainties can be introduced to DOAS, we will focus only on the most common examples.

8.7.1 Interferences

A number of problems are known to cause interferences between different trace gases in the DOAS analysis. The most common is the contamination of absorption cross-sections and reference spectra with other trace gas absorptions. At a first glance, this should be easily overcome by carrying out careful laboratory experiments to determine the absorption cross-sections of the respective gases. In a number of cases, however, it is difficult to prepare pure gases. For example, HONO made in the laboratory often contains NO_2, which is either a byproduct of the HONO formation process, or is formed through secondary reactions in the cell. It is also difficult to make pure NO_2, since it is partially transformed to HONO on the walls of the cell.

A simple calculation can show what effect a contamination of a trace gas reference spectrum will have on the results of a DOAS analysis. We will simplify (8.14) to provide a mathematical description of the influence of such an interference by only considering two absorbers:

$$J(i) = J_0(i) + a'_1 \times S'_1(i) + a'_2 \times S'_2(i). \tag{8.42}$$

We analyse this spectrum with the linear function described earlier (8.15), where the reference spectrum of absorber 2 is contaminated by absorptions from compounds 1: $S_2(i) = S_2^*(i) + \varepsilon \times S_1(i)$, where $S_2^*(i)$ is the reference spectrum of the pure absorber 2. The model function, omitting the polynomial and the shift and squeeze of the reference spectra, is then:

$$F(i) = 1 + a_1 \times S_1(i) + a_2 \times S_2(i) = 1 + a_1 \times S_1(i) + a_2 \times \left(S_2^*(i) + \varepsilon \times S_1(i)\right). \tag{8.43}$$

The least squares analysis of this equation will then result in the correct scaling factor for a'_2, but the result for absorber 1 would be $a'_2 = (a'_1 - a'_2 \times \varepsilon)$. The true absorption would thus be underestimated, depending on the amount of contamination and the scaling factor for trace gas 2. In certain cases, i.e. NO_2 and HONO, this interference may be difficult to recognise because the two trace gases are also linked to each other through their chemistry in the atmosphere. It is then basically impossible to separate spectral interference from correlation from the atmospheric chemistry of these compounds.

The only solution to this problem is thus to ensure that the trace gas absorption cross-section or reference spectra are not contaminated. This can often be done by comparing different measurements or different published absorption cross-sections.

8.7.2 Spectrometer Stray Light and Offsets

A problem common to all grating spectrometers, which leads to systematic uncertainties, is stray light or unremoved offsets in the spectra. Stray light originates from light entering the spectrometer that is not directly projected onto the detector, i.e. other orders of the diffraction grating or light of undetected wavelengths. This light has to be destroyed through absorption on the spectrometer walls. Because a part of this light will eventually fall onto the detector, a small amount of non-dispersed light will be measured. Offsets can originate from drifting electronics or light that enters the spectrometer accidentally, for example due to scattered solar light in active DOAS applications. We will assume here that this light has a smooth spectral characteristic. One can then describe the intensity at a certain wavelength as follows, where we use I_{stray} as the intensity from all effects:

$$I_{\text{meas}}(\lambda) = I(\lambda) + I_{\text{stray}}(\lambda). \tag{8.44}$$

To illustrate the consequences of stray light, we will use (6.8) and change it according to (8.44), for both $I(\lambda)$ and $I'(\lambda)$:

$$\ln\left(\frac{I_{\text{meas}}(\lambda)}{I'_{0\ \text{meas}}(\lambda)}\right) = \ln\left(\frac{I(\lambda) + I_{\text{stray}}(\lambda)}{I'_0(\lambda) + I_{\text{stray}}(\lambda)}\right) = -\sigma'(\lambda) \times c \times L. \tag{8.45}$$

It is difficult to assess the overall effect of the stray light contribution to the change in optical density, and thus the number concentration that would be derived through (8.45). In typical DOAS applications, the contribution of the stray light is, however, small as compared to $I(\lambda)$ or $I'_0(\lambda)$. One can, therefore, simplify (8.45) to gain further insight of the effect stray light has on DOAS. In the following derivation, we use the fact that $\ln(1+x) \approx x$ if x is close to zero:

$$\begin{aligned}
\ln\left(\frac{I(\lambda) + I_{\text{stray}}(\lambda)}{I_0'(\lambda) + I_{\text{stray}}(\lambda)}\right) &= \ln\left(\frac{I(\lambda)}{I_0'(\lambda)}\right) + \ln\left(1 + \frac{I_{\text{stray}}(\lambda)}{I(\lambda)}\right) - \ln\left(1 + \frac{I_{\text{stray}}(\lambda)}{I_0'(\lambda)}\right) \\
&\approx \ln\left(\frac{I(\lambda)}{I_0'(\lambda)}\right) + \frac{I_{\text{stray}}(\lambda)}{I(\lambda)} - \frac{I_{\text{stray}}(\lambda)}{I_0'(\lambda)} \\
&= \ln\left(\frac{I(\lambda)}{I_0'(\lambda)}\right) + \frac{I_{\text{stray}}(\lambda)}{I_0'(\lambda)} \times \left(\frac{I_0'(\lambda)}{I(\lambda)} - 1\right) \\
&\approx \ln\left(\frac{I(\lambda)}{I_0'(\lambda)}\right) + \frac{I_{\text{stray}}(\lambda)}{I_0'(\lambda)} \times \ln\left(\frac{I_0'(\lambda)}{I(\lambda)}\right) \\
&= \ln\left(\frac{I(\lambda)}{I_0'(\lambda)}\right) \times \left(1 - \frac{I_{\text{stray}}(\lambda)}{I_0'(\lambda)}\right) = -\sigma'(\lambda) \times c \times L\,.
\end{aligned} \tag{8.46}$$

The optical density derived based on the DOAS approach would thus be reduced by the ratio $I_{\text{stray}}(\lambda)/I'_0(\lambda)$. Because this ratio is often less than 1%, the error imposed by stray light is small. However, a number of cases are known, especially the measurements below the wavelength of 300 nm, where $I_{\text{stray}}(\lambda)/I_0(\lambda)$ can reach 10% or more. In this case, the stray light has to be considered.

Besides the influence on the derived optical densities, stray light can also influence the shape of an absorption spectrum. We can distinguish two cases where this will occur. First, if the stray light intensity varies over the spectrum, absorption bands at different wavelengths will be changed differently through the stray light, i.e. the optical density of bands with higher stray light will be reduced more than those with lower stray light. Consequently, the relative sizes of the absorption bands, and thus the overall shape of the spectrum, will change. The second case is found when the spectrum has strong intensity variations.

If the shape of the absorption spectrum is changed, residual structures can appear. It is, in principle, possible to correct the stray light effect during the analysis procedure. However, in most cases, it is easier to reduce the stray light already during the measurement, for example, by using filters or adequate light sources.

9

Scattered-light DOAS Measurements

The absorption spectroscopic analysis of sunlight scattered by air molecules and particles as a tool for probing the atmospheric composition has a long tradition. Götz et al. (1934) introduced the 'Umkehr' technique, which is based on the observation of a few select wavelengths of scattered sunlight. The analysis of strong absorption in the ultraviolet allowed the retrieval of ozone concentrations in several atmospheric layers, which yielded the first remotely measured vertical profiles of (stratospheric) ozone. The COSPEC technique developed in late 1960s was the first attempt to study tropospheric species by analysing scattered sunlight in a wider spectral range with the help of an optomechanical correlator (Millan et al. 1969; Davies 1970), see Sect. 5.7. It has been applied over three decades for measurements of total emissions of SO_2 and NO_2 from various sources, e.g. industrial emissions (Hoff and Millan, 1981) and volcanic plumes (Stoiber and Jepsen, 1973; Hoff et al. 1992). Scattered sunlight was also used to study stratospheric and tropospheric NO_2, as well as other stratospheric species by ground-based differential optical absorption spectroscopy (DOAS) (Noxon, 1975; Noxon et al., 1979; Pommereau, 1982; McKenzie et al., 1982; Solomon et al., 1987). An overview of different scattered light techniques is given in Table 9.1.

Scattered sunlight DOAS is an experimentally simple and very effective technique for the measurement of atmospheric trace gases and aerosols. Since scattered light DOAS instruments analyse radiation from the sun, rather than relying on artificial sources, they are categorised as **passive** DOAS instruments (see also Chap. 6).

All passive DOAS instruments are similar in their optical setup, which essentially consists of a telescope to collect light, coupled to a spectrometer–detector combination (see Chap. 7). However, different types of passive DOAS instruments employ a wide variety of observation geometries for different platforms and measurement objectives.

The earliest scattered light DOAS applications were ground-based and predominately observed light from the zenith (Noxon, 1975; Syed and Harrison, 1980; McMahon and Simmons, 1980; Pommereau, 1982, 1994;

U. Platt and J. Stutz, *Scattered-light DOAS Measurements*. In: U. Platt and J. Stutz, Differential Optical Absorption Spectroscopy, Physics of Earth and Space Environments, pp. 329–377 (2008)
DOI 10.1007/978-3-540-75776-4_9

Table 9.1. Overview and history of the different scattered light passive DOAS applications

Method	Measured quantity	No. of axes, technique	References
COSPEC	NO_2, SO_2, I_2	1, (S)	Millan et al., 1969; Davies, 1970; Hoff and Millan, 1981; Stoiber and Jepsen, 1973; Hoff et al., 1992
Zenith scattered light DOAS	Stratospheric NO_2, O_3, OClO, BrO, IO	1	Noxon, 1975; Noxon et al., 1979; Harrison, 1979; McKenzie and Johnston, 1982; Solomon et al., 1987a; Solomon et al., 1987b, McKenzie et al., 1991; Fiedler et al., 1993; Pommereau and Piquard, 1994a,b; Kreher et al., 1997; Wittrock et al., 2000a
Zenith sky + Off-axis DOAS	Stratospheric OClO	2	Sanders et al., 1993
Off-axis DOAS	Stratospheric BrO profile	1	Arpaq et al., 1994
Zenith scattered light DOAS	Tropospheric IO, BrO	1	Friess et al., 2001, 2004
Off-axis DOAS	Tropospheric BrO	1	Miller et al., 1997
Sunrise Off-axis DOAS + direct moonlight	NO_3 profiles	2, S	Weaver et al., 1996; Smith and Solomon, 1990; Smith et al., 1993
Sunrise Off-axis DOAS	Tropospheric NO_3 profiles	1	Kaiser, 1997; von Friedeburg et al., 2002
Aircraft-DOAS	Tropospheric BrO	2	McElroy et al., 1999
Aircraft zenith sky + Off-axis DOAS	"near in-situ" Stratospheric O_3	3	Petritoli et al., 2002
AMAX-DOAS	Trace gas profiles	8+, M	Wagner et al., 2002; Wang et al., 2003; Heue et al., 2003
Multi Axis DOAS	Tropospheric BrO profiles	4, S	Hönninger and Platt, 2002
Multi Axis DOAS	Tropospheric BrO profiles	4, S	Hönninger et al., 2003b

Table 9.1. (continued)

Multi Axis DOAS	Trace gas profiles	2-4, M	Löwe et al., 2002; Oetjen, 2002; Heckel, 2003
Multi Axis DOAS	NO_2 plume	8, M	V. Friedeburg, 2003
Multi Axis DOAS	BrO in the marine boundary layer	6, S/M	Leser et al., 2003; Bossmeyer, 2002
Multi Axis DOAS	BrO and SO_2 fluxes from volcanoes	10, S	Bobrowski et al., 2003
Multi Axis DOAS	BrO emissions from a Salt Lake	4, S	Hönninger et al., 2003a
Multi Axis DOAS	IO emissions from a Salt Lake	6, S	Zingler et al., 2005

S = Scanning instrument, M = Multiple telescopes.

McKenzie et al., 1982, 1991; Solomon et al., 1987, 1988, 1993; Perner et al., 1994; Van Roozendael et al., 1994; Slusser et al., 1996). This Zenith Scattered Light–DOAS (ZSL-DOAS) geometry is particularly useful for the observation of stratospheric trace gases, and has made major contributions to the understanding of the chemistry of stratospheric ozone, in particular through the measurement of stratospheric NO_2, OClO, BrO, and O_3 (Pommereau, 1982, 1994; McKenzie et al., 1982, 1991; Solomon et al., 1987, 1988, 1993; Perner et al., 1994; Van Roozendael et al., 1994a,b,c; Slusser et al., 1996; Sanders, 1996; Sanders et al., 1997).

The next development in scattered light DOAS employed an off-axis geometry (Sanders et al., 1993) and observed the sky at one low-elevation angle to improve the sensitivity of the instrument. Recently, this idea was expanded by employing multiple viewing geometries. This Multi-Axis DOAS (MAX-DOAS) method typically employs 3–10 different viewing elevations (Winterrath et al., 1999; Friess et al., 2001; Hönniger and Platt, 2003; Wagner et al., 2004). In contrast to the earlier instruments, MAX-DOAS is more sensitive to tropospheric trace gases, and thus offers a large number of possible applications. It should be noted that at the time of writing this book, MAX-DOAS was still very much a method in development, and many of the possible applications have not been extensively explored. The most recent ground-based passive DOAS application makes use of modern solid-state array detectors, expanding the number of viewing channels to hundreds. This imaging DOAS can provide a spectroscopic 'photo' of the composition of the atmosphere, e.g. of the emissions from a smoke stack.

Early on in the development and use of scattered light DOAS, platforms other than the ground were explored. Schiller et al. (1990) report ZSL-DOAS measurements on-board the NASA DC8 research aircraft. Similar measurements were reported by McElroy et al. (1999) and Pfeilsticker and Platt

(1994). The use of scattered light DOAS on mobile platforms allows the access to remote areas that can only be reached through air, e.g. the remote ocean and polar regions. In the recent years, MAX-DOAS has also been adapted to airborne platforms. While ground-based MAX-DOAS typically uses viewing elevations from the zenith to the very low elevations, the range of airborne MAX-DOAS extends from zenith viewing to nadir (downwards) viewing, thus covering a whole 180°. This viewing direction arrangement allows the measurement of trace gases below and above the aircraft (Wang et al., 2003).

One of the most exciting developments of passive DOAS in the last decade was the launch of various spaceborne DOAS instruments (see Chap. 11). The instruments typically operate in a nadir viewing mode to provide global coverage of the distribution of trace gases such as NO_2 and HCHO. Instruments such as SCHIAMACHY allow the limb-observations of scattered sunlight, with the goal of deriving vertical trace gas profiles.

A common characteristic, that distinguishes scattered light absorption spectroscopy measurements from active DOAS (for examples see Chap. 10) or direct sunlight DOAS is the lack of a clearly defined light path. Considerable effort thus has to be invested in converting the observed trace gas absorption strength to a quantity that is useful for the interpretation of observations. This usually involves modelling the radiation transport in the atmosphere (see Chap. 3) to determine an effective light path length in the atmosphere.

This chapter provides a general introduction into the methods required to interpret scattered light DOAS measurements. We will begin by introducing the basic concepts that are needed to understand scattered light DOAS, and then discuss the details of radiative transfer calculations. Since the techniques to analyse the absorption spectra were already discussed in Chap. 8, we will concentrate on the interpretation of trace gas abundances prevailing along the atmospheric light path.

9.1 Air Mass Factors (AMF)

The classical concept of absorption spectroscopy as an analytical method is based on the knowledge of absorption path length and the assumption that the conditions along the light do not vary (Chap. 6). For scattered and direct sunlight DOAS measurements, in which the light crosses the vertical extent of the atmosphere, this assumption is usually not valid. New concepts thus have to be introduced to interpret these measurements. In this section we will introduce these concepts and the quantities that are necessary to quantitatively analyse DOAS observations. We will use an approach that loosely follows the history and development of the interpretation of spectroscopy observations beginning with direct sun observations, followed by zenith and off-axis scattered sunlight applications.

Before discussing the individual aspects of these observational strategies, it is useful to introduce the quantity that is commonly the final result of passive DOAS observations, the vertical column density (VCD). Historically, the vertical column density (V) has been defined as the concentration of a trace gas vertically integrated over the entire extent of the atmosphere:

$$V = \int_0^\infty c(z)\, dz \,. \tag{9.1}$$

In recent years, this concept has been expanded by varying the limits of integration to cover the stratosphere, troposphere, or height intervals of the atmosphere. We will, therefore, expand this equation by introducing partial columns:

$$V(z_1, z_2) = \int_{z_1}^{z_2} c(z)\, dz \,. \tag{9.2}$$

9.1.1 Direct Light AMF

The earliest applications of absorption spectroscopy in the atmosphere relied on the measurement of direct sun or moonlight. As a consequence of the movement of the solar or lunar disk in the sky (Fig. 9.1), the path length used in Lambert–Beer's law to convert the observed trace gas absorption changes as a function of the solar or lunar position. It is common to use the angle between the zenith and the sun or moon to quantify this position. This Solar (or Lunar)-Zenith-Angle (SZA, LZA), ϑ, is 0° when the sun or moon is in the zenith, and 90° when they are on the horizon. In addition, the Solar (Lunar)-Azimuth-Angle (SAZ, LAZ) is used to define the horizontal position. The SAZ (LAZ) is zero by definition when the sun or moon is in northern direction and increases clockwise. We can, however, see that the azimuth angle does not play an important role when interpreting direct solar or lunar measurements.

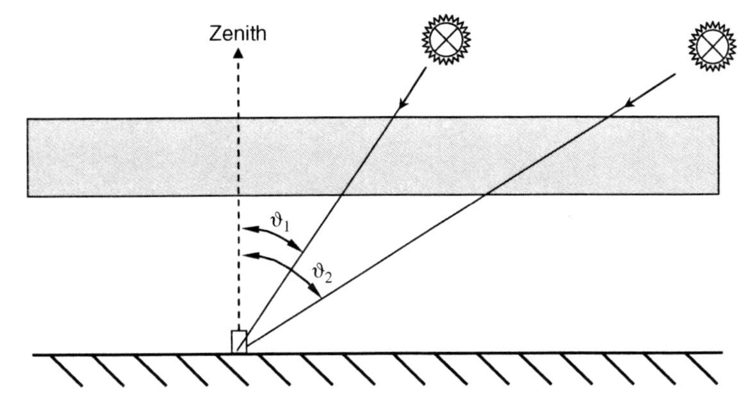

Fig. 9.1. Sketch of direct light observation geometries. In first approximation, the light path through a trace gas layer varies with $1/\cos\vartheta$ (ϑ = zenith angle of celestial body observed)

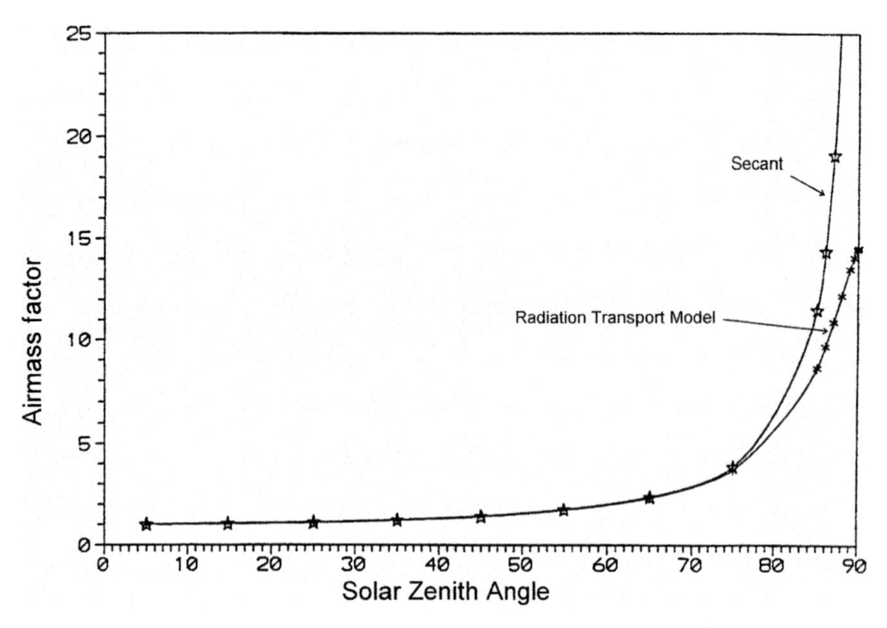

Fig. 9.2. Direct light air mass factors. Model calculations including curvature of earth as well as refraction inside the atmosphere are compared to the simple secants $\vartheta = 1/\cos\vartheta$ approximation. Deviations become apparent at $\vartheta > 70°$ (from Frank, 1991)

To describe the observations of trace gases, we introduce the 'slant column density' (SCD), S. Historically, SCD has been defined by the concentration integrated over the light path in the atmosphere.

$$S = \int_0^\infty c(s)\, ds\ . \tag{9.3}$$

In contrast to the definition of VCD in (9.1), the element of path ds does not need to be vertical. In the case sketched in Fig. 9.1, the SCD can be determined by the geometrical enhancement of slanted light path in the atmosphere, i.e. $ds = 1/\cos\vartheta\, dz$ for small SZA. This concept of the SCD will, however, lead to problems in the interpretation of scattered light observations, since the column seen by the instrument is an 'apparent' column, which is intensity weighted over an infinite number of different light paths through the atmosphere. We will, therefore, define SCD more generally from the observed column density as the ratio of measured differential optical density D' and known differential absorption cross-sections σ', i.e. $S = D'/\sigma'$ (see Chaps. 6 and 8).

To relate the observed SCD to the desired result of the measurement, i.e. the vertical column density, we now introduce the airmass factor (AMF), A, as:

$$A = \frac{S}{V}\ . \tag{9.4}$$

The AMF is the proportionality factor between the observed column density and the VCD (see Noxon et al., 1979). The most basic example of an AMF is the direct light AMF for an observation geometry where the instrument looks directly towards a celestial body (e.g. sun, moon, star), which is assumed to be point-like. Neglecting the curvature of earth and refraction in the atmosphere (i.e. for sufficiently small zenith angles), we obtain for the AMF as A_D (Fig. 9.1):

$$A_D = \frac{\text{length of slant path}}{\text{length of vertical path}} \approx \frac{1}{\cos \vartheta} . \tag{9.5}$$

Up to an SZA of $\approx 75°$, (9.5) is a good approximation for the direct AMF (Fig. 9.2). Above 75°, effects such as the earth's curvature and atmospheric refraction have to be considered. Refraction in the atmosphere is caused by the dependence of the refractive index of air on temperature, pressure and thus its change with altitude.

9.1.2 Scattered Zenith Light AMF

A multitude of passive DOAS applications use scattered sunlight to measure trace gas absorptions. The telescopes in these scattered sunlight DOAS instruments are aimed at a point in the sky other than the sun or moon. Consequently, we need to consider the viewing direction of the DOAS telescope in addition to the solar position. This viewing direction is again characterised by two angles: the elevation, which gives the angle in the vertical between the horizon (for ground based instruments) and the viewing direction. The zenith in this case is at an elevation of $\alpha = 90°$. For downward-looking geometries, such as from airborne or satellite instruments, we will use negative values, i.e.

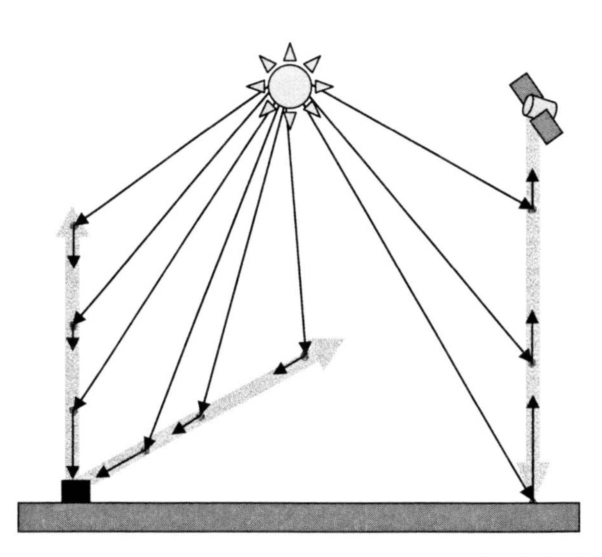

Fig. 9.3. Sketch of ground-based and satellite-borne passive DOAS observations

the nadir is at $\alpha = -90°$. The viewing direction can also be measured from the zenith, in which case viewing parallel to the ground is at a zenith angle of $\vartheta = 90°$ and the nadir is at $\vartheta = 180°$. The second angle that is important is the viewing azimuth angle, defined in the same way as for the solar position.

The zenith viewing geometry, i.e. $\alpha = 90°$ has historically been one of the most successful applications of passive DOAS. Many of the basic concepts of AMF calculations have been determined for this viewing geometry, mostly in the context of studying stratospheric trace gases and the chemistry leading to the Antarctic ozone hole.

As illustrated in Fig. 9.3, in zenith scattered light DOAS applications the irradiance received by the detector originates from light scattered by the air molecules and particles that are located along the viewing direction of the telescope. Assuming, for now, that only one scattering process occurs between the sun and the detector, one can gain a basic understanding of the zenith sky observations.

Figure 9.4 illustrates that two processes have to be taken into account to understand the measurement of radiances at the detector. First, one has to consider the efficiency by which solar light is scattered from its original direction towards the detector. Secondly, one needs to consider the extinction, either by trace gas absorption or by Rayleigh scattering, along the different light paths. The main process changing the direction of light under clear sky conditions is Rayleigh scattering in the zenith, which depends primarily on air density, i.e. the number of scattering air molecules. The scattering efficiency will thus be highest close to the ground and decrease exponentially

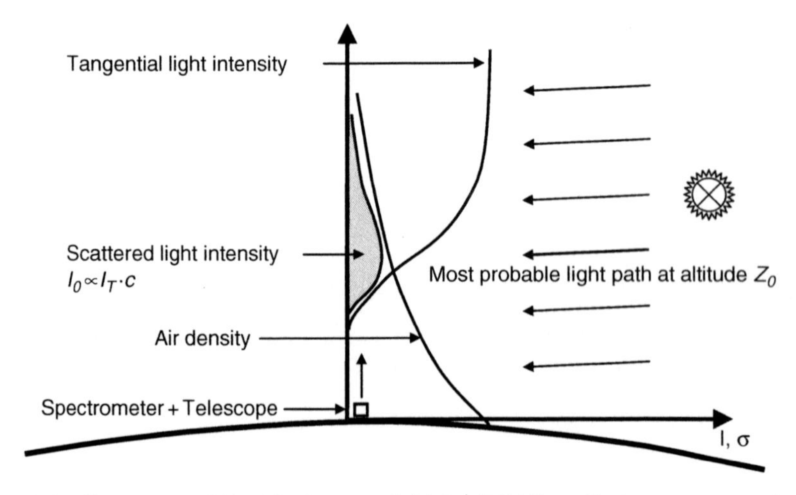

Fig. 9.4. Geometry of Zenith Scattered Light DOAS–radiation transport in the atmosphere. There is an infinite number of possible light paths. However, at solar zenith angles around 90° the light seen by the spectrometer most likely originates from a certain altitude range around z_i

with altitude. In a thin layer at height z', the intensity of light scattered towards the detector depends on the intensity reaching the scattering point, I_z, the Rayleigh scattering cross-section, σ_R, and the air density, $\rho(z)$ (Solomon et al., 1987):

$$I_S(\lambda, z') = I_0(\lambda, z') \cdot \sigma_R(\lambda) \cdot \rho(z')\, dz' \,. \tag{9.6}$$

The extinction along the light path will depend on the length of each light path in the atmosphere, the concentration of air molecules and, in the case of strong absorbers, the concentration of these gases.

$$I_S(\lambda, z) = I_0(\lambda, z) \times \exp\left(-\sigma_R(\lambda) \int_z^{\infty} \rho(z') \cdot A(z', \vartheta)\, dz' \right) , \tag{9.7}$$

where $A(z)$ is the direct sun AMF for the light before scattering at altitude z'. Because the AMF depends on the solar zenith angle, the length of each light path will also depend on the SZA, as can be seen in Fig. 9.1. This leads to a decrease in intensities I(z) with SZA. The dependence of Rayleigh scattering on air density indicates that the intensity reaching the zenith point at which it is scattered will be lowest close to the ground and increase with altitude.

Combining the two effects, which show opposite altitude dependences, gives rise to a distribution of scattered light intensity that is small at the ground, increases to a maximum, and then decreases again with altitude. The maximum of this scattered light distribution represents the most probable height that the observed light originates from. The contribution of scattering in a thin layer at altitude z to the intensity observed by the detector is then:

$$\begin{aligned} I_S(\lambda, z) = \sigma_R(\lambda) \cdot \rho(z) \cdot I_0(\lambda) \cdot \exp\left(-\sigma_R(\lambda) \int_z^{\infty} \rho(z') \cdot A(z', \vartheta)\, dz \right) \\ \times \exp\left(-\sigma_R(\lambda) \int_h^{z} \rho(z')\, dz' \right) . \end{aligned} \tag{9.8}$$

The last term in this equation is the Rayleigh extinction on the light path from the scattering height to the detector, which is assumed to be at an altitude h.

This equation allows the discussion of the dependence of scattering properties with the SZA. At large SZA, the light paths in the atmosphere become large, and at lower altitudes the intensity reaching the scattering height is reduced more than at higher altitudes (Fig. 9.5). The most probable scattering height, therefore, moves upwards as the SZA increases. The most probable scattering height also depends on the wavelength, due to the wavelength dependence of Rayleigh scattering. Z_o typically varies from about 26 km (327 nm) to 11 km (505 nm) at 90 deg SZA.

In the presence of an absorbing trace gas, (9.8) is expanded by including the absorption cross-section and the trace gas concentration at each height.

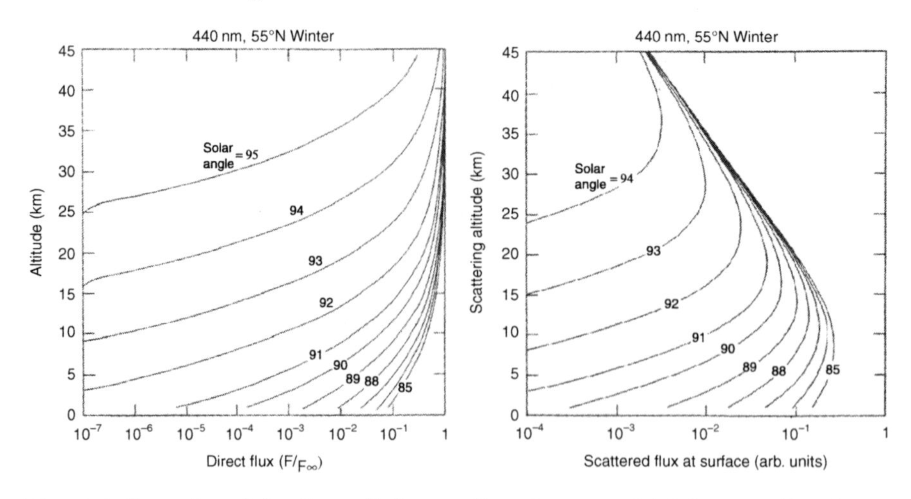

Fig. 9.5. Intensity of the direct (**left panel**) and scattered (**right panel**) radiative flux as a function of altitude for various SZA (from Solomon et al, 1987, Copyright by American Geophysical Union (AGU), reproduced by permission of AGU)

$$I_S^A(\lambda, z) = I_S(\lambda, z) \cdot \exp\left(-\sigma(\lambda) \int_z^\infty C(z') \cdot A(z', \vartheta)\, dz\right) \cdot \exp\left(-\sigma(\lambda) \int_h^z C(z')\, dz'\right) . \tag{9.9}$$

Because the large direct sunlight AMFs, $A(z', \vartheta)$, are approximately equal to $1/\cos\vartheta$, the absorption is largest at large SZAs. The slant light paths through the atmosphere are quite long under twilight conditions: at 90° zenith angle and 327 nm, the horizontal light path length through the stratosphere is 600 km. The AMFs for the path below the scattering height are, i.e. the second exponential term in (9.9), equal to unity.

Our discussion shows that the light reaching the detector is an average over a multitude of rays, each of which takes a somewhat different route through the atmosphere. The detector, therefore, measures the intensity-weighted average of the absorptions along the different light paths arriving at the telescope. This 'apparent' column density S is, for historic reasons, also called SCD, although it has no resemblance to the slanted column of direct solar measurements.

Based on our discussion above, we can now use the definition of SCD (9.3) to write down a simplified expression for SCD for scattered sunlight:

$$S(\vartheta) = \frac{1}{\sigma(\lambda)} \ln\left[\frac{\int_h^\infty I_S^A(\lambda, z) dz}{\int_h^\infty I_S(\lambda, z) dz}\right] . \tag{9.10}$$

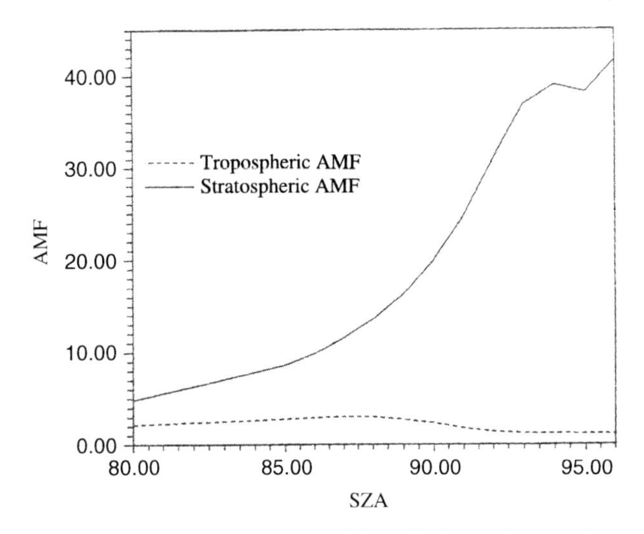

Fig. 9.6. Stratospheric and tropospheric AMF of NO_2 at 445 nm determined using a single scattering radiative transfer model (from Stutz, 1992)

This SCD can now be used in (9.4) to calculate the AMF for zenith scattered light. Figure 9.6 shows such an AMF for a stratospheric absorber at 450 nm. As expected, the AMF increases with SZA as the light path in the stratosphere becomes longer and the most probable scattering height moves upwards. The decrease at very large SZA occurs when Z_0 moves above the absorption height.

While our simplified description illustrates the principles of radiative transfer and AMF calculations, it is insufficient to make accurate AMF calculations. Other physical processes such as scattering by aerosol particles, refraction, and multiple scattering need to be considered in the AMF calculations. Interpretation of the SCD, therefore, requires radiation transport modelling. More details on atmospheric radiation transport are available in Chap. 4, and, for example, Solomon et al. (1987), Frank and Platt (1990), or Marquard et al. (2000). To put it simply, the detected light from the zenith can be represented by a most probable light path through the atmosphere defined by the most likely scattering height Z_o in the zenith.

9.1.3 Scattered Off-axis and Multi-axis AMF

Scattered-light DOAS viewing geometries other than the zenith have become increasingly popular in recent years. The motivation for using smaller elevation angles is twofold. With respect to stratospheric measurements, lower viewing angles can improve the detection limits by increasing the light intensity reaching the detector. The stronger motivation is the ability to achieve larger AMFs for tropospheric trace gases. To understand these motivations, the underlying radiative transfer principles are discussed here.

Our argument follows very closely the approach we have adopted for ZSL-DOAS in Sect. 9.1.2. The light of the detector originates from scattering processes within the line of view of the detector. Because the detector aims at lower elevations than in the ZSL case, i.e. the viewing path crosses through layers with a higher air density, scattering events closer to the ground will contribute more to the detected intensity. Consequently, the most probable scattering height will move downwards in the atmosphere as the viewing elevation angle decreases. It is typically somewhere in the troposphere for all wavelengths. We can now expand (9.8) and (9.9) by including an AMF, $A_T(z, \alpha)$, for the path between the scattering event and the detector. In an approximation based on purely geometrical arguments, $A_T(z, \alpha)$ is equal to $1/\cos\alpha$, i.e. it increases with decreasing viewing elevation angle. This will change the distribution of the scattering term in (9.9), as well as increase the Rayleigh extinction between the scattering event and the detector. It should be noted here that both the SZA and the elevation angle also influence σ_R (see Chap. 4).

$$
\begin{aligned}
I_S(\lambda, z, \alpha) = \sigma_R(\lambda) \cdot \rho(z) \cdot A(z', \alpha) \cdot I_0(\lambda) \\
\cdot \exp\left(-\sigma_R(\lambda) \int_z^\infty \rho(z') \cdot A_S(z', \vartheta)\, dz\right) \\
\cdot \exp\left(-\sigma_R(\lambda) \int_h^z \rho(z') \cdot A_T(z', \alpha)\, dz'\right) . \qquad (9.11)
\end{aligned}
$$

The smaller elevation angles also influence the absorption of trace gases. Equation (9.9) thus has to be expanded to include $A_T(z, \alpha)$ in the second integral:

$$
\begin{aligned}
I_S^A(\lambda, z, \alpha) = I_S(\lambda, z, \alpha) \cdot \exp\left(-\sigma(\lambda) \cdot \int_z^\infty C(z') \times A_S(z', \vartheta)\, dz\right) \\
\cdot \exp\left(-\sigma(\lambda) \cdot \int_h^z C(z') \cdot A_T(z', \alpha)\, dz'\right) . \qquad (9.12)
\end{aligned}
$$

One can see that the lower elevation in our simplified model does not change the first integral that describes the light path before the scattering event. The behaviour of stratospheric trace gases, for example, does not depend on the elevation angle, while that of tropospheric trace gases will. The SCD for the off-axis case can now be calculated according to (9.10).

To address the separation of stratospheric and tropospheric absorbers in more detail, we will further simplify our discussion by concentrating on the most probable light path. This eliminates the integrations in (9.10) to calculate the SCD and the AMF. Figure 9.7 illustrates this simplified view of off-axis viewing geometries. In this simplified picture, the AMFs $A_S(z, \vartheta)$ and $A_T(z, \alpha)$ are independent of altitude z and can be approximated as $1/\cos\vartheta$

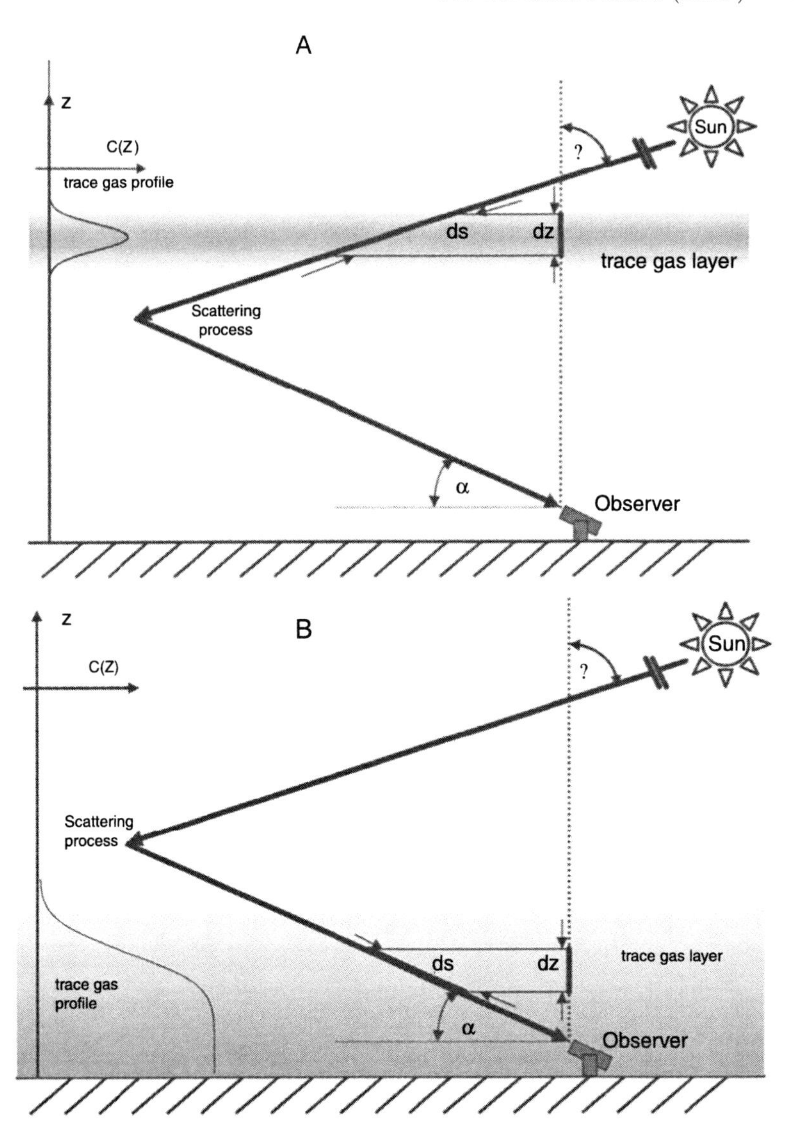

Fig. 9.7. Geometry of Off-axis DOAS and a sketch of the associated radiation transport in the atmosphere. Like in the case of ZSL-DOAS, there is an infinite number of possible light paths (from Hönninger, 1991)

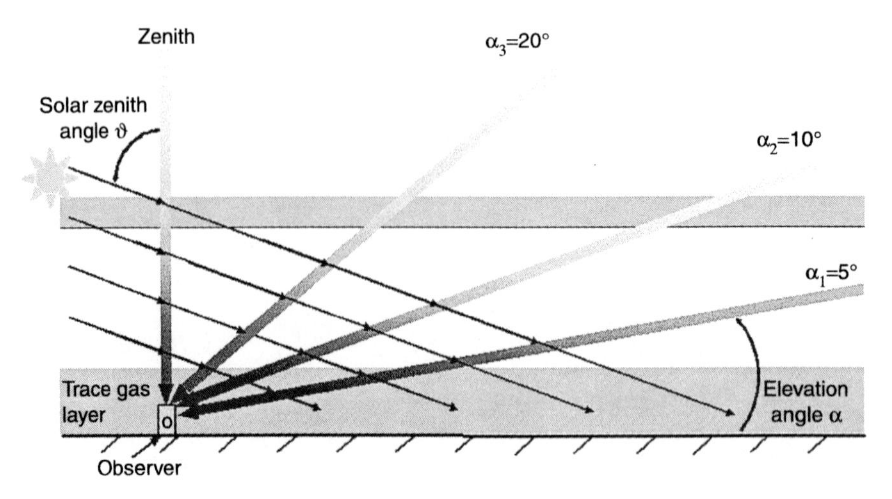

Fig. 9.8. Geometry of Multi-axis DOAS (MAX-DOAS) and a sketch of the associated radiation transport in the atmosphere. As in the case of ZSL-DOAS, there is an infinite number of possible light paths

and $1/\cos\alpha$, respectively. We further introduce the tropospheric and stratospheric vertical column densities, $\mathrm{VCD_T}$ and $\mathrm{VCD_S}$, which integrate the vertical trace gas concentration profile from the ground to the scattering altitude and from the scattering altitude to the edge of the atmosphere, respectively. After these simplifications, we find that the SCD can be described by:

$$S(\alpha, \vartheta) = V_T \cdot A_T(\alpha) + V_S \cdot A_S(\vartheta) = \frac{V_T}{\cos\alpha} + \frac{V_S}{\cos\vartheta}\,. \qquad (9.13)$$

The SCD, therefore, depends both on the SZA, which controls the contribution of the stratospheric column, and the elevation angle, which controls the contribution of the tropospheric column.

While this equation illustrates the dependence of the SCD and the total AMF on the SZA, the elevation angle, and the vertical trace gas profile, it is highly simplified. In the case of low elevations, which are often used to increase the tropospheric path length, multiple scattering events, curvature of the earth, and refraction become significant (see Fig. 9.8). In addition, the higher levels of aerosols in the troposphere make Mie scattering an important process that must be included in the determination of AMF. To consider all these effects, a detailed radiative transfer model is required. A short description of such models will be given in Sect. 9.2.

9.1.4 AMFs for Airborne and Satellite Measurements

Airborne and satellite DOAS measurements have become an important tool to study atmospheric composition on larger scales. These measurements are based on scattered sunlight detection, and AMFs have to be calculated to interpret the observations. In principle, the approach is very similar to that shown for other scattered light applications.

As with the multi-axis approach, the SZA and the viewing angle have to be considered. In this case, however, the instruments look downwards and can, at least at higher wavelengths, see the ground. Besides the scattering on air molecules and aerosol particles, clouds and the albedo of the ground have to be considered. The ground is typically parameterised by a wavelength-dependent albedo and the assumption that the surface is a Lambertian reflector, or by a bi-directional reflectivity function (BDRF), which parameterises the reflection based on incoming and outgoing reflection angles. Clouds seen from an airplane or a satellite can be parameterised by introducing parallel layers of optically thick scatterers in a multiple scattering model or, in the case of thick clouds, by parameterising them as non-Lambertian reflectors in the model at a certain altitude (Kurosu et al., 1997).

An additional problem, which we will not discuss here in detail, is the fact that downward-looking airborne or satellite instruments often observe areas of the earth's ground that may also be partially covered by clouds. The spatial averaging over the earth surface together with clouds is a challenge for any radiative transfer model.

9.1.5 Correction of Fraunhofer Structures Based on AMFs

A challenge in applying DOAS to the measurement of atmospheric trace gases is the solar Fraunhofer structure, which manifests itself as a strong modulation of $I_0(\lambda)$ due to absorption in the solar atmosphere (see Chap. 6). We will discuss two approaches to overcome this problem, one for the measurement of stratospheric trace gases and one for the measurement of tropospheric gases. Both techniques are based on the choice of suitable Fraunhofer reference spectra that can be used in the analysis procedures described in Chap. 8, or in simple terms by which the observed spectrum is divided. Ideally, one would like to choose a spectrum without any absorption of the respective atmospheric trace gas. Since this is not possible, both techniques rely on the choice of a Fraunhofer reference spectrum where the trace gas absorptions are small. Based on our discussion in Sect. 9.1.5, this is equivalent to a small AMF in the reference spectrum as compared to the actual observation.

In the case of ZSL measurements of stratospheric trace gases, our earlier discussions revealed that the AMF increases with the solar zenith angle. The obvious choice for a Fraunhofer reference spectrum is a spectrum measured at small SZA. Ideally, the ratio of a spectrum taken at a higher SZA, for example at sunset, and the Fraunhofer spectrum will consist of pure trace gas absorptions. The ratio of (9.9) for two different SZAs would eliminate the solar intensity, i.e. $I_S(\lambda, z)$. However, this approach poses another challenge since the division also eliminates the absorptions that are originally in the Fraunhofer reference spectrum, which are not known. The analysis of the ratio between the two spectra results in the so-called differential slant column density (DSCD), S', which is the difference between the SCDs of high SZA, ϑ_2, and the reference spectrum measured at low SZA ϑ_1:

$$S' = S(\vartheta_2) - S(\vartheta_1) . \tag{9.14}$$

A determination of the VCD is not directly possible from S'. A solution to this problem presents itself if the vertical column density of the absorbing trace gas remains constant with time, i.e. same at ϑ_2 and ϑ_1. Equation (9.14) can then be written as:

$$S' = S'(\mathrm{AMF}) = \mathrm{V} \times \mathrm{A}(\vartheta_2) - \mathrm{S}(\vartheta_1) \tag{9.15}$$

This linear equation can then be used to determine the VCD and $\mathrm{S}(\vartheta_1)$ by plotting the DSCD against the AMF and applying a linear fit to the resulting curve. An example of this so-called **Langley Plot** is shown in Fig. 9.9. The slope of the curve gives the VCD, while the extrapolation to $A = 0$ results in an ordinate intersection at $-S_0$.

In the case where the interest is more in tropospheric trace gases, the above method will not lead to the desired result since the tropospheric AMF is only weakly dependent on the SZA. However, as shown in Sect. 9.1.3, changing the

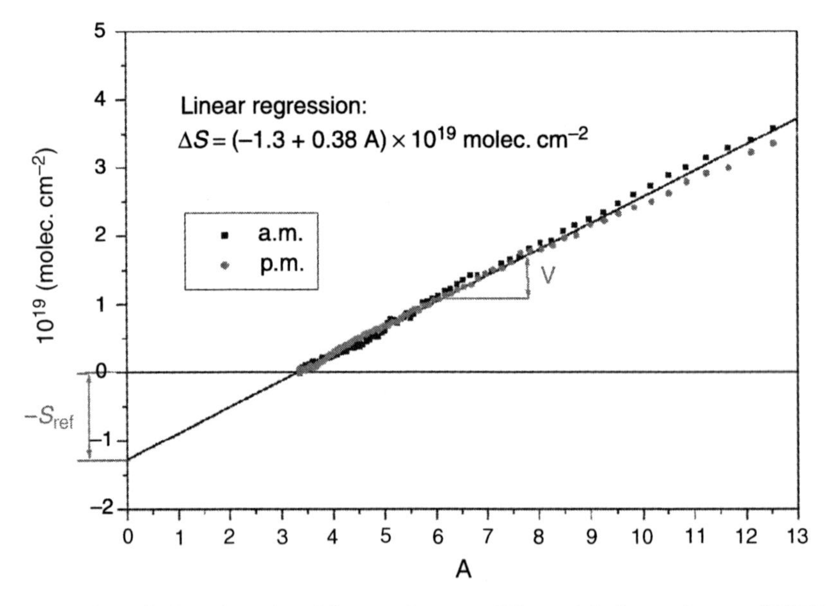

Fig. 9.9. Sample Langley plot: Measured ozone differential slant columns (DSCDs) are plotted as a function of the airmass factor A calculated for the solar zenith angle of the measurement. The slope of the plot indicates the vertical column density V. The ordinate section indicates the slant column density in the solar reference spectrum, S_{ref}

viewing elevation will considerably influence the tropospheric AMF (9.13). The approach to measure tropospheric gases is thus to use a zenith spectrum ($\alpha = 90°$) as the Fraunhofer reference to analyse spectra measured at lower elevations. To guarantee that the stratospheric AMF is the same for these two spectra, one has to measure the zenith and the low elevation spectrum simultaneously, or at least temporally close together. Using (9.13), this approach describes:

$$S' = S(\alpha, \vartheta) - S(90, \vartheta) = V_T \times A_T(\alpha) . \tag{9.16}$$

It should be noted that, despite the fact that we have called V_T the tropospheric vertical column density, the height interval over which the vertical concentration profile is integrated in V_T depends on the radiative transfer. Typically V_T does not cover the entire troposphere, but rather the boundary layer and the free troposphere. A more detailed discussion of this will be given in Sect. 9.3.

The dependencies described in (9.13) gave rise to a new method that uses simultaneous measurements of one or more low-viewing elevations, together with a zenith viewing channel to measure tropospheric trace gases. This multi-axis DOAS method is described in detail in Sect. 9.3.3.

9.1.6 The Influence of Rotational Raman scattering, the 'Ring Effect'

Named after Grainger and Ring [1962], the Ring effect, it manifests itself by reducing the optical density of Fraunhofer lines observed at large solar zenith angles (SZA), compared to those at small SZAs. This reduction is on the order of a few percent. However, because atmospheric trace gas absorptions can be more than an order of magnitude smaller than the Ring effect, an accurate correction is required.

Several processes, such as rotational and vibrational Raman scattering, aerosol fluorescence, etc. have been suggested as explanations of the Ring effect. Recent investigations [Bussemer 1993, Fish and Jones 1995, Burrows et al. 1995, Joiner et al. 1995, Aben et al. 2001] show convincingly that rotational Raman scattering is the primary cause of the Ring effect.

In short, light intensity scattered into a passive DOAS instrument can be expressed as:

$$I_{meas} = I_{Rayleigh} + I_{Mie} + I_{Raman} = I_{elastic} + I_{Raman}$$

The accurate determination of $I_{elastic}$ and I_{Raman} requires detailed radiative transfer calculations for each observation. However, Schmeltekopf et al. [1987] proposed an approximation which is based on the inclusion of a "Ring Spectrum" in the spectral analysis of the observations (see Chap. 8).

Based on the logarithm of the measured spectrum, $\ln(I_{meas})$, and the equation above the following approximation can be made.

$$\ln(I_{meas}) = \ln\left(I_{elastic} \cdot \frac{I_{elastic} + I_{Raman}}{I_{elastic}}\right)$$
$$= \ln(I_{elastic}) + \ln\left(1 + \frac{I_{Raman}}{I_{elastic}}\right) \approx \ln(I_{elastic}) + \frac{I_{Raman}}{I_{elastic}},$$

where the ratio of the Raman and the elastic part of the intensity is considered the Ring spectrum:

$$I_{Ring} = \frac{I_{Raman}}{I_{elastic}}$$

This approximation has been proven to be simple and effective. An experimental and a numerical approach exist to determine a Ring spectrum.

1) The first approach is based on the different polarization properties of atmospheric scattering processes. Rayleigh scattering by air molecules is highly polarized for a scattering angles near 90° (see Chap. 4, Sect. 4.2.2). Light scattered by rotational Raman scattering, on the other hand, is only weakly polarized (see Sect. 4.2.3). By measuring the intensity of light polarized perpendicular and parallel to the scattering plane, the rotational

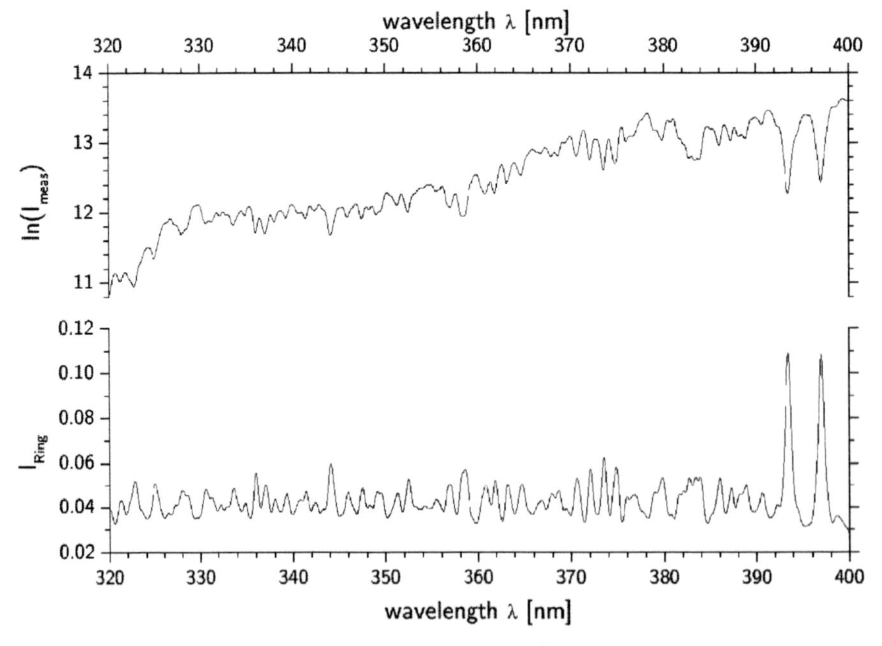

Fig. 9.10. Sample Ring spectrum (I($_{Ring}$)) calculated for the evaluation of UV spectra taken during ALERT2000. Shown is also the logarithm of the Fraunhofer reference spectrum(I_{meas}) used for the calculation. The spectrum was taken on April 22, 2000 at 15:41 UT at a local solar zenith angle of 70° and zenith observation direction (from Hönninger, 2001)

Raman component, and therefore a "Ring spectrum", can be determined [Schmeltekopf et al. 1987, Solomon et al. 1987].

This approach faces a number of challenges. Mie scattering also contributes to the fraction of non-polarized light in the scattered solar radiation. The presence of aerosol or clouds therefore makes the determination of the Ring spectrum difficult. Also, the atmospheric light paths for different polarizations may be different, and can thus contain different trace gas absorptions, which can affect the DOAS fit of these gases.

2) The second approach uses the known energies of the rotational states of the two main constituents of the atmosphere, O_2 and N_2, to calculate the cross section for rotational Raman scattering. This can be realized, either by including Raman scattering into radiative transfer models [e.g. Bussemer 1993; Fish and Jones 1995; Funk 2000], or by calculating the pure ratio of the cross sections for Raman and Rayleigh scattering. In many realizations this calculation is based on measured Fraunhofer spectra (e.g. MFC [Gomer et al. 1993]) leading to a Raman cross section which has all the spectral characteristics of the respective instrument. The rotational Raman spectrum is then divided by the measured Fraunhofer spectrum to determine the Ring spectrum (Fig. 9.10). Note, the Fraunhofer spectrum must be corrected for rotational Raman scattering to represent pure elastic scattering. In most cases, this approach leads to an excellent correction of the Ring effect.

9.2 AMF Calculations

The detailed calculation of AMFs requires the consideration of different physical processes influencing the radiative transfer (RT) in the atmosphere. Scattering processes, either Rayleigh or Mie scattering, reflection on the earth's surface, refraction, the curvature of the earth, and the vertical distribution of trace gases play a role in the transfer of solar radiation. Consequently, sophisticated computer models are employed to calculate the RT in the atmosphere and the AMFs needed to retrieve vertical column densities from DOAS observations of scattered sun-light.

It is beyond the scope of this book to describe the details of RT models that are currently in use for AMF calculations. The interested readers can find details of RT modelling in Solomon et al. (1987), Perliski and Solomon (1992), Perliski and Solomon (1993), Stamnes et al. (1988), Dahlback and Stamnes (1991), Rozanov et al. (1997), Marquard (1998), Marquard et al. (2000), Rozanov et al. (2000), Rozanov (2001), Spurr (2001), and v. Friedeburg (2003).

However, we give a short overview of the most commonly used RT methods and the input data required to run these models.

The traditional computation procedure for the calculation of AMFs for a certain absorber is straightforward (e.g. Frank, 1991; Perliski and Solomon, 1993). For this purpose, the following computation steps are performed:

1. The radiances $I_S(\lambda, 0)$ and $I_S{}^A(\lambda, \sigma)$ are calculated by an appropriate RT model. This means that two model simulations must be performed: one simulation where the absorber is included and one simulation where the absorber is omitted from the model atmosphere.
2. The slant (or apparent) column density S is calculated according to the DOAS method, i.e. (9.1).
3. The vertical column density is calculated by integrating the number density of the considered absorber, which has been used as an input parameter for the modelling of $I_S{}^A(\lambda, \sigma)$ in step 1, in the vertical direction over the spatial extension of the model atmosphere.
4. The AMF is then derived according to (9.4).

This computational procedure is used to calculate AMFs when multiple scattering is taken into consideration and was also used in some single-scattering RT models. It was first described and applied by Perliski and Solomon (1993). At present, it appears to be implemented in all existing RT models that consider multiple scattering, such as discrete ordinates radiative transfer (DISORT) models (Stamnes et al., 1988; Dahlback and Stamnes, 1991), GOMETRAN/SCIATRAN (Rozanov et al., 1997), AMFTRAN (Marquard, 1998; Marquard et al., 2000), the integral equation method RT models (Anderson and Lloyd, 1990), and the backward Monte Carlo RT models (e.g. Perliski and Solomon, 1992; Marquard et al., 2000). We refer to this method as the 'traditional' (linear) AMF computation method.

There is a second method to calculate AMFs, which was often used for single-scattering radiative transport models. In this approximation, it is assumed that the radiation propagating through the atmosphere is scattered once before it is detected. It is evident that this scattering process must occur along the detector's viewing direction. Mainly, some of the earlier single-scattering RT models apply this method (see Sarkissian et al., 1995; and references therein). The method is based on a linear weighting scheme (see e.g. Solomon et al., 1987), and thus there is a second assumption entering into this method, namely, that the total optical density at the detector position is equivalent to the sum of the optical densities along the single paths weighted by their probabilities. However, this is, in general, not valid if there are several paths through the atmosphere, which is the case for measurements of scattered radiation.

9.2.1 Single-scattering RT Models

Earlier RT models considered only one scattering event in the atmosphere, and were thus termed 'single-scattering models'. Figure 9.11 illustrates the approach most often taken in these simple RT models. The curved atmosphere

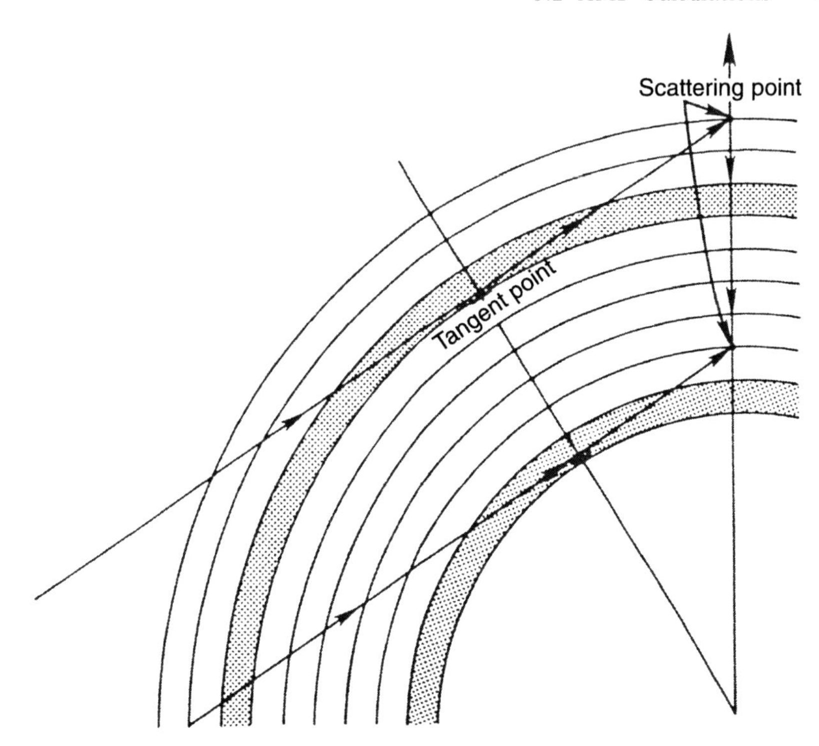

Fig. 9.11. Definition of the tangent point in spherical geometry (from Perliski and Solomon, 1993, Copyright by American Geophysical Union (AGU), reproduced by permission of AGU)

is subdivided in distinct layers. Light travels from the direction of the sun, defined by the SZA, through the atmosphere until it reaches the zenith over the DOAS instrument. At the zenith, a parameterisation of Rayleigh scattering and Mie scattering is used to calculate the fraction of light scattered towards the earth surface from this scattering point (Fig. 9.12). The vertical profiles of air density and aerosol concentration which are needed both for the calculation of extinction along the light path and the scattering efficiency at the scattering point, are input parameters of the model. Refraction is considered whenever a ray enters a new layer following Snell's law. Absorption is calculated from the path length in each layer, vertical concentration profile, and absorption cross-section supplied to the model. The model then performs a numeric integration of (9.10) to derive the SCD for each SZA and wavelength. The ratio of this SCD with the VCD calculated from the vertical trace profile is then the desired AMF.

A number of single-scattering models have been developed over the years (Frank, 1991; Perliski and Solomon, 1993; Schofield et al., 2004). The advantage of these models is their simplicity and the lesser use of computer resources. Perliski and Solomon (1993) discuss the disadvantages of these models. In particular, the omission of multiple scattering events poses a

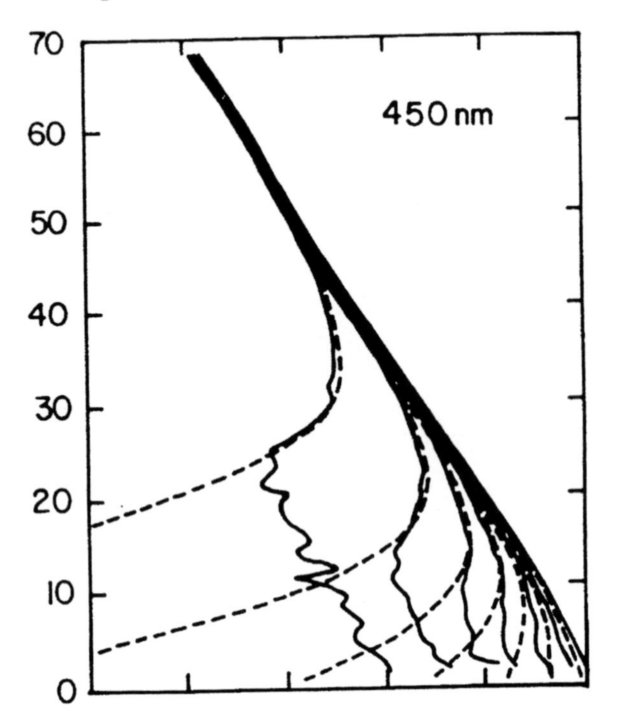

Fig. 9.12. Contribution of different altitudes to the intensity detected by a ground-based ZSL-DOAS instrument for different SZA. The dashed and solid lines show results from a single and a multiple scattering model, respectively. (from Perliski and Solomon, 1993, Copyright by American Geophysical Union (AGU), reproduced by permission of AGU)

serious challenge. Figure 9.12 illustrates the change in the contribution of each altitude to the detected intensity for different SZA. In particular in the lower atmosphere, the contribution of multiple scattering cannot be ignored. This effect is even more severe if off-axis geometries are used, where the most probable scattering height is in the troposphere. Multiple scattering also has to be considered when high aerosol levels are encountered. For these reasons, single-scattering models are rarely used.

9.2.2 Multiple-scattering RT Models

Radiative transfer models that consider multiple scattering events are the standard in current AMF calculations. A common approach to these calculations is to solve the RT equations described in Chap. 3 for direct and diffuse radiation. Several implementations for the numerical solution have been brought forward over the past years. Rozanov et al. (2000, 2001) describe a combined differential-integral (CDI) approach that solves the RT equation in its integral form in a pseudo-spherical atmosphere using the characteristics method. The model considers single scattering, which is calculated truly spherical and multiple scattering, which is initialised by the output of the pseudo-spherical model. This model is known as SCIATRAN (http://www.iup.uni-bremen.de/sciatran/).

Other models are based on the discrete ordinate method to solve the RT equation. In this approach, the radiation field is expressed as a Fourier cosine series in azimuth. A numerical quadrature scheme is then used to replace the integrals in the RT equation by sums. The RTE equation is, therefore, reduced to a set of coupled linear first-order differential equations, which are consequently solved (Lenoble, 1985; Stamnes et al., 1988; Spurr, 2001). This model also employs a pseudo-spherical geometry for multiple scattering in the atmosphere.

A number of RT models rely on the Monte Carlo method, which is based on statistical sampling experiments on a computer. In short, RT is quantified by following a large number of photons as they travel through the atmosphere. While in the atmosphere, the photons can undergo random processes such as absorption, Rayleigh scattering, Mie scattering, reflection on the ground, etc. For each of these processes, a probability is determined, which is then used to randomly determine if, at any point in the atmosphere, the photon undergoes a certain process. A statistical analysis of the fate of the photon ensemble provides the desired RT results. More information on Monte Carlo methods can be found in Lenoble (1985). While the most logical approach to implement Monte Carlo methods for AMF calculations is to follow photons entering the atmosphere and counting them as they arrive at the detector, this 'forward Monte Carlo' method is slow and numerically inefficient. Consequently, it is not used for this specific application. However, the 'backward Monte Carlo method', in which photons leave the detector and are then traced through the atmosphere, has proven to be a reliable approach to calculate AMFs (Perliski and Solomon, 1993; Marquard et al., 2000; von Friedeburg et al., 2003). The advantage of Monte Carlo methods is their precise modelling of RT without the need for complex numerical solutions or the simplifications of the underlying RT equation (Chap. 4). The disadvantage, however, is that Monte Carlo models are inherently slow due to the large number of single-photon simulations that are needed to determine a statistically significant result.

9.2.3 Applications and Limitations of the 'Traditional' DOAS Method for Scattered Light Applications

The classical DOAS approach has been widely and successfully applied to measurements of scattered sun-light. It uses the same tools as described in Chaps. 6 and 8. In short, based on Lambert–Beer's law:

$$I(\lambda) = I_0(\lambda) \cdot \exp\left(-\sigma(\lambda) \cdot c \cdot L\right) , \tag{9.17}$$

one can calculate the product of path length and concentration of a trace gas, i.e. the column density, by measuring $I(\lambda)$ and $I_0(\lambda)$ and using the known absorption cross-section of the gas $\sigma(\lambda)$.

$$c \cdot L = -\frac{1}{\sigma(\lambda)} \ln\left(\frac{I(\lambda)}{I_0(\lambda)}\right) . \tag{9.18}$$

Most applications of scattered light DOAS follow this approach by using the measurement in the zenith or at a low elevation angle as $I(\lambda)$ and, as described above, a spectrum with small SZA or different elevation angle for $I_0(\lambda)$. In this case, the fitting of one or more absorption cross-sections modified to the instrument resolution will yield the SCD, or more precisely the DSCD. It is also clear that the SCD is independent of the wavelength. This appears to be trivial here. However, we will see below that this may not be the case for certain scattered light measurements.

In this section we argue that this classical DOAS approach can lead to problems in the analysis of scattered light applications. To simplify the discussion, we assume that $I_0(\lambda)$ is the solar scattered light without the presence of an absorber as already assumed in Sect. 9.1. The discussion can easily be expanded to the case where a different solar reference spectrum is used (see Sect. 9.1.5).

We will use a simplified version of (9.10), where we replace the integrals over height with integrals over all possible light paths that reach the detector. In addition, we simplify this equation further by replacing the integrals with a SCD for each path, S'. As assumed in the classical approach, the left side of the equation is the desired VCD times the AMF, the SCD determined by DOAS measurements.

$$V \cdot A = \frac{1}{\sigma(\lambda)} \ln\left[\frac{\int_{all\,paths} I_S(\lambda, z) \cdot \exp\left(-\sigma(\lambda) \cdot S'\right) dz}{\int_{all\,paths} I_S(\lambda, z) dz}\right] . \tag{9.19}$$

The comparison with (9.10) reveals that the application of the logarithm is not as straightforward as in the case of Lambert–Beer's law. The logarithm has to be applied on the integrals over the intensities of the light passing the atmosphere on different light paths. We can now distinguish three cases to further interpret this equation:

1. In the case that SCD S' is the same for all paths, i.e. in case of direct solar measurements, the exponential function in the numerator can be moved in front of the integral and we have the classical Lambert–Beer's law. In this case, the methods described in Chap. 8, i.e. fitting of absorption cross-section, can be applied.
2. If we assume that the exponential function in the numerator of (9.19) can be approximated using $\exp(x) \approx 1 + x$, for $-\varepsilon < x < \varepsilon$, which is the case for a weak absorber, we can perform the following approximation:

$$\int_{all\ paths} I_S(\lambda) \cdot \exp(-\sigma(\lambda) \cdot S') \approx \int_{all\ paths} I_S(\lambda) \cdot (1 - \sigma(\lambda) \cdot S') =$$

$$= \int_{all\ paths} I_S(\lambda) - \sigma(\lambda) \cdot \int_{all\ paths} I_S(\lambda) \cdot S'$$

$$= \left[\int_{all\ paths} I_S(\lambda) \right] \cdot \left(1 - \sigma(\lambda) \cdot \bar{S}\right) \quad with \quad \bar{S} = \frac{\int_{all\ paths} I_S(\lambda) \cdot S'}{\int_{all\ paths} I_S(\lambda)}.$$

$$\approx \left[\int_{all\ paths} I_S(\lambda) \right] \cdot \exp\left(-\sigma(\lambda) \cdot \bar{S}\right)$$

In the case of a weak absorber, i.e. typically with an optical density below 0.1, the classical approach can still be employed since the absorption cross-section $\sigma(\lambda)$ is now outside of the integral. The SCD $\bar{S}$ is the intensity-weighted average of all slant columns on different paths. One can, therefore, use the fitting of an absorption cross-section and a classical AMF for the analysis of the data.

3. In the case of a strong absorber (OD > 0.1), such as ozone in the ultraviolet wavelength region, the approximation used above cannot be employed. The RT in the atmosphere cannot be separated from the trace gas absorption. The integral in the numerator of (9.19) now becomes dependent on $\sigma(\lambda)$ through absorptions along each path, as well as the weighing of each path during the integration, i.e. paths with stronger absorptions have a smaller intensity and thus contribute less to the integral than paths that have weaker absorptions.

Richter (1997) investigated this effect and found that the classical approach introduces small VCD errors of ~2% for ozone absorptions in the ultraviolet for SZAs below 90° in ZSL applications. However, the VCD error can reach 15% if the classical DOAS approach is used for SZAs above 90° in this wavelength region. For ozone in the visible and NO_2, the error generally remains below 2% for all SZA. Richter (1997) also showed that in the UV above 90° SZA, the residual of the fit increases due to this effect. He proposes an extended DOAS approach, which instead of using absorption cross-sections in the fitting procedure uses wavelength-dependent slant column optical densities extracted from a RT model.

An additional problem in the use of the DOAS approach for solar measurements is the temperature dependence of absorption sections. Because the light reaching the detector crosses the entire atmosphere, absorption occurs at different temperatures found at different altitudes. Equation (9.19), therefore, needs to be expanded by introducing a temperature dependent $\sigma(\lambda, T)$. The light path through the atmosphere, i.e. the RT, now plays an important role since regions with different temperatures are weighted differently. Again, the solution of this problem lies in the combination of RT and absorption spectroscopy (Marquard et al., 2000).

9.3 AMFs for Scattered Light Ground-Based DOAS Measurements

The analysis of scattered light DOAS observations relies on the principles outlined earlier. However, the different observational setups require analysis strategies adapted to the particular geometry of each application. Central problems include the dependence of the radiation transport, and thus the AMFs, on the – a-priori unknown – amount of Mie scattering and the location of trace gases in the atmosphere. Here we discuss how AMFs depend on various atmospheric parameters, such as the solar position, the trace gas profile, etc., and how this impacts the analysis of DOAS observations.

9.3.1 ZSL-DOAS Measurements

Zenith scattered light applications have been and still are widely used to study stratospheric chemistry. The AMF can vary widely as a function of wavelength, trace gas absorption, vertical trace gas profile, and stratospheric aerosol loading (e.g. Solomon et al., 1987; Perliski and Solomon, 1993; Fiedler et al., 1994). On the other hand, tropospheric clouds are of comparatively little influence, thus making ground-based measurements by this technique possible when the SZA is smaller than 95°.

Dependence on SZA

For SZAs smaller than 75°, the AMF can be approximated by $1/\cos\vartheta$. At larger SZA, a RT model yields results as shown in Fig. 9.13 for ozone. The AMF increases continuously, reaching a value of 6–20 at 90° SZA, depending on the wavelength.

A small dip around 93° is caused by the most likely scattering height passing above the altitude of the stratospheric absorption layer, leading to a reduction of the effective path length. Wavelengths above 550 nm do not show this effect since the influence of scattering and absorption processes in the atmosphere, which also contribute, decrease as the wavelength increases.

Dependence on Solar Azimuth

The dependence of the AMF on the solar azimuth is generally small for ZSL appications. The only exception is when the stratospheric trace gas or aerosol is not homogeneously distributed, and the changing effective viewing direction causes a change in the parameters influencing the RT.

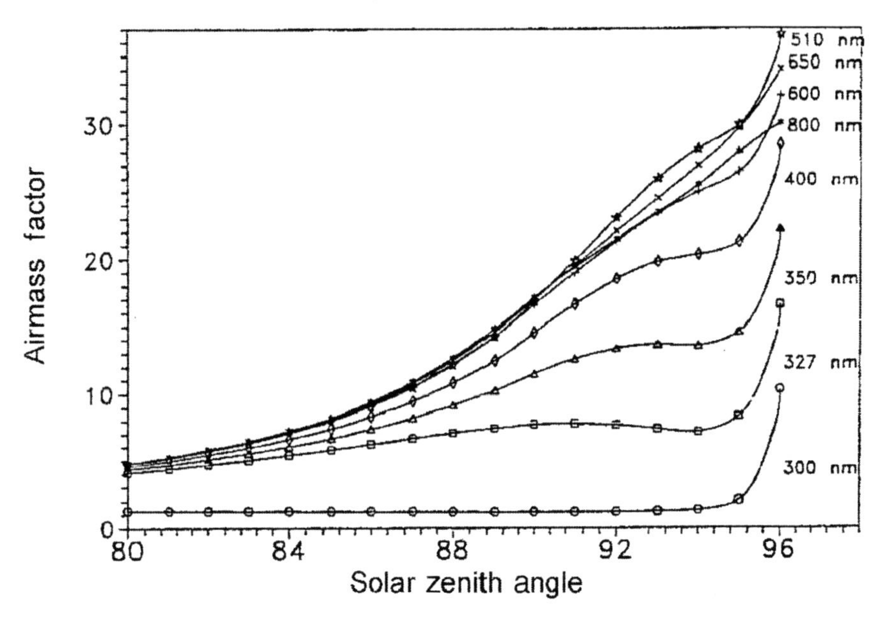

Fig. 9.13. Examples of Zenith Scattered Light (ZSL) airmass factors (AMFs) for stratospheric ozone and different wavelength as a function of solar zenith angles (in degrees) (from Frank, 1991)

Dependence on Wavelength

The ZSL-AMF depends on wavelength in the same way as Rayleigh and Mie scattering, as well as certain trace gas absorptions, are wavelength dependent. Figure 9.14 shows the dependence of the ozone AMF on the wavelength. The most notable feature is a steep increase of the AMF at lower wavelength. This is predominantely caused by the wavelength dependence of Raleigh scattering, which influences the weighing of the absorption layer.

A minimum around 570 nm is caused by the Chappuis ozone absorption band, which increases in strength at increasing SZA.

Dependence on Trace Gas Profile

The AMF also depends on the vertical profile of the respective trace gas. As illustrated in Fig. 9.4, the light collected at the ground is weighted towards a distribution around the most probable scattering angle. A trace gas profile with a maximum at this altitude will lead to a larger AMF than a profile with a maximum above or below the most probable scattering altitude, because the maximum of the profile is more heavily weighted. An extreme example for the dependence of the vertical profile is a comparison of the ZSL AMF of a trace gas located in the troposphere and the stratosphere. Figure 9.6 shows that the stratospheric AMF, for example for NO_2 around 445 nm, increases

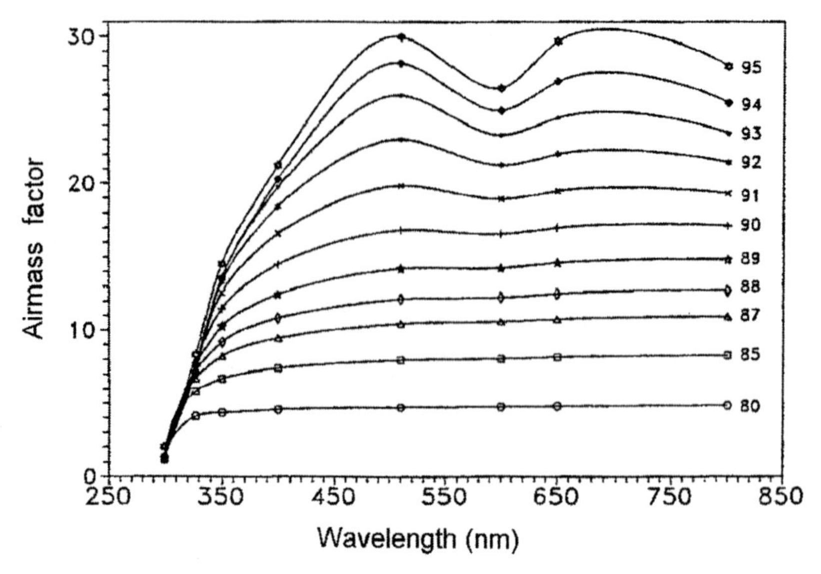

Fig. 9.14. Examples of Zenith Scattered Light (ZSL) airmass factors (AMFs) for stratospheric ozone and different solar zenith angles (in degrees) as a function of wavelength (from Frank 1991)

from 6 at 80° SZA to 20 at 90° SZA. In contrast, the tropospheric AMF is much smaller with a value of 2 at 80° SZA and ~1 at 90° SZA. The much smaller values of the tropospheric AMF are due to the relatively small amount of light being scattered in the troposphere and the short path on which light scattered in the stratosphere passes through the troposphere. Above 90° SZA, less and less light is scattered in the troposphere, and the tropospheric AMF approaches 1 as stratospheric light passes the troposphere vertically.

Dependence on the Aerosol Profile

Aerosol particles have an influence on the RT since they efficiently scatter solar light into the receiving instrument. The impact on the AMF depends on the vertical distribution of aerosol and its scattering coefficient. For example, a stratospheric aerosol layer located below the maximum concentration of a stratospheric trace gas will lead to a reduction of the AMF since the most probable scattering altitude is shifted downwards. On the other hand, the AMF can, theoretically, be increased if an aerosol layer and a trace gas layer are collocated. The most prominent example of the impact of aerosol on AMF occured during the 1992 volcanic eruption of Mount Pinatubo. Since the volcanic aerosol was located below the ozone layer, the AMF were changed up to 40% relative to the pre-eruption case (e.g. Dahlback et al., 1994).

Tropospheric aerosol has little influence on the ZSL-AMF of a stratospheric trace gas, in particular at high SZA, since most of the scattering events occur in the stratosphere. Similarly, tropospheric clouds have little influence.

Chemical Enhancement

An additional problem in the AMF determination is found when measuring photoreactive species that change their concentration according to solar radiation (e.g. Roscoe and Pyle, 1987). Examples of such species are NO_2 and BrO. Because their concentration and vertical profile change during sunrise and sunset, the temporal change in these parameters have to be considered when calculating the AMF dependence on the SZA. This is typically achieved by using correction parameters derived from a photchemical model of stratospheric chemistry.

Accuracy of AMF Determinations

The accuracy of the AMF calculations directly affects the accuracy of the VCDs derived by ZSL–DOAS instruments. Much effort thus has gone into comparing RT models (Sarkissian et al., 1995; Hendrick et al., 2004, 2006).

Hendrick et al. (2006), for example, compared six RT models to determine the systematic difference between different solutions to the ZSL-DOAS retrieval. The models included single-scattering models, multiple-scattering models based on DISORT, CDIPI, or similar analytical approaches, and one Monte Carlo model. All models were constrained with the same boundary conditions, which included time-dependent profiles of the photoreactive trace gases BrO, NO_2, and OClO to describe chemical enhancement. Figure 9.15 shows a comparison of the models run in single scattering (SS) and multiple scattering (MS) mode. Note that the figure shows the SCD calculated by the models rather than the AMF. For BrO and OClO, the different models agreed better than $\pm5\%$. The agreement for NO_2 is $\pm2\%$ for all, except one model. There is a systematic difference between SS and MS models, in particular for OClO, for which the altitude of the aerosol is similar to that of the OClO layer. These results show the typical systematic uncertainty of current RT models for ZSL-DOAS interpretation. It should be noted that this intercomparison does not take into account uncertainties introduced by the errors in the aerosol profile, trace gas profile, and chemical enhancement used for the retrieval of real ZSL-DOAS observations.

9.3.2 Off-axis-DOAS Measurements

As discussed in Sect. 9.1.3, a change of viewing direction can be beneficial for scattered light DOAS measurements in various respects. For example, Sanders et al. (1993) observed stratospheric OClO over Antarctica during twilight using an 'off-axis' geometry. Because the sky is substantially brighter towards the horizon in the direction of the sun at large SZAs, the signal-to-noise ratio of the measurements can be considerably improved as compared to zenith geometry. Sanders et al. (1993) also pointed out that the off-axis geometry increases the sensitivity for lower absorption layers. They found

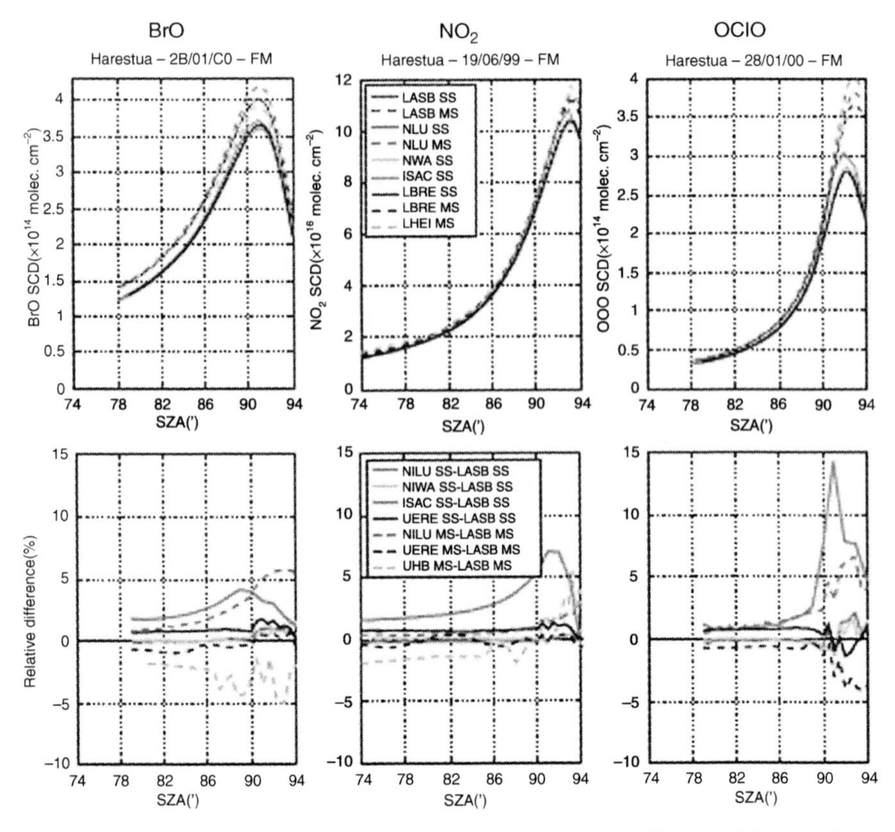

Fig. 9.15. Comparison of stratospheric and tropospheric SCDs of NO_2 at 445 nm from various RT models (from Hendrick et al., 2006)

that absorption by tropospheric species (e.g. O_4) is greatly enhanced in the off-axis viewing mode, whereas for an absorber in the stratosphere (e.g. NO_2) the absorptions for zenith and off axis geometries are comparable. One of the challenges of the 'off-axis' measurements is the increased complexity of the RT calculations. We will discuss the general implications of a non-zenith viewing geometry in the following section in more detail.

9.3.3 MAX-DOAS Measurements

The main difference between ZSL–DOAS and 'off-axis' or 'multi-axis'-DOAS is a viewing elevation angle different from 90°. This leads to an increased effective path length in the troposphere and only lesser changes in stratospheric trace gases AMF as compared to the zenith viewing geometry. Consequently, tropospheric absorbers are more heavily weighed in low elevation observations. This property is one of the main motivations to use low elevation viewing angles. However, it also increases the number of parameters that have to be considered in the RT calculations. In addition to the parameters, as we have already discussed for the ZSL case, one now also has to consider the vertical profiles of tropospheric trace gases and aerosol. The RT is also more dependent on the albedo and the solar azimuth.

The behaviour of the AMF under various conditions has been discussed by Hönninger et al. (2003) using the Monte Carlo radiative transfer model "Tracy" (v. Friedeburg, 2003), which includes multiple Rayleigh and Mie scattering, the effect of surface albedo, refraction, and full spherical geometry. To investigate the dependence of the AMF on the vertical distribution of an absorbing trace gas, Hönninger et al. (2003) considered a number of artificial profiles (Fig. 9.16). These were used together with a number of different aerosol extinction profiles and phase functions (Fig. 9.16). Calculations were performed at a wavelength of $\lambda = 352$ nm, and a standard atmospheric scenario for temperature, pressure and ozone. The vertical grid size in the horizontal was 100 m in the lowest 3 km of the atmosphere, 500 m between 3 and 5 km, and 1 km from 5 km up to the top of the model atmosphere at 70 km.

SZA Dependence of the AMF/Stratospheric AMF

The change of the AMF with SZA depends strongly on the vertical distribution of the trace gas. As in the ZSL geometry, the AMF for a stratospheric trace gas depends strongly on the SZA, in particular, at large SZA (left panel in Fig. 9.17). The dependence of tropospheric AMF is much smaller and only becomes significant above a SZA of ~75° (middle panel in Fig. 9.17). Above 75°, a small dependence on SZA can be observed. For trace gases that are

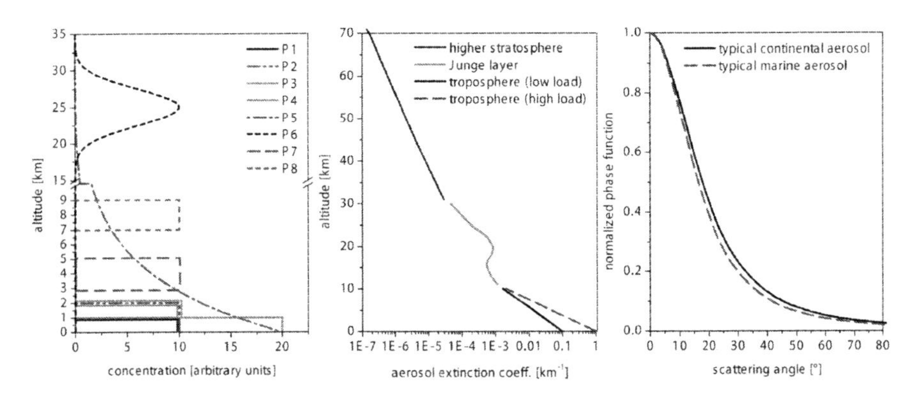

Fig. 9.16. Profile shapes of an atmospheric absorber (**left**), atmospheric aerosol (**middle**) and the aerosol scattering phase functions (**right**) used for MAX-DOAS radiation transport studies. The profiles P1–P4 assume a constant trace gas concentration in the 0–1 km and 0–2 km layers of the atmosphere. Profile P5 is that of the oxygen dimer, O_4. P5 is a purely a stratospheric profile centred at 25 km with a FWHM of 10 km (from Hönninger et al., 2003)

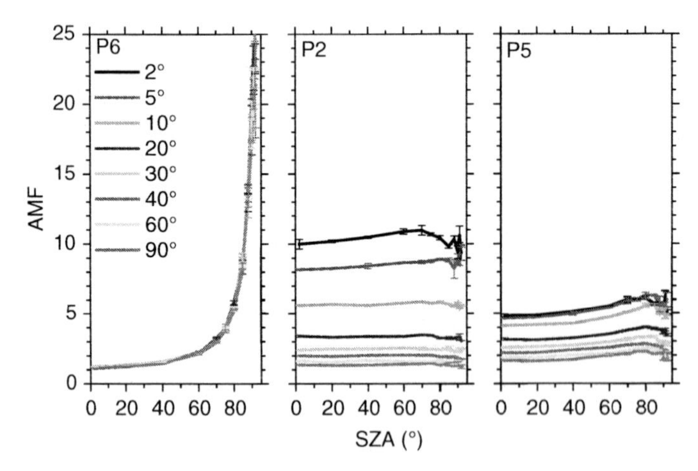

Fig. 9.17. SZA-dependence of the AMF for the typical stratospheric profile P6 and the boundary layer profile P2 as well as the O_4 profile P5 for comparison (for description of profiles see caption of Fig. 9.16). The expected strong SZA dependence is observed for the stratospheric absorber, with no significant dependence on the viewing direction. In contrast, for the tropospheric profiles P2 and P5 significant differences for the various viewing directions can be seen, while the SZA dependence is significant only at higher SZA (from v. Friedeburg, 2003; Hönninger et al., 2003)

present both in the troposphere and the stratosphere the total AMF will be a mixture of the terms.

The functional dependence of the AMF on the SZA can be best understood by analysing the altitude of the first and last scattering events between the sun and the detector (Fig. 9.18). In the model atmosphere investigated by Hönninger et al. (2003), the first scattering altitude (FSA) for $\alpha = 2°$ and SZA $< 75°$ is approximately 6 km, while the last scattering altitude, i.e. the altitude of the last scattering event before a photon reaches the MAX-DOAS instrument, is ~0.6 km.

At larger SZA, the FSA slowly moves upwards in the atmosphere into the stratosphere. This is in agreement with the concept of an upward-moving most probable light path as the sun sets (see Sect. 9.1.2), and explains the SZA dependence of the stratospheric AMF. The LSA is largely independent on the SZA. Therefore, the tropospheric AMF changes little.

Dependence of AMF on Viewing Elevation

The dependence of the AMF on the viewing elevation angle, α, is strongly influenced by the vertical profile of the trace gas. Stratospheric AMFs show little dependence on α (left panel in Fig. 9.17) at low SZA because the first scattering event occurs below the stratosphere, and the light path in the stratosphere is approximately geometric, i.e. only proportional to $1/\cos(\vartheta)$. Only at larger SZA does the stratospheric AMF show a weak dependence on α.

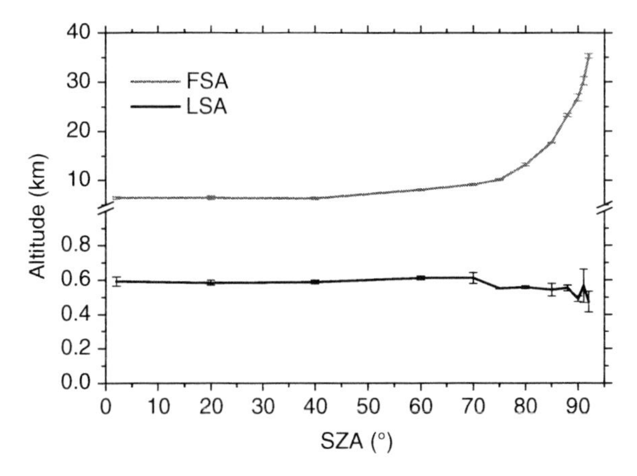

Fig. 9.18. Average altitude of first and last scattering event (FSA, LSA) for observation at 2° elevation angle. The altitude of the FSA strongly depends on the SZA, whereas the LSA altitude is largely independent of the SZA. Note the axis break at 1 km altitude and the expanded y-scale below (from Hönninger et al., 2003)

For a trace gas located in the lowest kilometre of the atmosphere, the dependence of the AMF on α is strong (middle panel in Fig. 9.17). Since most of the scattering events occur above the trace gas absorptions, the dependence is close to geometric, i.e. proportional to $1/\sin(\alpha)$ (see also Sect. 9.1.3). Deviations from this dependence may only be observed at SZA larger than 75°.

For trace gases extending above 1 km or located in the free troposphere, the dependence on α is more complicated than for the lower tropospheric case. In general, one finds that the dependence on α decreases as the altitude of the trace gas increases above the last scattering altitude. The dependence nearly disappears at the altitude of the first scattering event.

The dependence of the AMF α is the basis of MAX-DOAS. If simultaneous (or temporally close) measurements are made at different elevation angles α, there is essentially no change in ϑ and thus in the stratospheric part of the AMF. Thus, the stratospheric contribution to the total absorption can be regarded essentially a constant offset to the observed SCD.

Influence of the Trace Gas Profile Shapes

In the previous section, we indicated that the AMF depends on the vertical profile of the trace gas. Hönninger et al. (2003) calculated AMFs for six different profiles, P1–P6 in Fig. 9.19, for a pure Rayleigh scattering atmosphere, i.e. no aerosol. The dependence on the elevation angle is strongest for trace gases located close to the ground (profile P1) and decreases as the gases are located higher in the atmosphere, reaching AMFs of about 15 for very small α. A comparison with the geometric AMF shows how well $1/\sin(\alpha)$ approximates the AMF in this case. As the profiles extend further aloft

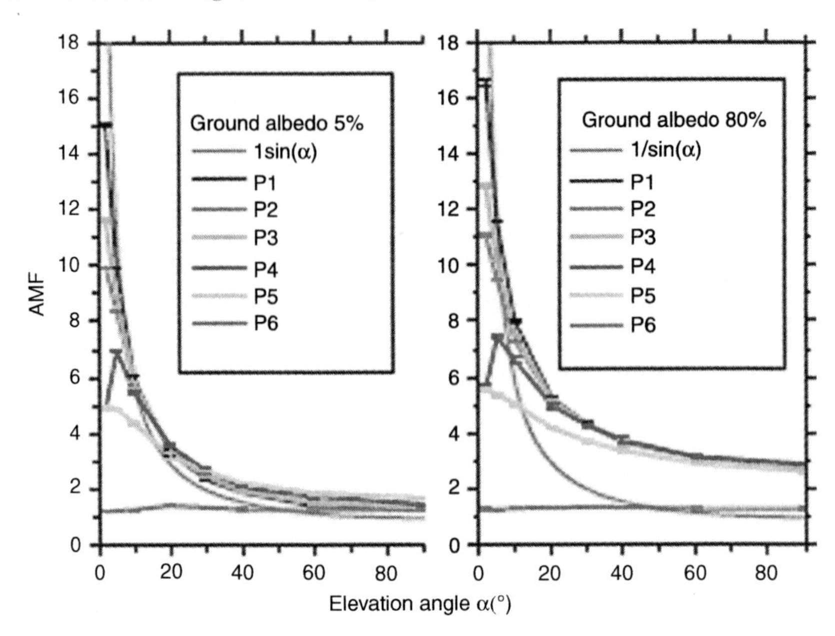

Fig. 9.19. AMF dependence on the viewing direction (elevation angle α) for the profiles P1–P6 (for description of profiles see caption of Fig. 9.16) calculated for 5% ground albedo (**left**) and 80%, albedo (**right**), respectively. A pure Rayleigh case was assumed (from v. Friedeburg, 2003; Hönninger et al., 2003)

(P2, P3, P5), the AMF decreases. Profile P5 deserves special attention since it describes the exponential decreasing concentration of atmospheric O_4. Because the O_4 levels and the vertical profile of O_4 do not change, they can be used to validate RT calculations (see Sect. 9.3.4).

For elevated trace gas layers, i.e. profile P4, the AMF peaks at $\alpha = 5$, because at lower viewing elevation angles a DOAS instrument would predominately see the air below the layer. The AMF is almost independent of the viewing direction for the stratospheric profile P6.

Dependence on Surface Albedo

Figure 9.20 shows that tropospheric AMFs (P1 – P5) are also influenced by the surface albedo. In general, higher albedos lead to larger AMF, because the reflection at the ground and the upwelling radiation will increase the effective absorption path length in the troposphere. Light that enters a DOAS instrument after being reflected at the ground may have passed parts through an absorption layer in the lower troposphere twice, thus increasing the trace gas absorption as it enters the detector. Since the effect of higher albedo increases all AMFs similarly, it has no significant effect on the elevation angle dependencies.

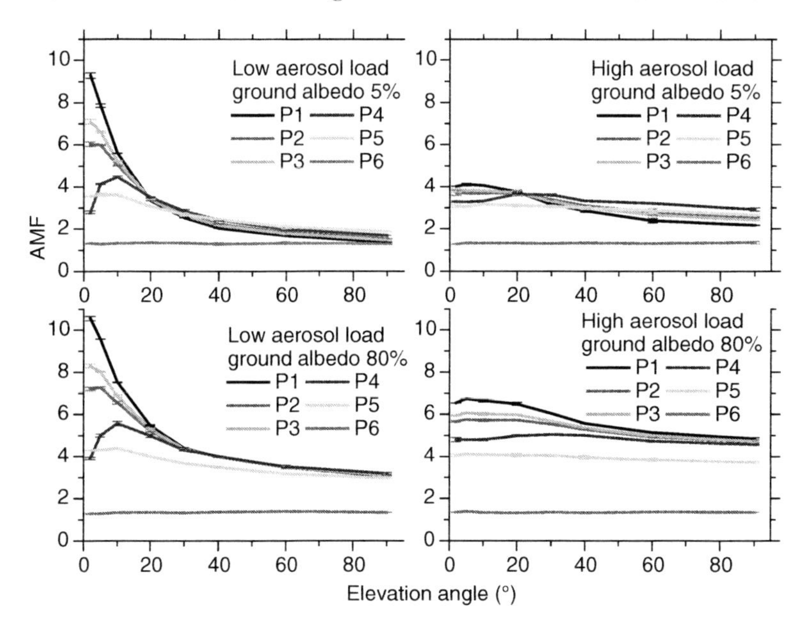

Fig. 9.20. AMF as a function of viewing direction, i.e. elevation angle α for the profiles P1–P6 (see caption of Fig. 9.16) calculated for low (**left column**) and high (**right column**) aerosol load (Fig. 9.16) and for 5% (**top**) and 80% (**bottom**) ground albedo, respectively (from v. Friedeburg, 2003)

Dependence on Aerosol Profile

The presence of aerosol particles in the atmosphere significantly enhances the scattering of radiation. Figure 9.20 illustrates how the AMFs for different viewing directions change for two different aerosol scenarios (high and low tropospheric aerosol load, Fig. 9.20) as well as for large and small surface albedo (see above). The AMF for a low aerosol load shows a similar shape to the Rayleigh case (Fig. 9.19). However, the absolute values of the AMFs at elevation angles below 20° are considerably reduced as compared to the Rayleigh case. Consequently, the geometric approximation is not good at low elevation angles, even at low aerosol loads. At high tropospheric aerosol loads, the absolute AMFs are further reduced and the dependence on α becomes much weaker.

The comparison of Rayleigh, low aerosol, and high aerosol cases also show how the dependence of the AMF on α changes with aerosol load. For trace gases in the lower part of the troposphere (P1–P3), the difference between AMF at low and high elevations slowly decreases. An extreme case of this dependence is dense fog, for which the α dependence of the AMF completely disappears. The high aerosol load and high albedo case in Fig. 9.20 also shows that in this case the AMF at very small α, i.e. 2°, can indeed be smaller than those at slightly larger α, e.g. 5°. This effect can be explained by the reduction of effective light path length at low elevation angles under these conditions (see below). The shift of the AMF maximum to larger α is even more pronounced for an elevated trace gas layer (P4), for which the maximum is at $\alpha = 5°$ for the Rayleigh case, $\alpha = 10°$ for a low aerosol load, and $\alpha = 40°$ for a high aerosol load.

The optical properties of aerosol, i.e. the scattering phase function (Fig. 9.16) and the single scattering albedo, influence the AMF little as shown by the comparison of a continental and marine aerosol in Fig. 9.21.

The fact that the tropospheric AMF decreases with increasing tropospheric aerosol load can be explained by a shorter mean free path of photons due to aerosol extinction. In particular, at low α, the light path in the lowest atmospheric layers is shorter, and a less slanted path in the higher layers is probable as illustrated in Fig. 9.22.

The last scattering altitude, which we introduced earlier, is a key parameter to understand the sensitivity of MAX-DOAS measurements towards different vertical profiles of gases and aerosol. The LSA dependence on the viewing elevation angle for different aerosol profiles is shown in Fig. 9.23. The LSA moves downward in the atmosphere as α decreases. In addition, the LSA decreases for increasing aerosol load and increasing albedo. In general, the geometric approximation for the AMF can only hold for trace gases below the LSA. Once the LSA reaches the trace gas layer, the AMF begins to decrease. Due to the α dependence of the LSA, this decrease occurs first at small α, leading to smaller AMF at very low elevation angles.

Dependence of the AMF on the Solar Azimuth Angle

In contrast to the ZSL measurements, low elevation angle measurements can show a dependence of the AMF on the SZA. In general, this dependence is

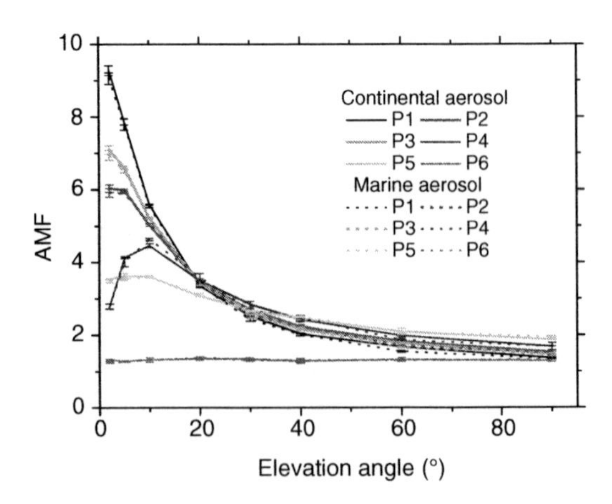

Fig. 9.21. Effect of different aerosol types on the AMF. Only small differences result from different scattering phase functions (from v. Friedeburg, 2003)

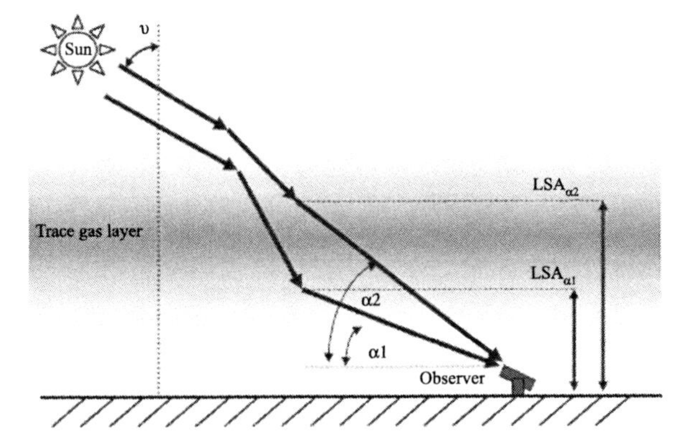

Fig. 9.22. The last scattering altitude (LSA): for low elevation angles, the mean free path in the viewing direction is shorter due to higher density and/or aerosol load. This can result in the slant path through absorbing layers at higher levels being shorter for lower elevation angles than for higher ones (from v. Friedeburg, 2003; Hönninger et al., 2003)

determined by the relative azimuth angle between the sun and the viewing direction. The effect arises due to the shape of scattering phase functions for the atmospheric scattering processes. Light paths taken by photons at different relative azimuth angles may be different, thus impacting the AMF. Figure 9.24 shows that, overall, the relative azimuth angle, here calculated for a SZA of 30° and the profile P4, has only a small influence on the AMF and

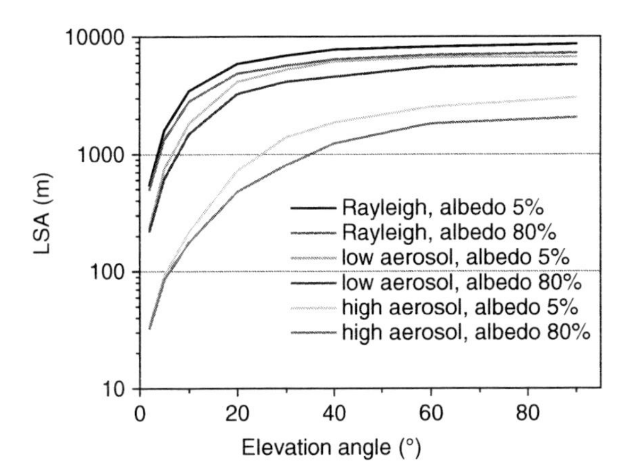

Fig. 9.23. Last scattering altitude (LSA) for pure Rayleigh, low and high aerosol load scenarios and both 5% and 80% albedo. The LSA is generally below 1 km for the lowest elevation angle and above 1 km for the highest elevation angles, especially for zenith viewing direction (from v. Friedeburg, 2003; Hönninger et al., 2003)

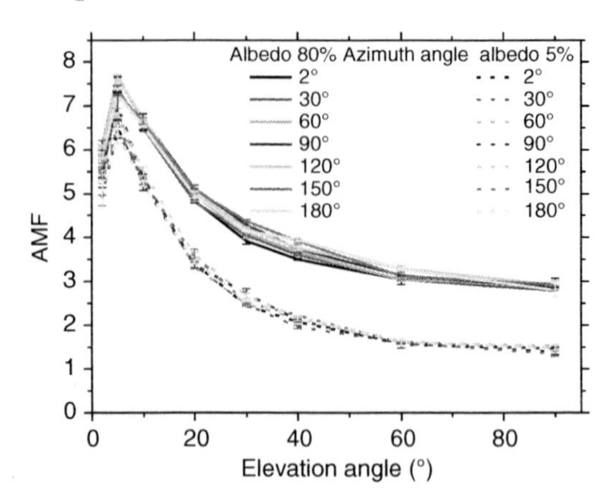

Fig. 9.24. Azimuth-dependence of the AMF for 30° SZA and profile P4 (trace gas layer at 1–2 km altitude). For relative azimuth angles between 2° (looking almost towards the sun) and 180° (looking away from the sun) only a small effect can be seen (from v. Friedeburg, 2003; Hönninger et al., 2003)

its α dependence. However, this effect is found to increase with ground albedo and tropospheric aerosol load.

9.3.4 Accuracy of MAX-DOAS AMF Calculations

Determining the accuracy of RT calculations of AMFs is a challenging task. As in the ZSL case, intercomparisons of different RT models allow the estimate of the uncertainties introduced by numerical calculations in the models. We will discuss such an intercomparison in this section. In addition, MAX-DOAS measurements of tropospheric trace gases offer another opportunity to constrain or validate RT calculations: the use of observations of stable atmospheric gases such as O_2 and O_4. This will be the topic of discussion in the second part of this section.

Model Intercomparison/Model Accuracy

The accurate calculation of AMF for low viewing elevations has only recently been implemented in RT models. Successful models include spherical geometry, refraction, and multiple-scattering. Figure 9.25 illustrates an intercomparison of four RT models, which were previously described in Sect. 9.3.1 (Hendrick et al., 2006). All models used the same boundary conditions, i.e. trace gas and aerosol profiles. Comparisons were made for viewing elevation angles of 5°, 10°, and 20°. In general, the agreement between the models was better than ±5% for the two trace gases investigated: NO_2 (Fig. 9.25) and HCHO (not shown). Hendrick et al. (2006) conclude that the largest discrepancies between the models are caused by the different treatments of aerosol scattering in each model.

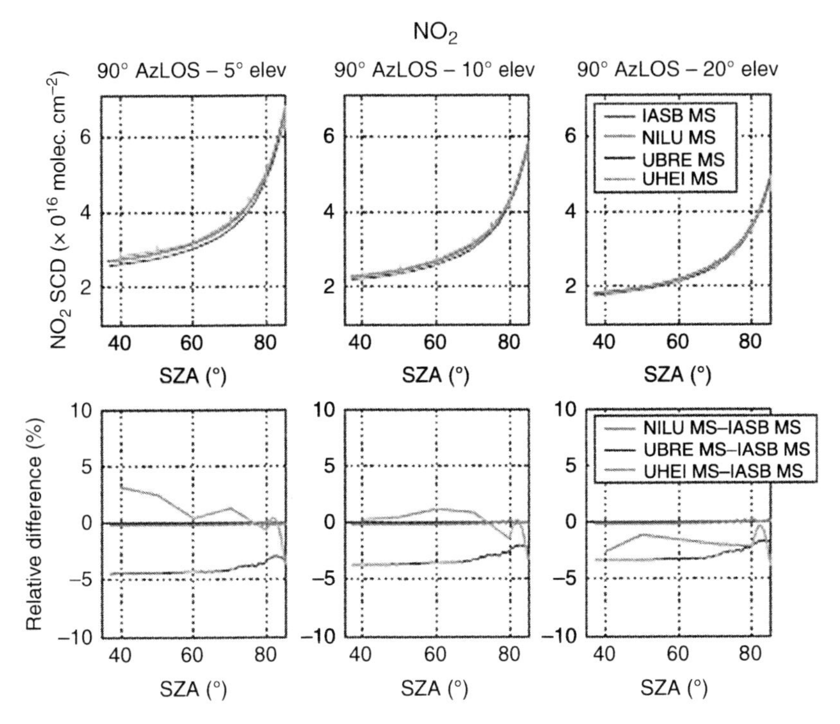

Fig. 9.25. Intercomparison of low viewing elevation slant column densities for NO_2 calculated by four different multiple scattering models (from Hendrick et al., 2006)

Validation of RT Calculations with O_2 and O_4 Measurements

One of the most challenging aspects of RT calculations is to determine the accuracies of AMFs for atmospheric measurements. The observation of the absorptions of O_2 or the oxygen dimer, O_4, offers an excellent tool to validate RT calculations and various input parameters, such as aerosol profiles, optical properties, etc. Alternatively, these measurements can be used to determine the aerosol load of the atmosphere with the goal to improving the RT calculations (Friess et al., 2006). The atmospheric O_2 and O_4 concentration profiles are well known, i.e. the O_4 profile $c_{O4}(z) = (0.21 \cdot c_{air}(Z))^2$, and are fairly constant in time (see the O_4 profile P5 in Fig. 9.16). Only small changes of the air density c_{air} on temperature and barometric pressure have to be taken into account.

The measured O_2 and/or O_4 SCDs for a series of elevation angles, e.g. from 2° to 90°, can be compared to a series of calculated O_2 and O_4 SCDs (for the temperature and pressure as recorded during the measurement). The comparison between observed and measured O_2 or O_4 SCDs can give valuable insight into the quality of the RT calculations. Furthermore, the aerosol profile can be varied until the best agreement is reached. Figure 9.26 shows

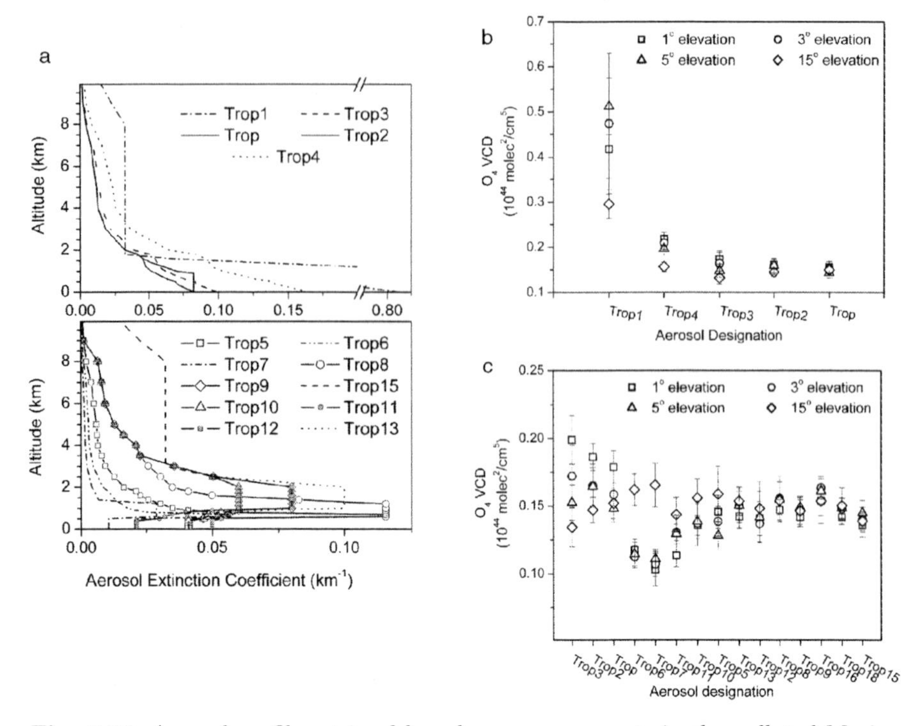

Fig. 9.26. Aerosol profile retrieval based on measurements in the polluted Marine boundary layer using observations of O_4 SCD and radiative transfer calculations based on the aerosol profiles shown in panel (**a**). The O_4 vertical column densities retrieved based on the observations and RT calculations are shown in panel (**b**) and (**c**) for two different days. The O_4 VCD with the best agreement between the four viewing elevation angles indicates the best estimate for the aerosol profile for each day (from Pikelnaya et al., 2007, Copyright by American Geophysical Union (AGU), reproduced by permission of AGU)

an example of such an optimisation for O_4 observations (Pikelnaya et al., 2007). Panels (a) in Fig. 9.26 shows different aerosol profiles that were used in a RT model to calculate AMFs of O_4. Panel (b) and (c) of Fig. 9.26 show the VCDs derived from these AMFs for O_4 SCD observation on two different days. On the first day, the profile TROP, which described a fairly constant aerosol extinction in a marine boundary layer of 700 m depth and an exponential decay above this altitude, led to the best agreement between the VCDs calculated for elevation angles of 1°, 3°, 5°, and 15°. On the second day [Panel (c) in Fig. 9.26], the profile TROP9, which consisted mainly of an aerosol layer between 1 km and 2 km altitude, led to the best agreement between the observed O_4 VCDs. This example exemplifies the sensitivity of O_4 observations on the aerosol profile and the potential of this approach to retrieve these profiles.

9.3.5 The Box-AMF Concept

With the knowledge of the aerosol profile, either through a retrieval based on O_4 slant column densities and intensities (Friess et al., 2006) or through other means, the main unknown in the MAX-DOAS RT is the vertical profile of the trace gases.

To understand the sensitivity of MAX-DOAS measurements and RT calculations on the vertical distribution, the **Box Air Mass Factor** was introduced. The Box-AMF represents the contribution of a trace gas located at a certain altitude interval, or vertical 'box', to the overall AMF under given atmospheric conditions. It is thus a measure of the sensitivity of a particular viewing direction towards an absorber being present in a specific altitude. The SCD of a trace gas, and thus indirectly the AMF, at a given elevation viewing angle, α_i, is related to the Box-AMF through the following equation:

$$S(\alpha_i) = \sum_{j=1}^{m} \left[(A_{Box})_{ij} \cdot \Delta h_j \cdot c_j \right] . \qquad (9.20)$$

In this equation, Δh_j is the height of altitude interval j, c_j the concentration, and $(A_{\mathrm{Box}})_{ij}$ the Box-AMF in this interval for elevation viewing angle, α_{i}.

Figure 9.27 shows modelled Box-AMFs for trace gas layers of $\Delta h_j = 100\,\mathrm{m}$ thickness from the ground up to 2 km altitude. The Box-AMF is largest near the surface for very low elevation angles. In particular, for α smaller than 10°, the Box-AMFs vary strongly with altitude. Consequently, MAX-DOAS observations at small α are very sensitive to the shape of the vertical profile in the lower troposphere.

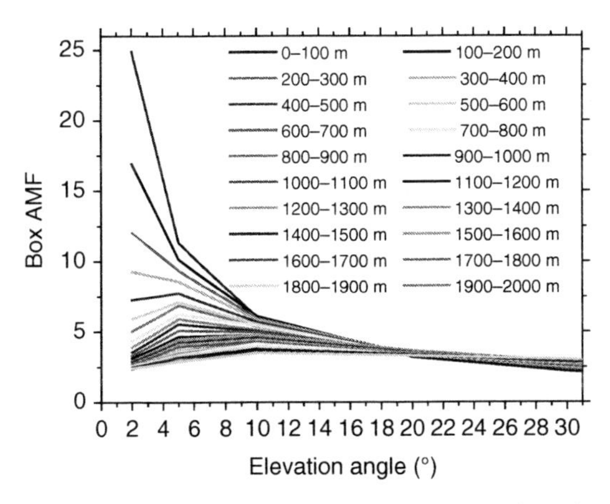

Fig. 9.27. Box AMF for the model layers below 2 km altitude. The sensitivity towards the lowest layers changes strongly for elevation angles below 10°, for elevation angles smaller than 5° the sensitivity already decreases for layers above 400 m altitude (from v. Friedeburg, 2003; Hönninger et al., 2003)

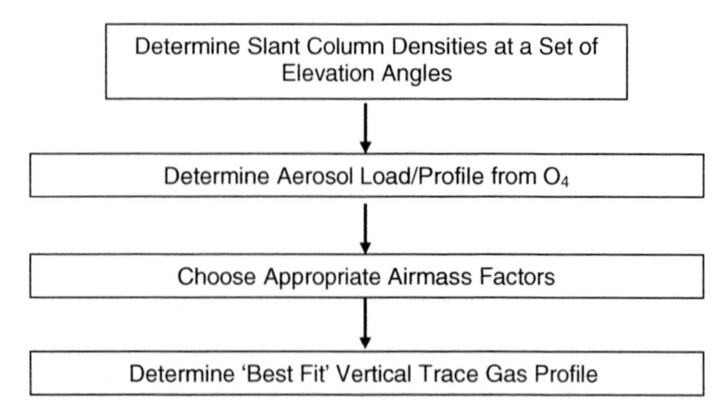

Fig. 9.28. Diagram of an AMAX-DOAS evaluation procedure including the determination of the aerosol load from O_4 observations

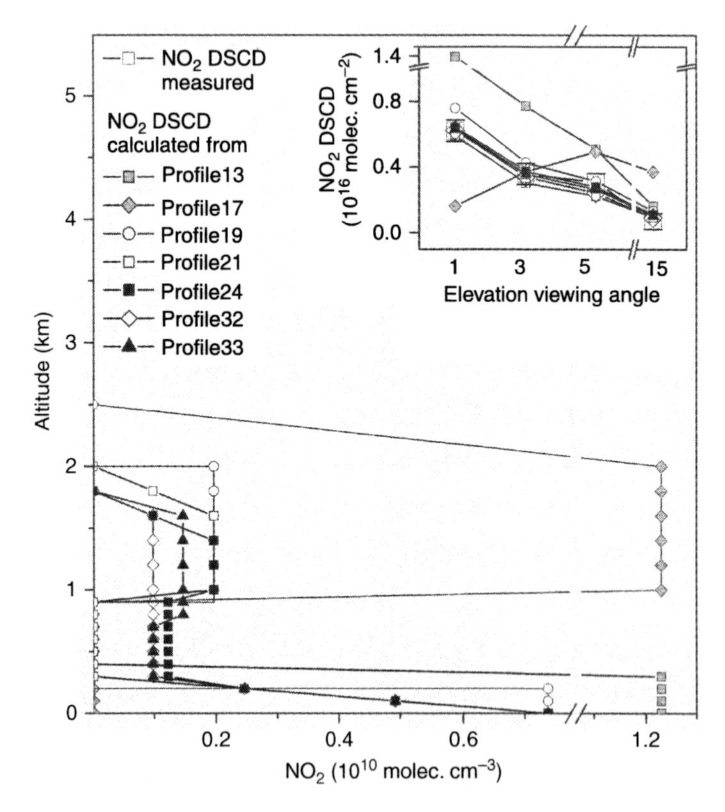

Fig. 9.29. Retrieval of the vertical profile of NO_2 in the polluted marine boundary layer of the Gulf of Maine (Pikelnaya et al., 2005). Shown are different vertical profiles of NO_2 used in the retrieval and a comparison between the calculated SCD and the observed SCD (see insert). The best agreement was found for Profile 24 (from Pikelnaya et al., 2007, Copyright by American Geophysical Union (AGU), reproduced by permission of AGU)

This sensitivity can be used to retrieve trace gas profiles from MAX-DOAS observations. By measuring trace gas SCDs at different elevation angles below 20°, one can set up a linear equation system based on (9.20), which can then be solved for the concentrations in each layer.

Figure 9.28 illustrates the steps necessary to derive trace gas profiles from MAX-DOAS measurements. The first step is measurement of O_4 and other trace gases at various viewing elevation angles, followed by the aerosol profile retrieval as shown in Fig. 9.26. In a final step, a vertical trace gas profile can be retrieved. Such a retrieval is illustrated in Fig. 9.29 for the example of a layer of NO_2 close to the ground and another layer from 1 to 2 km altitude (Pikelnaya et al., 2005). The figure shows different NO_2 profiles and a comparison between the measured and the calculated DSCDs for each profile for different viewing elevation angles. The TROP9 aerosol profile from Fig. 9.26 was used in the retrieval of the NO_2 profile for this day. The best agreement between the observed and modelled SCD was achieved for Profile 24. This example illustrates the capabilities of combined MAX-DOAS observations and RT calculations to retrieve vertical trace gas profiles.

9.4 Aircraft Observed Scattered Light (AMAX-DOAS)

Airborne platforms give high mobility to DOAS applications, allowing rapid mapping of trace gas distributions over a relatively large area. They also allow an additional degree of altitude separation by using upward and downward viewing directions (Wahner et al., 1989a,b, 1990a,b; Schiller et al., 1990; Brandjen et al., 1994; Pfeilsticker and Platt, 1994, 1997; Erle et al., 1998; McElroy et al., 1999; Petritoli et al., 2002; Melamed et al., 2003; Bruns et al., 2004).

The radiation transport for an airborne instrument will, in addition to the factors discussed earlier, depend on the flight altitude. One might expect that an upward-looking instrument will only observe trace gases located above the flight altitude. However, the RT calculations in Fig. 9.30 for viewing angles larger than 90° of an aircraft flying at 10 km altitude show that gases below the aircraft still have a considerable influence on upward-looking observations. As the viewing angle approaches the zenith, this contribution becomes smaller, but it never completely disappears. The fact that gases below the aircraft can be seen in upward-viewing direction is explained by the scattering of light originating below the flight altitude, i.e. through reflection at the ground, clouds or upward Rayleigh scattering into the detector. The contribution of this portion of the signal depends on the surface albedo, which increases this portion and the scattering properties above the aircraft, i.e. aerosol, clouds.

In downward viewing directions and viewing directions smaller than 90°, light is observed from both above, because solar radiation first has to cross this region, and below the aircraft (Fig. 9.30). The decrease of the weighing functions towards the ground in Fig. 9.30 is caused by the decreasing intensity

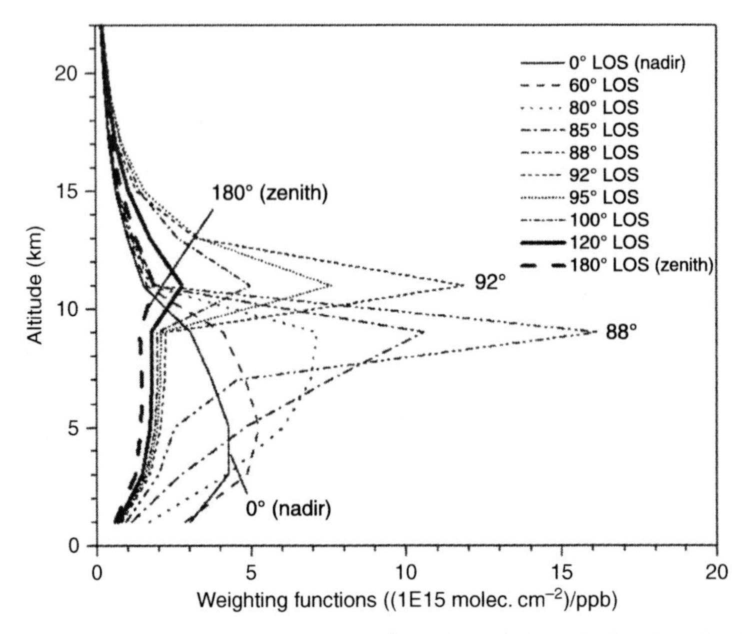

Fig. 9.30. Calculated weighting functions (Box AMFs) for a flight altitude of 10 km and NO_2 profiles anticipated for mid latitudes on the northern hemisphere spring at 51.6°SZA. Each weighting function corresponds to a different line of sight (LOS). Nadir direction is 0°. The magnitude of the weighting functions is small at the surface and above 15 km, revealing that the slant columns are not very sensitive to NO_2 in these regions (from Bruns et al., 2004)

scattered into the detector at lower altitudes due to higher extinction near the ground. The weighing functions in the downward-viewing direction strongly depend on surface albedo and wavelength.

It should be noted here that the presence of clouds below or above the aircraft considerably complicate the RT calculations and the interpretation of airborne DOAS observations.

9.5 Satellite Observed Scattered Light

Atmospheric trace gases can also be observed by DOAS from satellite platforms by measuring the light backscattered from the atmosphere and the surface. Several viewing geometries are possible. The most popular viewing geometries are Nadir View and Limb View. Many aspects of the RT for satellite observations are similar to those for the ground-based and aircraft-based observations discussed earlier. This section will thus concentrate on the particular problems arising in spaceborne observations.

9.5.1 Radiative Transfer in Nadir Geometry – the Role of Clouds

In near-nadir geometry (see Fig. 9.31), the instrument looks down to a point on earth below its current position. The radiation observed by the satellite instrument originates as solar light, which is backscattered or (diffuse) reflected from the atmosphere and the earth's surface, respectively.

The AMF for the case of pure backscattering from the earth's surface (Fig. 9.31a) can be approximated by geometric considerations. Neglecting the curvature of the earth, the AMF for an SZA ϑ and the observation angle (angle between nadir and actual observation direction), α, is given by:

$$A_{\text{surface}} = 1/\cos\vartheta + 1/\cos\alpha \, .$$

Since the radiation passes the atmosphere twice, A_{surface} is always greater than 2 (see Fig. 9.32).

In reality, however, a fraction of the radiation will be backscattered in the atmosphere (Fig. 9.31b). Figure 9.32 shows a comparison between the geometric AMFs and those calculated with a RT model for a stratospheric and tropospheric absorber. The relative ratio of surface and atmosphere backscattered radiation depends on various factors. The contribution of Rayleigh scattering increases towards shorter wavelengths. Consequently, the contribution of the lower atmosphere decreases for lower wavelength, and satellite observations in the UV are not very sensitive to trace gases in the atmospheric boundary layer. Similarly, trace gas absorptions, in particular those of ozone, can change the contribution of the lower atmosphere. Below 300 nm, where the ozone layer absorbs most radiation, the troposphere cannot be observed. On the other hand, this wavelength dependence can be used for vertical trace gas

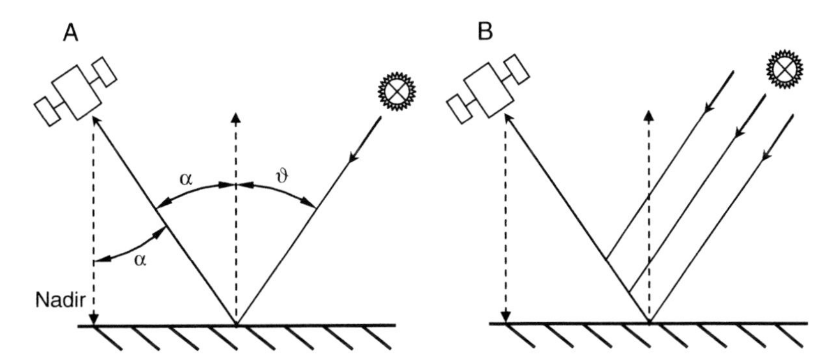

Fig. 9.31. Geometry of a (near) nadir viewing satellite instrument and a sketch of the associated radiation transport in the atmosphere: (**a**) Without scattering in the atmosphere the instrument would only record radiation scattered back from the surface, the AMF would always exceed 2. (**b**) In reality, in addition to reflection from the surface, as in the case of other Scattered-Light DOAS geometries, there is an infinite number of possible paths of scattered radiation between the light source (i.e. the sun) and the instrument

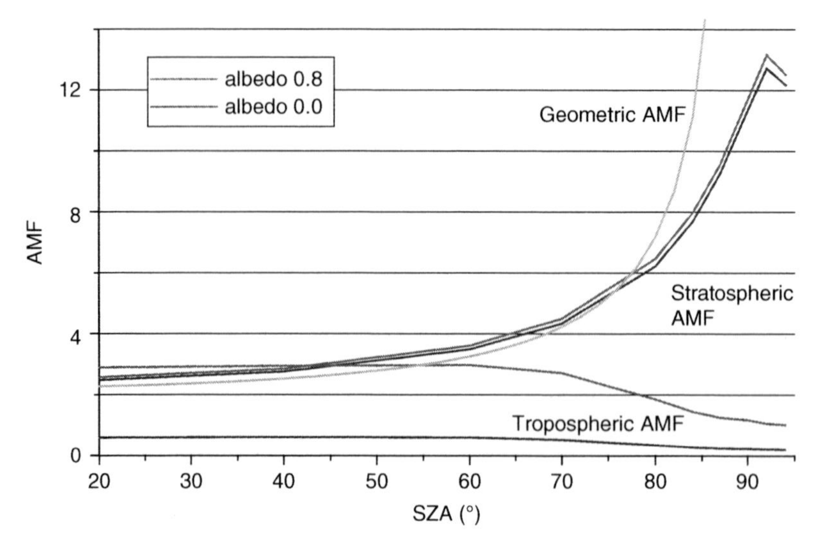

Fig. 9.32. Airmass factors for a nadir viewing satellite instrument as a function of solar zenith angle ϑ for two ground albedo values. Three cases are shown: Geometric AMF ($1/\cos\vartheta$), AMF for a tropospheric trace gas profile, AMF for a stratospheric trace gas profile

profiling if a gas is measured at different wavelengths. Mie scattering in the atmosphere will enhance the fraction of backscattered light above an aerosol layer while decreasing the contribution of the trace gases below the layer. An extreme example is clouds that basically block the contribution of the atmosphere to the SCD below the cloud. However, tropospheric aerosol can also considerably change the weighing of different altitudes in the atmosphere in satellite observations. Backscatter and absorption in the atmosphere will generally reduce the AMF to values below A_{surface} for a given set of angles ϑ and α (Fig. 9.32). The surface albedo will predominantly impact the tropospheric AMF (Fig. 9.32). Higher albedos lead to considerably higher AMF.

Another important consequence of part of the radiation being scattered in the atmosphere rather than traversing it entirely is a reduction in sensitivity towards lower altitudes. Thus, a trace gas layer located close to the surface will give a lower apparent column density than a layer of the same VCD higher up in the atmosphere. This is illustrated in Fig. 9.33 showing the Box AMFs or sensitivity as a function of altitude for high and low albedo at UV and visible wavelengths.

Another RT problem, particular to satellite observations, arises from the spatial averaging of satellite observations. Typically, a satellite observes an area of the earth surface in the range of 100–1000 km^2. The large extent of this ground pixel leads to an averaging of parameters such as the surface albedo. A particular problem in this context is the possibility of partially cloudy ground pixels. Clouds will mask the fraction of the total trace gas column

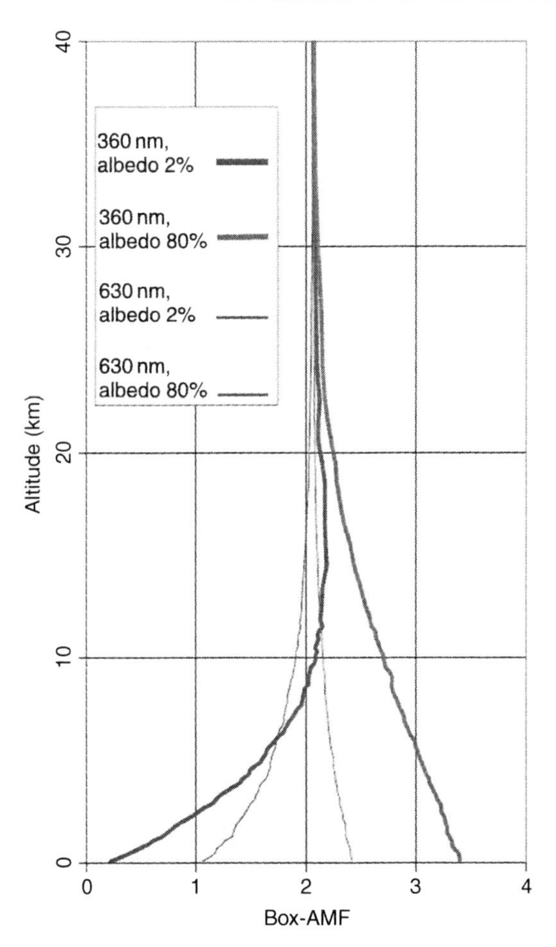

Fig. 9.33. Box airmass factors for a nadir viewing satellite – instrument as a function of box (trace gas layer) altitude for 360 nm and 630 nm and for two ground albedo values (2% and 80%), respectively (Figure courtesy of T. Wagner)

below the cloud. The total intensity received by the satellite instrument (neglecting radiation reflected from the surface below the cloud) is given by:

$$I_S = I_0 \cdot (f \cdot B_{\text{Cloud}} + (1 - f) \cdot B_{\text{Surface}}) \ ,$$

where B_{Cloud} and B_{Surface} denote the albedo of cloud and surface, respectively, and f is the fraction of the pixel covered by the cloud. Unfortunately, the effect of clouds is larger than expected at the first glance: For instance, assuming $B_{\text{Cloud}} = 0.8$, $B_{\text{Surface}} = 0.05$ and $f = 0.5$, only about 6% of the signal comes from the cloud-free part of the pixel. Accordingly, the derived column density will be dominated from the cloud-covered part of the pixel. In reality, the situation will be somewhat better, since some of the light penetrating the cloud towards the surface will actually return through the cloud to the instrument. Also, correction of the cloud effect is possible, but requires a very precise determination of the cloud fraction f.

In view of the importance of cloud correction for the analysis of tropospheric trace gases, an accurate and reliable cloud detection algorithm is essential for the correct determination of tropospheric trace gas column densities. While detection of clouds from satellite, e.g. through thermal IR radiometry, has been in use for a long time, it turns out that DOAS-type satellite instruments are better served by cloud detection using the same instrument's data.

A series of algorithms were developed for the retrieval of cloud parameters. For instance, for the Global Ozone Monitoring Experiment (GOME), there is the official GOME cloud product' ICFA (Initial Cloud Fitting Algorithm) (Kuze and Chance, 1994) and the FRESCO algorithm (Fast REtrieval Scheme for Clouds from the Oxygen-A-Band) (Koelemeijer et al., 2001), both using the GOME channels with moderate spectral resolution. In addition, there are several algorithms using broad radiometers of GOME with higher spatial resolution, the Polarization Monitoring devices (PMD).

Two different quantities are usually applied for cloud retrieval: (1) The absorption of the O_2-A-Band. Clouds reduce the penetration of light down to low layers of the atmosphere, thus the retrieved O_2 column density is reduced for a cloudy pixel as compared to a cloud-free measurement, where the absorption mainly depends on cloud coverage, cloud albedo, and cloud top height. This approach is used by ICFA and FRESCO, but cannot be applied to the PMD instruments because of their insufficient spectral resolution. (2) The main idea of a second class of algorithms is that clouds can also be identified through the overall intensity of reflected light, which is hardly affected by trace

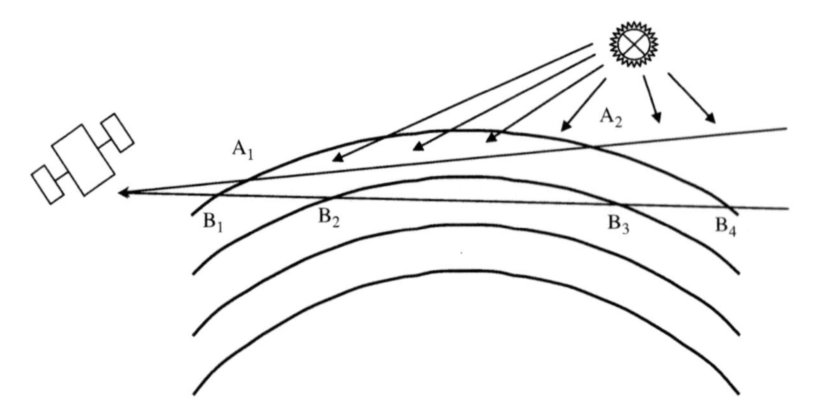

Fig. 9.34. Geometry of a limb viewing satellite instrument and a sketch of the associated radiation transport in the atmosphere. As in the case of other Scattered-light DOAS geometries there are an infinite number of possible light paths between the light source (i.e. the sun) and the instrument

gas absorptions, because clouds are usually brighter than the surface. These intensities are mainly independent of cloud top height, but they also depend on cloud coverage and cloud albedo. This approach is applied using small spectral windows of the detectors with moderate spectral resolution (FRESCO) and by the algorithms using the PMD instruments. All these algorithms retrieve an effective cloud fraction, a parameter that combines cloud coverage (cloud abundance) of the pixel and cloud albedo (Grzegorski et al., 2006).

9.5.2 The Analysis of Satellite-limb Scattered Light Observations

In limb-view geometry, the radiation scattered from the edge of the earth's disk (the limb) is observed by a suitable telescope. This mode of observation allows the derivation of vertical trace gas concentration profiles at relatively high accuracy and vertical resolution. The RT in this case is dominated by the scattering events at different altitudes within the atmosphere. Synthetic limb measurements and weighting functions (WFs) with several orders of scattering and surface reflection were computed by Kaiser and Burrows (2003). Comparisons reveal the wavelength-dependent contributions of single scattering and the second orders of scattering and surface reflection, showing that the single-scattering approximation is sufficient for the calculation of the weighing functions during the retrieval process. Models such as those described for the MAX-DOAS evaluations are thus also usable for the AMF calculation for LIMB geometry.

10

Sample Application of 'Active' DOAS with Artificial Light Sources

Over the past 25 years, active DOAS measurements have been made for a multitude of purposes. From the initial studies of the atmospheric composition in the mid 1970s, a wide variety of designs and applications in the open atmosphere, as well as in the laboratory, have been developed.

For instance, broadband lasers were used as a light source (Rothe et al., 1974; Perner et al., 1976; Perner et al., 1991; Amerding et al., 1991; Dorn et al., 1993, 1996, 1998; Brauers et al., 1996, 1999; Brandenburger et al., 1998). However, the majority of the designs rely on arc-lamps as light sources (e.g. Bonafe et al., 1976; Kuznetzov and Nigmatullina, 1977; Noxon et al., 1978, 1980; Platt, 1977; Evangelisti et al., 1978; Platt et al., 1979; Platt and Perner, 1980, 1983; Johnston and McKenzie, 1984; Dorn and Platt, 1986; Edner et al., 1986, 1990, 1992, 1993a,b, 1994a; Axelsson et al., 1990a,b, 1995; Axelsson and Lauber, 1992; Amerding et al., 1991; Wagner, 1990; Hallstadius et al., 1991; Galle et al., 1991; Biermann et al., 1991; Stevens and Vossler, 1991; Gall et al., 1991; Fayt et al., 1992; Plane and Nien, 1992; Vandaele et al., 1992; Hausmann et al., 1992, 1997; Stevens et al., 1993; Hausmann and Platt, 1994; Martini et al., 1994; Plane and Smith, 1995; Evangelisti et al., 1995; Heintz et al., 1996; Harder et al., 1997; Flentje et al., 1997; Lamp et al., 1998; Reisinger et al., 1998; Russwurm, 1999; Allan et al., 1999, 2000; Hebestreit et al., 1999; Alicke, 2000; Gölz et al., 2001; Kim and Kim, 2001; Geyer et al., 1999, 2001a,b,c, 2002, 2003; Matveev et al., 2001; Volkamer et al., 2001; Alicke et al., 2002; Veitel et al., 2002; Lee et al., 2002; Yu et al., 2004; Saiz-Lopez et al., 2004a,b; Lee et al., 2005; Zingler and Platt, 2005). In addition there is a small number of recent studies using artificial 'lights of opportunity' to perform DOAS measurements (Yoshii et al., 2003; Fuqi et al., 2005). It is impossible to review the entire breadth of DOAS studies over the past three decades. However, it is instructive to consider examples for some of the most valuable uses of active DOAS. The intention of this chapter is to provide an overview of the abilities DOAS offers to research in atmospheric chemistry and physics. We hope that this overview will inspire the development of even more applications in the future.

U. Platt and J. Stutz, *Sample Application of 'Active' DOAS with Artificial Light Sources*. In: U. Platt and J. Stutz, Differential Optical Absorption Spectroscopy, Physics of Earth and Space Environments, pp. 379–427 (2008)
DOI 10.1007/978-3-540-75776-4_10

10.1 Air Pollution Studies and Monitoring Applications

Motivated by the desire to understand and mitigate air quality problems in urban areas such as Los Angeles, one of the earliest uses for active DOAS was the investigation of the processes leading to the formation of urban air pollution. Today, DOAS is used as a tool to monitor air quality and also as a benchmark for other techniques. DOAS is an Environmental Protection Agency (EPA) approved method to monitor primary pollutants such as O_3, NO_2, and SO_2. In addition, the Association of German Engineers (Verein Deutscher Ingenieure, VDI) issued a draft guideline for the testing of DOAS instruments for air pollution monitoring (VDI, 2005). A number of commercial DOAS instruments are available today. All these instruments work according to the principles described in earlier chapters. The advantages of DOAS for monitoring purposes are high accuracy and the ability to measure all three pollutants simultaneously.

In the following section, we will concentrate on state-of-the-art research applications to illustrate the abilities that DOAS has to offer today's research community.

10.1.1 Measurement of Urban Pollutants

The measurement of pollutants is an important undertaking, both for air pollution-related research and the monitoring of air quality.

Ozone, *nitrogen dioxide*, and *sulphur dioxide* are some of the most important species connected to air pollution. SO_2 and NO_2 (in the form of NO which is rapidly converted to NO_2; see Chap. 2) are released from combustion sources. Ozone is formed throughout the day by photochemical processes, and is the main component of the so-called Los Angeles Smog (see Chap. 2). All three species are regulated in most industrialized countries (e.g. by the US Environmental Protection Agency), and considerable effort is put into their continuous measurement for air quality monitoring.

Ozone, NO_2, and SO_2 are, in most cases, the strongest absorbers in DOAS spectra acquired in polluted areas. These three trace gases are thus the most commonly measured species in active DOAS applications. Figure 10.1 shows an example of an analysis of O_3, NO_2, and SO_2 and in addition CH_2O in the UV made in Heidelberg, Germany in 1994 by a long-path DOAS instrument, such as those described in Chap. 7. The figure illustrates again how the analysis procedure outlined in Chap. 8 is used to separate the different absorption structures in an atmospheric absorption spectrum. The wavelength range displayed in Fig. 10.1 is typically chosen for the detection of O_3, SO_2, and HCHO in long-path DOAS applications. An example of the measurement of NO_2 was already given in Chap. 8.

Figures 10.2 and 10.3 show examples of measured time series of O_3, NO_2, SO_2, and a number of other trace gases during field experiments in Milan,

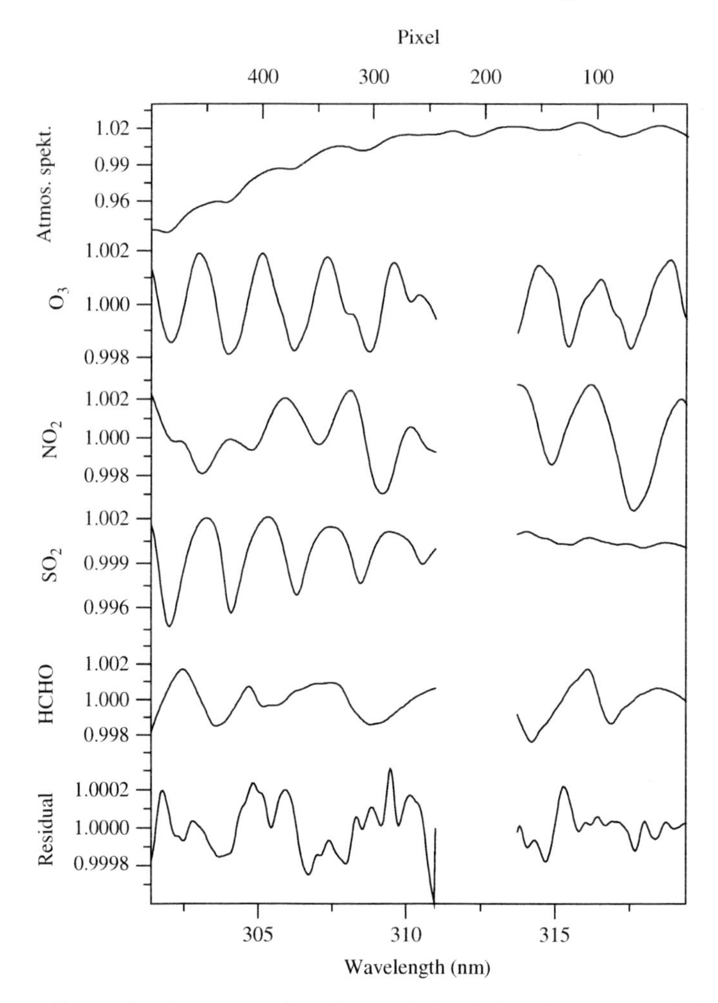

Fig. 10.1. Example of a spectral analysis of O_3, NO_2, SO_2, and HCHO in the polluted air of Heidelberg. The absorptions identified in the atmospheric spectrum (*top trace*) were: O_3, 21.1 ± 0.5 ppb; SO_2, 0.64 ± 0.01 ppb; and HCHO, 3.7 ± 0.1 ppb. The NO_2 mixing ratio was not determined since this wavelength interval is not optimal for its analysis. The missing area around 312 nm was excluded from the analysis procedure

Italy, and Houston, USA, respectively. The goal of both, the Pianura Padania Produzione di Ozono (Po-Valley Ozone Production, PIPAPO) study in Milan and the Texas Air Quality Study (TEXAQS) in Houston, was to investigate the formation of ozone and particles in the polluted environment of these cities.

The measurements in Milan were made with a long-path DOAS instrument that employed retro-reflectors to fold the light path once. This instrument

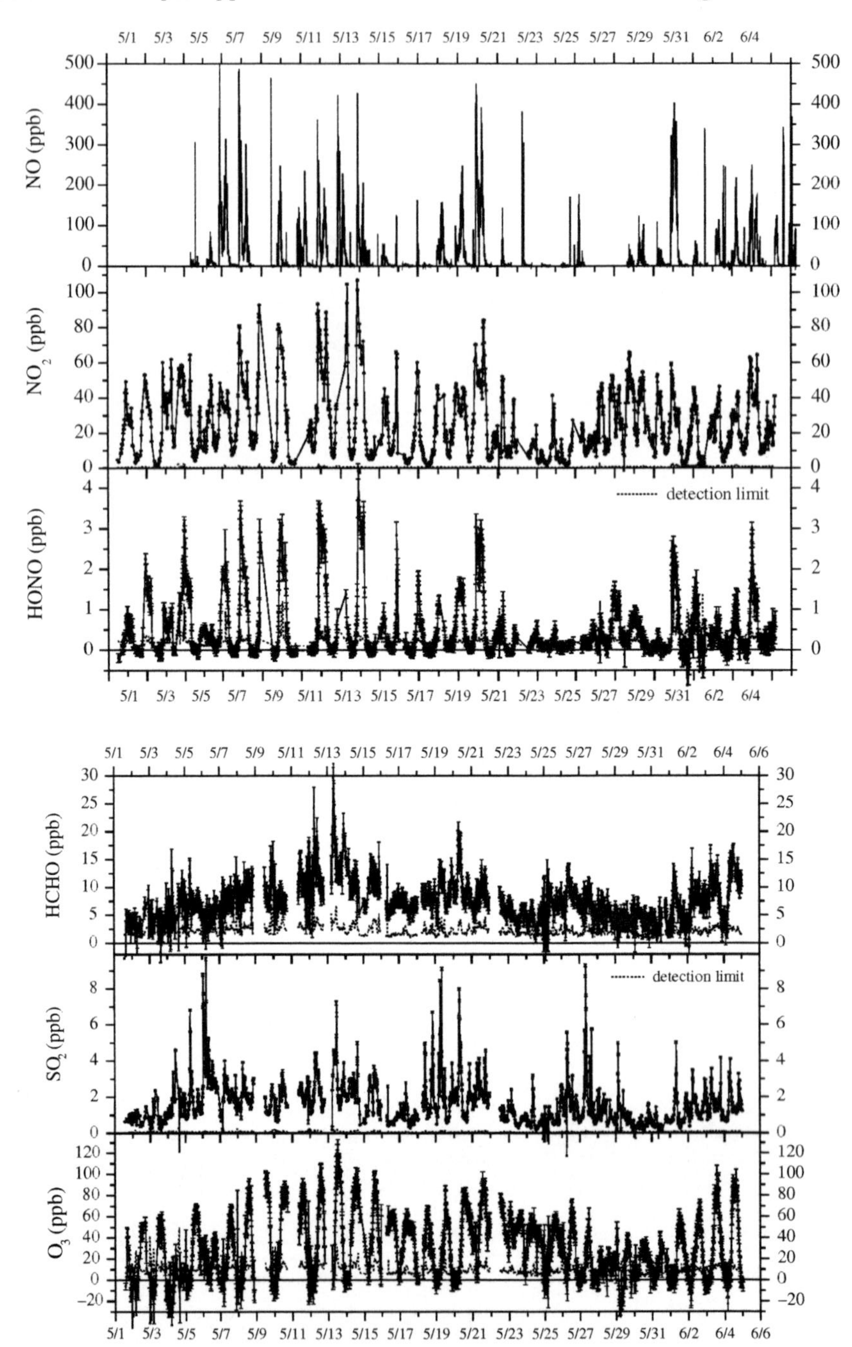

Fig. 10.2. Time series of O_3, NO_2, SO_2 HCHO, and HONO mixing ratios measured during the PIPAPO study in Milan, Italy. In addition, NO (top trace) was measured by chemoluminescence (from Alicke et al., 2002, Copyright by American Geophysical Union (AGU), reproduced by permission of AGU)

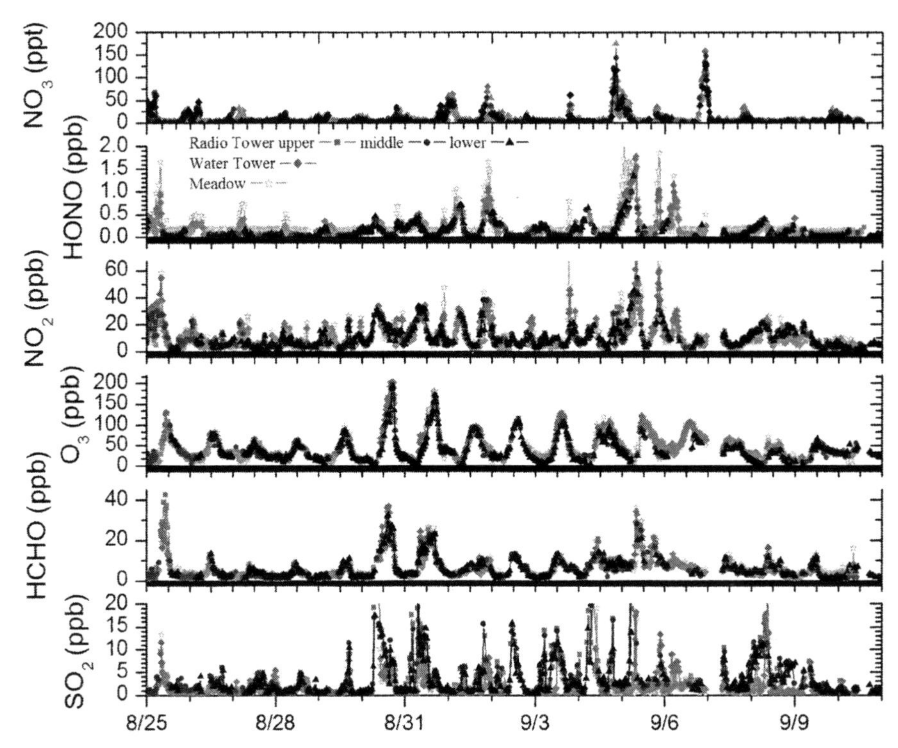

Fig. 10.3. O_3, NO_2, and SO_2, HONO, and NO_3 mixing ratios during the Texas Air Quality Study (TEXAQS) in summer 2000.The measurements in Houston, Texas (USA) were performed with two long-path DOAS instruments on five different light paths. A more detailed description of the set-up is given in Stutz, et al., (2004b)

was set up about 8 km north of the city centre of Milan at a small airport. The instrument could measure alternating on three light paths of different elevation (Fig. 10.6). Only the data from the upper light path are shown in Fig. 10.2. The mixing ratios of ozone in Milan show strong diurnal variations, with low levels at night when emission of NO converts O_3 to NO_2, and high levels after noon when ozone is photolytically formed (see Chap. 2). Mixing ratios of ozone exceeded 100 ppb on several occasions. Nitrogen dioxide shows a diurnal variation opposite to O_3, with high values at night and low values during the day. The NO_2 mixing ratios reached up to 100 ppb, showing that Milan is heavily polluted. The NO data, measured by a chemiluminescence instrument (see Chap. 5), also show extremely high mixing ratios. While the O_3 and NO_2 data are closely related to each other, the SO_2 data in Milan behaves quite differently. This can be explained by the different sources and chemistry of SO_2 which shows a higher short-term variability. Levels of SO_2 reached up to 8 ppb.

The observations in Houston (Fig. 10.3) were made using two DOAS instruments measuring on five different light paths (Fig. 10.15). A discussion of this set-up will be given in Sect. 10.1.2. The instruments were located 8 km south-east of downtown Houston, close to an industrial area with a large number of refineries. In general, the behaviour of O_3 and NO_2 is similar to that in Milan, and is typical for a polluted urban area. The levels of NO_2 are lower in Houston. However, ozone reached mixing ratios above 200 ppb on two occasions. SO_2 levels were also higher in Houston than in Milan, most likely due to emissions from the nearby refineries.

The time series from these two field experiments (Figs. 10.2 and 10.3) illustrate the ability of modern long-path DOAS instruments to continuously measure O_3, NO_2, and SO_2 at a time resolution of 5–15 min. The detection limits reached during these two campaigns are shown in Tables 10.1 and 10.2. One will notice that the detection limits vary with the light path length. This is caused by the presence of residual structures in the spectra of both experiments (see Chap. 8). The detection limits in Tables 10.1 and 10.2 give an indication of the performance of modern DOAS instruments.

Formaldehyde plays an important role in the chemistry of the troposphere. It is formed in the oxidation of most hydrocarbons, and also directly emitted from combustion sources. The photolysis of HCHO forms HO_2 radicals, which in the presence of NO are rapidly converted into OH. HCHO is thus an important HO_X precursor. DOAS measurements for the PIPAPO and TEXAQS field experiments are shown in Figs. 10.2 and 10.3. Typical levels in polluted urban areas are $\sim$ 10 ppb. Only in heavily polluted environments with a fast photochemistry, such as Houston, are higher HCHO levels observed. The HCHO levels in Houston (Fig. 10.3) roughly follow the O_3 mixing ratios. This

Table 10.1. Overview of the species measured in Milan, Italy and the observed mixing ratios (adapted from Alicke et al., 2002)

Species	Wavelength window (nm)	Absorption cross-section	Maximum mixing ratio (ppb/date)	Detection limit (ppb)
HONO	375–336	(Stutz et al., 2000)	4.4 ± 0.2, 13.5.98 21:25	0.2
NO_2	375–336	(Harder et al., 1997)	115.2 ± 0.6, 13.5.98 21:08	0.6
NO	–		480, 7.5.98 21:30	0.25
SO_2	322–303	(Vandaele et al., 1994)	9.1 ± 0.22, 19.5.98 8:47	0.18
O_3	322–303	(Bass and Paur, 1985)	123 ± 9, 13.5.98 12:37	8.5
HCHO	322–303	(Cantrell et al., 1990)	33.6 ± 1.8, 13.5.98 8:32	1.6

The distance between telescope and retro-reflectors was 1.25 km.
*All times GMT (local – 2 h)

Table 10.2. Summary of data coverage, average and best detection limits for the DOAS results in Houston, Texas (USA), see details in Stutz et al., 2004b

	Lightpath 'meadow'	Lightpath 'water tower'	Lightpaths 'radio tower'
Distance (km)	0.75	1.9	6.1
Data coverage	08/20–09/10	08/18–09/10	08/24–09/12
Detection limit NO_3 (ppt)			
Average	23	8	2.6
Best	2	1.5	0.8
Detection limit HONO (ppt)			
Average	320	90	38
Best	120	50	16
Detection limit NO_2 (ppb)			
Average	0.88	0.4	0.13
Best	0.24	0.2	0.05
Detection limit O_3 (ppb)			
Average	9	4	2.5
Best	4.7	1.5	0.9
Detection limit HCHO (ppt)			
Average	1	0.46	0.34
Best	0.5	0.15	0.11
Detection limit SO_2 (ppt)			
Average	260	130	180
Best	140	40	40

The last column refers to all three light paths ending at the radio tower (red, blue, and black colour in Fig. 10.3)

is due to the link of ozone photochemical formation with an increased rate of hydrocarbon oxidation, and thus HCHO production.

HCHO mixing ratios are an important parameter for the study of urban air quality. The importance of HCHO as a HO_X radical precursor is shown in Fig. 10.4. The DOAS data during PIPAPO (Fig. 10.2) were used to determine the HO_X formation due to the photolysis of O_3, HCHO, and HONO, and the ozonolysis of alkenes. In the polluted environment of Milan, HCHO photolysis is the most important HO_X precursor. The correct representation of HCHO in air quality models, as well as the accurate determination of HCHO concentrations in the field, is thus an important aspect of atmospheric chemistry research.

Finally, it should be noted here that the measurement of HCHO in the field is still a major challenge (e.g. Cardenas et al., 2000; Hak et al., 2005). The capabilities that DOAS offers for the investigation of HCHO chemistry have thus far not been entirely explored.

The photolysis of *Nitrous Acid* (HONO) can be the dominant HO_x source in the early morning, as shown in Fig. 10.4. Nitrous acid was an elusive

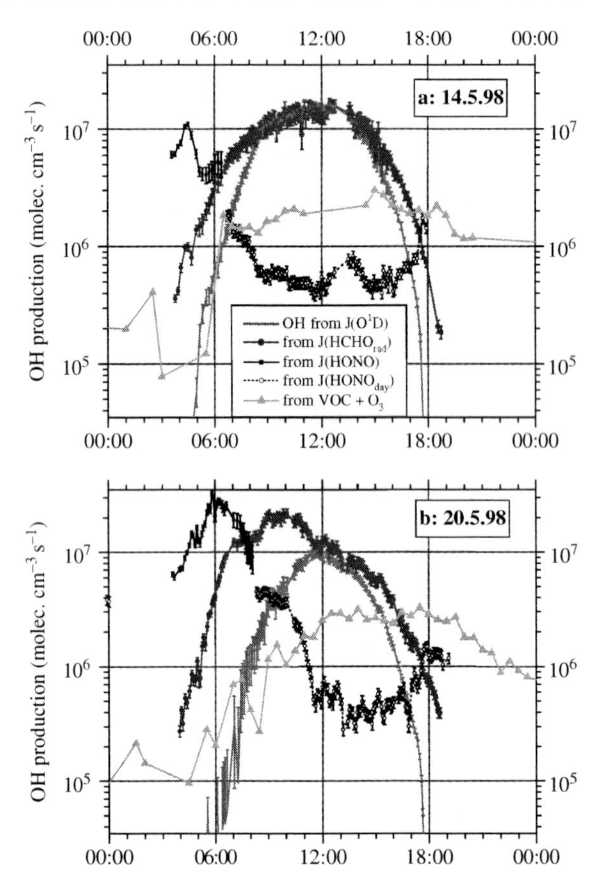

Fig. 10.4. Comparison of three different OH radical producing species for 14.5.1998 (*graph a*) and 20.5.98 (*graph b*). The data are plotted on a logarithmic scale for a better overview. The impact of HONO photolysis on the OH formation in the early morning hours can be easily seen. The estimated daytime OH production from HONO (see text) is shown with a *dotted line* (from Alicke et al., 2002, Copyright by American Geophysical Union (AGU), reproduced by permission of AGU)

compound until its discovery in the atmosphere by DOAS in 1979 (Perner and Platt, 1979). Figure 10.5 shows the first identification of HONO absorptions in a DOAS spectrum. Since the original measurement of HONO, DOAS has become the most reliable technique to measure HONO in the atmosphere. The high selectivity and the absence of interference with other trace gases, in this case primarily NO_2, make DOAS an excellent choice to measure HONO. Although the importance of HONO as a HO_X precursor has been known for decades, there are still major uncertainties about the mechanisms of its formation. It is known that HONO is formed through the heterogeneous conversion of NO_2 on various surfaces, as described in Chap. 2. Details about the nature of the surfaces or the mechanisms of the conversion are unclear. The investigation of HONO formation is thus an active field of research, where DOAS is playing an important role.

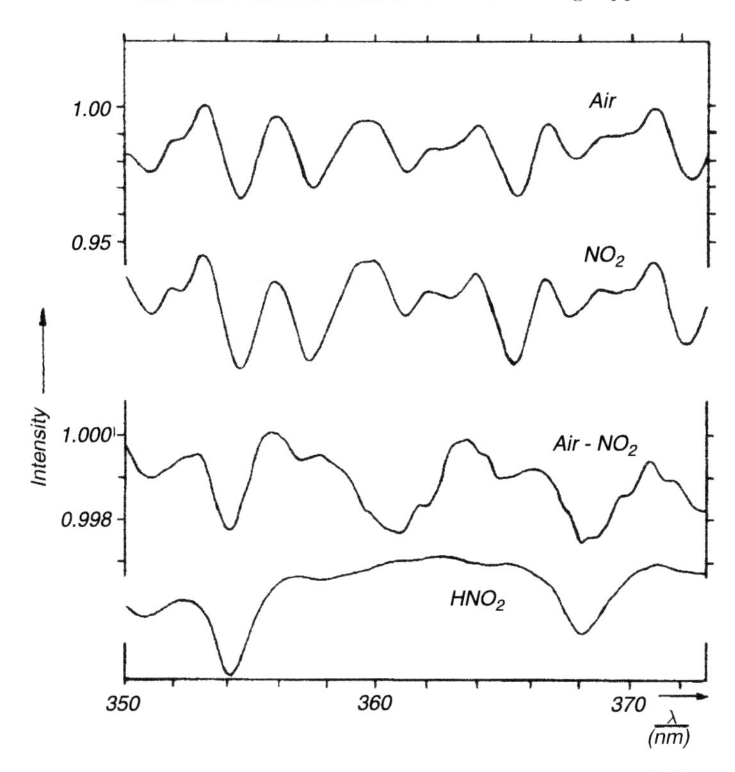

Fig. 10.5. First atmospheric absorption spectrum of nitrous acid (from Perner and Platt, 1979, Copyright by American Geophysical Union (AGU), reproduced by permission of AGU)

Modern DOAS applications have moved away from the simple measurement of HONO to more process-oriented studies of the formation of HONO. As mentioned earlier, it is suspected that the source of HONO in the atmosphere is the heterogeneous conversion of NO_2. Because DOAS measures both NO_2 and HONO in a single absorption spectrum, it offers exciting opportunities to study this chemistry.

Following the hypothesis that the formation of HONO on the ground (e.g. Perner and Platt, 1979; Stutz et al., 2004a) leads to deposition of its precursor NO_2, and release of HONO from the ground one would, therefore, expect a positive NO_2 gradient (smaller concentration at the ground than above), and a negative HONO gradient above the ground. Based on this hypothesis, Stutz et al. (2002a) set up an experiment to measure gradients and fluxes of these two species during the PIPAPO experiment. Figure 10.6 shows the set-up, which was based on a modern long-path DOAS instrument (see Chap. 7), that sequentially aimed at three retro-reflectors mounted at different heights on a 4-m-high tower. The distance between telescope and retro-reflectors was 1.25 km.

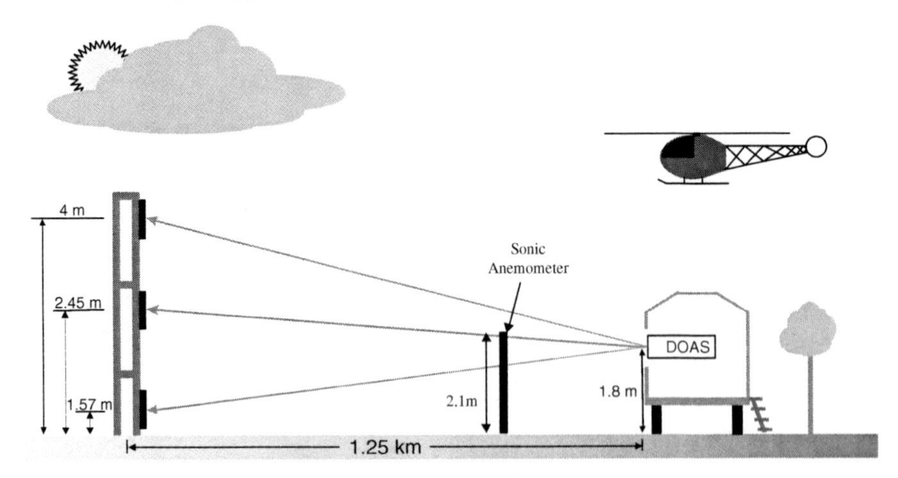

Fig. 10.6. Set-up of the experiment to measure gradients and fluxes of NO_2 and HONO during the PIPAPO experiment. The DOAS instrument aimed sequentially at the three retro-reflectors mounted on the tower at 1.25 km distance (from Stutz et al., 2002a, Copyright by American Geophysical Union (AGU), reproduced by permission of AGU)

The top panels of Figs. 10.7 and 10.8 show the NO_2 and HONO data acquired during one night. Figure 10.7, Panel b and Fig. 10.8, Panel b show the gradients calculated based on the original absorptions, where the uncertainty is determined from the error propagation of the original observations. The bottom panels then describe the fluxes and the deposition velocities of the two trace gases. The data show that during this night a deposition of both gases was observed. Based on this observation, Stutz et al. (2002a) suggest that the loss of HONO on surfaces is an important process in the atmosphere. They conclude, based on data from other nights, that the HONO loss outweighs its formation if the $HONO/NO_2$ concentration ratio exceeds $\sim 3\%$. This result shows that there is a maximum amount of HONO formation in the atmosphere. However, there are also clear indications for HONO emission from cars, in particular from DOAS studies of HONO and $HONO/NO_X$ ratios in automobile exhaust (Perner and Platt, 1980; Kessler, 1984; Kirchstetter et al., 1996; Lammel and Cape, 1996).

The example shown here illustrates how modern DOAS instruments can be used in dedicated set-ups to answer specific questions on atmospheric chemistry.

Aromatic hydrocarbon measurements are another example of successful DOAS applications. Aromatics have strong absorption features in the wavelength range 250–300 nm (Fig. 10.9). While, in principle, it should be easy to measure these absorptions in the atmosphere, several problems have to be overcome to provide accurate measurements of these species. First, the short wavelengths increase Rayleigh and Mie extinction (see Chap. 3), decreasing the maximum path length that can be achieved. In addition, strong temperature-dependent ozone absorptions are also located in this wavelength

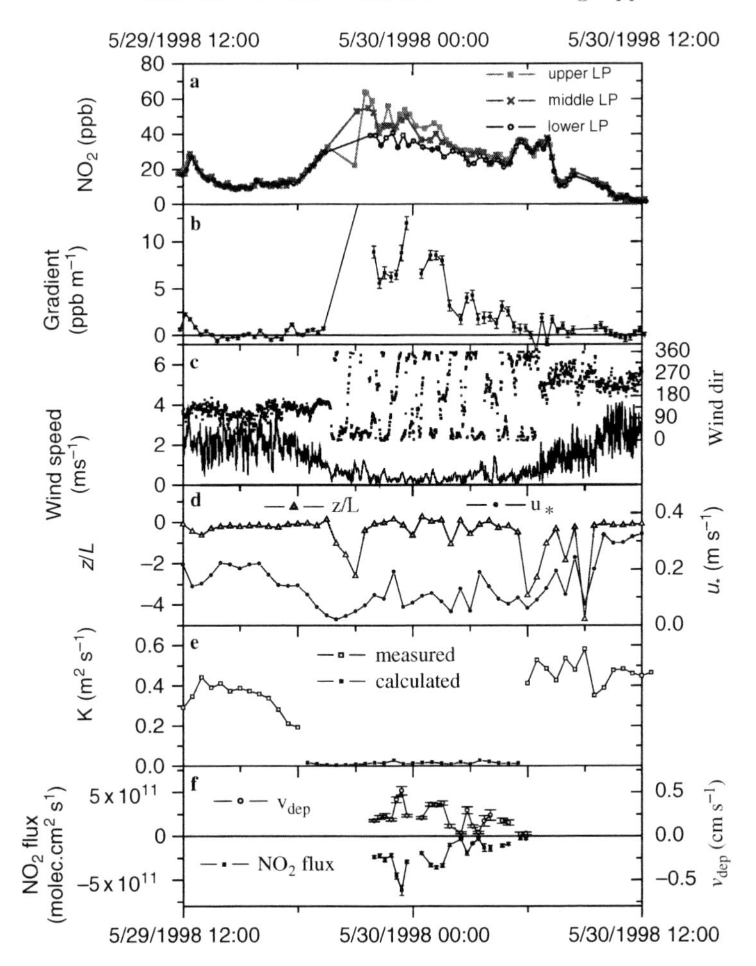

Fig. 10.7. NO_2 gradients during the night of May 29, 1998 in Milan, Italy. Gradients well above the detection limit were observed continuously for many hours during this night (from Stutz et al., 2002a, Copyright by American Geophysical Union (AGU), reproduced by permission of AGU)

window. During polluted daytime conditions, the absolute absorptions can severely reduce the transmitivity of the atmosphere. The temperature dependence of the differential absorption cross-sections is often challenging to incorporate in a DOAS analysis. Furthermore, molecular oxygen and O_2–O_2, as well as O_2–N_2 dimers, also have strong structured absorptions between 250 and 280 nm. The O_2 absorptions are in saturation at long light paths. At the limited resolution of most DOAS instruments, the shape of these absorptions thus depends mainly on the path-length in the atmosphere. Details of this effect are discussed in Chap. 6 (Sect. 6.6). Because the modelling of these absorptions based on literature cross-section is difficult, other solutions have to be found. Finally, spectrometer stray light poses a challenge for measurement

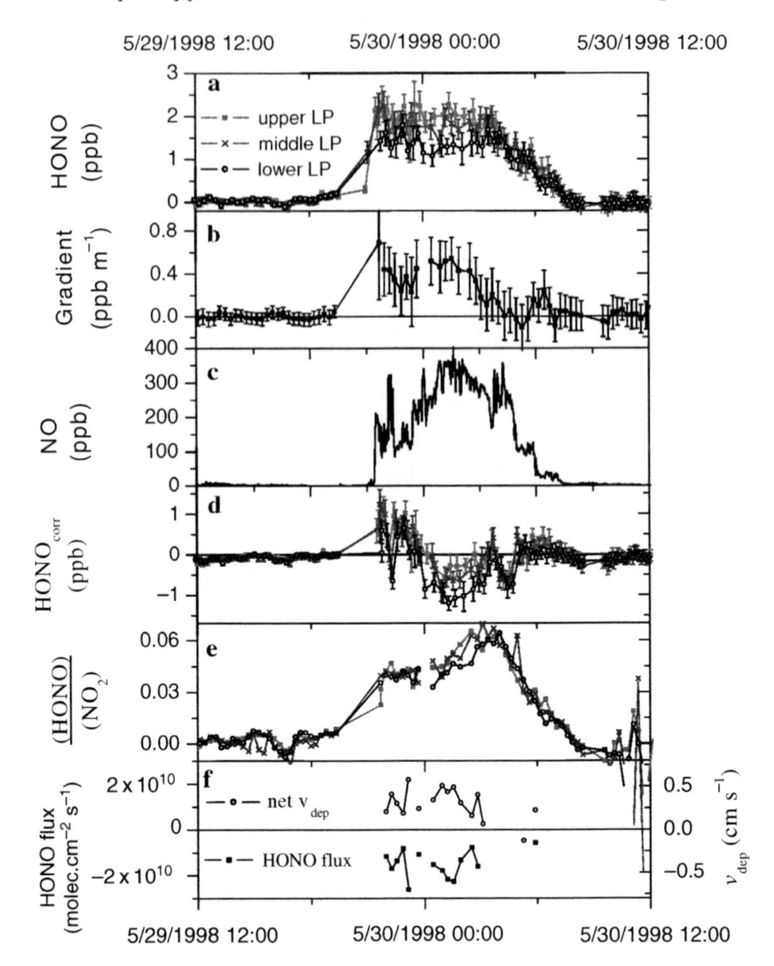

Fig. 10.8. HONO gradients during the night of May 29: (**a**) shows the mixing ratios on the individual light paths (LP). The gradient is displayed in (**b**). To remove the influence of direct HONO emissions, we calculated $HONO_{corr}$ by subtracting 0.65% of the NO_x (not shown here) from the HONO mixing ratios (**d**). The ratio of HONO to NO_2 mixing ratios is displayed in (**e**). The fluxes and net deposition velocities is shown in (**f**) (from Stutz et al., 2002a, Copyright by American Geophysical Union (AGU), reproduced by permission of AGU)

in the UV. Many light sources used in active DOAS have their emission maximum in the visible range and thus provide low intensity in the UV. The imperfect suppression of visible light in spectrometers leads to stray light levels that can be 10% and more of the signal between 250 and 300 nm.

Long-path DOAS measurements in the atmosphere, both for research and for air quality monitoring purposes with commercial instruments, have been reported during a number of occasions. Frequently, the problems with the O_2 and O_2–O_2/O_2–N_2 dimer absorptions are solved by ratioing the measured spectrum to a spectrum recorded on the same light path. The difficulty with

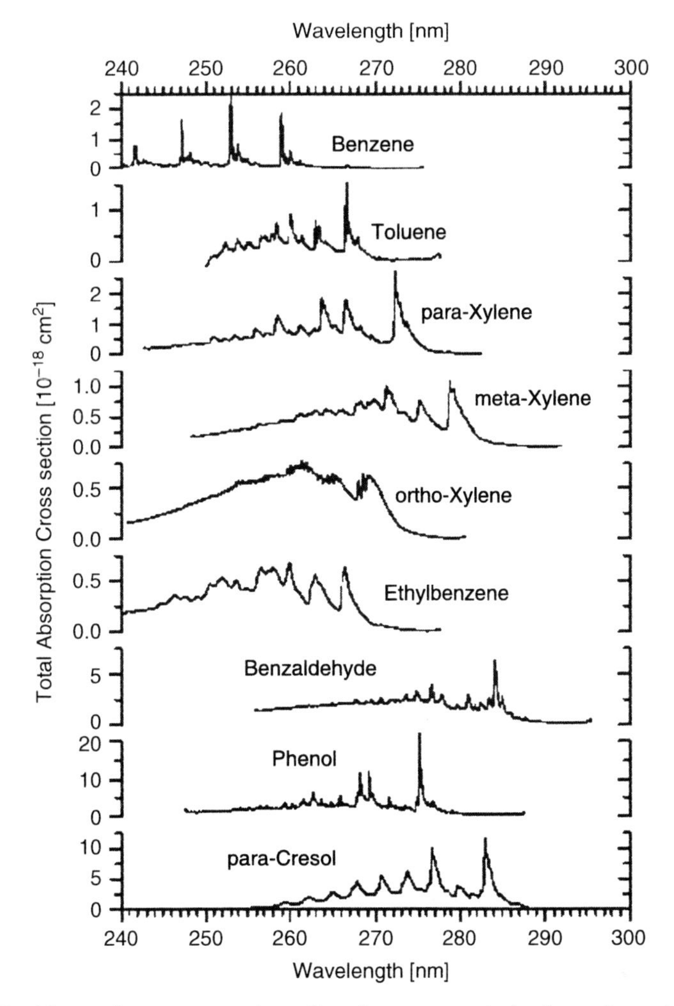

Fig. 10.9. Absorption cross-section of various aromatic hydrocarbons in the UV wavelength range

this approach is to determine the atmospheric concentration of aromatics at the time this reference spectrum is recorded. If the level of aromatics is not known, only relative measurements can be performed. The temperature-dependent ozone absorption can, for example, be modelled by including ozone reference spectra for different temperatures in the fit. Spectrometer stray light is reduced by either using UV band-pass filters or by correcting the stray light

The determination of aromatics in the atmosphere is a promising application of DOAS, in fact a series of measurements have been reported (e.g. Löfgren, 1992; Senzig, 1995; Axelsson et al., 1995; Barrefors, 1996; Volkamer et al., 1998; Volkamer, 2001; Kurtenbach et al., 2001). However, we will discuss here another application: the measurement of aromatic hydrocarbons in a traffic tunnel to determine the emission ratios of these species from an average car fleet. The measurements were made by Kurtenbach et al. (2001) with a open-path multi-reflection DOAS instrument with a base-path length of 15 m inside a tunnel in Wuppertal, Germany. The cell was operated on a total path-length of 720 m. Details of the White cell set-up can be found in Chap. 7. The time resolution of the measurements was ~ 1 min, much faster than typical gas chromatographs (GCs).

The multi-reflection cell offers a number of advantages with respect to the measurement of aromatics. The most important advantage is the ability to enclose the optics and flush it with clean air to derive a reference spectrum of O_2 and O_2–O_2/O_2–N_2 dimer absorptions. If this is not feasible, one can derive such a spectrum from open-path measurements with a collocated GC. The determination of the aromatics from the GC is more representative for a multi-reflection cell that only probes a small volume. For the multi-reflection cell measurements by Kurtenbach et al. (2001) and Ackermann (2000), both approaches were used.

To ensure that small temporal changes in the instrument function do not influence the analysis of the spectra, in particular for the strong O_2 absorptions, a cross-convolution technique is applied. In this method, the atmospheric spectrum is convoluted with the instrument function of the O_2 reference spectrum (and if measured at the same time also of the reference spectra of the aromatics), while the reference spectra are convoluted with the instrument function of the atmospheric spectra (Volkamer et al., 1998). After this procedure, both sets of spectra have the same characteristics and can be analysed with the DOAS method.

To overcome challenges with spectrometer stray light, two measurements were performed. The first measurement was performed without any additional filters in the light beam. During the second measurement, a long-pass filter (Schott WG305) blocked light of wavelengths below 300 nm. This spectrum thus determines the spectrometer stray light originating from wavelengths above 300 nm. Because the maximum intensity of the Xe-arc lamp employed during the measurements is around 500 nm, this is the major portion of the spectrometer stray light. The measurement with the filter was then subtracted from the first measurement to derive the true spectrum between 250 and 300 nm. This spectrum was analysed using the methods of Chap. 8 and the cross-convoluted spectra. All aromatic reference spectra in the study of Kurtenbach et al. (2001) were measured with the instrument in the field and then later calibrated with the absorption cross-sections by Etzkorn et al. (2000), because no high resolution absorption cross-sections for aromatics are currently available. The O_2 reference spectrum was measured at exactly the same path length that was later used in the tunnel measurements by filling a Teflon tube around the multi-reflection cell with synthetic air. The top panel of Fig. 10.10 shows an example of a tunnel measurement of various aromatic hydrocarbons. The absorptions are clearly identified.

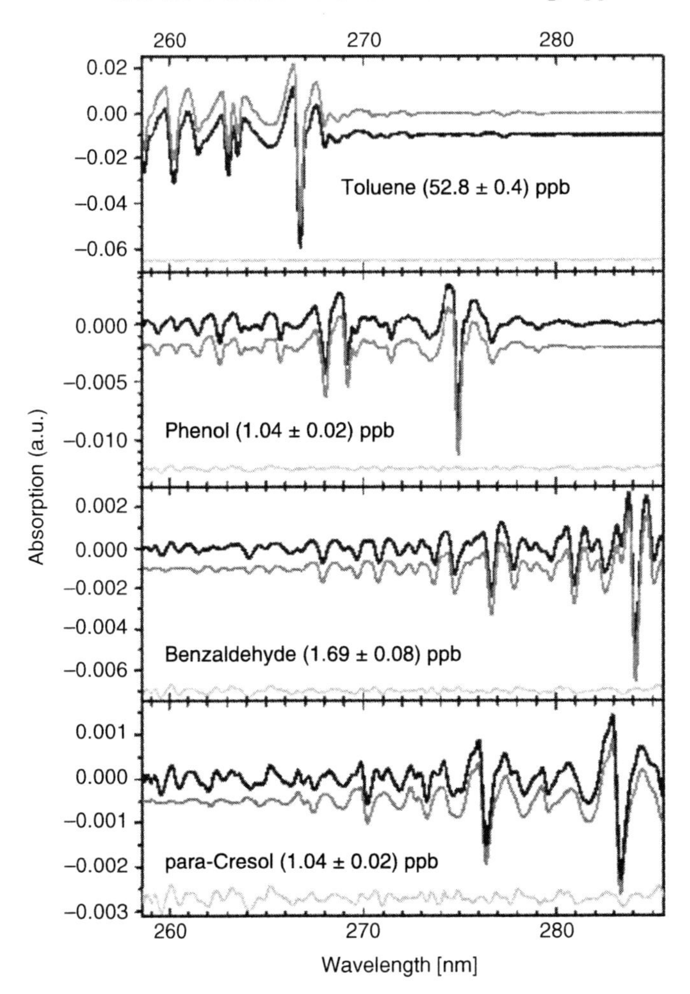

Fig. 10.10. Example of an analysis of toluene, phenol, benzaldehyde, and p-cresol measured in a traffic tunnel. The figure compares the scaled reference spectrum with the sum of this spectrum and the residual. This comparison gives a visualisation of the quality of the fit. Also shown are the residual specta (thin lines near the bottom of each panel), which in this case were much smaller than the absorptions of the trace gases (from Ackermann, 2000)

Table 10.3 lists the maximum mixing ratio observed from the various aromatics in the tunnel, as well as the detection limits of the measurements. For benzene and toluene, the detection limit is on the order of 1 ppb. For other aromatics, such as benzaldehyde and p-cresol, the detection limits are even lower. While the detection limits are greater than those from GC measurements, the DOAS system is much faster (1 min compared with 20 min) and needs no calibration.

Table 10.3. Detection limit of various aromatic hydrocarbons during the tunnel measurements in Wuppertal (from Ackermann, 2000)

Species	Highest mixing ratio observed (ppb)	Detection limit (ppb)
Benzene	36.8	0.6
Toluene	83.9	0.7
Ethyl-benzene	18.1	0.8
Benzaldehyde	2.2	0.07
para-Xylene	13.0	0.3
meta-Xylene	11.0	1.0
ortho-Xylene	20.0	2.0
Phenol	2.80	0.15
para-Cresol	1.07	0.03
SO_2	33.0	0.5

The time traces of 1 week of measurements (Fig. 10.11) show that the levels of aromatics in the tunnel depend strongly on the traffic density (top trace in Fig. 10.11), as one would expect from pollutants that originate from vehicular emissions.

Ackermann (2000) used the data from the tunnel measurements to derive the emission factors for the various aromatic hydrocarbons (Table 10.4). These parameters are essential for the conditions in urban air quality models that are used to develop mitigation strategies for urban pollution. This example shows that DOAS can contribute in other ways to solve today's air pollution problems than the simple monitoring of trace gas concentrations.

Ammonia plays an important role as a reducing agent in the atmosphere. Through reaction with acids, such as H_2SO_4 and HNO_3, it contributes to the formation of particular matter, and is thus an important component of smog. High numbers of particles are often found in areas where urban pollution is mixed with rural air containing elevated levels of NH_3. The most important sources of ammonia are the bacterial activity in the decomposition of animal waste, in the soil, and biomass burning.

Measurements of ammonia with DOAS have, for example, been reported for flue gas analysis (Mellqvist and Rosén 1996) as well as in ambient air (Neftel et al., 1990). Here, we show an example for the measurement of ammonia concentrations at a dairy farm. Mount et al. (2002) used a coaxial long-path DOAS system with an array of retro-reflectors located at a distance of up to 150 m (Fig. 10.12). The main difference to the set-ups described in Chap. 7 is the introduction of a double spectrograph, for improved stray light suppression in the UV. Because NH_3 absorbs between 190 and 220 nm (Fig. 10.13, right panel), at even smaller wavelengths than aromatics, stray light suppression is critical for this application, as explained.

The left panel of Fig. 10.13 shows an example of a raw spectrum recorded with the instrument shown in Fig. 10.12. The intensity rapidly decreases

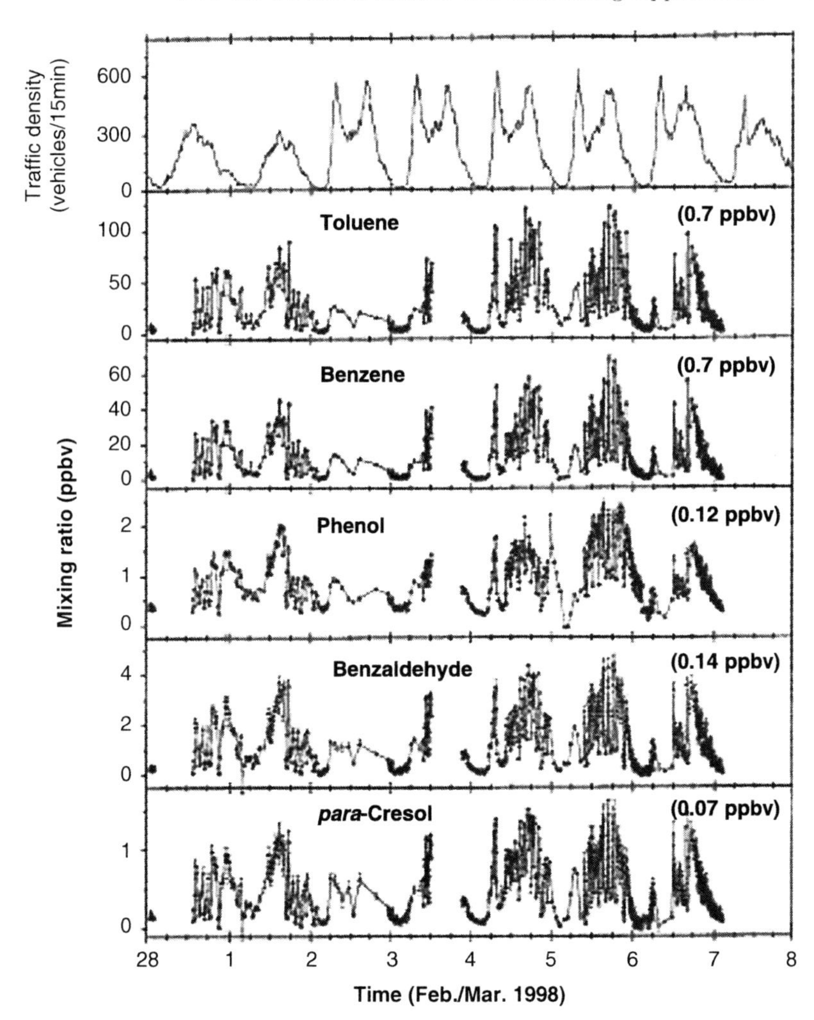

Fig. 10.11. Diurnal variations of the mixing ratios of various aromatic hydrocarbons during the first week of February 1998. Measurements were made with an open multi-reflection system inside a traffic tunnel in Wuppertal, Germany (Ackermann, 2000; Kurtenbach et al., 2001). The *top panel* shows the traffic density in the tunnel. The *numbers* in parenthesis denote the detection limits of the respective aromatics (from Kurtenbach et al., 2001)

towards shorter wavelengths. In general, the light intensity is low compared with long-path instruments operating in the visible range. However, the intensity is sufficient to allow the identification of ammonia in ambient air. The right panel of Fig. 10.13 shows a comparison of a spectrum measured with the light path passing over a NH_3 source compared with a differential absorption cross-section of NH_3. Taking into account saturation effects in the raw ratio, the agreement between the two curves is excellent.

Table 10.4. Emission factors for aromatic hydrocarbons derived from DOAS measurements in a traffic tunnel (from Ackermann, 2000)

Species	Emission factors (milligram per vehicle kilometre driven)		
	Weekday	Weekend	Total
Toluene	56.7 ± 4.5	60.5 ± 8.5	58.0 ± 5.0
Benzene	24.7 ± 1.9	27.8 ± 7.2	25.4 ± 3.2
Ethyl-benzene	18.5 ± 1.5	24.3 ± 4.1	19.2 ± 2.3
Benzaldehyde	1.62 ± 0.12	3.4 ± 0.5	1.1 ± 0.2
Para-xylene	7.4 ± 0.6	9.1 ± 1.4	7.3 ± 0.8
Meta-xylene	8.9 ± 0.8	10.4 ± 1.6	9.2 ± 1.1
Ortho-xylene	17.7 ± 1.5	25.8 ± 5.6	18.4 ± 2.1
Phenol	1.44 ± 0.12	1.76 ± 0.35	1.5 ± 0.2
P-cresol	0.85 ± 0.10	1.25 ± 0.21	0.89 ± 0.10

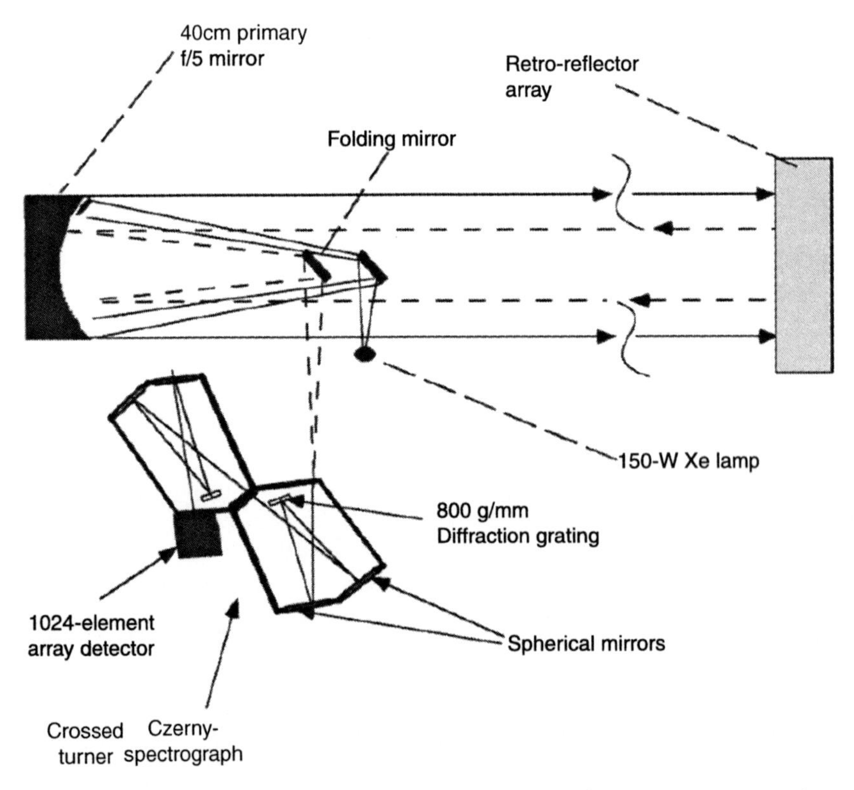

Fig. 10.12. Instrumental set-up for the measurement of ammonia emission from a dairy farm (from Mount et al., 2002, Copyright Elsevier, 2002)

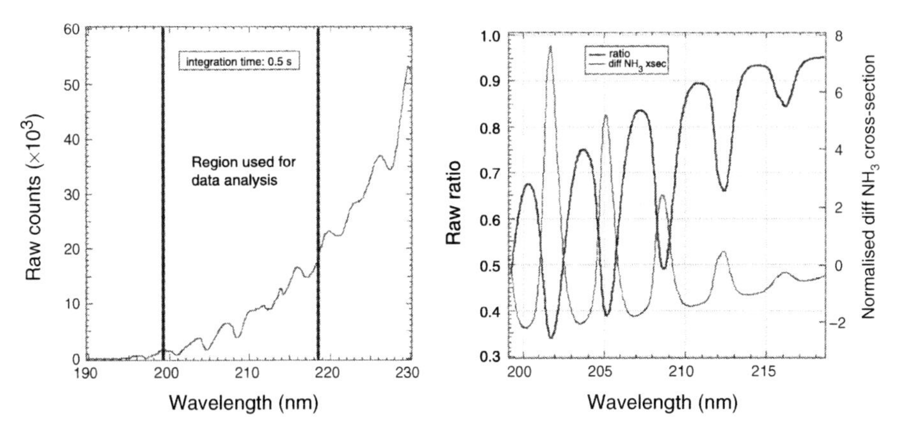

Fig. 10.13. Left panel: Raw spectrum measured on a 68 m path with an integration time of 0.5 s (Mount et al., 2002).The decrease of the intensity towards shorter wavelengths can be clearly seen. **Right panel**: Comparison of the ratio of two spectra with and without NH_3 in the light beam, to the absorption cross-section of NH_3. The difference in shape is caused by saturation effects in the raw ratio (from Mount et al., 2002, Copyright Elsevier, 2002)

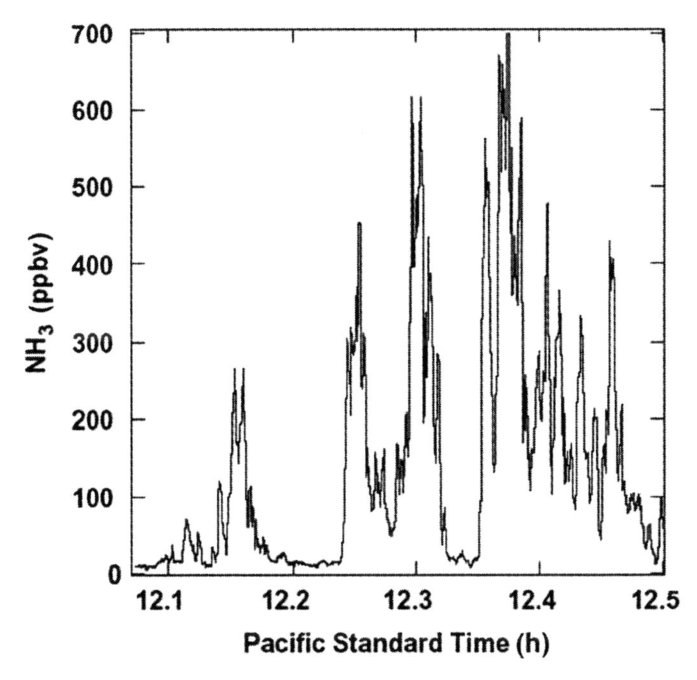

Fig. 10.14. Example of ammonia mixing ratios measured downwind of a slurry lagoon (Mount et al., 2002). The integration time for one measurement was 0.6 s (from Mount et al., 2002, Copyright Elsevier, 2002)

Mount et al. (2002) used this instrument to monitor the release of ammonia on a diary farm. Figure 10.14 shows 1 h of high frequency measurement downwind of a slurry lagoon. The rapid variations are caused by shifting winds and turbulent cells moving into the light path.

The observations by Mount et al. (2002) illustrate how active DOAS instruments can be applied to fence-line monitoring, in this case of the release of ammonia from a dairy farm. The potential of DOAS for the measurement of NH_3 has thus far not been extensively explored and there is great potential for further developments.

10.1.2 Vertical Profiles of Air Pollution by Multiple DOAS Light Beams

As we already described above, active DOAS instruments can be used to measure the vertical distribution of trace gases. This was first explored by Platt in 1977, with the goal of measuring SO_2 deposition velocities. While this technique was not used for nearly two decades, it was revived recently to study nocturnal chemical processes. The investigation of NO_2 to HONO conversion was a first example of vertical profiling. Here, we want to discuss another example that was set up to study the vertical variation of nocturnal chemistry.

Two long-path DOAS instruments were set up in Houston, Texas, during TEXAQS 2000 (see also Sect. 10.1.1 above). The measurements were made on five different light paths, as illustrated in Fig. 10.15. The first DOAS system measured on the two short light paths taking $\sim$ 5 min for one series of measurements. The second system took $\sim$ 15 min to measure all three long light paths. The raw data of these measurements are shown in Fig. 10.3.

A simple deconvolution process was then used to derive trace gas mixing ratios averaged in the boxes indicated in Fig. 10.15. Because the measurements were made sequentially, all data were linearly interpolated to the time of the upper long path measurement. The path integrated concentration along the ith light path is denoted as S_i, and the concentrations in a specific height interval as C_i, where the light paths are numbered from the ground upwards.

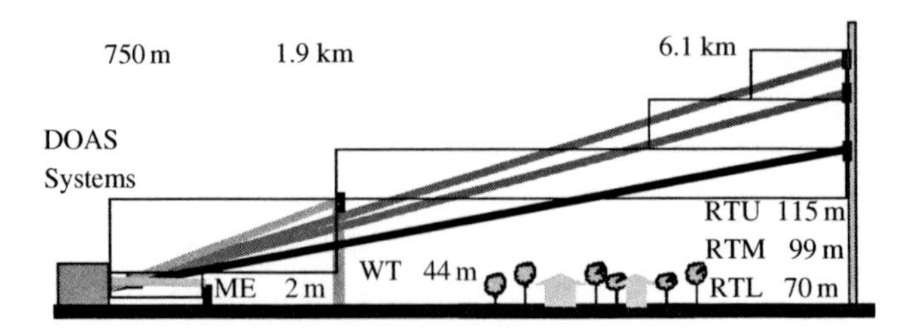

Fig. 10.15. Set-up during the TEXAQS 2000 experiment. Five retro-reflector arrays were mounted at different distances and altitudes. The measurements were performed by two DOAS instruments (from Stutz et al., 2004b, Copyright by American Geophysical Union (AGU), reproduced by permission of AGU)

The data of the two lowest paths directly represent the concentrations at 2-m altitude and in the height interval (2–44 m). The other concentrations C_3, C_4, and C_5 in the height intervals, respectively, (44–70 m), (70–99 m), and (99–115 m) are calculated from the ground upwards with the following equation:

$$C_i = \frac{h_i - H}{h_i - h_{i-1}} S_i - \frac{h_{i-1} - H}{h_i - h_{i-1}} S_{i-1} .$$

An important assumption, which underlies the application of this equation, is that the trace gases are homogeneously distributed in the horizontal direction. This can be tested by using a number of techniques outlined by Stutz et al., 2004b. It is also essential to determine the errors of the concentrations averaged over the height intervals. Since differences of the measurements are formed and the effective absorptions path in the boxes is shorter than the initial path length, the errors of the box averaged concentrations increase. One has to be careful to assess whether the measured concentrations are statistically meaningful. Four days of box-averaged mixing ratios during times when horizontal uniformity was encountered and statistically significant trace gas levels were found are shown in Fig. 10.16. During a number of instances vertical profiles of the various trace gases were observed. Clearest are the

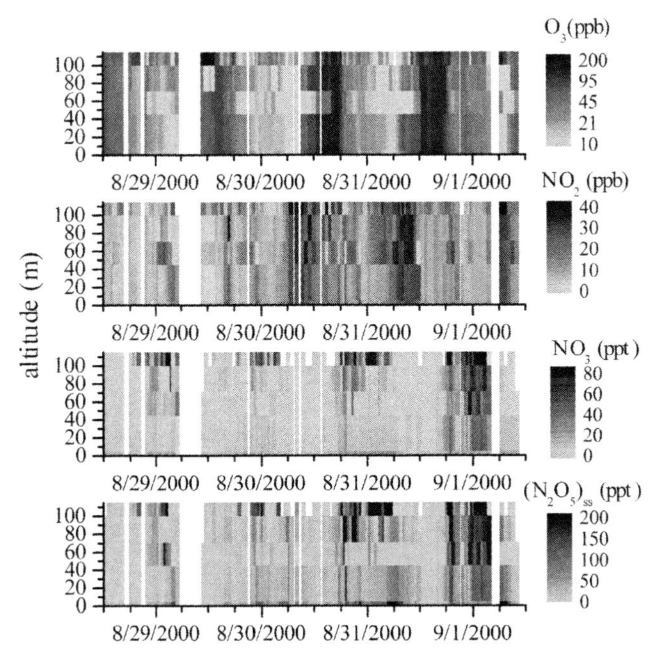

Fig. 10.16. Observations of the spatial distribution of trace gas mixing ratios during 4 days in Houston, TX. The N_2O_5 mixing ratios are calculated from the steady state of measured NO_2, NO_3, and N_2O_5. The O_3 mixing ratios are displayed in a logarithmic colour coding (from Stutz et al., 2004b, Copyright by American Geophysical Union (AGU), reproduced by permission of AGU)

nocturnal maxima of NO_3 in the upper nocturnal boundary layer each night. On several occasions, vertically non-uniform distributions of O_3 and NO_2 are observed. For example, on August 31, a plume of NO_2 between 50 and 90 m was encountered.

The nocturnal profiles of the various gases are more clearly seen in Fig. 10.17. The profiles show that the levels of all gases vary with altitude. This variation, and in particular that of NO_3, the dominant nocturnal radical, leads to an altitude-dependent nocturnal chemistry. Similar experiments

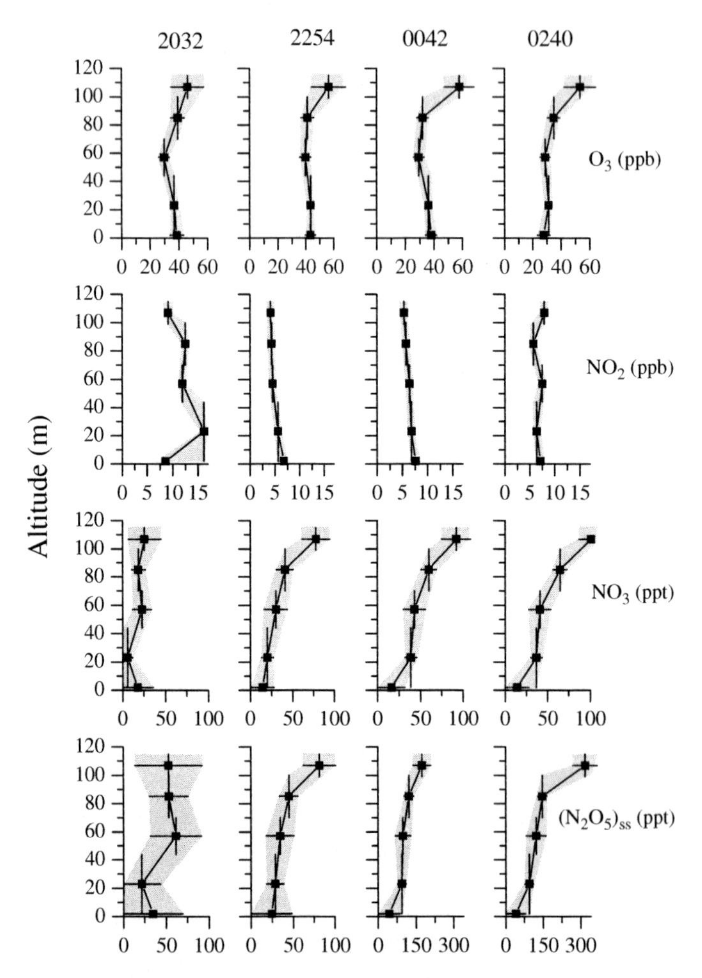

Fig. 10.17. Vertical mixing ratio profiles of O_3, NO_2, NO_3, and N_2O_5 near Houston, Texas (USA) during the night of 8/31-9/1, 2000 at four different times (noted on *top of the graphs*). The N_2O_5 mixing ratios shown are calculated from the steady state of measured NO_2, NO_3, and N_2O_5 (from Stutz et al., 2004b, Copyright by American Geophysical Union (AGU), reproduced by permission of AGU)

have confirmed these results. Figures 10.16 and 10.17 show examples of the abilities of modern DOAS instruments to perform process studies in the atmosphere. Active DOAS continues to contribute to the advancement of our understanding of pollution in the atmosphere.

10.2 Investigation of Free Radical Processes in the Atmosphere

As outlined in Chap. 2, free radicals play a central role in many chemical systems, such as flames, living cells, or in the atmosphere. Due to their high chemical reactivity, radicals are the driving force for most chemical processes in the atmosphere. In particular, they initiate and carry reaction chains. Thus, knowledge of the concentration of those species in the atmosphere is a key requirement for the investigation of atmospheric chemistry. For instance, degradation of most oxidisable trace gases in the atmosphere, such as hydrocarbons, carbon monoxide, or H-CFC's, is initiated by free radicals.

As a consequence of their high reactivity, the steady state concentration of free radicals in the atmosphere is generally quite low, even compared with atmospheric trace gas standards. For instance, the atmospheric lifetime of the OH radical never exceeds 1 s. A consequence of this short lifetime is that transport processes can be largely neglected when considering the budgets of most free radicals. Therefore, measurement of the instantaneous concentration of free radicals together with a comprehensive set of other trace gases linked with the radical's chemical cycles allows the study of in situ chemical reactions in the atmosphere (Perner et al., 1976, 1987a; Platt et al., 1987, 1988, 2002;

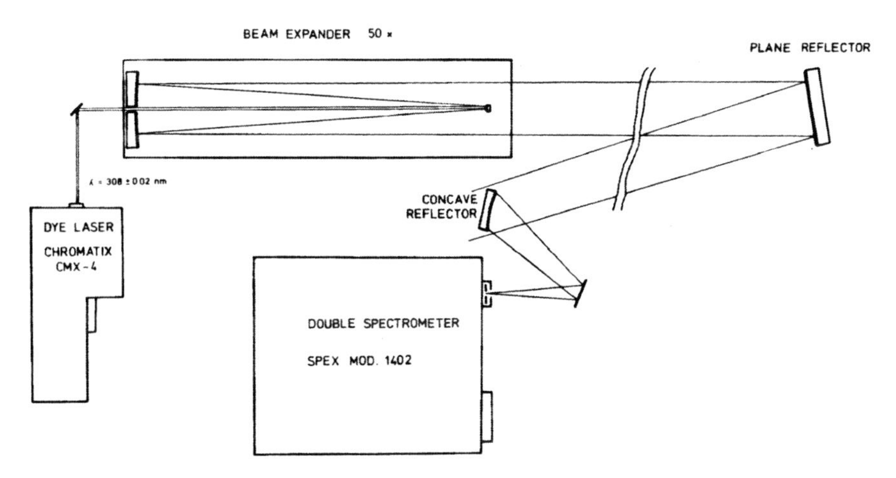

Fig. 10.18. Optical set-up of the high-resolution DOAS instrument used for the first reliable detection of OH in the atmosphere (Perner et al., 1976). A broadband, frequency-doubled dye laser serves as light source

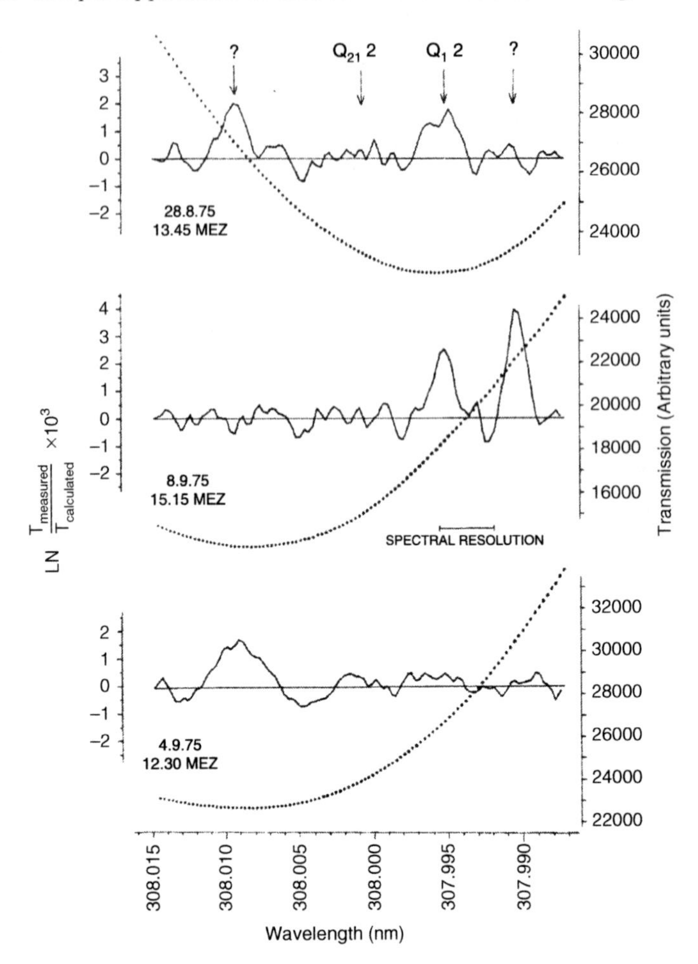

Fig. 10.19. High-resolution spectrum of OH. The DOAS instrument was used for the first reliable detection of OH in the atmosphere (from Perner et al., 1976, Copyright by American Geophysical Union (AGU), reproduced by permission of AGU)

Dorn et al., 1988, 1993, 1996; Mount 1992; Mount and Eisele, 1992; Eisele et al., 1994; Brauers et al. 1996, 1999; Hausmann et al., 1997; Hofzumahaus et al., 1998).

Several key free radicals of interest to atmospheric chemistry can be measured by DOAS with very high (sub – ppt) sensitivity. To date, the following radicals have been measured: Hydroxyl radicals (OH), nitrate radicals (NO_3), and halogen oxide radicals (ClO, OClO, BrO, OBrO, IO, and OIO). In fact, most of them were detected for the first time in the atmosphere by DOAS.

In this section, we will discuss some examples of the measurement of these radical species. Again, we can only show examples of the many measurements that have been made of radical species.

10.2.1 Measurement of OH Radicals by DOAS

A series of OH measurement campaigns were performed with the DOAS technique, initially using optomechanical scanning of the spectra (Perner et al., 1976, 1987a, Platt et al., 1988), and in later instruments employing diode array detectors (Dorn et al., 1988; Hofzumahaus et al., 1991), see Sect. 7.9.2. Observed OH radicals levels ranged from the detection limit ($1–2 \times 10^6$ molec. cm^{-3}) to about 10^7 molec. cm^{-3}. Figures 10.18, 10.19, and 10.20 show the optical set-up using a frequency doubled dye laser as light source, examples of a spectrum and a diurnal profile of the OH concentration.

The most recent implementations of DOAS OH instruments are based on multi-reflection cells with base path lengths of 20–40 m (Brauers et al., 1996). This instrument reaches detection limits down to 1.5×10^6 molec. cm^{-3} in an integration time of about 200 s. Figure 10.21 shows the different absorption features that can be found in the wavelength window used for the measurement of OH. In particular, interferences of SO_2, HCHO, and naphthalene are observed.

Figure 10.22 shows the identification of the OH absorptions in an atmospheric spectrum. The spectrum was measured using the multi-channel scanning technique (MCST), which is explained in detail in Sect. 7.14.2. It should be noted that the amplitude of the residual is extremely low in this measurement, with a peak-to-peak equivalent optical density of $\sim 10^{-4}$.

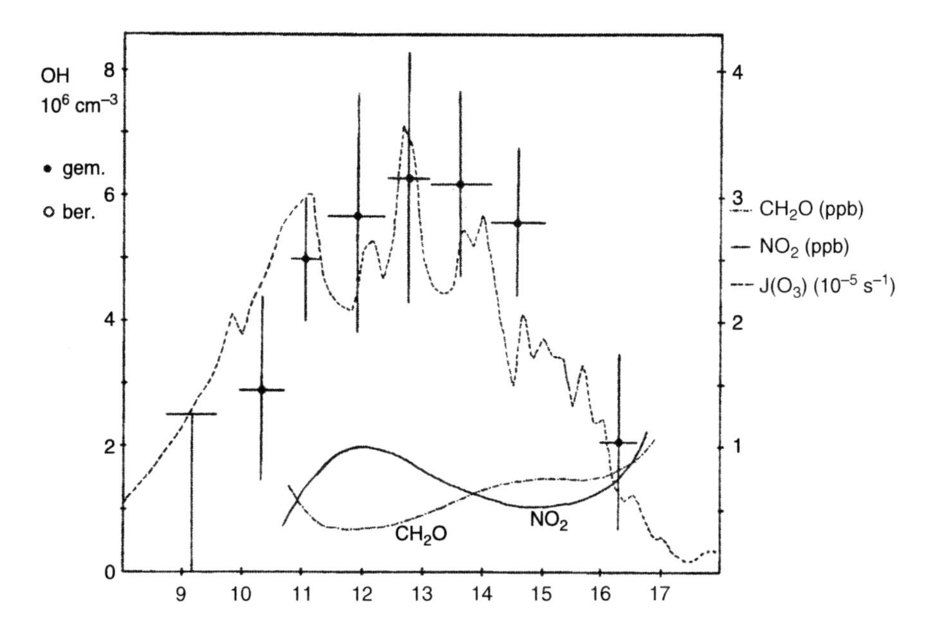

Fig. 10.20. Diurnal OH profile measured with a later version of the OH DOAS instrument (see Fig. 7.48) (from Platt et al., 1987)

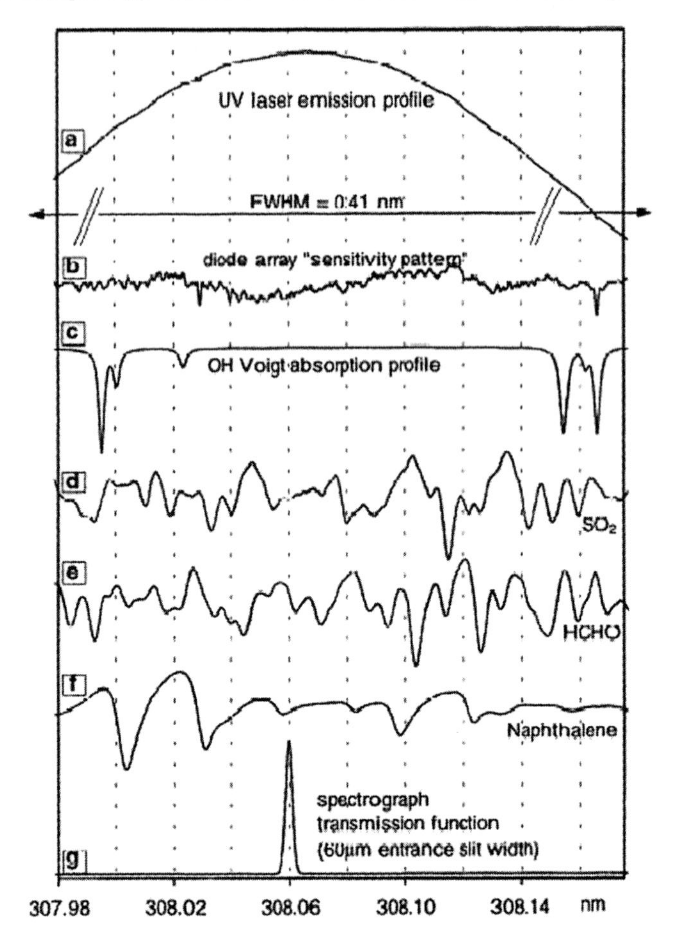

Fig. 10.21. OH absorption spectrum (*trace c*) together with the spectra of several interfering species SO_2, CH_2O, and naphthalene (*traces d–f*). Also shown are the laser intensity profile (*trace a*) and the diode array sensitivity pattern (from Brauers et al., 1996, Copyright by American Geophysical Union (AGU), reproduced by Permission of AGU)

The measurements in Figs. 10.20 and 10.23 show the diurnal profile of OH with a maximum around noon. The comparison of the OH concentration with $J(O^1D)$ in Fig. 10.23, confirms the strong influence of solar radiation on OH levels.

10.2.2 Measurement of NO_3 Radicals

Nitrate radicals, NO_3 and OH, are in a sense complementary to each other (see Chap. 2): OH radicals are directly produced by photochemical reactions, thus their concentration is highest at noon and low at night. Conversely, nitrate radicals are effectively destroyed during the day, thus they will reach high

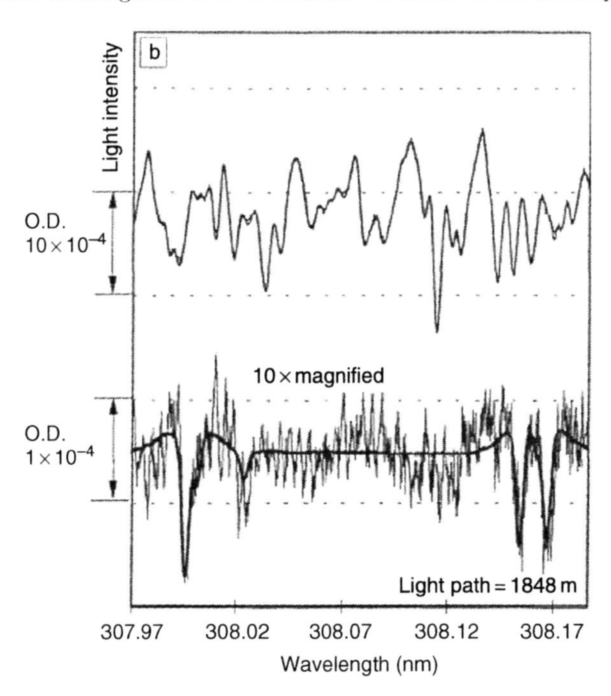

Fig. 10.22. Ambient OH measurement in Pennewitt, Germany, August 16, 1994. The atmospheric spectrum (*top trace*) is dominated by the absorptions of SO_2 as shown in *trace d* of Fig. 10.21. After the removal of this absorption, the structures of OH can be clearly identified in the spectrum (*bottom trace*). The absorptions correspond to an OH concentration of $8.8 \cdot \times 10^6$ molec. cm^{-3} (from Brauers et al., 1996, Copyright by American Geophysical Union (AGU), reproduced by permission of AGU)

concentrations predominately at night. Both species are responsible for the removal of hydrocarbons and NO_X from the atmosphere. The product of NO_3 and OH reactions with NO_2 is nitric acid. We already discussed the role DOAS played in the detection of OH radicals. Here, we will concentrate on the importance of DOAS measurements for nocturnal radical chemistry, i.e. NO_3 chemistry.

NO_3 was first identified in the troposphere spectroscopically by Platt et al. (1979) and Noxon et al. (1980). Figure 10.24 shows the first reported NO_3 measurement by active DOAS. Due to the width of the NO_3 absorption bands and location in the visible part of the spectrum NO_3 can readily be measured by active DOAS using a variety of thermal light sources (incandescent lamps, Xe-arc lamps). Since then a large number of NO_3 measurements in the atmosphere have been reported (Platt et al., 1980a, 1981, 1984, 1990; Harris et al., 1983; Brauers et al., 1990; Platt, 1991; Plane and Nien, 1991; Winer and Biermann, 1991; Platt and Heintz, 1994; Platt and Hausmann, 1994; Plane and Smith, 1995; Platt and Janssen, 1996; Heintz et al., 1996; Carslaw et al., 1997; Aliwell

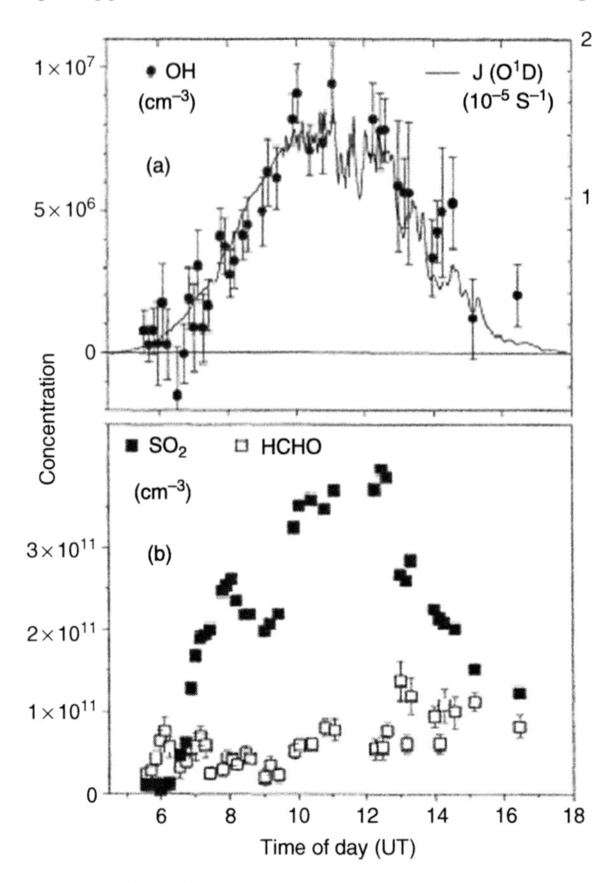

Fig. 10.23. Diurnal profiles of OH concentration and ozone photolysis frequency at a rural site in north-eastern Germany (**top panel**). Also shown are the concentrations of HCHO and SO_2 that are determined in the measurement (**bottom panel**) (from Dorn et al., 1996, Copyright by American Geophysical Union (AGU), reproduced by permission of AGU)

and Jones, 1998; Allan et al., 1999, 2000; Geyer et al., 1999, 2001a,b,c; Gölz et al., 2001; Ball et al., 2001; Geyer and Platt, 2002; Stutz et al., 2002; Kurtenbach et al., 2002; Geyer et al., 2003; Kern et al., 2006). A challenge in the measurement of NO_3 is the interference with a water-absorption band system around 650 nm (see, e.g. Fig. 10.26). Because the spectral structure of this band system is much narrower than that of NO_3 filtering procedures are often used to reduce the effect of water (Geyer et al., 1999). Another method to remove this water band is the use of daytime atmospheric spectra as water references. This technique has been employed in most NO_3 measurements in the past (see top panel in Fig. 10.27). However, as pointed out recently by Geyer et al. (2003), NO_3 levels can be elevated under conditions with high O_3 mixing ratios ($[O_3] > 100$ ppb) during the day. In this case, more care has

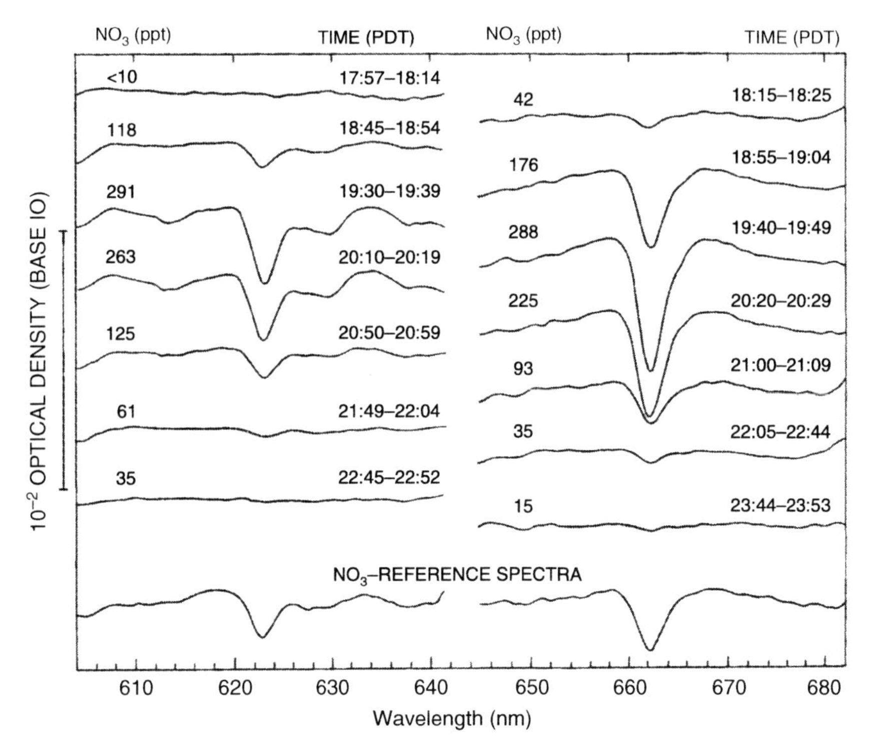

Fig. 10.24. Example of NO_3 radical spectra recorded in an urban area (Riverside, Los Angeles, USA). The absorption structures due to H_2O and O_2 are already removed (from Platt et al., 1979, Copyright by American Geophysical Union (AGU), reproduced by permission of AGU)

to be taken in the choice of water reference spectra, i.e. Geyer et al. (2003) used daytime spectra taken during periods of high NO mixing ratios as water references.

Because DOAS is the most commonly used technique to measure the atmospheric concentration of NO_3, there are many examples of its application. Figures 10.24 and 10.26 show NO_3 spectra in an urban and rural environment, respectively. Figure 10.25 presents the corresponding time series of NO_3 after sunset in an urban environment. Because NO_3 also plays an important role in marine areas, we also show an example taken in the remote marine boundary layer (Fig. 10.27). This spectrum illustrates that very low NO_3 mixing ratios, close to 1 ppt, can be measured accurately.

Observations in the Marine Boundary Layer

A number of NO_3 DOAS observations in the marine boundary layer have been reported (e.g. Heintz et al., 1996; Martinez et al., 2002). Here, we will review data measured at Mace Head, Ireland and Weybourne, England (Allan et al., 2000; Carslaw, 1997).

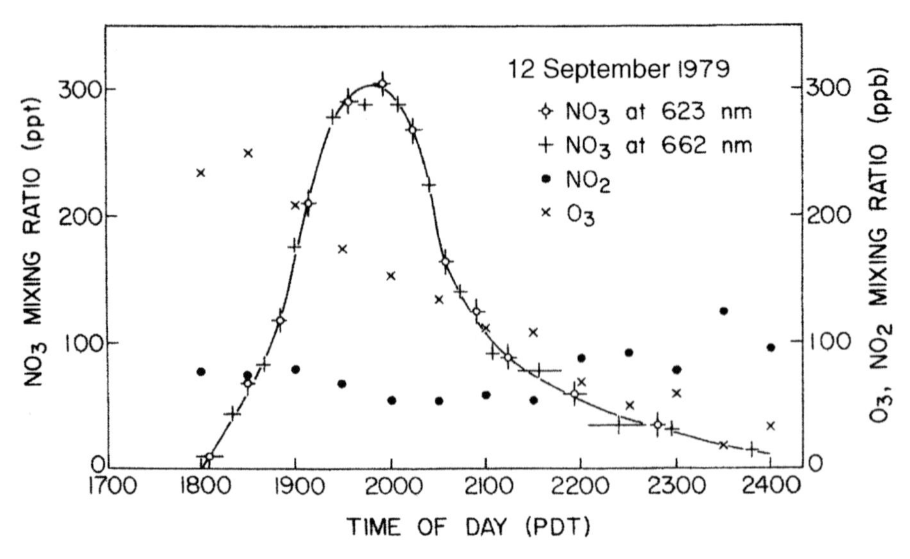

Fig. 10.25. Example of NO_3 radical time series in an urban area (Riverside, Los Angeles, USA) after sunset. Data are derived from the series of spectra shown in Fig. 10.24. (from Platt et al., 1979, Copyright by American Geophysical Union (AGU), reproduced by permission of AGU)

Figure 10.28 shows an overview of a time series of NO_3 measurements. The maximum NO_3 mixing ratios observed at Mace Head are up to 40 ppt. A clear dependence on the origin of the air is observed. Clean atlantic or polar air has low NO_3 levels, while continental air shows much higher NO_3. This can be explained by the much higher NO_x levels in these air masses, which increases the formation rate of NO_3.

The data in Fig. 10.28 can be used to derive the removal rate of NO_X at night through the calculation of N_2O_5 and its uptake on the aerosol. In addition, it can be used to determine the nocturnal oxidation of DMS. Figure 10.29 shows another application of these measurements by showing a comparison of NO_3 and HO_2+RO_2 levels. This correlation is due to the oxidation of organic species by NO_3 at night (see Sect. 2.5.2).

Observations in Rural and Forested Areas

Measurements of NO_3 in rural and forested areas are of particular significance since emissions of mono-terpenes, which rapidly react with NO_3, are often highest there. A continental long-term study was reported at the continental site Lindenberg near Berlin, Germany by Geyer et al. (2001) (Fig. 10.30). Observations of this type offer the possibility of statistical investigations of the night-time NO_3 and N_2O_5 concentrations and the NO_3 production and

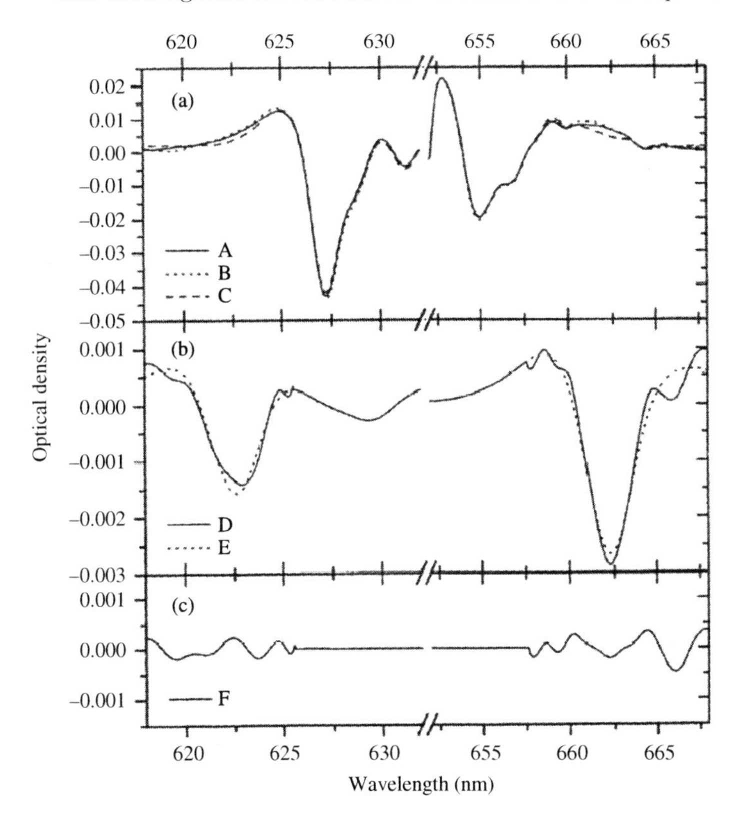

Fig. 10.26. Example of night-time NO_3 radical absorption in a rural area (August 5, Pabsthum, Germany). (Geyer et al., 1999). The *top panel (a)* shows the atmospheric spectrum at 2150 UT **(trace A)** and two daytime references (**traces B, C**), which are dominated by water absorptions. The *middle panel (b)* shows a comparison of the scaled NO_3 reference with added residual (**D**) and the NO_3 reference (E)). The optical density of NO_3 absorption is $\sim$ 0.35% which, in this case, was equivalent to 9.8 ± 1.1 ppt. The *bottom panel (c)* shows the residual of the analysis (**trace F**). Details about the analysis procedure can be found in (from Geyer et al., 1999, Copyright by American Geophysical Union (AGU), reproduced by permission of AGU)

degradation frequencies, see Sect. 2.5. In addition, the mean contribution and the seasonal variation of the different sink mechanisms of NO_3, which provides an essential tool for quantifying the average effect of NO_3 on the oxidation capacity of the atmosphere, can be studied.

Observations in Urban Areas

Figure 10.24 showed examples of urban spectra of NO_3. Surprisingly, few measurements of NO_3 have been performed in this environment. This is partially

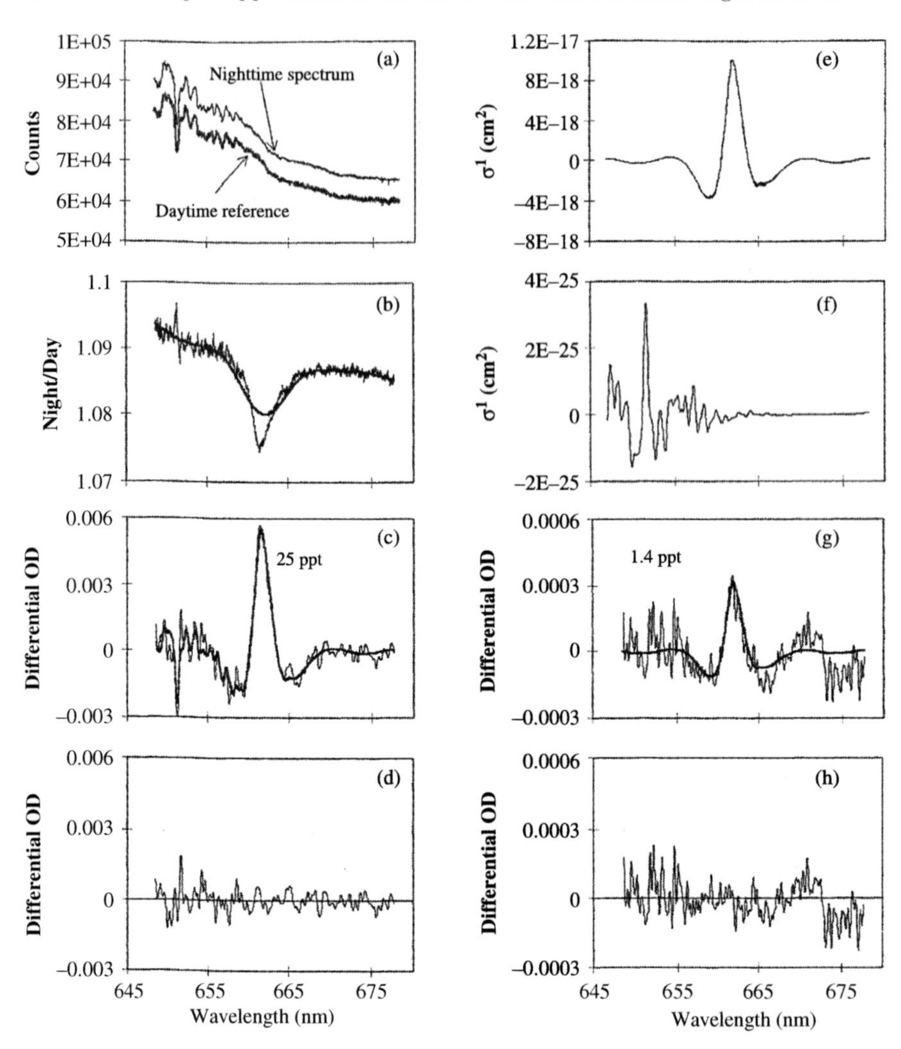

Fig. 10.27. Example of night-time NO_3 radical absorption in the marine boundary layer (Allan et al., 2000): (**a**) shows the daytime reference and the nocturnal spectrum; (**b**) is the ratio of the two spectra with a low-pass filter; (**c**) and (**g**) show the fit of the H_2O and NO_2, cross-sections (**e and f**) and (**d**) is the residual. The result of the fit gave a NO_3-mixing ratio of 24.7 ± 0.7 ppt. An example with 1.9 ± 0.2 ppt of NO_3 is shown in (**g**). The measurements were made with a long-path DOAS instrument on a 8.4-km-long light path at Mace Head, Ireland. Details on the instrument and the analysis procedure can be found in (from Allan et al., 2000, Copyright by American Geophysical Union (AGU), reproduced by permission of AGU)

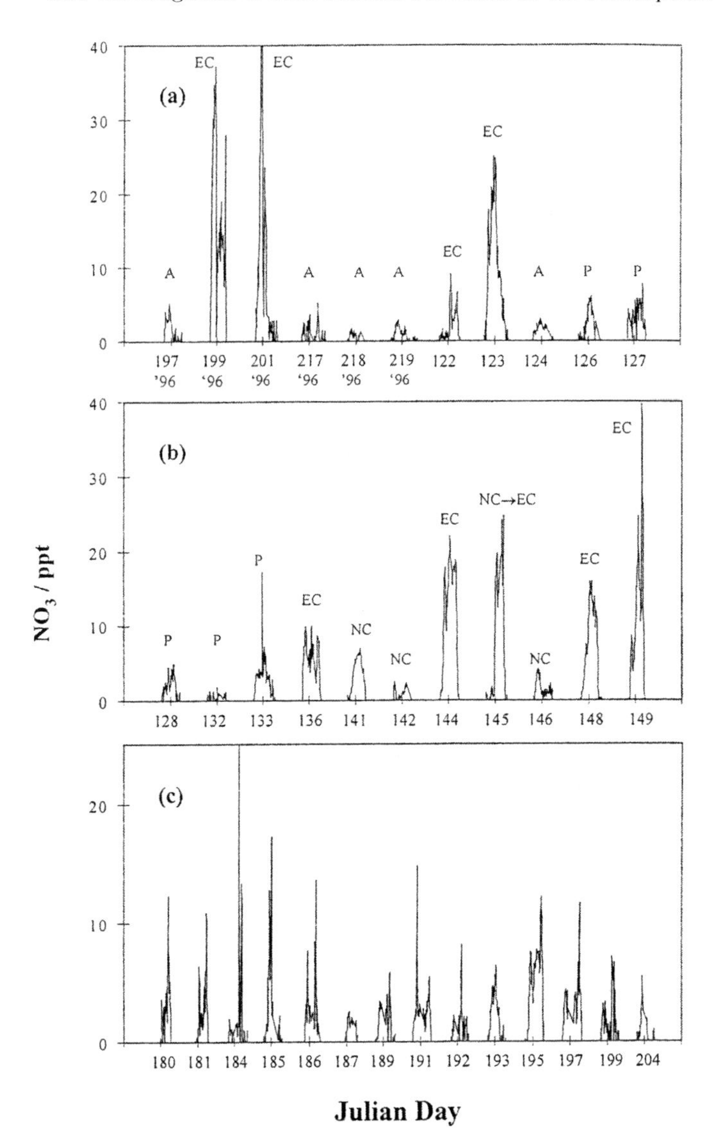

Fig. 10.28. NO_3 time series collected in the marine boundary layer of Mace Head, Ireland (**a** and **b**). The letters denote the origin of the observed air masses: A, atlantic; P, polar marine; EC, easterly continental; NC, northerly continental (from Allan et al., 2000, Copyright by American Geophysical Union (AGU), reproduced by permission of AGU)

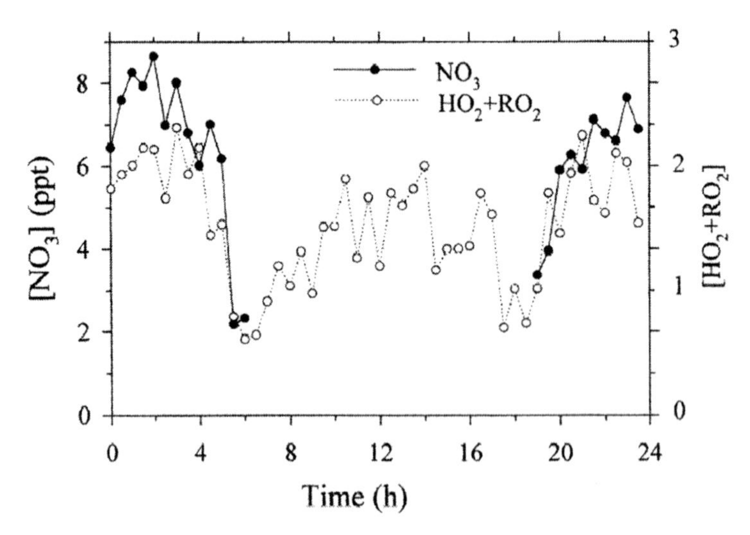

Fig. 10.29. Diurnal variation of NO_3 and $HO_2 + RO_2$ produced by averaging half-hourly data taken during 5 days (from Carslaw et al., 1997, Copyright by American Geophysical Union (AGU), reproduced by permission of AGU)

explained by the often high NO levels at the ground in cities, which efficiently destroy NO_3. As shown in Figs. 10.16 and 10.17, the levels of NO_3 are quite high in the upper nocturnal boundary layer. Mixing ratios above 100 ppt are frequently observed in cities, in which both precursors of NO_3, O_3, and NO_2, are elevated. Studies by Geyer and Stutz (Geyer et al., 2003; Stutz et al., 2004) showed that NO_3 can play an important role for the removal of NO_x and VOCs emitted by traffic in cities.

10.2.3 Measurement of Halogen Oxides

The discovery of boundary-layer ozone depletion events in the Arctic (Oltmans and Komhyr, 1986; Barrie et al., 1988; Bottenheim et al., 1990; Niki and Becker, 1993) has increased the interest in the chemistry of reactive halogens in the troposphere. DOAS measurements by Hausmann and Platt (1994) showed for the first time that these depletion events, during which ozone drops from ~ 40 ppb to unmeasurable levels (<2 ppb), are associated with high concentrations of reactive bromine, in particular BrO. Sources of gas phase Br are most likely heterogeneous reactions of oxides of nitrogen with sea-salt surfaces (Barrie et al., 1988; Finlayson-Pitts et al., 1990; Fan and Jacob 1992). Details of this chemistry are discussed in Chap. 2.

Here, we will present a number of examples of the measurement of halogen oxides and point out how these observations have advanced our understanding of halogen chemistry.

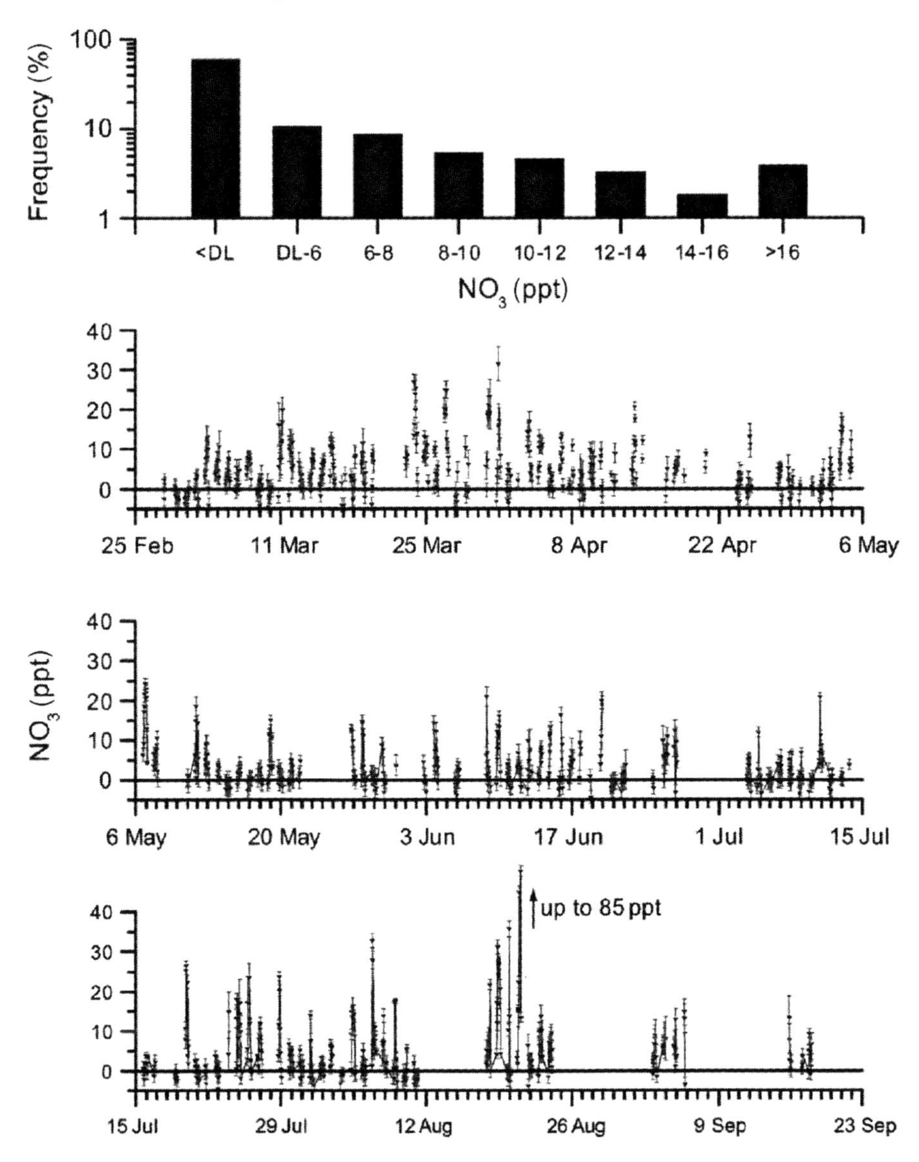

Fig. 10.30. Night-time NO_3 radical concentration profiles observed during a long-term campaign in a semi-rural area (Lindenberg near Berlin, Germany). Time series and frequency distribution (top panel, in 2-ppt intervals centred at the number given, DL = detection limit) of NO_3 observed by DOAS measurements averaged over a 10-km light path from February 27 to September 18, 1998. During the night from August 20–21, 1998, NO_3 reached peak mixing ratios up to 85 ppt. (from Geyer et al., 2001, Copyright Elsevier, 2001)

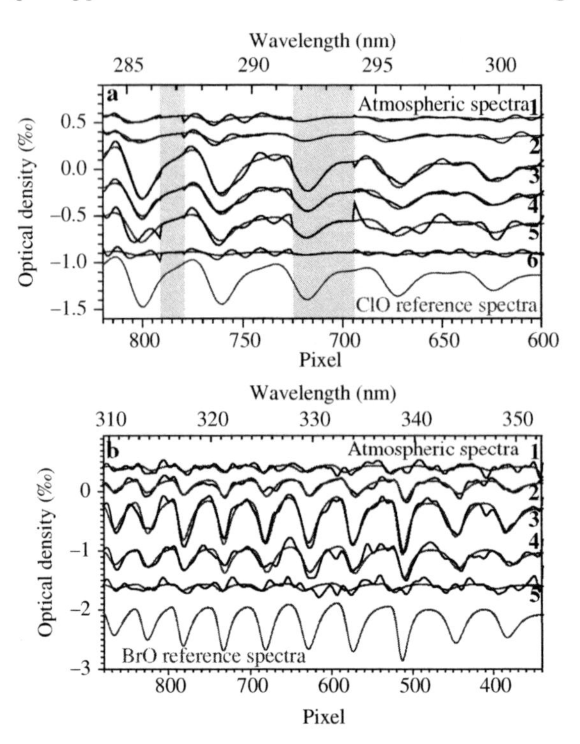

Fig. 10.31. ClO (**a**) and BrO (**b**) absorption structures recorded at the Great Salt Lake during October 14, 2000. The *black lines* are the atmospheric spectra after removing all absorptions except those by the respective halogen oxides. The *blue lines* are the respective fitted reference spectra that were calculated from a literature absorption cross-section (Orphal et al., in press; Sander and Friedl, 1989). Spectra are offset to each other to simplify the graph. The *grey intervals* in (a) mark spectral regions that were excluded from the fit (from Stutz et al., 2002b, Copyright by American Geophysical Union (AGU), reproduced by permission of AGU)

First, we show examples of the identification of the various halogen oxides in the atmosphere. Figures 10.31 and 10.35 show the spectra of ClO and BrO at the Great Salt Lake in Utah (Stutz et al., 2002). Similar examples for the detection of BrO can be found for the Dead Sea (Hebestreit et al., 1999) and polar regions (e.g. Hausmann and Platt, 1994; Tuckermann et al., 1997). The detection of ClO is more challenging due to the shorter wavelengths and the weaker absorptions.

Figures 10.32 and 10.33 show the identification of IO (Alicke et al., 1999) and OIO (Allan et al., 2001) in the atmosphere of Mace Head. Measurements of IO have been made at other locations and there is increasing evidence that IO is more ubiquitous than thus far assumed (Platt and Hönninger, 2003; Saiz-Lopez et al., 2004b; Zingler and Platt, 2005; Peters, 2004; Peters et al., 2005).

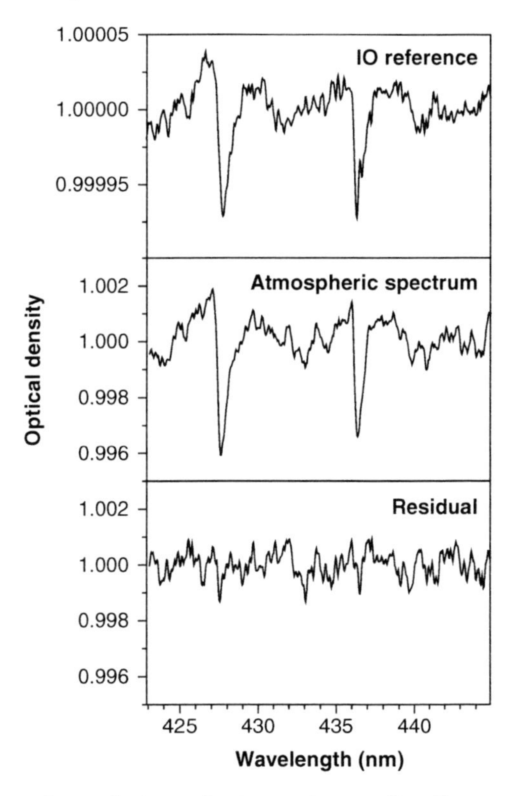

Fig. 10.32. Comparison of atmospheric spectrum after the removal of NO_2 and H_2O absorptions (centre panel) with the IO spectrum (top panel) clearly shows the presence of IO (from Alicke et al., 1999)

The study of the chemistry of BrO in polar regions was the first scientific application for the capability of DOAS to measure halogen oxides. Figure 10.34 shows data from a field experiment in Ny Alesund, Spitsbergen, during Spring 1996. The onset of ozone decay coincides with elevated BrO levels, providing evidence for the catalytic O_3 destruction cycles described in Chap. 2. The observation at the Dead Sea, Israel, (Fig. 10.35) confirm these results. The Dead Sea has the highest BrO levels thus far reported. With up to 100 ppt of BrO, the complete destruction of O_3 proceeds very fast and influences wide areas in the Dead Sea valley (Hebestreit et al., 1999; Matveev et al., 2001).

DOAS has provided evidence for the presence of four halogen oxides and is the most reliable method to measure these compounds today. Since tropospheric halogen chemistry is an active research field, DOAS will continue making contributions by providing accurate measurements of these species.

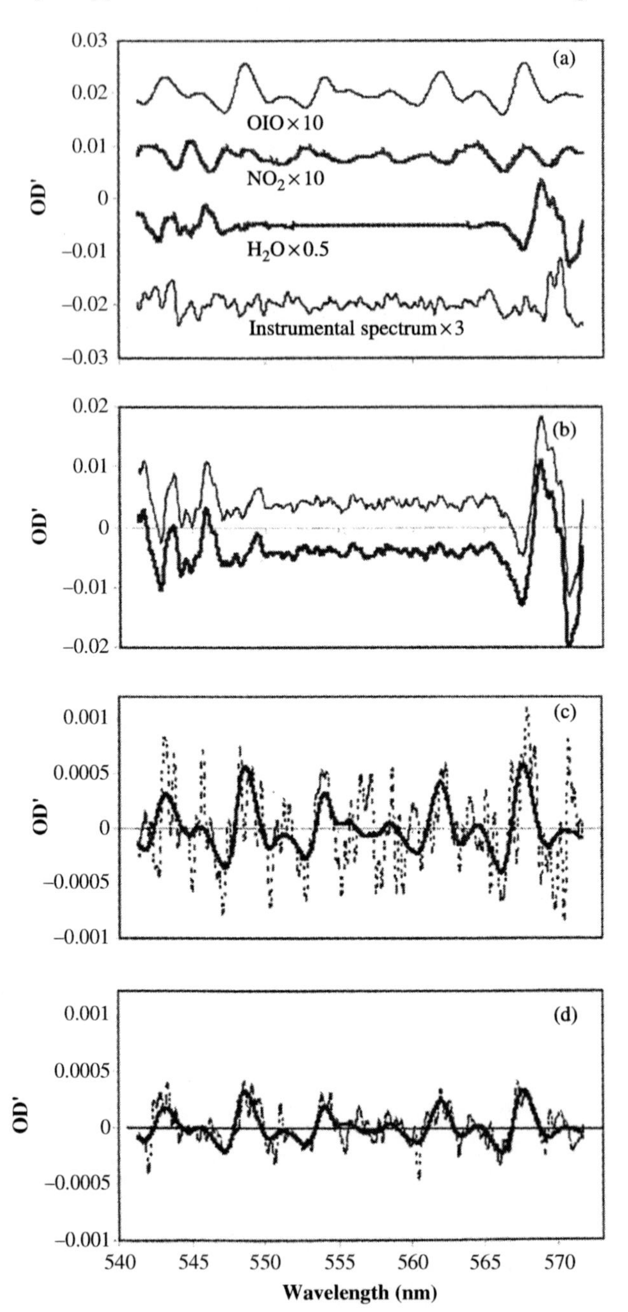

Fig. 10.33. OIO analysis from an atmospheric spectrum. *Panel* (**a**) shows the reference spectra used in the analysis; *Panel* (**b**) depicts the atmospheric spectrum containing all the absorbers. In *panel* (**c**) the absorptions of NO_2 and H2O have been removed. The OIO absorptions in this spectrum correspond to a mixing ratio of 1.8 ± 0.3 ppt. *Panel* (**d**) shows the residual of the analysis (from Allan et al., 2001, Copyright by American Geophysical Union (AGU), reproduced by permission of AGU)

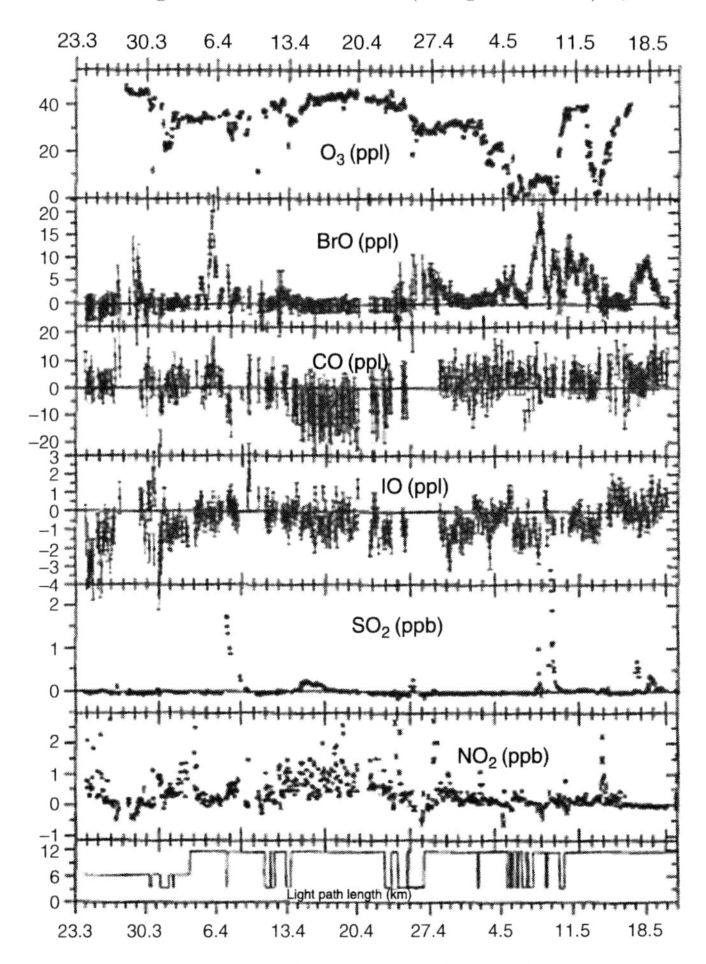

Fig. 10.34. DOAS measurements during Spring 1996 in Ny Alesung, Spitsbergen. The anti-correlation between BrO and O_3 can clearly be seen. Errors are 1σ uncertainties (adapted from Tuckermann 1996)

10.3 Investigation in Photoreactors (Smog Chambers) by DOAS

Since its development in the late-1970s DOAS has also been used in aerosol and smog chambers. To our knowledge the first measurements were made in Riverside in 1979 to investigate the formation of nitrous acid.

More recently, comprehensive DOAS measurements were performed in the EUROpean PHOto REactor (EUPHORE) in Valencia, Spain (Becker, 1996). The set-up is shown in Fig. 10.37. In particular, kinetic and mechanistic studies of the degradation of monocyclic aromatic species were followed by DOAS (Volkamer, 2001; Volkamer et al., 2001, 2002). A sample time series

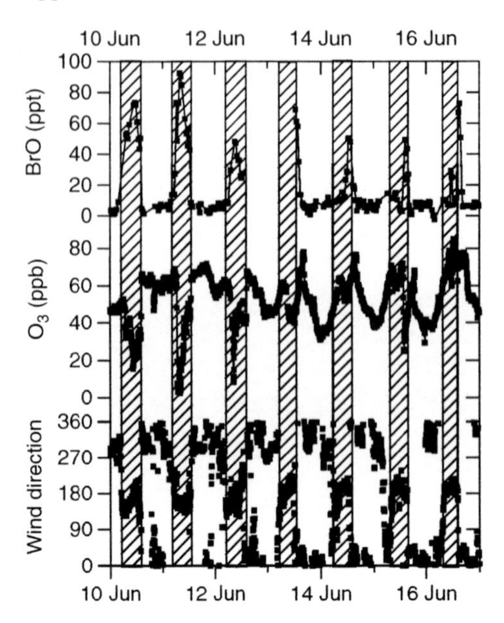

Fig. 10.35. DOAS measurements at the Dead Sea, Israel. The regular pattern in O_3 and BrO is caused by the changing wind direction and the chemical destruction of ozone (from Hebestreit et al., 1999)

of photochemical p-xylene degradation and corresponding formation of the degradation products p-tolualdehyde, 2,5-dimethylphenol, and glyoxal – all measured by DOAS – is shown in Fig. 10.38.

10.4 Validation of Active DOAS

An important aspect for all measurement techniques is the determination of the observational accuracy. In particular, for measurements in the open atmosphere only one method is available to determine the accuracy of a measurement, the comparison of different analytical techniques. The statistical approach, i.e. repeating a measurement several times, is not available due to the ever changing conditions in the atmosphere.

As described in Chap. 6, DOAS is an absolute analytical method. Calibrations are not needed and the accuracy is determined by the accuracy of the absorption cross-sections used in the analysis of the DOAS spectra. However, as in all analytical techniques, the possibility of unknown systematic errors always has to be considered. An intercomparison between DOAS and other techniques is thus a valuable exercise. Often intercomparisons between DOAS and other techniques are used for the benefit of the other techniques which may have calibration problems, for example OH-LIF systems.

Shown here are a number of examples for intercomparisons for the most important trace gases. Measurements of O_3 and SO_2 were made by a long-path

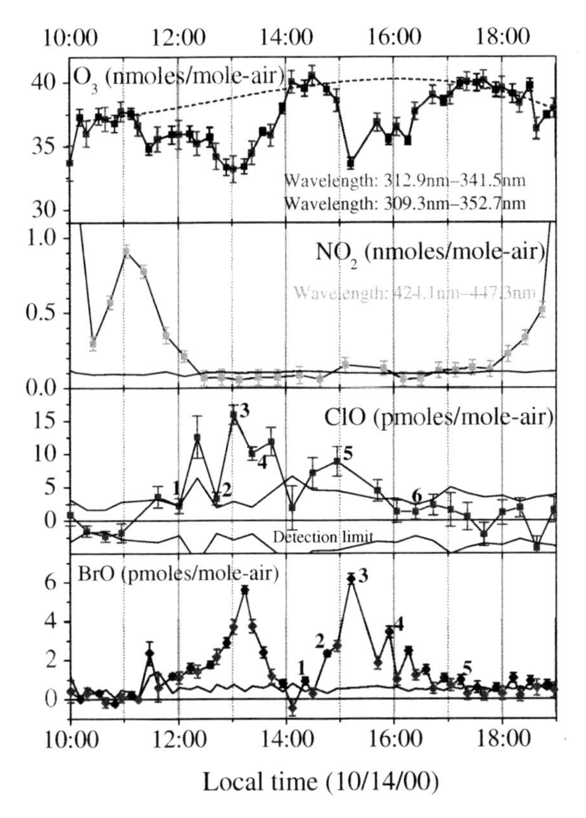

Fig. 10.36. Mixing ratios of O_3, NO_2, BrO, and ClO measured at ground level on October 14, 2000. The different symbols in the ozone and BrO traces indicate the different grating position used in the measurements. The *solid brown lines* show the detection limit determined for each individual spectrum (from Stutz et al., 2002, Copyright by American Geophysical Union (AGU), reproduced by permission of AGU)

DOAS instrument, a UV ozone monitor, and a commercial SO_2 monitor at the University of East Anglia Atmospheric Observatory at Weybourne, England (Stutz, 1996). Weybourne is located at the East coast of England, in an area with few pollution sources. The observed air masses were thus fairly well mixed, facilitating the comparison of the in situ technique with the DOAS data which were averaged over a 3.5-km-long light path. Figure 10.39 shows the excellent agreement between the methods. The difference between the DOAS and the in situ O_3 measurement is $\sim$ 5%, which is in the range of the accuracy of both the methods. While the slope of the SO_2 correlation is 1.01, the data is considerably more scattered than that of O_3. This is mainly caused by localised SO_2 sources close to the observatory, for example ships cruising off the coast. In both cases, small offsets were found. The reason for these offsets is unclear.

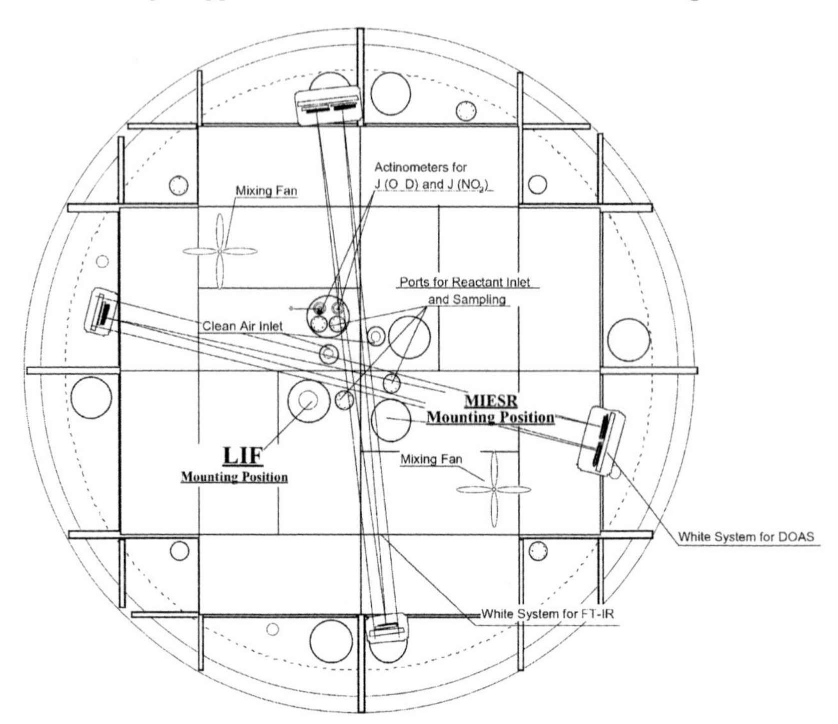

Fig. 10.37. Schematic of the DOAS multi-reflection system (White system) in the EUPHORE chamber in Valencia, Spain (from Volkamer et al., 2002, reproduced by permission of the PCCP Owner Societies)

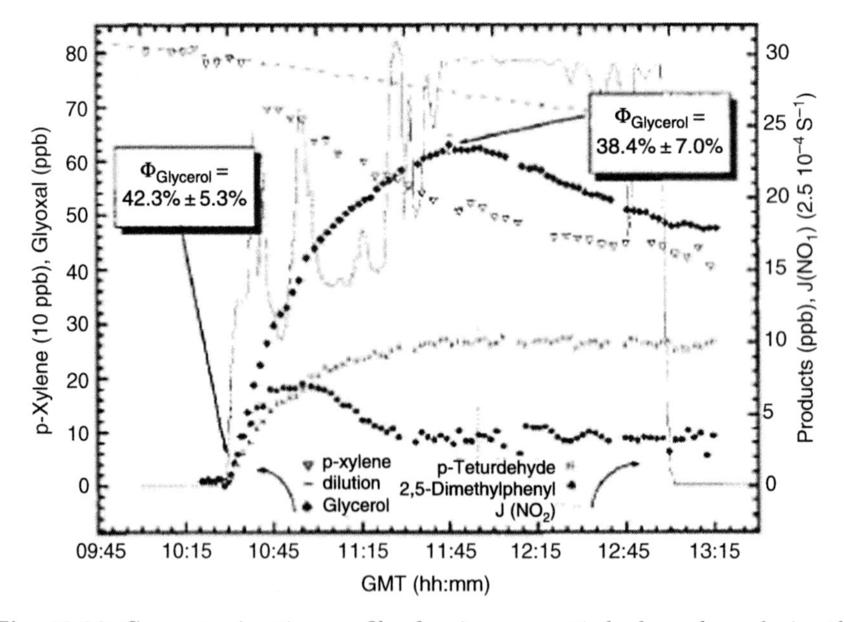

Fig. 10.38. Concentration time profile of various aromatic hydrocarbons during the oxidation of p-xylene (from Volkamer et al., 2002, reproduced by permission of the PCCP Owner Societies)

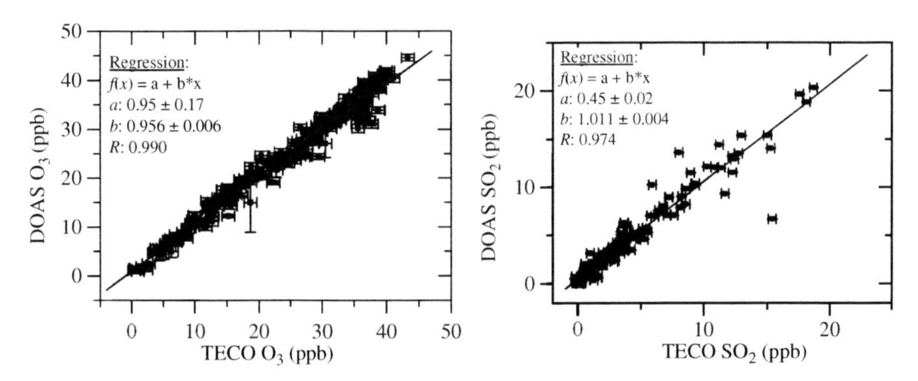

Fig. 10.39. Intercomparison between in situ O_3 and SO_2 instrument and a long-path DOAS system in Weybourne, England (from Stutz, 1996)

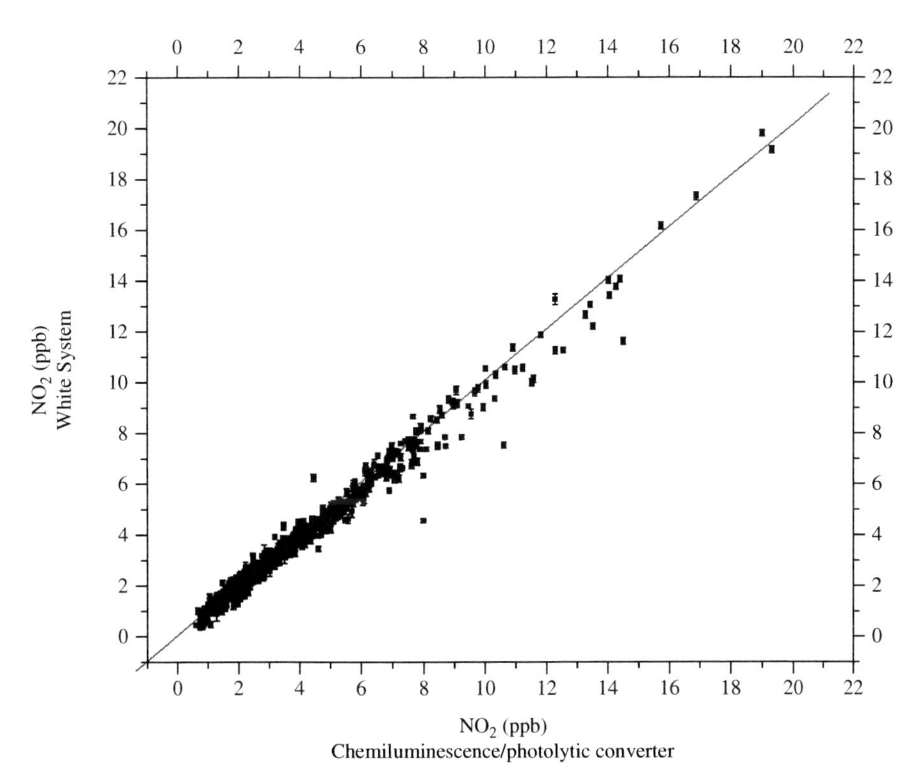

Fig. 10.40. Intercomparison of a multi-reflection DOAS instrument and an in situ NO_2 monitor. Both instruments were set up side-by-side at an altitude of 8 m above the ground on a meadow 50 km southeast of Berlin, Germany, during the BERLIOZ experiment (from Alicke et al., 2003, Copyright by American Geophysical Union (AGU), reproduced by permission of AGU)

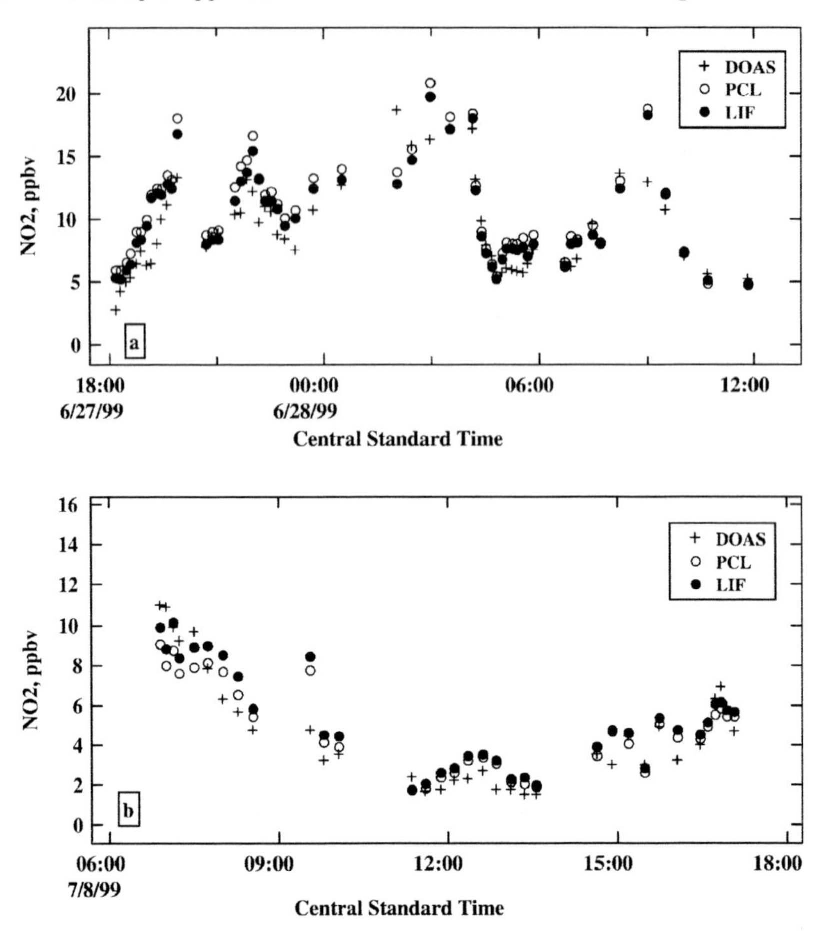

Fig. 10.41. Comparison of three NO_2 measurement methods: DOAS, Photolytic Converter/chemoluminescence, PCL, and Laser Induced Fluorescence, LIF (from Thornton et al. 2000); during the Southern Oxidant Study (SOS) 1999, in Nashville, TN, USA. The measurement was performed in a shallow river valley. The DOAS light beam crossed the entire extent (vertical and horizontal) of the valley, while the in situ instruments sampled 10 m above the ground (from Thornton et al., 2003, Copyright by American Geophysical Union (AGU), reproduced by permission of AGU)

Figure 10.40 shows the intercomparison of a multi-reflection DOAS system with 15m base path length and a NO_2 in situ instrument based on the photolytic conversion of NO_2 to NO followed by the detection of NO through chemoluminescence (Alicke et al., 2003). This intercomparison was performed during the BERLIOZ campaign in 1998, ~ 50 km southeast of the city centre of Berlin, Germany. A linear regression weighted by the error of the DOAS system yields a slope of 1.006 ± 0.005 (White system against chemoluminescence system). The

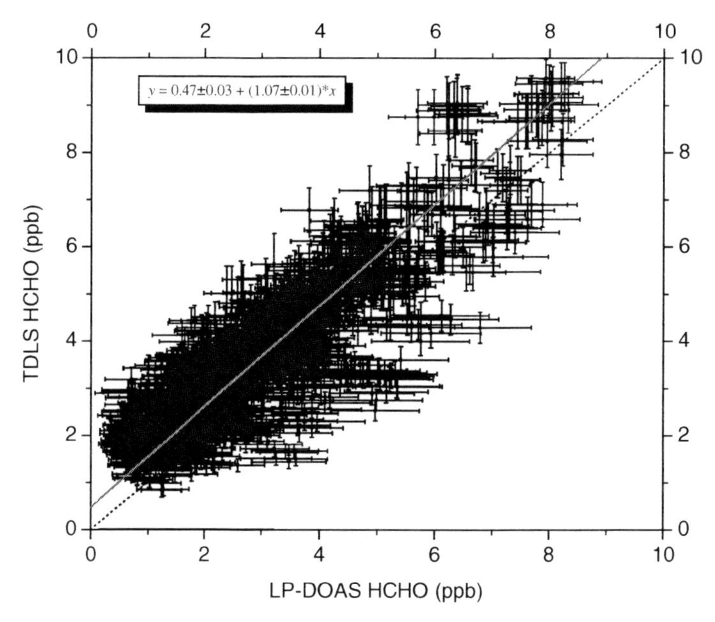

Fig. 10.42. Intercomparison of simultaneous formaldehyde measurements by a Tuneable Diode Laser Spectrometer (TDLS) in situ instrument and a long-path DOAS system during the Southern Oxidant Study (SOS), 1999 in Nashville, TN, USA. Shown is the entire dataset during 4 weeks of this experiment. (TDLS HCHO data courtesy of A. Fried, NCAR)

intercept of (36 ± 19) ppt is statistically insignificant. The excellent agreement between the two methods shows the accuracy of both methods.

Another intercomparison between three different methods is shown in Fig. 10.41 (Thornton et al., 2003). Here, a long-path DOAS instrument (path-length 2×1.35 km) was compared with two co-located point measurements by a Photolytic Converter/chemoluminescence instrument (PCL), and a laser induced fluorescence (LIF) system. The comparison shows that, in many situations, the agreement between the two in-situ instruments is better than with the DOAS instrument. This apparent disagreement is caused by the spatial averaging of the DOAS system. While an in situ instrument observes only a small air volume, the DOAS data represent an average over an extended air mass.

Under certain meteorological conditions, this can lead to a disagreement between DOAS and other methods. Whenever data from a long-path DOAS instrument is used, this averaging has to be considered. An example for this effect can also be observed in Fig. 10.16, where the spatial inhomogeneity of certain trace gases is illustrated.

During the same field experiment, intercomparison between a TDLS HCHO instrument (Fried A, personal communication) and the long-path DOAS was performed. While the statistical errors of both the methods were $\sim$ 0.2–0.5 ppb, Fig. 10.42 shows an excellent agreement between the two

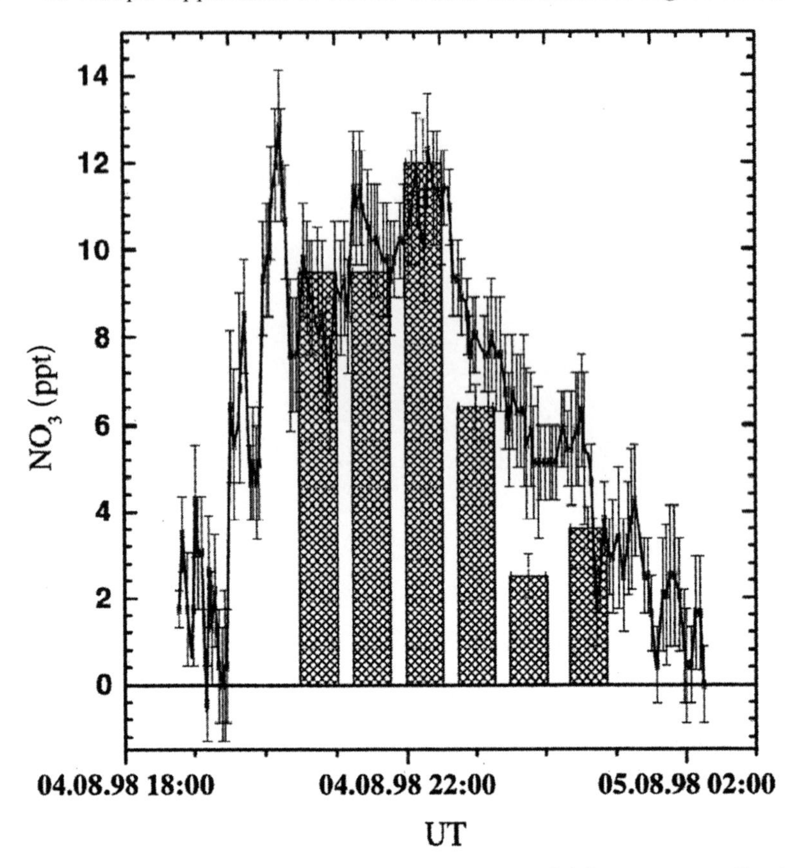

Fig. 10.43. Observation performed by a long-path DOAS instrument (*line*) and a matrix isolation electron spin resonance method (*bars*) during the BERLIOZ experiment, near Berlin, Germany (from Geyer et al., 1999, Copyright by American Geophysical Union (AGU), reproduced by permission of AGU)

methods. The slope of 1.07 ± 0.01 is within the range of the accuracy of both instruments.

The various intercomparisons show the accuracy of DOAS when compared with other methods. In general, the agreement is within the accuracies determined by the different absorption cross-sections. However, spatial inhomogenities can have a considerable influence on the quality of the intercomparisons.

While many intercomparisons for pollutants have been made, few examples exist for radical species. Geyer et al. (1999), showed a comparison between a long-path DOAS instrument and a matrix isolation electron spin resonance (MIESR) method. These measurements were performed during the BERLIOZ experiment in 1998. The authors also included data from an older field experiment in Deuselbach, Germany in 1983. The comparison of the

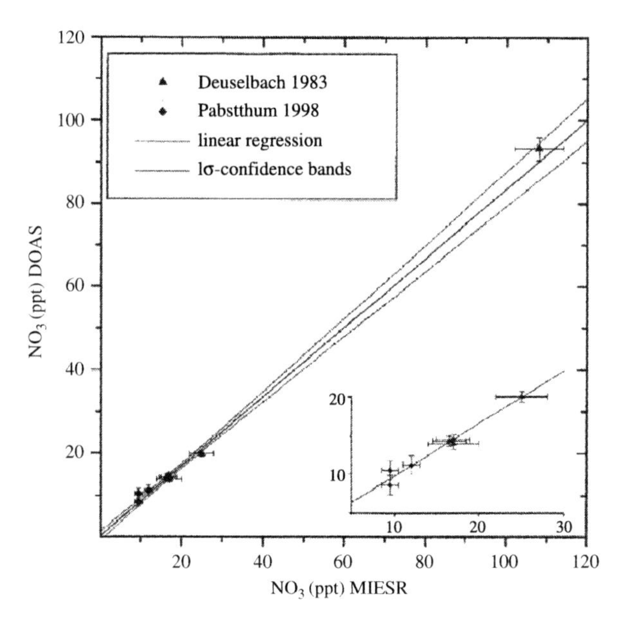

Fig. 10.44. Correlation between DOAS and MIESR measurements from two field experiments. Data marked with *triangles* were measured in Deuselbach in May/June 1983, while data denoted by *diamonds* were obtained during BERLIOZ in summer 1998 (from Geyer et al., 1999, Copyright by American Geophysical Union (AGU), reproduced by permission of AGU)

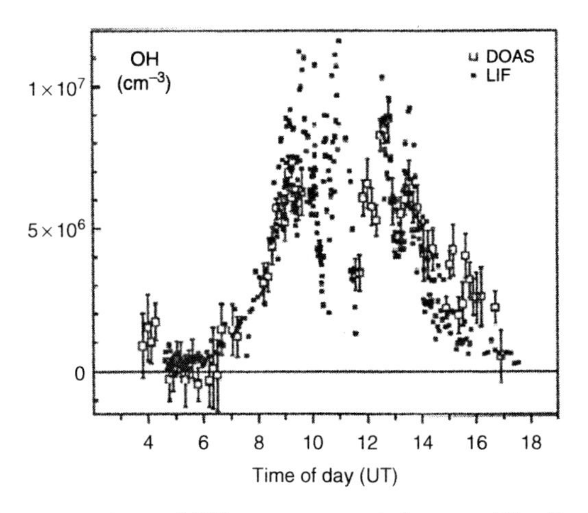

Fig. 10.45. Intercomparison of OH measurements by a multi-reflection DOAS system and a laser-induced fluorescence instrument (from Brauers et al., 1996, Copyright by American Geophysical Union (AGU), reproduced by permission of AGU)

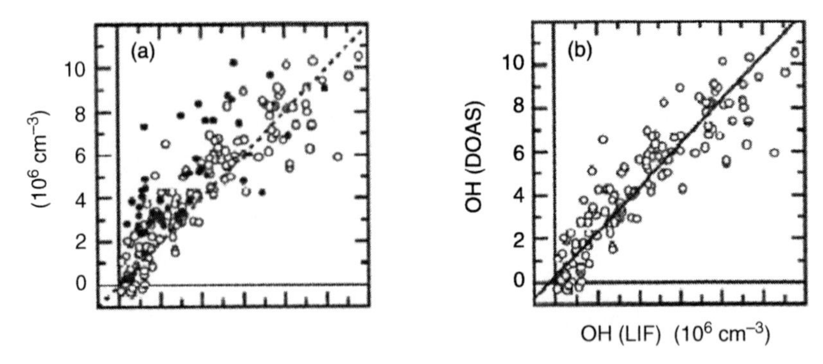

Fig. 10.46. Correlation of DOAS and laser-induced fluorescence data. *Panel* (a) shows the entire dataset for 11 days in August 1994, while *Panel* (b) presents a reduced dataset, where data from a windsector from 285–20 was excluded. The correlation coefficient was 0.9. The *solid line* represent the weighted linear fit given by $[OH]_{DOAS} = (1.01 \pm 0.04) \times [OH]_{LIF} + (0.28 \pm 0.15)$ (from Brauers et al., 1996, Copyright by American Geophysical Union (AGU), reproduced by permission of AGU)

measured data (Fig. 10.43) shows an excellent agreement, in particular considering that the DOAS are measured over a 3.5-km-long path while the MIESR data were sampled at one point. In addition, the MIESR data are averaged over 1 h, while the DOAS time resolution was ~ 5 min.

The correlation plot of the BERLIOZ and Deuselbach data also illustrates the excellent agreement between the two methods (Fig. 10.44). It is remarkable

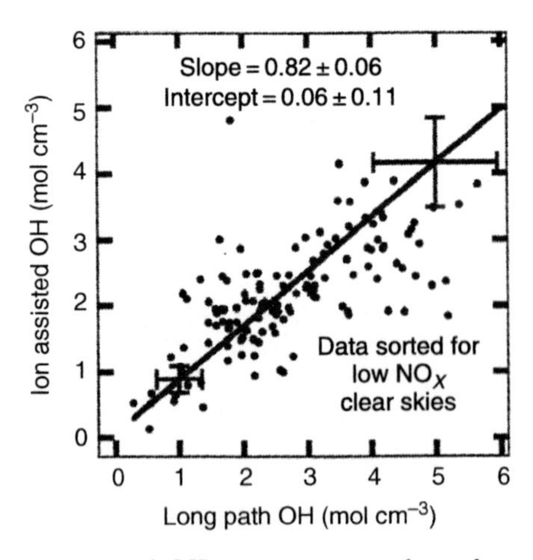

Fig. 10.47. Intercomparison of OH measurements by a long-path DOAS system and a chemical ionisation mass spectrometer in the mountains near Boulder, CO (Mount et al., 1997). Only data from clear days with NO_X levels below 500 ppt are shown (from Mount et al., 1997, Copyright by American Geophysical Union (AGU), reproduced by permission of AGU)

that this agreement is achieved using two datasets that were acquired 15 years apart.

The study by Geyer et al. (1999) leaves little doubt about the accuracy of both DOAS and MIESR data.

One of the most challenging measurements is that of the OH radical. Because of its high reactivity and low concentrations, atmospheric measurements typically have large uncertainties. It is thus essential to compare different analytical methods of this elusive radical to assess how well we can determine atmospheric OH concentration. Brauers et al. (1996) showed one of the first intercomparisons between a DOAS and a LIF instrument. The DOAS instrument was based on a multi-reflection system with 20 m base path length (see also above). Figure 10.45 shows the comparison between the data of both instruments during a diurnal cycle. The agreement is quite good, as can also be seen in the correlation of data over an 11-day period (Fig. 10.46).

Mount et al. (1997), showed a similar OH comparison between a long-path DOAS instrument and a chemical ionisation mass spectrometer, confirming the accuracy of the measurements (Fig. 10.47).

11

Sample Application of 'Passive' DOAS

For almost a century, atmospheric trace gas abundances have been measured by their absorption of sunlight. There are several possible measurement geometries, which have specific applications, advantages, and drawbacks:

- Direct sunlight (or moon and starlight)
- Scattered sunlight in zenith view geometry (ZSL-DOAS)
- Scattered sunlight in 'off-axis' view geometry
- Scattered sunlight in 'multi-axis' view geometry (MAX-DOAS)

In all passive DOAS applications, determination of the (effective) length of the light path is more difficult than for active DOAS (see Chap. 10). The methods used to overcome this challenge through the calculation of air mass factors (AMF) were introduced in Chap. 9.

Direct sunlight is (during daylight hours) readily available and allows a relatively simple calculation of the optical path (see Chap. 9). The practical realization of this approach requires a manual or automatic mechanism to point the instrument to the sun (or moon or star). A famous application of direct light spectroscopy is the determination of the O_3 total column density, which has been routinely performed at many stations worldwide by UV absorption spectroscopy since the technique was introduced by S. Dobson during the 1920s (Dobson and Harrison, 1926; see Chap. 6). Clearly, all sunlight reaching the surface of earth has to traverse the entire atmosphere, picking up the spectral signature of its constituents. Thus, it is straightforward to determine the total column density of a trace gas from direct light measurements. However, there is little or no information about the altitude distribution of the species. The nighttime composition of the atmosphere can be probed by using direct moonlight or starlight. Nevertheless, there has been surprisingly little use of direct sunlight spectroscopy in the UV or visible range for the determination of atmospheric species other than ozone (some recent measurements are described in Sect. 11.1).

Determination of atmospheric trace gases by **scattered** sunlight spectroscopy has been used to probe the atmosphere for three quarters of a century.

U. Platt and J. Stutz, *Sample Application of 'Passive' DOAS*. In: U. Platt and J. Stutz, Differential Optical Absorption Spectroscopy, Physics of Earth and Space Environments, pp. 429–494 (2008)
DOI 10.1007/978-3-540-75776-4_11

Initially most of the studies concentrated on stratospheric species, in particular ozone (see Chap. 9), almost invariably using the zenith view geometry. The influence of tropospheric gases and aerosol was seen as a nuisance, merely reducing the accuracy of the measurements. To determination of stratospheric composition by passive, ground-based instrumentation makes the best possible use of the available technology. This might have been the reason why the analysis of tropospheric species by passive differential absorption spectroscopy with other geometries such as MAX-DOAS came only recently into widespread use. During the last few years, a multitude of innovative techniques and observation geometries for probing the atmosphere close to the instrument emerged. These include the aforementioned MAX-DOAS approach (see Sects. 11.3.2 through 11.3.3), as well as also plume scanning schemes (see Sect. 11.3.8), imaging DOAS (IDOAS; see Sect. 11.3.9), and topographic target DOAS.

Another area of rapid development is the observation of atmospheric composition from space. While this technology started in the 1960s with the mapping of the stratospheric ozone column (TOMS), the recent decade has seen enormous development, marked by several pioneering instruments, which record complete spectra at sufficient spectral resolution ($\Delta\lambda < 1$nm) to allow DOAS retrieval of trace gas columns (e.g. Global Ozone Monitoring Experiment, GOME; Scanning Imaging Absorption Spectrometer for Atmospheric Cartography, SCIAMACHY; Ozone Monitoring Instrument, OMI; GOME-2; and Optical Spectrograph and Infrared Imager System, OSIRIS).

11.1 Atmospheric Measurements by Direct Light Spectroscopy

Direct light spectroscopy using radiation from celestial bodies (sun, moon, or stars) allows the direct determination of total trace gas column densities. There are essentially three advantages over scattered light schemes. First, the extension of the light path in comparison with the vertical path, the AMF (see Chap. 9), is given by $1/\cos\vartheta$, with ϑ denoting the zenith angle of the celestial body. Secondly, spectroscopic analysis of direct sunlight benefits from the high intensity. Lastly, the ring effect is absent. In particular, moonlight or starlight is so weak that, with present technology, there is no alternative to spectroscopy with direct light. Nevertheless, moonlight or starlight allows the study of nighttime chemistry, for instance of molecules that are destroyed by sunlight, such as chlorine dioxide (OClO) or the nitrate radical (NO_3).

One disadvantage of direct light DOAS is that Fraunhofer signatures vary when imaging different parts of the sun, as discussed in Chap. 9. Moreover, some manual or automatic mechanism to point the instrument to the sun (or moon/star) is required, which can make direct light measurements from mobile carriers difficult. Nevertheless, direct light DOAS measurements from aircraft and balloon were reported as outlined in Sect. 11.1.2.

11.1.1 Ground-based Measurement of Atmospheric Species

Some of the most scientifically successful moonlight DOAS measurements were those of the OClO column density, as for example performed at the McMurdo Antarctic station in 1986 by Solomon et al. (1987b) (Figs. 11.1 and 11.2). A clear increase of OClO with increasing direct light AMF (dashed line) is observed. (Note that evening twilight measurements using scattered sunlight were performed at the same time, as shown in Fig. 11.12.

Direct moonlight measurements were also made in Kiruna, Sweden, by Wagner et al. (2002b) in order to obtain improved absorption cross-sections of the oxygen dimer $(O_2)_2$ or O_4. The investigated O_4 bands were found to show an increase of the peak absorption with decreasing temperature, ranging from $\approx$ 13%/100 K at 477.3 and 532.2 nm, $\approx$ 20% at 360.5 and 577.2 nm to $\approx$ 33% at 380.2 and 630.0 nm. Moreover, with the exception of the band at 380.2 nm, the O_4 absorption cross-sections were found to be somewhat larger than previous measurements (see Appendix B).

For the separation of stratospheric and tropospheric column densities of BrO, a combination of direct sunlight and ZSL-DOAS was employed in a

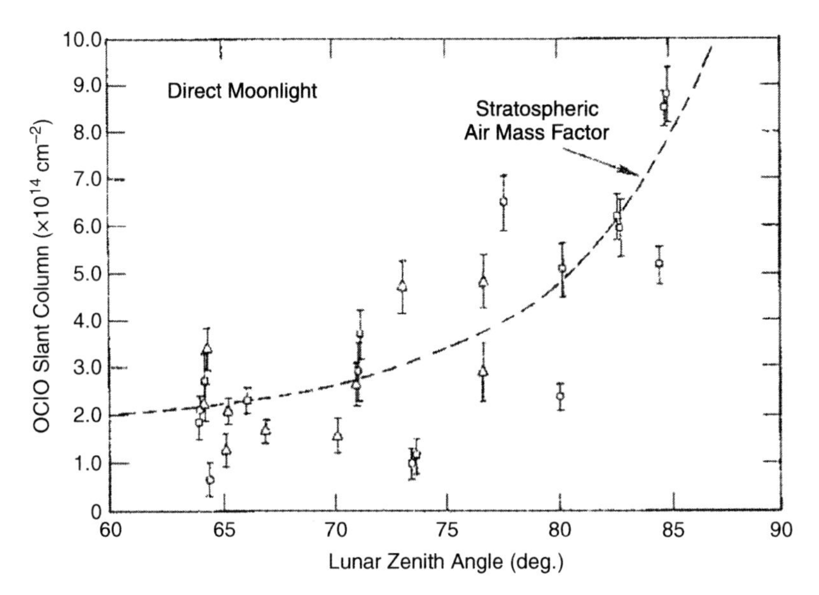

Fig. 11.1. Variation of the OClO SCD at McMurdo, Antarctica during the nights of 16–19 September 1986 as a function of lunar zenith angle (from Solomon et al., 1987b, Copyright by American Geophysical Union (AGU), reproduced by permission of AGU)

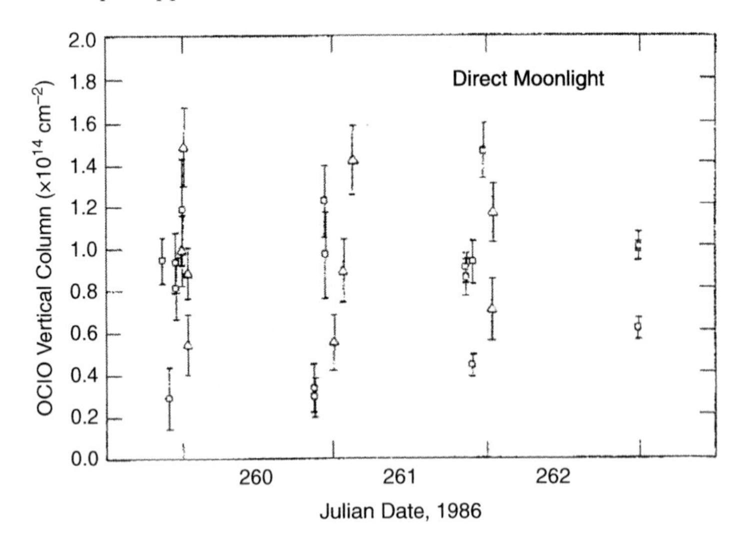

Fig. 11.2. Variation of the OClO VCDs at McMurdo, Antarctica, during the nights of 16–19 September 1986 (from Solomon et al., 1987b, Copyright by American Geophysical Union (AGU), reproduced by permission of AGU)

measurement campaign at the Arrival Heights Antarctic station during Austral spring of 2002 (Schofield et al., 2004, 2006).

11.1.2 Balloon- and Aircraft-borne Measurement of Stratospheric Species

Direct light balloon-borne DOAS spectroscopy allows the determination of vertical profiles of several trace gases in the atmosphere (Rigaud et al., 1983; Pommereau and Piquard, 1994a,b; Renard et al., 1996, 1997a,b, 1998, 2000a,b; Harder et al., 1998, 2000; Ferlemann et al., 1998, 2000; Fitzenberger et al., 2000; Pundt et al., 2002; Bösch et al., 2003; Weidner et al., 2005; Butz et al., 2006; Dorf et al., 2006). The precise imaging of the sun from an unmanned, airborne platform usually requires sophisticated hardware that relies on active orientation control of the balloon payload and mirror steering, as for instance described by Hawat et al. (1998). However, some simple, passive entrance optics are also in use, which rely on the radiation scattered from a diffuser plate close to the entrance optics (see below).

Basically there are two types of observation geometries, which both allow the retrieval of vertical trace gas profiles. Both geometries are shown in Fig. 11.3: (1) Observation during ascent (Pos. 1) and (2) Occultation during constant payload altitude (the 'float period', Pos. 2). The analysis of the geometry during ascent is straightforward, with the air mass factor given by $A(\vartheta) \approx 1/\cos\vartheta$, (with ϑ denoting the SZA). Small corrections to the AMF due to the earth's curvature are also necessary. The average trace gas

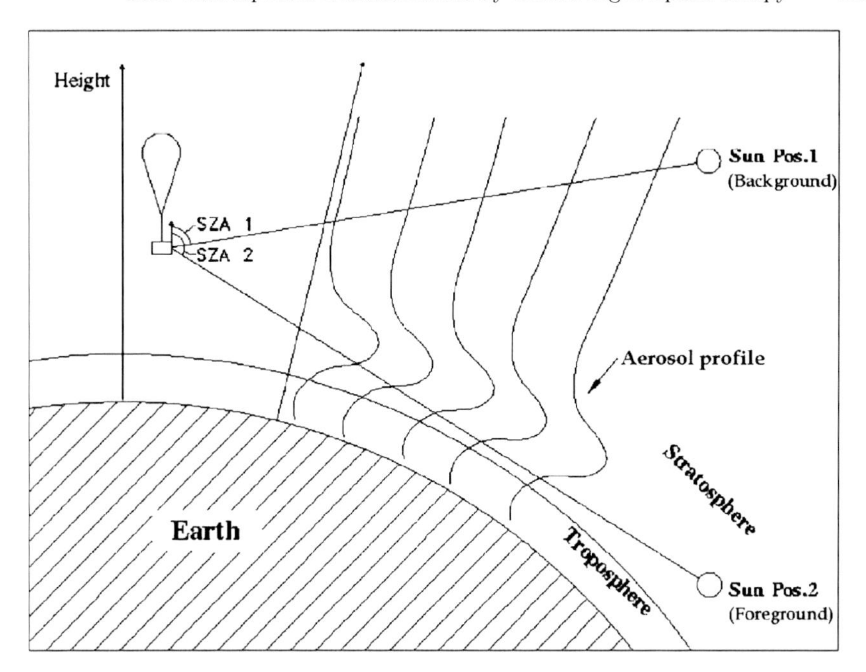

Fig. 11.3. Schematic of direct sunlight DOAS balloon measurements. During balloon ascent, typically at high solar elevations (Sun Pos. 1), differentiation of the consecutive series of trace gas column measurements yields a vertical profile. At constant balloon altitude (float level), another profile is derived during sunset or sunrise (from Osterkamp, 1997)

concentration between two altitude levels of the instrument z_{i} and $z_{\mathrm{i+1}}$ is then determined by taking the difference between the associated vertical column densities (VCDs) above the instrument:

$$\bar{c}\left(z_i, z_{i+1}\right) = V\left(z_{i+1}\right) - V\left(z_i\right) = \frac{S\left(z_{i+1}, \vartheta\right)}{A\left(\vartheta\right)} - \frac{S\left(z_i, \vartheta\right)}{A\left(\vartheta\right)},$$

where V and S denote the vertical and slant trace gas columns, respectively. Occultation experiments during the float period require more sophisticated de-convolution techniques, as described by Ferlemann et al. (1998).

Figure 11.4 shows an example of the trajectory of a balloon, and the 'line of sight' as a function of altitude (pressure) and solar zenith angle (SZA) for ascent and float period. The BrO slant column density is given by the integral of the BrO concentration (contour lines) times the path length of the beam traversing the respective atmospheric layer. Note that, in these pressure–SZA plots, the rays are not straight lines. Figure 11.5 compares several BrO profiles determined by in-situ techniques and balloon-borne direct sunlight DOAS measurements.

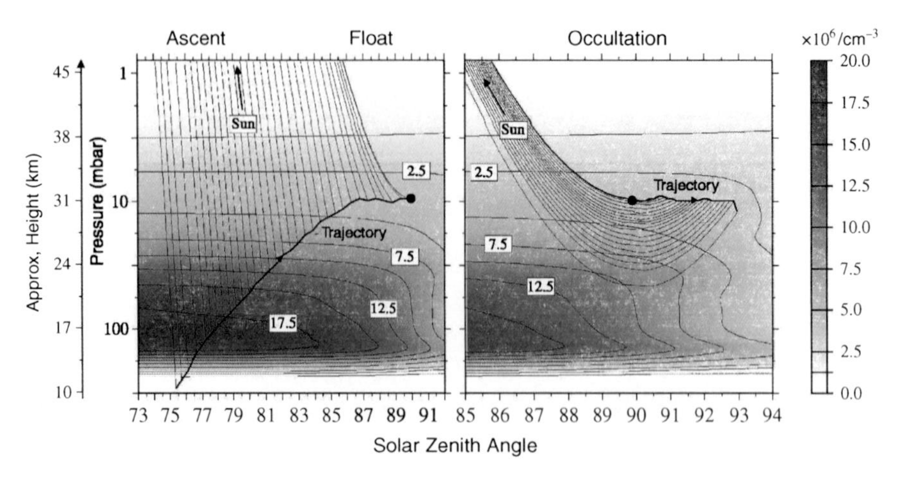

Fig. 11.4. Model simulations of stratospheric BrO as a function of SZA and pressure (altitude) at sunset for a balloon position at 40.5°N, 7.5°W on 23 November, 1996. Superimposed is the observation geometry of the balloon direct sunlight DOAS BrO measurements. **Left panel**: Balloon trajectory and 'line of sight' of each 10th spectrum. **Right panel**: Same for solar occultation (sunset). The amount of BrO along the line of sight is given by the integral of the local BrO concentration (*contour lines*) times the path element of each light beam traversing the atmospheric layer (figure from Harder et al., 2000, Copyright by American Geophysical Union (AGU), reproduced by permission of AGU)

An interesting example of a balloon borne direct sunlight design is the Systéme D'Analyse par Observations Zénithales (SAOZ) (Pommereau and Piquard, 1994a,b), which consists of a small, lightweight spectrometer (spectral range 270–620 nm) capable of determining NO_2 and O_3 absorption spectra. Passive optics accept direct sunlight and reflect it through a set of diffuser plates into the spectrometer. This setup is similar to the SAOZ-BrO shown in Fig. 11.6. While the setup is simple, and has the great advantage of not requiring moving parts or control electronics, its drawback is the large acceptance angle. A fraction of scattered light is also recorded along with the direct sunlight, which can complicate the evaluation of the data. On the other hand, the design makes the instrument very simple and lightweight requiring only small balloons. In fact, the instrument has been successfully flown more than a hundred times (Pundt et al., 2002). In addition, an improved version of the instrument – the 'SAOZ-BrO' (spectral range 350–420 nm) was developed, which had more light throughput and better sampling of the spectrum. SAOZ-BrO was added to the payload in a series of 16 balloon flights, which took place between late winter of 1997 and spring of 2000, mostly from an Arctic site (Kiruna, Sweden, 68°N), but also from mid-latitude sites (southern France, 43/44°N).

From the flights in the winters of 1998, 1999, and 2000, average profiles of O_3, NO_2, OClO, and BrO are shown in Fig. 11.7. During the first two

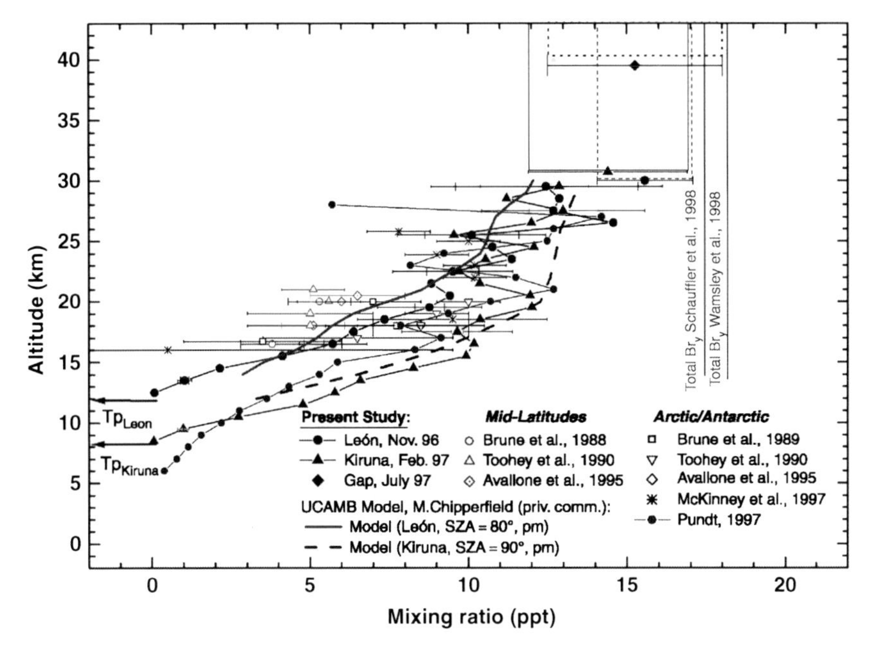

Fig. 11.5. Comparison of BrO profiles measured by several groups and techniques, including balloon-borne direct sunlight DOAS spectroscopy (from Harder et al. 1998, 2000, Copyright by American Geophysical Union (AGU), reproduced by permission of AGU)

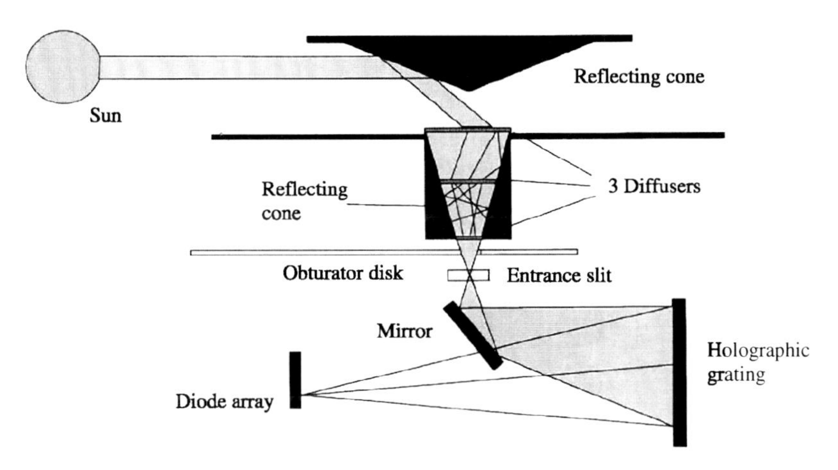

Fig. 11.6. Layout of the 'SAOZ-BrO' balloon spectrometer and entrance optics (Pundt, 1997). Radiation is accepted from a −5 to +15° elevation range and all azimuth directions by a downward pointing, reflecting cone and by a set of three diffuser plates mounted inside a second reflecting cone in front of the spectrometer (from Pundt et al., 2002, Copyright by American Geophysical Union (AGU), reproduced by permission of AGU)

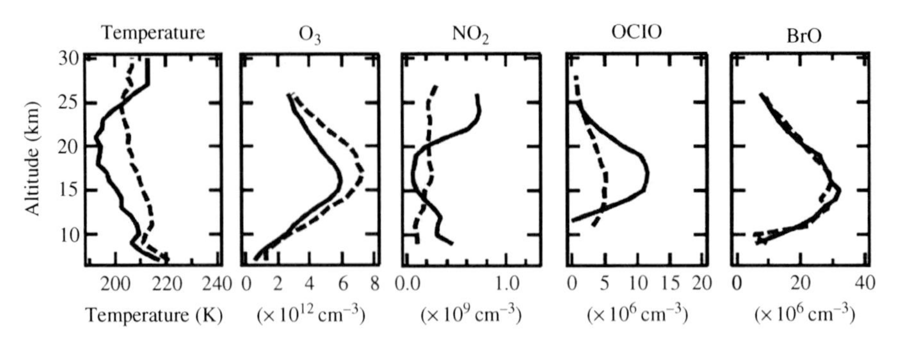

Fig. 11.7. Average vertical distribution of temperature, sunset ozone, NO_2 and OClO and daytime BrO observed by SAOZ and the SAOZ-BrO instruments flown on the same balloon in the cold, chlorine activated winter vortex of 2000 (*solid lines*) and in warmer conditions in 1998 and 1999 (*dashed lines*). (from Pundt et al., 2002, Copyright by American Geophysical Union (AGU), reproduced by permission of AGU)

winters, relatively 'warm' stratospheric conditions prevailed, whereas in 2000 a cold polar vortex showed signs of chlorine activation. While there appears to be little difference in BrO between the different states of the stratosphere, OClO is greatly enhanced in the cold vortex.

Aircraft-based direct light measurements also require sophisticated entrance optics, to image the light source into the spectrometer. To our knowledge, to date only direct moonlight (but no direct sunlight) measurements have been performed from aircrafts mostly to study nighttime levels of OClO, NO_2, and NO_3 in the Arctic (Solomon et al., 1988; Schiller et al., 1990; Wahner et al., 1990a,b; Pfeilsticker et al., 1997; Erle, 1999) and Antarctic (Wahner et al., 1989a) stratosphere. For instance, inside the winter polar vortices, nighttime OClO VCDs of $(1\text{–}2.5) \cdot 10^{14}\,\mathrm{cm}^{-2}$ were found, with the higher values prevailing in Antarctica.

11.2 Stratospheric Measurements by Ground-based Scattered Light DOAS

Since the 1930s, scattered light spectroscopy has been used to determine vertical profiles of stratospheric ozone (Götz et al., 1934). Plans to develop a fleet of supersonic aircraft in the 1970s led to an increased interest in stratospheric NO_2. However, it was only after the discovery of the ozone hole in 1985 that the list of observed stratospheric trace species expanded further to include OClO, BrO, O_4, NO_3, and HCHO. Scattered light DOAS measurements played an important role in the study of these species. Compared to direct light observations (see Sect. 11.1), scattered light observations can typically use much simpler entrance optics. However, scattered light spectroscopy requires the determination of the AMF, which is typically much more complicated than in the case of direct light measurements (see Chap. 9).

11.2.1 Determination of Stratospheric NO_2 and O_3 from the Ground

Ozone measurements by the Umkehr technique are not the topic of this book, so we shall only give a very brief account, and mention O_3 measurements only as part of other observations. The Umkehr technique inverts measurements of the intensity ratio of zenith scattered sunlight at a pair of UV wavelengths (e.g. 311.0 and 329.0 nm) recorded at a series of SZAs between about 70° and 90° (Götz et al., 1934; Mateer et al., 1996) to determine O_3 vertical profiles from the upper troposphere through the stratosphere. The measurements are usually made with Dobson spectrometers (Dobson and Harrison, 1926) operated in zenith scattered light mode. Sample results are given in Fig. 11.8

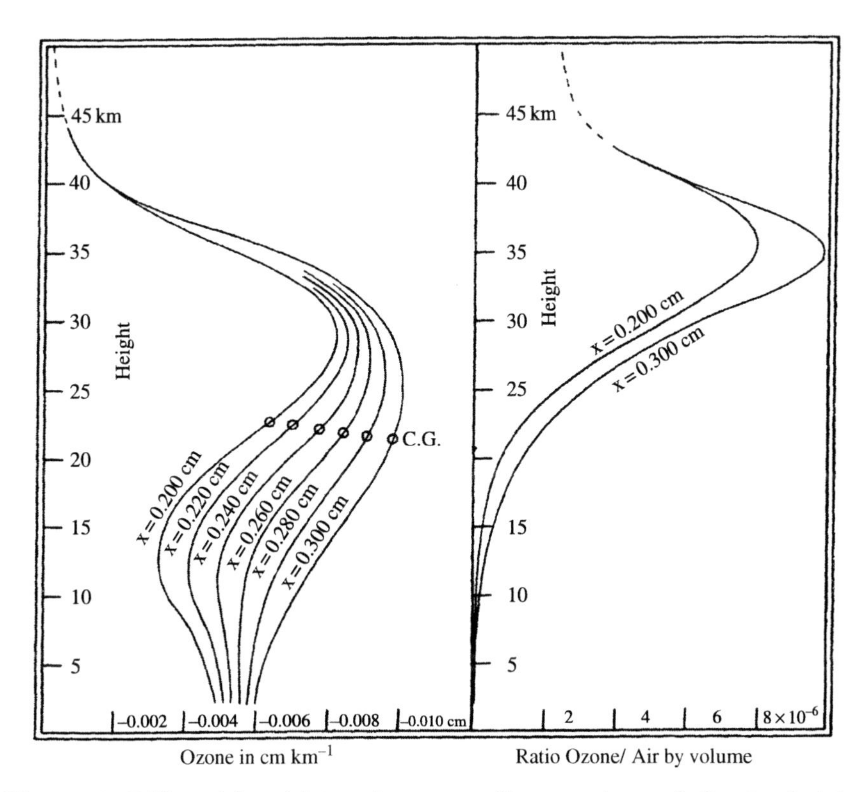

Fig. 11.8. Differential and integral ozone profiles over Arosa, Switzerland, determined in 1932 by the Umkehr technique. The data are derived from an inversion of measurements of the intensity ratio at a pair of UV wavelengths recorded in a series of SZAs from about 70° to 90° (figure from Götz et al., 1934)

showing differential and integral ozone profiles over Arosa, Switzerland (Götz et al., 1934).

The first species other than ozone to be measured by a scattered light DOAS-type technique was NO_2. The total column of stratospheric NO_2 was first determined from zenith scattered sunlight spectra by Brewer using the structured absorption of this molecule around 450 nm (Brewer et al., 1973). Brewer et al. (1973) also noted the increase of the NO_2 column density during the course of the day. Later, J. Noxon discovered the sharp decline of the wintertime NO_2 column at about 50°N in northern Canada (Fig. 11.9). The phenomenon, now referred to as 'Noxon Cliff' is due to conversion of most of the NO_x into nitric acid (via NO_3–N_2O_5) in the absence of sunlight (Noxon, 1975, 1979; Noxon et al., 1978, 1979).

In the following years, a large number of measurements of stratospheric O_3 and NO_2 by passive DOAS were reported. These measurements were mostly made from ground-based stations, but aircrafts, balloons, and ships were also used. These measurements are summarised in Table 11.1. Here, we present a small selection of ground-based observations of stratospheric NO_2 using zenith scattered sunlight.

A time series of stratospheric NO_2 columns measured by zenith scattered sunlight DOAS during morning and evening twilight at Lauder, New Zealand between 1981 and 1999, is shown in Fig. 11.10. The NO_2 columns at morning and evening twilight show a pronounced seasonal cycle, and the

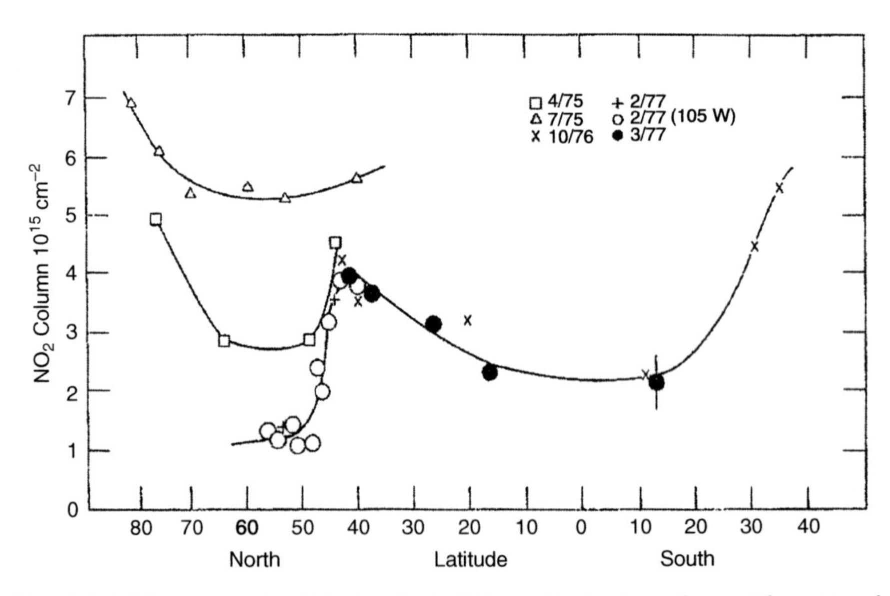

Fig. 11.9. Measurements of stratospheric NO_2 vertical column by zenith scattered sunlight during evening twilight at various latitudes (from Noxon, 1979, Copyright by American Geophysical Union (AGU), reproduced by permission of AGU)

Table 11.1. Overview and history of the different scattered light passive DOAS applications

Method	Measured quantity	No. of axes, techn.	References
COSPEC	NO_2, SO_2, I_2	1, S	Millan et al. (1969), Davies (1970), Hoff and Millan (1981), Stoiber and Jepsen (1973), and Hoff et al. (1992).
Zenith scattered light DOAS	Stratospheric NO_2, O_3, OClO, BrO, IO	1	Noxon (1975), Noxon et al. (1979), Harrison (1979), Syed and Harrison (1980), McMahon and Simmons (1980), McKenzie and Johnston (1982, 1983, 1984), Pommereau et al. (1982, 1984), Solomon et al. (1987a,b, 1989b, 1993b, 1994a,b), Mount et al. (1987), Pommereau and Goutail (1988a,b), Johnston and McKenzie (1989), Sanders et al. (1989, 1999), Wahner et al. (1990a), McKenzie et al. (1991), Kreher (1991), Johnston et al. (1992), Fiedler et al. (1993), Kioke et al. (1993, 1994, 1999), Elokhov and Gruzdev (1993, 1995), Pommereau and Piquard (1994b), Pfeilsticker and Platt (1994), Perner et al. (1994), Van Roozendael et al. (1994a,b,c, 1997), Kondo et al. (1994), Kreher et al. (1995, 1997, 1999), Hofmann et al. (1995), Gil et al. (1996, 2000), Senne et al. (1996), Vaughan et al. (1997), Pfeilsticker et al. (1997a,1999a,b), Slusser et al. (1997), Sarkissian et al. (1997), Goutail et al. (1999), Eisinger et al. (1997), Aliwell et al. (1997, 2002), De Maziere et al. (1998), Roscoe et al. (1999, 2001), Richter et al. (1999), Liley et al. (2000), Fish et al. (2000), Wittrock et al. (2000), Sinnhuber et al. (2002), Tørnkvist et al. (2002), Hendrick et al. (2004), and Schofield et al. (2004, 2006)
Zenith sky + off-axis DOAS	Stratospheric OClO	2	Sanders et al. (1993)
Off-axis DOAS	Stratospheric BrO profile	1	Arpaq et al. (1994)
Zenith scattered light DOAS	Tropospheric IO, BrO	1	Friess et al. (2001, 2003)

(continued)

Table 11.1. (continued)

Off-axis DOAS	Tropospheric BrO	1	Miller et al. (1997)
Sunrise off-axis DOAS + direct moonlight	NO_3 profiles	2, S	Weaver et al. (1996), Smith and Solomon (1990) and Smith et al. (1993)
Sunrise off-axis DOAS	Tropospheric NO_3 profiles	1	Kaiser (1997) and von Friedeburg et al. (2002)
Aircraft-DOAS	Tropospheric BrO	2	McElroy et al. (1999)
Aircraft zenith sky + off-axis DOAS	'Near in-situ' strato-spheric O_3	3	Petritoli et al. (2002)
AMAX-DOAS	Trace gas profiles	8+, M	Wagner et al. (2002a), Wang et al. (2003), and Heue et al. (2003)
MAX-DOAS	Tropospheric BrO profiles	4, S	Hönninger and Platt (2002)
MAX-DOAS	Tropospheric BrO profiles	4, S	Hönninger et al. (2003b)
MAX-DOAS	Trace gas profiles	2–4, M	Löwe et al. (2002), Oetjen (2002), and Heckel (2003)
MAX-DOAS	NO_2 plume	8, M	von Friedeburg (2003)
MAX-DOAS	BrO in the marine boundary layer	6, S/M	Leser et al. (2003) and Bossmeyer (2002)
MAX-DOAS	BrO and SO_2 fluxes from volcanoes	10–90, S	Bobrowski et al. (2003), Galle et al. (2003), McGonigle et al. (2005), Lee et al. (2005), and Oppenheimer et al. (2006)
MAX-DOAS	BrO emissions from a Salt Lake	4, S	Hönninger et al. (2003a)
MAX-DOAS	BrO and IO emissions from a Salt Lake	6, S	Zingler et al. (2005)

S: Scanning instrument; M: Multiple telescopes

increase of the NO_2 column during daytime is clearly seen. Distinct variations of the annual cycle are due to the combined effects of the solar cycle, the quasi-biennial oscillation QBO, and the El Nino southern oscillation ENSO. The effects of volcanic eruptions in 1982 and 1992 are also visible. In addition, a trend in the NO_2 column of +5% per decade was found. The morning/evening ratio has a clear seasonal cycle, but appears to be unaffected by the other periodic variables and volcanic events.

Figure 11.11 shows plots of the NO_2 and O_3 VCDs derived from ZSL-DOAS measurements during two cruises of the research vessel Polarstern in 1990 and 1993 from Bremerhaven, Germany to Ushuaia, Argentina (1990) (Kreher et al., 1995) and Cape Town, South Africa (1993) (Senne et al., 1996). Spectra recorded at 90° SZA were evaluated around 445 ± 7 and 493 ± 23 nm for NO_2 and O_3, respectively. The minimum around the equator in both species, as well as the increase of the NO_2 column during daylight hours, is clearly visible. In addition, the height of the NO_2 maximum is calculated (Fig. 11.11c).

11.2.2 Observation of Halogen Radicals in the Polar Stratosphere

During the 1990s, improved technology allowed the study of weaker atmospheric absorbers than O_3 and NO_2, in particular halogen oxide radicals. Of the halogen monoxide radicals FO, ClO, BrO, and IO, only the latter two have structured absorption features that are at sufficiently long wavelengths to be detected by scattered light absorption spectroscopy. The structured spectrum of ClO is centred around 290 nm, with the longest wavelength band near 308 nm. The low intensities in this wavelength range make it very difficult to detect CIO with ground-based or aircraft-based instruments, even at the relatively high levels reached under ozone hole conditions (see Sect 2.12). Stratospheric ClO might be detectable with future, improved instrumentation or in direct sunlight. However, some problems might arise from the overlapping structured absorption of ozone. Many observations of polar stratospheric (and partly tropospheric, see Sect. 11.3) IO (Pundt et al., 1998; Wittrock et al., 2000a; Frieß et al., 2001) and BrO (Solomon et al., 1989b, 1993a; Wahner et al., 1989, 1990b; Arpag et al., 1994; Fish et al., 1995; Kreher et al., 1997; Frieß, 2001; Frieß et al., 1999, 2004, 2005; Van Roozendael et al., 2002, 2003; Eisinger et al., 1997; Aliwell et al., 1997, 2002; Richter et al., 1999; Sinnhuber et al., 2002; Tørnkvist et al., 2002; Schofield et al., 2004, 2006) by ground-based instruments have been reported.

As outlined in Sect. (2.12), the symmetric OClO molecule is a good indicator of stratospheric chlorine activation. It is also much easier to measure than ClO, since its prominent absorption bands are around 360 nm. The detection of OClO in the Antarctic stratosphere during the ozone hole period in 1986 by Solomon et al. (1987b) provided important evidence for perturbed chlorine chemistry as the primary cause for the dramatic ozone loss (see also Chap. 2). Since then a series of further measurements of OClO by ground and aircraft,

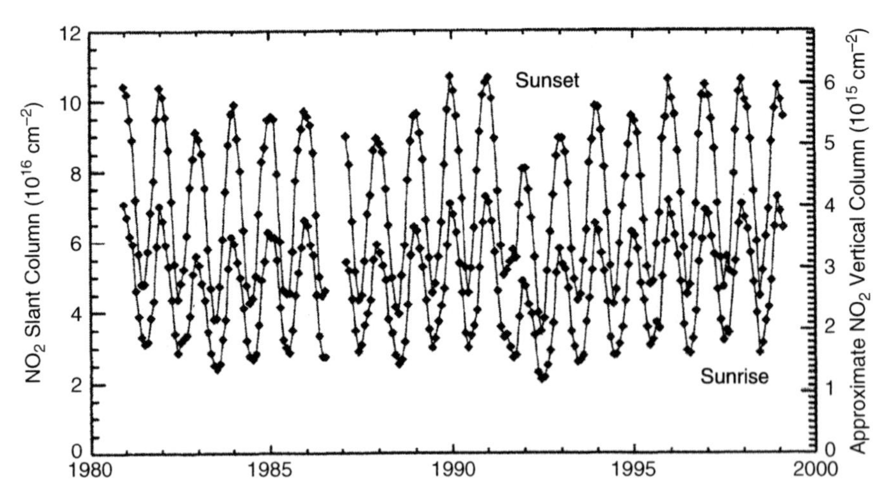

Fig. 11.10. Monthly means of stratospheric NO_2 slant columns (*left scale*) measured by zenith scattered sunlight DOAS during morning and evening twilight (labelled as 'sunrise' and 'sunset') at Lauder, New Zealand, between 1981 and 1999. Approximate vertical columns are indicated at the *right-hand scale*, assuming a fixed AMF of 17.5 (from Liley et al., 2000, Copyright by American Geophysical Union (AGU), reproduced by permission of AGU)

as well as satellite-based passive DOAS, have been made (Solomon et al., 1987b, 1988, 1989b,d, 1990, 1993a; Schiller et al., 1990; Perner et al., 1991, 1994; Sanders et al., 1993, 1996; Fiedler et al., 1993; Pommereau and Piquard, 1994b; Kreher et al., 1996; Gil et al., 1996; Renard et al., 1997a,b; Pfeilsticker et al., 1997a; Sanders et al., 1999; Tørnkvist et al., 2002). Figure 11.12 shows OClO column data from Solomon et al. (1987b) at the McMurdo Antarctic station by evening twilight measurements. Direct moonlight measurements were made at the same time (see Sect. 11.1.1, Figs. 11.1 and 11.2).

Some results encompassing simultaneous ozone, NO_2, BrO, and OClO time series recorded in Kiruna at the Arctic circle during spring of 1995 are shown in Fig. 11.13 (Otten, 1997; Otten et al., 1998). The measurements were made by ZSL-DOAS during morning and evening twilight, respectively. The OClO data show a clear correlation, with periods of high potential vorticity (PV) in the stratosphere and low stratospheric temperatures.

11.2.3 Halogen Radical Observation in the Mid-latitude Stratosphere

Ground-based measurements of stratospheric BrO have been made at a number of mid-latitude sites (Arpag et al., 1994; Aliwell et al., 1997, 2002; Fish et al., 1997; Eisinger et al., 1997; Richter et al., 1999; Sinnhuber et al., 2002; Van Roozendael et al., 2002). Figure 11.14 shows as an example the measured

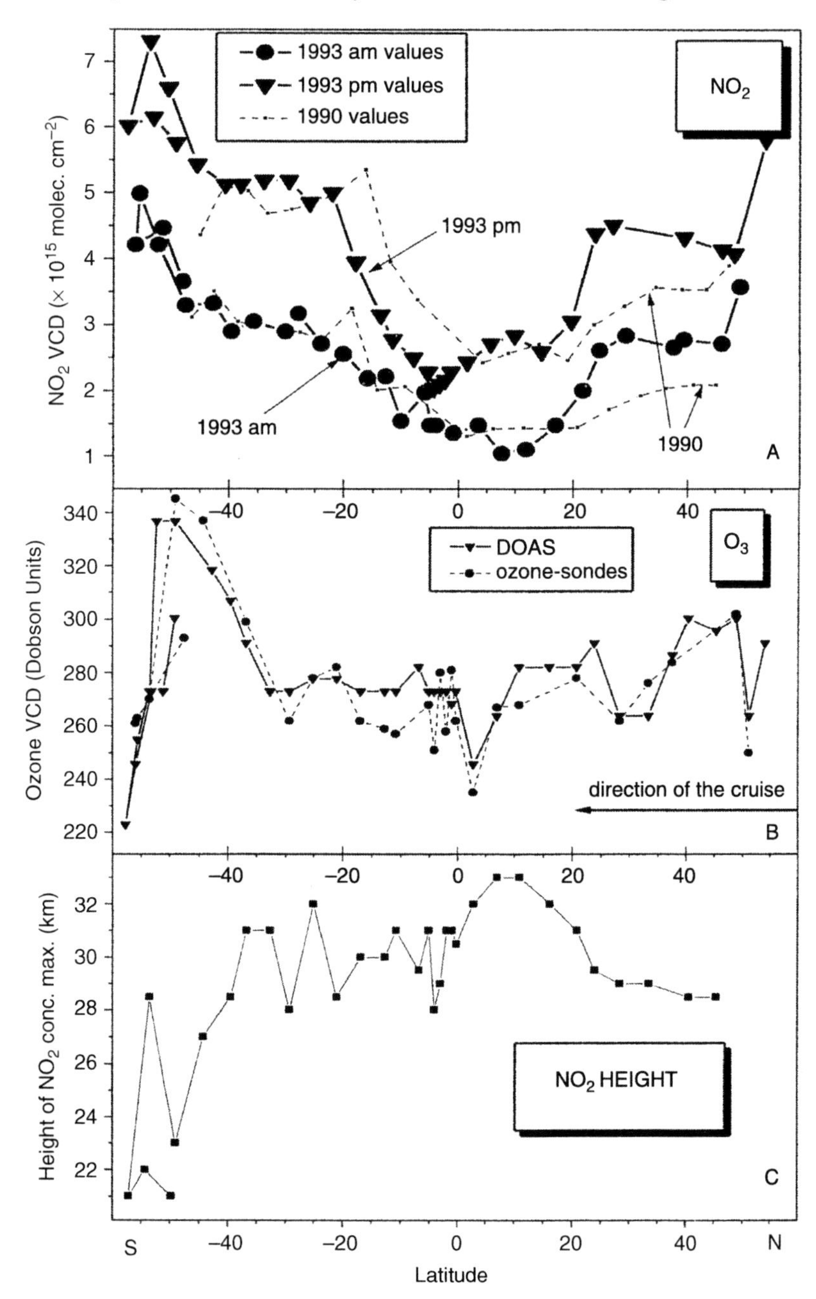

Fig. 11.11. Latitudinal cross-sections of NO_2 (**a**) and O_3 (**b**) as recorded by zenith scattered light DOAS during cruises of the research vessel Polarstern from Bremerhaven, Germany, to Ushuaia, Argentina (1990), and Cape Town, South Africa (1993). In addition, the calculated height of the NO_2 maximum is shown in (**c**) (from Senne et al., 1996, Copyright by American Geophysical Union (AGU), reproduced by permission of AGU)

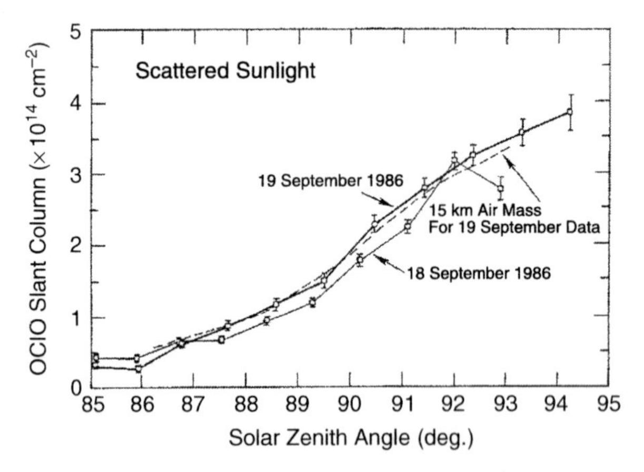

Fig. 11.12. Variation of the OClO SCDs at McMurdo, Antarctica during twilight of 18 and 19 September, 1986. *Dashed line*: calculated AMFs for 19 September (from Solomon et al., 1987b, Copyright by American Geophysical Union (AGU), reproduced by permission of AGU)

and modelled difference in the morning–evening slant column densities (SCDs) of BrO at mid-latitudes in the northern and southern hemisphere (Sinnhuber et al., 2002). BrO PM–AM differences show a distinct annual cycle, with a maximum in winter.

11.2.4 Observation of Stratospheric Trace Gas Profiles

Since the 1930s, scattered light spectroscopy has been used to determine vertical profiles of stratospheric ozone by the Umkehr technique (see Sect. 11.2; Götz et al. (1934)), rather than the total column density as described in Sect. 11.2.1. Despite significant improvements of the inversion algorithms since then (Mateer et al., 1996; Rodgers, 1990, 2000) surprisingly few attempts to derive vertical distributions of trace gases from ground-based zenith scattered light UV-vis spectroscopy have been reported (McKenzie et al., 1991; Preston et al., 1997; Denis et al., 2003; Schofield et al., 2004, 2006; Hendrick et al., 2004). However, the vertical resolution of all these measurements is on the order of the atmospheric scale height. Preston et al. (1997) and Denis et al. (2003) retrieved the NO_2 vertical distribution in the stratosphere from zenith sky observations using the optimal estimation method (Rodgers, 1990, 2000). In contrast, Schofield et al. (2004, 2006) applied the retrieval algorithm to a combination of ground-based zenith-sky and direct-sun measurements of BrO. Due to the light path geometry, zenith-sky and direct-sun observations are dominantly sensitive to the stratosphere and the troposphere, respectively. Therefore, combining both observation geometries in a single retrieval provides information on both stratospheric and tropospheric absorbers.

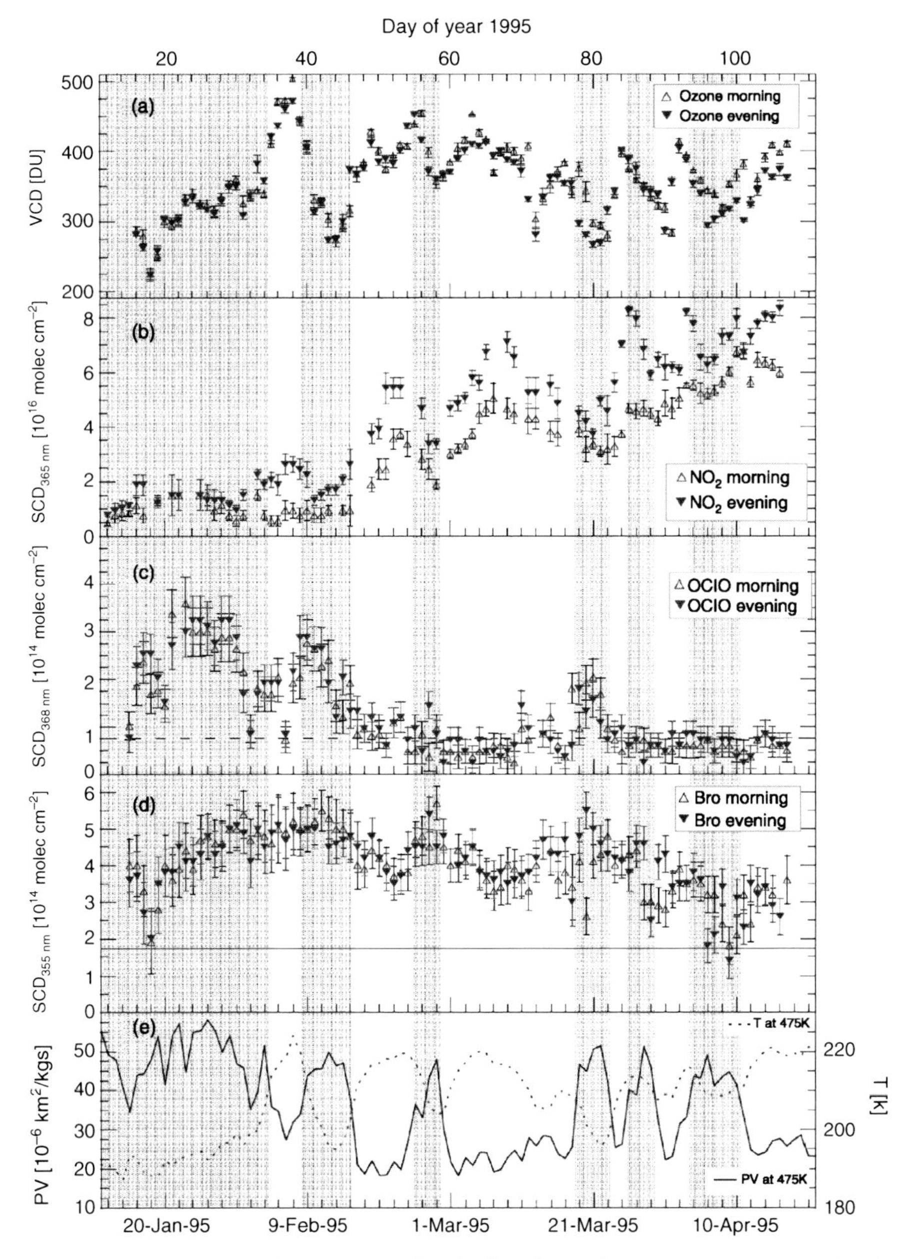

Fig. 11.13. Variation of the ozone VCD (86°–89° SZA) and NO_2, BrO, and OClO SCDs at Kiruna, Sweden during morning and evening twilight (90° SZA) January through April, 1995. *Bottom panel*: Temperature at 475 K isentropic surface (*dashed line*), potential vorticity, PV (ECMWF 1996) (from Otten, 1997)

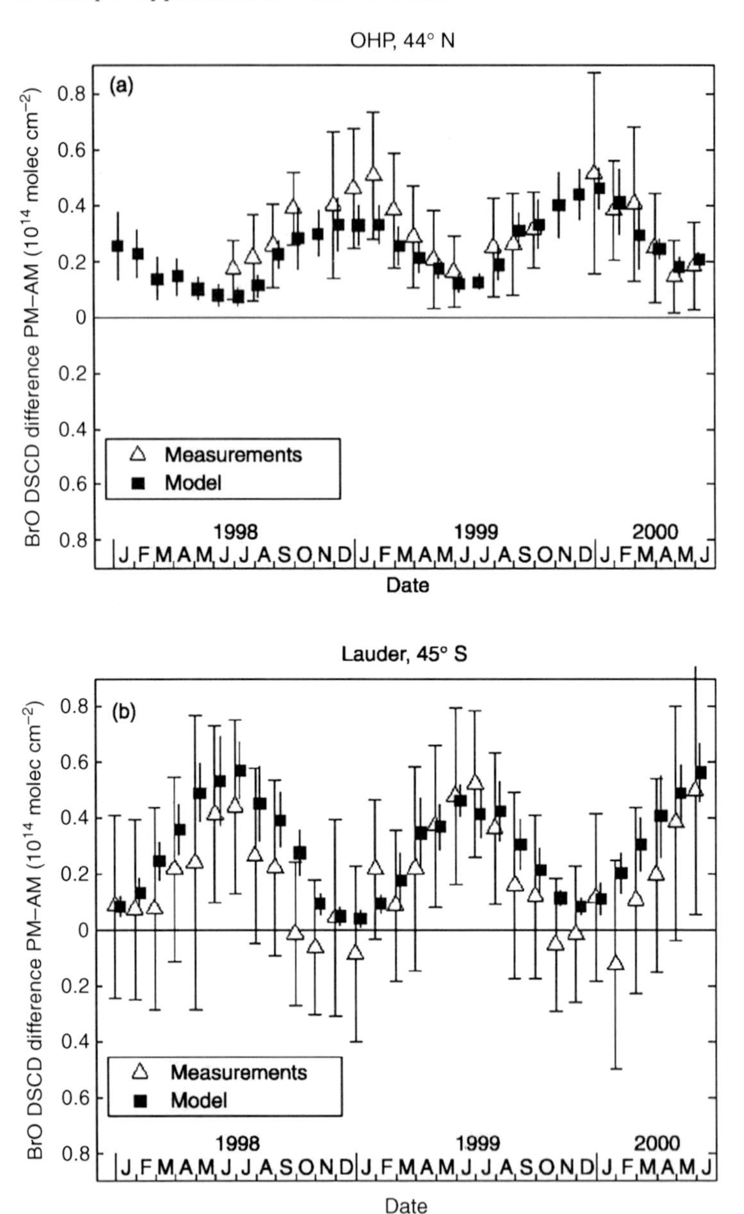

Fig. 11.14. Comparison of the measured and modelled difference between the evening and morning BrO DSCD for Observatoire de Haute-Provence (OHP, 44°N) and Lauder (45°S). The error bars represent monthly averages and their standard deviation (1σ) (from Sinnhuber et al., 2002, Copyright by American Geophysical Union (AGU), reproduced by permission of AGU)

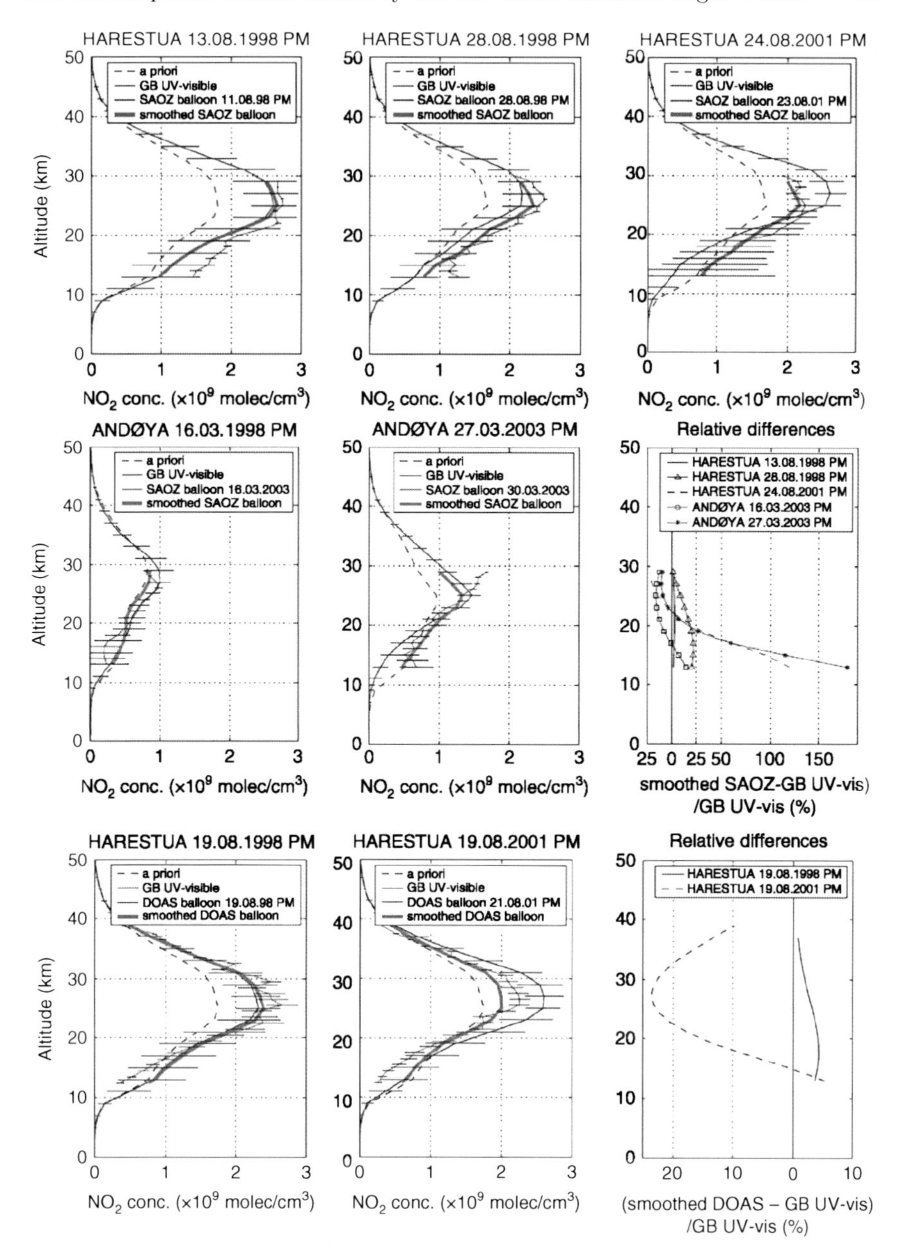

Fig. 11.15. Comparison between NO_2 profiles derived from ground-based UV-vis measurements at Harestua (sunset, summer conditions) and Andoya, Norway (sunset, late winter–early spring conditions) and SAOZ and DOAS balloon profiles (all balloons were launched from Kiruna, Sweden). The SAOZ/DOAS balloon profiles are smoothed by convolution with the ground-based UV-vis averaging kernels. The relative differences appear in the *right lower plot* (from Hendrick et al., 2004)

The study by Hendrick et al. (2004) presents a comprehensive comparison of NO_2 profiles retrieved from zenith scattered UV-vis DOAS measurements, with balloon and satellite observations (see Sects. 11.1.2 and 11.7) showing agreement at the $<25\%$ level, see Fig. 11.15.

All recent studies stressed the impact of photochemistry on the retrieved profile information. Many relevant trace species (e.g. NO_2 and BrO) display a strong diurnal variation, which considerably complicates the retrieval since the observed variation of the measurements with the SZA depends not only on the scattering geometry, but also on photochemistry (see e.g. Roscoe and Pyle, 1987). For instance, the concentration of NO_2 and BrO increases and decreases with increasing SZA, respectively.

11.3 Measurement of Tropospheric Species by Ground-based DOAS

Early applications of passive DOAS focussed on the detection of stratospheric trace gases. In recent years, novel "off-axis" observational geometries, i.e. viewing elevations different than the zenith, have allowed the observation of tropospheric trace gases.

A multitude of different observation geometries have been introduced. In particular, MAX-DOAS has been shown to be a powerful technique for the study of tropospheric constituents [see e.g. Hönninger et al. (2004a), Wagner et al. (2004), Bruns et al. (2004), and the discussion of MAX-DOAS AMFs in Chap. 9].

As illustrated in Fig. 11.16, two extreme cases of the spatial distribution of trace gases have to be considered for DOAS applications:

1. A localised trace gas plume (e.g. a stack plume) with very low trace gas concentrations outside the plume. In this case, reference spectra can be taken by pointing the instrument away from the plume. A scan through the plume will thus reveal the column densities inside the plume.
2. An infinitely extended trace gas layer, where the trace gas concentration only depends on altitude. Some elements of the vertical trace gas distribution can be derived from observation of column densities seen at different observation angles, as described in Chap. 9.

Clearly, real conditions will be in between the two extreme cases. Nevertheless, they are a useful tool in considering the principle of the different passive DOAS techniques.

Scanning of trace gas plumes (in Fig. 11.16a) emitted by various sources such as volcanoes, power plants, or cities is now systematically used as a tool for studying the emission and transformation of gases in the atmosphere. In addition, novel instrumentation allows the visualisation of trace gas distributions by I-DOAS. If extended layers exist (in Fig. 11.16b), the trace gas concentrations and profiles can be derived.

Another direction of development includes the use of 'known' gas distributions to study other properties of the atmosphere, such as photon path lengths or properties of the atmospheric aerosol. Further applications, like observation of polarisation properties, Raman scattering, or azimuth effects offer exciting opportunities for future research. In this section, we give a few examples of what has been achieved to date.

11.3.1 MAX-DOAS Observations in Polluted Regions

As described earlier, MAX-DOAS instruments are relatively simple and easy to operate. However, one drawback of MAX-DOAS is the need for precise knowledge of the atmospheric radiation transport, where the dominating uncertainty is the contribution of Mie scattering, i.e. the aerosol load. Therefore, algorithms have been developed, which determine the properties of atmospheric aerosol scattering from MAX-DOAS measurements of the O_4 SCDs by means of radiative transfer modelling (see Chap. 9), as described by Wagner et al. (2004).

An example of MAX-DOAS measurements of O_4 and urban NO_2 performed in Heidelberg, Germany during July of 2003, is presented in Fig. 11.17.

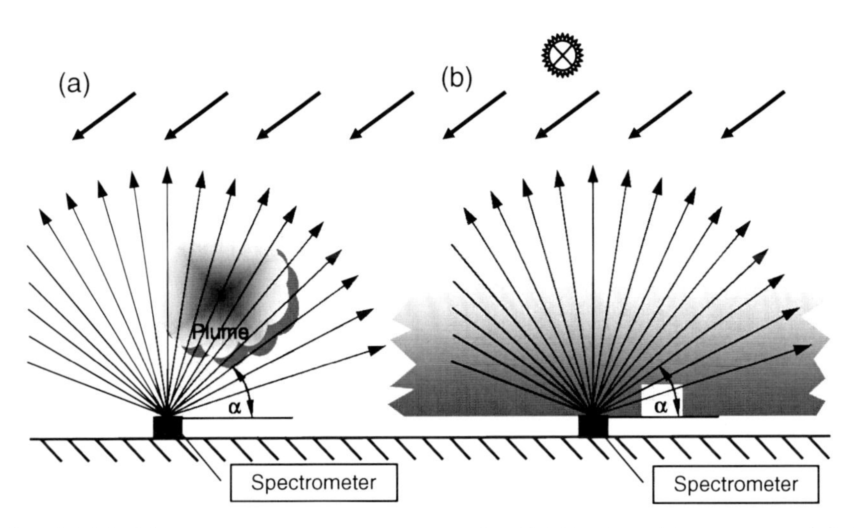

Fig. 11.16. There are two extreme cases in the distribution of trace gases. (**a**) A localised trace gas plume (e.g. a stack plume); (**b**) An (infinitely) extended trace gas layer

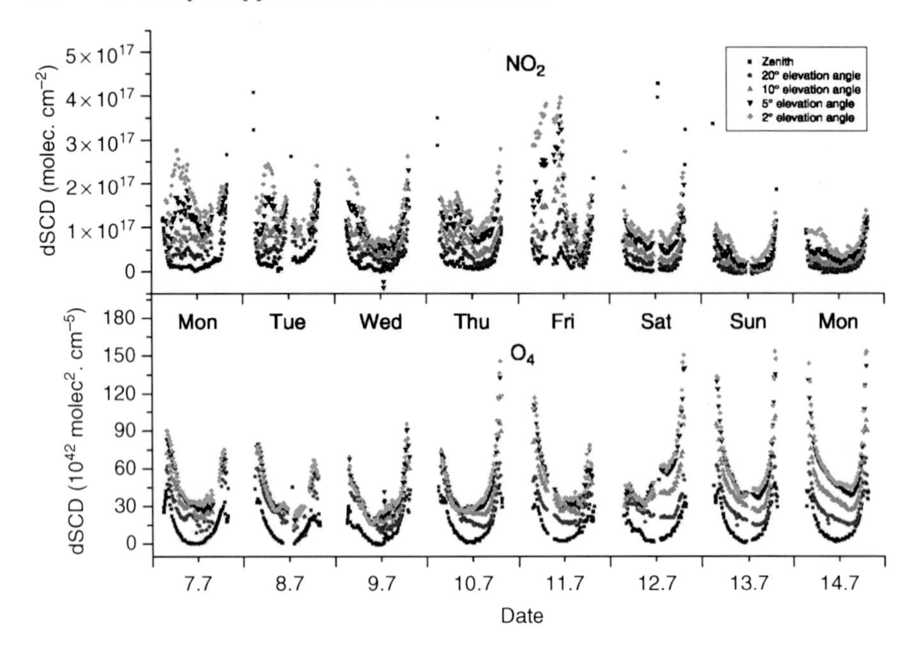

Fig. 11.17. NO_2 (*upper panel*) and O_4 (*lower panel*) SCDs recorded in Heidelberg, Germany during 8 days in July 2003. For evaluation, a reference spectrum of the 6th of July at noon was used. The elevation angle is indicated by the symbol colour as denoted in the legend (from Sinreich, 2004)

The retrieval of NO_2 concentrations from this data is split into three steps. First, sets of O_4 SCDs for a variety of aerosol scenarios (e.g. combinations of profile and optical densities) are calculated using a radiative transfer model (Chap. 9), and then compared to the O_4 SCDs measured for a series of elevation angles. From the best match, the adequate aerosol scenario is determined. Second, this best-matching aerosol scenario is taken as input to calculate the AMFs of other trace gases (here NO_2) using a radiation transport model to derive their height profiles. The flowchart of this method is sketched in Fig. 11.19 (Sinreich, 2004; Sinreich et al., 2005).

Figure 11.18 shows the application of the technique used to derive trace gas concentrations and mixing-layer height of glyoxal (CHOCHO) and NO_2 together with the vertical aerosol optical density and layer height. The measurements were made by passive MAX-DOAS at Massachusetts Institute of Technology (MIT), Cambridge, MA, USA (Sinreich et al., 2007).

11.3.2 MAX-DOAS Observations of Halogen Oxides at Mid-latitudes

An emerging application of MAX-DOAS is the measurement of halogen oxides in the troposphere. In Sects. 11.3.2–11.3.5 we will show examples of this

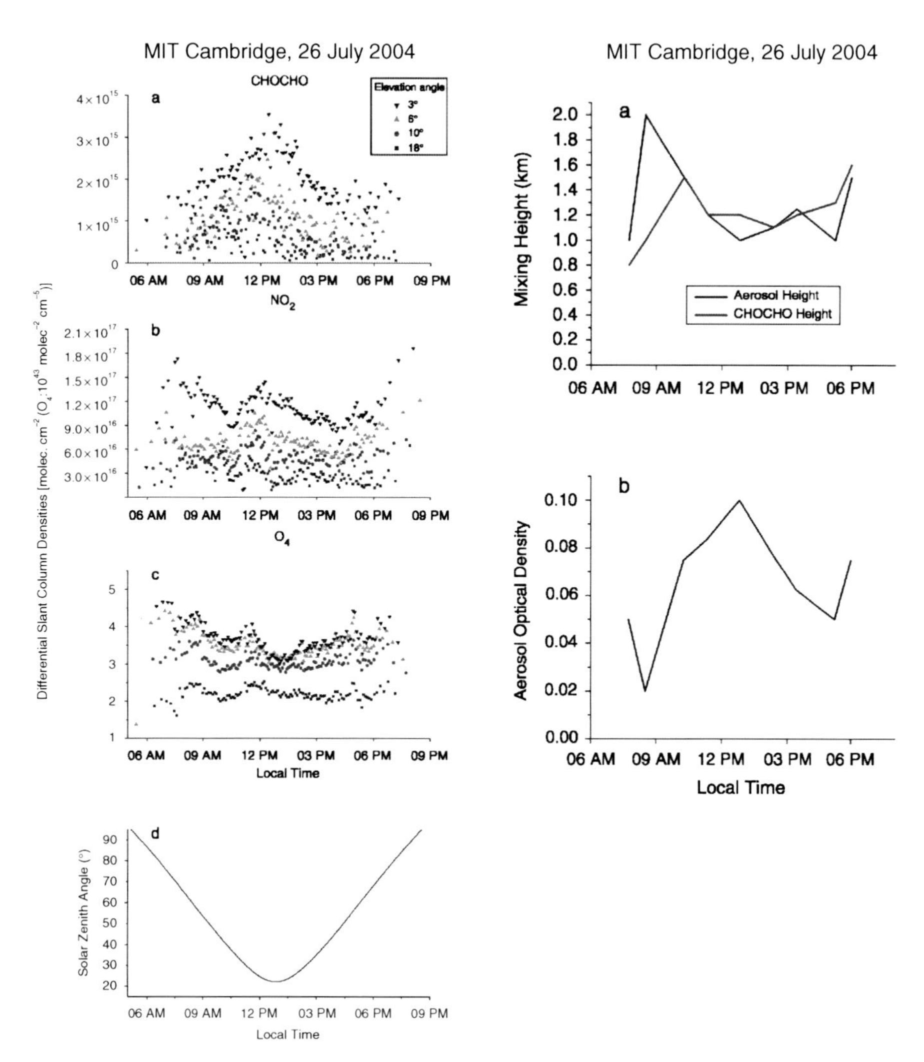

Fig. 11.18. Left panel: Differential SCDs of CHOCHO, NO_2, and O_4 measured by passive MAX-DOAS at MIT, Cambridge, MA, USA on 26 July 2004 (**a**–**c**) at different elevation angles. In (**d**) the SZA is plotted. **Right panel**: Estimated mixing height for CHOCHO and aerosol (**a**) as well as aerosol optical density (**b**) retrieved by radiative transfer modelling (from Sinreich et al., 2007)

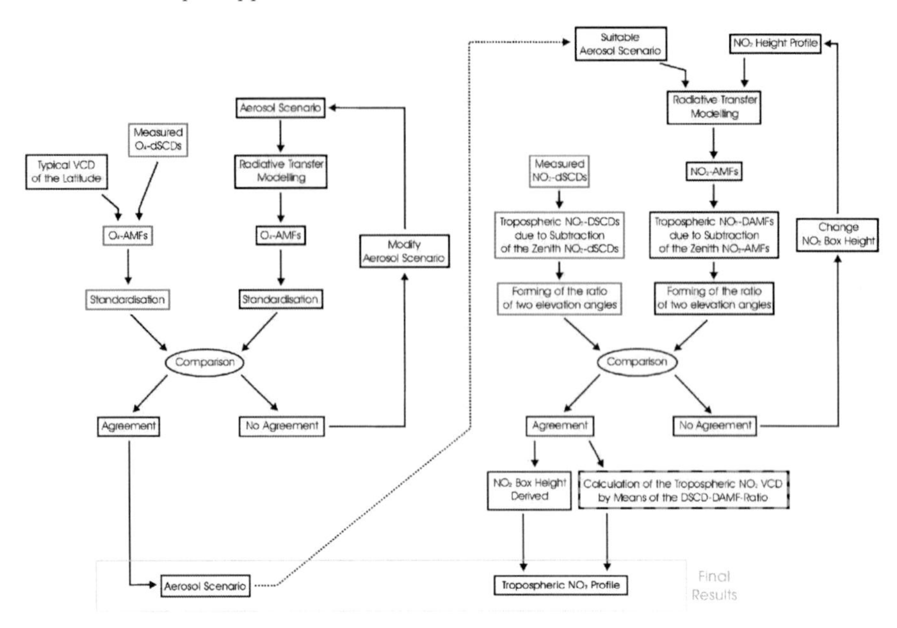

Fig. 11.19. Flowchart of the MAX-DOAS retrieval process for trace gas concentrations. **Left**: retrieval of the aerosol scenario; **Right**: derivation of the tropospheric NO_2 profile based on the aerosol profile (*dotted line*). (from Sinreich et al., 2005, reproduced by permission of The Royal Society of Chemistry (RSC))

technique at different geographical locations. Field measurements of reactive halogen and sulphur gas emissions were performed at the world's largest salt pan (area $\approx$ 10000 km^2), the Salar de Uyuni, Southern Altiplano, Bolivia (19.8°–20.7°S, 67.0°–68.2°W, >3600 m above sea level, see Fig. 11.20), during the dry season in October/November 2002 by Hönninger et al. (2004b). Bromine monoxide (BrO) and sulphur dioxide (SO_2) were studied by ground-based scattered light MAX-DOAS at various locations upwind of the Salar and on the surface of the salt pan.

Significant amounts of BrO were found at all locations, with the lowest levels at the northern edge of the Salar (upwind), while BrO SCDs (differential with respect to zenith) of up to 3.7×10^{14} molec. cm^{-2} were observed inside the Salar, see Fig. 11.21. Using the MAX-DOAS vertical profile information, BrO mixing ratios of >20 ppt were derived, which can have significant impact on tropospheric ozone chemistry both locally (Fig. 11.21, lower panel) and regionally. In addition, sulphur dioxide (SO_2) SCDs of $>5 \times 10^{16}$ molec. cm^{-2} were found around the Salar by the same technique, suggesting emissions from activity of nearby volcanoes.

Fig. 11.20. Map of the Salar de Uyuni and surrounding area. Measurement sites are marked 'Salar 1' through 'Salar 4' (from Hönninger et al., 2004b, Copyright by American Geophysical Union (AGU), reproduced by permission of AGU)

11.3.3 Halogen Oxide Radicals in the Polar Troposphere

In the Arctic, passive DOAS measurements of BrO have been performed near Kiruna, Sweden (Frieß et al., 1999), as well as at Alert and the Hudson Bay, Canada (Hönninger and Platt, 2001; Hönninger et al., 2004c). Figure 11.22 shows a time series of simultaneous BrO measurements by active long-path DOAS and passive MAX-DOAS. A comparison of the data in Fig. 11.23 shows good agreement between both techniques. Tropospheric IO was measured by Wittrock et al. (2000a, 2004) in Ny Alesund on the island of Spitsbergen.

In Antarctica, several studies at Neumayer Station targeted IO and BrO abundances (Frieß, 2001; Frieß et al., 2001, 2004, 2005). Tropospheric BrO was also observed at Arrival Heights, Antarctica (Kreher et al., 1997; Schofield et al., 2006).

11.3.4 Halogen Oxide Radicals in the Free Troposphere

The free troposphere is a region which is difficult to access with ground-based, spectroscopic instruments (but not with airborne instruments, see Sects. 11.1.2 and 11.4), since zenith observations at large (around 90°) SZAs are most sensitive to stratospheric absorbers, while multi-axis observations

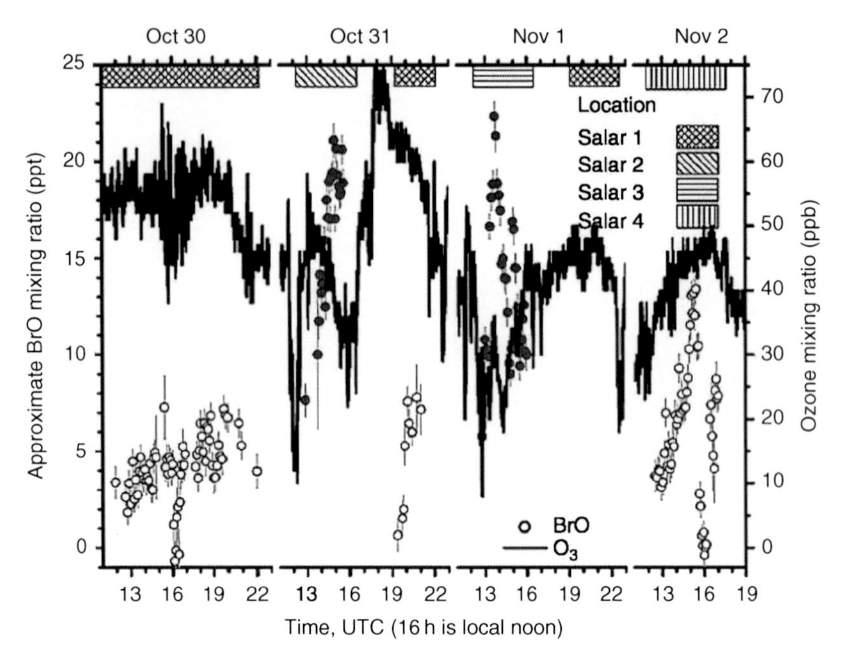

Fig. 11.21. Top panel: MAX-DOAS time series of BrO (differential) SCDs recorded 30 October to 2 November, 2002. Measurement locations are indicated at the *top*. Error bars (1σ) are shown in gray. **Bottom panel**: Approximate BrO mixing ratio derived from the MAX-DOAS data and in-situ ozone mixing ratios. Note that the ozone time series consists of measurements at different locations including the transects, as indicated at the top. BrO data from Salar2 and Salar3 are shown in *colour/solid* symbols (from Hönninger et al., 2004b, Copyright by American Geophysical Union (AGU), reproduced by permission of AGU)

are most sensitive to the lowest few kilometres of the atmosphere. Nevertheless, an attempt was made to estimate the BrO vertical column using Langley plots (Van Roozendael et al., 2002, 2003). In this method, a linear relationship between AMF and slant column is assumed, and the vertical column is derived from the slope of the correlation curve. Figure 11.24 shows BrO vertical columns derived from Langley plot based evaluation of data from stations of the Network for the Detection of Stratospheric Change (Andoya, Harestua, Arrival Heights, and Observatoire de Haute-Provence, OHP) and GOME – over-flight measurements. Langley plot evaluations are given in two ranges of SZAs. BrO columns obtained during twilight, when the method is mostly sensitive to the stratosphere, are systematically smaller than columns derived at lower SZAs. Langley plot 'noon' evaluations are qualitatively consistent with Gome which, together with previous results, is a clear an indication for the presence of BrO in the free troposphere.

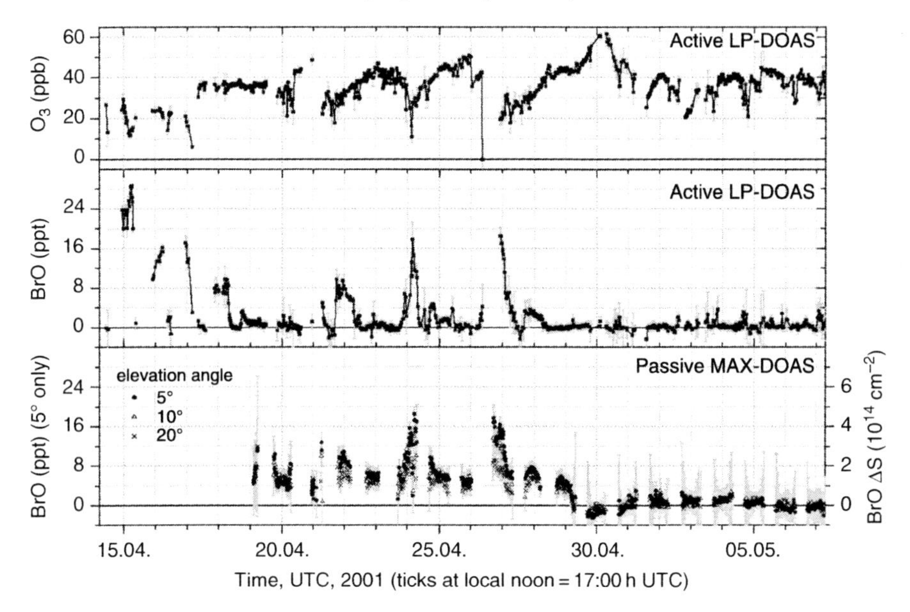

Fig. 11.22. Time series of LP-DOAS O_3 (*top*), LP-DOAS BrO (*centre*) and MAX-DOAS BrO (*bottom*) recorded at Kuujjuarapik, Nunavik, Quebec, Canada in the Hudson Bay (55°N and 75°W) during April 2001. Data for different elevation angles are shown in different colours/symbols. MAX-DOAS column densities are indicated on the right scale. The MAX-DOAS mixing ratio scale is valid for the 5° data points only (from Hönninger et al., 2004c)

11.3.5 Trace Gases in the Marine Environment

Ship-borne MAX-DOAS observations of mid-latitude tropospheric BrO and NO_2 abundances were performed during a cruise of the German research vessel Polarstern from Bremerhaven, Germany to Cape Town, South Africa in October, 2000 by (Leser et al., 2003). O_4 columns were also determined. The SCDs of the gases were sequentially measured at different elevation angles above the horizon (5°–90°). From the SCDs, the concentration of the investigated absorbers in the marine boundary layer was derived. During a period of two days in the region north of the Canary Islands (around 35°N and 13°W), significant levels of boundary layer BrO (on the order of 1 ppt) were found (see Fig. 11.25), while during the remaining time, BrO data ranged below the detection limit. Boundary layer NO_2 was found near Europe and the Canary Islands, with mixing ratios in the range of several ppb. In the remote marine area Southwest of Africa, upper limits ranging from 24 to 100 ppt were derived.

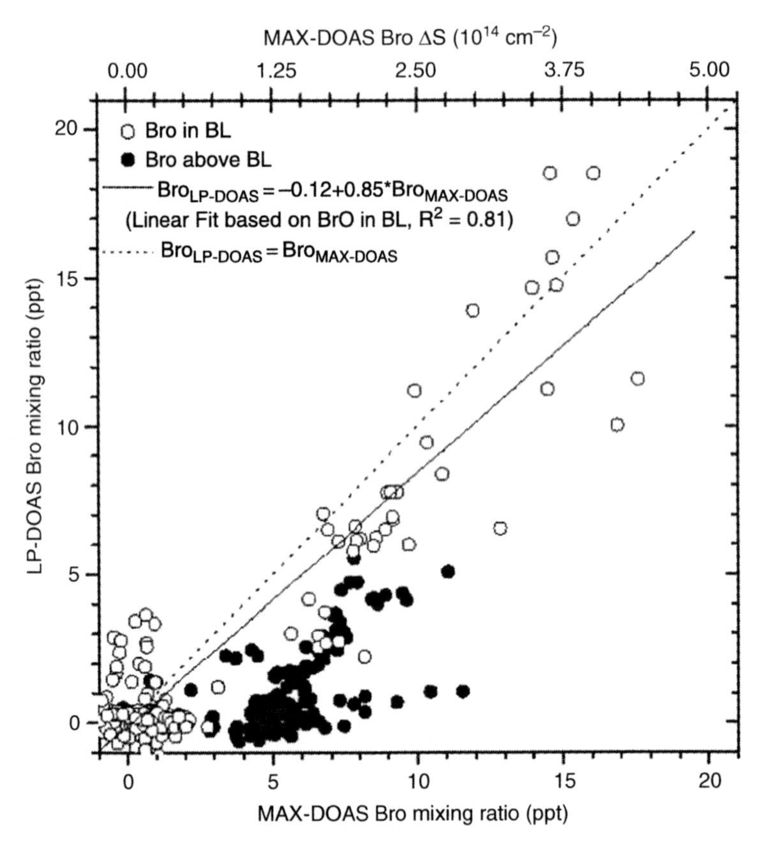

Fig. 11.23. Scatter plot of BrO mixing ratios derived by active LP-DOAS BrO versus passive MAX-DOAS. *Filled symbols* indicate MAX-DOAS signatures of BrO located above the boundary layer. The linear fit based on BrO in the boundary layer (*open symbols*) shows good agreement between both techniques (from Hönninger et al., 2004c, Copyright by American Geophysical Union (AGU), reproduced by permission of AGU)

11.3.6 Determination of Aerosol Properties from MAX-DOAS Observations

An interesting capability of DOAS measurements is the determination of elements of radiation transport in the atmosphere by studying the structured absorption due to molecules with known abundance such as O_2, O_4, or water vapour (see Sects. 11.3.1 and 11.9). These data can be used to infer the aerosol optical density, as described by Sinreich (2004), Bruns et al. (2004), Sinreich et al. (2005), and Frieß et al. (2006). The retrieval procedure for aerosol data and trace gas concentrations is briefly outlined in Sect. 11.3.1. Figure 11.26 shows the calculated relationship between aerosol extinction coefficient and the ratio of the O_4 AMFs – and thus measured O_4 column densities – for 2° and 90° observation elevation angles.

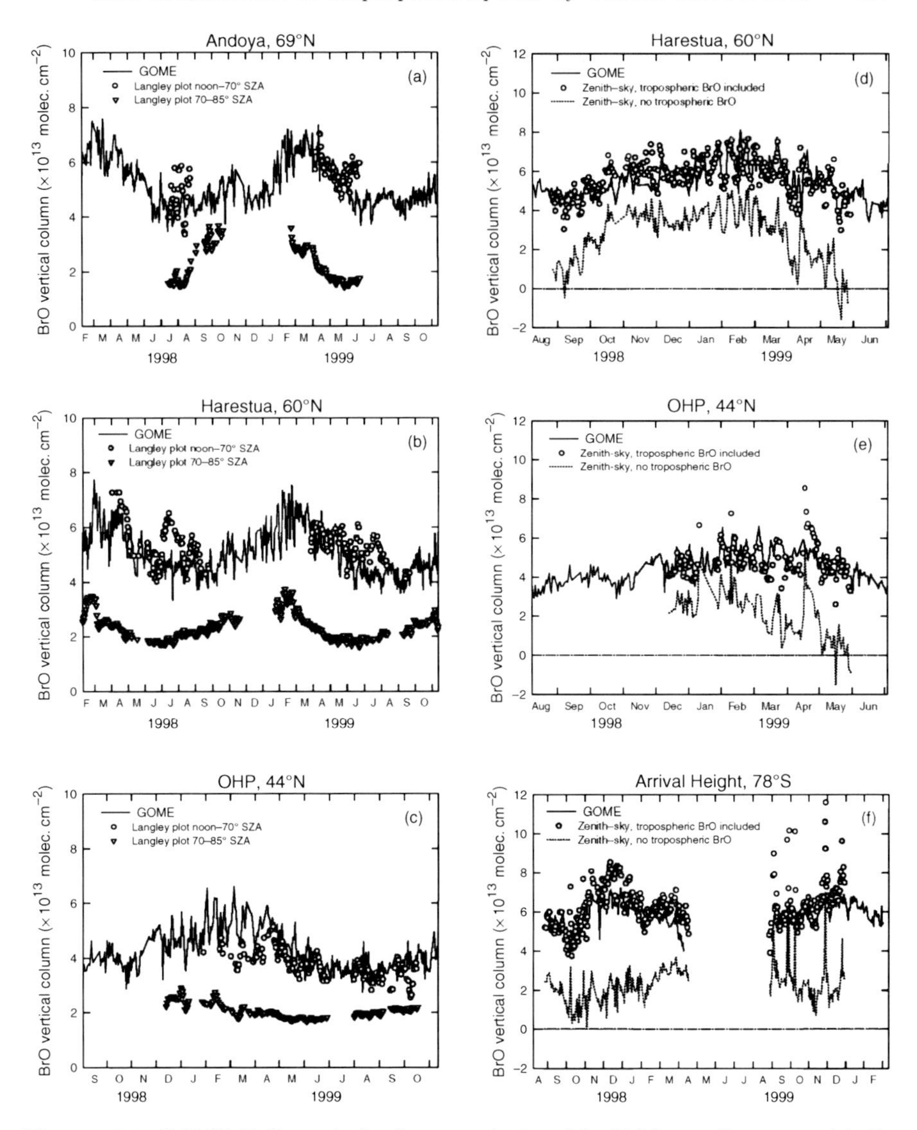

Fig. 11.24. GOME BrO vertical columns calculated in 500-km radius around indicated ground-based stations, compared to vertical columns derived from zenith sky data by a Langley plot method (*left panel*) or from noon 'GOME overpass' analyses using a seasonal reference spectrum (*right panel*). (from Van Roozendael et al., 2002)

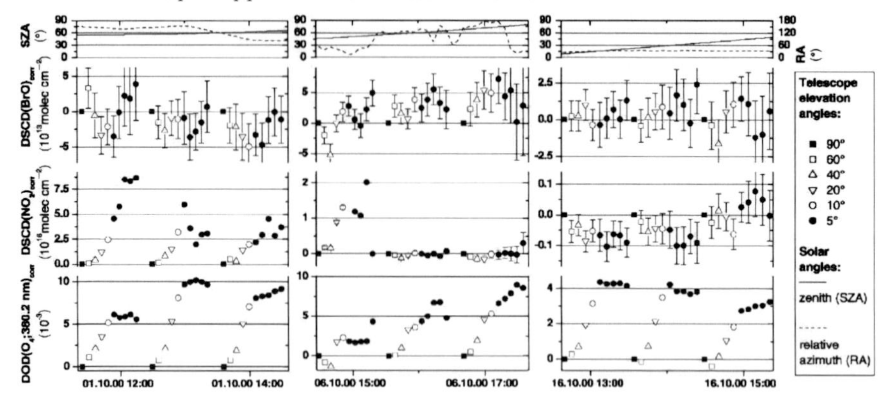

Fig. 11.25. Typical time series of (differential) MAX-DOAS SCDs of BrO (*top*), NO_2 (*centre*), and O_4 (*bottom*) at various observation elevation angles, as indicated by different *symbols*, during a trip of the research vessel Polarstern in October, 2000. Also given are SZA and relative azimuth RA (azimuth angle between viewing direction and sun). Error bars indicate 1σ fit errors. **Left panel**: On 1 October in the English Channel high NO_2 SCDs at low elevation angles indicate elevated NO_2. **Centre panel**: The pattern of the BrO series during the afternoon of 6 October, north of the Canary Islands, is typical for BrO in the marine BL. Boundary layer NO_2 levels are below the detection limit after 15:30 h. **Right panel**: On 16 October, in the remote Atlantic Southwest of Africa, both NO_2 and BrO are below the detection limits (note expanded scale). (from Leser et al., 2003, Copyright by American Geophysical Union (AGU), reproduced by permission of AGU)

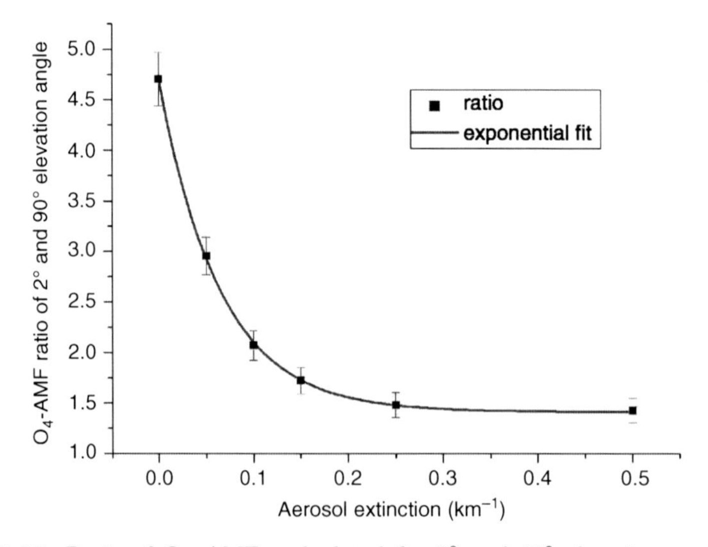

Fig. 11.26. Ratio of O_4 AMFs calculated for 2° and 90° elevation angles as a function of the aerosol extinction coefficient (assuming a constant aerosol extinction within the lowest 2 km of the atmosphere). The method is sensitive to aerosol extinctions between about 0.25 and 0.01 km^{-1} (corresponding to visibility ranges of about 40–16 km) (from Sinreich et al., 2005, reproduced by permission of The Royal Society of Chemistry (RSC))

11.3.7 Determination of NO_3 Vertical Profiles

From morning twilight scattered radiation measurements, it is possible to determine nitrate radical (NO_3) vertical concentration profiles. The technique relies on the rapid photolysis of the NO_3 molecule in direct sunlight. The basic idea is illustrated in Fig. 11.27. The technique uses the light scattered into the detector while the sun is still below the horizon. The instrument's line of sight is shown as a dotted line; it crosses the solar terminator (straight line in the top panel). At a high SZA 1, the scattered light traverses the distance between B and C in the not yet illuminated part of the atmosphere still containing the NO_3 accumulated there over night. At a smaller SZA 2 (bottom panel), the direct sunlit zone has extended to lower altitudes, and the light path B/C through the non-illuminated layers is much shorter, leading to a smaller NO_3 SCD. A sequence of measurements allows the derivation of a vertical NO_3 profile. The technique was originally suggested by Smith and Solomon (1990), who derived a profile from the stratosphere down to 3 km. Smith et al. (1993) repeated the experiment in the Antarctic region, finding negligible levels of NO_3 in the troposphere. Fish et al. (1999) modelled near-ground NO_3 concentrations of $<1.3 \times 10^9\,\mathrm{cm}^{-3}$ ($\approx$50 ppt) and an increase to higher concentrations of $(1.75 - 2.5) \times 10^9\,\mathrm{cm}^{-3}$ ($\approx$65–90 ppt) within a layer between 300 and 1700 m altitude, followed by a sharp decrease above 1700 m. Povey et al. (1998) found average nocturnal NO_3 abundances of about 100–200 ppt, in boxes centred at 1.5 and 1.2 km, respectively. An experimental verification of these NO_3 profile structures was reported by two groups (von Friedeburg et al., 2002; Allan et al., 2002; Coe et al., 2002). An example is shown in Fig. 11.28.

11.3.8 Emission from Point Sources

A new application of passive absorption spectroscopy is the quantification of plume emissions from localised sources such as smoke stacks, industrial plants, urban centres, biomass fires, and volcanoes. The approach is similar to the MAX-DOAS principle (see Sects. 11.3.1 through 11.3.6), in that scattered light is observed under a variety of angles and (differential) SCDs are derived from the recorded spectra. In contrast to MAX-DOAS, which assumes, at least in its simplest application, horizontally homogenous trace gas distribution, scanning emission plumes makes use of the inhomogeneity of the trace gas distribution under investigation. Some early measurements were reported by, for example, Milan (1980) and Edner et al. (1994b).

In this application, column densities are measured in a series of angles approximately perpendicular to the propagation direction of the plume (Fig. 11.29). From a simple geometric approach, the approximate extent of the plume can be inferred, based solely on the DOAS measurements and wind direction. Each column density $S(\alpha)$ is an integral of the trace gas concentration along the line of sight through the plume (arrows Fig. 11.29a). Integrating

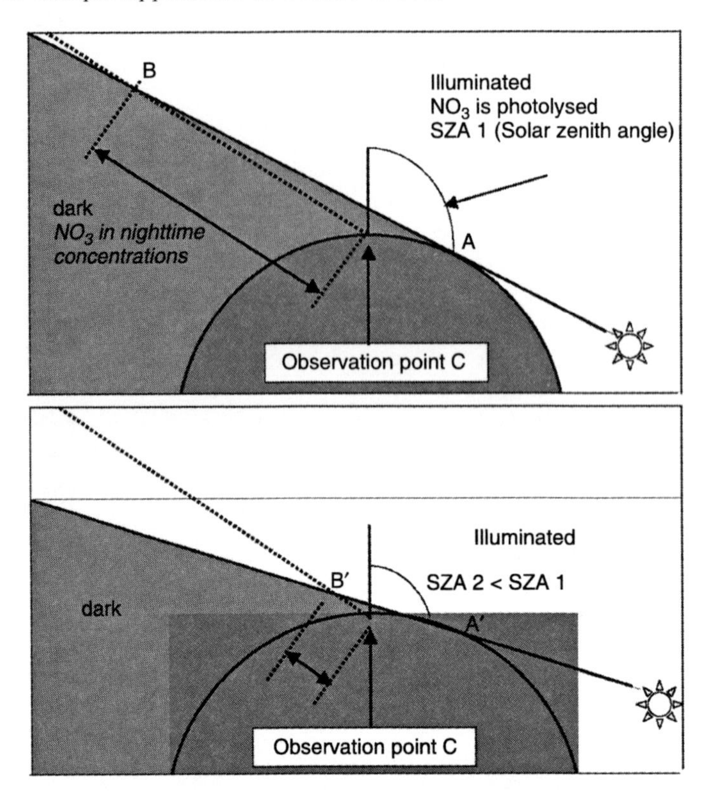

Fig. 11.27. Viewing geometry of off-axis spectroscopy for morning-twilight NO_3 measurements (from von Friedeburg et al., 2002, Copyright by American Geophysical Union (AGU), reproduced by permission of AGU)

over the spatial coordinate $s \approx R \cdot \alpha$ associated with the observation elevation angle α (see Fig. 11.29), where R denotes the distance from the spectrometer to the centre of the plume, the amount of trace gas Q (in molec. cm^{-1}) per unit length of plume is derived:

$$Q = R \int S\,(\alpha)\, d\alpha \approx R \int \int c\,(s) ds\, d\alpha = \int\limits_{A} c\,(\vec{x})\, dA'.$$

By using two spectrometers, the plume can be triangulated from the elevation angles α and β, under which the centre of the plume is seen. The trace gas flux J from the point source can be calculated if the wind speed v_P at the position of the plume is known, as $J = Q \cdot v_P$. The wind speed can either be derived from meteorological data, or it can be determined by observing the temporal variation $S(t)$ in the plume at different downwind distances from the source. From the spatial separation and the temporal correlation of the fluctuations in $S(t)$, the wind speed can be calculated (see e.g. Galle et al., 2004; McGonigle et al., 2005).

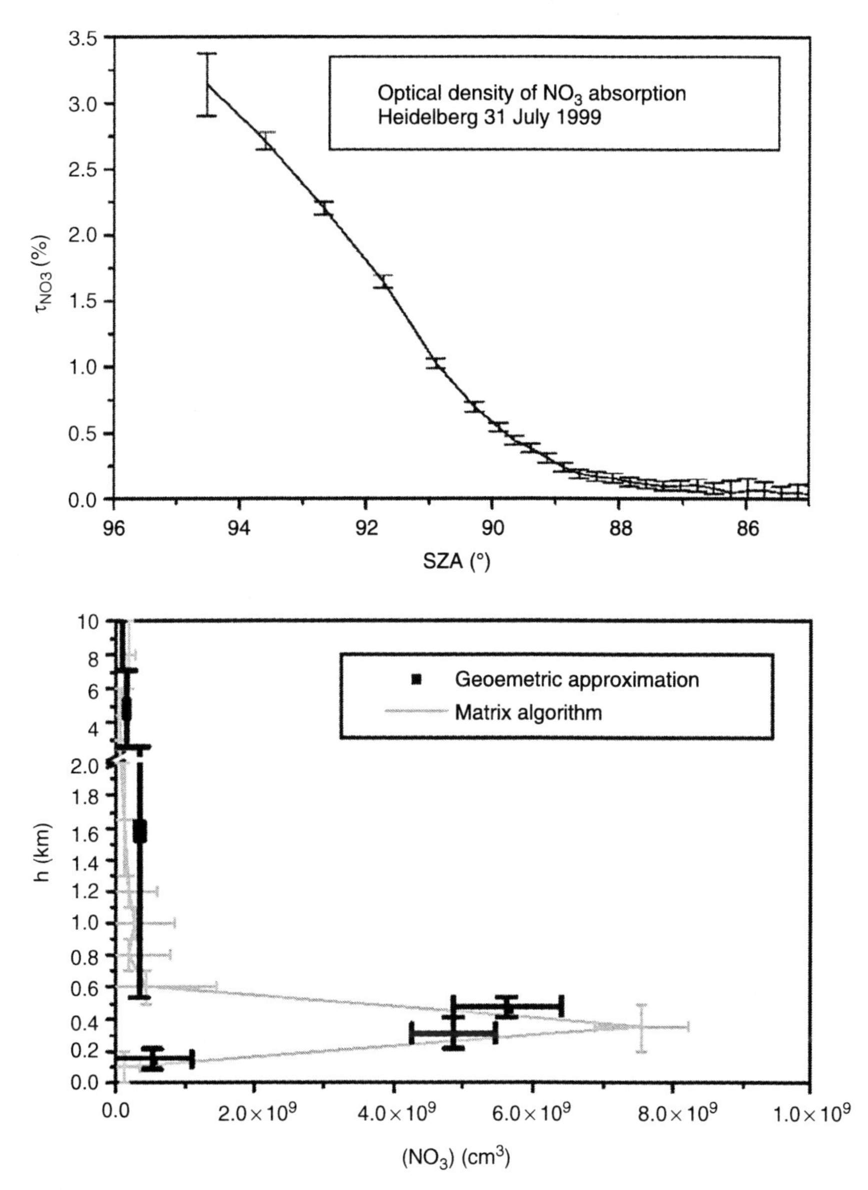

Fig. 11.28. Off-axis spectroscopy for NO_3 profile determination. **Upper panel**: Typical decrease of the NO_3 slant-path optical density as a function of decreasing SZA during sunrise, observed at Heidelberg on 31 July 1999. Error bars denote the 2σ error. **Lower panel**: Comparison between the corresponding NO_3 profiles obtained by geometric approximation (*black*) and the modified matrix technique (*gray*). The *vertical bars* refer to the layer for which a given concentration value is valid. The concentration data derived with the geometric approximation are averaged over the lowest 400 m to allow for a better comparison with the matrix method (from von Friedeburg et al., 2002, Copyright by American Geophysical Union (AGU), reproduced by permission of AGU)

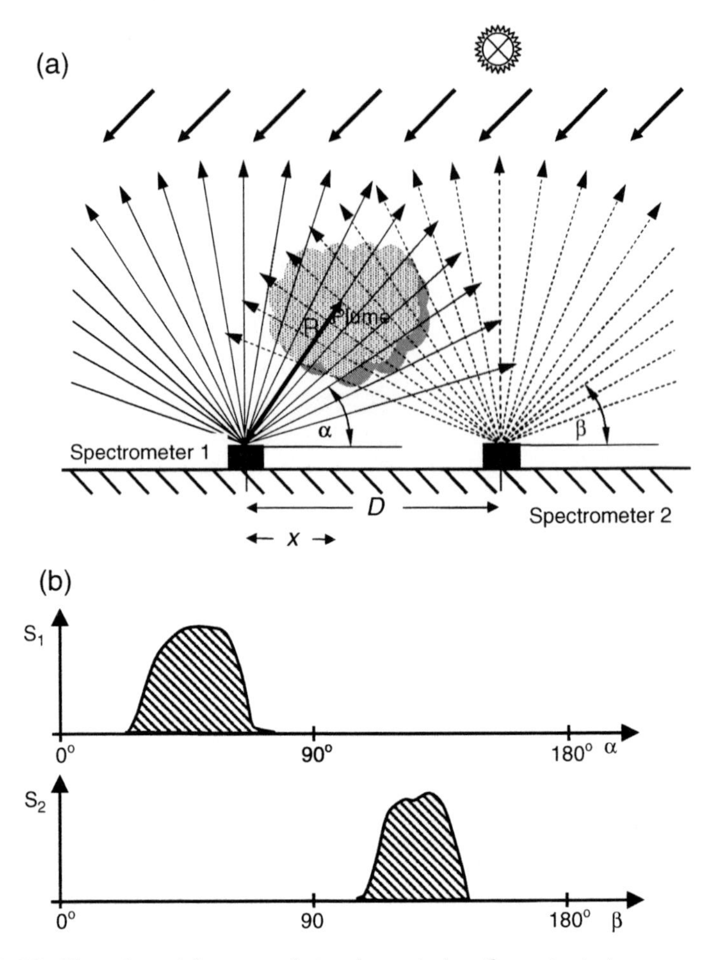

Fig. 11.29. Experimental approach to characterize the extent, trace gas content, and trace gas flux in a plume (**a**). The column densities S_1 and S_2 seen by spectrometers 1 and 2 when scanning in a plane (approximately) perpendicular to the propagation direction of the plume is sketched in (**b**). One spectrometer is enough to locate the plume. Two or more spectrometers allow triangulation of the plume.

Many measurements of SO_2, BrO (Galle et al., 2003; Bobrowski et al., 2003, 2007; McGonigle et al., 2005; Bobrowski, 2005; Oppenheimer et al., 2006), ClO (Bobrowski, 2005; Lee et al., 2005) and OClO (Bobrowski, 2005) in volcanic plumes have been performed by passive DOAS techniques. Figure 11.30 shows measurements of the Soufrier Hills volcano plume on Montserrat. The BrO plume width (FWHM) seen from the instrument location ranged over 34°. The distance between the plume centre and the observation point was about 1 km. The corresponding plume width is $\approx$760 m (2σ). Combined with the measured SCDs (BrO $\approx 2 \times 10^{15}$, $SO_2 \approx 1.75 \times$

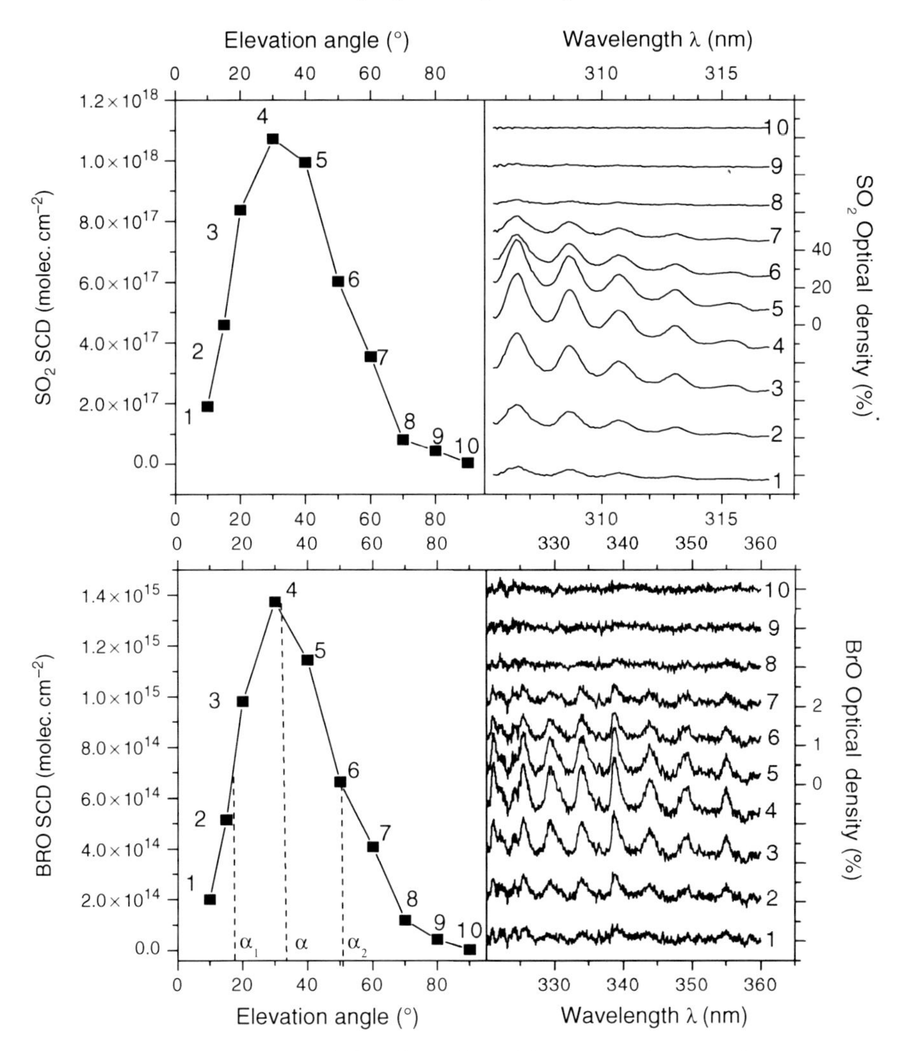

Fig. 11.30. Cross-sections of the plume from Monserat volcano. The observation point was about 4 km from the summit, on 25 May, 2002 where the plume was observed over an angular range of 34°. Measured BrO and SO_2 spectra corresponding to the individual viewing directions are shown on the *right*. (In both panels, subsequent spectra, except spectrum 5, are vertically shifted for better illustration.) (from Bobrowski et al., 2003)

10^{18} molec. cm^{-2}), mixing ratios in the plume of ≈1 ppb for BrO and ≈1 ppm SO_2, respectively, were derived (Bobrowski et al., 2003).

The second example in Fig. 11.31 shows repeated scans through the plume of a large power station (Alcanitz, Spain) yielding simultaneously measured SCDs of NO_2 and SO_2 (Bobrowski, 2005).

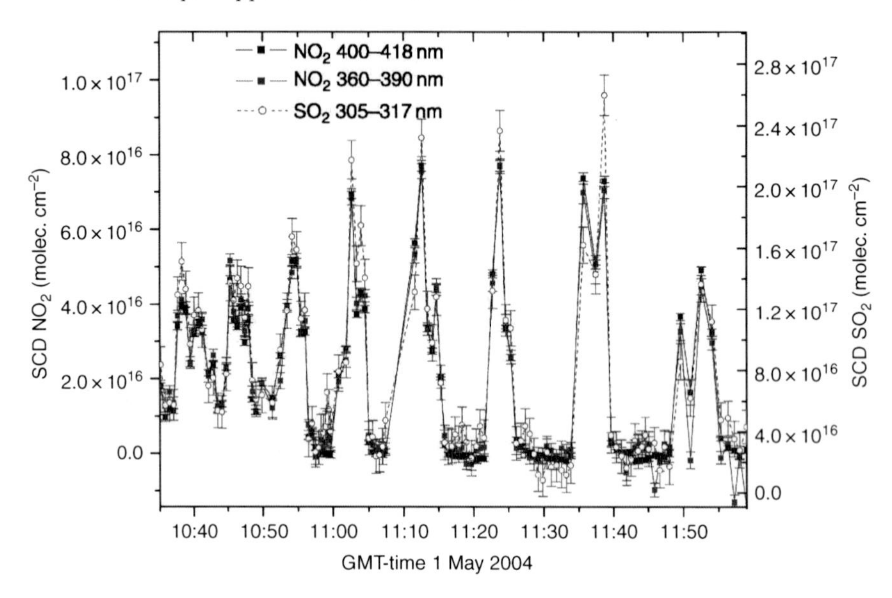

Fig. 11.31. Repeated scans through the plume of the Alcanitz power station in northern Spain. The evaluation yields simultaneous SCDs of NO_2 and SO_2 (from Bobrowski, 2005)

11.3.9 Imaging Trace Gas Distributions (I-DOAS)

Spatially resolved remote identification and quantification of trace gases in the atmosphere is desirable in various fields of scientific research, as well as in public security and industrial contexts. A ground-based remote sensing instrument permitting spatially resolved measurement of atmospheric trace gas distributions by passive DOAS has been developed by Lohberger et al. (2004) based on a 'pushbroom' imaging spectrometer. The instrument permits spatially resolved simultaneous measurement of different trace gas concentrations with high selectivity and sensitivity. Typically, imaging of an object is understood as taking a two-dimensional "picture", i.e. a 2D-array of pixel with intensity information. Imaging spectroscopy aquires electromagnetic spectra for each spatial pixel. This additional spectral dimension contains information about wavelength-dependent radiation intensities. Thus, the measurement results in a three-dimensional data product (with two spatial dimensions and one wavelength dimension). This could be called a 'super colour' or 'hyperspectral' image, in contrast to an ordinary colour image containing intensity information of only three wavelength ranges. Two techniques to obtain this information, which are distinguished by the temporal sequence of acquiring data of the three dimensions are currently used. 'Whiskbroom' imaging sensors acquire the spectrum of one spatial pixel at a time. The disadvantage of the whiskbroom scanning principle is the long time needed to acquire an image. 'Pushbroom' sensors simultaneously image one spatial direction (for example the vertical in Fig. 11.32) and spectrally disperse the light of all pixel in this dimension simultaneously, using a two-dimensional detector. Thus, scanning is needed only in the one spatial direction (horizontal in Fig. 11.32), greatly reducing the total measurement time. Instruments using the pushbroom technique are typically more complex and expensive. In order to take advantage of the reduced measurement time, light throughput and detector photon efficiency need to be higher than those of whiskbroom instruments.

Lohberger et al. (2004) employed the pushbroom technique, where the spectrum of each pixel was analysed by the DOAS technique in order to determine trace gas column densities. The result is a two-dimensional image showing spatially resolved atmospheric trace gas column densities. Radiative transfer is simplified by the small distances between the instrument and the object of investigation, compared to lengths by atmospheric Rayleigh and Mie scattering of 22 and 25 km at 400 nm, respectively. The typical distance to a visualized gas plume is below 1 km. Consequently, additional geometric considerations for optically thin plumes, the contribution of photons scattered into the instrument without passing the gas cloud is essentially negligible. A sample result showing the NO_2 plume from a heating plant is shown in Fig. 11.33. While a conventional photograph of the stack shows no visible plume; the DOAS evaluation of the spectra recorded with the pushbroom spectrometer clearly reveals enhanced NO_2 column densities around the top of the stack. The same technique was used to visualise volcanic SO_2 and BrO emissions as described by Bobrowski et al. (2007).

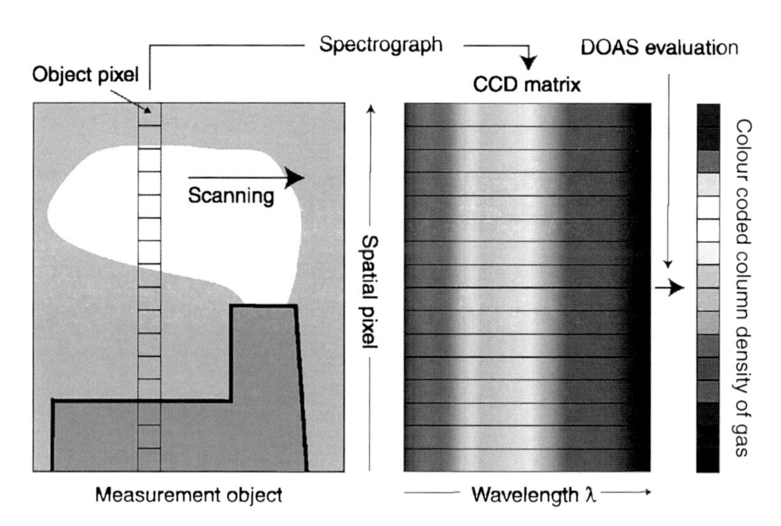

Fig. 11.32. I-DOAS measurement principle. The instrument used by Lohberger et al. (2004) utilises pushbroom imaging (see text) to simultaneously acquire spectra of one spatial direction (i.e. *vertical*). Spectral information along the second spatial dimension (*horizontal*) is obtained by scanning the scene using a mirror. The spectra are evaluated using the DOAS technique (from Lohberger et al., 2004)

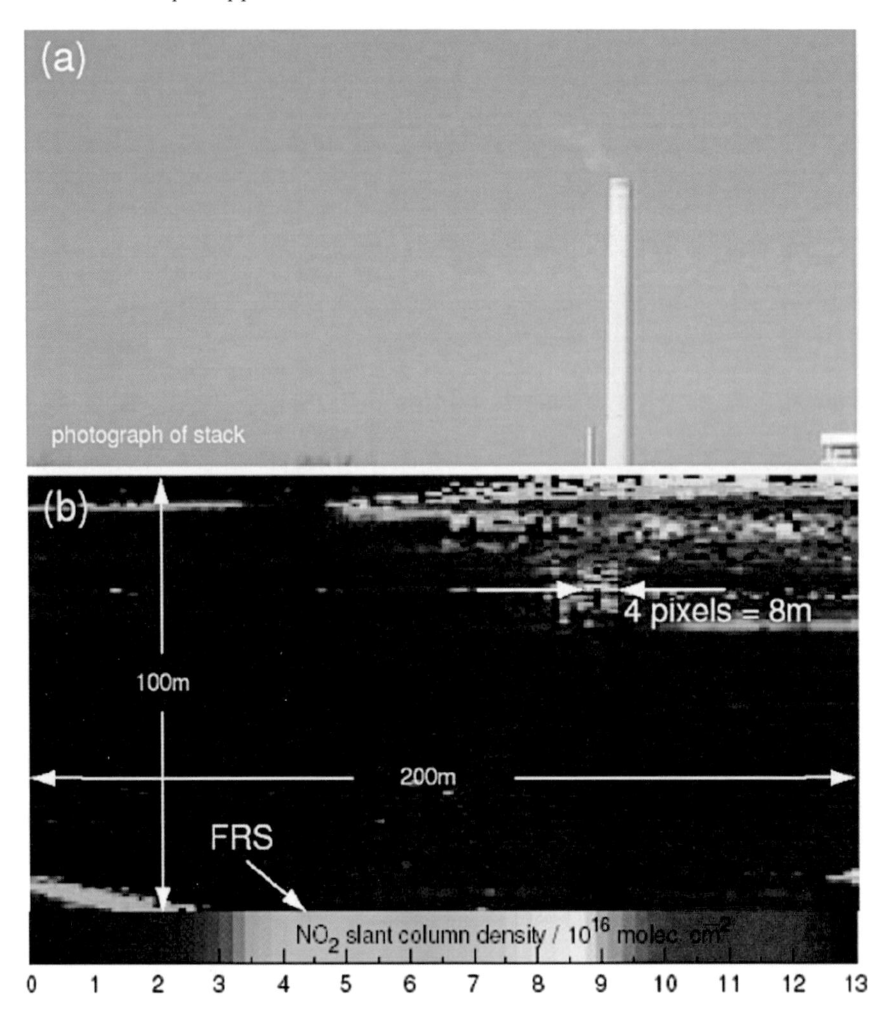

Fig. 11.33. Visualisation of the NO_2 plume of a power station's exhaust fume stack (**a**) under **calm** wind conditions, 10 December, 2002. (**b**) clearly shows the NO_2 plume with an FRS taken from the *left side* (*arrow*). (from Lohberger et al., 2004)

11.4 Scattered Light Aircraft Measurements of Stratospheric Species

Scattered light DOAS measurements of stratospheric species from airborne platforms offer several advantages over ground-based observations. In upward-viewing geometries, the influence of the atmosphere below the platform is usually very small. Thus the fraction of the atmosphere above the platform can be probed. This can be used, for example, to selectively probe the stratosphere.

In addition, during ascent or descent of the platform, vertical profiles can be recorded (see Sect. 9.4).

Early studies employed the then customary zenith scattered light geometries on aircraft platforms to map stratospheric O_3, NO_2, BrO, and OClO. Aircraft campaigns in both, the Antarctic and Arctic were performed on a DC-8 (Wahner et al., 1989a,b, 1990a,b; Schiller et al., 1990), in the northern hemisphere on a Transall transport aircraft (Pfeilsticker and Platt, 1994; Pfeilsticker et al., 1997, 1999; Erle et al., 1998), and other, smaller aircraft (McElroy et al., 1999; Petritoli et al., 2002; Melamed et al., 2003).

An example of a comparison of measured and modelled differential SCDs of BrO and OClO is given in Fig. 11.34. The dependence of BrO and OCIO on aerosol surface area and temperature was studied using a sensitivity analysis with the model (Erle et al., 1998).

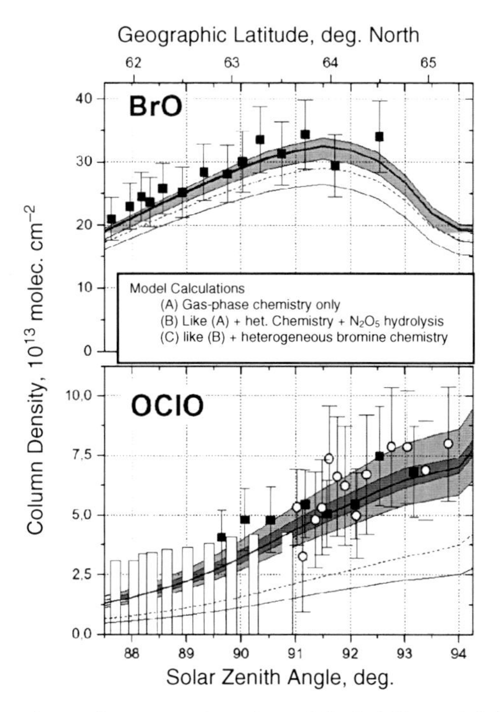

Fig. 11.34. Comparison of measured and modelled differential SCDs of BrO and OClO as a function of the SZA for 13 December, 1994. Measurements, indicated by symbols with error bars, were made by zenith scattered light DOAS on board the Transall research aircraft of the German air force. The geographical latitude of the measurements is shown at the *top* of the plot. The area shaded in *light gray* indicates the effect of multiplying or dividing the aerosol surface by a factor of 2, respectively. The area shaded in *dark gray* indicates the effect of reducing or raising the temperature of ± 3 K, respectively (from Erle et al., 1998, Copyright by American Geophysical Union (AGU), reproduced by permission of AGU)

11.5 Scattered Light Aircraft Measurements of Tropospheric Species

More recently, the MAX-DOAS technique was adapted to aircraft platforms, where it is known as Airborne Multi AXis DOAS (AMAX-DOAS). As outlined in Sect. 7.10.5, a large number of viewing directions are possible. In a recent implementation (Heue et al., 2004a,b; Bruns et al., 2004, 2006), two groups of telescopes were mounted at the top and bottom of the aircraft body. Besides a nadir, and zenith viewing telescope, telescopes collecting radiation from several angles (e.g. $\pm 2°$) above and below the aircraft were installed. Thus, some degree of vertical resolution was obtained. Results were reported by Wang et al. (2004), Fix et al. (2005), Heue et al. (2005), and Bruns et al. (2006).

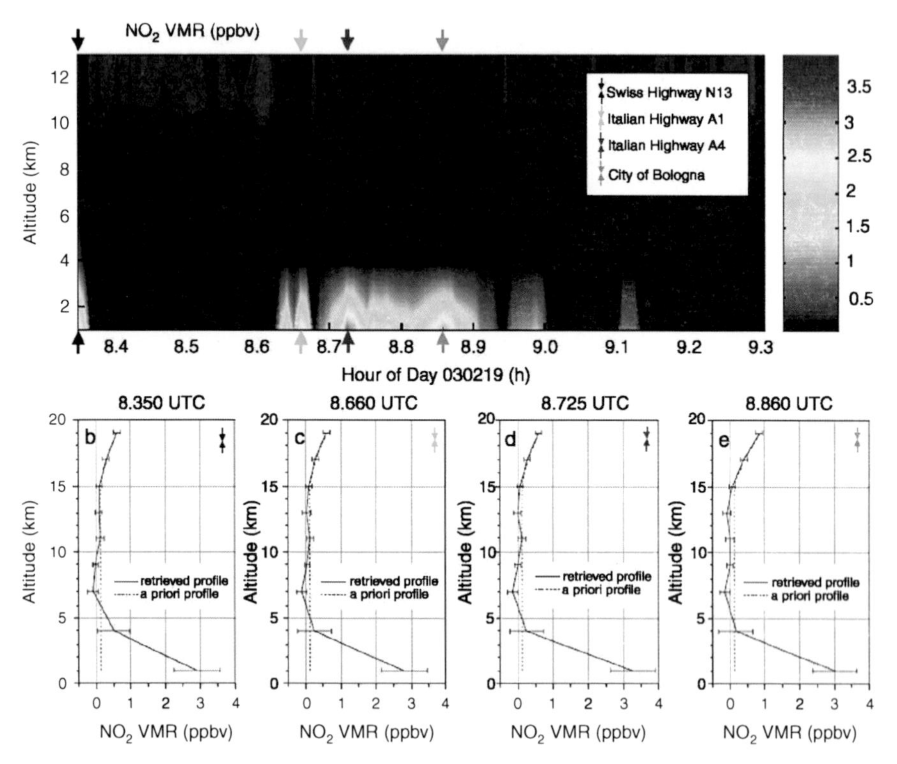

Fig. 11.35. Retrieved NO_2 profiles of a Falcon flight from Basel, Switzerland, to Tozeur, Tunesia, crossing the Alps and northern Italy on 19 February 2003. The top panel (a) shows a colour contour of all retrieved NO_2 profiles. Plots (b–d) in the lower panel show the retrieved profiles of tropospheric NO_2 mixing ratios at 8.350, 8.660, 8.725, and 8.860 UTC, respectively. The *dark blue* and *green arrows* in panel (**a**) mark the spatial positions of the profiles shown in plots (**b**) and (**c**), the *red* and *light blue arrows* mark the positions of the profiles represented by plots (**d**) and (**e**). (from Bruns et al., 2006)

Figure 11.35 shows some results of the two-dimensional NO_2 distribution retrieved from AMAX-DOAS measurements performed during a flight of the research aircraft Falcon from Basel, Switzerland, to Tozeur, Tunesia, crossing the Alps and northern Italy in essentially north–south direction on 19 February, 2003. Enhanced levels of NO_2 in the boundary layer are clearly discernible when the aircraft traverses several highways and the city of Bologna.

11.6 Satellite Observations Using DOAS Techniques

Optical absorption spectroscopy, and in particular DOAS, is an important and rapidly growing application for the remote sensing of atmospheric composition from space. In Chap. 6, several modes of observation were presented. These include:

- Nadir observations of sunlight reflected by the earth's surface or the atmosphere (earthshine)
- Limb observations of sunlight scattered at the horizon
- Solar occultation i.e. observation of sunrise/sunset through the atmosphere
- Lunar occultation
- Stellar occultation

In addition to the observation mode, the orbit of the satellite largely determines its performance in observing earth. The orbits can be categorised as Low Earth Orbits (LEO) or GEOstationary Orbits (GEO). Presently, all satellites capable of DOAS-type measurements are in LEOs.

A particular LEO is the (nearly) polar orbit, which allows observation of the entire surface of earth. Among the polar orbits, the sun synchronous orbit has the special advantage of observing earth always at (nearly) the same local time as shown in Fig. 11.36. In this orbit, the non-sphericity of earth, in combination with a small tilt of the orbit (inclination with respect to the equator $>90°$), leads to a precession keeping the plane of the orbit at a fixed angle relative to the line sun–earth. Presently, the instruments GOME, SCIAMACHY, OMI, and GOME-2 (Table 11.2) are on polar, sun synchronous orbits. Figure 11.37 shows the scanning scheme of the GOME instrument.

GOME consists of four spectrometers of moderate resolution that cover the spectral range from 290 to 790 nm with 1024 channels each. Analysing the spectrum of sunlight reflected by the earth (earthshine) using the DOAS technique yields column densities of a series of trace gases in the atmosphere (e.g. NO_2, BrO, OClO, SO_2, H_2O, O_3, and O_4). The GOME instrument monitors the earth in nadir view (looking downward approximately perpendicular to the earth's surface) and records a spectrum every 1.5 s. Using a scanning mirror, the instrument records three pixels perpendicular to the direction of flight and one 'backscan' pixel at the end of each scan. The

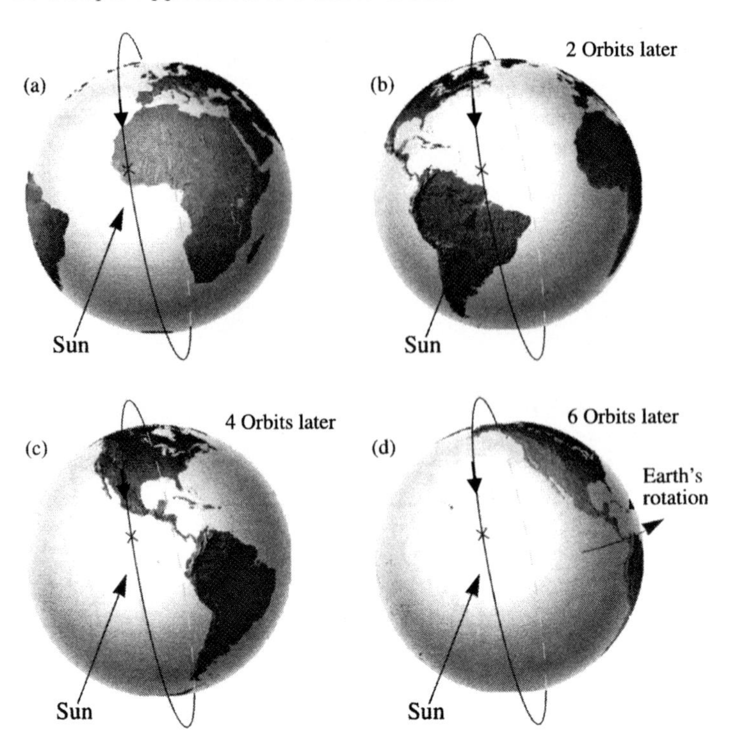

Fig. 11.36. Illustration of a sun synchronous polar satellite orbit, for example with an altitude of 800 km and an orbiting time of 100 minutes. The satellite crosses the equator always at the same local time. Examples for such satellites are: ERS-1, ERS-2 (GOME): 10:30 equator crossing time; ENVISAT (SCIAMACHY) 11:00 equator crossing time

size of each 'ground pixel', i.e. the footprint on the ground, is normally 320 by 40 km (see Fig. 11.37). Additionally, a polarisation monitoring device (PMD) (Bednarz, 1995) simultaneously scans the earth with a higher spatial but lower spectral resolution. It measures integrated intensities in three spectral bands between 295 and 745 nm. Each GOME pixel is spatially resolved by 16 subpixels of 20×40 km each. This subpixel information can be used for the determination of the cloud fraction and the ground albedo, as described in Guzzi et al. (1998), Wenig (1998) and Wenig et al. (1999), and Grzegorski et al. (2006). More recent instruments, e.g. GOME-2, SCIAMACHY, and OMI, offer smaller ground pixels or additional infrared (IR) channels (SCIAMACHY), or employ simultaneous observations of all cross-track spatial pixels, i.e. pushbroom scanning instead of whiskbroom scanning (OMI).

Table 11.2. Satellite instruments, species measured, satellite platform, and orbit. (The list is not intended to be complete, but merely to illustrate past and current instrumentation.)

Name	Target species	Satellite platform	Orbit
Atmospheric trace molecule spectroscopy (ATMOS)	O_3, NO_x, N_2O_5, ClO, NO_2, HCl, HF, CH_4, CFCs, etc. (upper troposphere)	Space Shuttle Spacelab-3 (1985), ATLAS-1,2 and 3 (1992,1993, 1994)	inclined
Backscatter ultraviolet (BUV) ozone experiment	O_3 (profiles)	Nimbus-4 (1970–1974)	Polar
Global ozone monitoring experiment (GOME)	O_3, NO_2, H_2O, BrO, OClO, SO_2, HCHO, clouds, aerosol	ESA-ERS-2 (1995–present)	Polar, Sun Sync.
Global ozone monitoring experiment-2 (GOME-2)	O_3, NO_2, H_2O, BrO, OClO, SO_2, HCHO, clouds, aerosol	METOP-1 to METOP-3 (2006, 2010/11, 2015/16)	Polar, Sun Sync.
Global ozone monitoring by occultation of stars (GOMOS)	O_3, NO_2, upper troposphere	ESA ENVISAT (2002)	Polar, Sun Sync.
Imaging atmospheric sounding instrument (IASI)	O_3, CO, CH_4, N_2O, SO_2	METOP-1 (2006)	Polar, Sun Sync.
Ozone dynamics ultraviolet spectrometer (ODUS)	SO_2, NO_2, BrO, OClO	GCOM-A1 Prog, Japan (2005)	inclined
Ozone monitoring instrument (OMI)	O_3, SO_2, NO_2, BrO	NASA-EOS-CHEM (2004)	Polar, Sun Sync.
Optical Spectrograph and Infrared Imager System (OSIRIS)	O_3, NO_2	ODIN, Sweden (2001–present)	Polar, Sun Sync.
Polar Ozone and Aerosol Measurement (POAM-II, III)	O_3, NO_2, H_2O, aerosol in the upper troposphere	SPOT-3 (1993–1996), SPOT-4 (1998–	Polar, Sun Sync.

Table 11.2. (continued)

Stratospheric Aerosol and Gas Experiment 1 (SAGE I)	O_3, NO_2, H_2O, aerosol in the upper troposphere	NASA Atm. Explorer Mission (1979–1981)	inclined
Stratospheric Aerosol and Gas Experiment 2 (SAGE II)	O_3, NO_2, H_2O, aerosol in the upper troposphere	Earth Radiation Budget Sat. (1984–present)	inclined
Stratospheric Aerosol and Gas Experiment 3 (SAGE III)	O_3, OClO, BrO, NO_2, NO_3, aerosols	Meteor 3M (2001); Inter. Space Stat.	inclined
Solar backscatter ultraviolet (SBUV) ozone experiment	O_3 profiles	Nimbus-7 (1979–1990)	Polar
Scanning imaging absorption spectrometer for atmospheric cartography (SCIAMACHY)	O_2, O_3, O_4, NO, NO_2, N_2O, BrO, OClO H_2CO, H_2O, SO_2, HCHO, CO, CO_2, CH_4, clouds, aerosols, p, T, col. and profiles	ESA-ENVISAT (2001–present)	Polar, Sun Sync.
Total ozone monitoring spectrometer (TOMS)	O_3	Nimbus 7 (1979–1992) ADEOS (1996–1997) Earth Probe (1996–present) Meteor (1992–1994)	Polar

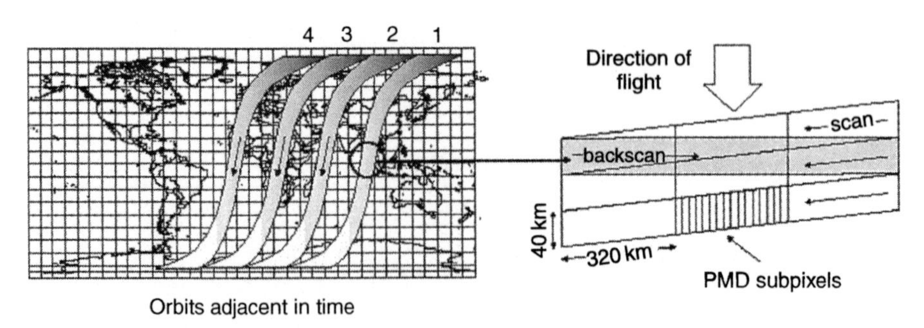

Fig. 11.37. Whiskbroom scanning scheme of the GOME instrument (from Leue et al., 2001a, Copyright by American Geophysical Union (AGU), reproduced by permission of AGU)

11.7 Satellite Observations of Stratospheric Species

Mapping of stratospheric species, in particular ozone (starting in 1970 with BUV on Nimbus 4) and NO_2 (starting 1979 with SAGE I), is the oldest application of satellite sensors. An example is satellite generated maps of the Antarctic ozone hole (although it was not originally discovered by satellite sensors), well known to the public from numerous presentations in the media.

11.7.1 Stratospheric O_3

The DOAS technique offers a number of well-known advantages over the Dobson-type evaluation of ozone absorption. Thus, evaluation of O_3 from GOME and SCIAMACHY data is a standard procedure (e.g. Lambert et al., 2001; Weber et al., 2003). Figure 11.38 shows as an example the daily minimum found polewards from 50°S during the Antarctic winters 1995–2001 (shaded area) in contrast to the winter of 2002, when a vortex-split occurred.

11.7.2 Stratospheric NO_2

One specific advantage of UV-vis satellite sensors is that, in contrast to other satellite sensors, they are sensitive to the total stratospheric NO_2 column, including the lower stratosphere. In particular, at polar latitudes a substantial fraction of the stratospheric NO_2 VCD is located in the lower stratosphere regions. The earth is usually covered daily at the equator. Global coverage is achieved after one to six days, depending on the sensor. From satellite NO_2 observations, it is possible to investigate the variation of the stratospheric NO_2

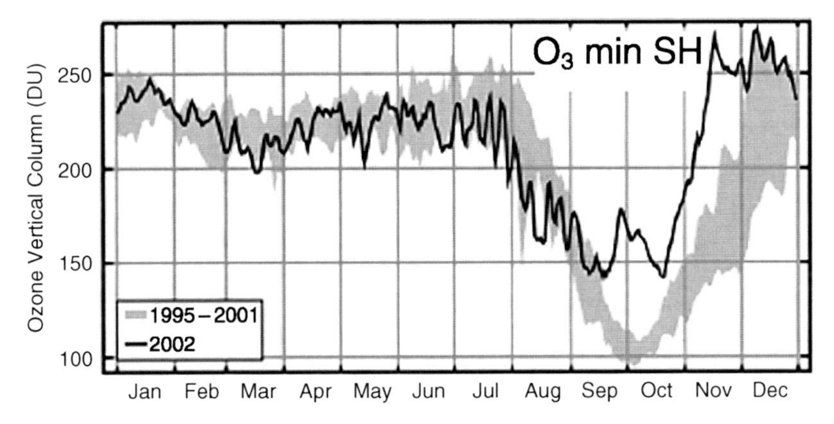

Fig. 11.38. Daily minimum GOME total ozone observed polewards of 50°S during the Austral winter of 2002 (*solid line*). *Shadings* indicate the range observed during the winters of 1995–2001 (from Richter et al., 2005)

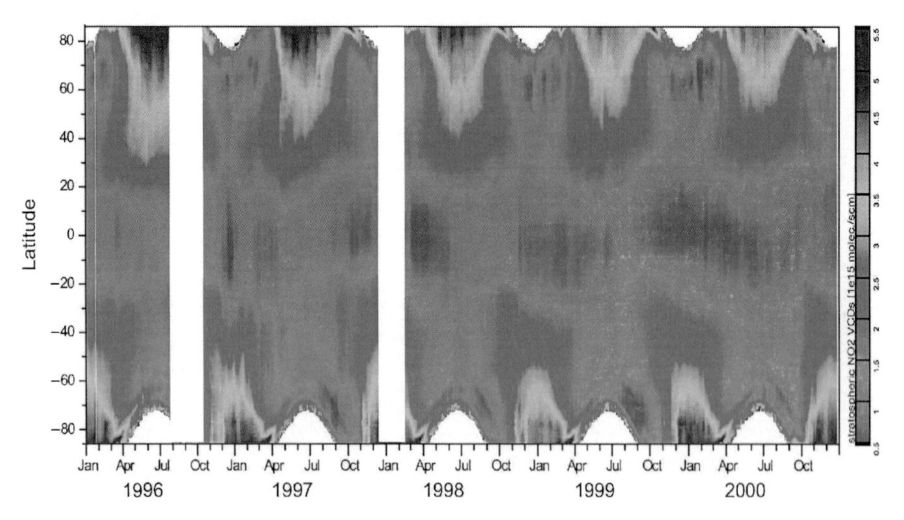

Fig. 11.39. Time series of the zonally averaged stratospheric NO_2 VCD (scale: $0 \ldots 5.5 \times 10^{15}$ molec. cm^{-2}) as a function of latitude and time from 1996 to 2000. (from Wenig et al., 2004, Copyright by American Geophysical Union (AGU), reproduced by permission of AGU)

VCD as a function of latitude and season, and thus to monitor and investigate several important aspects of stratospheric chemistry and dynamics, e.g. the Noxon-Cliff, zonal symmetry, or inter-hemispheric differences.

The GOME instrument made measurements over the globe for nearly 8 years. These data were used by Richter and Borrows (2002) and Wenig et al. (2004) to construct global maps of stratospheric NO_2. Figure 11.39 shows the time series of the zonally averaged stratospheric NO_2 VCD as a function of latitude and time from 1996 to 2000. The seasonal variability growing from mid- to high latitudes with maximum NO_2 columns during polar summer is clearly visible. Time series for average NO_2 VCDs over Europe, USA, and Brazil are shown in Fig. 11.40.

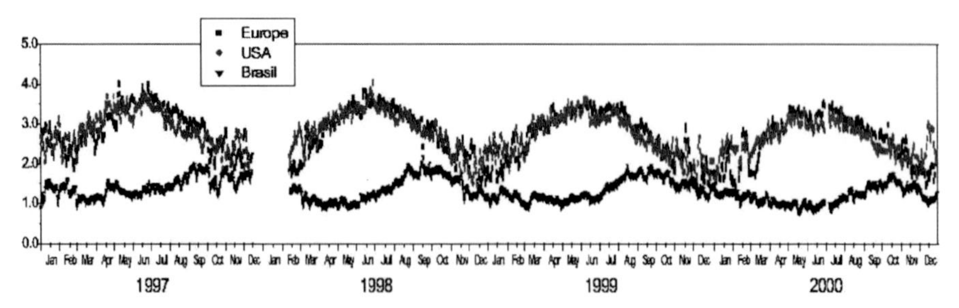

Fig. 11.40. Time series of the NO_2 VCD in units of 10^{15} molec. cm^{-2} (January 1997 to December 2000). DOAS evaluation using a fixed Fraunhofer reference spectrum (from 1 June, 1997) (from Wenig et al., 2004, Copyright by American Geophysical Union (AGU), reproduced by permission of AGU)

11.7.3 Stratospheric OClO

Several studies demonstrated that the observation of OClO is a good indicator for stratospheric chlorine activation. The GOME instrument allows daily observation of this species in polar regions. The first global data set of OClO and continuous time series of its occurrence in both winter stratospheres was reported by Wagner et al. (2001a, 2002). Figure 11.41 shows that OClO regularly occurs over Antarctica with little variation in total amount and temporal variation during the different winters, while the OClO occurrence is much more variable in the Arctic winter stratosphere. The primary reason is the larger dynamic activity resulting in warmer temperatures during the Arctic winter. About 40% larger OClO column amounts were found in the Antarctic polar stratosphere than in its northern counterpart, a further indication for a significantly more efficient chlorine activation in the Antarctic late winter and spring stratosphere.

An overview of the possibilities of mapping of stratospheric species by satellite-based DOAS observations is given in Fig. 11.42, which shows OClO SCDs, O_3 vertical columns, and NO_2 vertical columns (in molec. cm^{-1}) during the Antarctic vortex split in Austral spring of 2002. The results show (see Richter et al., 2004) that chlorine activation in the vortex, as indicated by the OClO abundance, was similar to previous years until the major warming on

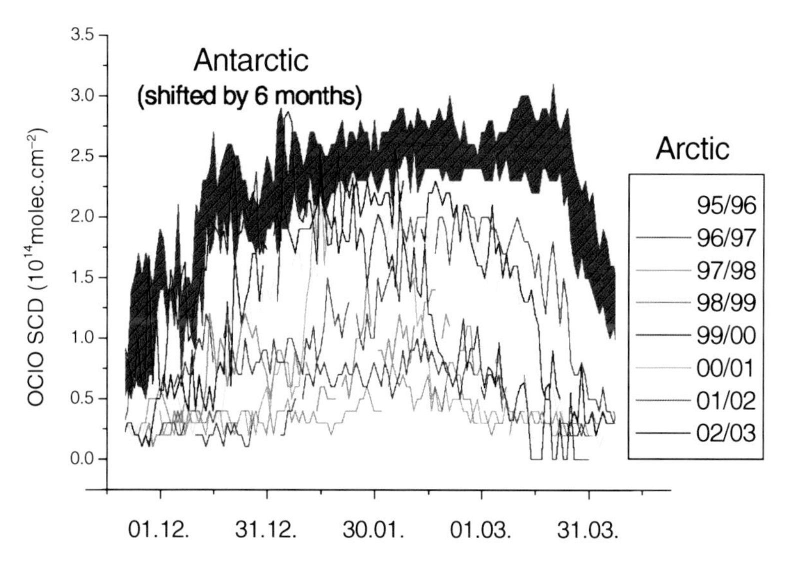

Fig. 11.41. Maximum daily OClO SCD measured by GOME over the Arctic (*thin coloured lines*) and the Antarctic (*red hatched area*) in the winters 1995 through 2003. For comparison with the measurements over the Arctic, the Antarctic OClO SCD measurements are shifted by 6 months. Note the small year-to-year variability of the chlorine activation in the Antarctic compared to the Arctic (from Wagner et al., 2002, Copyright by American Geophysical Union (AGU), reproduced by permission of AGU)

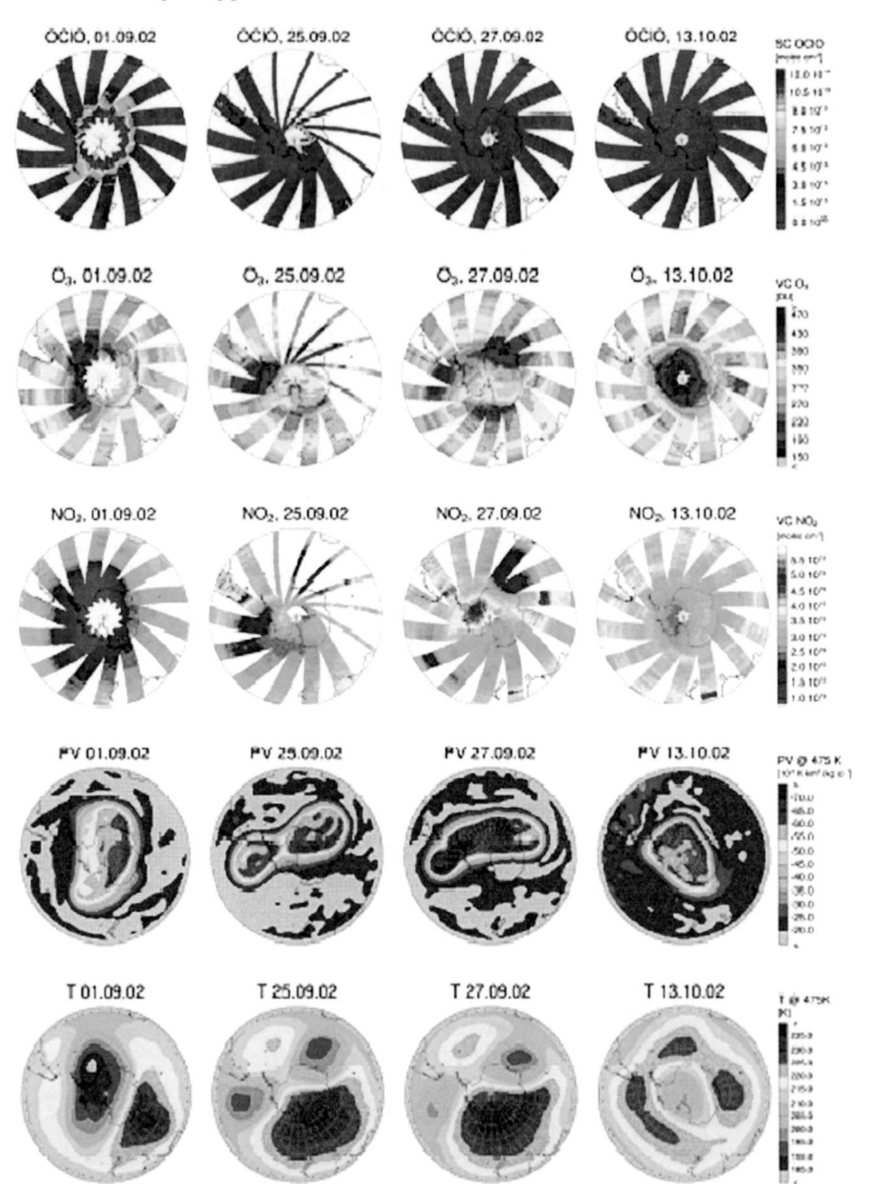

Fig. 11.42. The Antarctic vortex split in Austral spring of 2002: OClO (slant columns in molec. cm^{-1}, not corrected for light path enhancement), O_3 vertical columns (in Dobson units), and NO_2 vertical columns (in molec. cm^{-1}) in comparison to PV and temperature from ECMWF (at 475 K potential temperature surface). The data before the major warming event (9 and 25 September) on 26 September and after the warming (26 September and 31 October) are shown. (from Richter et al., 2005)

26 September, 2002, after which it decreased rapidly. Similarly, NO_2 columns were only slightly larger than in previous years before the warming, indicating strong denoxification and probably also denitrification.

After the warming, very large NO_2 columns were observed for a few days, which then decreased again as the vortex re-established until the final warming. Ozone columns were much larger than in any previous year from September onwards (see Fig. 11.42), mainly as a result of the unusual dynamical situation. The analysis of the global long-term time series of GOME measurements since 1996 provided the opportunity to set the austral winter 2002 into perspective. The data reveal the large difference in variability of chlorine activation between the two hemispheres, whereas denoxification shows surprisingly little variation from year to year in both hemispheres. However, in Antarctica, depletion of stratospheric NO_2 is usually sustained for about one month longer than in the Arctic as a result of the stable vortex. Compared to the observations in the northern hemisphere, the austral winter 2002 was stable, cold, and had a high potential for chemical ozone destruction.

11.8 Satellite Observations of Tropospheric Species

The inherent difficulty in probing the troposphere from satellite platforms is the retrieval of absorption features of tropospheric species from reflected earthshine, which also crosses the stratosphere. Some tropospheric trace gases of interest (e.g. O_3 and NO_2) are also abundant in the stratosphere, while others (such as SO_2, HCHO, or water vapour) are not. Various procedures have been developed to separate the tropospheric signal from stratospheric contribution:

1. The comparison of signals from known NO_2-free regions with those from polluted tropospheric regions gives the NO_2 in the boundary later. This technique is known as 'reference sector method'.
2. The difference between columns measured above clouds and those in cloud-free situations can give concentrations at levels depending on the cloud height, the so-called 'cloud slicing'.
3. The differences in the spectroscopic features (e.g. due to pressure and temperature) can be used to obtain trace gas profiles (e.g. of ozone) in the troposphere.

A considerable problem in observing tropospheric species from space is the presence of clouds (see Sect. 9.5), thus cloud detection and quantification schemes have to be used. If high precision is required, as for instance in the case of long-lived greenhouse gases (CH_4 and CO_2), then reduction of the column due to surface elevation must also be considered.

Use of the data ranges from the simple scrutiny of global or regional concentration maps, which provides confirmation of, or provokes questions about what was not previously observable. The data also provide a direct comparison with the output from chemical transport models (CTM) on global and regional scales and are used for realistic validation and sensitivity assessment of these models. This leads to substantial improvement of accuracy and reliability of CTMs. Satellite measurements are also useful in providing real boundary conditions for operational models. In addition, source strengths of trace gases can be derived (Leue et al., 2001; Palmer et al., 2003). For field campaigns, the knowledge of the actual concentrations of appropriate species in the vicinity of the campaign area is available. In short, satellite measurement of tropospheric trace gas abundances has become an essential element of tropospheric research.

11.8.1 Tropospheric O_3

Despite ozone's strong absorption features in the UV and visible spectral ranges, which led to its early spectroscopic detection in the atmosphere (see Sect. 11.2), the ozone in the troposphere is among the most difficult species to derive from satellite platforms. The challenge is to separate the tropospheric partial O_3 column from the ten times larger stratospheric column (see Fig. 11.43).

Several DOAS-based retrieval algorithms for tropospheric O_3 have been developed and successfully applied (Lambert et al., 2001, Edwards et al., 2003).

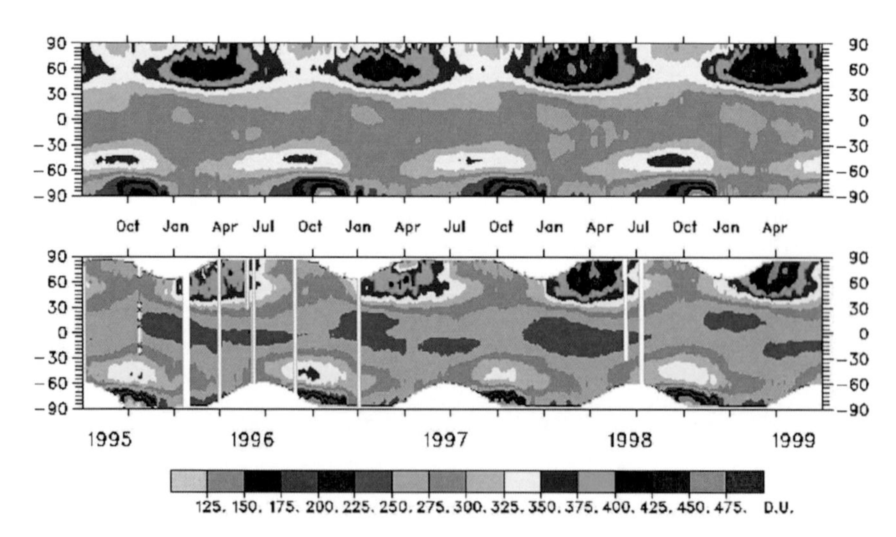

Fig. 11.43. Global ozone, 1995–1997: The lower part of the figure shows four years of observation of the total ozone column with GOME (1 July 1995–30 June 1999) The upper part shows four individual years of a multi-year simulation with the chemistry-climate model ECHAM4.L39(DLR)/CHEM (from Borrell et al., 2001)

11.8.2 Tropospheric NO_2

From the total column densities of NO_2, the tropospheric residual can be deduced by a multi-step 'reference sector' technique described by Leue et al. (2001). A flowchart of the complete procedure is given in Fig. 11.44. Figure 11.45 shows the mean global tropospheric NO_2 VCD for the year of 1997 (Leue et al., 1998, 2001). There is a clear correlation between areas of high NO_2 and industrialised regions. Similar techniques were developed by Richter and Borrows (2002) and Martin et al. (2002).

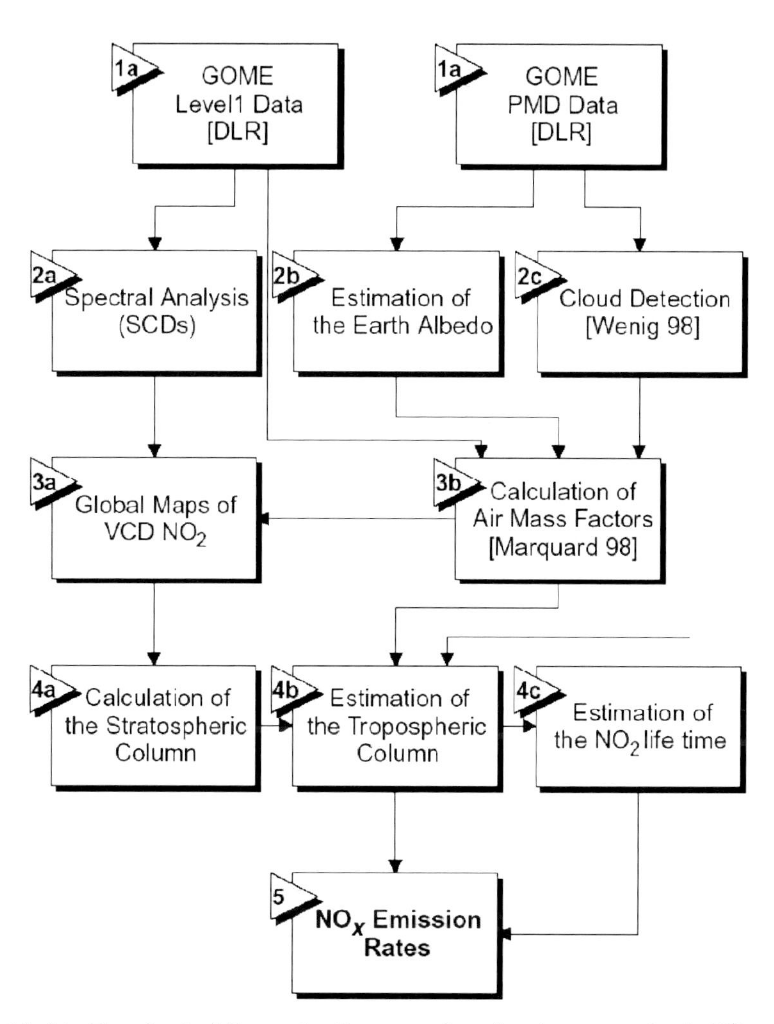

Fig. 11.44. Flowchart of the evaluation procedure for the tropospheric NO_2 residual (from Leue et al., 2001a, Copyright by American Geophysical Union (AGU), reproduced by permission of AGU)

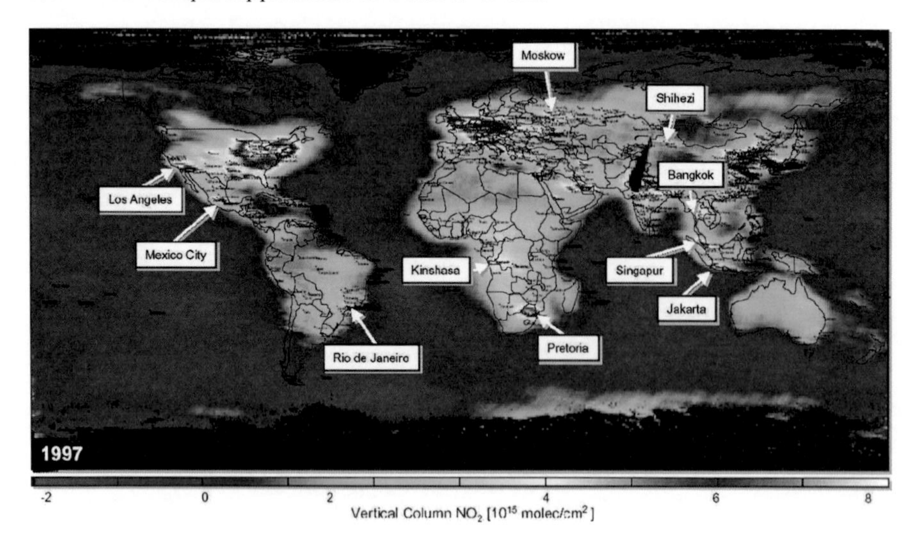

Fig. 11.45. Mean tropospheric residual of the NO_2 VCD for the year of 1997 measured by GOME (from Leue et al., 2001a, Copyright by American Geophysical Union (AGU), reproduced by permission of AGU)

Much higher spatial resolution (30×60 km) is obtained by the SCIAMACHY instrument (see example in Fig. 11.46). This map is averaged over all available data, spanning 18 months, which reduces the effects of seasonal variations in biomass burning, and human activity. The high NO_2 VCDs are associated with major cities across North America and Europe, along with other sites such as Mexico City in Central America, and coal-fired power plants in the eastern Highveld plateau of South Africa. The highest NO_2 levels anywhere on earth are found above north eastern China. This is the most prominent change from previous satellite surveys, and reflects the country's rapid pace of industrialisation (Richter et al., 2005; van der A et al., 2006). Across Southeast Asia and much of Africa, NO_2 produced by biomass

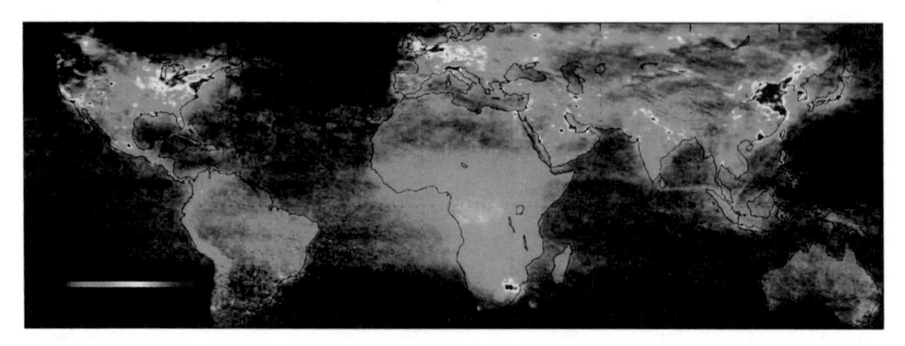

Fig. 11.46. Mean tropospheric NO_2 VCD for 2003–2004 measured by SCIAMACHY (from Beirle 2004)

burning is seen. Ship tracks (NO_2 from ship exhaust stacks; see also Beirle et al., 2004b; Beirle 2004) are also visible in some locations, for instance at the Red Sea and the Indian Ocean between the southern tip of India and Indonesia.

GOME satellite data was also used to derive the weekly variation of the NO_2 column densities of populated or industrial areas of the world (Beirle et al., 2003a). Figure 11.47 shows the normalised amplitude of the NO_2 variation as a function of the day of week averaged over the period from 1996 to 2001. There is a pronounced weekly cycle in Western Europe, North America, and parts of Asia (Korea and Japan). The cycle is absent in Northeast China.

Biomass burning events have been studied using multiple satellite sensors by Edwards et al. (2003). Another area of research where satellite measurements are an important tool is studies of the intercontinental transport of air pollution (Spichtinger et al., 2001; Wenig et al., 2003). New instruments such as SCIAMACHY also allow the study of regional air quality (Fig. 11.48).

11.8.3 Tropospheric Formaldehyde

Formaldehyde (CH_2O) can also be measured by GOME. It serves as an indicator of anthropogenic or natural combustion (biomass burning) and degradation of hydrocarbons (Fig. 11.49). Elevated HCHO levels are found in highly forested areas and in regions with biomass burning (Wittrock et al., 2000b). Figure 11.50 shows the distribution of formaldehyde over North America

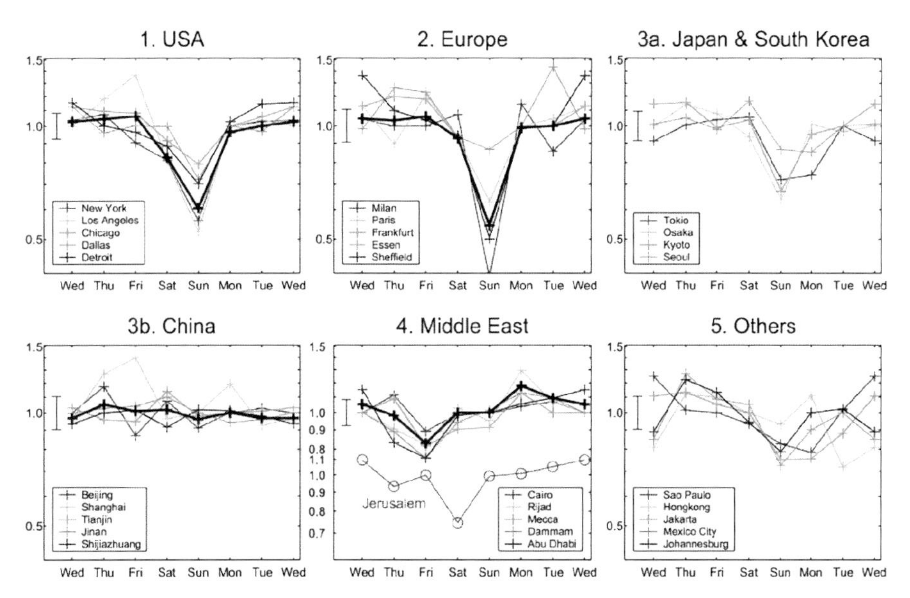

Fig. 11.47. Six-year mean (1996–2001) of GOME tropospheric NO_2 VCDs as a function of day of week (from Beirle et al., 2003a)

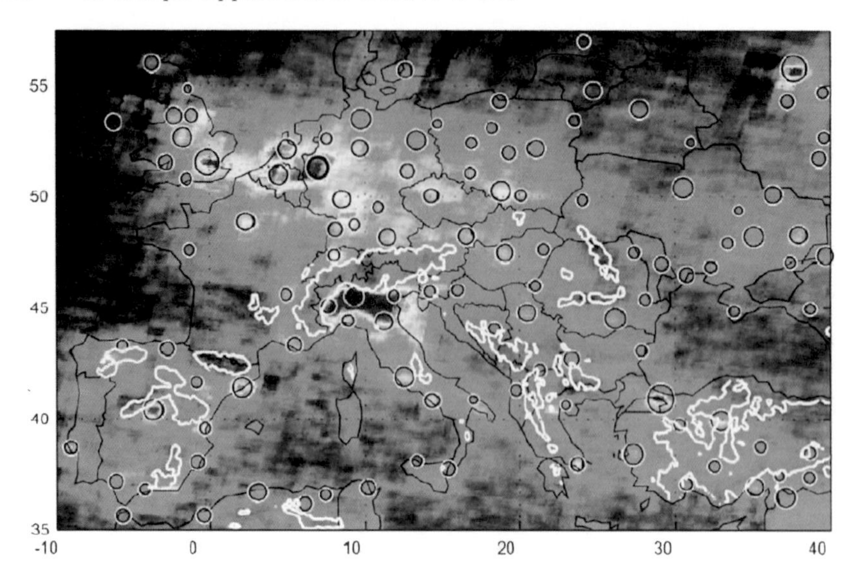

Fig. 11.48. High resolution (80 × 40 km) mode NO_2 map of Europe. Cities over 500,000 inhabitants are encircled (from Beirle et al., 2004a)

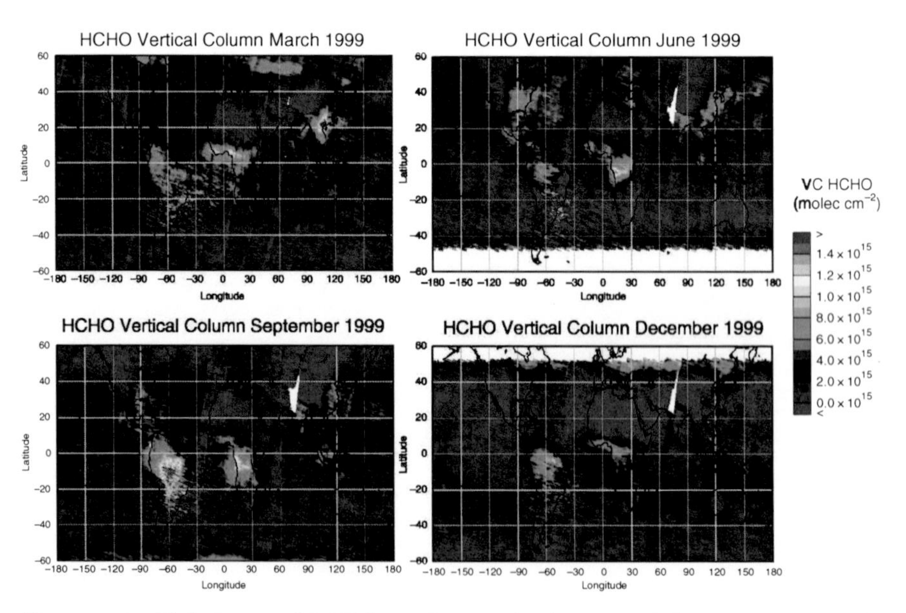

Fig. 11.49. Global view of HCHO total vertical columns in different seasons of the year 1999 (from Wittrock et al., 2000b)

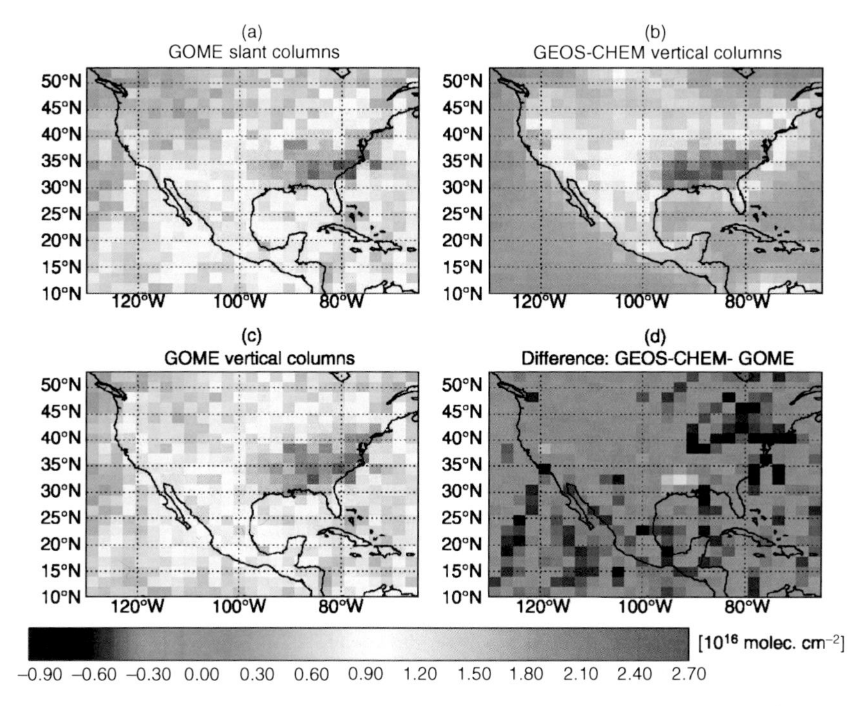

Fig. 11.50. HCHO total vertical columns over the US for July 1996 on a 2° by 2.5° grid. Panel (**a**) shows the observed slant columns from GOME, panel (**b**) shows the total vertical columns from the GEOS-CHEM global 3-D model, panel (**c**) shows the total vertical columns from GOME after application of the AMF, and panel (**d**) shows the differences between the modelled and observed total vertical columns. Both model and observations are for 10–12 local time and for cloud cover <40%. The high formaldehyde levels in the Southeast USA are attributed to degradation of biogenic hydrocarbons (from Palmer et al., 2003, Copyright by American Geophysical Union (AGU), reproduced by permission of AGU)

(Palmer et al., 2003). The high formaldehyde levels in the Southeast USA are attributed to degradation of biogenic hydrocarbons.

11.8.4 Tropospheric SO_2

In the GOME spectra, the characteristic absorption structures of SO_2 can be readily identified in the spectral range of about 315–327 nm. Using DOAS, integrated columns of SO_2 can be determined. However, due to the increased Rayleigh scattering at these small wavelengths, a large fraction of light observed from space originates in the mid-troposphere. Therefore, the sensitivity to the lowest part of the troposphere is strongly reduced compared to the upper troposphere. This effect can be accounted for by making a priori assumptions on the shape of the vertical SO_2 distribution. Details of the assumed SO_2 profile have a significant impact on the final result.

While the global background concentration of SO_2 is difficult to quantify with GOME measurements, volcanic eruptions can readily be observed, and the emission plumes can be monitored over several days (Eisinger and Burrows, 1998). This is true also holds for minor explosions and continuous out gassing processes, which are difficult to monitor from the ground. The retrieval of SO_2 column densities from GOME is strongly affected by the spectral interference due to some instrumental effects, the so-called diffuser plate structures. These are artificial structures induced in the measured GOME spectra due to variation in the position of sun with respect to the instrument throughout the year. Efforts are made to eliminate the impact of these diffuser plate structures on the retrieval of SO_2 column densities by using some simple techniques (Richter and Wagner, 2001). A region from 60°E to 95°E and 15°S to 30°S in the remote Indian Ocean was selected, where the SO_2 column density was presumed to be negligible. The whole data set was normalised with respect to this region. While it may be possible that some information was lost, the results show a very good consistency as can be seen in Fig. 11.51.

Under favourable conditions (no clouds and strong inversion), significantly enhanced SO_2 columns can be observed in regions with intense coal burning, in particular during winter time. The detection of anthropogenic SO_2 emissions from space is another demonstration of the up-to-now unparalleled sensitivity of the GOME instrument towards tropospheric constituents (Khokhar et al., 2005).

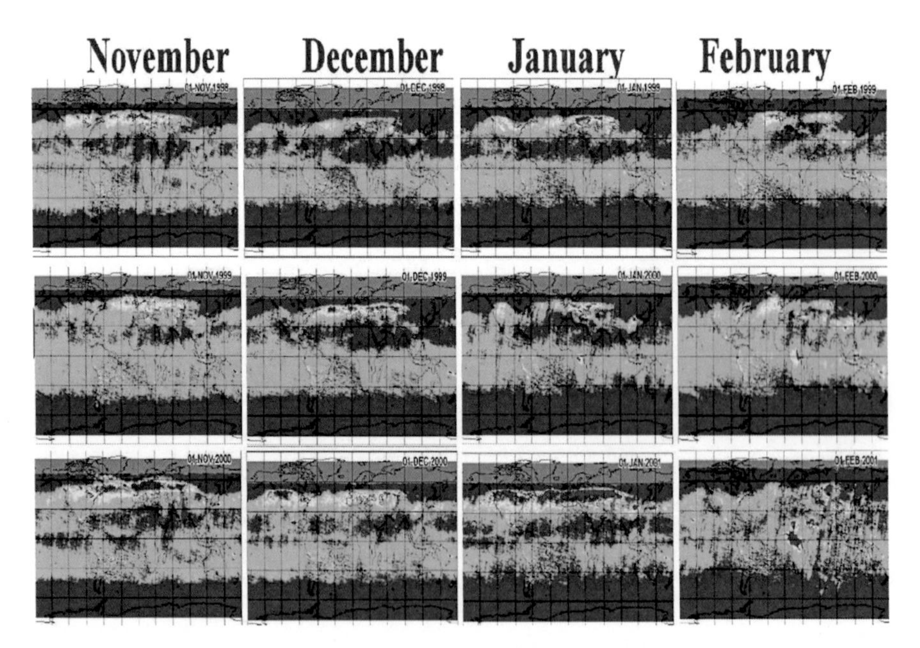

Fig. 11.51. Monthly mean SO_2 total vertical columns over Western Europe for the months November through February (*columns*) of the years 1998/1999 (*top row*), 1999/2000 (*centre row*), and 2000/2002 (*bottom row*). (from Khokhar et al., 2005)

11.8.5 Tropospheric BrO

Polar BrO

From satellite observations, it was for the first time possible to monitor the spatial and temporal evolution of enhanced BrO concentrations in the boundary layer on a global scale (see Fig. 11.52; Wagner and Platt, 1998; Hegels et al., 1998; Wagner et al., 2001b; Richter et al., 1998, 2002; Hollwedel et al., 2004). Wagner et al. (2001b) concluded that the BrO concentrations in the boundary layer measured by GOME are in agreement with those from previous ground-based observations, and that elevated boundary layer BrO coincides with ozone depletion events (observed by ground-based instruments). As shown in Fig. 11.53, the events start to appear in January and July for the northern and southern hemisphere, respectively. The onset of the events is most probably determined by the end of polar night over regions where sea ice surfaces are present. The latest events were observed during June and December, probably determined by the melting of the sea ice surfaces. The highest boundary layer BrO concentrations and the largest spatial extension

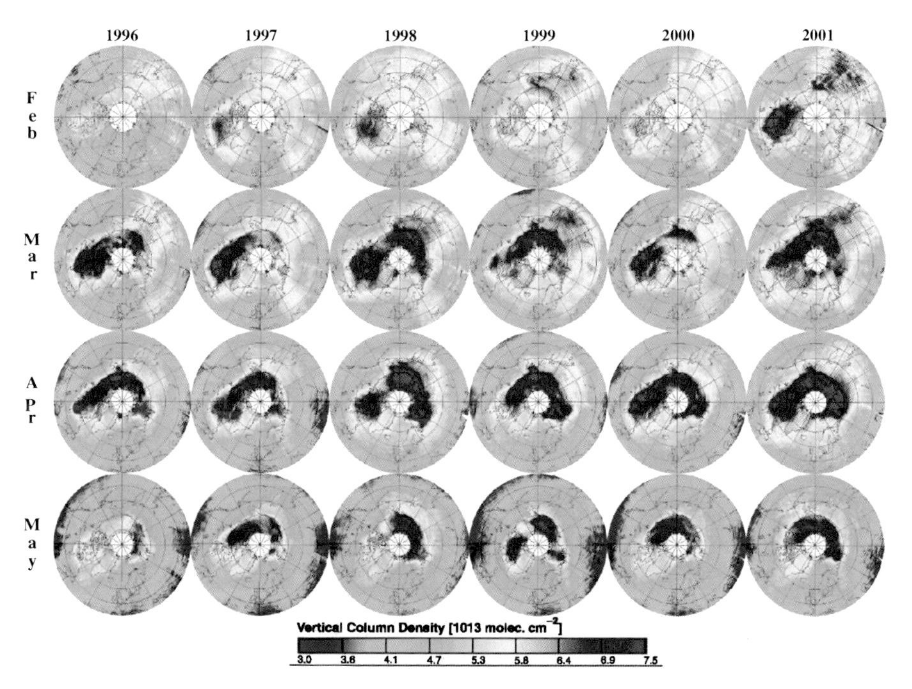

Fig. 11.52. Monthly means of the tropospheric BrO VCDs observed by GOME for the northern hemisphere. (SZA 90°, AMF for standard stratospheric profile) during the spring months February–May (*from top to bottom*) of the years 1996–2001 (*from left to right*). The colour scale ranges from 3×10^{13} to 7.5×10^{13} molec. cm^{-2}. (from Hollwedel et al., 2004)

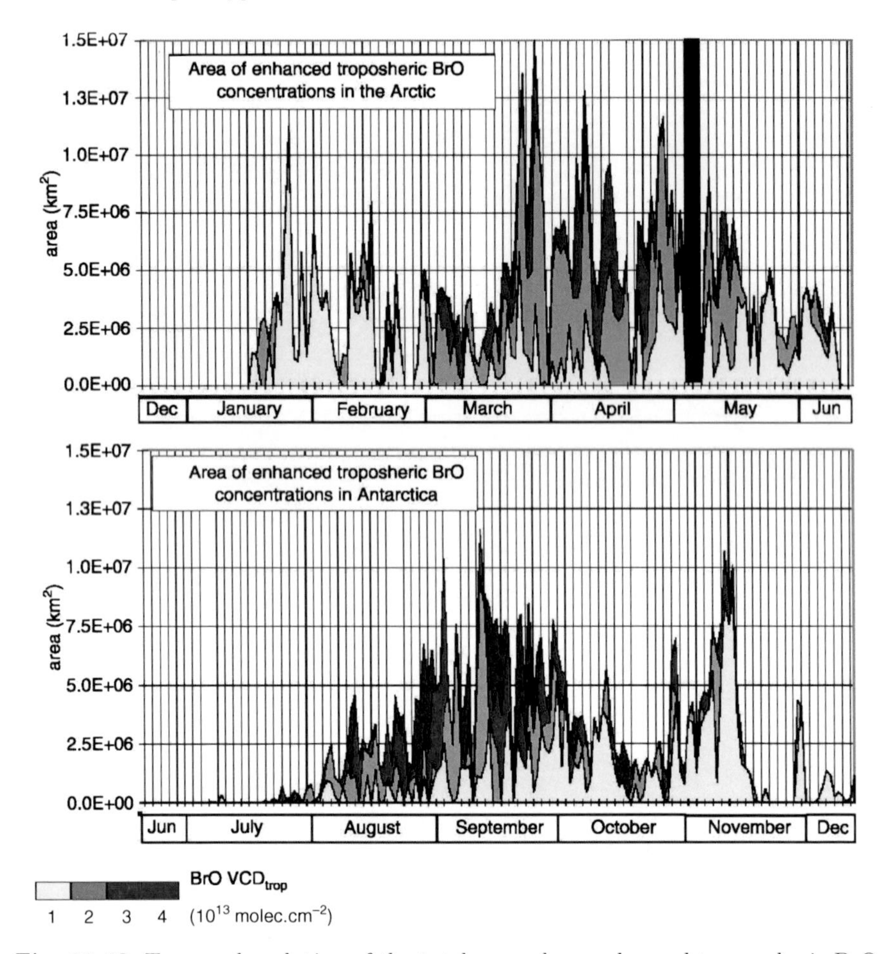

Fig. 11.53. Temporal evolution of the total area where enhanced tropospheric BrO VCDs are observed by GOME for the northern (*upper panel*) and the southern hemisphere (*bottom panel*). The magnitude of the tropospheric BrO VCD is indicated by different colours. It was determined using the stratospheric AMF and can thus serve only as a rough indicator. The *black box* in the *upper panel* indicates a period with no data (from Wagner et al., 2001b, Copyright by American Geophysical Union (AGU), reproduced by permission of AGU)

of the 'bromine clouds' occur during March/April and August/September in the northern and southern hemisphere, respectively. During this time, about 8×10^{30}–10×10^{30} BrO molecules (≈1000–1300 t of Br) are present in the polar boundary layer. The total area reaches 1.5×10^7 km^2, about 3% of the surface of earth. The average BrO load (from the beginning of August to the middle of November in the Antarctic, and from the middle of January to the beginning of June in the Arctic) is about 3×10^{30}–4×10^{30} molecules (equivalent to 400–550 t of bromine). In the northern hemisphere, the areas are up to 20% larger compared with the southern hemisphere; however, in the southern hemisphere, the peak BrO concentrations are higher. Events of enhanced boundary layer BrO concentrations were also sometimes found at mid-latitudes (e.g. around the Caspian Sea). High BrO appears to be related to the occurrence of freezing sea water. Enhanced marine boundary layer BrO concentrations might have gone undetected due to the small ground albedo of the open oceans. Thus, the total area where enhanced boundary layer BrO concentrations occur is probably underestimated. The release of reactive bromine compounds without the presence of (one year old) sea ice is not very probable because: (1) no enhanced BrO concentrations in coastal regions are observed during the summer periods and (2) enhanced BrO concentrations above the Antarctic continent or above Greenland are found only very seldom and are probably related to transport processes. This, in particular, confirms the assumptions of Lehrer (1999). In summary, it appears that whenever the conditions for the autocatalytic amplification mechanisms are fulfilled, high concentrations of reactive bromine compounds are indeed present. This might indicate that small concentrations of reactive bromine (possibly not measurable with current instrumentation) are nearly always present in the troposphere. Further data were presented by Hollwedel et al. (2004).

11.8.6 Tropospheric Carbon Monoxide

Carbon monoxide (CO) has relatively strong absorption bands in the near IR. For its retrieval, a fit window has to be selected as a compromise between the higher instrument sensitivity at shorter wavelengths and a better signal-to-noise ratio at longer wavelengths. Frankenberg et al. (2005) chose a window in channel 8, between 2324 and 2335 nm. CO is a relatively weak absorber whose absorption lines are overlaid by strong absorptions by CH_4 and H_2O.

In addition, in contrast to the UV/vis spectral region, the near IR exhibits peculiarities that make it somewhat difficult to apply simple, 'classical' DOAS algorithms; this is mostly due to the strong sensitivity of the line shape to temperature and pressure in the IR, and the presence of unresolved absorption lines. To overcome these problems, Buchwitz et al. (2000) introduced the concept of weighting function modified DOAS (WFM-DOAS). Other research groups, like Schrijver (2004), also developed modified algorithms for the near-IR spectral region; some preliminary results are shown by Buchwitz et al. (2004) and Gloudemans et al. (2004). Frankenberg et al. (2004) developed the 'Iterative Maximum A Posteriori'-DOAS (IMAP-DOAS), which was applied to obtain the global CO and CH_4 column density maps shown in Figs. 11.54 and 11.55. In addition to these modifications of the classical DOAS algorithm, instrumental problems of higher dark currents of the near-IR detectors are an important issue and have to be analysed and corrected in detail.

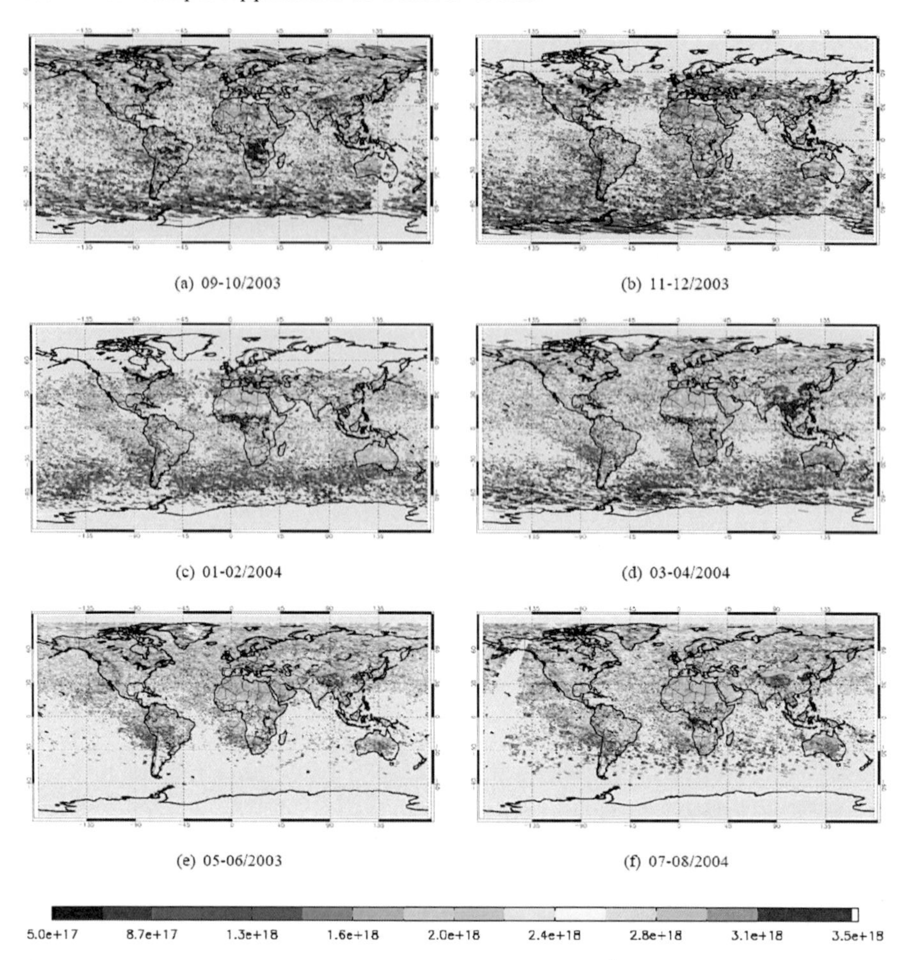

Fig. 11.54. Maps of maximum CO VCDs (molec. cm^{-2}) within a two-month period from September 2003 to August 2004, as derived from SCIAMACHY near-IR spectra. The most striking enhancements are seen in equatorial America and Africa, where the high CO levels are probably due to biomass burning (from Frankenberg et al., 2005)

11.8.7 Tropospheric Methane

Like CO, the greenhouse gas methane (CH_4) has relatively strong absorption bands in the near IR. For its retrieval, similar problems have to be overcome as in the case of CO (see Sect. 11.8.6). An additional difficulty arises from the fact that the CH_4 lifetime in the atmosphere is around 10 years (compared to a few months in the case of CO), and therefore the CH_4 mixing ratio is quite uniform throughout the global troposphere. In order to resolve structures in the CH_4 distribution due to sources and sinks, the column density has to be measured with a precision of about 1%. Frankenberg et al. (2005) solved the problem

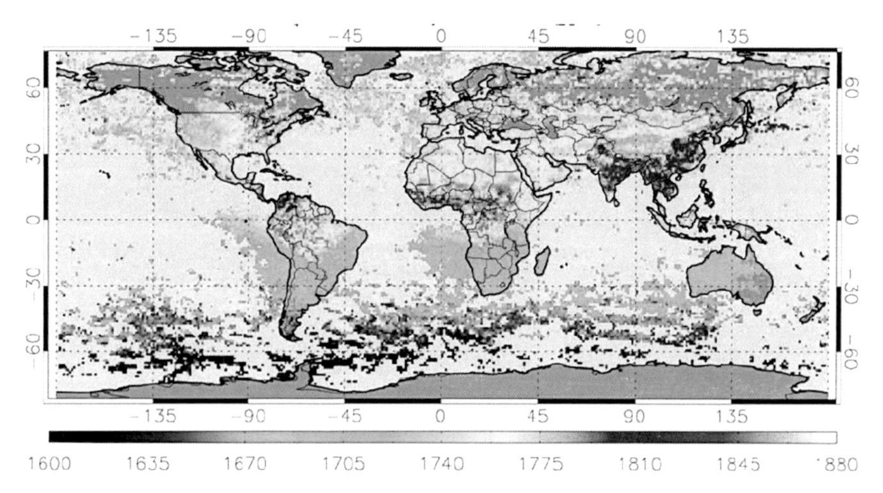

Fig. 11.55. Global map of column-averaged methane mixing-ratios (in ppb). Data are averaged over the time period of August through November 2003 on a 1° ×° 1 horizontal grid. At least five measurements (and up to 150) are taken for each grid cell. Few observations are available over the ocean due to the low ocean reflectivity substantially reducing the signal-to-noise ratio there. Occasionally, sun glint or clouds at low altitudes allow measurement over the ocean. Note that the scale only spans a concentration range of 1740 ± 140 ppb (from Frankenberg et al., 2005)

by ratioing the CH_4 columns derived from satellite near-IR measurements (SCIAMACHY on ENVISAT; see Table 11.2) by CO_2 columns derived from IR data from the same instrument.

Two problems affecting the retrieval in the near IR have to be considered. First, the measurement of the VCD of a well-mixed gas scales linearly with the total column of air, and thus with surface pressure. Hence, a proxy for surface pressure is required in order to convert column density data into mixing ratios. Second, clouds and aerosols can substantially alter the light path of the recorded photons. Carbon dioxide exhibits only very small variations in its total columns and the retrieval window (1562–1585 nm) is spectrally close to the CH_4 retrieval window (1630–1670 nm). Thus, it is well suited as a proxy for both surface pressure (or, in the presence of clouds, cloud top pressure) and changes in the light path due to aerosols and partial cloud cover. Consequently, it is possible to derive the column-averaged dry (no water vapour included) volume mixing ratio (VMR) of CH_4 in the atmosphere from the ratio of the CH_4 VCD and the CO_2 VCD scaled to the averaged atmospheric CO_2 mixing ratio (Frankenberg et al., 2005, 2006).

11.8.8 Tropospheric Water Vapour

Water vapour has structured absorption features in the red part of the spectrum, which can be used to retrieve the water vapour total column (Noël

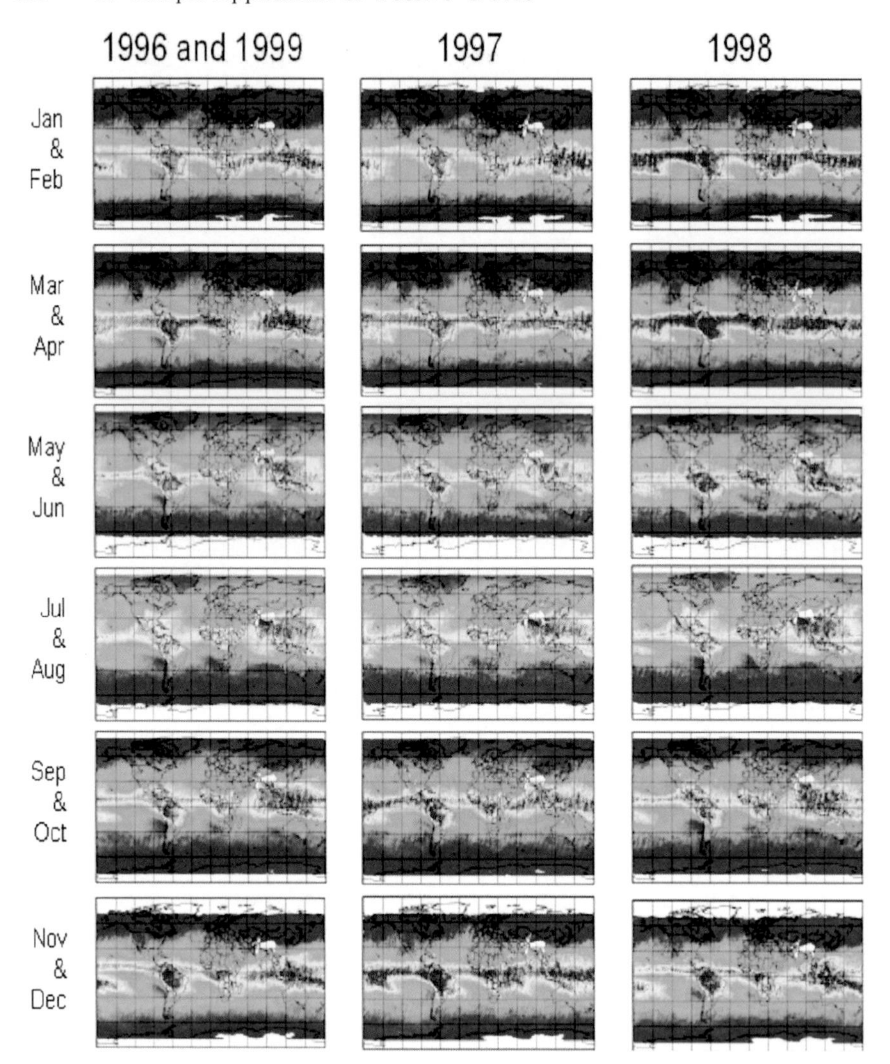

Fig. 11.56. Time series of two-month averages of the H_2O VCD. *Left*: average for normal years (1996 and 1999); *middle and right*: average for the El-Niño years (1997 and 1998). Strong El-Niño induced anomalies are found from September/October 1997 to March/April 1998 (from Wagner et al., 2005, Copyright by American Geophysical Union (AGU), reproduced by permission of AGU)

et al., 1999, 2002; Casadio et al., 2000; Maurellis et al., 2000b; Wagner et al., 2003, 2005; Lang and Lawrence, 2004). Following the customary DOAS evaluation, the H_2O SCD has to be corrected for the non-linearity arising from the fact that the fine structured H_2O absorption lines are not spectrally resolved by the GOME instrument (see Sect. 6.7). Then, the corrected H_2O SCD is divided by a 'measured' AMF, which is derived from the simultaneously evaluated O_4 absorption. Thus, the effects of radiation transport and clouds are eliminated. Compared to 'traditional' water vapour measurements from satellite, the DOAS data have essentially uniform sensitivity to H_2O throughout the atmosphere, thus they see the bulk of the H_2O close to the surface.

An example, where the DOAS fit was performed in the wavelength interval 611–673 nm is shown in Fig. 11.56 (Wagner et al., 2005). A band of high H_2O VCDs always surrounds the tropics according to the high surface temperatures. In most months, the local maxima are located over the continents, e.g. Brazil or Indonesia. Local minima at the equator are found at the east coast of Africa. As expected, the belt of maximum H_2O VCD follows the latitudinal variation of the ITCZ location with season; this variation is especially pronounced above the continents. During El-Niño periods, the global pattern of the H_2O VCD changes significantly. In addition, a very strong increase of H_2O in the equatorial Pacific, and a shift southwards can be recognised. This shift is most obvious at the end of 1997 and the beginning of 1998.

11.9 Determination of Photon Path Lengths by 'Reversed DOAS'

A novel application of the DOAS 'reverses' the customary DOAS approach. It measures the lengths of photon paths in clouds by making use of the known concentrations of atmospheric constituents such as molecular oxygen (O_2), tropospheric ozone, or the oxygen dimer $(O_2)_2$, O_4.

11.9.1 Average Path Lengths from Low Resolution Measurement of Weak Absorbers

Oxygen dimers have very wide spectral features (>1 nm), and they can therefore be monitored by simple, low-resolution instruments (Erle et al., 1995). Time series indicating average photon path lengths in clouds of almost 30 km at Kiruna (Sweden) were reported by Wagner et al. (1998).

11.9.2 Path Length Distributions from High Resolution Measurement of Strong Absorbers

By analysing the absorption of individual rotational lines (e.g. of O_2), it is possible to infer not only the average light path length in clouds, but also elements of the path-length distribution function (photon PDF) of solar photons transmitted to the ground (Pfeilsticker et al., 1995, 1998a,b; Pfeilsticker, 1999;

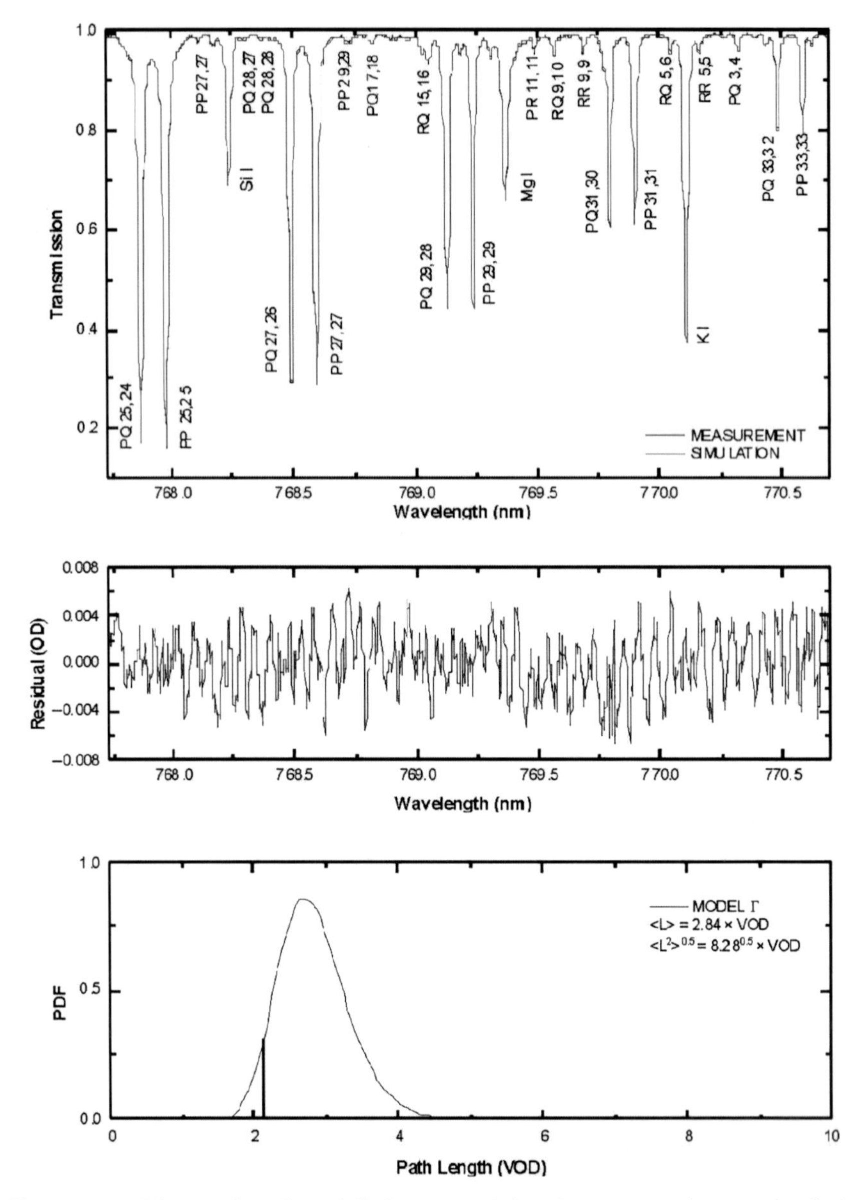

Fig. 11.57. Measured and modelled oxygen A-band spectrum (*upper panel*), inferred residual spectrum (*middle panel*) and inferred photon PDF for the observation between 12:32 and 12:33 UT over Cabauw/NL on 23 September, 2001. The *black vertical line* indicates the optical path for the direct sunlight. Photon path lengths are given in units of vertical atmosphere (VOD) (from Scholl 2006)

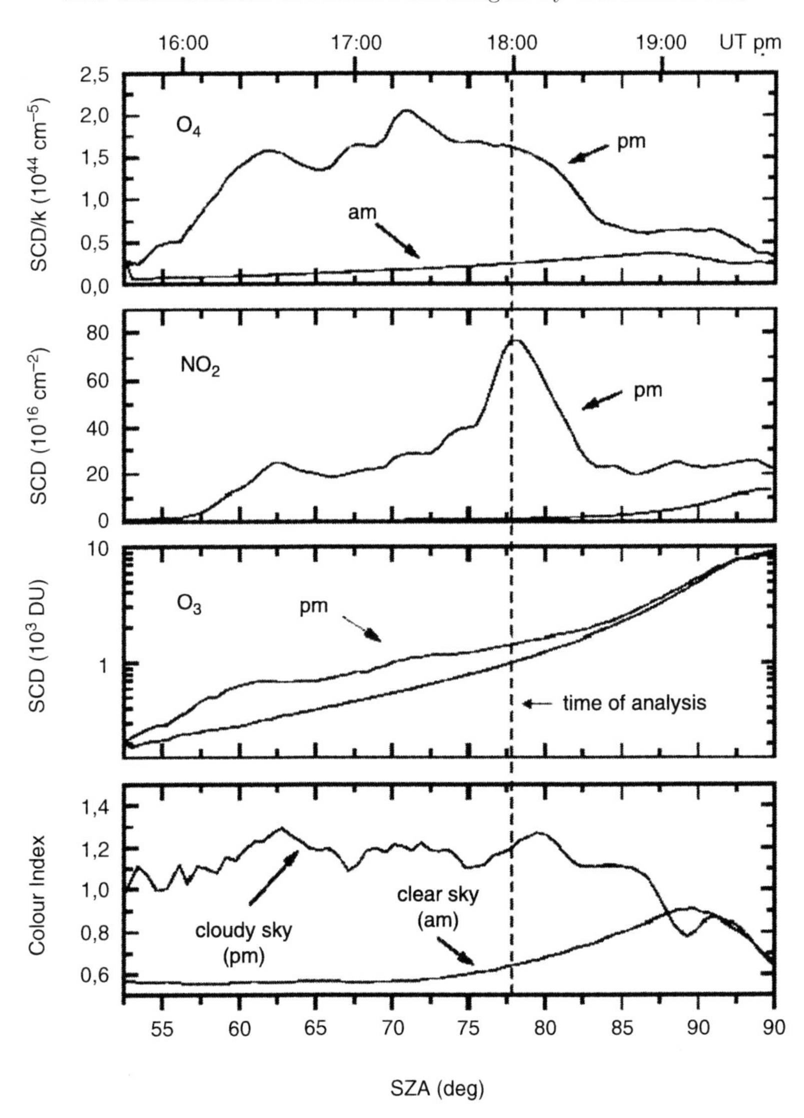

Fig. 11.58. The time development of the observed SCDs of various trace gases and the cloud index. The evening values during the cloudy period are clearly enhanced. Note the logarithmic scale for O_3. The *dashed line* indicates the time of analysis (figure from Winterrath et al., 1999, Copyright by American Geophysical Union (AGU), reproduced by permission of AGU)

Veitel et al., 1998; Wagner et al., 1998; Min et al., 2001; Funk, 2000; Funk and Pfeilsticker, 2003). The knowledge of photon PDFs is of primary interest for atmospheric radiative transfer, and thus for the climate. The method relies on the analysis of highly resolved oxygen A-band (762–775 nm) spectra

(Fig. 11.57) observed with a small field of view ($-1°$) telescopes in zenith scattered skylight. It uses the fact that solar photons are randomly scattered by molecules, aerosols, and cloud droplets while traversing the atmosphere. Since a fraction of the solar photons is absorbed in the atmosphere, they on average travel different distances, depending on opacity. Spectral intervals with a largely changing atmospheric opacity but known absorber concentration and spectroscopic constants, such as in the spectral region of the oxygen A-band, thus contain information on the photon PDF. Recent studies have shown that the photon PDF is largely correlated with the total amount of liquid water in the atmosphere, the spatial arrangement of clouds, cloud inhomogeneity, and solar illumination (Pfeilsticker et al., 1998; Pfeilsticker 1999; Veitel et al., 1998; Min et al., 2001; Funk and Pfeilsticker, 2003). For example, for thunderstorm cumulus nimbus (Cb) clouds, average photon path lengths of up to 100 km have been observed; whereas for stratus (Sc) clouds, photon paths are only 50–100% larger than for the direct and slant path of the sun's rays. Moreover, it appears that even though the spatial distribution of cloud droplet is inhomogeneous, with the moments of the density behaving like multi-fractals, photon PDFs tend to be mono-fractal, mostly due to the so-called radiative smoothing for optically thick clouds (Savigny et al., 1999).

11.9.3 Measurement of Trace Gases Inside Clouds

The phenomenon of light path extension inside clouds described in Sect. 11.9 allows determination of trace gas concentrations inside clouds. By analysing the radiation originating from the bottom of a cloud, the light path extension inside the cloud can be determined (e.g. from O_4), while the column density of trace gases (e.g. NO_2) can be simultaneously determined. By comparing to a situation without cloud at a different time or point at the sky in the case of patch cloudiness, the excess trace gas column due to the cloud can be determined. By dividing the excess trace gas column density by the extra light path in the cloud, the average trace gas concentration inside the cloud can be derived. This technique was used by Noxon (1976), Franzblau and Popp (1989), Winterrath et al. (1999), and Langford et al. (2004) to derive the evolution of NO_2 during thunderstorm events and thus to derive the NO_x production by lightning. Figure 11.58 shows the evolution of the absorption due to O_4, NO_2, and O_3 in a thunderstorm cloud.

12

DOAS: Yesterday, Today, and Tomorrow

The preceding two chapters presented examples of the technological development and the scientific contributions of the DOAS technique. This final chapter will give a historical overview of this method in order to provide an outlook on the future of DOAS, with respect to both its further technological development, and upcoming scientific and monitoring applications. Following the approach taken throughout this book, we will distinguish between active DOAS, which employs artificial light sources, and passive DOAS, which uses sunlight.

12.1 Passive DOAS Applications

Figure 12.1 illustrates how, beginning with the measurements of total ozone in 1930s by Dobson through direct sunlight, the passive DOAS method has branched out in a multitude of different applications. Direct sun- or moonlight applications are quite rare at the present time. Among these, the most important are balloon borne DOAS measurements (Ferlemann et al., 1998; Harder et al., 1998), which have given important insights into the vertical distribution of trace gases such as NO_2 and BrO. A few direct moon measurements of photolabile species, such as NO_3, have also been reported. While not employed frequently in the past, direct sunlight observations have been revived through the occultation measurements of the SCIAMACHY (Scanning Imaging Spectrometer for Atmospheric Chartography) instrument.

The measurement of scattered sunlight, which was initiated by Noxon in 1975 by aiming his instrument at the zenith, led to a better understanding of both ozone and NO_2 in the stratosphere. Driven by the discovery of the Antarctic ozone hole in 1985, DOAS rapidly found its place as a key tool to study the halogen-catalysed ozone destruction mechanisms and the budget of nitrogen species. The measurements of BrO and OClO by Solomon et al. (1987b, 1989d) revealed that chlorine and bromine chemistry played a crucial role in the formation of the ozone hole. Followed by this success, variations

U. Platt and J. Stutz, *DOAS: Yesterday, Today, and Tomorrow.* In: U. Platt and J. Stutz, Differential Optical Absorption Spectroscopy, Physics of Earth and Space Environments, pp. 495–504 (2008)
DOI 10.1007/978-3-540-75776-4_12

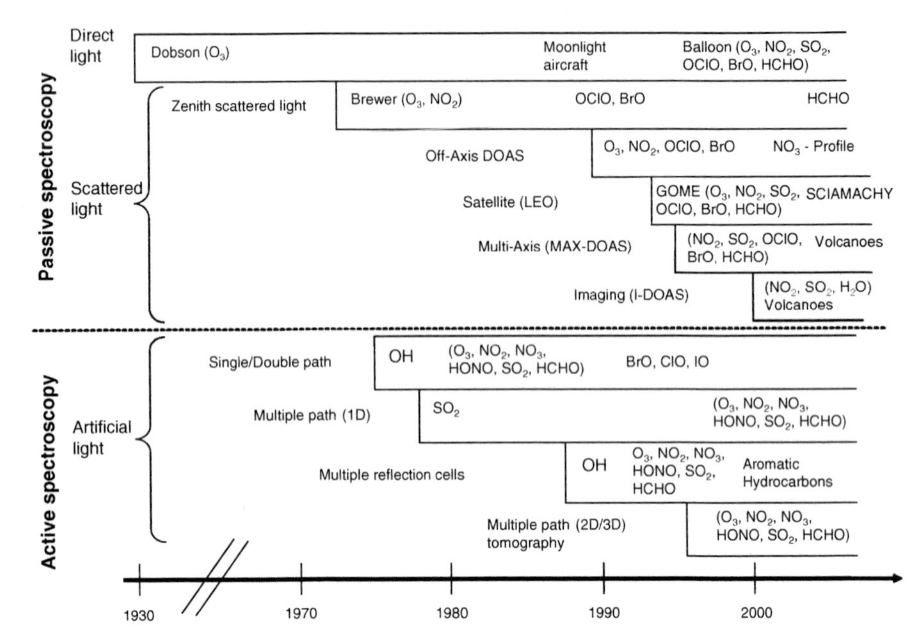

Fig. 12.1. Development of differential optical spectroscopy since 1930

of the zenith-light technique were developed by aiming the telescope to lower elevations and increasing the sensitivity, for example for NO_3 (e.g. Sanders et al., 1987). This advance culminated in the recent development of Multi-Axis DOAS, which uses multiple viewing angles. MAX-DOAS also expands the use of passive DOAS systems from mostly stratospheric applications, to the measurements of tropospheric trace gases.

One of the most exciting developments over the past decade was the expansion of DOAS to satellite-based measurements. With the launch of GOME (Global Ozone Monitoring Instrument) on board ERS-2 (Earth Research Satellite 2 by ESA) in 1995, a new era of DOAS measurements began. GOME's unique ability to measure multiple trace gases (despite its name only hinting ozone), such as O_3, NO_2, BrO, HCHO, and SO_2, on a global scale has expanded the scale of DOAS applications from local and regional applications, to global coverage. Moreover, for the first time species in the troposphere, and even in the boundary layer, can be routinely monitored from space. The next generation of satellite instruments, SCIAMACHY, was launched in 2002 and is now operational.

Passive DOAS techniques today encompass a wide variety of applications. Balloon-borne occultation experiments are used in the study of upper tropospheric and stratospheric chemistry and the validation of satellites. The measurement of NO_2 and other species using zenith-scattered sunlight is used in the monitoring of the stratosphere in the Network for the Detection of Stratospheric Change (NDSC) and the Global Atmospheric Watch (GAW) network.

Currently, most active efforts concentrate on the application of MAX-DOAS. A number of uses have emerged in recent years. In particular, the fact that MAX-DOAS instruments can be built small and inexpensively makes them ideal to use in remote and difficult to access areas, such as polar regions, volcanoes, and the ocean. Several applications have shown the value of this new method, for example to measure halogen oxides in the remote Arctic and Salt Lakes, measure NO_2 and BrO on board ships, and monitor emissions from volcanoes (Bobrowski et al., 2003). MAX-DOAS has become a powerful new method. While its applications show rapid growth, it is undergoing further refinement at this moment. Much of the effort of the DOAS community in recent years has also focused on the use of satellite borne DOAS instruments, in particular the GOME instrument.

The advances of passive DOAS over the past few years points to a number of developments that are likely to occur in the future. These developments are, in part, technical. However, a number of interesting applications are expected to emerge.

12.1.1 MAX-DOAS

The rapid development of MAX-DOAS over the past few years offers a large number of practical applications. A number of these applications have been demonstrated. Others are in the development phase, and some are currently just ideas that may or may not be developed. One of the most promising uses for MAX-DOAS is as an inexpensive and easy to use automatic technique to monitor both the level of air pollution, and the emissions from individual sources. One should, however, point out that, in contrast to current air pollution monitoring stations, MAX-DOAS leads to an altitude-averaged measurement in the lower atmosphere. This is, however, a tremendous advantage if one wants to monitor total levels of gases such as NO_2 or HCHO in the troposphere. Due to this averaging, MAX-DOAS measurements are only weakly influenced by local effects and/or small-scale transport phenomena. In addition, the monitoring of trace gas transport above the boundary layer is particularly helpful in the monitoring of long-rage pollution transport.

A recent example of the monitoring capabilities by MAX-DOAS is the remote measurement of SO_2 emitted from volcanoes. Small spectrometers have made it possible to develop miniaturised instruments that can easily be carried to volcano craters. Continuous measurements of SO_2 and other gases are often used in the monitoring of the state of a volcano, with the goal of predicting its next eruption. While few MAX-DOAS instruments are currently dedicated to this task, the advantages of these instruments make it likely that this will become a widespread application. A similar application that has thus far not been attempted is the monitoring of emissions from forest fires.

Another promising application of MAX-DOAS is the monitoring of pollutant emissions, both from industrial point sources, as well as from area sources such as traffic. Again, the path averaging capabilities of MAX-DOAS will allow the determination of the total emissions. It is, for example, straightforward to set up a DOAS system downwind of a power plant to monitor the composition of the plume at some distance from the smoke stack. Setting up an instrument close to a road or in a city will allow the measurement of averaged emissions from cars. Many other applications are possible.

12.1.2 Aerosol and Cloud Monitoring

While DOAS has primarily been used for the measurement of atmospheric trace gases, the ability to measure gases with constant mixing ratios in the atmosphere, i.e. O_2 and O_4, allows for new applications in cloud and aerosol monitoring. This technique has been applied in the study of radiative properties of clouds, and is based on the fact that aerosol scattering in the atmosphere changes the light path length, and thus the optical density of the O_2 and O_4 absorptions. Using radiative transfer calculations, one can therefore extract information about the vertical distribution of the aerosol and its optical properties.

12.1.3 Imaging DOAS

An expansion of the plume monitoring technique uses MAX-DOAS to quantitatively 'visualise' emission plumes. By using two-dimensional detectors, it is possible to acquire a one-dimensional (1D) spectral image of a plume. An additional fast scanning mirror then moves the horizontal viewing angle along the plume, resulting in a data-set of absorption spectra in a 2D field of viewing direction. This can then be converted into a trace gas concentration image with typically 100×100 pixels, clearly showing the plume of a specific trace gas. This truly 'multiple-axis' approach is a unique tool for the visualisation of trace gas distribution, and a large number of applications are conceivable.

12.1.4 Tomography

Another promising, application of MAX-DOAS, the testing of which has just begun, is the tomographic observation of 2D or 3D trace gas distributions. The basic idea here is to deploy many simple MAX-DOAS systems throughout an urban area. The instruments are then consecutively aimed in many different directions, so that viewing geometries cross at multiple locations (Frins et al., 2006). The tomographic analysis of these observations results in 2D or even 3D concentration fields that can, for example, be used to identify pollution sources, or to initialise regional air pollution models.

12.1.5 Satellite Instruments

One of the most exciting aspects of DOAS in recent years has been the success of the space-borne GOME instrument. Results from GOME have shown the usefulness of DOAS observations from space for many applications. The next step in this development was taken in 2002 with the launch of SCIAMACHY. While at the time of writing this book only examples of data have been available, the smaller ground-pixel size, the expansion of the useable wavelength range in the IR, and the ability to make limb and occultation measurements adds a new aspect to space-borne DOAS.

The most useful application for DOAS from space may be the continuous monitoring of tropospheric pollution levels from space which, like today's meteorological satellites, would provide maps of the 'chemical weather'. To achieve this goal, the time frequency of measurements over any point of earth has to be increased from every 1–3 days to perhaps every 2 h. One way in which this could be achieved would be the deployment of a number of smaller satellites in polar orbit that pass over any point on earth, following each other in a determined temporal pattern. Ultimately, one would wish to deploy DOAS instruments in geostationary orbit, which would allow the recording of the diurnal variation of the atmospheric composition. First proposals of such an instrument have been made. In summary, it is very likely that a network of 'chemical weather' satellites will available in the next two decades.

12.2 Active DOAS Applications

The first active DOAS instruments were developed in the 1970s as research tools for the measurements of various radicals, in particular OH. To develop systems capable of measuring O_3, NO_2, and SO_2, as well as NO_3 and HONO, the 'broadband' laser used in these initial systems was soon replaced by thermal light sources (e.g. Xe arc lamps), which give a much broader spectrum and are technologically simpler. These early instruments, which were based on opto-mechanical 'slotted disk' detectors, are the basis of most commercial air pollution monitoring instruments available today. In addition, the desire to fold the long light path in a small volume led to the development of multi-reflection DOAS systems. Improvements of the various active DOAS instruments were more subtle. The use of new photodiode array detectors, first for OH measurements in 1985, and later for broadband active DOAS in 1995, increased the measurement frequency and allowed the use of longer light paths, thus improving the detection limits. The introduction of the coaxial telescope/retroreflector set-up greatly simplified the deployment of the instruments in the field. In addition, the number of trace gases that are measured with DOAS has expanded over the past decade, and now includes all tropospheric halogen oxides and many aromatics.

12.2.1 New Trace Gases

The increase in the number of species measured by active DOAS has been driven in large part by the enhancement of detection limits and new applications for this technique. Despite these improvements, there is still about an order of magnitude in sensitivity to be gained in active DOAS applications. An extrapolation of the success in the past indicates that it is very likely that new trace gases will enter the pool of species that can be observed. Possible candidates for such species are, for example, polycyclic aromatics, and oxygenated hydrocarbons such as acrolein or ketones.

12.2.2 Infrared Measurements

A number of trace gases that play an important role in atmospheric chemistry and climate research, such as CH_4 and CO, absorb in the near infrared wavelength region. By expanding DOAS to longer wavelengths, new species can be measured, opening opportunities for new applications. It is well possible that more trace gases will be targeted in the near infrared wavelength region in the future.

12.2.3 Hydrocarbons

Possible target species in the near IR region between 1 and 2 μm, wavelength region are hydrocarbons. The overtone spectra of the C–H stretch are located in this spectral region, thus allowing the measurement of these compounds. While the absorption may be too small to allow for ambient monitoring, DOAS could be used for emission or fence-line monitoring (see below).

12.2.4 Air Pollution Monitoring

A number of commercial active DOAS instruments are available on the market to monitor pollutants such as O_3, SO_2, and NO_2. While some instruments have been employed by air quality agencies, the potential of DOAS for monitoring purposes has not been fully exploited. DOAS offers a number of advantages over other methods that are often not recognised. Due to the simultaneous measurement capability, one DOAS instrument can replace a whole set of instruments, which are individually dedicated to the measurement of O_3, SO_2, NO_2, or BTX. DOAS is an absolute method that does not need to be calibrated. The cost of calibrations is thus absent. DOAS instruments are also easy to run automatically, and remote control can be implemented. Finally, taking into account the legal implications of air quality measurements, DOAS measurements provide a 'photograph' of the chemical composition of the atmosphere which can be used as incontestable evidence in court.

12.2.5 BTX Monitoring

The measurement of the ambient concentrations of aromatic hydrocarbons, such as benzene, toluene, and xylenes, has not been extensively explored, and is a promising expansion of the pollution monitoring capabilities of DOAS.

Research applications have illustrated that, by overcoming challenges with overlaying O_2–O_2, and O_2–N_2 dimers, and O_3, the accurate measurement of a wide variety of aromatic hydrocarbons is possible. It should be noted that DOAS is, in contrast to most other techniques, able to distinguish different isomers of aromatic compounds.

12.2.6 Fence-Line Monitoring

A thus far unexplored application of DOAS is the monitoring of emissions downwind from large sources. This so-called fence-line monitoring is often difficult to achieve in a quantitative manner, since dilution will reduce the pollutant concentrations. In a long-path application, however, the integral over an entire plume can be measured, thus making the result less dependent on dilution. Active DOAS systems set up near such a source can thus provide continuous data on the total amount of a species emitted.

12.2.7 Tomography

Initial experiments have shown that DOAS can be used to make tomographic measurements. A set-up of a large number of active DOAS instruments with multiple crossing light beams, together with established tomographic methods, will allow the determination of 2D or even 3D concentration fields. This is a promising approach for pollution monitoring in urban areas, as well as in industrial complexes, that desire spatially resolved emission control. The installation of such a tomographic system would be challenging, since a large number of instruments and reflectors have to be deployed and maintained. The extensive effort in setting up the system would, however, be rewarded by the unique information that such a network would provide. It is thus only a question of time until a sponsor for such work can be found and DOAS tomography is implemented.

12.2.8 Range Resolved Technology/Broadband LIDAR

Another application that has not been extensively explored is the use of the DOAS approach in range resolved techniques, i.e. LIDAR. In contrast to classical LIDARs, which are based on one or multiple lasers, a DOAS LIDAR would require a high-intensity broadband light source that allows the detection of an entire spectrum. While the detection capabilities for time resolved measurements of the returning light are available with modern CCD detectors, a suitable light source has not yet been found. The ability to observe emissions, for example, of NO_x, with a high spatial resolution is, however, a very promising application, both for research and for monitoring purposes.

12.3 Development of the Underlying Technology

In the past, the advancement of DOAS has often gone hand in hand with the availability of new technology. For example, the multiplex advantage of new solid-state detectors, such as photodiode arrays, has reduced the noise in the spectroscopic detection, thus allowing for longer light paths in active DOAS applications, and shorter exposure times in passive DOAS measurements. The availability of inexpensive miniaturised spectrometers has made it possible to build small and light MAX-DOAS instruments that can be carried to a volcano crater for SO_2 monitoring. Many such examples can be found in the history of DOAS. It is thus valuable to speculate how new technology will influence the development of DOAS in the future.

12.3.1 New Light-Sources

Illumination technology in recent years has seen an increased use of light emitting diodes. Today they are, for example, used in traffic lights, and flash lights. Initial applications for home illumination are also becoming available. The high brightness of modern LEDs makes them a natural choice for DOAS applications. Advantages include the low power consumption, longevity, and high stability of these sources. Initial tests have shown that LED-based active DOAS instruments are feasible and the technical implementation is only a question of time.

Tunable diode lasers have also seen a rapid development over the past years. The introduction of the quantum cascade lasers, which can be operated at ambient temperatures, makes these systems easier to use and also less expensive. The scanning ability results in absorption spectra that can be analysed using DOAS techniques. The combination of TDLS and DOAS would open novel applications in the infrared wavelength region.

The development of a broadband LIDAR that is based on DOAS depends on the availability of suitable light sources. Further development of lasers, flash lamps, and other sources could provide such a source, making this exciting instrument possible.

12.3.2 New Detectors

The development of photodiode arrays, and more recently CCD and CMOS detectors, have considerably influenced DOAS. Both expensive high-quality and low-cost CCD detectors are available today. The development of this technology is, however, still advancing, and new detectors become available every year. In particular, for active DOAS, high capacity (large full well capacity) detectors are not yet available, thus limiting the use of these detectors mostly to passive DOAS applications. An increase readout speed would also be highly desirable.

DOAS has mostly been restricted to the UV and visible wavelength range. The upper wavelength limits were determined in the beginning by the availability of suitable photomultiplier tubes, and later by the bandgap limit of silicon used in photodiode arrays and CCDs. New semiconductor technology has led to the development of solid-state detector arrays based on gallium-arsenide that also work above wavelength of 1 μm. These detectors, which work similar to CCDs, open a new spectral window for DOAS. As the development of these detectors advances, DOAS will also expand its IR capabilities.

12.3.3 New Software

A goal for future DOAS applications is to further advance the automation of DOAS instruments. While advances have been made in the control of instruments and the long-term operation, improvements in the analysis software with respect to the fully automated extraction of trace gas concentrations from the absorption spectra, is needed. To reduce the need of initial manual spectral evaluation, and to simplify the process of data reduction, methods have to be found that automatically recognise changes in the atmosphere and the instrument, and provide corrections for these effects. In addition, algorithms for automatic quality control and improved error calculations are needed. These improvements will particularly benefit the use of DOAS as a monitoring technique, where ease of operation and reliability is a key component.

Advances in software development are also expected for MAX-DOAS applications. While radiative transfer models are available to interpret MAX-DOAS data, the algorithms to invert the observations in order to derive vertical profiles of trace gases are still under development. Within the next few years, one can expect considerable advancement in these methods.

12.3.4 Improved System Design

The design of DOAS instruments will proceed in two directions: First, the improvement of current detection limits and the expansion to new wavelength regions. Second, the development of smaller, more automated, and less expensive DOAS systems. The detection limits of current active DOAS instruments are about one order of magnitude above the physical limits imposed, by photon noise. There is thus great potential in improving these instruments to measure lower trace gas levels, improve the accuracy of the measurements, and expand the pool of observable species. While it is currently not clear how these instruments can be further improved, it is very likely that a breakthrough will be achieved during the next decade.

The development of miniaturised spectrometers has allowed the construction of smaller DOAS instruments. The lower power consumption of the systems also allows for operation in remote locations. It is expected that this trend will continue in the future and that DOAS instruments will decrease in size. The use of these systems in highly mobile platforms, such as cars, ships, and aircrafts, is thus on the horizon. The low price of these instruments also makes it very likely that the number of DOAS instruments will increase, and that observation networks will be established. This will most likely occur first with respect to the monitoring of volcanoes, but later expand to air pollution and emission monitoring. Ultimately, one can imagine that further miniaturised 'personal' DOAS instruments may become a tool for trace gas monitoring for the interested scientist and personnel working in air quality control.

There is little doubt that DOAS will be an increasingly valuable tool in our ongoing efforts to understand the atmosphere and the chemical and physical processes occurring in it.

13

Literature

Internet Addresses

Keller-Rudek, H. and Moortgat, G.K. (2005), MPI-Mainz-UV-VIS Spectral Atlas of Gaseous Molecules. URL: www.atmosphere.mpg.de/spectral-atlas-mainz

Smith, P.L., Heise, C., Esmond, J.R., Kurucz, R.L. (1995), Atomic spectral line database from CD-ROM 23 of R.L. Kurucz

Kurucz, R.L. and Bell, B. (1995), Atomic Line Data Kurucz CD-ROM No. 23. Cambridge, Mass.: Smithsonian Astrophysical Observatory. URL: http://cfa-www.harvard.edu/amdata/ampdata/kurucz23/sekur.html

Science-softCon UV/Vis Spectra Data Base URL: http:// www. science-softcon.de/

Rothman, L.S., Barbe, A., Chris Benner, D., Brown, L.R., Camy-Peyret, C., Carleer, M.R., Chance, K., Clerbaux, C., Dana, V., Devi, V.M., Fayt, A., Flaud, J.-M., Gamache, R.R., Goldman, A., Jacquemart, D., Jucks, K.W., Lafferty, W.J., Mandin, J.-Y., Massie, S.T., Nemtchinov, V., Newnham, D.A., Perrin, A., Rinsland, C.P., Schroeder, J., Smith, K.M., Smith, M.A.H., Tang, K., Toth, R.A., Vander Auwera, J., Varanasi, P., Yoshino, K. (2003), The HITRAN molecular spectroscopic database: edition of 2000 including updates through 2001, Journal of Quantitative Spectroscopy & Radiative Transfer **82**, 5–44. http://www.hitran.com

References

Aben, I., Stam, D.M., Helderman, F.: The ring effect in skylight polarisation. Geophys. Res. Lett. **28**(3), 519–522 (2001)

Ahmed, S.M., Kumar, V.: Measurement of photoabsorption and fluorescence cross sections for carbon disulfide at 188–213 nm and 287.5–339.5 nm. J. Phys. **39**, 367–380 (1992)

Albritton, D.L., Schmeltekopf, A.L., Zare, R.N.: An introduction to the least-squares fitting of spectroscopic data, In: Narahar, R.K., Weldon, M.W. (eds.) Molecular Spectroscopy: Modern Research. Academic, Orlando (1976)

Aldener, M., Brown, S.S., Stark, H., Daniel, J.S., Ravishankara, A.R.: Near-IR absorption of water vapor: pressure dependence of line strengths and an upper limit for continuum absorption. J. Mol. Spectrosc. **232**, 223–230 (2005)

Alicke, B.: Messung von troposphärischen Halogenoxidradikalen in mittleren Breiten. Diplomarbeit, Ruprecht-Karls-Universität Heidelberg, Germany (1997)

Alicke, B., Hebestreit, K., Stutz, J., Platt, U.: Detection of iodine oxide in the marine boundary layer. Nature **397**, 572–573 (1999)

Alicke, B.: The role of nitrous acid in the boundary layer. Ph.D. thesis, University of Heidelberg (2000)

Alicke, B., Platt, U., Stutz, J.: Impact of nitrous acid photolysis on the total hydroxyl radical budget during the limitation of oxidant production/Pianura Padana Produzione di Ozono study in Milan. J. Geophys. Res. **107**(D22), 8196 (2002). doi: 10.1029/2000JD000075

Alicke, B., Geyer, A., Hofzumahaus, A., Holland, F., Konrad, S., Pätz, H.W., Schäfer, J., Stutz, J., Volz-Thomas, A., Platt, U.: OH formation by HONO photolysis during the BERLIOZ experiment. J. Geophys. Res. **108**(D4), 8247 (2003). doi:10.1029/2001JD000579, (PHOEBE: BERLIOZ special section)

Aliwell, S.R., Jones, R.L., Fish, D.J.: Mid-latitude observations of the seasonal variation of BrO 1. Zenith-sky measurements. Geophys. Res. Lett. **24**, 1195–1198 (1997)

Aliwell, S.R., Jones, R.L.: Measurements of tropospheric NO_3 at midlatitude. J. Geophys. Res. **103**, 5719–5727 (1998)

Aliwell, S.R., Van Roozendael, M., Johnston, P.V., Richter, A., Wagner, T., Arlander, D.W., Burrows, J.P., Fish, D.J., Jones, R.L., Tørnkvist, K.K., Lambert, J.-C., Pfeilsticker, K., Pundt, I.: Analysis for BrO in zenith-sky spectra—an intercomparison exercise for analysis improvement. J. Geophys. Res. **107**(D14), 4199 (2002). doi: 10.1029/2001JD000329

Allan, B.J., Carlslaw, N., Coe, H., Burgess, R.A., Plane, J.M.C.: Observation of the nitrate radical in the marine boundary layer. J. Atmos. Chem. **33**, 129–154 (1999)

Allan, B.J., McFiggans, G., Plane, J.M.C., Coe, H.: Observation of iodine monoxide in the remote marine boundary layer. J. Geophys. Res. **105**, 14363–14369 (2000a)

Allan, B.J., McFiggans, G., Coe, H., Plane, J.M.C., McFayden, G.G.: The nitrate radical in the remote marine boundary layer. J. Geophys. Res. **105**(D19), 24191–24204 (2000b)

Allan, B.J., Plane, J.M.C., McFiggans, G.: Observations of OIO in the remote marine boundary layer. Geophys. Res. Lett. **28**, 1945–1948 (2001)

Allan, B.J., Plane, J.M.C., Coe, H., Shillito, J.: Observation of NO_3 concentration profiles in the troposphere. J. Geophys. Res. **107**(D21), 4588 (2002). doi:10.1029/2002JD002112

Amerding, W., Herbert, A., Schindler, T., Spiekermann, M., Comes, F.J.: In situ measurements of tropospheric OH radicals—a challenge for the experimentalist. Ber. Bunsenges. Phys. Chem. **94**, 776–781 (1990)

Amerding, W., Herbert, A., Spiekermann, M., Walter, J., Comes, F.J.: Fast scanning DOAS—A very promising technique for monitoring OH and other tropospheric trace gases. Fresenius. J. Anal. Chem. **340**, 654–660 (1991)

Amerding, W., Herbert, A., Spiekermann, M., Walter, J., Comes, F.J.: Ein schnell durchstimmbares Laserspektrometer. Ber. Bunsenges. Phys. Chem. **96**, 314 (1992)

Amerding, W., Spiekermann, M., Grigonis, R., Walter, J., Herbert, A., Comes, F.J.: Fast scanning laser DOAS for local monitoring of trace gases, in particular tropospheric OH radicals. Ber. Bunsenges. Phys. Chem. **96**, 314–318 (1992)

Ammann, M., Kalberer, M., Jost, D.T., Tobler, L., Rössler, E., Piguet, D., Gäggeler, H.W., Baltensperger, U.: Heterogeneous production of nitrous acid on soot in polluted air masses. Nature **395**, 157–160 (1998)

Anderson, D.E., Lloyd, S.A.: Polar twilight UV-visible radiation field: perturbations due to multiple scattering, ozone depletion, stratospheric clouds, and surface albedo. J. Geophys. Res. **95**, 7429–7434 (1990)

Anderson, S.M., Mauersberger, K.: Laser measurements of ozone absorption cross sections in the Chappuis band. Geophys. Res. Lett. **19**(9), 933–936 (1992)

Andreae, M.O., Ferek, R.J., Bermond, F., Byrd, K.P., Chatfield, R.B., Engstrom, R.T., Hardin, S., Houmere, P.D, LeMarrec, F., Raemdonck, H.: Dimethyl sulfide in the marine atmosphere. J. Geophys. Res. **90**, 12891–12900 (1985)

Andres-Hernandez, M.D., Notholt, J., Hjorth, J., Schrems, O.: A DOAS study on the origin of nitrous acid at urban and non-urban sites. Atmos. Environ. **30**(2), 175–180 (1996)

Andrews, L.C.: Field guide to atmospheric optics, SPIE field guides, Vol. FG02. SPIE Press, Bellingham (2004)

Angström, A.: On the atmospheric transmission of sun radiation and on dust in the air. Geogr. Ann. Stockholm **11**, 156–166 (1929)

Angström, A.: On the atmospheric transmission of sun radiation, II. Geogr. Ann. Stockholn **12**, 130–159 (1930)

Appel, B.R., Winer, A.M., Tokiwa, Y., Biermann, H.W.: Comparison of atmospheric nitrous acid measurements by annular denuder and differential optical absorption systems. Atmos. Environ. **24A**, 611–616 (1990)

Arpag, K.A., Johnston, P.V., Miller, H.L., Sanders, R.W., Solomon, S.: Observations of the stratospheric BrO column over Colorado, 40N. J. Geophys. Res. **99**, 8175–8181 (1994)

Arrhenius, S.: On the influence of carbonic acid in the air upon the temperature of the ground. The London, Edinburgh, and Dublin Philosophical Magazine and Journal of Science **41**, 237–276 (1896)

Ashworth, S.H., Allan, B.J., Plane, J.M.C.: High resolution spectroscopy of the OIO radical: Implications for the ozone depleting potential of iodine. Geophys. Res. Lett. **29**(10), 1456–1459 (2002)

Atkinson, R., Carter, W.P.L., Pitts, J.N., Jr, Winer, A.M.: Measurements of nitrous acid in an urban area. Atmos. Environ. **20**, 408–409 (1986)

Atkinson, R., Baulch, D.L., Cox, R.A., Hampson, R.F., Kerr, J.A., Troe, J.: Evaluated kinetic and photochemical data for atmospheric chemistry: Supplement III, J. Phys. Chem. Ref. Data **18**, 881–1095 (1989)

Atkinson, R.: Gas-phase tropospheric chemistry of organic compounds: a review. Atmos. Environ. **24A**, 1–41 (1990)

Aumont, B., Chervier, F., Laval, S.: Contribution of HONO sources to the $NO_X/HO_X/O_3$ chemistry in the polluted boundary layer. Atmos. Environ. **37**, 487–498 (2003)

Axelsson, H., Galle, B., Gustavsson, K., Regnarsson, P., Rudin, M.: A transmitting/receiving telescope for DOAS-measurements using retroreflector technique. In: Digest of topical meeting on optical remote sensing of the atmosphere. OSA **4**, 641–644 (1990a)

Axelsson, H., Edner, H., Galle, B., Ragnarson, P., Rudin, M.: Differential optical absorption spectroscopy (DOAS) measurements of ozone in the 280–290 nm wavelength region. Appl. Spectrosc. **44**, 1654–1658 (1990b)

Axelsson, L., Lauber, A.: Measurement of sulfur dioxide with the differential optical absorption technique combined with Fourier transformation. Appl. Spectrosc. **46**, 1832–1836 (1992)

Axelsson, H., Eilard, A., Emanuelsson, A., Galle, B., Edner, H., Ragnarson, P., Kloo, H.: Measurement of aromatic hydrocarbons with the DOAS technique. Appl. Spectrosc. **49**, 1254–1260 (1995)

Ball, S.M., Povey, I.M., Norton, E.G., Jones, R.L.: Broadband cavity ringdown spectroscopy of the NO_3 radical. Chem. Phys. Lett. **342**, 113–120 (2001)

Ball, S.M., Jones, R.L.: Broadband cavity ringdown spectroscopy. Chem. Rev. **103**, 5239 (2003)

Ball, S.M., Langridge, J.M., Jones, R.L.: Broadband cavity enhanced absorption spectroscopy using light emitting diodes. Chem. Phys. Lett. **398**, 68–74 (2004)

Barrefors, G.: Monitoring of benzene, toluene and p-xylene in urban air with differential optical absorption spectroscopy technique. Sci. Total Environ. **189/190**, 287–292 (1996)

Barrie, L.A., Bottenheim, J.W., Shnell, R.C., Crutzen, P.J., Rasmussen, R.A.: Ozone destruction and photochemical reactions at polar sunrise in the lower Arctic atmosphere. Nature **334**, 138–141 (1988)

Barrie, L., Platt, U.: Arctic tropospheric chemistry: an overview. Tellus B **49**, 449–454 (1997)

Barringer, A.R., Davies, J.H., Moffat, A.J.: The problems and potential in monitoring pollution from satellites, AIAA paper, presented at the AIAA earth resources observations and information systems meeting, Annapolis, MD, 2–4 March 1970, A1AA Paper No. 70–305, pp. 62–75 (1970)

Bass, A.M., Paur, R.J.: The ultraviolet cross section of ozone. I. The measurements. In: Zerefos, C., Ghazy, A (eds.) Proceedings of the Quadrennial Ozone Symposium, Chalkidiki, pp. 606–616 (1985)

Bates, D.R., Nicolet, M.: The photochemistry of the atmospheric water vapor. J. Geophys. Res. **55**, 301 (1950)

Bates, D.R., Witherspoon, A.E.: The photochemistry of some minor constituents of the earth's atmosphere (CO_2, CO, CH_4, N_2O). Mon. Not. R. Astron. Soc. **112**, 101 (1952)

Bates, T.S., Charlson, R.J., Gammon, R.H.: Evidence for the climatic role of marine biogenic sulfur. Nature **329**, 319–321 (1987)

Becker, K.H., Schurath, U., Tatarczyk, T.: Fluorescence determination of low formaldehyde concentrations in air by dye laser excitation. Appl. Opt. **14**, 310–313 (1975a)

Becker, K.H., Inocennio, A., Schurath, U.: The reaction of ozone with hydrogen sulfide and its organic derivatives. Int. J. Chem. Kinet. **7**, 205–220 (1975b)

Becker, K.H. (ed.): The European Photoreactor EUPHORE, Final Report of the EC-Project, Contract EV5V-CT92-0059. Department of Physical Chemistry, Bergische Universität Wuppertal (1996)

Beckett, W.S., Russi, M.B., Haber, A.D., Rivkin, R.M., Sullivan, J.R., Tameroglu, Z., Mohsenin, V., Leaderer, B.P.: Effect of nitrous acid on lung function in asthmatics: a chamber study. Environ. Health. Perspect. **103**, 372–375 (1995)

Bednarz, F. (ed.): GOME, Global Ozone Monitoring Experiment. Users manual, European Space Research and Technology Centre (ESTEC), Frascati, Italy (1995)

Beek de, R., Vountas, M., Rozanov, V.V., Richter, A., Burrows, J.P.: The ring effect in the cloudy atmosphere. Geophys. Res. Lett. **28**(4), 721–724 (2001)

Beilke, S., Markusch, H., Jost, D.: Measurements of NO-oxidation in power plant plumes by correlation spectroscopy. In: Versino, B., Ott, H. (eds.) Proceedings of 2nd Europion Symposium on the Physico-Chemical Behaviour of Atmospheric Pollutants, Varese, 29 September to 1 October 1981, D. Reidel Publishing Company, Dordrecht, Holland, pp. 448–459 (1981)

Beirle, S., Platt, U., Wenig, M., Wagner, T.: Weekly cycle of NO_2 by GOME measurements: a signature of anthropogenic sources. Atmos. Chem. Phys. **3**, 2225–2232 (2003)

Beirle, S.: Estimating source strengths and lifetime of nitrogen oxides from satellite data. Ph.D. thesis, University of Heidelberg, Heidelberg (2004)

Beirle, S., Platt, U., Wenig, M., Wagner, T.: Highly resolved global distribution of tropospheric NO_2 using GOME narrow swath mode data. Atmos. Chem. Phys. **4**, 1913–1924 (2004a)

Beirle, S., Platt, U., von Glasow, R., Wenig, M., Wagner, T.: Estimate of nitrogen oxide emission from shipping by satellite remote sensing. Geophys. Res. Lett. **31**, L18102 (2004b). doi:10.1029/2004GL020312

Beirle, S., Platt, U., Wenig, M., Wagner, T.: NO_X production by lightning estimated with GOME. Adv. Space Res. **34**, 793–797 (2004c)

Belmiloud, D., Schermaul, R., Smith, K.M., Zobov, N.F., Brault, J.W., Learner, R.C.M., Newnham, D.A., Tennyson, J.: New studies of the visible and near infra-red absorption by water vapour and some problems with the database. Geophys. Res. Lett. **27**(22), 3703–3706 (2000)

Benedick, R.E.: Ozone diplomacy. Harvard University Press, Cambridge (1991)

Bergman, S..: Lehrbuch der Experimentalphysik, Bd. III, Optik,. F. Matossi. Walter de Gruyter & Co., Berlin (1966)

Berresheim, H., Elste, T., Plass-Dülmer, C., Eisele, F.L., Tanner, D.J.: Chemical ionization mass spectrometer for long-term measurements of atmospheric OH and H_2SO_4. Int. J. Mass Spectrom. **202**, 91–109 (2000)

Bevington, P.R.: Data reduction and error analysis for the physical sciences. McGraw-Hill, New York (1969)

Biermann, H.W., Green, M., Neider, J.N.: Long-pathlength DOAS (differential optical absorption spectrometer) system for the in situ measurement of xylene in indoor air. In: Schiff, H.I. (ed.) Measurements of Atmospheric Gases, Vol. 1433, pp. 2–7. SPIE-The International Society of Optical Engineering, Bellingham (1991)

Biermann, H.W., Tuazon, E.C., Winer, A.M., Wallington, T.J., Pitts, J.N.: Simultaneous absolute measurements of gaseous nitrogen species in urban ambient air by long pathlength infrared and ultraviolet-visible spectroscopy. Atmos. Environ. **22**, 1545–1554 (1988)

Bloss, W., Gravestock, T., Heard, D.E., Ingham, T., Johnson, G.P., Lee, J.D.L.: Application of a compact all solid-state laser system to the in situ detection of atmospheric OH, HO_2, NO and IO by laser-induced fluorescence. J. Environ. Monit. **5**, 21–28 (2003)

Bobrowski, N., Hönninger, G., Galle, B., Platt, U.: Detection of bromine monoxide in a volcanic plume. Nature **423**, 273–276 (2003)

Bobrowski, N.: Volcanic gas studies by multi axis differential optical absorption spectroscopy. Ph.D. thesis, University of Heidelberg, Heidelberg (2005)

Bobrowski, N., Hönninger, G., Lohberger, F., Platt, U.: I-DOAS: a new monitoring technique to study the 2D distribution of volcanic gas emissions. J. Volcanol. Geotherm. Res. **150**, 329–338 (2006)

Bobrowski, N., Glasow, R.v., Aiuppa, A., Inguaggiato, S., Louban, I., Ibrahim, O.W., Platt, U.: Reactive halogen chemistry in volcanic plumes. J. Geophys. Res. **112**, DO6311, doi: 10.1029/2006JD007206 (2007)

Bodeker, G.E., Shiona, H., Eskes, H.: Indicators of Antarctic ozone depletion. Atmos. Chem. Phys. **5**, 2603–2615 (2005)

Bonafe, U., Cesari, G., Giovanelli, G., Tirabassi, T., Vittori, O.: Mask correlation spectrophotometry advanced methodology for atmospheric measurements. Atmos. Environ. **10**, 469–474 (1976)

Bongartz, A., Kamens, J., Welter, F., Schurath, U.: Near-UV absorption cross sections and trans/cis equilibrium of nitrous acid. J. Phys. Chem. **95**, 1076–1082 (1991)

Borrell, P., Burrows, J.P., Platt, U., Zehner C.: Determining Tropospheric Concentrations of Trace Gases from Space. ESA Bulletin **107**, 72–81 (2001)

Borrell, P., Burrows, J.P., Richter, A., Platt, U., Wagner, T.: New directions: new developments in satellite capabilities for probing the chemistry of the troposphere. Atmos. Environ. **37**, 2567–2570 (2003)

Bossmeyer, J.: Ship-based multi-axis differential optical absorption spectroscopy measurements of tropospheric trace gases over the Atlantic Ocean: new measurement concepts. Diploma Thesis, University of Heidelberg, Germany (2002)

Bösch, H.: Studies of the stratospheric nitrogen and iodine chemistry by balloon-borne DOAS measurements and model calculations. Ph.D. thesis, University of Heidelberg (2002)

Bösch, H., Camy-Peyret, C., Chipperfield, M.P., Fitzenberger, R., Harder, H., Platt, U., Pfeilsticker, K.: Upper limits of stratospheric IO and OIO inferred from center-to-limb-darkening-corrected balloon-borne solar occultation visible spectra: implications for total gaseous iodine and stratospheric ozone. J. Geophys. Res. **108**, 445 (2003). doi: 10.1029/2002JD003078

Bottenheim, J.W., Gallant, A.G., Brice, K.A.: Measurements of NO_y species and O_3 at 82°N latitude. Geophys. Res. Lett. **13**, 113–116 (1986)

Bottenheim, J.W., Barrie, L. W., Atlas, E., Heidt, L.E., Niki, H., Rasmussen, R.A., Shepson, P.B.: Depletion of lower tropospheric ozone during Arctic spring: the polar sunrise experiment 1988. J. Geophys. Res. **95**, 18555–18568 (1990)

Brand, J.C.D., Srikameswaran, K.: The Π^*–Π (2350 A) band system of sulphur dioxide. Chem. Phys. Lett. **15**, 130–132 (1972)

Brand, J.C.D., Jones, V.T., DiLauro, C.: The $^3B_1$–$^1A_1$ band system of sulphur dioxide: rotational analysis of the (010), (100), and (110) bands. J. Mol Spectrosc **45**, 404–411 (1973)

Brand, J.C.D., Nanes, R.: The 3400–3000 A absorption of sulphur dioxide. J. Mol Spectrosc **46**, 194–199 (1973)

Brandenburger, U., Brauers, T., Dorn, H.-P., Hausmann, M., Ehhalt, D.H.: In-situ measurement of tropospheric hydroxyl radicals by folded long-path laser absorption during the field campaign POPCORN in 1994. J. Atmos. Chem. **31**, 181–204 (1998)

Brandtjen, R., Klüpfel, T., Perner, D.: Airborne measurements during the European Arctic stratospheric ozone experiment: observation of OClO. Geophys. Res. Lett. **21**, 1363–1366 (1994)

Brasseur, G., Solomon, S.: Aeronomy of the middle atmosphere. D. Reidel Publishing Company, Dordrecht, 2nd edition (1986)

Brassington, D.J.: Sulphur dioxide absorption cross-section measurements from 290 to 317 nm. Appl. Opt. **20**, 3774–3779 (1981)

Brassington, D.J., Felton, R.C., Jolliffe, B.W., Marx, B.R., Moncrieff, J.T.M.: Errors in spectroscopic measurements of SO_2 due to nonexponential absorption of laser radiation, with application to the remote monitoring of atmospheric pollution. Appl. Opt. **23**, 469–475 (1984)

Brauers, T., Dorn, H.-P., Platt, U.: Spectroscopic measurements of NO_2, O_3, SO_2, IO and NO_3 in maritime air. In: Restelli, G., Angeletti, G. (eds.) Physico-Chemical Behaviour of Atmospheric Pollutants, Proceedings of the 5th European Symposium, Varese, Italia, pp. 237–242. Kluwer Academic Publishers, Dordrecht (1990)

Brauers, T.: FZ-Jülich, Germany (1992) Unpublished results

Brauers, T., Hausmann, M., Brandenburger, U., Dorn, H.-P.: Improvement of differential optical absorption spectroscopy using multichannel scanning technique. Appl. Opt. **34**(21), 4472–4479 (1995)

Brauers, T., Aschmutat, U., Brandenburger, U., Dorn, H.-P., Hausmann, M., Heßling, M., Hofzumahaus, A., Holland, F., Plass-Dülmer, C., Ehhalt, D.H.: Intercomparison of tropospheric OH radical measurements by multiple folded long- path laser absorption and laser induced fluorescence. Geophys. Res. Lett. **23**, 2545–2548 (1996)

Brauers, T., Hausmann, M., Bister, A., Kraus, A., Dorn, H.-P.: OH radicals in the boundary layer of the Atlantic Ocean: 1. Measurements by long-path laser absorption spectroscopy. J. Geophys. Res. **106**, 7399ff (1999)

Brenninkmeijer, C.A.M., Manning, M.R., Lowe, D.C., Wallace, G.A., Sparks, R.J., Volz-Thomas, A.: Interhemispheric asymmetry in OH abundance inferred from measurements of atmospheric ^{14}CO. Nature **356**, 50–52 (1992)

Brewer, A.W., McElroy, C.T., Kerr, J.B.: Nitrogen dioxide concentrations in the atmosphere. Nature **246**, 129–133 (1973)

Brimblecombe, P., Heymann, M.: TRAP-45—analysis of tropospheric air pollution problems and air pollution abatement strategies in Europe since 1945, A Subproject in EUROTRAC-2, International Scientific Secretariat, GSF-Forschungszentrum für Umwelt and Gesundheit GmbH, Munich, Germany (1998)

Bröske, R., Kleffmann, J., Wiesen, P.: Heterogeneous conversion of NO_2 on secondary organic aerosol surfaces: a possible source of nitrous acid (HONO) in the atmosphere? J. Atmos. Chem. Phys. **3**, 469–474 (2003)

Brown, S.S., Stark, H., Ciciora, S.J., Ravishankara, A.R.: In-situ measurement of atmospheric NO_3 and N_2O_5 via cavity ring-down spectroscopy. Geophys. Res. Lett. **28**(17), 3227–3230 (2001)

Brown, S.S., Stark, H., Ciciora, S.J., McLaughlin, R.J., Ravishankara, A.R.: Simultaneous in situ detection of atmospheric NO_3 and N_2O_5 via cavity ring-down spectroscopy. Rev. Sci. Instrum. **73**(9), 3291–3301 (2002)

Brown, S.S.: Absorption spectroscopy in high-finesse cavities for atmospheric studies. Chem. Rev. **103**(12), 5219–5238 (2003)

Brown, S.S., Stark, H., Ryerson, T.B., Williams, E.J., Nicks, D.K., Jr, Trainer, M., Fehsenfeld, F.C., Ravishankara, A.R.: Nitrogen oxides in the nocturnal boundary layer: Simultaneous in situ measurements of NO_3, N_2O_5, NO_2, NO, and O_3. J. Geophys. Res. **108**(D9), 4299 (2003). doi:10.1029/2002JD002917

Brune, W.H., Stevens, P.S., Mather, J.H.: Measuring OH and HO_2 in the troposphere by laser induced fluorescence at low pressure. J. Atmos. Sci. (1995)

Brunekreef, B., Holgate, S.T.: Air pollution and health. Lancet **360**, 1233–1242 (2002)

Bruns, M., Buehler, S.A., Burrows, J.P., Heue, K.-P., Platt, U., Pundt, I., Richter, A., Rozanov, A., Wagner, T., Wang, P.: Retrieval of profile information from airborne multi axis UV/visible skylight absorption measurements. Appl. Opt. **43**(22), 4415–4426 (2004)

Bruns, M., Buehler, S.A., Burrows, J.P., Richter, A., Rozanov, A., Wang, P., Heue, K.P., Platt, U., Pundt, I., Wagner, T.: NO_2 profile retrieval using airborne multi axis UV-visible skylight absorption measurements over central Europe. Atmos. Chem. Phys. **6**, 3049–3058 (2006)

Buchwitz, M., Rozanov, V., Burrows, J.: A near-infrared optimised DOAS method for the fast global retrieval of atmospheric CH_4, CO, CO_2, H_2O, and N_2O total column amounts from SCIAMACHY Envisat-1 nadir radiances. J. Geophys. Res. **105**, 15231–15245 (2000)

Burkholder, J.B., Talukdar, R.K.: Temperature dependence of the ozone absorption cross section over the wavelength range 410 to 760 nm. Geophys. Res. Lett. **21**, 581–584 (1994)

Burkholder, J.B., Curtius, J., Ravishankara, A.R., Lovejoy, E.R.: Laboratory studies of the homogeneous nucleation of iodine oxides. Atmos. Chem. Phys. **4**, 19–34 (2004)

Burrows, J.P., Tyndall, G.S., Moortgat, G.K.: Absorption spectrum of NO_3 and kinetics of the reactions of NO_3 with NO_2, Cl, and several stable atmospheric species at 298 K. J. Phys. Chem. **89**, 4848–4856 (1985)

Burrows, J.P., Chance, K.V., Crutzen, P.J., Fishman, J., Fredericks, J.E., Geary, J.C., Johnson, T.J., Harris, G.W., Isaksen, I.S.A., Kelder, H., Moortgat, G.K., Muller, C., Perner, D., Platt, U., Pommereau, J.-P., Rodhe, H., Roeckner, E., Schneider, W., Simon, P., Sundqvist, H., Vercheval, J.: SCIAMACHY phase A study—scientific requirements specification, report to European space Agency (1991)

Burrows, J., Vountas, M., Haug, H., Chance, K., Marquard, L., Muirhead, K., Platt, U., Richter, A., Rozanov, V.: Study of the ring effect, final report for ESA contract 109996/94/NL/CN, ESA ITT No. AO/1-2778/94/NL/CN (1995)

Burrows, J., Platt, U., Chance, K., Vountas, M., Rozanov, V., Richter, A., Haug, H., Marquard, L.: Study of the ring effect. European Space Agency, Noordiwijk (1996)

Burrows, J.P., Dehn, A., Deters, B., Himmelmann, S., Richter, A., Voigt, S., Orphal, J.: Atmospheric remote-sensing reference data from GOME: part I. Temperature-dependent absorption cross-sections of NO_2 in the 231–794 nm range. J. Quant. Spectrosc. Radiat. Transf. **60**, 1025–1031 (1998)

Burrows, J.P., Weber, M., Buchwitz, M., Rozanov, V., Ladstätter-Weißenmayer, A., Richter, A., DeBeek, R., Hoogen, R., Bramstedt, K., Eichmann, K.-U.,

Eisinger, M., Perner, D.: The global ozone monitoring experiment (GOME): mission concept and first scientific results. J. Atmos. Sci. **56**, 151–171 (1999a)

Burrows, J.P., Richter, A., Dehn, A., Deters, B., Himmelmann, S.: Atmospheric remote-sensing reference data from GOME: part 2. Temperature-dependent absorption cross sections of O_3 in the 231–794 nm range. J. Quant. Spectrosc. Radiat. Transf. **61**, 509–517 (1999b)

Bussemer, M.: Der Ring-Effekt: Ursachen und Einfluß auf die spektroskopische Messung stratosphärischer Spurenstoffe. Diploma thesis in Physics, University Heidelberg (1993)

Butkovskaya, N.I., Le Bras, G.: Mechanism of the NO_3 + DMS reaction by discharge flow mass spectrometry. J. Phys. Chem. **98**, 2582–2591 (1994)

Butz, A., Bösch, H., Camy-Peyret, C., Chipperfield, M., Dorf, M., Dufour, G., Grunow, K., Jeseck, P., Kühl, S., Payan, S., Pepin, I., Pukite, J., Rozanov, A., von Savigny, C., Sioris, C., Wagner, T., Weidner, F., Pfeilsticker, K.: Intercomparison of stratospheric O_3 and NO_2 abundances retrieved from balloon borne direct sun observations and Envisat/SCIAMACHY limb measurements. J. Atmos. Chem. Phys. **6**, 1293–1314 (2006)

Calvert, J.G.: Hydrocarbon involvement in photochemical smog formation in Los Angeles atmosphere. Environ. Sci. Technol. **10**, 256–262 (1976)

Calvert, J.G., Lazrus, A., Kok, G.L., Heikes, B.G., Walega, J.G., Lind, J., Cantrell, C.A.: Chemical mechanisms of acid generation in the troposphere. Nature **317**, 27–35 (1985)

Calvert, J.G., Yarwood, G., Dunker, A.M.: An evaluation of the mechanism of nitrous acid formation in the urban atmosphere. Res. Chem. Intermed. **20**(3–5), 463–502 (1994)

Campbell, M.J., Sheppard, J.C.,. Au, B.F.: Measurement of hydroxyl concentration in boundary – layer air by monitoring CO oxidation. Geophys. Res. Lett. **6**, 175–178 (1979)

Campbell, M.J., Farmer, J.C., Fitzner, C.A., Henry, M.N., Sheppard, J.C., Hardy, R.J., Hopper, J.F., Muralidhar, V.: Radiocarbon tracer measurements of atmospheric hydroxyl radical concentrations. J. Atmos. Chem. **4**, 413–427 (1986)

Camy-Peyret, C., Jeseck, P., Hawat, T., Durry, G., Berubeé, Rochette, L., Huguenin, D.: The LPMA balloon borne FTIR spectrometer for remote sensing of the atmospheric constituents. In: Proceedings of 12th ESA Symposium on Rocket and Balloon Programmes and Related Research (1995)

Camy-Peyret, C., Bergqvist, B., Galle, B., Carleer, M., Clerbaux, C., Colin, R., Fayt, C., Goutail, F., Nunes-Pinharanda, M., Pommereau, J.P., Hausmann, H., Platt, U., Pundt, I., Rudolph, T., Hermans, C., Simon, P.C., Vandaele, A.C., Plane, J.M.C., Smith, N.: Intercomparison of instruments for tropospheric measurements using differential optical absorption spectroscopy. J. Atmos. Chem. **23**, 51–80 (1996)

Camy-Peyret, C., Payan, S., Jeseck, P., Té, Y., Hawat, T., Pfeilsticker, K., Harder, H., Fitzenberger, R., Bösch, H.: Recent results obtained with the LPMA and DOAS balloon-borne instruments during the ILAS, SABINE and THESEO campaigns. In: Proceedings of 14th ESA Symposium on Rocket and Balloon Programmes and Related Research (1999)

Canrad-Hanovia.: Compact arc lamps, data sheet. Canrad-Hanovia Inc., Newark (1986)

Cantrell, C.A., Stedman, D.H., Wendel, G.J.: Measurement of atmospheric peroxy radicals by chemical amplification. Anal. Chem. **56**, 1496–1502 (1984)

Cantrell, C.A., Davidson, J.A., McDaniel, A.H., Shetter, R.E., Calvert, J.G.: Temperature-dependent formaldehyde cross section in the near-ultraviolet spectral region. J. Phys. Chem. **94**, 3902–3908 (1990)

Cantrell, C.A., Shetter, R.E., Lind, J.A., Mcdaniel, A.H., Calvert, J.G., Parrish, D.D., Fehsenfeld, F.C., Buhr, M.P., Trainer, M.: An improved chemical amplifier technique for peroxy radical measurements. J. Geophys. Res. **98**, 2897–2909 (1993)

Cardenas, L.M., Brassington, D.J., Allan, B.J., Coe, H., Alicke, B., Platt, U., Wilson, K.M., Plane, J.M.C., Penkett, S.A.: Intercomparison of formaldehyde measurements in clean and polluted atmospheres. J. Atmos. Chem. **37**, 53–80 (2000)

Carpenter, L.J., Sturges, W.T., Penkett, S.A., Liss, P.S., Alicke, B., Hebestreit, K., Platt, U.: Short-lived alkyl iodides and bromides at Mace Head, Ireland: links to biogenic sources and halogen oxide production. J. Geophys. Res. **104**, 1679–1689 (1999)

Carpenter, L.J., Hebestreit, K., Platt, U., Liss, P.S.: Coastal zone production of IO precursors: a 2-dimensional study. Atmos. Chem. Phys. **1**, 9–18 (2001)

Carroll, M.A., Sanders, R.W., Solomon, S., Schmeltekopf, A.L.: Visible and near-ultraviolet spectroscopy at McMurdo Station, Antarctica, 6. Observations of BrO. J. Geophys. Res. **94**(D14), 16633–16638 (1989)

Carslaw, N., Carpenter, L., Plane, J.M.C., Allan, B.J., Burgess, R.A., Clemitshaw, K., Coe, C.H., Penkett, S.A.: Simultaneous observations of nitrate and peroxy radicals in the marine boundary layer. J. Geophys. Res. **102**, 18917–18933 (1997)

Carslaw, N., Creasey, D.J., Heard, D.E., Lewis, A.C., McQuaid, J.B., Pilling, M.J., Monks, P.S., Bandy, B.J., Penkett, S.A.: Modeling OH, HO_2, and RO_2 radicals in the marine boundary layer, 1, Model construction and comparison with field measurements. J. Geophys. Res. **104**, 30241–30255 (1999)

Casadio, S., Zehner, C., Pisacane, G., Putz, E.: Empirical retrieval of the atmospheric air mass factor (ERA) for the measurement of water vapour vertical content using GOME data. Geophys. Res. Lett. **27**(10), 1483–1486 (2000)

Cauer, H.: Schwankungen der Jodmenge der Luft in Mitteleuropa, deren Ursachen und deren Bedeutung für den Jodgehalt unserer Nahrung (Auszug). Angewandte Chemie **52**(11), 625–628 (1939)

Chance, K., Palmer, P.I., Spurr, R.J.D., Martin, R.V., Kurosu, T.P., Jacob, D.J., et al.: Satellite observations of formaldehyde over North America from GOME. Geophy. Res. Lett. **27**(21), 3461–3464 (2000)

Chance, K., Kurosu, T.P., Sioris, C.E.: Undersampling correction for arraydetector-based satellite spectrometers. Appl. Opt. **44**(7), 1296–1304 (2005)

Chapman, S.: On ozone and atomic oxygen in the upper atmosphere. Philos. Mag. **7**, 369–383 (1930)

Chappuis, J.: Sur le spectre d'absorption de l'ozone. C.R. Acad. Sci. **91**, 985 (1880)

Charlson, R.J., Rhode, H.: Factors controlling the acidity of natural rainwater. Nature **295**, 683–685 (1982)

Charlson, R.J., Lovelock, J.E., Andreae, M.O., Warren, S.G.: Oceanic phytoplankton, atmospheric sulphur, cloud albedo and climate. Nature **326**, 655–661 (1987)

Chen, Y.Q., Zhu, L.: Wavelength-dependent photolysis of glyoxal in the 290–420 nm region. J. Phys. Chem. A **107**(23), 4643–4651 (2003)

Chernin, S.M., Barskaya, E.G.: Optical multipass matrix systems. Appl. Opt. **30**, 51–57 (1991)

Chernin, S.M.: New generation of multipass systems in high resolution spectroscopy. Spectrochim. Acta A **52**, 1009–1022 (1996)

Chin, M., Davis, D.D.: Global sources and sinks of OCS and CS_2 and their distribution. Global Biogeochem. Cycles **7**, 321–337 (1993)

Chin, M., Jacob, D.J.: Anthropogenic and natural contributions to tropospheric sulphate: a global model analysis. J. Geophys. Res. **101**, 18691–18699 (1996)

Clarke, D., Basurah, H.: Polarization measurements of the ring effect in the daytime sky. Planet. Space Sci. **37**, 627–630 (1989)

Clemitshaw, K.C., Carpenter, L.J., Penkett, S.A., Jenkin, M.E.: A calibrated peroxy radical chemical amplifier for ground-based tropospheric measurements. J. Geophys. Res. **102**, 25405–25416 (1997)

Clemitshaw, K.C.: A Review of instrumentation and measurement techniques for ground-based and airborne field studies of gas-phase tropospheric chemistry. Crit. Rev. Environ. Sci. Technol. **34**, 1–108 (2004) ISSN: 1064-3389, doi: 10.1080/10643380490265117

Coe, H., Jones, R.L., Colin, R., Carleer, M., Harrison, R.M., Peak, J., Plane, J.M.C., Smith, N., Allan, B., Clemitshaw, K.C., Burgess, R.A., Platt, U., Etzkorn, T., Stutz, J., Pommereau, J.-P., Goutail, F., Nunes-Pinharanda, M., Simon, P., Hermans, C., Vandaele, A.-C.: A comparison of differential optical absorption spectrometers for measurement of NO_2, O_3, SO_2 and HONO. In: Borrell, P.M., Borrell, P., Cvitaš, T., Kelly, K., Seiler, W. (eds.) Proceedings of EUROTRAC Symposium 1996: Transport and Transformation of Pollutants, pp. 757–762. Computational Mechanics Publications, Southampton (1997) ISBN 1 85312 498 2

Coe, H., Allan, B.J., Plane, J.M.C.: Retrievals of vertical profiles of NO_3 from zenith sky measurements using an optimal estimation method. J. Geophys. Res. **107**(D21), 4587 (2002). doi:10.1029/2002JD002111

Coheur, P.-F., Fally, S., Carleer, M., Clerbaux, C., Colin, R., Jenouvrier, A., Mérienne, M.-F., Hermans, C., Vandaele, A.C.: New water vapour line parameters in the 26000–13000 cm^{-1} region. J. Quant. Spectrosc. Radiat. Transf **74**(4), 493–510 (2002)

Colin, R., Carleer, M., Simon, P.C., Vandaele, A.C., Dufour, P., Fayt, C.: Atmospheric absorption measurement by Fourier transform DOAS, EUROTRAC annual report 1991 (TOPAS subproject), pp 14–46 (1991)

Coquart, B., Jenouvrier, A., Merienne, M.F.: The NO_2 absorption spectrum. II. Absorption cross section at low temperatures in the 400–500 nm region. J. Atmos. Chem. **21**, 251–261 (1995)

Cornu, A.: Observation de la limite ultraviolette du spectre solaire a diverses altitudes. C.R. Acad. Sci. **89**, 808 (1879)

Cox, R.A., Sheppard, D.W., Stevens, M.P.: Absorption coefficients and kinetics of the BrO radical using molecular modulation. J. Photochem. **19**, 189–207 (1982)

Cox, R.A., Bloss, W.J., Jones, R.L., Rowley, D.M.: OIO and the atmsopheric cycle of iodine. Geophys. Res. Lett **26**, 1857–1860 (1999)

CRC Handbook of Chemistry and Physics, 88th Edition (Internet Version 2008), In: David R.L., (ed.) CRC Press/Taylor and Francis, Boca Raton, FL. (2008)

Crutzen, P.J.: The influence of nitrogen oxides on the atmospheric ozone content. Qart. J. Roy. Met. Soc. **96**, 320–325 (1970)

Crutzen, P.J.: A discussion of the chemistry of some minor constituents in the stratosphere and troposphere. Pure Appl. Geophys. **106–108**, 1385–1399 (1973)

Crutzen, P.J.: Photochemical reactions initiated by and influencing ozone in unpolluted tropospheric air. Tellus **26**, 47–57 (1974)

Crutzen, P.J., Müller, R., Brühl, Ch., Peter, T.: On the potential importance of the gas phase reaction $CH_3O_2 + ClO \longrightarrow ClOO + CH_3O$ and the heterogeneous reaction $HOCl + HCl \longrightarrow H_2O + Cl_2$ in "ozone hole" chemistry. Geophys. Res. Lett. **19**, 1113–1116 (1992)

Cunningham, R.W.: Comparison of three methods for determining fit parameter uncertainties for the Marquardt compromise. Comput. phys. **7**(5), 570 (1993)

Dahlback, A., Stamnes, K.: A New spherical model for computing the radiation-field available for photolysis and heating at twilight. Plan. Spa. Sci. **39**(5), 671–683 (1991)

Dahlback, A., Rairoux, P., Stein, B., Del Guasta, M., Kyrö, E., Stefanutti, L., Larsen, N., Braathen, G.: Effects of stratospheric aerosols from the Mt. Pinatubo eruption on ozone measuremnts at Sodankylá Finland in 1991/92. Geophys. Res. Lett. **21**(13), 1399–1402 (1994)

Dave, J.V., Mateer, C.L.: A preliminary study on the possibility of estimating total atmospheric ozone from satellite measurements. J. Atmos. Sci. **24**, 414–427 (1967)

Davidson, J.A., Cantrell, C.A., McDaniel, A.H., Shetters, R.E., Madronich, S., Calvert, J.G.: Visible-ultraviolet absorption cross sections for NO_2 as a function of Temperature. J. Geophys. Res. **93**, 7105–7112 (1988)

Davies, J.H.: Correlation spectroscopy. Anal. Chem. **42**, 101A–112A (1970)

Davies, J.H., van Egmond, N.D., Wiens, R., Zwick, H.: Recent developments in environmental sensing with the barringer correlation spectrometer. Can. J. Remote Sens. **1**, 85–94 (1975)

Davis, D.D., Rodgers, M.O., Fischer, S.D., Asai, K.: An experimental assessment of the O_3/H_2O interference problem the detection of natural levels of OH via laser induced fluorescence. Geophys. Res. Lett. **8**, 69–72 (1981)

Davis, D., Crawford, J., Liu, S., McKeen, S., Bandy, A., Thornton, D., Rowland, F., Blake, D.: Potential impact of iodine on tropospheric levels of ozone and other critical oxidants. J. Geophys. Res. **101**, 2135–2147 (1996)

Davies, W.E., Vaughan, G., O'Connor, F.M.: Observations of near-zero ozone concentrations in the upper troposphere at mid-latitudes. Geophys. Res. Lett. **25**, 1173–1176 (1998)

De Maziere, M., Van Roozendael, M., Hermans, C., Simon, P.C., Demoulin, P., Roland, G., Zander, R.: Quantitative evaluation of the post-Pinatubo NO_2 reduction and recovery, based on 10 years of FTIR and UV-Visible. J. Geophys. Res. **103**, 10849–10858 (1998)

Dieke, D.R., Crosswhite, H.M. The ultraviolet bands of OH, Fundamental data. J. Quant. Spectrosc. Radiat. Transfer **2**, 97–199 (1961)

Dillon, T.J., Tucceri, M.E., Hölscher, D., Crowley, J.N.: Absorption cross-section of IO at 427.2 nm and 298 K. J. Photochem. Photobiol. A. Chem. **176**, 3–14

Dimpfl, W.L., Kinsey, J.L: Radiative lifetimes of OH (A $^2\Sigma$) and Einstein coefficients of the A-X system of OH and OD. J. Quant. Spectrosc. Radiat. Transfer **21**, 233–241 (1979)

Dobson, G.M.B., Harrison, D.N.: Measurements of the amount of ozone in the earth's atmosphere and its relation to other geophysical conditions, Part 1. Proc. R. Soc. Lond. A **110**, 660–693 (1926)

Dobson, G.M.B.: A photoelectric spectrophotometer for measuring the amount of atmospheric ozone. Proc. Phys. Soc. **43**, 324–339 (1931)

Dobson, G.M.B.: Forty years' research on atmospheric ozone at Oxford: a history. Appl. Opt. **7**, 387–406 (1968)

Dorf, M., Bösch, H., Butz, A., Camy-Peyret, C., Chipperfield, M.P., Engel, A., Goutail, F., Grunow, K., Hendrick, F., Hrechanyy, S., Naujokat, B., Pommereau, J.-P., Van Roozendael, M., Sioris, C., Stroh, F., Weidner, F., Pfeilsticker, K.: Balloon-borne stratospheric BrO measurements: comparison with Envisat/SCIAMACHY BrO limb profiles. Atmos. Chem. Phys. **6**, 2483–2501 (2006)

Dorn, H.-P., Platt, U.: Eine empfindliche optische Nachweismethode für Spurenstoffe in der Atmosphäre. Elektrizitätswirtschaft **24**, 967–970 (1986)

Dorn, H.-P., Callies, J., Platt, U., Ehhalt, D.H.: Measurement of tropospheric OH concentrations by laser long-path absorption spectroscopy. Tellus B **40**, 437–445 (1988)

Dorn, H.-P., Neuroth, R., Brauers, T., Brandenburger, U., Ehhalt, D.H.: Measurement of tropospheric OH radical concentrations by differential UV laser long-path absorption. In: Schiff, H.I., Platt, U. (eds.) Proceedings of SPIE Symposium on Optical Methods in Atmospheric chemistry, pp 361–366 (1993)

Dorn, H.-P., Neuroth, R., Hofzumahaus, A.: Investigation of OH absorption cross sections of rotational transitions in the $A^2\Sigma^+$, v$' = 0 < -X^2\Pi$, $v'' = 0$ band under atmospheric conditions: implications for tropospheric long-path absorption measurements. J. Geophys. Res. **100**, 7397–7409 (1995)

Dorn, H.P., Brandenburger, U., Brauers, T., Hausmann, M., Ehhalt, D.H.: In- situ detection of tropospheric OH radicals by folded long- path laser absorption. results from the POPCORN field campaign in August 1994. Geophys. Res. Lett. **23**, 2537–2540 (1996)

Durry, G., Megie, G.: Atmospheric CH_4 and H_2O monitoring with near-infrared InGaAs laser diodes by the SDLA, a balloonborne spectrometer for tropospheric and stratospheric in situ measurements. Appl. Opt. **38**(36), 7342–7354 (1999)

Dvortsov, V.L., Geller, M.A., Solomon, S., Schauffler, S.M., Atlas, E.L., Blake, D.R.: Rethinking reactive halogen budgets in the midlatitude lower stratosphere. Geophys. Res. Lett. **26**, 1699–1702 (1999)

Eckhardt, H.D.: Simple model of corner reflector phenomena. Appl. Opt. **10**, 1559–1566

Edner, H., Sunesson, A., Svanberg, S., Uneus, L., Wallin, S.: Differential optical absorption spectroscopy system used for atmospheric mercury monitoring. Appl. Opt. **25**, 403–409 (1986)

Edner, H., Amer, R., Ragnarson, P., Rudin, M., Svanberg, S.: Atmospheric NH_3 monitoring by long-path UV absorption spectroscopy. In: Proceedings of International Congress on Optical Science and Engineering. Environment and Pollution Measurement Sensors and Systems, Den Haag (1990)

Edner, H., Ragnarson, P., Svanberg, S.: A multi-path DOAS system for large area pollution monitoring. In: Borrell, P.M., et al. (eds.) Proceedings of EUROTRAC Symposium 1992, pp. 220–223, SPB Academic Publishing BV, Den Haag (1992)

Edner, H., Ragnarson, P., Spännare, S., Svanberg, S.: Differential optical absorption spectroscopy (DOAS) system for urban atmospheric pollution monitoring. Appl. Opt. **32**, 327–333 (1993a)

Edner, H., Ragnarson, P., Spännare, S., Svanberg, S.: Differential optical absorption spectroscopy (DOAS) system for urban atmospheric pollution monitoring. Appl. Opt. **32**, 327 (1993b)

Edner, H., Ragnarsson, P., Svanberg, S., Wallinder, E.: Simultaneous tropospheric ozone monitoring using lidar and DOAS systems. Lund Reports on Atomic Physics LRAP – 155 (1994a)

Edner, H., Ragnarson, P., Svanberg, S., Wallinder, E., Ferrara, E., Cioni, R., Raco, B., Taddeucci, G.: Total fluxes of sulfur-dioxide from the Italian volcanoes Etna, Stromboli and Vulcano measured by differential absorption LIDAR and passive differential optical-absorption spectroscopy. J. Geophys. Res. **99**(D9), 18827–18838 (1994b)

Edwards, D.P., Lamarque, J.-F., Attié, J.-L., Emmons, L.K., Richter, A., Cammas, J.-P., Gille, J.C., Francis, G.L., Deeter, M.N., Warner, J., Ziskin, D.C., Lyjak, L.V., Drummond, J.R., Burrows, J.P.: Tropospheric ozone over the tropical Atlantic: a satellite perspective. J. Geophys. Res. **108**(D8), 4237 (2003). doi:10.1029/2002JD002927

Ehhalt, D.H., Drummond, J. W.: The tropospheric cycle of NO_X. In: Chemistry of the Unpolluted and Polluted Troposphere. D. Reidel Publishing, p. 219 (1982)

Ehhalt, D.H., Dorn, H.-P., Poppe, D.: The chemistry of the hydroxyl radical in surface air. Proc. Roy. Soc. Edinburgh **97B**, 17–34 (1991)

Ehhalt, D.H.: Gas phase chemistry of the troposphere In: Zellner, R. (ed.) Global Aspects of Atmospheric Chemistry. Deutsch Bunsen-Gesellschaft für Physikalische Chemie, Steinkopf, Darmstadt, Springer, New York (1999)

Eisberg-Resnik, R.: Quantum physics. John Wiley & Sons, Inc. (1985)

Eisele, F.L., Tanner, D.J.: Ion-assisted tropospheric OH measurements. J. Geophys. Res. **D5**, 9295–9308 (1991)

Eisele, F.L., Mount, G.H., Fehsenfeld, F.C., Harder, J., Madronicj, E., Parrish, D.D., Roberts, J., Trainer, M., Tanner, D.: Intercomparison of tropospheric OH and ancillary trace gas measurements at Fritz Peak Observatory, Colorado. J. Geophys. Res. **D99**, 18605–18626 (1994)

Eisele, F.L., Tanner, D.J., Cantrell, C.A., Calvert, J.G.: Measurements and steady state calculations of OH concentrations at Mauna Loa Observatory. J. Geophys. Res. **101**, 14665–14679 (1996)

Eisinger, M., Richter, A., Ladstätter-Weißmayer, A., Burrows, J.P.: DOAS zenith sky observations: 1. BrO measurements over Bremen (53°N) 1993–1994. J. Atmos. Chem. **26**, 93–108 (1997)

Eisinger, M., Burrows, J.P.: Tropospheric sulfur dioxide observed by the ERS-2 GOME instrument. Geophys. Res. Lett. **25**, 4177–4180 (1998)

Elokhov, A.S., Gruzdev, A.N.: Spectrometric measurements of of total NO_2 in different regions of the globe. SPIE **2107**, 111–121 (1993)

Elokhov, A.S., Gruzdev, A.N.: Estimation od tropospheric and stratospheric NO_2 from spectrometric measurements of column NO_2 abundances. SPIE **2506**, 444–455 (1995)

Engeln, R., Berden, G., Peeters, R., Meijer, G.: Cavity enhanced absorption and cavity enhanced magnetic rotation spectroscopy. Rev. Sci. Instrum. **69**, 3763–3769 (1998)

Erle, F., Pfeilsticker, K., Platt, U.: On the influence of tropospheric clouds on zenith-scattered-light measurements of stratospheric species. Geophys. Res. Lett. **22**, 2725–2728 (1995)

Erle, F., Grendel, A., Perner, D., Platt, U., Pfeilsticker, K.: Evidence of heterogeneous bromine chemistry on cold stratospheric sulphate aerosols. Geophys. Res. Lett. **25**, 4329–4332 (1998)

Erle, F.: Untersuchungen zur Halogenaktivierung der winterlichen arktischen Stratosphäre anhand flugzeuggestützter spektroskopischer Messungen. Ph.D. thesis, University of Heidelberg (1999)

ESA.: The GOME users manual. In: Bednarz, F. (ed.) ESA Publication Division, ESTEC, Noordwijk, The Netherlands (1995)

Etzkorn, T., Klotz, B., Sörensen, S., Patroescu, I.V., Barnes, I., Becker, K.H., Platt, U.: Gas-phase absorption cross sections of 24 monocyclic aromatic hydrocarbons in the UV and IR spectral ranges. Atmos. Environ. **33**, 525–540 (1999)

EUROTRAC Final Report.: Vol. 8: Instrument development for atmospheric research and monitoring. In: Bösenberg, J., Brassington, D., Simon, P.C. (eds.) Chapter 11, Springer Verlag, Berlin, Heidelberg, New York, ISBN 3-540-62516-X (1997)

Evangelisti, F., Giovanelli, G., Orsi, G., Tirabassi, T., Vittori, O.: Application features of mask correlation spectrophotometry to long horizontal paths. Atmos. Environ. **12**, 1125–1131 (1978)

Evangelisti, F., Baroncelli, A., Bonasoni, P., Giovanelli, G., Ravegnani, F.: Differential optical absorption spectrometer for measurement of tropospheric pollutants. Appl. Opt. **34**, 2737–2744 (1995)

Fabry, C.: L'absorption de l'ultraviolet par l'ozone et la limite du spectre solaire. J. Phys. Radium **3**, 196–206 (1913)

Famy, O.G., Famy, M.J.: Mutagenicity of N-α-Acetoxyethyl-N-ethylnitrosamine and N,N-Diethylnitrosamine in relation to the mechanism of metabolic activation of Dialkylnitrosamines. Cancer Res. **36**, 4504–4512 (1976)

Fan, S.-M., Jacob, D.J.: Surface ozone depletion in the Arctic spring sustained by bromine reactions on aerosols. Nature **359**, 522–524 (1992)

Farman, J.C., Gardiner, B.G., Shanklin, J.D.: Large losses of total ozone in Antarctica reveal seasonal ClO_X/NO_X interaction. Nature **315**, 207–210 (1985)

Fastie, W.G.: Ebert spectrometer reflections. Phys. Today **4**(1), 37–43 (1991)

Fayt, C., Dufour, P., Hermans, C., van Roozendael, M., Simon, P.C.: Instrument and software development for DOAS measurements of atmospheric constituents. In: Borrell, P.M., et al. (eds.) Proceedings of EUROTRAC Symposium 1992, pp. 231–233.SPB Academic Publishing BV, Den Haag (1992)

Febo, A., Perrino, C., Allegrini, I.: Measurement of nitrous acid in Milan, Italy, by DOAS and diffusion denuders. Atmos. Environ. **30**(21), 3599–3609 (1996)

Feigelson, M.E.: Radiation in a cloudy atmosphere. D. Reidel Publishing Company, Dordrecht (1981)

Felton, C.C., Sheppard, J.C., Campbell, M.J.: The radiochemical hydroxyl radical measurement method. Environ. Sci. Technol. **24**, 1841–1847 (1990)

Fenger, J.: Urban air quality. Atmos. Environ. **33**(29), 4877–4900 (1999)

Ferlemann, F., Camy-Peyret, C., Fitzenberger, R., Harder, H., Hawat, T., Osterkamp, H., Perner, D., Platt, U., Schneider, M., Vradelis, P., Pfeilsticker, K.: Stratospheric BrO profile measured at different latitudes and seasons: measurement technique. Geophys. Res. Lett. **25**, 3847–3850 (1998)

Ferlemann, F., Bauer, N., Fitzenberger, R., Harder, H., Osterkamp, H., Perner, D., Platt, U., Schneider, M., Vradelis, P., Pfeilsticker, K.: Differential optical absorption spectroscopy instrument for stratospheric balloon-borne trace gas studies. Appl. Opt. **39**, 2377–2386 (2000)

Fickert, S., Adams, J.W., Crowley, J.N.: Activation of Br_2 and BrCl via uptake of HOBr onto aqueous salt solutions. J. Geophys. Res. **104**, 23719–23728 (1999)

Fiedler, M., Frank, H., Gomer, T., Hausmann, M., Pfeilsticker, K., Platt, U.: Groundbased spectroscopic measurements of stratospheric NO_2 and OClO in Arctic winter 1989/90. Geophys. Res. Lett. **20**, 963–966 (1993)

Fiedler, S.E.: Incoherent broad-band cavity-enhanced absorption spectroscopy. Ph.D. thesis, D83, Faculty II—Mathematics and Sciences, Technical University of Berlin (TU) (2005)

Filsinger, F.: MAX-DOAS measurements of tropospheric BrO at the Hudson Bay. Diploma thesis, University of Heidelberg (2004)

Finlayson-Pitts, B.J., Johnson, S.N.: The reaction of NO_2 with NaBr: Possible source of BrNO in polluted marine atmospheres. Atmos. Environ. **22**, 1107–1112 (1988)

Finlayson-Pitts, B.J., Ezell, M.J., Pitts, J.N.: Formation of chemically active chlorine compounds by reactions of atmospheric NaCl particles with gaseous N_2O_5 and $ClONO_2$. Nature **337**, 241–244 (1989)

Finlayson-Pitts, B.J., Livingston, F.E., Berko, H.N.: Ozone destruction and bromine photo chemistry in the Arctic spring. Nature **343**, 622–625 (1990)

Finlayson-Pitts, B.J., Pitts, J.N.: Chemistry of the Upper and Lower Atmosphere: Theory, Experiments and Applications, Vol. xxii, 969 pp. Academic, San Diego (2000)

Finlayson-Pitts, B.J., Wingen, L.M., Sumner, A.L., Syomin, D., Ramazan, K.A.: The heterogeneous hydrolysis of NO_2 in laboratory systems and in outdoor and indoor atmospheres. An integrated mechanism. Phys. Chem. Chem. Phys. **5**, 223–242 (2003)

Fischer.: Modelling of low-power high-pressure discharge lamps. Philips J. Res. **42**, 58–85 (1987)

Fischer, H.: Remote sensing of atmospheric trace gases. Interdisc. Sci. Rev. **18**(3), 185–191 (1993)

Fischer, H., Oelhaf, H.: Remote sensing of vertical profiles of atmospheric trace constituents with MIPAS limb emission spectrometers. Appl. Opt. **35**(16), 2787–2796 (1996)

Fish, D.J., Aliwell, S.R., Jones, R.L.: Mid-latitude observations of the seasonal variation of BrO: 2. interpretation and modelling study. Geophys. Res. Lett. **24**, 1199–1202 (1997)

Fish, D.J., Roscoe, H.K., Johnston, P.V.: Possible causes of stratospheric NO_2 trends observed at Lauder, New Zealand. Geophys. Res. Lett. **27**, 3313–3316 (2000)

Fitzenberger, R., Bösch, H., Camy-Peyret, C., Chipperfield, M.P., Harder, H., et al.: First profile measurements of tropospheric BrO. Geophys. Res. Lett. **27**, 2921–2924 (2000)

Fix, A., Ehret, G., Flentje, H., Poberaj, G., Gottwald, M., Finkenzeller, H., Bremer, H., Bruns, M., Burrows, J.P., Kleinböhl, A., Küllmann, H., Kuttippurath, J., Richter, A., Wang, P., Heue, K.-P., Platt, U., Pundt, I., Wagner, T.: SCIAMACHY validation by aircraft remote measurements: design, execution, and first results of the SCIA-VALUE mission. Atmos. Chem. Phys. **5**, 1273–1289 (2005)

Fleischmann, O.C., Burrows, J.P.: University of Bremen, Germany, unpublished results (2002)

Fleischmann, O.C., Orphal, J., Burrows, J.P.: New ultraviolet absorption cross-sections of BrO at atmospheric temperatures measured by time-windowing Fourier transform spectroscopy. J. Photochem. Photobiol. A Chem. **168**, 117–132 (2004)

Flentje, H., Dubois, R., Heintzenberg, J., Karbach, H.J.: Retrieval of aerosol properties from boundary layer extinction measurements with a DOAS system. Geophys. Res. Lett. **24**, 2019–2022 (1997)

Foster, K.L., Plastridge, R.A., Bottenheim, J.W., Shepson, P.B., Finlayson-Pitts, B.J., Spicer, C.W.: The role of Br_2 and BrCl in surface ozone destruction at polar sunrise. Science **291**, 471–474 (2001)

Francis, P., Burton, M.R., Oppenheimer, C.: Remote measurements of volcanic gas compositions by solar occultation spectroscopy. Nature **396**, 567–569 (1998)

Frank, H., Platt, U.: Advanced calculation procedures for the interpretation of skylight measurements. In: Proceedings of First European Ozone Workshop, pp. 65–68. Schliersee (1990)

Frank, H.: Ein Strahlungstransportmodell zur Interpretation von spektroskopischen Spurenstoffmessungen in der Erdatmosphäre. Diploma thesis, University of Heidelberg (1991)

Frankenberg, C., Platt, U., Wagner, T.: Iterative maximum a posteriori (IMAP)-DOAS for trace gas retrieval of strong absorbers: model studies for CH_4 and CO_2 retrieval from near infrared spectra of SCIAMACHY onboard ENVISAT. Atmos. Chem. Phys. **4**, 6067–6106 (2004)

Frankenberg, C., Platt, U., Wagner, T.: Retrieval of CO from SCIAMACHY onboard ENVISAT detection of strongly polluted areas and seasonal patterns in global CO abundances. Atmos. Chem. Phys. **4**, 8425–8438 (2004)

Frankenberg, C., Meirink, J.F., van Weele, M., Platt, U., Wagner, T.: Assessing methane emissions from global space-borne observations. Science **308**, 1010–1014 (2005)

Frankenberg, C., Meirink, J.F., Bergamaschi, P., Goede, A., Heimann, M., Körner, S., Platt, U., van Weele, M., Wagner, T.: Satellite chartography of atmospheric methane from SCIAMACHY onboard ENVISAT: (I) Analysis of the years 2003 and 2004. J. Geophys. Res. **111**, D07303 (2006). doi: 10.1029/2005JD006235

Franzblau, E., Popp, C.J.: Nitrogen oxides produced from lightning. J. Geophys. Res. **84**(D8), 11089–11104

Fricke, W., Beilke, S.: Changing concentrations and deposition of sulfur and nitrogen compounds in Central Europe between 1980 and 1992. In: Slanina, J., Angeletti, G., Beilke, S (eds.) Air Pollution Research Report 47, CEC, Proceedings of Joint Workshop CEC/BIATEX of Eurotrac, 4–7 May 1993, Aveiro, Portugal, pp. 9–30 (1993)

Fricke, W., Uhse, K.: Anteile von Witterung und Emissionsminderung am SO_2-Rückgang in Deutschland, Staub-Reinhaltung der Luft **54**, 289–296 (1994)

Friedeburg, V.C., Wagner, T., Geyer, A., Kaiser, N., Vogel, B., Vogel, H., Platt, U.: Derivation of troposphere NO_3 profiles using Off-axis-DOAS measurements during sunrise and comparison with simulations. J. Geophys. Res. **107**, D13 (2002). doi: 10.1029/2001JD000481

Friedeburg, V.C.: Derivation of trace gas information combining differential optical absorption spectroscopy with radiative transfer modelling. Ph.D. thesis, University of Heidelberg (2003)

Frieß, U., Chipperfield, M., Otten, C., Platt, U., Pyle, J., Wagner, T., Pfeilsticker, K.: Intercomparison of measured and modelled BrO slant column amounts for the Arctic winter and spring 1994/95. Geophys. Res. Lett. **26**, 1861–1864 (1999)

Frieß, U.: Spectroscopic measurements of atmospheric trace gases at Neumayer-station, Antarctica. Ph.D. thesis, University of Heidelberg, Heidelberg (2001)

Frieß, U., Wagner, T., Pundt, I., Pfeilsticker, K., Platt, U.: Spectroscopic measurements of tropospheric iodine oxide at Neumayer station, Antarctica. Geophys. Res. Lett. **28**, 1941–1944 (2001)

Frieß, U., Hollwedel, J., König-Langlo, G., Wagner, T., Platt, U.: Dynamics and chemistry of tropospheric bromine explosion events in the Antarctic coastal region. J. Geophys. Res. **109**, D06305 (2004). doi:10.1029/2003JD004133

Frieß, U., Kreher, K., Johnston, P.V., Platt, U.: Ground-based DOAS measurements of stratospheric trace gases at two Antarctic stations during the 2002 ozone hole period. J. Atmos. Sci. **62**(3), 765–777 (2005). doi: 10.1175/JAS-3319.1 (JAS-1076)

Frieß, U., Monks, P.S., Remedios, J.J., Rozanov, A., Sinreich, R., Wagner, T., Platt, U.: MAX-DOAS O_4 measurements: a new technique to derive information on atmospheric aerosols. (II) Modelling studies. J. Geophys. Res. **111**, D14203 (2006). doi:10.1029/2005JD006618

Frins, E., Bobrowski, N., Platt, U., Wagner, T.: Tomographic MAX-DOAS observations of sun illuminated targets: a new technique providing well defined absorption paths in the boundary layer. Appl. Opt. **45**(24), 6227–6240 (2006)

Fung, K., Grosjean, D.: Determination of nanogram amounts of carbonyls as 2,4-dinitrophenylhydrazine by high performance liquid chromatography. Aerosol Sci. Technol. **53**, 168–171 (1981)

Funk, O.: Photon path length distributions for cloudy skies; oxygen a-band measurements and radiative transfer calculations. Ph.D. thesis, University of Heidelberg (2000)

Funk, O., Pfeilsticker, K.: Photon path lengths distributions for cloudy skies: oxygen A-Band measurements and model calculations. Annal. Geophys. **21**, 615–626 (2003)

Fuqi, S., Kuze, H., Yoshii, Y., Nemoto, M., Takeuchi, N., Kimura, T., Umekawa, T., Yoshida, T., Hioki, T., Tsutsui, T., Kawasaki, M.: Measurement of regional distribution of atmospheric NO_2 and aerosol particles with flashlight long-path optical monitoring. Atmos. Environ. **39**, 4959–4968 (2005)

Gall, R., Perner, D., Ladstätter-Weissenmayer, A.: Simultaneous determination of NH_3, SO_2, NO and NO_2 by direct UV-absorption in ambient air. Fresenius J. Anal. Chem. **340**, 646–649 (1991)

Galle, B., Axelsson, H., Edner, H., Eilard, A., Mellqvist, J., Ragnarson, P., Svanberg, S., Zetterberg, L.: Development of DOAS for atmospheric trace species monitoring, EUROTRAC Annual Report 1991 (1991)

Galle, B., Klemedtsson, L., Griffith, D.W.: Application of an FTIR for measurements of N_2O fluxes using micrometeorological methods, an ultralarge chamber system and conventional field chambers. J. Geophys. Res. **99**, 16575–16583 (1994)

Galle, B., Oppenheimer, C., Geyer, A., McGonigle, A, Edmonds, M., Horrocks, L.: A miniaturised ultraviolet spectrometer for remote sensing of SO_2 fluxes: a new tool for volcano surveillance. J. Volcanol. Geotherm. Res. **119**, 214–254 (2003)

Galle, B., Platt, U, Van Roozendael, M., Oppenheimer, C., Hansteen, T., Boudon, G., Burton, M., Delgado, H., Strauch, W., Malavassi, E., Garzon, G.,

Pullinger, C., Kasereka, M., Molina, M., Molina, L., Carn, S.: NOVAC, network for observation of volcanic and atmospheric change. Project proposal to the European Union (2004)

Gamache, R.R., Goldman, A., Rothman, L.S.: Improved spectral parameters for the three most abundant isotopomers of the oxygen molecule. J. Quant. Spectrosc. Radiat. Transf. **59**(3–5), 495–509 (1998)

George, G.A., Morris, G.C.: The intensity of absorption of naphthalene from $30000\,cm^{-1}$ to $53000\,cm^{-1}$. J. Mol. Spectrosc. **26**, 67–71 (1968)

George, L.A., Hard, T.M., O'Brien, R.J.: Measurement of free radicals OH and HO_2 in Los Angeles smog. J. Geophys. Res. **104**, 11643–11655 (1999)

German, K.R.: Direct measurement of the of the radiative lifetime of the A $^2\Sigma^+$(v'=0) states of OH and OD. J. chem. Phys. **62**, 2584–2587 (1975)

Gershenzon, M.Y., Il'in, S., Fedetov, N.G., Gershenzon, Y.M.: The mechanism of reactive NO_3 uptake on dry NaX (X=Cl, Br). J. Atmos. Chem. **34**, 119–135 (1999)

Geyer, A., Alicke, B., Mihelcic, D., Stutz, J., Platt, U.: Comparison of tropospheric NO_3 radical measurements by differential optical absorption spectroscopy and matrix isolation electron spin resonance. J. Geophys. Res. **104**, 26097–26105 (1999)

Geyer, A.: The role of the nitrate radical in the boundary layer – observations and modeling studies. Doctoral Thesis, University of Heidelberg, Germany (2000)

Geyer, A., Ackermann, R., Dubois, R., Lohrmann, B., Müller, T., Platt, U.: Long-term observation of nitrate radicals in the continental boundary layer near Berlin. Atmos. Environ. **35**, 3619–3631 (2001a)

Geyer, A., Alicke, B., Konrad, S., Schmitz, T., Stutz, J., Platt, U.: Chemistry and oxidation capacity of the nitrate radical in the continental boundary layer near Berlin. J. Geophys. Res. **106**, 8013–8025 (2001b)

Geyer, A., Platt, U.: The temperature dependence of the NO_3 degradation frequency—a new indicator for the contribution of NO_3 to VOC oxidation and NO_X removal in the atmosphere. J. Geophys. Res. **107**, 4431–4442 (2002). doi: 10.1029/2001JD001215

Geyer, A., Hofzumahaus, A., Holland, F., Konrad, S., Klüpfel, T., Pätz, H.-W., Perner, D., Schäfer, H.-J., Volz-Thomas, A., Platt, U.: Nighttime production of peroxy and hydroxyl radicals during the BERLIOZ campaign. Observations and modeling studies. J. Geophys. Res. **108**(D4), 8249 (2003a). doi:10.1029/2001JD000656, (PHOEBE: BERLIOZ special section)

Geyer, A., Alicke, B., Ackermann, R., Martinez, M., Harder, H., Brune, W., Piero di Carlo, Williams, E., Jobson, T., Hall, S., Shetter, R., Stutz, J.: Direct observations of daytime NO_3: implications for urban boundary layer chemistry. J. Geophys. Res. **108**(D12), 4368 (2003b). doi:10.1029/2002JD002967

Geyer, A., Stutz, J.: The Vertical structure of OH-HO_2-RO_2 Chemistry in the nocturnal boundary layer: A one-dimensional study. J. Geophys. Res. **109**, D16301, doi: 10.1029/2003JD004425 (2004)

Gil, M., Puentedura, O., Yela, M., Parrondo, C., Jadhav, D., Thorkelsson, B.: OClO, NO_2 and O_3 total columns observations over Iceland during the winter 1993/94. Geophys. Res. Lett. **23**, 3337–3340 (1996)

Gil, M., Puentedura, O., Yela, E., Cuevas, M.: Behavior of NO_2 and O_3 columns during the eclipse of February 26, 1998, as measured by visible spectroscopy. J. Geophys. Res. **105**, 3583 (2000)

Giovanelli, G., Tirabassi, T., Sandroni, S.: Sulphur dioxide plume structure by mask correlation spectroscopy. Atmos. Environ. **13**, 1311–1318 (1979)

Gleason, W.A., Dunker, A.M.: Investigation of background radical sources in a teflon-film irradiated chamber. Environ. Sci. Technol. **23**, 970–978 (1989)

Goldman, A., Gillis, J.R.: Spectral line parameters of the A $^2\Sigma^+ \leftarrow$ X $^2\Pi$ (0, 0) band of OH for atmospheric and high temperatures. J. Quant. Spectrosc. Radiat. Transfer **25**, 111–135 (1981)

Goodman, A.L., Underwood, G.M., Grassian, V.H.: Heterogeneous reaction of NO_2: Characterization of gas-phase and adsorbed products from the reaction, 2 NO_2(g) + H_2O(a) $\longrightarrow$ HONO(g) + NO_3(a) on hydrated silica particles. J. Phys. Chem. A **103**, 7217–7223 (1999)

Gölz, C., Senzig, J., Platt, U.: NO_3 initiated oxidation of biogenic hydrocarbons, CHEMOSPHERE. Glob. Change Sci. **3**, 339–352 (2001)

GOME Users Manual.: SP-1182, European Space Agency, Publications Division, ES-TEC, Noordwijk, The Netherlands, F. Bednarz (ed.), ISBN 92-9092-327-x (1995)

Gomer, T., Stutz, J., Heintz, F., Platt, U.: MFC Handbook. University of Heidelberg (1995)

Götz, P.F.W., Meetham, A.R., Dobson, G.M.B.: The vertical distribution of ozone in the atmosphere. Proc. R. Soc. Lond. A **145**, 416–446 (1934)

Goutail, F., Pommereau, J.-P., Phillips, C., Deniel, C., Sarkissian, A., Lefévre, F., Kyrö, E., Rummukainen, M., Eriksen, P., Andersen, S.B., Kaastad-Hoiskar, B.-A., Braathen, G., Dorokhov, V., Khatatov, V.U.: Depletion of column ozone in the Arctic during the winters of 1993–94 and 1994–95. J. Atmos. Chem. **32**, 1–34 (1999)

Goy, C.A., Pritchard, C.A.: Pressure dependence of the visible iodine bands. J. Mol. Spectrosc. **12**, 38–44 (1964)

Grainger, J.F., Ring, J.: Anomalous fraunhofer line profiles. Nature **193**, 762 (1962)

Grant, W.B., Menzies, R.T.: A survey of laser and selected optical systems for remote measurement of pollutant gas concentrations. J. Air Pollut. Control Assoc. **33**, 187–194 (1983)

Grassi, L., Guzzi, R.: Theoretical and practical consideration on the construction of a zero geometrical loss multi-pass cell based on the use of monolithic multiple-face retro-reflectors. Appl. Opt. **40**(33), 6062–6071 (2001)

Greenblatt, G.D., Orlando, J.J., Burkholder, J.B., Ravishankara, A.R.: Absorption measurements of oxygen between 330 and 1140 nm. J. Geophys. Res. **95**, 18577–18582 (1990)

Grzegorski, M., Wenig, M., Platt, U., Stammes, P., Fournier, N., Wagner, T.: The Heidelberg iterative cloud retrieval utilities (HICRU) and its application to GOME data. Atmos. Chem. Phys. **6**, 4461–4476 (2006)

Guenther, A., Hewitt, C.N., Erickson, D., Fall, R., Geron, C., Graedel, T., Harley, P., Klinger, L., Lerdau, M., McKay, W.A., Pierce, T., Scholes, B., Steinbrecher, R., Tallamraju, R., Taylor, J., Zimmerman, P.: A global model of natural volatile organic compound emissions. J. Geophys. Res. **100**(D5), 8873–8892 (1995). doi: 10.1029/94JD02950

Gurlit, W., Giesemann, C., Ebert, V., Zimmermann, R., Platt, U., Wolfrum, J., Burrows, J.P.: Lightweight diode laser spectrometer “CHILD” for balloon-borne measurements of water vapor and methane. Appl. Opt. **44**(1), 91–101 (2005)

Gutzwiller, L., Arens, F., Baltensperger, U., Gäggeler, H.W., Ammann, M.: Significance of semivolatile diesel exhaust organics for secondary HONO formation. Environ. Sci. Technol. **36**, 677–682 (2002)

Guzzi, R., Burrows, J., Cervino, M., Levoni, C., Cattani, E., Kurosu, T., Torricella, T.: GOME cloud and aerosol data products algorithms development. Report, ESA Contract 11572/95/NL/CN (1998)

Guzzi, R., Zoffoli, S., Corradini, S., Chiarini, M.: Information content of the radiative transfer theory. Agenzia Spaziale Italiana, Rome Report (2003)

Haagen-Smit, A.J.: Chemistry and physiology of Los Angeles smog. Ind. Eng. Chem. **44**, 1342–1346 (1952)

Haagen-Smit, A.J., Fox, M.M.: Photochemical ozone formation with hydrocarbons and automobile exhaust. J. Air Pollut. Control Assoc. **4**, 105–109 (1954)

Hak, C., Pundt, I., Trick, S., Kern, C., Platt, U., Dommen, J., Ordóñez, C., Prévôt, A.S.H., Junkermann, W., Astorga-Lloréns, C., Larsen, B.R., Mellqvist, J., Strandberg, A., Yu, Y., Galle, B., Kleffmann, J., Lörzer, J.C., Braathen, G.O., Volkamer, R.: Intercomparison of four different in-situ techniques for ambient formaldehyde measurements in urban air. J. Atmos. Chem. Phys. **5**, 2881–2900 (2005)

Hall, C.T., Blacet, F.E.: Separation of the absorption spectra of NO_2 and N_2O_4 in the range of 2400–5000 A. J. Chem. Phys. **20**, 1745–1749 (1952)

Hallstadius, H., Unéus, L., Wallin, S.: System for evaluation of trace gas concentration in the atmosphere based on the differential optical absorption spectroscopy technique. Proc. Soc. Photo. Opt. Instrum. Eng. **1433**, 36–43 (1991)

Hallquist, M., Stewart, D.J., Stephenson, S.K., Cox, R.A.: Hydrolysis of N_2O_5 on sub-micron sulfate aerosols. Phys. Chem. Chem. Phys. **5**(16), 3453–3463 (2003)

Hamada, Y., Merer, A.J.: Rotational structure at the long wavelength end of the 2900 A system of SO_2. Can. J. Phys. **52**, 1443–1457 (1974)

Hanst, P.L., Lefohn, A.S., Gay, B.W.: Detection of atmospheric pollutants at parts-per-billion levels by infrared spectroscopy. Appl. Spectrosc. **27**, 188–198 (1973)

Hanst, P.L.: Air pollution measurement by Fourier transform spectroscopy. Appl. Opt. **17**, 1360–1366 (1978)

Hanst, P.L., Hanst, S.T.: Gas measurement in the fundamental infrared region. In: Sigrist, M.W. (ed.) Air Monitoring by Spectroscopic Techniques, Chemical Analysis Series, Vol. 127, pp. 335–470. Wiley, New York (1994)

Hard, T.M., Chan, C.Y., Mehrabzadeh, A.A., Obrien, R.J.: Diurnal HO_2 cycles at clean air and urban sites in the troposphere. J. Geophys. Res. **97**, 9785–9794 (1992)

Hard, T.M., Mehrabzadeh, A.A., Chan, C.Y., O'Brien, R.J.: FAGE measurements of tropospheric HO with measurements and model interferences. J. Geophys. Res. **97**, 9795–9817 (1992)

Hard, T.M., George, L.A., O'Brien, R.J.: FAGE determination of tropospheric HO and HO_2. J. Atmos. Sci. **52**, 3354–3372 (1995)

Hard, T.M., George, L.A., O'Brien, R.J.: An absolute calibration for gas-phase hydroxyl measurements. Environ. Sci. Technol. **36**, 1783–1790 (2002)

Harder, J.W., Brault, J.W., Johnston, P.V., Mount, G.H.: Temperature dependent NO_2 cross section at high spectral resolution. J. Geophys. Res. **102**(D3), 3861–3879 (1997)

Harder, J.W., Jakoubek, R.O., Mount, G.H.: Measurement of tropospheric trace gases by long-path differential absorption spectroscopy during the 1993 OH photochemistry experiment. J. Geophys. Res. **102**, 6215–6226 (1997)

Harder, H., Camy-Peyret, C., Ferlemann, F., Fitzenberger, R., Hawat, T., Osterkamp, H., Perner, D., Platt, U., Schneider, M., Vradelis, P., Pfeilsticker, K.: Stratospheric BrO profile measured at different latitudes and seasons. Atmospheric observations. Geophys. Res. Lett. **25**, 3843–3846 (1998)

Harder, H., Bösch, H., Camy-Peyret, C., Chipperfield, M., Fitzenberger, R., Payan, S., Perner, D., Platt, U., Sinnhuber, B.-M., Pfeilsticker, K.: Comparison of measured and modelled stratospheric BrO: implications for the total amount of stratospheric bromine. Geophys. Res. Lett. **27**, 3695–3698 (2000)

Harris, G.W., Carter, W.P.L., Winer, A.M., Pitts, J.N., Platt, U., Perner, D.: Observations of nitrous acid in the Los Angeles atmosphere and implications for the predictions of ozone-precursor relationships. Environ. Sci. Technol. **16**, 414–419 (1982)

Harris, G.W., Winer, A.M., Pitts, J.N., Platt, U., Perner, D.: Measurement of HONO, NO_3 and NO_2 by long-path differential optical absorption spectroscopy in the Los Angeles basin: In: Killinger, D.K., Mooradian, A. (eds.) Optical and Laser Remote Sensing, Vol. 39, pp. 106–113. Springer, New York (1983)

Harris, G.W., Mackay, G.I., Iguchi, T., Mayne, L.K., Schiff, H.I.: Measurements of formaldehyde in the troposphere by tunable diode laser absorption spectroscopy. J. Atmos. Chem. **8**, 119–137 (1989)

Harrison, A.W.: Midsummer stratospheric NO_2 at latitude 45°S. Can. J. Phys. **57**, 1110–1117 (1979)

Harrison, R.M., Peak, J.D., Collins, G.M.: Tropospheric cycle of nitrous acid. J. Geophys. Res. **101**, 14429–14439 (1996)

Hartley, W.N.: On the probable absorption of solar radiation by atmospheric ozone. Chem. News **42**, 268 (1880)

Hartley, W.N.: On the absorption spectrum of ozone. J. Chem. Soc. **39**, 57–60 (1881)

Harwood, M.H., Jones, R.L.: Temperature dependent ultraviolet-visible absorption cross sections of NO_2 and N_2O_4: low-temperature measurements of the equilibrium constant for 2 NO_2 to N_2O_4. J. Geophys. Res. **99**, 22.955–22.964 (1994)

Harwood, M., Burkholder, J., Hunter, M., Fox, R., Ravishankara, A.: Absorption cross sections and self-reaction kinetics of the IO radical. J. Phys. Chem. A **101**, 853–863 (1997)

Hashmonay, R.A., Yost, M.G., Wu, C.-F.: Computed tomography of air pollutants using radial scanning path-integrated optical remote sensing. Atmos. Environ. **33**, 267–274 (1999)

Hastie, D.R., Weißenmayer, M., Burrows, J.P., Harris, G.W.: Calibrated chemical amplifier for atmospheric RO_X measurements. Anal. Chem. **63**, 2048–2057 (1991)

Haug, H.: Raman-Streuung von sonnenlicht in der Erdatmosphäre, Diploma thesis, University of Heidelberg (1996)

Haug, H., Pfeilsticker, K., Platt, U.: Vibrational Raman scattering in the atmosphere, University of Heidelberg, unpublished manuscript (1996)

Haugen (ed.): Workshop on micrometeorology. American Meteorological Society Science Press, Ephrata (1973)

Hausmann, M., Ritz, D., Platt, U.: New coaxial "long-path-DOAS" system: first application to BrO measurement in the Arctic troposphere. In: Schiff, H.I.,

Platt, U. (eds.) Proceedings Europto Series. Optical Methods in Atmospheric Chemistry, Vol. **1715**, pp. 341–352. (1992)

Hausmann, M., Rudolf, T., Platt, U.: Spectroscopic Measurement of Bromine Oxide, Ozone, and Nitrous acid in Alert, NATO—ASI Series Subseries I "Global Environmental Change", Vol. 7, pp. 189–203. Springer-Verlag (1993)

Hausmann, M., Platt, U.: Spectroscopic measurement of bromine oxide and ozone in the high Arctic during polar sunrise experiment 1992. J. Geophys. Res. **99**, 25399–25413 (1994)

Hausmann, M., Brandenburger, U., Brauers, T., Dorn, H.-P.: Detection of tropospheric OH radicals by long-path differential-optical-absorption spectroscopy: experimental setup, accuracy, and precision. J. Geophys. Res. **102**, 16011–16022 (1997)

Hausmann, M., Brandenburger, U., Brauers, T., Dorn, H.-P.: Simple Monte Carlo methods to estimate the spectra evaluation error in differential-optical-absorption spectroscopy. Appl. Opt. **38**(3), 462–475 (1999)

Hawat, T.M., Camy-Peyret, C., Torguet, R.J.: Suntracker for atmospheric remote sensing. Opt. Eng. **37**(05), 1633–1642 (1998)

Heard, D.E., Pilling, M. J.: Measurement of OH and HO_2 in the troposphere. Chem Rev **103**(12), 5163–5198 (2003)

Hebestreit, K., Stutz, J., Rosen, D., Matveev, V., Peleg, M., Luria, M., Platt, U.: First DOAS measurements of tropospheric BrO in mid latitudes. Science **283**, 55–57 (1999)

Hecht, E.: Optics, 4th edn. Adison Wesley, New York. ISBN 0-8053-8566-5 (2002)

Heckel, A.: Messungen troposphärischer Spurengase mit einem MAXDOAS-Instrument Nachweis von troposphärischem Formaldehyd in Norditalien während der Format Kampagne. Diploma thesis, University of Bremen (2003)

Heckel, A., Richter, A., Tarsu, T., Wittrock, F., Hak, C., Pundt, I., Junkermann, W., Burrows, J.P.: MAX-DOAS measurements of formaldehyde in the Po-Valley. Atmos. Chem. Phys. **5**, 909–918 (2005)

Hegels, E., Crutzen, P.J., Klüpfel, T., Perner, D., Burrows, P.J.: Globale distribution of atmospheric bromine Monoxide from GOME on Earth-observing satellite ERS 2. Geophys. Res. Lett. **25**, 3127–3130 (1998)

Heintz, F., Flentje, H., Dubois, R., Platt, U.: Long-term observation of nitrate radicals at the TOR-Station Kap Arkona (Rügen). J. Geophys. Res. **101**, 22891–22910 (1996)

Heismann, B.: Eine CCD-Kamera zur Messung atmosphärischer Spurenstoffe. Diploma thesis, University of Heidelberg (1996)

Helleis, F., Crowley, J., Moortgat, G.: Temperature dependent rate constants and production branching ratios for the gas phase reaction between CH_3O_2 and ClO. J. Phys. Chem. **97**, 11464–11473 (1993)

Helleis, F., Crowley, J., Moortgat, G.: Temperature dependent CH_3OCl formation in the reaction between CH_3O_2 and ClO. Geophys. Res. Lett. **21**(17), 1795–1798 (1994)

Hendrick, F., Barret, B., Van Roozendael, M., Boesch, H., Butz, A., De Mazière, M., Goutail, F., Hermans, C., Lambert, J.-C., Pfeilsticker, K., Pommereau, J.-P.: Retrieval of nitrogen dioxide stratospheric profiles from ground-based zenith-sky UV-visible observations: validation of the technique through correlative comparisons. Atmos. Chem. Phys. **4**, 2091–2106 (2004) SRef-ID: 1680-7324/acp/2004-4-2091

Hendrick, F., Van Roozendael, M., Kylling, A., Petritoli, A., Rozanov, A., Sanghavi, S., Schofield, R., von Friedeburg, C., Wagner, T., Wittrock, F., Fonteyn, D., De Mazière, M.: Intercomparison exercise between different radiative transfer models used for the interpretation of ground-based zenith-sky and multi-axis DOAS observations. Atmos. Chem. Phys. **6**, 93–108 (2006)

Hermes, Th.: Lichtquellen und Optik für die Differentielle Optische Absorptionsspektroskopie. Diploma thesis in physics, University of Heidelberg (1999)

Herriott, D., Kogelnik, H., Kompfner, R.: Off-axis paths in spherical mirror interferometers. Appl. Opt. **3**, 523–526 (1964)

Herriott, D.R., Schulte, H.J.: Folded optical delay lines. Appl. Opt. **4**, 883–889 (1965)

Heue, K.-P., Bruns, M., Burrows, J.P., Lee, W.-D., Platt, U., Pundt, I., Richter, A., Wagner, T., Wang, P.: NO_2 over the tropics and the arctic measured by the AMAXDOAS in September 2002. Proceedings of the 16th ESA symposium on rocket and balloon program and related research, St. Gallen, 02–05 June 2003, ESA SP-530 (2003)

Heue, K.-P., Bruns, M., Burrows, J.P., Friedeburg, V.C., Lee, W.-D., Platt, U., Pundt, I., Richter, A., Wagner, T., Wang, P.: Validation of scientific NO_2-SCIAMACHY data using the AMAXDOAS instrument. Atmos. Chem. Phys. **5**, 1039–1051 (2005)

Himmelmann, S., Orphal, J., Bovensmann, H., Richter, A., Ladstätter-Weißenmayer, A., Burrows, J.P.: First observation of the OIO molecule by time-resolved flash photolysis absorption spectroscopy. Chem. Phys. Lett. **251**, 330–334 (1996)

Hinkley, E.D. (ed.): Laser Monitoring of the Atmosphere, Topics in Applied Physics, Vol. 14. Springer, Berlin (1976)

Hirokawa, J., Onaka, K., Kajii, Y., Akimoto, H.: Heterogeneous processes involving sodium halide particles and ozone: molecular bromine release in the marine boundary layer in the absence of nitrogen oxides. Geophys. Res. Lett. **25**, 2449–2452 (1998)

Hoff, R.M., Millan, M.M.: Remote SO_2 mass flux measurements using Cospec. J. Air Poll. Cont. Assoc. **31**(4) 381–384 (1981)

Hoff, R.M.: Differential SO_2 column measurements of the Mt. Pinatubo volcanic plume. Geophys. Res. Lett. **19**(2), 175–178 (1992)

Hofmann, D., Bonasoni, P., De Maziere, M., Evangelisti, F., Giovanelli, G., Goldman, A., Goutail, F., Harder, J., Jakoubek, R., Johnston, P., Kerr, J., McElroy, Tom., McKenzie, R., Mount, G., Platt, U., Pommereau, J.-P., Sarkissian, A., Simon, P., Solomon, S., Stutz, J., Thomas, A., Van Roosendael, M., Wu, E.: Intercomparison of UV/visible spectrometers for measurement of stratospheric NO_2 for the network for the detection of stratospheric change. J. Geophys. Res. **100**, 16765–16791 (1995)

Hoffmann, T., O'Dowd, C.D., Seinfeld, J.H.: IO homogeneous nucleation. An explanation for coastal new particle formation. Geophys. Res. Lett. **28**(10), 1949–1952 (2001)

Hofzumahaus, A., Dorn, H.-P., Callies, J., Platt, U., Ehhalt, D.H.: Tropospheric OH concentration measurements by laser long-path absorption spectroscopy. Atmos. Environ. **25A**, 2017–2022 (1991)

Hofzumahaus, A., Aschmutat, U., Heßling, M., Ehhalt, F., Holland, D.H.: The measurement of tropospheric OH radicals by laser- induced fluorescence spectroscopy during the POPCORN field campaign. Geophys. Res. Lett. **23**, 2541–2544 (1996)

Hofzumahaus, A., Brauers, T., Aschmutat, U., Brandenburger, U., Dorn, H.-P., Hausmann, M., Heßling, M., Holland, F., Plass-Dülmer, C., Sedlacek, M., Weber, M., Ehhalt, D.H.: Reply to comment by Lanzendorf et al. Geophys. Res. Lett. **24**, 3039–3040 (1997)

Hofzumahaus, A., Aschmutat, U., Brandenburger, U., Brauers, T., Dorn, H.-P., Hausmann, M., Hessling, M., Holland, F., Plass-Dulmer, C., Ehhalt, D.H.: Intercomparison of tropospheric OH measurements by different laser techniques during the POPCORN campaign 1994. J. Atmos. Chem. **31**(1–2), 227–246 (1998)

Hoiskar, B.A.K., Dahlbak, A., Vaughan, G., Braathen, G.O., Goutail, F., Pommereau, P., Kivi, R.: Interpretation of ozone measurements by ground-based visible spectroscopy—a study of seasonal dependence of airmass factors for ozone on climatology data. J. Quant. Spectrosc. Radiat. Transf. **57**, 569–579 (1997)

Holland, F., Aschmutat, U., Heßling, M., Hofzumahaus, A., Ehhalt, D.H.: Highly time resolved measurements of OH during POPCORN using laser-induced fluorescence spectroscopy. J. Atmos. Chem. **31**, 205–225 (1998)

Holland, F., Hessling, M., Hofzumahaus, A.: In-situ measurement of tropospheric OH radicals by laser-induced fluorescence: A description of the KFA instrument. J. Atmos. Sci. **52**, 3393–3401 (1995)

Holland, F., Hofzumahaus, A., Schäfer, J., Kraus, A., Pätz, H.: Measurements of OH and HO_2 radical concentrations and photolysis frequencies during BERLIOZ. J. Geophys. Res. **108**(D4), 8246 (2003). doi:10.1029/2001JD001393 (2003)

Hollwedel, J., Wenig, M., Beirle, S., Kraus, S., Kühl, S., Wilms-Grabe, W., Platt, U., Wagner, T.: Year-to- year variability of polar tropospheric BrO as seen by GOME, (Proc. COSPAR 2002). Adv. Space Res. 804–808 (2004)

Hoogen, R., Rozanov, V.V., Burrows, J.P.: Ozone profiles from GOME satellite data: Algorithm description and first validation. J. Geophys. Res. **104**(D7), 8263–8280 (1999)

Hönninger, G.: Referenzspektren reaktiver Halogenverbindungen für DOAS Messungen. Diploma thesis, Institut für Umweltphysik, University of Heidelberg (1999)

Hönninger, G.: Halogen oxide studies in the boundary layer by multi axis differential optical absorption spectroscopy and active longpath-DOAS. Ph.D. thesis, University of Heidelberg (2002)

Hönninger, G., Platt, U.: The role of BrO and its vertical distribution during surface ozone depletion at Alert. Atmos. Environ. **36**, 2481–2489 (2002)

Hönninger, G., Friedeburg, C.V., Platt, U.: Multi axis differential absorption spectroscopy (MAX-DOAS). Atmos. Chem. Phys. **4**, 231–254 (2004a)

Hönninger, G., Bobrowski, N., Palenque, E.R., Torrez, R., Platt, U.: Reactive bromine and sulfur emissions at salar de uyuni, Bolivia. J. Geophys. Res. **31**, L04101 (2004b). doi:10.1029/2003GL018818

Hönninger, G., Leser, H., Sebastian, O., Platt, U.: Ground-based measurements of halogen oxides at the Hudson Bay by active long path DOAS and passive MAX-DOAS. Geophys. Res. Lett. **31**, L04111 (2004c). doi:10.1029/2003GL018982

Horn, D., Pimentel, G.C.: 2.5 km low-temperature multiple-reflection cell. Appl. Opt. **10**, 1892–1898 (1971)

Horowitz, A., Meller, R., Moortgat, G.K.: The UV/Visible absorption cross section of the a-dicarbonyl compounds: pyruvic acid, biacetyl and glyoxal. J. Photochem. Photobiol. A Chem. **146**, 19–27 (2001)

Howard, C.J.: Kinetics of the reaction of HO_2 with NO_2. J. Chem. Phys. **67**, 5258–5263 (1977)

Howie, W.H., Lane, I.C., Newman, S.M., Johnson, D.A., Orr-Ewing, A.J.: The UV absorption of ClO. Phys. Chem. Chem. Phys. **1**, 3079–3085 (1999)

Hübler, G., Perner, D., Platt, U., Tönnissen, A., Ehhalt, D.H.: Groundlevel OH radical concentration: new measurements by optical absorption. J. Geophys. Res. **89**, 1309–1319

Hutley, M.C.: Diffraction Gratings. Academic, London (1982) (ISSN 0308-5392; 6)

Impey, G.A., Shepson, P.B., Hastie, D.R., Barrie, L.A., Anlauf, K.G.: Measurements of photolyzable chlorine and bromine during the polar sunrise experiment 1995. J. Geophys. Res. **102**(D13), 16005–16010 (1997)

IPCC, Climate Change.: The IPCC Scientific Assessment, Intergovernmental Panel on Climate Change. Cambridge University Press, Cambridge (1992)

IPCC: Climate change 2001. Third assessment report of the intergovernmental panel on climate change. Cambridge University Press, Cambridge (2002)

Isaacs, R.G., Wang, W.-C., Worsham, R.D., Goldenberg, S.: Multiple scattering LOWTRAN and FASCODE models. Appl. Opt. **26**, 1272–1281 (1987)

Jaeglé, L., Jacob, D.J., Brune, W.H., Faloona, I., Tan, D., Heikes, B.G., Kondo, Y., Sachse, G.W., Anderson, B., Gregory, G.L., Singh, H.B., Pueschel, R., Ferry, G., Blake, D.R., Shetter, R.E.: Photochemistry of HO_X in the upper troposphere at northern midlatitudes. J. Geophys. Res. **105**, 3877–3892 (2000)

Janssen, J.: Analyse spectrale des e'le'ments de l'atmosphe're terrestre. C. R. Hebd. Seances Acad. Sci. **101**, 649–651 (1885)

Janssen, J.: Sur les spectres d'absorption de l'oxyge'ne. C. R. Hebd. Seances Acad. Sci. **102**, 1352–1353 (1886)

Janssen, M.A.: An Introduction to the Passive Remote Atmospheric Remote Sensing by Microwave Radiometry, pp. 1–36. Wiley, New York (1993)

Jenkin, M.I., Cox, R.A., Williams, D.J.: Laboratory studies of the kinetics of formation of nitrous acid from the thermal reaction of nitrogen dioxide and water vapour. Atmos. Environ. **22**, 487–498 (1988)

Jimenez, J.L., Bahreini, R., Cocker, D.R. III, Zhuang, H., Varutbangkul, V., Flagan, R.C., Seinfeld, J.H., O'Dowd, C.D., Hoffmann, T.: New particle formation from photooxidation of diiodomethane (CH_2I_2). J. Geophys. Res. **108**(D10), 4318 (2003). doi: 10.1029/2002JD002452

Jimenez, R., Taslakov, M., Simeonov, V., Calpini, B., Jeanneret, F., Hofstetter, D., Beck, M., Faist, J., van den Bergh, H.: Ozone detection by differential absorption spectroscopy at ambient pressure with a 9.6 μm pulsed quantum-cascade laser. Appl. Phys. B **78**, 249 (2004)

Jobson, B.T., Niki, H., Yokouchi, Y., Bottenheim, J., Hopper, F., Leaitch, R.: Measurements of C_2-C_6 hydrocarbons during polar sunrise experiment 1992. J. Geophys. Res. **99**, 25355–25368 (1994)

Johnston, H.S., Graham, R.: Photochemistry of NO_x and HNO_x Compounds. Can. J. Chem. **52**(8), 1415–1423, doi: 10.1139/cjc-52-8, 1415 (1974)

Johnston, H.S., Cantrell, C.A., Calvert, J.G.: Unimolecular decomposition of NO_3 to form NO and O_2 and a review of N_2O_5/NO_3 kinetics. J. Geophys. Res. **91**, 5159–5172 (1986)

Johnston, H.S., Morris, E.D., Van den Bogaerde, J.: Molecular modulation kinetic spectrometry. ClOO and ClO_2 radicals in the photolysis of clorine in oxygen. J. Am. Chem. Soc. **91**, 7712–7727 (1969)

Johnston, H.S.: Reduction of stratospheric ozone by nitrogen oxide catalyst from supersonic transport exhaust. Science **173**, 517–522 (1971)

Johnston, P.V., McKenzie, R.L.: Long-path absorption measurements of NO_2 in rural New Zealand. Geophys. Res. Lett. **11**, 69–72 (1984)

Johnston, P.V., McKenzie, R.L.: NO_2 observations at 45S during the decreasing phase of solar cycle 21, from 1980 to 1987. J. Geophys. Res. **94**, 3473–3486 (1989)

Johnston, P.V., McKenzie, R.L., Keys, J.G., Matthews, W.A.: Observations of depleted stratospheric NO_2 following the Pinatubo volcanic eruption. Geophys. Res. Lett. **19**, 211–213 (1992)

Johnston, P.V.: Making UV/Vis Cross sections, reference Fraunhofer and synthetic spectra, Unpublished Manuscript, NIWA, Lauder, Oct. 1996 (1996)

Joiner, J., Bhartia, P.K.: The determination of cloud pressures from rotational Raman scattering in satellite backscatter ultraviolet measurements. J. Geophys. Res. **100**, 23019–23026 (1995)

Joiner, J., Bhartia, P.K., Cebula, R.P., Hilsenrath, E., McPeters, R.D.: Rotational-Raman scattering (ring effect) in satellite backscatter ultraviolet measurements. Appl. Opt. **34**(21), 4513–4525 (1995)

Jones, D.G.: Photodiode array detectors in UV-VIS spectroscopy: part I. Anal. Chem. **57**, 1057–1073 (1985a)

Jones, D.G.: Photodiode array detectors in UV-VIS spectroscopy: part II. Anal. Chem. **57**, 1207–1214 (1985b)

Joseph, D.M., Ashworth, S.H., Plane, J.M.C.: The absorption cross-section and photochemistry of OIO. J. Photochem. Photobiol. A Chem. **176**, 68–77 (2005)

Jourdain, J.L., LeBras, G., Poulet, G., Combourieu, J., Rigaud, R., Leroy, B.: UV absorption spectrum of $ClO(A^2\Pi\text{-}X^2\Pi)$ up to the (1,0) band. Chem. Phys. Lett. **57**, 109–112 (1978)

Junge, C, Chagnon, C.W., Manson, J.E.: A world-wide stratospheric aerosol layer. Science **133**(3463), 1478–1479 (1961)

Junge, C.E.: Air chemistry and radioactivity, Vol. 4. (International Geophysics). Academic, New York (1963)

Junkermann, W., Ibusuki, T.: FTIR spectroscopic measurements of surface bond products of nitrogen oxides on aerosol surfaces: Implications for heterogeneous HNO_2 production. Atmos. Environ. **26**, 3099–3103 (1992)

Junkermann, W., Platt, U., Volz-Thomas, A.: A photoelectric detector for the measurement of photolysis frequencies of ozone and other atmospheric molecules. J. Atmos. Chem. **8**, 203–227 (1989)

Kaiser, N.: Off-axis-Messungen von troposphärischem NO_3. Diploma thesis, University of Heidelberg (1997)

Kaiser, J.W., Burrows, J.P.: Fast weighting functions for retrievals from limb scattering measurements. J. Quant. Spectrosc. Radiat. Transf. **77**(3), 273–283 (2003). doi: 10.1016/S0022-4073(02)00125-5

Kalberer, M., Ammann, M., Arens, F., Gäggeler, H.W., Baltensperger, U.: Heterogeneous formation of nitrous acid (HONO) on soot aerosol particles. J. Geophys. Res. **104**, 13825–13832 (1999)

Kanaya, Y., Sadanaga, Y., Matsumoto, J., Sharma, U.K., Hirokawa, J., Kajii, Y., Akimoto, H.: Daytime HO_2 concentrations at Oki Island, Japan, in summer 1998: Comparison between measurement and theory. J. Geophys. Res. **105**, 24205–24222 (2000)

Kanaya, Y., Sadanaga, Y., Hirokawa, J., Kajii, Y., Akimoto, H.: Development of a ground-based LIF instrument for measuring HOx radicals: Instrumentation and calibrations. J. Atmos. Chem. **38**, 73–110 (2001)

Kasparian, J., Rodriguez, M., Méjean, G., Yu, J., Salmon, E., Wille, H., Bourayou, R., Frey, S., André, Y.-B., Mysyrowicz, A., Sauerbrey, R., Wolf, J.-P., Wöste, L.: White-light filaments for atmospheric analysis. Science **301**, 61–64 (2003)

Keeling, C.D., Barcastow, R.B., Bainbridge, A.E., Ekdahl, C.A., Guenther, P.R., Waterman, L.S.: Atmospheric carbon dioxide variations at Mauna Loa observatory, Hawaii. Tellus **28**, 538–551 (1976)

Keller-Rudek, H., Moortgat, G.K.: MPI-Mainz-UV-VIS spectral atlas of gaseous molecules. (2005) URL: www.atmosphere.mpg.de/spectral-atlas-mainz

Kern, C.: Applicability of light-emitting diodes as light sources for active long path DOAS measurements: a feasibility study. Diploma thesis in physics, Institut für Umweltphysik, University of Heidelberg (2004)

Kern, C., Trick, S., Rippel, B., Platt, U.: Applicability of light-emitting diodes as light sources for active DOAS measurements. Appl. Opt. **45**, 2077–2088 (2006)

Kerr, J.B., McElroy, C.T., Evans, W.F.: Mid-latitude summertime measurements of stratospheric NO_2. Can. J. Phys. **60**, 196–200 (1982)

Kessler, C., Perner, D., Platt, U.: Spectroscopic measurements of nitrous acid and formaldehyde—implications for urban photochemistry. In: Versino, B., Ott, H. (eds.) Proceedings of the 2nd European Symposium on Physico-Chemical Behavior of Atmospheric Pollutants, September 29 to October 1, pp. 393–400, Varese (1981)

Kessler, C.: Gasförmige Salpetrige Säure (HNO_2) in der belasteten Atmosphäre. Ph.D. thesis, University of Cologne (1984)

Kessler, C., Platt, U.: Nitrous acid in polluted air masses: Sources and formation pathways. Proceeding on the 3rd European Symposium on PhysioChemical Behaviour of Atmospheric Pollutants, Varese, Italia, 10–12 Apr., pp. 412–422, D. Reidel, Norwell, Mass (1984)

Khalil, M.A.K., Rasmussen, R.A., Gundwardena, A.: Atmospheric methyl bromide: Trends and global mass balance. J. Geophys. Res. **98**, 2887–2896 (1993)

Khokhar, M.F., Frankenberg, C., Van Roozendael, M., Beirle, S., Kühl, S., Richter, A., Platt, U., Wagner, T.: Satellite observations of atmospheric SO_2 from volcanic eruptions during the time period of 1996 to 2002. Adv. Space Res. **36**(5), 879–887 (2005)

Kim, K.-H., Kim, M.-Y.: Comparison of an open path differential optical absorption spectroscopy system and a conventional in situ monitoring system on the basis of long term measurements of SO_2, NO_2, and O_3. Atmos. Environ. **35**, 4059–4072 (2001)

King, M.D., Dick, E.M., Simpson, W.R.: A new method for the atmospheric detection of the nitrate radical (NO_3). Atmos. Environ. **34**, 685–688 (2000)

Kirchstetter, T.W., Harley, R.A., Littlejohn, D.: Measurement of nitrous acid in motor vehicle exhaust. Environ. Sci. Technol. **30**(9), 2843–2849 (1996)

Kleffmann, J., Becker, K.H., Wiesen, P.: Heterogeneous NO_2 conversion processes on acid surfaces: possible atmospheric implications. Atmos. Environ. **32**, 2721–2729 (1998)

Kleffmann, J., Gavriloaiei, T., Hofzumahaus, A., Holland, F., Koppmann, R., Rupp, L., Schlosser, E., Siese, M., Wahner, A.: Daytime formation of nitrous acid: a major source of OH radicals in a forest. Geophy. Res. Lett. **32**, L05818 (2005). doi:10.1029/2005GL022524

Klein, U., Wohltmann, I., Lindner, K., Künzi, K.F.: Ozone depletion and chlorine activation in the Arctic winter 1999/2000 observed in Ny-Ålesund. J. Geophys. Res. **107**(D20), 8288 (2002). doi:10.1029/2001JD000543

Kley, D.: Tropospheric chemistry and transport. Science **276**, 1043–1045 (1997)

Knight, G., Ravishankara, A.R., Burkholder, J.B.: Laboratory studies of OBrO. J. Phys. Chem. A **104**, 11121–11125 (2000)

Knoll, P., Singer, R., Kiefer, W.: Improving spectroscopic techniques by a scanning multichannel method. Appl. Spectrosc. **44**, 776–782 (1990)

Koelemeijer, R.B.A., Stammes, P., Hovenier, J.W., de Haan, J.F.: A fast method for retrieval of cloud parameters using oxygen A band measurements from the global ozone monitoring experiment. J. Geophys. Res. **106D**, 3475–3490 (2001)

Koike, M., Kondo, Y., Matthews, W.A., Johnston, P.V., Yamazaki, K.: Decrease of stratospheric NO_2 at 44° N caused by Pinatubo volcanic aerosols. Geophys. Res. Lett. **20**, 1975–1978 (1993)

Koike, M., Jones, N.B., Matthews, W.A., Johnston, P.V., McKenzie, R.L., Kinnison, R. L., Rodriguez, J.: Impact of Pinatubo aerosols on the partitioning between NO_2 and HNO_3. Geophys. Res. Lett. **21**, 597–600 (1994)

Koike, M., Kondo, Y., Matthews, W.A., Johnston, P.V., Nakajima, P.V., Kawaguchi, A., Nakane, H., Murata, I., Budiyono, A., et al.: Assessment of the uncertainties in the NO_2 and O_3 measurements by visible spectrometers. J. Atmos. Chem. **32**, 121–145 (1999)

Kolmogorov, A.N.: The local structure of turbulence in incompressible viscous fluid for very large Reynolds number. Dokl. Akad. Nauk SSSR **30**, 229–303 (1941). Reprinted in Proc. R. Soc Lond. A **434**, 15–17 (1991)

Kondo, Y., Matthews, W.A., Solomon, S., Koike, M., Hayashi, M., Yamazaki, K., Nakajima, H., Tsukui, K.: Ground based measurements of column amounts of NO_2 over Syowa Station, Antarctica. J. Geophys. Res. **99**, 14535–14548 (1994)

Kosterev, A.A., Tittel, F.K.: Chemical sensors based on quantum cascade lasers. IEEE J. Quantum Electron. **38**, 582 (2002)

Kraus, S.: DOASIS A Framework Design for DOAS. Dissertation, University of Mannheim, Germany (2005)

Kreher, K.: Messung der Breitenverteilung (50°N–70°S) von stratosphärischem Ozon und Stickstoffdioxid mittels optischer Absorptionsspektroskopie. Diploma thesis, University of Heidelberg (1991)

Kreher, K., Fiedler, M, Gomer, T, Stutz, J., Platt, U.: The latitudinal distribution (50°N–50°S) of NO_2 and O_3 in October/November 1990. Geophys. Res. Lett. **22**, 1217–1220 (1995)

Kreher, K., Keys, J.G., Johnston, P.V., Platt, U., Liu, X.: Ground-based measurements of OClO and HCl in austral spring 1993 at arrival heights, Antarctica. Geophys. Res. Lett. **23**, 1545–1548 (1996)

Kreher, K., Johnston, P.V., Wood, S.W., Platt, U.: Ground-based measurements of tropospheric and stratospheric BrO at arrival heights (78°S), Antarctica. Geophys. Res. Lett. **24**, 3021–3024 (1997)

Kreher, K., Bodeker, G.E., Kanzawa, H., Nakane, H., Sasano, H.: Ozone and temperature profiles measured above Kiruna inside, at the edge of, and outside the Arctic polar vortex in February and March 1997. Geophys. Res. Lett. **26**, 715–718 (1999)

Kromminga, H., Orphal, J., Spietz, P., Voigt, S., Burrows, J.P.: The temperature dependence (213–293 K) of the absorption cross-sections of OClO in the

340–450 nm region measured by Fourier-transform spectroscopy. J. Photochem. Photobiol. A Chem. **157**, 149–160 (2003)

Künzli, N., Kaiser, R., Medina, S., et al.: Public-health impact of outdoor and traffic-related air pollution: a European assessment. Lancet **356**, 795–801 (2000)

Kurosu, T., Rozanov, V.V., Burrows, J.P.: Parameterization schemes for terrestrial water clouds in the radiative transfer model GOMETRAN. J. Geophys. Res. **102**(D18), 21809–21823 (1997)

Kurtenbach, R., Becker, K.H., Gomes, J.A.G., Kleffmann, J., Lörzer, J.C., Spittler, M., Wiesen, P., Ackermann, R., Geyer, A., Platt, U.: Investigations of emissions and heterogeneous formation of HONO in a road traffic tunnel. Atmos. Environ. **35**, 3385–3394 (2001)

Kurtenbach, R., Ackermann, R., Becker, K.H., Geyer, A., Gomes, J.A.G., Lörzer, J.C., Platt, U., Wiesen, P.: Verification of the contribution of vehicular traffic to the total NMVOC emissions in Germany and the importance of NO_3 chemistry in the city air. J. Atmos. Chem. **42**, 395–411 (2002)

Kurucz, R.L., Bell, B.: Atomic Line Data, Kurucz CD-ROM No. 23. Cambridge, Mass.: Smithsonian Astrophysical Observatory. Available in the internet: http://cfa-www.harvard.edu/amdata/ampdata/kurucz23/sekur.html 1995

Kuze, A., Chance, K.V.: Analysis of cloud-top height and cloud coverage from satellites using the O_2 A and B bands. J. Geophys. Res. **99**, 14482–14491 (1994)

Kuznetzov, B.I., Nigmatullina, K.S.: Optical determination of the nitrogen dioxide content in the atmosphere. Izv. Atmos. Oceanic Phys. **13**, 614–617 (1977)

Laan, E., de Vries, J., Kruizinga, B., Visser, H., Levelt, P., van den Oord, G.H.J., Maelkki, A., Leppelmeier, G., Hilsenrath, E.: Ozone monitoring with the OMI instrument. In: Proceedings of SPIE 45th Annual Meeting (Imaging Spectrometry VI: Sensor Applications). The International Symposium on Optical Science and Technology, pp. 334–343. San Diego (2000)

Lambert, J.C., Van Roozendael, M., Simon, P.C., Pommereau, J.P., Goutail, F., Gleason, J.F., Andersen, S.B., Arlander, D.W., Buivan, N.A., Claude, H., De La Noe, J., De Maziere, M., Dorokhov, V., Eriksen, P., Green, A., Karlsen Tornqvist, K., Kastadt Hoiskar, B.A., Kyro, E., Leveau, J., Merienne, M.F., Milinevsky, G., Roscoe, H.K., Sarkissian, A., Shanklin, J.D., Staehelin, J., Wahlstrom Tellefsen, C., Vaughan, G.: Combined characterization of GOME and TOMS total ozone measurements from space using ground-based observations from the NDSC. Adv. Space Res. **26**, 1931–1940 (2001)

Lammel, G., Cape, J.N.: Nitrous acid and nitrite in the atmosphere. Chem. Soc. Rev. **25**, 361–369 (1996)

Lammel, G., Perner, D.: The atmospheric aerosol as a source of nitrous acid in the polluted atmosphere. J. Aerosol Sci. **19**, 1199–1202 (1988)

Lamp, T., Ropertz, A., Weber, K., van Haaren, G.: First results of ambient air measurements with different remote sensing systems over a lake in Germany. Proc. Soc. Photo. Opt. Instrum. Eng. **3534**, 162–172 (1998)

Lang, R., Lawrence, M.G.: Evaluation of the hydrological cycle of MATCH driven by NCEP reanalysis data: comparison with GOME water vapor field measurements. Atmos. Chem. Phys. Discuss. **4**, 7917–7984 (2004)

Langford, A.O., Portmann, R.W., Daniel, J.S., Miller, H.L., Solomon, S.: Spectroscopic measurement of NO_2 in a Colorado thunderstorm: determination of the mean production by cloud-to-ground lightning flashes. J. Geophys. Res. **109**, D11304 (2004). doi: 10.1029/203JD004158

Larche, K.: Die Strahlung des Xenon—Hochdruckbogens hoher Leistungsaufnahme. Z. Phys. **136**, 74–86 (1953)

Lary, D.J., Chipperfield, M.P., Toumi, R., Lenton, T.: Heterogeneous atmospheric bromine chemistry. J. Geophys. Res. **101**, 1489–1504 (1996)

Lary, D.J.: Halogens and the chemistry of the free troposphere. Atmos. Chem. Phys. **5**, 227–237 (2005)

Lauer, A., Dameris, M., Richter, A., Burrows, J.P.: Tropospheric NO_2 columns: a comparison between model and retrieved data from GOME measurements. Atmos. Chem. Phys. **2**, 67–78 (2002)

Lazlo, B., Kurylo, M.J., Huie, R.E.: Absorption cross section, kinetics of formation, and self-reaction of the IO radical via laser photolysis of $N_2O/I_2/N_2$ mixtures. J. Phys. Chem. **99**, 11701–11707

Le Bras, G., Gölz, C., Platt, U.: Production of peroxy-radicals in the DMS oxidation during night-time. In: Restelli, G., Angeletti, G. (eds.) Dimethylsulphide: Oceans, atmosphere and climate: Proceedings of the International Symposium, Belgirate, Italy, 13–15 October 1992, Kluwer Academic Publishers, pp. 251–260 (1993)

Le Bras, G., Platt, U.: A possible mechanism for combined chlorine and bromine catalysed destruction of tropospheric ozone in the Arctic. Geophys. Res. Lett. **22**, 599–602 (1995)

Lee, D.S., Köhler, I., Grobler, E., Rohrer, F., Sausen, R., Gallardo-Klenner, L., Olivier, J.G.J., Dentener, F.J., Bouwman, A.F.: Estimates of global NO_X emissions and their uncertainties. Atmos. Environ. **31**, 1735–1749 (1997)

Lee, J.S., Kuk, B.J., Kim, Y.J.: Development of a differential optical absorption spectroscopy (DOAS) system for the detection of atmospheric trace gas species; NO_2, SO_2, and O_3. J. Korean Phys. Soc. **41**, 693–698 (2002)

Lee, C., Kim, Y.K., Tanimoto, H., Bobrowski, N., Platt, U., Mori, T., Yamamoto, K.: Remote measurement of volcanic halogen oxides and observation of surface ozone depletion. Geophys. Res. Lett. **32**, L21809 (2005). doi:10.1029/2005GL023785

Leighton, P.A.: Photochemistry of Air Pollution. Academic, New York (1961)

Lelieveld, J., Crutzen, P.J.: Influence of cloud and photochemical processes on tropospheric ozone. Nature **343**, 227–233 (1990)

Lelieveld, J., Crutzen, P.J.: The role of clouds in tropospheric photochemistry. J. Atmos. Chem. **12**, 229–267 (1991)

Lenoble, J.: Radiative Transfer in Scattering and Absorbing Atmospheres: Standard Computational Procedures. A. Deepak Publishing, Hampton (1985)

Lerner, J.M., Thevenon, A.: The optics of spectroscopy, Jobin-Yvon Optical Systems/Instrumentss SA (1988)

Leser, H., Hönninger, G., Platt, U.: MAX-DOAS measurements of BrO and NO_2 in the marine boundary layer. Geophys. Res. Lett. **30**(10), 1537 (2003). doi: 10.1029/2002GL015811

Leue, C., Wenig, M., Platt, U.: Retrieval of atmospheric trace gas concentrations. In: Jähne, B., Haußecker, H., Geißler, P. (eds.) Handbook of Computer Vision and Applications. Volume III: Systems and Applications, Academic Press, San Diego (1999)

Leue, C., Wenig, M., Wagner, T., Platt, U., Jähne, B.: Quantitative analysis of NO_X emission from global ozone monitoring experiment satellite image sequences. J. Geophys. Res. **106**, 5493–5505 (2001)

Levelt, P.F., van den Oord, B., Hilsenrath, E., Leppelmeier, G.W., Bhartia, P.K., Malkki, A., Kelder, H., van der, A.R.J., Brinksma, E.J., van Oss, R., Veefkind, P., van Weele, M., Noordhoek, R.: Science Objectives of EOS-Aura's Ozone Monitoring Instrument (OMI). In: Proceedings of Quadrennial Ozone Symposium, Sapporo, Japan, pp. 127–128 (2000)

Levenberg, K.:A method for the solution of certain non-linear problems in least squares. Quant. Appl. Math. **2**, 164–168 (1944)

Levy, H.: Normal atmosphere: large radical and formaldehyde concentrations predicted. Science **173**, 141–143 (1971)

Li, S.-M.: Equilibrium of particle nitrite with gas phase HONO: tropospheric measurements in the high Arctic during polar sunrise. J. Geophys. Res. **99**, 25469–25478 (1994)

Liley, J.B., Johnston, P.V., McKenzie, R.L., Thomas, A.J., Boyd, I.S.: Stratospheric NO_2 variations from a long time series at Lauder, New Zealand. J. Geophys. Res. **105**(D9), 11633–11640 (2000)

Lindberg, S., Brooks, S., Lin, C.-J., Scott, K.J., Landis, M.S., Stevens, R.K., Goodsite, M., Richter, A.: Dynamic oxidation of gaseous mercury in the Arctic troposphere at polar sunrise. Environ. Sci. Technol. **36**, 1245–1256 (2002)

Livesey, N.J., Read, W.G., Froidevaux, L., Waters, J.W., Santee, M.L., Pumphrey, H.C., Wu, D.L., Shippony, Z., Jarnot, R.F.: The UARS microwave limb sounder version 5 data set: theory, characterization, and validation. J. Geophys. Res. **108**(D13), 4378 (2003). doi:10.1029/2002JD002273

Löfgren, L.: Determination of benzene and toluene in urban air with differential optical absorption spectroscopy. Int. J. Environ. Anal. Chem. **47**, 69–74 (1992)

Logan, J.A., Prather, M.J., Wofsy, S.C., McElroy, M.B.: Tropospheric chemistry: a global perspective. J. Geophys. Res. **86**, 7210–7254 (1981)

Lohberger, F., Hönninger, G., Platt, U.: Ground based imaging differential optical absorption spectroscopy of atmospheric gases. Appl. Opt. **43**(24), 4711–4717 (2004)

Long, W.A.: Raman Spectroscopy. McGraw-Hill, New York (1977)

Longfellow, C.A., Imamura, T., Ravishankara, A.R., Hanson, D.R.: HONO solubility and heterogeneous reactivity on sulfuric acid surfaces. J. Phys. Chem. A **102**, 3323–3332 (1998)

Lovelock, J.E.: Gaia: A New Look at Life on Earth. Oxford University Press, Oxford (1979)

Löwe, A.G., Adukpo, D., Fietkau, S., Heckel, A., Ladstätter-Weißenmayer, A., Medeke, T., Oetjen, H., Richter, A., Wittrock, F., Burrows, J.P.: Multi-axis-DOAS observations of atmospheric trace gases at different latitudes by the global instrument network BREDOM. In: Proceedings of 10th Science Conference of IAMAS, CACGP and 7th Science Conference of IGAC, September 2002, Crete (2002)

MacManus, J.B., Kebabian, P.L., Zahniser, M.S.: Astigmatic mirror multipass absorption cells for long-path-length spectroscopy. Appl. Opt. **34**, 3336–3348 (1995)

Majewski, W., Meerts, W.L.: Near-UV spectra with fully resolved rotational structure of naphthalene and perdeuterated naphthalene. J. Mol. Spectrosc. **104**, 271–281 (1984)

Mandelman, M., Nicholls, R.W.: The absorption cross section and f-values for the $v'' = 0$ progression of bands and associated continuum for the ClO (A^2Pi-X^2Pi) system. J. Quant. Spectrosc. Radiat. Transf. **17**, 483–491 (1977)

Martinez, M., Perner, D., Hackenthal, E., Kultzer, S., Schultz, L.: NO_3 at Helgoland during the NORDEX campaign in October 1996. J. Geophys. Res. **105**(D18), 22685–22695 (2000)

Marquard, D.W.: An algorithm for least-squares estimation of nonlinear parameters. J. Soc. Indust. Appl. Math. **11**, 431–441 (1963)

Marquard, L.C.: Modellierung des Strahlungstransports in der Erdatmosphäre für absorptionsspektroskopische Messungen im ultravioletten und sichtbaren Spektralbereich, Doktoral thesis, University of Heidelberg (1998)

Marquard, L.C., Wagner, T., Platt, U.: Improved approaches for the calculation of air mass factors required for scattered light differential optical absorption spectroscopy. J. Geophys. Res. **105**, 1315–1327 (2000)

Martin, R.V., Chance, K., Jacob, D.J., Kurosu, T.P., Spurr, R.J.D., Bucsela, E., Gleason, J.F., Palmer, P.I., Bey, I., Fiore, A.M., Li, Q., Yantosca, R.M., Koelemeijer, R.B.A.: An improved retrieval of tropospheric nitrogen dioxide from GOME. J. Geophys. Res. **107**(D20), 4437 (2002). doi:10.1029/2001JD001027

Martin, R.V., Parrish, D.D., Ryerson, T.B., Nicks, D.K., Jr, Chance, K., Kurosu, T.P., Jacob, D.J., Sturges, E.D., Fried, A., Wert, B.P.: Evaluation of GOME satellite measurements of tropospheric NO2 and HCHO using regional data from aircraft campaigns in the southeastern United States. J. Geophys. Res. **109**, D24307 (2004). doi:10.1029/2004JD004869

Martinez, M., Arnold, T., Perner, D.: The role of bromine and chlorine chemistry for arctic ozone depletion events in Ny-Ålesund and comparison with model calculations. Ann. Geophys. **7**, 941–956 (1999)

Martini, L., Sladkovic, R., Slemr, f., Werle, P.: Monitoring of air pollutants: long term intercomparison of DOAS with conventional techniques. Proceedings Of 87th Annual Meeting & Exhibition of Air & Waste Management Association, Cincinnati, Ohio, 19–24 June 1994

Marx, B.R., Birch, K.P., Felton, R.C., Jolliffe, B.W., Rowley, W.R.C., Woods, P.T.: High-resolution spectroscopy of SO_2 using a frequency-doubled continuous-wave dye laser. Opt. Comm. **33**, 287–291 (1980)

Mateer, C.L., Dutsch, H.U., Staehelin, J.: Influence of a priori profiles on trend calculations from Umkehr data. J. Geophys. Res. **101**(D11), 16779–16787 (1996)

Matsumoto, J., Imai, H., Kosugi, N., Kajii, Y.: Development of a measurement system of nitrate radical and dinitrogen pentoxide using a thermal conversion/laser-induced fluorescence. Rev. Sci. Instrum. **76**, 064101 (2005). doi: 10.1063/1.1927098

Matveev, V., Peleg, M., Rosen, D., Tov-Alper, D.S., Stutz, J., Hebestreit, K., Platt, U., Blake, D., Luria, M.: Bromine oxide – ozone interaction over the dead sea. J. Geophys. Res. **106**, 10375–10378 (2001)

Maurellis, A.N., Lang, R., van der Zande, W.J.: A new DOAS parametrization for retrieval of trace gases with highly-structured absorption spectra. Geophys. Res. Lett. **27**, 4069–4072 (2000a)

Maurellis, A.N., Lang, R., van der Zande, W.J., Aben, I., Ubachs, W.: Precipitable water column retrieval from GOME data. Geophys. Res. Lett. **27**, 903–906 (2000b)

McConnell, J.C., Henderson, G.S., Barrie, L., Bottenheim, J., Niki, H., Langford, C.H., Templeton, E.M.J.: Photochemical bromine production implicated in Arctic boundary-layer ozone depletion. Nature **355**, 150–152 (1992)

McElroy, C.T., McLinden, C.A., McConnell, J.C.: Evidence for bromine monoxide in the free troposphere during Arctic polar sunrise. Nature **397**, 338–340 (1999)

McGonigle, A.J.S., Hilton, D.R., Fischer, T.P., Oppenheimer, C.: Plume velocity determination for volcanic SO_2 flux measurements. Geophys. Res. Lett. **32**, L11302 (2005). doi:10.1029/2005GL022470

McKeen, S.A., Trainer, M., Hsie, E.Y., Tallamraju, R.K., Liu, S.C.: On the indirect determination of atmospheric OH radical concentrations from reactive hydrocarbon measurements. J. Geophys. Res. **95**, 7493–7500 (1990)

McKenzie, R.L., Johnston, P.V.: Seasonal variations in stratospheric NO_2 at 45° S. Geophys. Res. Lett. **9**, 1255–1258 (1982)

McKenzie, R.L., Johnston, P.V.: Stratospheric nitrogen dioxide measurements at arrival heights, Antarctica, N.Z. Antarct. Rec. **5**, 12 (1983)

McKenzie, R.L., Johnston, P.V.:Springtime stratospheric NO_2 in Antarctica, Geophys. Res. Lett. **11**, 73–75 (1984)

McKenzie, R.L., Johnston, P.V., McElroy, C.T., Kerr, J.B., Solomon, S.: Altitude distributions of stratospheric constituents from ground-based measurements at twilight. J. Geophys. Res. **96**, 15499–15511 (1991)

McMahon, B.B., Simmons, E.L.: Ground based measurements of atmospheric NO_2 by differential optical absorption. Nature **287**, 710–711 (1980)

McManus, J.B., Kebabian, P.L.: Narrow optical interference fringes for certain setup conditions in multipass absorption cells of the Herriott type. Appl. Opt. **29**(7), 898–900 (1990)

McManus, J.B., Kebabian, P.L., Zahniser, M.S.: Astigmatic mirror multipass absorption cells for long-path-length spectroscopy. Appl. Opt.-LP **34**(18), 3336–3348 (1995)

Melamed, M.L., Solomon, S., Daniel, J.S., Langford, A.O., Portmann, R.W., Ryerson, T.B., Nicks, D.K., Jr, McKeeen, S.A.: Measuring reactive nitrogen emissions from point sources using visible spectroscopy from aircracft. J. Environ. Monit. **5**, 29–34 (2003)

Meller, R., Moortgart, G.K.: Temperature dependence of the absorption cross sections of formaldehyde between 223 and 323 K in the wavelength range 225–375 nm. J. Geophys. Res. **201**, 7089–7101 (2000)

Mellqvist, J., Rosén, A., Axelsson, H.: Temperature dependence of the absorption spectra of nitrogen oxide, nitrogen dioxide and sulfur dioxide in the application of differential optical. Analyst **117**, 417–418 (1992)

Mellqvist, J., Rosén, A.: DOAS for flue gas monitoring – I. Temperature in the U.V./visible absorption spectra of NO, NO_2, SO_2 and NH_3. J. Quant. Spectrosc. Radiat. Transf. **56**, 187–208 (1996)

Mentel, T.F., Bleilebens, D., Wahner, A.: A study of nighttime nitrogen oxide oxidation in a large reaction chamber – the fate of NO_2, N_2O_5, HNO_3, and O_3 at different humidities. Atmos. Environ. **30**, 4007–4020 (1996)

Mérienne, M.F., Jenouvrier, A., Coquart, B.: The NO_2 absorption spectrum. I. Absorption cross-sections at ambient temperature in the 300–500 nm region. J. Atmos. Chem. **20**, 281–297 (1995)

Mérienne, M.F., Jenouvrier, A., Hermans, C., Vandaele, A.C., Carleer, M., Clerbaux, C., Coheur, P.F., Colin, R., Fally, S., Bach, M.: Water vapor line parameters in the 13,000–9250 cm^{-1} region. JQSRT **82**, 99–117 (2003)

Mie, G.: Beiträge zur Optik trüber Medien, speziell kolloidaler Metallösungen, Annalen der Physik, Vierte Folge, Band **25**(3), 377–445 (1908)

Migeotte, M.V.: Lines of methane at 7.7μ in the solar spectrum. Phys. Rev. **74**, 112–113 (1948)
Migeotte, M.: The fundamental band of carbon monoxide at 4.7 μm in the solar spectrum. Phys. Rev. **75**, 1108–1109 (1949)
Mihelcic, D., Holland, F., Hofzumahaus, A., Hoppe, L., Konrad, S., Müsgen, P., Pätz, H.-W., Schmitz, T., Schäfer, H.-J., Schmitz, T., Volz-Thomas, A., Bächmann, K., Schlomski, S., Platt, U., Geyer, A., Alicke, B., Moortgat, G.: Peroxy radicals during BERLIOZ at Pabstthum: measurements, radical budgets, and ozone production. J. Geophys. Res. **108**(D4), 8254 (2003). doi:10.1029/2001JD001014, (PHOEBE: BERLIOZ special section)
Mihelcic, D., Muesgen, P., Ehhalt, D.H.: An improved method of measuring tropospheric NO_2 and RO_2 by matrix isolation and electron spin resonance. J. Atmos. Chem. **3**, 341–361 (1985)
Mihelcic, D., Klemp, D., Müsgen, P., Pätz, H.W., Volz-Thomas, A.: Simultaneous measurements of peroxy and nitrate radicals at Schauinsland. J. Atmos. Chem. **16**, 313–335 (1993)
Millan, M., Townsend, S., Davies, J.: Study of the Barringer refractor plate correlation spectrometer as a remote sensing instrument. Utias rpt. 146, m.a.sc. thesis, University of Toronto, Toronto, Ontario, Canada (1969)
Millán, M.M.: A study of the operational characteristics and optimization procedures of dispersive correlation spectrometers for the detection of trace gases in the atmosphere. Ph.D. thesis, Universidad de Toronto, Ontario (1972)
Millan, M.M., Hoff, R.M.: Dispersive correlation spectroscopy: a study of mask optimization procedures. Appl. Opt. **16**, 1609–1618 (1977)
Millan, M.: Recent advances in correlation spectroscopy for the remote sensing of SO_2. In: Proceedings of 4th Joint Conference on Sensing of Environmental Pollutants, pp. 40–43 (1978)
Millan, M.M.: Remote sensing of air pollutants. A study of some atmospheric scattering effects. Atmos. Environ. **14**, 1241–1253 (1980)
Miller, H.L., Weaver, A., Sanders, R.W., Arpag, K., Solomon, S.: Measurements of arctic sunrise surface ozone depletion events at Kangerlussuaq. Greenland (67 ° N to 51 ° W), Tellus B, **49**(5), 496–509 (1997)
Min, Q.-L., Harrison, L.C., Clothiaux, E.: Joint statistics of photon path length and cloud optical depth: case studies. J. Geophys. Res. **106**, 7375–7386 (2001)
Minato, A., Sugimoto, N., Sasano, Y.: Optical design of cube-corner retroreflectors having curved mirror surfaces. Appl. Opt. **31**, 6015–6020 (1992)
Minato, A., Sugimoto, N.: Design of a four-element, hollow-cube corner retroreflector for satellites by use of a genetic algorithm. Appl. Opt. **37**, 438–442 (1998)
Mohammed-Tahrin, N., South, A.M., Newnham, D.A., Jones, R.L.: An accurate wavelength calibration for the ozone absorption cross-section in the near-UV spectral region, and its effect on the retrieval of BrO from measurements of zenith-scattered sunlight. J. Geophys. Res. **106**(D9), 9897–9907 (2001)
Molina, L.T., Molina, M.J.: Production of the Cl_2O_2 from the Self-Reaction of the ClO Radical. J. Phys. Chem. **91**(2), 433–436 (1987)
Molina, M.J., Rowland, F.S.: Stratospheric sink for chlorofluoromethans. Chlorine atom catalyzed destruction of ozone. Nature **249**, 810 (1974)
Monks, P.S., Carpenter, L.J., Penkett, S.A., Ayers, G.P.: Night-time peroxy radical chemistry in the remote marine boundary layer over the southern ocean. Geophys. Res. Lett. **23**, 535–538 (1996)

Montzka, S., Butler, J., Myers, R.M.T., Swanson, T., Clarke, A., Lock, L., Elkins, J.: Decline in the tropospheric abundance of halogen from halocarbons: implications for stratospheric ozone depletion. Science **272**, 1318–1320 (1996)

Mount, G.H., Sanders, R.W., Schemltekopf, A.L., Solomon, S.: Visible spectroscopy at McMurdo Station, Antarctica, 1. Overview and daily variations of NO_2 and O_3, austral spring, 1986. J. Geophys. Res. **92**, 8320–8328 (1987)

Mount, G.H.: The measurement of tropospheric OH by long path absorption. 1. Instrumentation. J. Geophys. Res. **97**, 2427–2444 (1992)

Mount, G.H., Eisele, F.L.: An intercomparison of tropospheric OH. Measurements at Fritz Peak Observatory, Colorado. Science **256**, 1187–1190 (1992)

Mount, G.H., Rumburg, B., Havig, J., Lamb, B., Westberg, H., Yonge, D., Johnson, K., Kincaid, R.: Measurement of atmospheric ammonia at a dairy using differential optical absorption spectroscopy in the mid-ultraviolet. Atmos. Environ. **36**, 1799–1810 (2002)

Mount, G.H., Sanders, R.W., Brault, J.W.: Interference effects in reticon photodiode array detectors. Appl. Opt. **31**, 851 (1992)

Mozurkewich, M., Calvert, J.G.: Reaction probability of N_2O_5 on Aqueous aerosol. J. Geophys. Res. **93**, 15889–15896 (1988)

Murtagh, D., Frisk, U., Merino, F., Ridal, M., Jonsson, A., Stegman, J., Witt, G., Eriksson, P., Jiménez, C., Megie, G., Noë, de la J., Ricaud, P., Baron, P., Ramon Pardo, J., Hauchcorne, A., Llewellyn, E.J., Edward, J., Degenstein, E.J., Gattinger, R.L., Lloyd, N.D., Evans, W.F.J., McDade, I.C., Haley, C.S., Sioris, C., Savigny, C.V., Solheim, B.H., McConnell, J.C., Strong, K., Richardson, E.H., Leppelmeier, G.W., Kyrölä, E., Auvinen, H., Oikarinen, L.: An overview of the Odin atmospheric mission. Can J. Phys. **80**, 309–319 (2002)

Nardi, B., Bellon, W., Oolman, L.D., Deshler, T.: Spring 1996 and 1997 ozonesonde measurements over McMurdo Station, Antarctica. Geophys. Res. Lett. **26**, 723–726 (1999)

Nash, T.: The colorimetric estimation of formaldehyde by means of the Hantzsch reaction. Biochem. J. **55**, 416–421 (1953)

Naus, H., Ubachs, W.: Visible absorption bands of the $(O_2)_2$-collision complex at pressures below 760 Torr. Appl. Opt. **38**, 3423–3428 (1999)

Neftel, A., Blatter, A., Staffelbach, T.: Gas phase measurements of NH_3 and NH_4^+ with differential optical absorption spectroscopy and gas stripping scrubber in combination with flow injection analysis. In: Restelli G., Angeletti, G. (eds.) Proceedings of Symposium on Physico-Chemical Behaviour of Atmospheric Pollutants, pp. 83–91. Kluwer, Dordrecht (1990)

Neftel, A., Blatter, A., Hesterberg, R., Staffelbach, T.: Measurements of concentration gradients of HNO_2 and HNO_3 over a semi-natural ecosystem. Atmos. Environ. **30**, 3017–3025 (1996)

Nestlen, M., Platt, U., Flothmann, D.: A new instrument for measuring dry deposition velocities of SO_2 using the eddy-correlation method. Proceedings of Workshop "Trockene Deposition", Oberursel 1981

Neuroth, R., Dorn, H.-P., Platt, U.: High resolution spectral features of a series of aromatic hydrocarbons and BrO: potential interferences with OH measurements. J. Atmos. Chem. **12**, 12287–12298 (1991)

Newchurch, M.J., Yang, E.-S., Cunnold, D.M., Reinsel, G.C., Zawodny, J.M., Russell, J.M. III: Evidence for slowdown in stratospheric ozone loss:

First stage of ozone recovery. J. Geophys. Res. **108**(D16), 4507 (2003). doi:10.1029/2003JD003471

Newitt, D.M., Outridge, L.E.: The ultraviolet absorption bands ascribed to HNO_2. J. Chem. Phys. **6**, 752–754 (1938)

Newman, S.M., Lane, I.C., Orr-Ewing, A.J., Newnham, D.A., Ballard, J.: Integrated absorption intensity and einstein coefficients for the O_2 $^{a^1\Delta_g \leftarrow X^3\Sigma_g^-}$ transition: a comparison of cavity ringdown and high resolution fourier transform spectroscopy with a long-path absorption cell. J. Chem. Phys. **110**(22), 10749–10757 (1999)

Newman, S.M., Orr-Ewing, A.J., Newnham, D.A., Ballard, J.: Temperature and pressure dependence of linewidths and integrated absorption intensities for the $^{a^1\Delta_g \leftarrow X^3\Sigma_g^-}$ (0,0) transition. J. Phys. Chem. A, **104**(42), 9467–9480 (2000)

Newnham, D.A., Ballard, J.: Visible absorption cross section and integrated absorption intensities of molecular oxygen (O_2 and O_4). J. Geophys. Res. **103**, 28801–28816 (1998)

Nguyen, M.T., Sumathi, R., Sengupta, D., Peeters, J.: Theoretical analysis of reactions related to the HNO_2 energy surface: OH + NO and H + NO_2. Chem. Phys. **230**(1), 1–11 (1998)

Nicolet, M.: On the molecular scattering in the terrestrial atmosphere: an empirical formula for its calculation in the homosphere. Planet. Space Sci. **32**(11), 1467–1468 (1984)

Niki, H., Becker, K.H. (eds.): NATO Advanced Research Workshop: "The Tropospheric Chemistry of Ozone in the Polar Regions", NATO ASI Series, Subseries I "Global Environmental Change". Springer, Heidelberg (1993)

Noël, S., Bovensmann, H., Burrows, J.P., Frerick, J., Chance, K.V., Goede, A.H.P.: Global atmospheric monitoring with SCIAMACHY. Phys. Chem. Earth (C). **24**(5), 427–434 (1999)

Noel, S., Buchwitz, M., Bovensmann, H., Burrows, J.P.: Retrieval of total water vapour column amounts from GOME/ERS-2 data. Remote Sensing of Trace Constituents in the Lower Stratosphere, Troposphere and the Earth's Surface: Global Observations, Air Pollution and the Atmospheric Correction **29**(11), 1697–1702 (2002)

Notholt, J., Hjorth, J., Raes, F.: Long path field measurements of aerosol parameters and trace gas concentrations – formation of nitrous acid during foggy periods. J. Aerosol Sci. **22**(S1), S411–S414 (1991)

Notholt, J., Hjorth, J., Raes, F., Schrems, O.: Simultaneous long path field measurements of HNO_2, CH_2O and aerosol. Ber. Bunsenges. Phys. Chem. **96**, 290–293 (1992)

Notholt, J., Toon, G.C., Lehmann, R., Sen, B., Blavier, J.-F.: Comparison of Arctic and Antarctic trace gas column abundances from ground-based FTIR spectrometry. J. Geophys. Res. **102**, 12863–12869 (1997)

Noxon, J.F., Goody, R.: Noncoherent scattering of skylight. Atmos. Ocean. Phys. **1**, 257–281 (1965)

Noxon, J.F.: Nitrogen dioxide in the stratosphere and troposphere measured by ground-based absorption spectroscopy. Science **189**, 547–549 (1975)

Noxon, J.F., Norton, R.B., Henderson, W.R.: Observation of atmospheric NO_3. Geophys. Res. Lett. **5**, 675–678 (1978)

Noxon, J.F.: Stratospheric NO_2. 2. Global behavior, J. Geophys. Res. **84**(C8), 5067–5076 (1979)

Noxon, J.F., Whipple, E.C., Hyde, R.S.: Stratospheric NO_2. 1. Observational method and behavior at midlatitudes. J. Geophys. Res. **84**(C8), 5047–5065 (1979)

Noxon, J.F., Norton, R.B., Marovich, E.: NO_3 in the troposphere. Geophys. Res. Lett. **7**, 125–128 (1980)

Noxon, J.F.: NO_3 and NO_2 in the mid-Pacific troposphere. J. Geophys. Res. **88**, 11017–11021 (1983)

NRC: Review of the NARSTO Draft Report: An assessment of tropospheric ozone pollution –A North American perspective. The National Academy of Sciences (2000)

O'Dowd, C.D.: Biogenic coastal aerosol production and its influence on aerosole radiative properties. J. Geophys. Res. **106**(D2), 1545–1549 (2001)

O'Dowd, C., Jimenez, J.L., Bahreini, R., Flagan, R.C., Seinfeld, J.H., Hämeri, K., Pirjola, L., Kulmala, M., Jennings, S.G., Hoffmann, T.: Marine aerosol formation from biogenic iodine emissions. Nature **417**, 632–636 (2002)

Oetjen, H.: Messung atmosphärischer Spurengase in Ny-Alesund, Aufbau und Inbetriebnahme eines neuen DOAS-Meßsystems, Diploma thesis, University of Bremen (2002)

Ogryzlo, E.A., Thomas, G.E.: Pressure dependence of the visible absorption bands of molecular iodine. J. Mol. Spectrosc. **17**, 198–202 (1965)

Oltmans, S.J., Komhyr, W.D.: Surface ozone distributions and variations from 1973 to 1984 measurements at the NOAA geophysical monitoring for climate change baseline observatories. J. Geophys. Res. **91**, 5229–5236 (1986)

Oppenheimer, C., Tsanev, V.I., Braban, C.F., Cox, R.A., Adams, J.W., Aiuppa, A., Bobrowski, N., Delmelle, P., Barclay, J., McGonigle, A.J.: BrO formation in volcanic plumes. Geochim. Cosmochim. Acta. **70**, 2935–2941 (2006)

Orlando, J.J., Tyndall, G.S.: The atmospheric chemistry of the HC(O)CO. Radical. Int. J. Chem. Kinet. **33**, 149–156 (2001)

Orphal, J., Bogumil, K., Dehn, A., Deters, B., Dreher, S., Fleischmann, O.C., Hartmann, M., Himmelmann, S., Homann, T., Kromminga, H., Spietz, P., Türk, A., Vogel, A., Voigt, S., Burrows, J.P.: Laboratory spectroscopy in support of UV-visible remote-sensing of the atmosphere. In: Pandalai S.G. (ed.) Recent Research Developments in Physical Chemistry, Transworld Research, Trivandrum, pp. 15–34 (2002)

Orphal, J.: A critical review of the absorption cross-sections of O_3 and NO_2 in the ultraviolet and visible. J. Photochem. Photobiol. A: Chem. **157**, 185–209 (2003)

Orphal, J., Fellows, C.E., Flaud, P.-M.: The visible absorption spectrum of NO_3 measured by high-resolution Fourier transform spectroscopy. J. Geophys. Res. **108**(D3), 4077 (2003). doi:10.1029/2002JD002489

Ortgies, G., Gericke, K.H., Comes, F.J.: Is UV laser induced fluorescence a method to monitor tropospheric OH? Geophys. Res. Lett. **7**, 905–908 (1980)

Osterkamp, H.: Messung von atmosphärischen O_4-Profilen. Diploma Thesis, University of Heidelberg, Germany (1997)

Osterkamp, H., Ferlemann, F., Harder, H., Perner, D., Platt, U., Schneider, M., Pfeilsticker, K.: First measurement of the atmospheric O_4 profile, in polar stratospheric ozone 1997. In: Proceedings of the Fourth European Symposium,

22 to 26 September, Schliersee, Germany, Air Pollution Report 66, pp. 478–481, European Commission, Brussels (1998)
Otten, C.: Messung stratosphärischer Spurenstoffe in den Wintern 1992/93 bis 1994/95 über Kiruna in Nordschweden, Ph.D. thesis, University of Heidelberg (1997)
Otten, C., Ferlemann, F., Platt, U., Wagner, T., Pfeilsticker, K.: Groundbased DOAS UV/visible measurements at Kiruna (Sweden) during the SESAME winters 1993/94 and 1994/95, J. Atmos. Chem. **30**, 141–162 (1998)
Oum, K.W., Lakin, M.J., DeHaan, D.O., Brauers, T., Finlayson-Pitts, B.J.: Formation of molecular chlorine from the photolysis of ozone and aqueous Sea-Salt particles. Science **279**, 74–77 (1998a)
Oum, K.W., Lakin, M.J., Finlayson-Pitts, B.J.: Bromine activation in the troposphere by the dark reaction of O_3 with seawater ice. Geophys. Res. Lett. **25**, 3923–3926 (1998b)
Pagsberg, P., Bjergbakke, E., Ratajczak, E., Sillesen, A.: Kinetics of the gas phase reaction OH+NO(+M) $\longrightarrow$ HONO(+M) and the determination of the UV absorption cross sections of HONO. Chem. Phys. Lett. **272**(5–6), 383–390 (1997)
Paldus, B.A., Zare, R.N.: CRDS an historical perspective and introduction, In: Busch, K.W., Busch, M.A. (eds.) Cavity-Ringdown Spectroscopy: An Ultralow-Absorption Measurement Technique, ACS, Washington, DC (1999)
Palmer, P.I., Jacob, D.J., Fiore, A.M., Martin, R.V., Chance, K., Kurosu, T.P.: Mapping isoprene emissions over North America using formaldehyde column observations from space. J. Geophys. Res. **108**(D6), 4180 (2003). doi:10.1029/2002JD002153
Parrish, D.D., Trainer, M., Williams, E.J., Fahey, D.W., Huebler, G., Eubank, C.S., Liu, S.C., Murphy, P.C., Albritton, D.L., Fehsenfeld, F.C.: Measurements of the NO_x-O_3 photostationary state at Niwot Ridge, Colorado. J. Geophys. Res. **91**, 5361–5370 (1986)
Paulson, S.E., Sen, A.D., Ping, L., Fenske, J.D., Fox, M.J.: Evidence for formation of OH radicals from the reaction of O_3 with alkenes in the gas phase. Geophys. Res. Lett. **24**(24), 3193–3196 (1997)
Paulson, S.E., Chung, M., Hasson, A.: OH radical formation from the gas-phase reaction of ozone with terminal alkenes, and the relationship between structure and mechanism. (Invited Feature Article) J. Phys. Chem. **103**, 8125–8138 (1999)
Penkett, S.A., Blake, N.J., Lightman, P., Marsh, A.R.W., Anwyl, P., Butcher, G.: The seasonal variation of nonmethane hydrocarbons in the free troposphere over the North Atlantic Ocean: possible evidence for extensive reaction of hydrocarbons with the nitrate radical. J. Geophys. Res. **98**, 2865–2885 (1993)
Penndorf, R.: Tables of the refractive index for standard air and the Rayleigh scattering coefficient for the spectral region between 0.2 and 20.0 μ and their application to atmospheric optics. J. Opt.Soc. Amer. **47**, 176–182 (1957)
Penney, C.M., St. Peters, R.L., Lapp, M.: Absolute rotational Raman cross sections for N_2, O_2, and CO_2, J. Opt. Soc. Am. **64**, 712–716 (1974)
Perliski, L.M., Solomon, S.: Radiative influences of Pinatubo volcanic aerosols on twilight observations of NO_2 column abundances. Geophys. Res. Lett. **19**, 1923–1926 (1992)
Perliski, L.M., Solomon, S.: On the evaluation of air mass factors for atmospheric near-ultraviolet and visible absorption spectroscopy. J. Geophys. Res. **98**, 10363–10374 (1993)

Perner, D., Ehhalt, D.H., Pätz, H.W., Platt, U., Röth, E.P., Volz, A.: OH-radicals in the lower troposphere, Geophys. Res. Lett. **3**, 466–468 (1976)

Perner, D., Platt, U.: Detection of nitrous acid in the atmosphere by differential optical absorption. Geophys. Res. Lett. **6**, 917–920 (1979)

Perner, D., Platt, U.: Absorption of light in the atmosphere by collision pairs of oxygen $(O_2)_2$. Geophys. Res. Lett. **7**, 1053–1056 (1980)

Perner, D., Kessler, C., Platt, U.: HNO_2, NO_2, and NO measurements in automobile engine exhaust by optical absorption. In: Grisar, R., Preier, H., Schmidke, G., Restelli, G. (eds.) Monitoring of Gaseous Pollutants by Tunable Diode Laser, Reidel, Dordrecht pp. 116–119 (1987a)

Perner, D., Platt, U., Trainer, M., Hübler, G., Drummond, J.W., Junkermann, W., Rudolph, J., Schubert, B., Volz, A., Ehhalt, D.H., Rumpel, K.J., Helas, G.: Measurement of tropospheric OH concentrations: a comparison of field data with model predictions. J. Atmos. Chem. **5**, 185–216 (1987b)

Perner, D., Klüpfel, T., Parchatka, U., Roth, A., Jörgensen, T.: Ground-based UV-vis spectroscopy: diurnal OClO – profiles during January 1990 above Söndre Strømfjord, Greenland. Geophys. Res. Lett. **18**, 787–790 (1991)

Perner, D., Roth, A., Klüpfel, D.: Groundbased measurements of stratospheric OClO, NO_2, and O_3 at Söndre Strömfjord in winter 1991/92. Geophys. Res. Lett. **21**, 1367–1370 (1994)

Perner, D., Arnold, T., Crowley, J., Klüpfel, T., Martinez, M., Seuwen, R.: The measurements of active chlorine in the atmosphere by chemical amplification. J. Atmos. Chem. **34**, 9–20 (1999)

Peters, C.: Studies of reactive halogen species (RHS) in the marine and mid-latitudinal boundary layer by active longpath differential optical absorption spectroscopy, Ph.D. thesis, University of Heidelberg (2004)

Peters, C., Pechtl, S., Stutz, J., Hebestreit, K., Hönninger, G., Heumann, K.G., Schwarz, A., Winterlik, J., Platt, U.: Reactive and organic halogen species in three different European coastal environments. Atmos. Chem. Phys. **5**, 3357–3375 (2005)

Petritoli, A., Ravegnani, F., Giovanelli, G., Bortoli, D., Bonafe, U., Kostadinov, I., Oulanovsky, A.: Off-Axis measurements of atmospheric trace gases by use of an airborne ultraviolet-visible spectrometer. Appl. Opt. **27**, 5593–5599 (2002)

Petropavlovskikh, I., Bhartia, P.K., DeLuisi, J.: New Umkehr ozone profile retrieval algorithm optimized for climatological studies. Geophys. Res. Lett. **32**, L16808 (2005). doi:10.1029/2005GL023323

Pfeilsticker, K., Platt, U.: Airborne measurements during the Arctic stratospheric experiment: observation of O_3 and NO_2. Geophys. Res. Lett. **21**, 1375–1378 (1994)

Pfeilsticker, K., Erle, F., Funk, Senne, T., Wagner, T., Platt, U.: Can enhanced tropospheric photon path lengths explain the anomalous cloud absorption phenomenon? In: Poster, Third European Symposium on Polar Stratospheric Ozone, Schliersee, 18–22 September (1995)

Pfeilsticker, K., Blom, C.E., Brandtjen, R., Fischer, H., Glatthor, N., Grendel, A., Gulde, T., Hopfner, M., Perner, D., Piesch, C., Platt, U., Renger, W., Sessler, J., Wirth, M.: Aircraft-borne detection of stratospheric column amounts of O_3, NO_2, OClO, $ClNO_3$, HNO_3, and aerosols around the arctic vortex (79°N to 39°N) during spring 1993.1. Observational data. J. Geophys. Res. **102**(D9), 10801–10814 (1997a)

Pfeilsticker, K., Erle, F., Platt, U.: Absorption of solar radiation by atmospheric O_4. J. Atm. Sci. **54**, 933–939 (1997b)

Pfeilsticker, K., Erle, F., Funk, O., Marquard, L., Wagner, T., Platt, U.: Optical path modifications due to tropospheric clouds: implications for zenith sky measurements of stratospheric gases. J. Geophys,. Res. **103**, 25323–25335 (1998a)

Pfeilsticker, K. Erle, F., Funk, O., Veitel, H., Platt, U.: First geometrical path lengths probability density function derivation of the skylight from spectroscopically highly resolved oxygen A-band observations. 1. Measurement technique, atmospheric observations and model calculations. J. Geophys. Res. **103**, 11483–11504 (1998b)

Pfeilsticker, K.: First geometrical path lengths probability density function derivation of the skylight from high resolution oxygen A-band spectroscopy. 2. Derivation of the Lévy index for the skylight transmitted by mid-latitude clouds. J. Geophys. Res. **104**, 4101–4116 (1999)

Pierson, A., Goldstein, J.: Stray light in spectrometers: causes and cures. Lasers. Optronics. Sept. issue, 67–74 (1989)

Pfeilsticker, K., Arlander, D.W., Burrows, J.P., Erle, F., Gil, M., Goutail, F., Hermans, C., Lambert, J-C Platt, U., Pommereau, J-P., Richter, A., Sarkissian, A., Van Roozendael, M., Wagner, T., Winterrath, T.: Intercomparison of the influence of tropospheric clouds on UV-visible absorption detected during the NDSC intercomparison campaign at OHP in June 1996. Geophys. Res. Lett **26** 1169–1173 (1999a)

Pfeilsticker, K., Erle, F., Platt, U.: Observation of the stratospheric NO_2 latitudinal distribution in the northern winter hemisphere. J. Atmos. Chem. **32**, 101–120 (1999b)

Pikelnaya, O., Hurlock, S., Trick, S., Stutz, J.: Measurements of reactive iodine species on the Isles of Shoals. Gulf of Maine. Eor Trans. AGU, **86**, Abstract A14A-06 (2005)

Pikelnaya, O., Hurlock, S.H., Trick, S., Stutz, J.: Intercomparison of multiaxis and long-path differential optical absorption spectroscopy measurements in the marine boundary layer. J. Geophys. Res. **112**, D10S01, doi: 10.1029/2006JD007727, (2007)

Pitts, J.N., Finlayson, B.J., Winer, A.M.: Optical systems unravel smog chemistry. Environ. Sci. Technol. **11**, 568–573 (1977)

Pitts, J.N., Jr: Formation and fate of gaseous and particulate mutagens and carcinogens in real and simulated atmospheres. Environ. Health Perspect. **47**, 115–140 (1983)

Pitts, J.N., Jr, Biermann, H.W., Atkinson, R., Winer, A.M.: Atmospheric implications of simultaneous night-time measurements of NO_3 radicals and HONO. Geophys. Res. Lett. **11**, 557–560 (1984)

Pitts, J.N., Biermann, H.W., Winer, A.M., Tuazon, E.C.: Spectroscopic identification and measurement of gaseous nitrous acid in dilute auto exhaust. Atmos. Environ. **18**, 847–854 (1984a)

Pitts, J.N., Sanhueza, E., Atkinson, R., Carter, W.P.L., Winer, A.M., Harris, G.W., Plum, C.N.: An investigation of the dark formation of nitrous acid in environmental chambers. Int. J. Chem. Kinet. **16**, 919–939 (1984b)

Pitts, J.N., Wallington, T.J., Biermann, H.W., Winer, A.M.: Identification and measurement of nitrous acid in an indoor environment. Atmos. Environ. **19**, 763–767 (1985)

Plane, J.M.C., Nien, C.-F.: Study of night-time NO_3 chemistry by differential optical absorption spectroscopy. In: Schiff, H.I. (ed.) SPIE Proceedings, Vol. 1433, Measurement of Atmospheric Gases (The International Society for Optical Engineering) Bellingham, pp. 8–20 (1991)

Plane, J.M.C., Nien, C.F.: Differential optical absorption spectrometer for measuring atmospheric trace gases. Rev. Sci. Instr. **63**, 1867–1876 (1992)

Plane, J.M.C., Smith, N.: Atmospheric monitoring by differential optical absorption spectroscopy. In: Clark, R.J.H., Hester, R.E. (eds.) Spectroscopy in Environmental Sciences, pp. 223–262. John Wiley & Sons Ltd., (city) (1995)

Platt, U.: Mikrometeorologische Bestimmung der SO_2-Abscheidung am Boden Dissertation, Uni Heidelberg (1977)

Platt, U.: Dry deposition of SO_2. Atmos. Environ. **12**, 363–367 (1978)

Platt, U., Perner, D., Pätz, H.: Simultaneous measurement of atmospheric CH_2O, O_3, and NO_2 by differential optical absorption. J. Geophys. Res. **84**, 6329–6335 (1979)

Platt, U., Perner, D.: Direct measurements of atmospheric CH_2O, HNO_2, O_3, NO_2 and SO_2 by differential optical absorption in the near UV. J. Geophys. Res. **85**, 7453–7458 (1980)

Platt, U., Perner, D., Harris, G.W., Winer, A.M., Pitts, J.N.: Detection of NO_3 in the polluted troposphere by differential optical absorption. Geophys. Res. Lett. **7**, 89–92 (1980a)

Platt, U., Perner, D., Harris, G.W., Winer, A.M., Pitts, J.N.: Observations of nitrous acid in an urban atmosphere by differential optical absorption. Nature **285**, 312–314 (1980b)

Platt, U., Perner, D., Schröder, J., Kessler, C., Tönnissen, A.: The diurnal variation of NO_3. J. Geophys. Res. **86**, 11965–11970 (1981)

Platt, U., Perner, D., Kessler, C.: The importance of NO_3 for the atmospheric NOx cycle from experimental observations. In: Proceedings of the 2nd Symposium Composition of the Nonurban Troposphere, Williamsburg, May 82, 25–28 (1982)

Platt, U., Perner, D.: Measurements of atmospheric trace gases by long path differential UV/visible absorption spectroscopy. In: Killinger, D.K., Mooradien, A. (eds.) Optical and Laser Remote Sensing, Springer Series in Optical Science, Vol. 39, pp. 95–105. Springer-Verlag, Berlin (1983)

Platt, U., Perner, D.: Ein Instrument zur spektroskopischen Spurenstoffmessung in der Atmosphäre. Fresenius Z. Anal. Chem. **317**, 309–313 (1984)

Platt, U., Winer, A.M., Biermann, H.W., Atkinson, R., Pitts, J.N.: Measurement of nitrate radical concentrations in continental air. Environ. Sci. Techn. **18**, 365–369 (1984)

Platt, U.: The origin of nitrous and nitric acid in the atmosphere. W. Jäschke, (ed.) NATO ASI-series, Springer Verlag, Heidelberg, pp. 299–319 (1986)

Platt, U., Rateike, M., Junkermann, W., Hofzumahaus, A., Ehhalt, D.H.: Detection of atmospheric OH radicals. Free. Rad. Res. Comms. **3**, 165–172 (1987)

Platt, U., Rateike, M., Junkermann, W., Rudolph, J., Ehhalt, D.H.: New tropospheric OH measurements. J. Geophys. Res. **93**, 5159–5166 (1988)

Platt, U., Perner, D., Semke, S.: Observation of nitrate radical concentrations and lifetimes in tropospheric air. In: Bojkov, R.D., Fabian, P. (eds.) Ozone in the Atmosphere. Proceedings of the Quadrennial Ozone Symposium 1988, pp. 512–515. A. Deepak Publishing, Hampton (1989)

Platt, U., LeBras, G., Poulet, G., Burrows, J.P., Moortgat, G.: Peroxy radicals from night-time reaction of NO_3 with organic compounds. Nature **348**, 147–149 (1990)

Platt, U.: Spectroscopic measurement of free radicals (OH, NO_3) in the atmosphere. Fresenius J. Anal. Chem. **340**, 633–637 (1991)

Platt, U.: Differential optical absorption spectroscopy (DOAS). In: Sigrist, M.W. (ed.) Air Monitoring by Spectroscopic Techniques. Chemical Analysis Series, Vol. 127, John Wiley & Sons, Inc, New York, Chichester, Brisbane, Toronoto, Singapore, ISBN 0–471–55875–3, pp. 27–84 (1994)

Platt, U., Heintz, F.: Nitrate radicals in tropospheric chemistry. Israel. J. Chem. **34**, 289–300 (1994a)

Platt, U., Hausmann, M.: Spectroscopic measurement of the free radicals NO_3, BrO, IO, and OH in the troposphere. Res. Chem. Intermed. **20**, 557–578 (1994b)

Platt, U., Janssen, C.: Observation and role of the free radicals NO_3, ClO, BrO and IO in the troposphere. Faraday Discuss. **100**, 175–198 (1996a)

Platt, U., Lehrer, E. (eds.): Arctic Tropospheric Halogen Chemistry. Final Report to EU (1996b)

Platt, U., Lehrer, E.: Arctic Tropospheric Ozone Chemistry, ARCTOC, Final Report of the EU-Project EV5V-CT93-0318, Heidelberg (1997)

Platt, U., Marquard, L., Wagner, T., Perner, D.: Corrections for zenith scattered light DOAS. Geophys. Res. Lett. **24**, 1759–1762 (1997)

Platt, U., Le Bras, G.: Influence of DMS on the NO_X – NO_Y partitioning and the NO_X distribution in the marine background atmosphere. Geophys. Res. Lett. **24**, 1935–1938 (1997)

Platt, U.: Modern methods of the measurement of atmospheric trace gases. J. Physical Chemistry Chemical Physics 'PCCP' **1**, 5409–5415 (1999)

Platt, U.: Air Monitoring by Differential Optical Absorption Spectroscopy, In: Meyers, R.A. (ed.) Encyclopedia of Analytical Chemistry, pp. 1936–1959, John Wiley & Sons Ltd, Chichester (2000)

Platt, U., Alicke, B., Dubois, R., Geyer, A., Hofzumahaus, A., Holland, F., Mihelcic, D., Klüpfel, T., Lohrmann, B., Pätz, W., Perner, D., Rohrer, F., Schäfer, J., Stutz, J.: Free radicals and fast photochemistry during BERLIOZ. J. Atmos. Chem. **42**, 359–394 (2002)

Platt, U., Hönninger, G.: The role of halogen species in the troposphere. Chemosphere **52**(2), 325–338 (2003)

Platt, U. Allan, W. Lowe, D.: Hemispheric average Cl atom concentration from $^{13}C/^{12}C$ ratios in atmospheric methane. Atmos. Chem. Phys. **4**, 2393–2399 (2004)

Platt, U., Pfeilsticker, K., Vollmer, M.: Radiation and optics in the atmosphere, Ch. 19. In: Träger, F. (ed.) Springer Handbook of Lasers and Optics, Springer, Heidelberg, ISBN-10: 0–387–95579–8, pp. 1165–1203 (2007)

Plum, C.N., Sanhueza, E., Atkinson, R., Carter, W.P.L., Pitts, J.N., Jr: OH radical rate constants and photolysis rates of a-dicarbonyls. Environ. Sci. Technol **17**, 479–484 (1983)

Pommereau, J.P.: Observation of NO_2 diurnal variation in the stratosphere. Geophys. Res. Lett. **9**(8), 850–853 (1982)

Pommereau, J.P., Goutail, F.: An advanced visible-UV spectrometer for atmospheric composition measurements. ESA SP-**270**, 197–200 (1987)

Pommereau, J.P., Goutail, F.: O_3 and NO_2 ground-based measurements by visible spectrometry during the arctic winter and spring 1988. Geophys. Res. Lett. **15**, 891–894 (1988a)

Pommereau, J.P., Goutail, F.: Stratospheric O_3 and NO_2 Observations at the southern polar circle in summer and fall 1988. Geophys. Res. Lett. **15**, 895 (1988b)

Pommereau, J.-P., Piquard, J.: Ozone and nitrogen dioxide vertical distributions by UV–visible solar occultation from balloons. Geophys. Res. Lett. **21**, 1227–1230 (1994a)

Pommereau, J.-P., Piquard, J.: Observations of the vertical distribution of stratospheric OClO. Geophys. Res. Lett. **21**, 1231–1234 (1994b)

Poppe, D., Wallasch, M., Zimmermann, J.: The dependence of the concentration of OH on its precursors under moderately polluted conditions – a model study. J. Atmos. Chem. **16**, 61–78 (1993)

Pope, F.D., Smith, C.A., Davis, P.R., Shallcross, D.E., Ashfold, M.N.R., Orr-Ewing, A.J.: Photochemistry of formaldehyde under tropospheric conditions. Faraday Discuss. **130**, 59–72 (2005a)

Pope, F.D., Smith, C.A., Ashfold, M.N.R., Orr-Ewing, A.J.: High-resolution absorption cross sections of formaldehyde at wavelengths from 313 to 320 nm. Phys. Chem. Chem. Phys. **7**, 79–84 (2005b)

Portmann, R.W., Brown, S.S., Gierczak, T., Talukdar, R.K., Burkholder, J.B., Ravishankara, A.R.: Role of nitrogen oxides in the stratosphere: a reevaluation based on laboratory studies. Geophys. Res. Lett. **26**(15), 2387–2390 (1999)

Povey, I.M., South, A.M., t'Kint de Roodenbeke, A., Hill, C., Freshwater, R.A., Jones, R.L.: A broadband lidar for the measurement of tropospheric constituent profiles from the ground. J. Geophys. Res. **10**(D3), 3369–3380 (1998)

Press, W.H., Flannery, B.P., Teukolsky, S.A., Vettering, W.T.: Numerical Recipes in C. Cambridge University Press, Cambridge (1986)

Preston, K.E., Jones, R.L., Roscoe, H.K.: Retrieval of NO_2 vertical profiles from ground-based UV-visible measurements: method and validation. J. Geophys. Res. **102**(D15), 19089–19097 (1997)

Pribram, J.K., Penchina, C.M.: Stray light in czerny-turner and ebert spectrometers. Appl. Opt. **7**, 2005–2014 (1968)

Price, P.N.: Pollutant tomography using integrated concentration data from non-intersecting optical paths. Atmos. Environ. **33**, 275–280 (1999)

Prinn, R., Cunnold, D., Rasmussen, R., Simmonds, P., Alyea, F., Crawford, A., Fraser, P., Rosen, R.: Atmospheric trends in methylchloroform and the global average for the hydroxyl radical. Science **238**, 945–950 (1987)

Pszenny, A.A.P., Keene, W.C., Jacob, D.J., Fran, S., Maben, J.R., Zetwo, M.P., Sringer-Young, M., Galloway, J.N.: Evidence of inorganic chlorine gases other than hydrogen chloride in marine surface air. Geophys. Res. Lett. **20**(8), 699–702 (1993)

Pundt, I., Pommereau, J.P., Lefevre, F.: Investigation of stratospheric bromine and iodine oxides using the SAOZ balloon sonde. In: Bojkov, R.D., Visconti, D. (eds.) Proceedings of XVIII Quadrennial Ozone Symposium L'Aquila, International Ozone Commission Italy, 12–21 September, pp. 575–578 (1996)

Pundt, I., Pommereau, J.P., Phillips, C., Lateltin, E.: Upper limit of iodine oxide in the lower stratosphere. J. Atmos. Chem. **30**, 173–185 (1998)

Pundt, I., Roozendael van, M., Wagner, T., Richter, A., Chipperfield, M.P., Burrows, J.P., Fayt, C., Hendrick, F., Pfeilsticker, K., Platt, U., Pommereau,

J.P.: Simultaneous UV-vis measurements of BrO from balloon, satellite and ground: implications for tropospheric BrO, In: Harris, N.R.P., Guirlet, M., Amanatis, G.T. (eds.) Air Pollution Research Report 73, Proceedings 5th European Symposium on Polar Stratospheric Ozone 1999, European Commission, Directorate General for Research Unit D.I. – Environment and Sustainable Development Programme, EUR 1934. pp. 316–319 (2000)

Pundt, I., Pommereau, J.-P., Chipperfield, M.P., Van Roozendael, M., Goutail, F.: Climatology of the stratospheric BrO vertical distribution by balloon-borne UV–visible spectrometry. J. Geophys. Res. **107**(D24), 4806 (2002). doi:10.1029/2002JD002230 (2002)

Rairoux, P., Schillinger, H., Niedermeier, S., Rodriguez, M., Ronneberger, F., Sauerbrey, R., Stein, B., Waite, D., Wedekind, C., Wille, H., Wöste, L., Ziener, C.: Remote sensing of the atmosphere using ultrashort laser pulses. App. Phys. B (Lasers and Optics) **71**, 573–580 (2000). doi: 10.1007/s003400000375

Ramacher, B., Rudolph, J., Koppmann, R.: Hydrocarbon measurements in the spring Arctic troposphere during the ARCTOC 95 campaign. Tellus **49B**, 466–485 (1997)

Ramacher, B., Rudolph, J., Koppmann, R.: Hydrocarbon measurements during tropospheric ozone depletion events: evidence for halogen atom chemistry. J. Geophys. Res. **104**, 3633–3653 (1999)

Rayleigh, L.: On the transmission of light through an atmosphere containing many small particles in suspension, and on the origin of the blue of the sky, Phil. Mag. **41**, 447–454. Also: in the scientific papers of Lord Rayleigh, Vol. 4, Dover, New York, 1964 (1899)

Reader, J., Sansonetti, C.J., Bridges, J.M.: Irradiances of spectral lines in mercury pencil lamps. Appl. Opt. **35**, 78–83 (1996)

Redemann Fischer, H.E., Fergg, F., Rabus, D.: Measurements of stratospheric NO_2 profiles using a gas correlation radiometer in the solar occultation mode. J. Atmos. Chem. **3**, 203–231 (1985)

Reisinger, A.R., Fraser, G.J., Johnston, P.V., McKenzie, R.L., Matthews, W.A.: Slow-scanning DOAS system for urban air pollution monitoring. In: Bojkov, R.D., Visconti, G. (eds.) Atmospheric ozone, Proceedings of the XVIII Quadrennial Ozone Symposium, L'Aquila, Italy, September 1996, Vol. 2, pp. 959–962. Parco Scientifico e Tecnologico d'Abruzzo, L'Aquila (1998)

Ren, X., Harder, H., Martinez, M., Lesher, R.L., Oliger, A., Simpas, J.B., Brune, W.H., Schwab, J.J., Demerjian, K.L., He, Y., Zhou, X., Gao, H.: OH and HO_2 chemistry in urban atmosphere of New York city. Atmos. Environ. **37**, 3639–3651 (2003)

Renard, J.-B., Pirre, M., Robert, C., Moreau, G., Huguenin, D., Russell, III J.M.: Nocturnal vertical distribution of stratospheric O_3, NO_2 and NO_3 from balloon measurements. J. Geophys. Res. **101**, 28793–28804 (1996)

Renard, J.-B., Pirre, M., Robert, C., Lefèvre, F., Lateltin, E., Nozière, B., Huguenin, D.: Vertical distribution of nighttime stratospheric NO_2 from balloon measurements: comparison with models. Geophys. Res. Lett. **24**, 73–76 (1997a)

Renard, J.B., Lefèvre, F., Pierre, M., Huguenin, D.: Vertical profile of night-time stratospheric OClO. J. Atmos. Chem. **26**, 65–76 (1997b)

Renard, J.B., Pierre, M., Robert, C., Huguenin, D.: The possible detection of OBrO in the stratosphere. J. Geophys. Res. **103**, 25383–25395 (1998)

Renard, J.-B., Chartier, M., Robert, C., Chalumeau, G., Berthet, G., Pirre, M., Pommerreau, J.-P., Goutail, F.: SALOMON: a new, light balloonborne UV-visible spectrometer for nighttime observations of stratospheric trace-gas species. Appl. Opt. **39**(3), 386–392 (2000a)

Renard, J.-B., Taupin, F.G., Rivière, E.D., Pirre, M., Huret, N., Berthet, G., Robert, C., Chartier, M.: Measurements and simulation of stratospheric NO_3 at Mid- and High-latitudes in the Northern Hemisphere. J. Geophys. Res. **106**, 32387–32399 (2000b)

Rhode, H., Charlson, R., Crawford, E.: Svante Arrhenius and the greenhouse effect. AMBIO **26**, 2–5 (1997)

Richter, A.: Absorptionsspektroskopische Messungen atmosphärischer Spurengase über Bremen, 53°N, Doktoral thesis, University of Brmen (1997)

Richter, A., Wittrock, F., Eisinger, M., Burrows, J.P.: GOME observation of tropospheric BrO in northern hemispheric spring and summer 1997. Geophys. Res. Lett. **25**, 2683–2686 (1998)

Richter, A., Eisinger, M., Ladstätter, Weißenmayer, A. Wittrock, F., Burrows, J.P.: DOAS zenith-sky observations: seasonal variations of BrO over Bremen (53°N) 1994–1995. J. Atmos. Chem. **32**, 83–99 (1999)

Richter, A., Burrows, J.P.: Retrieval of tropospheric NO_2 from GOME measurements. Adv. Space Res. **29**(11), 1673–1683 (2002)

Richter, A., Wittrock, F., Ladstätter-Weißenmayer, A., Burrows, J.P.: GOME measurements of stratospheric and tropospheric BrO. Adv. Space. Res. **29**(11), 1667–1672 (2002)

Richter, A., Burrows, J.P., Nüß, H., Granier, C., Niemeier, U.: Increase in tropospheric nitrogen dioxide over China observed from space. Nature **437**, 129–132 (2005). doi:10.1038/nature04092

Richter, A., Wittrock, F., Weber, M, Beirle, S., Kühl, S., Platt, U., Wagner, T., Wilms-Grabe, W., Burrows, J.P.: GOME observations of stratospheric trace gas distributions during the splitting vortex event in the Antarctic winter 2002 Part I: measurements. J. Atmos. Sci. **62**(3), 778–785 (2005)

Rigaud, P., Leroy, B., Le Bras, G., Poulet, G., Jourdain, J.L., Combourieu, J.: About the identification of some UV atmospheric absorptions laboratory study of ClO. Chem. Phys. Lett. **46**, 161 (1977)

Rigaud, P., Naudet, J.-P., Huguenin, D.: Simultaneous measurements of vertical distribution of stratospheric NO_3 and O_3 at different periods of the night. J. Geophys. Res. **88**, 1463–1467 (1983)

Rinsland, C.P., Mahieu, E., Zander, R., Jones, N.B., Chipperfield, M.P., Goldman, A., Anderson, J., Russell, J.M. III, Demoulin, P., Notholt, J., Toon, G.C., Blavier, J.-F., Sen, B., Sussmann, R., Wood, S.W., Meier, A., Griffith, D.W.T., Chiou, L.S., Murcray, F.J., Stephen, T.M., Hase, F., Mikuteit, S., Schulz, A., Blumenstock, T.: Long-term trends of inorganic chlorine from ground-based infrared solar spectra: past increases and evidence for stabilization. J. Geophys. Res. **108**(D8), 4252 (2003). doi:4210.1029/2002JD003001

Rityn, N.E.: Optics of corner cube reflectors. Sov. J. Opt. Tech. **34**, 198–201 (1967)

Ritz, D., Hausmann, M., Platt, U.: An improved open-path multireflection cell for the measurement of NO_2 and NO_3. In: Schiff, H.I., Platt, U. (eds.) Proceedings Europto series, Optical Methods in Atmospheric Chemistry, SPIE **1715**, 200–211 (1992)

Röckmann, T., Brenninkmeijer, C.A.M., Crutzen, P.J., Platt, U.: Short term variations in the $^{13}C/^{12}C$ ratio of CO as a measure of Cl activation during tropospheric ozone depletion events in the arctic. J. Geophys. Res. 104, 16911697 (1999)

Rodgers, C.D.: Characterisation and error analysis of profiles retrieved from remote sounding measurements. J. Geophys. Res. **95**(D5), 5587–5595 (1990)

Rodgers, C.D.: Inverse Methods for Atmospheric Sounding, Theory and Practice. World Scientific Publishing, Singapore (2000)

Rohrer, F., Bohn, B., Brauers, T., Brüning, D., Johnen, F.-J., Wahner, A., Kleffmann, J.: Characterisation of the photolytic HONO-source in the atmosphere simulation chamber SAPHIR. Atmos. Chem. Phys. Discuss. **4**, 7881–7915 (2004). SRef-ID: 1680-7375/acpd/2004-4-7881

Roscoe, H.K., Pyle, J.A.: Measurements of solar occultation: the error in a naive retrieval if the constituent's concentration changes. J. Atmos. Chem. **5**, 323–341 (1987)

Roscoe, H.K., Fish, D.J., Jones, R.L.: Interpolation errors in UV-visible spectroscopy for stratospheric sensing: implications for sensitivity, spectral resolution, and spectral range. Appl. Opt. **35**, 427–432 (1996)

Roscoe, H.K., Clemitshaw, K.C.: Measurement techniques in gas-phase tropospheric chemistry: a selective view of the past, present, and future. Science **276**, 1065–1072 (1997)

Roscoe, H.K., Johnston, P.V., Van Roozendael, M., Richter, A., Sarkissian, A., Roscoe, J., Preston, K.E., Lambert, J.-C., Hermans, C., et al.: Slant column measurements of O_3 and NO_2 during the NDSC intercomparison of zenith-sky UV-visible spectrometers in June 1996. J. Atmos. Chem. **32**, 281–314 (1999)

Roscoe, H.K., Charlton, A.J., Fish, D.J., Hill, J.G.T.: Improvements to the accuracy of measurements of NO_2 by zenith-sky visible spectrometers II: errors in offset using a more complete chemical model. J. Quant. Spectrosc. Radiat. Trans. **68**, 337–349 (2001)

Roscoe, H.K., Hill, J.G.T., Jones, A.E., Sarkissian, A.: Improvements to the accuracy of zenith-sky measurements of total ozone by visible spectrometers II: use of daily air-mass factors. J. Quant. Spectrosc. Radiat. Trans. **68**, 327–336 (2001)

Rothe, K.W., Brinkmann, U., Walther, H.: Applications of tunable dye lasers to air pollution detection: measurements of atmospheric NO_2 concentrations by differential absorption. Appl. Phys. **3**, 115–119 (1974)

Rothman, L.S., Gamache, R.R., Tipping, R.H., Rinsland, C.P., Smith, M.A.H., Benner, D.C., Malathy Devi, V., Flaud, J.-M., Camy-Peyret, C., Perrin, A., Goldman, A., Massie, S.T., Brown, L.R., Toth, R.A.: The HITRAN molecular database: editions of 1991 and 1992. J. Quant. Spectrosc. Radiat. Transf. **48**(5/6), 469–508 (1992)

Rothman, L.S., Rinsland, C.P., Goldman, A., Massie, S.T., Edwards, D.P., Flaud, J.-M., Perrin, A., Camy-Peyret, C., Dana, V., Mandin, J.-Y., Schroeder, J., McCann, A., Gamache, R.R., Wattson, R.B., Yoshino, K., Chance, K.V., Jucks, K.W., Brown, L.R., Nemtchinov, V., Varanasi, P.: The HITRAN molecular spectroscopic database and HAWKS (HITRAN Atmospheric Workstation): 1996 edition. J. Quant. Spectrosc. Radiat. Transf. **60**(5), 665–710 (1998)

Rothman, L.S., Barbe, A., Chris Benner, D., Brown, L.R., Camy-Peyret, C., Carleer, M.R., Chance, K., Clerbaux, C., Dana, V., Devi, V.M., Fayt, A., Flaud, J.-M., Gamache, R.R., Goldman, A., Jacquemart, D., Jucks, K.W., Lafferty,

W.J., Mandin, J.-Y., Massie, S.T., Nemtchinov, V., Newnham, D.A., Perrin, A., Rinsland, C.P., Schroeder, J., Smith, K.M., Smith, M.A.H., Tang, K., Toth, R.A., Vander Auwera, J., Varanasi, P., Yoshino, K.: The HITRAN molecular spectroscopic database: edition of 2000 including updates through 2001. J. Quant. Spectros. Rad. Transf. **82**, 5–44 (2003)

Rothman, L.S., Jacquemart, D., Barbe, A., Chris Benner, D., Birk, M., Brown, L.R., Carleer, M.R., Chackerian, C., Chance, K., Dana, V., Devi, V.M., Flaud, J.-M., Gamache, R.R., Goldman, A., Hartmann, J.-M., Jucks, K.W., Maki, A.G., Mandin, J.-Y. Massie, S.T., Orphal, J., Perrin, A., Rinsland, C.P., Smith, M.A.H., Tennyson, J., Tolchenov, R.N., Toth, R.A., Vander Auwera, J., Varanasi, P., Wagner, G.: The HITRAN 2004 molecular spectroscopic database. J. Quant. Spectrosc. Radiat. Transf. **96**, 139–204 (2005)

Rotstayn, L.D., Lohmann, U.: Simulation of the tropospheric sulfur cycle in a global model with a physically based cloud scheme. J. Geophys. Res. **107**(D21), 4592 (2002). doi:10.1029/2002JD002128

Rozanov, V., Diebel, D., Spurr, R., Burrows, J.: GOMETRAN: A radiative transfer model for the satellite project GOME – the plane-parallel version. J. Geophys. Res. **102**(D14), 16683–16695 (1997)

Rozanov, A., Rozanov, V.-V., Burrows, J.-P.: Combined differential-integral approach for the radiation field computation in a spherical shell atmosphere: Nonlimb geometry. J. Geophys. Res. **105**(D18), 22937–22942 (2000)

Rozanov, A., Rozanov, V.-V., Burrows, J.-P.: A numerical radiative transfer model for a spherical planetary atmosphere: combined differential-integral approach involving the Picard iterative approximation. J. Quant. Spectrosc. Radiat. Transf. **69**(4), 491–512 (2001)

Rudich, Y., Talukdar, R., Ravishankara, A.R.: Reactive uptake of NO_3 on pure water and ionic solutions. J. Geophys. Res. **D101**, 21023–21031 (1996)

Rudolph, J., Fu, B.R., Anlauf, T.A.K., Bottenheim, J.: Halogen atom concentrations in the Arctic troposphere derived from hydrocarbon measurements: Impact on the budget of formaldehyde. Geophys. Res. Lett. **26**(19), 2941–2944 (1999)

Russell, A.G., Cass, G.R., Seinfeld, J.H.: On some aspects of nighttime atmospheric chemistry. Environ. Sci. Technol. **20**, 1167–1172 (1986)

Russwurm, G.: Differential optical absorption spectroscopy (DOAS)—A Status Report on the Instrumentation (1999)

Saiz-Lopez, A., Plane, J.M.C., Shillito, J.A.: Bromine oxide in the mid-latitude marine boundary layer. Geophys. Res. Lett. **31**, L03111.1–L03111.4 (2004a). doi:10.1029/2003GL018 956

Saiz-Lopez, A., Plane, J.M.C., Shillito, J.A.: Novel iodine chemistry in the marine boundary layer. Geophys. Res. Lett. **31**, L04112 (2004b). doi: 10.1029/2003GL019215

Saiz-Lopez, A., Saunders, R.W., Joseph, D.M., Ashworth, S.H., Plane, J.M.C.: Absolute absorption cross-section and photolysis rate of I_2. Atmos. Chem. Phys. **4**, 1443–1450 (2004c) SRef-ID: 1680-7324/acp/2004-4-1443

Saiz-Lopez, A., Shillito, J.A., Coe, H., Plane, J.M.C.: Measurements and modelling of I_2, IO, OIO, BrO and NO_3 in the mid-latitude marine boundary layer. Atmos. Chem. Phys. Discuss. **5**, 9731–9767 (2005). SRef-ID: 1680-7375/acpd/2005-5-9731

Sakamaki, F., Hatakeyama, S., Akimoto, H.: Formation of nitrous acid and nitric oxide in the heterogeneous dark reaction of nitrogen dioxide and water vapor in a smog chamber. Int. J. Chem. Kinet. **XV**, 1013–1029 (1983)

Saltzman, B.E.: Colorimetric microdetermination of nitrogen dioxide in the atmosphere. Anal. Chem. **26**, 1948–1955 (1954)

Sanders, R., Solomon, S., Mount, G., Bates, M.W., Schmeltekopf, A.: Visible spectroscopy at McMurdo station, Antarctica: 3. Observation of NO_3. J. Geophys. Res. **92**, 8339–8342 (1987)

Sander, S.P., Friedl, R.R.: Kinetics and product studies of the reaction ClO + BrO using flash photolysis-ultraviolet absorption. J. Phys. Chem. **93**, 4764–4771 (1989)

Sanders, R.W., Solomon, S., Carroll, M.A., Schmeltekopf, A.L.: Visible and near-ultraviolet spectroscopy at McMurdo station, Antarctica 4. Overview and daily measurements of NO_2, O_3, and OClO during 1987. J. Geophys. Res. **94**(D9), 11381–11391 (1989)

Sanders, R.W., Solomon, S., Smith, J.P., Perliski, L., Miller, H.L., Mount, G.H., Keys, J.G., Schmeltekopf, A.L.: Visible and near-UV spectroscopy at McMurdo station, Antarctica, 9. Observations of OClO from April to October 1991. J. Geophys. Res. **98**(D4), 7219–7228 (1993)

Sanders, R.W.: Improved analysis of atmospheric absorption spectra by including the temperature dependence of NO_2. J. Geophys. Res. **101**, 20945–20952 (1996)

Sanders, R.W., Solomon, S., Kreher, K., Johnston, P.V.: An intercomparison of NO_2 and OClO measurements at arrival heights, Antarctica during austral spring 1996. J. Atmos. Chem. **33**, 283–298 (1999)

Sander, R., Keene, W.C., Pszenny, A.A.P., Arimoto, R., Ayers, G.P., Baboukas, E., Cainey, J.M., Crutzen, P.J., Duce, R.A., Hönninger, G., Huebert, B.J., Maenhaut, W., Mihalopoulos, N., Turekian, V.C., Van Dingenen, R.: Inorganic bromine in the marine boundary layer: a critical review. Atmos. Chem. Phys. **3**, 1301–1336 (2003)

Sansonetti, C.J., Salit, M.L., Reader, J.: Wavelengths of spectral lines in mercury pencil lamps. Appl. Opt. **35**, 74–77 (1996)

Sarkissan, A., Pommereau, J.P, Goutail, F.: Identification of polar stratospheric clouds from the ground by visible spectrometry. Geophys. Res. Lett. **18**, 779–782 (1991)

Sarkissian, A., Fish, D., Van Roozendael, M., Gil, M., Chen, H.B., Wang, P., Pommereau, J.P., Lenoble, J.: Ozone and NO_2 air-mass factors for zenith sky spectrometers: intercomparison of calculations with different radiative transfer models. Geophys. Res. Lett. **21**, 1113–1116 (1995)

Sarkissian, A., Vaughan, G., Roscoe, H.K., Bartlett, L.M., O'Connor, F.M., Drew, D.G., Hughes, P.A., Moore, D.: Accuracy of measurements of total ozone by a SAOZ ground-based zenith-sky visible spectrometer. J. Geophys. Res. **102**, 1379–1390 (1997)

Savigny, C.V., Funk, O., Platt, U., Pfeilsticker, K.: Radiative smoothing in zenith-scattered skylight transmitted to the ground. Geophys. Res. Lett **26**, 2949–2952 (1999)

Savitzky, A., Golay, M.J.E.: Smoothing and differentiation of data by simplified least squares procedures. Anal. Chem. **36**, 1627–1639 (1964)

Schall, C., Heumann, K.G.: GC determination of volatile organoiodine and organobromine compounds in seawater and air samples. Fresenius J. Anal. Chem. **346**, 717–722 (1993)

Schauffler, S.M., Atlas, E.L., Flocke, F., Lueb, R.A., Stroud, V., Travnicek, W.: Measurement of bromine-containing organic compounds at the tropical tropopause. Geophys. Res. Lett. **25**, 317–320 (1998)

Schermaul, R., Brault, J.W., Canas, A.A.D., Learner, R.C.M., Polyansky, O.L., Zobov, N.F., Belmiloud, D., Tennyson, J.: Weak line water vapour spectrum in the regions 13,200–15,000 cm^{-1}. J. Mol. Spectrosc. **211**, 169–178 (2002)

Schiff, H.I., Karecki, D.R., Harris, G.W., Hastie, D.R., Mackay, G.I.: A tunable diode laser system for aircraft measurements of trace gases. J. Geophys. Res. **95**, 10147–10154

Schiller, C., Wahner, A., Platt, U., Dorn, H.-P., Callies, J., Ehhalt, D.H.: Near UV atmospheric absorption measurements of column abundances during airborne Arctic stratospheric expedition, January–February 1989: 2. OClO observations. Geophys. Res. Lett. **17**, 501–504 (1990)

Schmidt, M.: Von Christian Friedrich Schönbein bis zum Ozonloch, ein Abriß der Geschichte der Ozonforschung, Max-Planck-Insitut für Aeronomie, Katlenburg-Lindau bei Göttingen, Germany (1988)

Schmölling, J.: Uebersicht über regulatorische Massnahmen zur Luftreinhaltung und deren Auswirkungen. AGF- Tagung "Luftreinhaltung – Luftverschmutzung", Bonn, 3–4, Nov. (1983)

Schneider, W., Moortgat, G.K., Tyndall, G.S., Burrows, J.P.: Absorption cross-sections of NO_2 in the UV and visible region (200–700 nm) at 298 K.J. Photochem. Photobiol. **40**, 195–217 (1987)

Schofield, R., Connor, B.J., Kreher, K., Johnston, P.V., Rodgers, C.D.: The retrieval of profile and chemical information from ground-based UV-visible spectroscopic measurements. J. Quant. Spectr. Rad. Transf. **86**, 115–131 (2004)

Schofield, R., Johnston, P.V., Thomas, A., Kreher, K., Connor, B.J., Wood, S., Shooter, D., Chipperfield, M.P., Richter, A., von Glasow, R., Rodgers, C.D.: Tropospheric and stratospheric BrO columns over arrival heights, Antarctica, 2002. J. Geophys. Res. **111**, D22310 (2006). doi:10.1029/2005JD007022

Scholl, T.: Photon path length distributions for cloudy skies – Their first and second-order moments inferred from high resolution oxygen A-Band spectroscopy. PhD Thesis, University of Heidelberg, Germany (2006)

Schönbein, C.F.: Beobachtungen über den bei der Elektrolysation des Wassers and dem Ausströmen der gewöhnlichen Electrizität aus Spitzen sich entwickelnden Geruch. Ann. Phys. Chem. **50**, 616 (1840)

Schroeder, W.H., Anlauf, K.G., Barrie, L.A., Lu, J.Y., Steffen, A., Schneeberger, D.R., Berg, T.: Arctic springtime depletion of mercury. Nature **394**, 331–332 (1998)

Schrötter, H.W., Klöckner, H.W.: Raman scattering cross-sections in gases. In: Weber, A. (ed.) Topics in Current Physics: Raman Spectroscopy of Gases and Liquids, Springer, Verlag (1979)

Schultz, M., Heitlinger, M., Mihelcic, D., Volz-Thomas, A.: Calibration source for peroxy radicals with built-in actinometry using H_2O and O_2 photolysis at 185 nm. J. Geophys. Res. **100**, 18811–18816 (1995)

Schulz-DuBois, E.O.: Generation of square lattice of focal points by a modified white cell. Appl. Opt. **12**, 1391–1393 (1973)

Schwartz, S.E.: Are global cloud albedo and climate crontrolled by marine phytoplankton. Nature **336**, 441–445 (1988)

Schweitzer, F., Mirabel, P., George, C.: Heterogeneous chemistry of nitryl halides in relation to tropospheric halogen activation. J. Atmos. Chem. **34**, 101–117 (1999)

Seinfeld, J.H., Pandis, S.N.: , Atmospheric Chemistry and Physics. John Wiley & Sons, Inc., New York (1998)

Seisel, S., Caloz, F., Fenter, F.F., Van den Bergh, H., Rossi, M.J.: The heterogeneous reaction of NO_3 with NaCl and KBr: a nonphotolytic source of halogen atoms. Geophys. Res. Lett. **24**, 2757–2760 (1997)

Senne, T., Stutz, J., Platt, U.: Measurement of the latitudinal distribution of NO_2 column density and layer height in Oct./Nov. 1993. Geophys. Res. Lett. **23**, 805–808 (1996)

Senzig, J.: Troposphärische DOAS-Messungen stickstoffhaltiger und aromatischer Verbindungen in Heidelberg, (in German), Diploma thesis, University of Heidelberg (1995)

Shangavi, S.: An effcient Mie theory implementation to investigate the influence of aerosols on radiative transfer, Diploma thesis, University of Heidelberg (2003)

Shapley, D.: Nitrosamines: scientists on the trail of prime suspect in. urban cancer. Science **191**, 268–270 (1976)

Shirinzadeh, B., Wang, C.C., Deng, D.Q.: Pressure dependence of ozone interference in the laser fluorescence measurements of OH in the atmosphere. Appl. Opt. **26**, 2102–2105 (1987)

Sigrist, M.W. (ed.): Air Monitoring by Spectroscopic Techniques, Chemical Analysis Series, Vol. 127. Wiley, New York (1994a)

Sigrist, M.W.: Air monitoring by laser photoacoustic spectroscopy. In: Sigrist, M.W. (ed.) Air Monitoring by Spectroscopic Techniques, Chemical Analysis Series, Vol. 127, pp. 163–238. Wiley, New York (1994b)

Sillman, S., Logan, J.A., Wofsy, S.C.: The Sensitivity of ozone to nitrogen oxides and hydrocarbons in regional ozone episodes. J. Geophys. Res. **95**, 1837–1852 (1990)

Simon, F.G., Schneider, W., Moortgat, G.K., Burrows, J.P.: A study of the ClO absorption cross-section between 240 and 310 nm and the kinetics of the self-reaction at 300 K. J. Photochem. Photobiol. A Chem. **55**, 1–23 (1990)

Simpson, W.R.: Continuous wave cavity ring-down spectroscopy applied to in situ detection of dinitrogen pentoxide (N_2O_5). Rev. Sci. Instrum. **74**(7), (2003). doi: 10.1063/1.1578705

Singh, H.B., Gregory, G.L., Anderson, B., Browell, E., Sachse, G.W., Davis, D.D., Crawford, J., Bradshaw, J.D., Talbot, R., Blake, D.R., Thornton, D., Newell, R., Merrill, J.: Low ozone in the marine boundary layer of the tropical Pacific Ocean: photochemical loss, chlorine atoms, and entrainment. J. Geophys. Res. **101**, 1907–1917 (1996)

Sinreich, R., Filsinger, F., Friess, U., Platt, U., Sebastian, O., Wagner, T.: MAX-DOAS detection of glyoxal during ICARTT 2004. Atmos. Chem. Phys. **7**, 1293–1303 (2007)

Sinreich, R., Friess, U., Wagner, T., Platt, U.: Multi axis differential optical absorption spectroscopy (MAX-DOAS) of gas and aerosol distributions. Farad. Disc. **130**, 153–164 (2005)

Sinnhuber, B.-M., Arlander, D.W., Bovensmann, H., Burrows, J.P., Chipperfield, M.P., Ennel, C.-F., Frieß, U., Hendrick, F., Johnston, P.V., Jones, R.L., Kreher, K., Mohamed-Tahrin, N., Müller, R., Pfeilsticker, K., Platt, U., Pommereau,

J.P., Pundt, I., Richter, A., South, A., Toernkvist, K.K., Van Roozendael, M., Wagner, T., Wittrock, F.: Comparison of measurements and model calculations of stratospheric bromine monoxide. J. Geophys. Res. **107**(D19), 4398 (2002). doi:10.1029/2001JD000940

Sinreich, R., Frieß, U., Wagner, T., Platt, U.: Multi axis differential optical absorption spectroscopy (MAXDOAS) of gas and aerosol distributions. Faraday Discuss. **130**, 153–164 (2005). doi: 10.1039/B419274P

Sioris, C.E., Evans, W.F.J.: Filling in of Fraunhofer and gas-absorption lines in sky spectra as caused by rotational Raman scattering. Appl. Opt. **38**(12), 2706–2713 (1999)

Sjödin, A.: Studies of the diurnal variation of nitrous acid in urban air. Environ. Sci. Technol. **22**, 1086–1089 (1988)

Slusser, J., Hammond, K., Kylling, A., Stamnes, K., Perliski, L., Dahlback, A., Anderson, D., DeMajistre, R.: Comparison of air mass computations. J. Geophys. Res. **101**, 9315–9321 (1996)

Slusser, J.R., Fish, D.J., Strong, E.K., Jones, R.L., Roscoe, H.K., Sarkissian, A.: Five years of NO_2 vertical column measurements at Faraday (65°S): evidence for the hydrolysis of $BrONO_2$ on Pinatubo aerosols. J. Geophys. Res. **102**, 12987–12993 (1997)

Smith, J., Solomon, S.: Atmospheric NO_3: 3. Sunrise disappearance and the stratospheric profile. J. Geophys. Res. **95**, 13819–13827 (1990)

Smith, J., Solomon, S., Sanders, R., Miller, H., Perliski, L., Keys, J., Schmeltekopf, A.: Atmospheric NO_3: 4. Vertical profiles at middle and polar latitudes at sunrise. J. Geophys. Res. **98**, 8983–8989 (1993)

Smith, N., Plane, J.M.C., Nien, C., Solomon, O.A.: Nighttime radical chemistry in the San Joaquin Valley. Atmos. Environ. **29**, 2887–2897 (1995)

Smith, P.L., Heise, C., Esmond, J.R., Kurucz, R.L.: Atomic spectral line database from CD-ROM 23 (1995 Atomic Line Data (Kurucz, R.L., Bell, B.) Kurucz CD-ROM No. 23. Cambridge, Mass.: Smithsonian Astrophysical Observatory) (1995), Solberg, S., Schmidtbauer, N., Semb, A., Stordal, F.: Boundary-layer ozone depletion as seen in the Norwegian Arctic in spring. J. Atmos. Chem. **23**, 301–332 (1996)

Smith, K.M., Newnham, D.A.: Near-infrared absorption spectroscopy of oxygen and nitrogen gas mixtures. Chem. Phys. Lett. **308** (1–2) 1–6 (1999)

Smith, F.G., King, T.A.: Optics and Photonics. An Introduction. Wiley, New York (2000). ISBN 0-471-48924-7

Smith, K.M., Newnham, D.A.: Near-infrared absorption cross sections and integrated absorption intensities of molecular oxygen (O_2, O_2-O_2, and O_2-N_2). J. Geophys. Res. **105**(D6), 7383–7396 (2000)

Smith, K.M., Newnham, D.A., Williams, R.G.: Collision-induced absorption of solar radiation in the atmosphere by molecular oxygen at 1.27 µm: field observations and model calculations. J. Geophys. Res. **106**(D7), 7541–7552 (2001)

Solomon, S., Mount, G., Sanders, R.W., Schmeltekopf, A.: Visible spectroscopy at McMurdo station, Antarctica: 2. Observation of OClO. J. Geophys. Res. **92**, 8329–8338 (1987b)

Solomon, S., Mount, G.H., Sanders, R.W., Jakoubek, R.O., Schmeltekopf, A.L.: Observations of the nighttime abundance of OClO in the winter stratosphere above Thule, Greenland. Science **242**, 550–555 (1988)

Solomon, S., Miller, H.L., Smith, J.P., Sanders, R.W., Mount, G.H., Schmeltekopf, A.L., Noxon, J.F.: Atmospheric NO_3, 1. Measurement technique and the annual cycle at 40°N. J. Geophys. Res. **94**, 11041–11048 (1989a)

Solomon, S., Sanders, R.W., Carroll, M.A., Schmeltekopf, A.L.: Visible and near-ultraviolet spectroscopy at McMurdo Station, Antarctica, 5. Observations of the diurnal variations of BrO and OClO. J. Geophys. Res. **94**, 11393–11403 (1989b)

Solomon, S., Sanders, R.W., Mount, G.H., Carroll, M.A., Jakoubek, R.O., Schmeltekopf, A.L.: Atmospheric NO_3, 2. Observations in polar regions. J. Geophys. Res. **94**(D13), 16423–16427 (1989c)

Solomon, S., Sanders, R.W., Carroll, M.A., Schmeltekopf, A.L.: Visible and near-ultraviolet spectroscopy at McMurdo Station, Antarctica, 6. Observations of the diurnal variations of BrO and OClO. J. Geophys. Res. **94**(D9), 11393–11403 (1989d). 10.1029/88JD03127

Solomon, S., Sanders, R.W., Miller, H.L. Jr: Visible and near-ultraviolet spectroscopy at McMurdo Station, Antarctica, 7. OClO diurnal photochemistry and implications for ozone destruction. J. Geophys. Res. **95**, 13807 (1990)

Solomon, S., Sanders, R.W., Garcia, R.R., Keys, J.G.: Increased chlorine dioxide over Antarctica caused by volcanic aerosols from Mount-Pinatubo. Nature **363**, 245–248 (1993a)

Solomon, S., Smith, J.P., Sanders, R.W., Perliski, L., Miller, H.L., Mount, G.H., Keys, J.G., Schmeltekopf, A.L.: Visible and near-ultraviolet spectroscopy at McMurdo Station, Antarctica, 8. Observations of nighttime NO_2 and NO_3 from April to October 1991. J. Geophys. Res. **98**, 993–1000 (1993b)

Solomon, S., Sanders, R.W., Jakoubek, R.O., Arpag, K., Stephens, S.L., Keys, J.G., Garcia, R.R.: Visible and near-ultraviolet spectroscopy at McMurdo Station, Antarctica, 10. Reductions of NO_2 due to Pinatubo aerosols. J. Geophys. Res. **99**, 3509–3516 (1994a)

Solomon, S., Garcia, R.R., Ravishankara, A.R.: On the role of iodine in ozone depletion. J. Geophys. Res. **99**, 20491–20499 (1994b)

Solomon, S.: Stratospheric ozone depletion: a review of concepts and history. Rev. Geophys. **37**, 275–316 (1999)

South, A.M., Povey, I.M., Jones, R.L.: Broadband lidar measurements of tropospheric water vapour profiles. J. Geophys. Res. **103**, 31191–31202 (1998)

Spicer, C.W., Chapman, E.G., Finlayson-Pitts, B.J., Plastridge, R.A., Hubbe, J.M., Fast, J.D., Berkowitz, C.M.: Unexpectedly high concentrations of molecular chlorine in coastal air. Nature **394**, 353–356 (1998)

Spicer, C.W., Plastridge, R.A., Foster, K.L., Finlayson-Pitts, B.J., Bottenheim, J.W., Grannas, A.M., Shepson, P.B.: Molecular halogens before and during ozone depletion events in the Arctic at polar sunrise: concentrations and sources. Atmospheric Environment **36**, 2721–2731 (2002)

Spichtinger, N., Wenig, M., James, P., Wagner, T., Platt, U., Stohl, A.: Satellite detection of a continental-scale plume of nitrogen oxides from boreal forest fires. Geophys. Res. Lett. **28**, 4579–4582 (2001)

Spietz, P., Gómez Martín, J., Burrows, J.P.: Effects of column density on I2 spectroscopy and a determination of I2 absorption cross section at 500 nm. Atmos. Chem. Phys. **6**, 2177–2191 (2006)

Spurr, R.J.D.: Linearized Radiative Transfer Theory: A General Discrete Ordinate Approach to the Calculation of Radiances and Analytic Weighting Functions, with Application to Atmospheric Remote Sensing. Ph.D. thesis, Technical University of Eindhoven (2001)

Staehelin, J., Thudium, J., Bühler, R., Volz-Thomas, A., Graber, W.: Trends in surface ozone concentrations at Arosa (Switzerland). Atmos. Environ. **28**, 75–87 (1994)

Stamnes, K., Tsay, S.-C., Wiscombe, W., Jayaweera, K.: Numerically stable algorithm for discrete-ordinate-method radiative transfer in multiple scattering and emitting layered media. Appl. Opt. **27**, 2502–2509 (1988)

Stern, D.: Global sulphur emissions from 1850 to 2000, Chemosphere 58, 163–175 (2005) (update from November 2005, http://www.rpi.edu/~sternd/datasite.html)

Stevens, R.K., Vossler, T.L.: DOAS (differential optical absorption spectroscopy) urban pollution. In: Schiff, H.I. (ed.) Measurements of Atmospheric Gases, Vol. 1433, pp. 25–35. SPIE-The International Society of Optical Engineering, Bellingham (1991)

Stevens, R.K., Drago, R.J., Mamane, Y.: A long path differential optical absorption spectrometer and EPA-approved fixed-point methods intercomparison. Atmos. Environ. **27B**, 231–236 (1993)

Stockwell, R.W., Calvert, J.G.: The near ultraviolet absorption spectrum of gaseous HONO and N_2O_3. J. Photochem. **8**, 193–203 (1978)

Stockwell, W.R. and Calvert, J.G.: The mechanism of the OH–SO_2 reaction. Atmos. Environ. **17**, 2231–2235 (1983)

Stoiber, R.E., Jepsen, A.: Sulfur-dioxide contributions to atmosphere by volcanos. Science **182**(4112), 577–578 (1973)

Strutt, R.J.: Ultra-violet transparency of the lower atmosphere, and its relative poverty in ozone. Proc. R. Soc. A **94**, 260–269 (1918)

Stuhl, F., Niki, H.: Flash photochemical study of the reaction OH + NO + M using resonance fluorescent detection of OH. J. Chem. Phys. **57**, 3677–3679 (1972)

Stutz, J., Platt, U.: Problems in using diode arrays for open path DOAS measurements of atmospheric species. In: Schiff, H.I., Platt, U. (eds.) Proceedings of the Europto Series. Optical Methods in Atmospheric Chemistry. **1715**, 329–340 (1992)

Stutz, J.: Messung der Konzentration troposphärischer Spurenstoffe mittels Differentieller-Optischer Absorptionsspektroskopie: Eine neue Generation von Geräten und Algorithmen. Ph.D. thesis, University of Heidelberg (1996)

Stutz, J., Platt, U.: Numerical analysis and error estimation of differential optical absorption spectroscopy measurements with least squares methods. Appl. Opt. **35**, 6041–6053 (1996)

Stutz, J., Platt, U.: Improving long-path differential optical absorption spectroscopy (DOAS) with a quartz-fiber mode-mixer. Appl. Opt. **36**, 1105–1115 (1997a)

Stutz, J., Platt, U.: A new generation of DOAS instruments. In: Bösenberg, J., Brassington, D.J., Simon, P.C. (eds.) EUROTRAC Final Report Vol 8: Instrument Development for Atmospheric Research and Monitoring, pp. 370–378 (1997b)

Stutz, J., Alicke, B., Neftel, A.: Nitrous acid formation in the urban atmosphere: gradient measurements of NO_3 and HONO over grass in Milan, Italy. J. Geophys. Res. **107**(D22), 8192 (2002a). doi: 10.1029/2001JD000390

Stutz, J., Ackermann, R., Fast, J.D., Barrie, L.: Atmospheric reactive chlorine and bromine at the Great Salt Lake, Utah. Geophys. Res. Lett. **29**, (2002b). doi:10.1029/2002GL014812

Stutz, J., Alicke, B., Ackermann, R., Geyer, A., Wang, S., White, A.B., Williams, E.J., Spicer, C.W., Fast, J.D.: Relative humidity dependence of HONO chemistry in urban areas. J. Geophys. Res. **109**, D03307 (2004a). doi:10.1029/2003JD004135

Stutz, J., Alicke, B., Ackermann, R., Geyer, A., White, A., Williams, E.: Vertical profiles of NO_3, N_2O_5, O_3, and NO_X in the nocturnal boundary layer: 1. Observations during the Texas Air Quality Study 2000. J. Geophys. Res. **109**, D12306 (2004b). doi:10.1029/2003JD004209

Stutz, J., Kim, E.S., Platt, U., Bruno, P., Perrino, C., Febo A.: UV-visible absorption cross-sections of nitrous acid. J. Geophys. Res. **105**, 14585–14592 (2000)

Sugimoto, N. and Minato, A.: Retroreflector with acute dihedral angles. Opt. Lett. **19**, 1660–1662 (1994)

Svanberg, S.: Atomic and Molecular Spectroscopy, 2nd edn. Springer Series on Atoms and Plasmas, Springer Berlin, Heidelberg (1992)

Svensson, R., Ljungström, E., Lindqvist, O.: Kinetics of the reaction between nitrogen dioxide and water vapour. Atmos. Environ. **21**, 1529–1539 (1987)

Syed, M.Q., Harrison, A.W.: Ground based observations of stratospheric nitrogen dioxide. Can. J. Phys. **58**, 788–802 (1980)

Tajime, T., Saheki, T., Ito, K.: Absorption characteristics of the γ-0 band of nitric oxide. Appl. Opt. **17**, 1290–1294 (1978)

Talmi, Y., Simpson, R.W.: Self-scanned photodiode array: a multichannel spectrometric detector. Appl. Opt. **19**, 1401–1414 (1980)

Talukdar, R.K., Longfellow, C.A., Gilles, M.K., Ravishankara, A.R.: Quantum yields of $O(^1D)$ in the photolysis of ozone between 289 and 329 nm as a function of temperature. Geophys. Res. Lett. **25**, 143–146 (1998)

Tang, T., McConnel, J.C.: Autocatalytic release of bromine from arctic snow pack during polar sunrise. Geophys. Res. Lett. **23**, 2633–2636 (1996)

Tellinghuisen, J.: Resolution of the visible-infrared absorption spectrum of I_2 into three contributing transitions. J. Chem. Phys. **58**, 2821–2834 (1973)

Thomas, K., Volz-Thomas, A., Kley, D.: Zur Wechselwirkung von NO_3-Radikalen mit wässrigen Lösungen: Bestimmung des Henry- und des Massenakkommodationskoeffizienten. Berichte des Forschungszentrums Jülich (Dissertation K. Thomas Univ. Wuppertal D468) 2755 (1993)

Thomsen, O.: Messung des Absorptionswirkungsquerschnitts von Schwefeldioxid im Wellenlängenbereich von 265 bis 298 nm. Diplomarbeit, Fachber, Physik Univ. Hamburg/GKSS-Geesthacht (1990)

Thornton, J.A., Wooldridge, P.J., Cohen, R.C., Martinez, M., Harder, H., Brune, W.H., Williams, E.J., Roberts, J.M., Fehsenfeld, F.C., Hall, S.R., Shetter, R.E., Wert, B.P., Fried, A.: Ozone production rates as a function of NO_X abundances and HO_X production rates in the Nashville urban plume. J. Geophys. Res. **107**(D12), 4146, doi: 10.1029/2001JD000932 (2002)

Thornton, J.A., Wooldridge, P.J., Cohen, R.C., Williams, E.J., Hereid, D., Fehsenfeld, F.C., Stutz, J., Alicke, B.: Comparisons of in situ and long path measurements of NO_2 in urban plumes. J. Geophys. Res. **108**(D16), 4496, doi:10.1029/2003JD003559 (2003)

Tolchenov, R.N., Naumenko, O., Zobov, N.F., Shirin, S.V., Polyansky, O.L., Tennyson, J., Carleer, M., Coheur, P.-F., Fally, S., Jenouvrier, A., Vandaele, A.C.: Water vapour line assignments in the 9250–26000 cm^{-1} frequency range. J. Mol. Spectrosc. **233**, 68–76 (2005)

Tørnkvist, K.K., Arlander, D.W., Sinnhuber, B.-M.: Ground-based UV measurements of BrO and OClO over Ny-Ålesund during Winter 1996 and 1997 and Andøya during Winter 1998/1999. J. Atmos. Chem. **43**, 75–106 (2002)

Toumi, R., Bekki, S.: The importance of the reactions between OH and ClO for stratospheric ozone. Geophys. Res. Lett. **20**, 2447–2450 (1993)

Toumi, R.: BrO as a sink for dimethylsulphide in the marine atmosphere. Geophys. Res. Lett. **21**, 117–120 (1994)

Tremmel, H.G., Junkermann, W., Slemr, F., Platt, U.: On the distribution of hydrogen peroxide in the lower troposphere over the Northeast United States during late summer 1988. J. Geophys. Res. **98**, 1083–1099 (1993)

Trick, S.: The formation of nitrous acid on urban surfaces—a physical-chemical perspective. Ph. D. thesis, University of Heidelberg (2004)

Trolier, M., Mauldin R.L., III Ravishankara, A.R.: Rate coefficients for the termolecular channel of the self-reaction of ClO (263 K). J. Phys. Chem. **94**, 4896–4907 (1990)

Trost, B., Stutz, J., Platt, U.: UV- absorption cross sections of a series of monocyclic aromatic compounds. Atmos. Environ. **31**, 3999–4008 (1997)

Tuazon, E.C., Winer, A.M., Graham, R.A., Pitts, J.N.: Atmospheric measurements of trace pollutants by kilometer-pathlength FT-IR spectroscopy. Environ. Sci. Technol. **10**, 259–299 (1980)

Tuazon, E.C., Winer, A.M., Pitts, J.N.: Trace pollutant concentrations in a multiday smog episode in the California South Coast Air Basin by long path length Fourier transform IR spectroscopy. Environ. Sci. Technol. **15**, 1232–1237 (1981)

Tucceri, M.E., Hölscher, D., Rodriguez, A., Dillon, T.J., Crowley, J.N.: Absorption cross section and photolysis of OIO. Phys. Chem. Chem. Phys. **8**, 834–846 (2006)

Tuckermann, M.: Troposphärische DOAS-Messungen zum halogenkatalysierten Ozonabbau im arktischen Frühjahr (Ny-Ålesund, Svalbard). Diploma thesis in physics, University of Heidelberg (1996)

Tuckermann, M., Ackermann, R., Gölz, C., Lorenzen-Schmidt, H., Senne, T., Stutz, J., Trost, B., Unold, W., Platt, U.: DOAS-observation of halogen radical-catalyzed Arctic boundary layer ozone destruction during the ARCTOC-campaigns 1995 and 1996 in Ny-Ålesund, Spitsbergen. Tellus B **49**, 533–555 (1997)

Tyndall, G.S., Orlando, J.J., Calvert, J.G.: Upper limit for the rate coefficient for the reaction $HO_2 + NO_2 \rightarrow HONO + O_2$. Environ. Sci. Technol. **29**, 202–206 (1995)

Ulshöfer, V.S.: Photochemische Produktion von Carbonylsulfid im Oberflächenwasser der Ozeane und Gasaustausch mit der Atmosphäre. Ph.D. thesis, University of Heidelberg (1995)

Ulshöfer, V.S., Andreae, M.O.: Carbonyl sulfide (COS) in the surface ocean and the atmospheric COS budget. Aquat. Geochem. **3**, 283–303 (1998)

Unold, W.: Bodennahe Messungen von Halogenoxiden in der Arktis. Diploma thesis, University of Heidelberg (1995)

URL: http://cfa-www.harvard.edu/amdata/ampdata/kurucz23/sekur.html

van de Hulst, H.C.: Multiple Light Scattering, Tables, Formulas and Applications, Vol. 1 (ISBN 0-12-710701-0) and Vol. 2 (ISBN 0-12-710702-9). Academic, New York (1980)

van der, A.R.J., Peters, D.H.M.U., Eskes, H., Boersma, K.F., Van Roozendael, M., De Smedt, I., Kelder, H.M.: Detection of the trend and seasonal variation in tropospheric NO_2 over China. J. Geophys. Res. **111**, D12317 (2006). doi:10.1029/2005JD006594

Van Doren, J.M., Watson, L.R., Davidovits, P., Worsnop, D.R., Zahniser, M.S., Kolb, Ch.E.: Temperature dependence of the uptake coefficients of HNO_3, HCl and N_2O_5 by water droplets. J. Phys. Chem. **94**, 3265–3269 (1990)

Van Roozendael, M., Hermans, C., DeMaziere, M., Simon, P.C.: Stratospheric NO_2 observations at the Jungfraujoch Station between June 1990 and May 1992. Geophys. Res. Lett. **21**, 1383–1386 (1994a)

Van Roozendael, M., de Maziere, M., Simon, S.: Ground-based visible measurements at the Jungfraujoch station since 1990. J. Quant. Spectrosc. Radiat. Transf. **52**, 231–240 (1994b)

Van Roozendael, M., Fayt, C., Bolsee, D., Simon, P.C., Gil, M., Yela, M., Cacho, J.: Cacho, ground-based stratospheric NO_2 monitoring at Keflavik (Iceland) during EASOE. Geophys. Res. Lett. **21**, 1379–1382 (1994c)

Van Roozendael, M., De Mazière, M., Hermans, C., Simon, P.C., Pommereau, J.P., Goutail, F., Tie, X.X., Brasseur, G.P. and Granier, C.: Ground-based observations of stratospheric NO_2 at high and mid-latitudes in Europe after the Mount Pinatubo eruption. J. Geophys. Res. **102**, 19171–19176 (1997)

Van Roozendael, M., Wagner, T., Richter, A., Pundt, I., Arlander, D.W., Burrows, J.P., Chipperfield, M., Fayt, C., Johnston, P.V., Lambert, J.-C., Kreher, K., Pfeilsticker, K., Platt, U., Pommereau, J.-P., Sinnhuber, B.-M., Tørnkvist, K.K., Wittrock, F.: Intercomparison of BrO measurements from ERS-2 GOME, ground-based and balloon platforms. Adv. Space Res. **29**, 1661–1666 (2002)

Van Roozendael, M., Fayt, C., Post, P., Hermans, C., Lambert, J.-C.: Retrieval of BrO and NO_2 from UV-visible observations. In: Borell, P., et al. (eds.) Sounding the Troposphere from Space: A New Era for Atmospheric Chemistry. Springer-Verlag, ISBN 3-540-40873-8, (2003)

Vandaele, A.C., Carleer, M., Colin, R., Simon, P.C.: Detection of urban O_3, NO_2, H_2CO and SO_2 using Fourier transform spectroscopy. In: Proceedings of International Symposium on Environmental Sensing, European Optical Society, Berlin, 22–26 June, SPIE, Bellingham, USA, ISBN 0–8194–0880–8, pp. 288–292 (1992)

Vandaele, A.C., Simon, P.C., Guilmot, J.M., Carleer, M., Colin, R.: SO_2 absorption cross section measurement in the UV using a Fourier transform spectrometer. J. Geophys. Res. **99**, 25599–25605 (1994)

Vandaele, A., Hermans, C., Simon, P., Van Roozendael, M., Guilmot, J., Carleer, M., Colin, R.: Fourier transform measurements of NO_2 absorption cross-section in the visible range at room temperature. J. Atmos. Chem. **25**, 289–305 (1996)

Vandaele, A.C., Hermans, C., Simon, P.C., Carleer, M., Colin, R., Fally, S. Merienne, M.F. Jenouvrier, A., Coquart, B.: Measurements of the NO_2 absorption cross section from $42,000\,cm^{-1}$ to $10,000\,cm^{-1}$ (238–1,000 nm) at 220 K and 294 K. J. Quant. Spectrosc. Radiat. Transf. **59**, 171–184 (1998)

Vandaele, A.C., Hermans, C., Fally, S., Carleer, M., Mérienne, M.-F., Jenouvrier, A., Coquart, B., Colin, R.: Absorption cross-sections of NO_2: simulation of temperature and pressure effects. J. Quant. Spectrosc. Radiat. Transf. **76**, 373–391 (2003). doi: 10.1016/S0022-4073(02)00064-X

Vaughan, G., Roscoe, H.K., Bartlett, L.M., O'Connor, F.M., Sarkissian, A., Van Roozendael, M., Lambert, J.C., Simon, P.C., Karlsen, K., Kastad Hoiskar, B.A., Fish, D.J., Jones, R.L., Freshwater, R.A., Pommereau, J.P., Goutail, F., Andersen, S.B., Drew, D.G., Huges, P.A., Moore, D., Mellquist, J., Hegels, E., Klüpfel,

T., Erle, F., Pfeilsticker, K., Platt, U.: An intercomparison of ground-based UV-visible sensors of ozone and NO_2. J. Geophys. Res. **102**, 1411–1422 (1997)

VDI.: Fernmessverfahren—Messungen in der bodennahen Atmosphäre nach dem DOAS-Prinzip—Messen gasförmiger Emissionen und Immissionen—Grundlagen, VDI Richtlinie VDI 4212, Blatt 1. (Remote sensing—atmospheric measurements near ground with DOAS—measurements of emissions and ambient air—fundamentals, VDI Guideline VDI 4212, 1) (2005)

Veitel, V., Funk, O., Kurz, C., Platt, U., Pfeilsticker, K: Geometrical path length probability density function of the skylight transmitted by mid-latitude cloudy skies; some case studies. Geophys. Res. Lett **25**, 3355–3358 (1998)

Veitel, H.: Vertical profiles of NO_2 and HONO in the boundary layer. Ph.D. thesis, University of Heidelberg (2002)

Veitel, H., Kromer, B., Mößner, M., Platt, U.: New techniques for measurements of atmospheric vertical trace gas profiles using DOAS. Environ. Sci. Pollut. Res., special issue **4**, 17–26 (2002)

Vitushkin, A.L., Vitushkin, L.F.: Design of a multipass optical cell based on the use of a shift corner cubes and rightangle prisms. Appl. Opt. **37**, 162–165 (1998)

Vogt, S.S., Tull, R.G., Kelton, P.: Self-scanned photodiode array: high performance operation in high dispersion astronomical spectrophotometry. Appl. Opt. **17**, 574–592 (1978)

Vogt, R., Finlayson-Pitts, B.J.: Tropospheric HONO and reactions of oxides of nitrogen with NaCl. Geophys. Res. Lett. **21**(21), 2291–2294 (1994)

Vogt, R., Crutzen, P.J., Sander, R.: A mechanism for halogen release from sea-salt aerosol in the remote marine boundary layer. Nature **383**, 327–330 (1996)

Voigt, S., Orphal, J., Bogumil, K., Burrows, J.P.: The temperature dependence (203–293 K) of the absorption cross sections of O_3 in the 230–850 nm region measured by Fourier-transform spectroscopy. J. Photochem. Photobiol. A Chem. **143**, 1–9 (2001)

Voigt, S., Orphal, J., Burrows, J.P.: The temperature- and pressure-dependence of the absorption cross-section of NO_2 in the 250–800 nm region measured by Fourier-transform spectroscopy. J. Photochem. Photobiol. A Chem. **149**, 1–7 (2002)

Volkamer, R., Etzkorn, T., Geyer, A., Platt, U.: Correction of the oxygen interference with UV spectroscopic (DOAS) measurements of monocyclic aromatic hydrocarbons in the atmosphere. Atmos. Environ. **32**, 3731–3747 (1998)

Volkamer, R.: A DOAS study on the oxidation mechanism of aromatic hydrocarbons under simulated atmospheric conditions. Ph.D. thesis, University of Heidelberg (2001)

Volkamer, R., Platt, U., Wirtz, K.: Primary and secondary glyoxal formation from aromatics: experimental evidence for the bicycloalkyl-radical pathway from BTX. J. Phys. Chem. A **105**, 7865–7874 (2001)

Volkamer, R., Klotz, B., Barnes, I., Imamura, T., Wirtz, K., Washida, N., Becker, K.-H., Platt, U.: OH-initiated oxidation of benzene: I. Phenol formation under atmospheric conditions. Phys. Chem. Chem. Phys. **4**, 1598–1610 (2002)

Volkamer, R., Spietz, P., Burrows, J.P., Platt, U.: High-resolution absorption cross-section of glyoxal in the UV/vis and IR spectral ranges. J. Photochem. Photobiol. A Chem. **172**, 35–46 (2005a). doi: 10.1016/j.jphotochem.2004.11.011

Volkamer, R., Molina, L.T., Molina, M.J., Shirley, T., Brune, W.H.: DOAS measurement of glyoxal as an indicator for fast VOC chemistry in urban air. Geophys. Res. Lett. **32**, L08806 (2005b). doi:10.1029/2005GL022616

Volz, A., Ehhalt, D.H., Derwent, R.G.: Seasonal and latitudinal variation of ^{14}CO and the tropospheric concentration of OH radicals. J. Geophys. Res. **86**, 5163–5171 (1981)

Volz, A., Kley, D.: Ozone measurements in the 19th century: an evaluation of the Montsouris series. Nature **332**, 240–242 (1988)

von Friedeburg, C., Pundt, I., Mettendorf, K.U., Wagner, T., Platt, U.: Multi-axis-DOAS measurements of NO_2 during the BAB II motorway emission campaign. Atmos. Env. **39**(5), 977–985 (2005)

von Glasow, R., Sander, R., Bott, A., Crutzen, P.J.: Modeling halogen chemistry in the marine boundary layer. 1. Cloud-free MBL. J. Geophys. Res. **107D**, 4341 (2002). doi:10.1029/2001JD000942

von Glasow, R., Crutzen, P.J.: Tropospheric halogen chemistry. In: Keeling, R.F. (ed.) The Atmosphere Vol. 4 Treatise on Geochemistry Holland, H.D., Turekian, K.K. (eds.), pp. 21–64. Elsevier-Pergamon, Oxford (2003)

von Glasow, R., Crutzen, P.J.: Model study of multiphase DMS oxidation with a focus on halogens. Atmos. Chem. Phys. **4**, 589–608 (2004)

von Glasow, R., von Kuhlmann, R., Lawrence, M.G., Platt, U., Crutzen, P.J.: Impact of reactive bromine chemistry in the troposphere. Atmos. Chem. Phys. **4**, 2481–2497 (2004)

Vountas, M., Rozanov, V.V., Burrows, J.P.: Impact of Raman scattering on radiative transfer in earth's atmosphere. J. Quant. Spectrosc. Radiat. Transf. **60**(6), 943–961 (1998)

Vountas, M., Richter, A., Wittrock, F., Burrows, J.P.: Inelastic scattering in ocean water and its impact on trace gas retrievals from satellite data. Atmos. Chem. Phys. **3**, 1365–1375 (2003)

Wagner, H.E.: Bestimmung der Spurengaszusammensetzung in der Troposphäre mit Langpfad-DOAS (Differenzielle Optische Absorptionsspektroskopie), Diploa thesis in physics, Inst. f. Physikalische und Theoretische Chemie, TU München (1990)

Wagner, T., Senne, T., Erle, F., Otten, C., Stutz, J., Pfeilsticker, K., Platt, U.: Determination of cloud properties and cloud type with DOAS measurements, Poster, Third European Symposium on Polar Strat. Ozone, Schliersee, 18–22 September (1995)

Wagner, T., Erle, F., Marquard, L., Otten, C., Pfeilsticker, K., Senne, Th., Stutz, J., Platt, U.: Cloudy sky photon path lengths as derived from DOAS observations. J. Geophys. Res. **103**, 25307–25321 (1998)

Wagner, T., Platt, U.: Observation of tropospheric BrO from the GOME satellite. Nature **395**, 486–490 (1998)

Wagner, T., Otten, C., Platt, U.: Observation of atmospheric NO_3 in the Arctic winter. Geophys. Res. Lett. **27**(21), 3441–3444 (2000)

Wagner, T., Leue, C., Pfeilsticker, K., Platt, U.: Monitoring of the stratospheric chlorine activation by global ozone monitoring experiment (GOME) OClO measurements in the austral and boreal winters 1995 through 1999. J. Geophys. Res. **106**, 4971–4986 (2001a)

Wagner, T., Leue, C., Wenig, M., Pfeilsticker, K., Platt, U.: Spatial and temporal distribution of enhanced boundary layer BrO concentrations measured by the GOME instrument aboard ERS-2. J. Geophys. Res. **106**, 24225–24235 (2001b)

Wagner, T., Wittrock, F., Richter, A., Wenig, M., Burrows, J.P., Platt, U.: Continuous monitoring of the high and persistent chlorine activation during the Arctic winter 1999/2000 by the GOME instrument on ERS-2. J. Geophys. Res. **107**(D20), 8267 (2002a). doi: 10.1029/2001JD000466

Wagner, T., Friedeburg, v.C., Wenig, M., Otten, C., Platt, U.: UV/vis observation of atmospheric O_4 absorptions using direct moon light and zenith scattered sunlight under clear and cloudy sky conditions. J. Geophys. Res. **107**(D20), 4424 (2002b). doi: 10.1029/2001JD001026

Wagner, T., Chance, K., Frieß, U., Gil, M., Goutail, F., Hönninger, Johnston, P.V., Karlsen-Tørnkvist, K., Kostadinov, I., Leser, H., Petritoli, A., Richter, A., Van Roozendael, M., Platt, U.: Correction of the ring effect and I_0effect for DOAS observations of scattered sunlight, ESA Technical Report (2002c)

Wagner, T., Heland, J., Zöger, M., Platt, U.: A fast H_2O total column density product from GOME—validation with in-situ aircraft measurements. Atmos. Chem. Phys. **3**, 651–663 (2003)

Wahner, A., Jakoubek, R.O., Mount, G.H, Ravishankara, A.R., Schmeltekopf, A.L.: Remote sensing observations of nighttime OClO column during the airborne antarctic ozone experiment, September 8, 1987. J.Geophys. Res. **94**, 11405–11411 (1989a)

Wagner, T., Dix, B., Friedeburg, C.v., Frieß, U., Sanghavi, S., Sinreich, R., Platt, U.: MAX-DOAS O_4 measurements: a new technique to derive information on atmospheric aerosols–principles and information content. J. Geophys. Res. **109**, D22205 (2004). doi: 10.1029/2004JD004904

Wahner, A., Jakoubek, R.O., Mount, G.H, Ravishankara, A.R., Schmeltekopf, A.L.: Remote sensing observations of daytime column NO_2 during the airborne Antarctic ozone experiment, August 22 to October 2, 1987. J.Geophys. Res., **94**, 16619–16632 (1989b)

Wahner, A., Callies, J., Dorn, H.-P., Platt, U., Schiller, C.: Near UV atmospheric absorption measurements of column abundances during airborne Arctic stratospheric expedition, January–February 1989: 1. Technique and NO_2 observations. Geophys. Res. Lett. **17**,497–500 (1990a)

Wahner, A., Callies, J., Dorn, H.-P., Platt, U., Schiller, C.: Near UV atmospheric absorption measurements of column abundances during airborne Arctic stratospheric expedition, January–February 1989: 3. BrO observations. Geophys. Res. Lett. **17**, 517–520 (1990b)

Wagner, T., Burrows, J.P., Deutschmann, T., Dix, B., Hendrick, F., Friedeburg, C.v., Frieß, U., Heue, K.-P., Irie, H., Iwabuchi, H., Keller, J., McLinden, C., Oetjen, H., Palazzi, E., Petrotoli, A., Platt, U., Postylyakov, O., Pukite, J., Richter, A., van Roozendael, M., Rozanov, A., Rozanov, V., Sinreich, R., Sanghavi, S., Wittrock, F.: Comparison of box-air-mass-factors and radiances for MAX-DOAS-geometries calculated from different UV/visible radiative transfer models. Atmos. Chem. Phys. Discuss. **6**, 9823–9876 (2006)

Wahner, A. Schiller, C.: Twilight variation of vertical column abundances of OClO and BrO in the North polar region. J. Geophys. Res. **97**(D8), 8047–8055 (1992)

Wahner, A., Mentel, T.F., Sohn, M.: Gas-phase reaction of N_2O_5 with water vapor: importance of heterogeneous hydrolysis of N_2O_5 and surface desorption of HNO_3 in a large teflon chamber. Geophys. Res. Lett. **25**, 2169–2172 (1998a)

Wahner, A., Mentel, T.F., Sohn, M., Stier, J.: Heterogeneous reaction of N_2O_5 on sodium nitrate aerosol. J. Geophys. Res. **103**, 31103–31112 (1998b)

Wamsley, P.R., Elkins, J.W., Fahey, D.W., Dutton, G.S., Volk, C.M., Myers, R.C., Montzka, S.A., Butler, J.H., Clarke, A.D., Fraser, P.J., Steele, L.P., Lucarelli, M.P., Atlas, E.L., Schauffler, S.M., Blake, D.R., Rowland, F.S., Sturges, W.T., Lee, J.M., Penkett, S.A., Engel, A., Stimpfle, R.M., Chan, K.R., Weisenstein, D.K., Ko, M.K.W., Salawitch, R.J.: Distribution of halon-1211 in the upper troposphere and lower stratosphere and the 1994 total bromine budget. J. Geophys. Res. **103**(D1), 1513–1526, 10.1029/97JD02466 (1998)

Wang, P., Bruns, M., Richter, A., Burrows, J.P., Heue, K.-P., Pundt, I., Wagner, T., Platt, U.: Validation of SCIAMACHY with AMAXDOAS measurements from the DLR Falcon. Geophys. Res. **5**, 09341 (Abstracts) (2003)

Wang, P., Richter, A., Bruns, M., Rozanov, V., Burrows, J., Heue, K.-P., Wagner, T., Pundt, I., Platt, U.: Measurements of tropospheric NO_2 with an airborne multi-axis DOAS instrument. Atmos. Chem. Phys. **5**, 337–343 (2005)

Wängberg, I., Etzkorn, T., Barnes, I., Platt, U., Becker, K.H.: Absolute determination of the temperature behaviour of the $NO_2 + NO_3 + (M) \Leftrightarrow N_2O_5 + (M)$ equilibrium. J. Phys. Chem. A **101**(50), 9694–9698 (1997)

Watts, S.: The mass budgets of carbonyl sulfide, dimethyl sulfide, carbon disulfide and hydrogen sulfide. Atmos. Environ. **34**, 761–779 (2000)

Wayne, R.P., Barnes, I., Biggs, P., Burrows, J.P., Canosa-Mas, C., Hjorth, J., Le Bras, G., Moortgat, G., Perner, D., Poulet, G., Restelli, G., Sidebottom, H.: The nitrate radical: physics, chemistry and the atmosphere. Atmos. Environ. **25A**, 1–250 (1991)

Wayne, R.P., Poulet, G., Biggs, P., Burrows, J.P., Cox, R.A., Crutzen, P.J., Haymann, G.D., Jenkin, M.E., LeBras, G., Moortgat, G.K., Platt, U., Schindler, R.N.: Halogen oxides: radicals, sources and reservoirs in the laboratory and in the atmosphere. Atmos. Environ. **29**, 2675–2884 (1995)

Weaver, A., Solomon, S., Sanders, R.W., Arpag, K., Miller, H.L.: Atmospheric No 3.5. Off-axis measurements at sunrise: Estimates of tropospheric NO_3 at 40 degrees N. J. Geophys. Res. **101**(D13), 18605–18612 (1996)

Weber, A.: Applications. In: Anderson, A. (ed.) The Raman Effect, Vol. 1 and 2. Dekker Inc., New York (1973)

Weber, M., Dhomse, S., Wittrock, F., Richter, A., Sinnhuber, B.-M., Burrows, J.P.: Dynamical control of NH and SH winter/spring total ozone from GOME observations in 1995–2002. Geophys. Res. Lett. **30**, 37.1–37.4 (2003). doi 10.1029/2002GL016799

Weibring, P., EdnerH., Svanberg, S., Cecchi, G., Pantani, L., Ferrara, R., Caltabiano, T.: Monitoring of volcanic sulphur dioxide emissions using differential absorption lidar (DIAL), differential optical absorption spectroscopy (DOAS) and correlation spectroscopy (COSPEC). Appl. Phys. B **67**(4), 419–426 (1998)

Weidner, F., Bösch, H., Bovensmann, H., Burrows, J. P., Butz, A., Camy-Peyret, C., Dorf, M., Gerilowski, K., Gurlit, W., Platt, U., von Friedeburg, C., Wagner, T., Pfeilsticker, K.: Balloon-borne limb profiling of UV/vis skylight radiances,

O_3, NO_2, and BrO: technical set-up and validation of the method. Atmos. Chem. Phys. **5**, 1409–1422 (2005). SRef-ID: 1680-7324/acp/2005-5-1409

Weinstock, B.: Carbon monoxide: residence time in the atmosphere. Science **166**, 224–225 (1969)

Wenig, M.: Satellite measurement of long-term global tropospheric trace distributions and source strength—algorithm development and data analysis. Ph.D. thesis, University of Heidelberg (2001)

Wenig, M., Spichtinger, N., Stohl, A., Held, G., Beirle, S., Wagner, T., Jähne, B., Platt, U.: Intercontinental transport of nitrogen oxide pollution plumes. Atmos. Chem. Phys. **3**, 387–393 (2003)

Wenig, M., Kühl, S., Beirle, S., Bucsela, E., Jähne, B., Platt, U., Gleason, J., Wagner, T.: Retrieval and analysis of stratospheric NO_2 from the global ozone monitoring experiment. J. Geophys. Res. **109**, D04315 (2004). doi:10.1029/2003JD003652

Wenig, M., Jähne, B., Platt, U.: Operator representation as a new differential optical absorption spectroscopy formalism. Appl. Opt. **44**(16), 3246–3253 (2005)

Wennberg, P.O., Hanisco, T.F., Jaeglé, L., Jacob, D.J., Hintsa, E.J., Lanzendorf, E.J., Anderson, J.G., Gao, R.-S., Keim, E.R., Donnelly, S.G., Del Negro, L.A., Fahey, D.W., McKeen, S.A., Salawitch, R.J., Webster, C.R., May, R.D., Herman, R.L., Proffitt, M.H., Margitan, J.J., Atlas, E.L., Schauffler, S.M., Flocke, F., McElroy, C.T., Bui, T.P.: Hydrogen radicals, nitrogen radicals, and the production of O_3 in the upper troposphere. Science **279**, 49–53 (1998)

Wennberg, P.O.: Bromine explosion. Nature **397**, 299–301 (1999)

Werle, P.W., Josek, K., Slemr, F.: Application of FM spectroscopy in atmospheric trace gas monitoring: a study of some factors influencing the instrument design. In: Schiff, H.I. (ed.) Measurement of Atmospheric Gases Vol. 1433, pp. 128–135. SPIE, Bellingham (1991)

West, P.W., Gaeke, G.C.: Fixation of sulfur dioxide as disulfitomercurate(II) and subsequent colorimetric estimation. Anal. Chem. **28**, 1816–1819 (1956)

White, J.U.: Long optical paths of large aperture. J. Opt. Soc. Am. **32**, 285–288 (1942)

White, J.U.: Very long optical paths in air. J. Opt. Soc. Am. **66**, 411–416 (1976)

WHO (World Health Organization): Monitoring ambient air quality for health impact assessment. WHO Regional Publications, European Series, No. 85. WHO Regional Office for Europe, Copenhagen; ISBN 92 890 1351 6 (1999)

Wilmouth, D.M., Hanisco, T.F., Donahue, N.M., Anderson, J.G.: Fourier transform ultraviolet spectroscopy of the $A(^2\Pi_{3/2}) \leftarrow X(^2\Pi_{3/2})$ transition of BrO. J. Phys. Chem. **103**, 8935–8944 (1999)

Wine, P.H., Ravishankara, A.R., Philen, D.L., Davis, D.D., Watson, R.T.: High resolution absorption cross sections for the $A^2\Pi - X^2\Pi$ system of ClO. Chem. Phys. Lett. **50**, 101–106 (1977)

Winer, A.M., Atkinson, R., Pitts, J.N.: Gaseous nitrate radical: possible night-time atmospheric sink for biogenic organic compounds. Science **224**, 156–159 (1984)

Winer, A.M., Biermann, H.W.: Measurements of nitrous acid, nitrate radicals, formaldehyde, and nitrogen dioxide for the Southern California Air Quality Study by differential optical absorption spectroscopy. In: Schiff, H. (ed.) Measurement of Atmospheric Gases Vol. 1433, pp. 44–57. SPIE, Bellingham (1991)

Winer, A.M., Biermann, H.W.: Long pathlength differential optical absorption spectroscopy (DOAS) measurements of gaseous HONO, NO_2 and HCHO in the California south coast air basin. Res. Chem. Intermed. **20**(3–5), 423–455 (1994)

Wingenter, O.W., Kubo, M.K., Blake, N.J., Smith, T.W., Blake, D.R., Rowland, F.S.: Hydrocarbon and halocarbon measurements as photochemical and dynamical indicators of atmospheric hydroxyl, atomic chlorine, and vertical mixing obtained during Lagrangian flights. J. Geophys. Res. **101**, 4331–4340 (1996)

Winterrath, T., Kurosu, T., Richter, A., Burrows, J.P.: O_3 and NO_2 in thunderstorm clouds: convection or production? Geophys. Res. Lett. **26**, 1291–1294 (1999)

Wiscombe, W.J.: Improved mie scattering algorithms. Appl. Opt. **19**, 1505–1509 (1980)

Wittrock, F., Müller, R., Richter, A., Bovensmann, H., Burrows, J.P.: Measurement of iodine oxide (IO) above Spitsbergen. Geophys. Res. Lett. **27**(10), 1471–1474 (2000a)

Wittrock, F., Richter, A., Ladstätter-Weißenmayer, A., Burrows, J.P.: Global observations of formaldehyde. Proceedings of ERS-ENVISAT Symposium, Gothenburg, October 2000. ESA publication SP-461 (2000b)

Wittrock, F., Oetjen, H., Richter, A., Fietkau, S., Medeke, T., Rozanov, A., Burrows, J.P.: MAX-DOAS measurements of atmospheric trace gases in Ny-Ålesund. Atmos. Chem. Phys. **4**, 955–966 (2004)

Wittrock, F., Richter, A., Oetjen, H., Burrows, J.P., Kanakidou, M., Myriokefalitakis, S., Volkamer, R., Beirle, S., Platt, U., Wagner, T.: Simultaneous global observations of glyoxal and formaldehyde from Space. Geophys. Res. Lett. **33**, L16804 (2006). doi:10.1029/2006GL026310

WMO.: WDCGG Data Summary, WDCGG No. 29, GAW DATA, Vol. IV-Greenhouse Gases and Other Atmospheric Gases. Japan Meteorological Agency in Co-Operation with World Meteorological Organisation (2005)

Wolf de, D.A.: Optical propagation through turbulent air. Opt. Laser Technol. **11**, Issue 1, February, 29–36 (1979)

Wood, E.C., Wooldridge, P.J., Freese, J.H., Albrecht, T., Cohen, R.C.: Prototype for in situ detection of atmospheric NO_3 and N_2O_5 via laser-induced fluorescence. Environ. Sci. Technol. **37**(24), 5732–5738 (2003)

Wu, C.Y.R., Judge, D.L.: SO_2 and CS_2 cross section data in the ultraviolet region. Geophys. Res. Lett. **8**, 769–771 (1981)

Yates, D.A., Kuwana, T.: Evaluation of a self-contained linear diode array detector for rapid scanning spectrophotometry. Anal. Chem. **48**, 510–514 (1976)

Yoshii, Y., Kuze, H., Takeuchi, N.: Long-path measurement of atmospheric NO_2 with an obstruction flashlight and a charge-coupled-device spectrometer. Appl. Opt. **42**(21), 4362–4368 (2003)

Young, A.T.: On the Rayleigh-scattering optical depth of the atmosphere. J. Appl. Meteorol. **20**, 328–330 (1981)

Yu Y., Geyer, A., Xie, P., Galle, B., Limin, C., Platt, U.: Observations of carbon disulfide by differential optical absorption spectroscopy in Shanghai. Geophys. Res. Lett. **31**, L11107 (2004). doi: 10.1029/2004GL019543

Yvon, S.A., Butler, J.H.: An improved estimate of the oceanic lifetime of atmospheric CH_3Br. Geophys. Res. Lett. **23**, 53–56 (1996)

Zabarnick, S.: Kinetics of the reaction OH + NO + M $\rightarrow$ HONO + M as a function of temperature and pressure in the presence of argon, SF_6, and N_2 bath gas. Chem. Phys. **171**(1–2), 265–273 (1993)

Zellner, R.,nd Hägele, J.: A double-beam UV-laser differential method for monitoring tropospheric trace gases. Opt. Laser Technol. **4**, 79–82 (1985)

Zender, C.S.: Global climatology of abundance and solar absorption of oxygen collision complexes. J. Geophys. Res. **104**, 24471–24484 (1999)

Zhou, X., Beine, H.J., Honrath, R.E., Fuentes, J.D., Simpson, W., Shepson, P.B., Bottenheim, J.W.: Snowpack photochemical production of HONO: a major source of OH in the Arctic boundary layer in springtime. Geophys. Res. Lett. **28**, 4087–4090 (2001)

Zhou, X., Civerolo, K., Dai, H., Huang, G., Schwab, J., Demerjian, K.: Summertime nitrous acid chemistry in the atmospheric boundary layer at a rural site in New York State. J. Geophys. Res. **107**, 4590 (2002). doi:10.1029/2001JD001539

Zhou, X., Gao, H., He, Y., Huang, G., Bertman, S., Civerolo, K., Schwab, J.: Nitric acid photolysis on surfaces in low-NO_X environments: significant atmospheric implications. Geophys. Res. Lett. **30**, 2217 (2003). doi:10.1029/2003GL018620

Zhu, L., Kellis, D., Ding, C.F.: Photolysis of glyoxal at 193, 248, 308 and 351 nm. Chem. Phys. Lett. **257**(5–6), 487–491 (1996)

Zimmermann, R.: Entwicklung eines miniaturisierten ballongetragenen Diodenlaser-Spektrometers zur Messung von stratosphärischen Methan- und Wasserdampfkonzentrationen. Ph.D. thesis, University of Heidelberg (2003)

Zingler, J., Platt, U.: Iodine oxide in the dead sea valley: evidence for inorganic sources of boundary layer IO. J. Geophys. Res. **110**, D07307 (2005). doi:10.1029/2004JD004993

Appendix A

Spectral Positions of Emission Lines from Calibration Lamps and Lasers

Emission lines are taken from the following internet site:
http://cfa-www.harvard.edu/amdata/ampdata/kurucz23/sekur.html

A.1 Cadmium Lines

Vaccum wavelength /nm	Air wavelength /nm	Log_gf	Element (name)	E_lower_lev. /cm^{-1})	J upper
214.5068	214.4393	-0.108	Cd II	0.000	1.5
226.5720	226.5019	-0.340	Cd II	0.000	0.5
228.8724	228.8018	0.336	Cd I	0.000	1.0
230.7334	230.6624	-0.980	Cd I	30656.130	0.0
231.3464	231.2753	-0.468	Cd II	46618.550	2.5
233.0003	232.9288	-0.960	Cd I	31826.996	1.0
257.3699	257.2928	-0.194	Cd II	44136.080	0.5
283.7726	283.6892	-0.980	Cd I	30114.017	1.0
288.1581	288.0736	-0.760	Cd I	30656.130	2.0
298.1492	298.0622	-0.410	Cd I	31826.996	3.0
340.4629	340.3652	-0.350	Cd I	30114.017	1.0
346.7195	346.6202	0.140	Cd I	30656.130	2.0
346.8650	346.7657	-0.360	Cd I	30656.130	1.0
361.1540	361.0510	0.410	Cd I	31826.996	3.0
361.3905	361.2875	-0.300	Cd I	31826.996	2.0
467.9458	467.8149	-0.970	Cd I	30114.017	1.0
480.1256	479.9914	-0.400	Cd I	30656.130	1.0
508.7242	508.5824	-0.230	Cd I	31826.996	1.0
644.0249	643.8470	-0.010	Cd I	43692.474	2.0

A.2 Mercury Lines

Wl_vac/nm	Wl_air/nm	log_gf	Element (name)	E_lower_lev ./cm^{-1}
226.0960	226.0260	-0.532	Hg II	51485.000
253.7279	253.6517	-1.382	Hg I	0.000
257.7057	257.6285	-1.714	Hg I	39412.300
275.3591	275.2777	-1.712	Hg I	37645.080
284.8516	284.7679	-0.101	Hg II	60608.000
289.4440	289.3592	-1.267	Hg I	39412.300
292.6262	292.5406	-1.481	Hg I	44042.977
296.8145	296.7278	0.470	Hg I	37645.080
302.2375	302.1495	-0.270	Hg I	44042.977
312.6571	312.5665	0.060	Hg I	39412.300
313.2453	313.1545	-0.040	Hg I	39412.300
313.2746	313.1838	-0.040	Hg I	39412.300
334.2434	334.1473	-1.100	Hg I	44042.977
365.5873	365.4832	0.140	Hg I	44042.977
366.3918	366.2875	-0.430	Hg I	44042.977
366.4318	366.3275	0.040	Hg I	44042.977
398.5017	398.3890	-1.730	Hg II	35514.000
404.7702	404.6559	-0.863	Hg I	37645.080
407.8978	407.7827	-1.470	Hg I	39412.300
435.9548	435.8323	-0.383	Hg I	39412.300
491.7435	491.6062	-1.506	Hg I	54068.781
546.2249	546.0731	-0.137	Hg I	44042.977
577.1193	576.9593	0.630	Hg I	54068.781
579.2266	579.0660	0.710	Hg I	54068.781

A.3 Hydrogen Lines

Wl_vac/nm	Wl_air /nm	Log_gf	Element (name)	E_lower_lev. ref./cm^{-1}
102.5722		-0.801	H I	0.000
121.5671		-0.080	H I	0.000
397.1195	397.0072	-0.993	H I	82259.105
410.2892	410.1734	-0.753	H I	82259.105
434.1682	434.0462	-0.447	H I	82259.105
486.2681	486.1323	-0.020	H I	82259.105
656.2822	656.1010	0.710	H I	82281.662
656.4610	656.2797	0.710	H I	82259.105

(continued)

Wl_vac/nm	Wl_air /nm	Log_gf	Element (name)	E_lower_lev. ref./cm^{-1}
901.7388	901.4913	-0.901	H I	97492.302
923.1550	922.9017	-0.735	H I	97492.302
954.8592	954.5973	-0.540	H I	97492.302
1005.2128	1004.9373	-0.303	H I	97492.302
1094.1089	1093.8093	0.002	H I	97492.302

A.4 Neon I Lines

Wl_vac/nm	Wl_air/nm	Log_gf	Element (name)	E_lower_lev ./cm^(-1)
576.6018	576.4419	-0.370	NE I	149657.042
585.4110	585.2488	-0.460	NE I	135888.717
590.4098	590.2462	-0.423	NE I	150858.508
594.6481	594.4834	-0.120	NE I	134041.840
607.6019	607.4337	-0.470	NE I	134459.287
609.7851	609.6163	-0.270	NE I	134459.287
614.4763	614.3063	-0.350	NE I	134041.840
633.6180	633.4428	-0.310	NE I	134041.840
638.4756	638.2991	-0.260	NE I	134459.287
640.4016	640.2246	0.360	NE I	134041.840
650.8326	650.6528	0.030	NE I	134459.287
653.4687	653.2882	-0.440	NE I	134818.640
660.0776	659.8953	-0.360	NE I	135888.717
668.0120	667.8276	-0.400	NE I	135888.717
671.8897	671.7043	-0.310	NE I	135888.717
693.1379	692.9467	0.030	NE I	135888.717

A.5 Zinc Lines

Wl_vac/nm	Wl_air/nm	Log_gf	Element (name)	E_lower_lev./ cm^{-1}
202.6136	202.5483	-0.086	Zn II	0.000
206.2663	206.2004	-0.377	Zn II	0.000
210.0604	209.9937	-0.032	Zn II	49355.040
213.9247	213.8573	0.161	Zn I	0.000
250.2743	250.1989	-0.495	Zn II	48481.000
255.8715	255.7948	-0.194	Zn II	49355.040
277.1671	277.0853	-0.576	Zn I	32501.421
280.1688	280.0862	-0.303	Zn I	32890.352
328.3274	328.2328	-0.377	Zn I	32311.350
330.3535	330.2584	-0.057	Zn I	32501.421
330.3892	330.2941	-0.534	Zn I	32501.421
334.5977	334.5015	0.246	Zn I	32890.352
334.6532	334.5570	-0.502	Zn I	32890.352
468.1444	468.0134	-0.815	Zn I	32311.350
472.3474	472.2153	-0.338	Zn I	32501.421
481.1872	481.0528	-0.137	Zn I	32890.352
636.4097	636.2338	0.150	Zn I	46745.413

Appendix B

Absorption Spectra of Molecules Measurable by DOAS

In Appendix B, we summarise the spectral features and approximate wavelengths of strong absorption structures of the most important species that are observed by DOAS (see Table B.1 for an overview). Frequently absorption cross-sections were measured during several studies; note that we give no recommendation as to which one to use.

B.1 Nitric Oxide, NO

There are four UV transitions of the NO molecule:

γ-band: $A^2\Sigma^+ \leftarrow X^2\Pi$
β-band: $B^2\Pi^+ \leftarrow X^2\Pi$
δ-band: $C^2\Sigma^+ \leftarrow X^2\Pi$
ε-band: $D^2\Pi^+ \leftarrow X^2\Pi$

The energy difference of the γ-band corresponds to wavelengths near 214 and 226 nm (e.g. Tajime et al., 1978; Mellqvist and Rosén, 1996); the wavelengths of the other bands are even shorter, in particular they are in a region where the O_2 absorption does not permit absorption paths exceeding a few metres in the atmosphere. However, the γ-band has been used successfully to measure the abundance of NO in the atmosphere (Gall et al., 1991; Edner et al., 1992; Edner et al., 1990; Mount et al. 2002), in smoke stacks (Gall et al., 1991; Mellqvist et al., 1992; Mellqvist and Rosén, 1996), and in car exhaust (Perner et al., 1987). It should be noted that strong atmospheric absorption and scattering does not allow light paths longer than about 1 km in this wavelength range.

B.2 Nitrogen Dioxide, NO_2

The NO_2 molecule has a continuous absorption below 200 nm (with a maximum at 170 nm), a short-wave UV system with band-heads at 235.0 and 249.1 nm ($B^2B_2 \leftarrow X^2A_1$ transition), and an extended system with absorptions

Table B.1. Substances detectable by active UV/vis absorption spectroscopy. Detection limits were calculated for a minimum detectable optical density of 5×10^{-4}

Species	Wavelength interval (nm)	Approx. diff. absorption cross-section 10^{-19} cm^2/molec.	Transition	Detection limit (L = 5 km, 20°C) ppb
SO_2	200–230	65	$c^1B_2 \leftarrow X^1A_1$	0.03*
	290–310	5.7	$A^1B_1 \leftarrow X^1A_1$	0.06
	340–390		$a^3B_2 \leftarrow X^1A_1$	
CS_2	320–340	0.4		0.9
NO	200–230	24	$A^2\Sigma^+ \leftarrow X^2\Pi$,	0.08*
NO_2	330–500	2.5	$A^2B_1 \leftarrow X^2A_1$	0.15
NO_3	600–670	200		0.002
NH_3	200–230	180	$A^1A_2 \leftarrow X^1A_1$	0.009*
HNO_2	330–380	5.1		0.07
O_3	300–330	0.1	$C \leftarrow X^1A_1$	3,7
CH_2O	300–360	0.48	$A^1A_2 \leftarrow X^1A_1$	0.8
CHOCHO	400–480		$A^1A_u - X^1A_g$	
ClO	260–300	35	$X^2\Pi \leftarrow A$	0.01
BrO	300–360	104	$X^2\Pi \leftarrow A$	0.004
IO	400–470	170	$X^2\Pi \leftarrow A$	0.002
OClO	325–435	116	$A^2A_2 \leftarrow X^2B_1$	
OBrO				
OIO	470–670	270		
Benzene	240–270	21.9	$A^1B_{2u} \leftarrow X^1A_{1g}$	0.017
Toluene	250–280	12.8		0.03
Xylene (o/m/p)	250–280	2.1/6.6/20.3		0.18/0.06/0.02
Phenol	260–290	198		0.002
Cresol (o/m/p)	250–280	20.1/31.8/87.2		0.018/0.011/0.004
Benzaldehyde	280–290	44		0.008

* Due to strong absorption/scattering only L = 1 km assumed.

from the UV (about 320 nm) through the green spectral region ($A^2B_1 \leftarrow X^2A_1$ transition). This latter system is responsible for the brown colour of gaseous NO_2; it is almost exclusively used for the DOAS measurement of NO_2 in the atmosphere and allows detection limits on the order of several 10 ppt at 10 km light path length.

There are numerous measurements of the NO_2 absorption cross-section (e.g. Hall and Blacet, 1952; Schneider et al., 1987; Davidson et al., 1988; Harwood and Jones, 1994; Coquart et al., 1995; Merienne et al., 1995; Vandaele et al., 1996, 1998, 2003; Harder et al., 1997; Burrows et al., 1998; Voigt et al., 2002; Vandaele et al., 2003). The temperature dependence of the cross-section was studied by Davidson et al. (1988), Harder et al. (1997), Burrows et al. (1998), Voigt et al. (2002), and Vandaele et al. (2003).

Points to watch when determining NO_2 absorption cross-sections are: (1) the formation of N_2O_4 in equilibrium with NO_2 (e.g. Harwood and Jones, 1994):

$$NO_2 + NO_2 \Leftrightarrow N_2O_4 .$$

While N_2O_4 is of no importance in the atmosphere, and has a smooth spectrum [e.g. Hall and Blacet 1952]. At high concentrations typically used in laboratory determinations of the NO_2 cross-sections, it might form in appreciable amounts and thus lower the apparent NO_2 cross-section, if not corrected for. (2) There might be heterogeneous formation of HONO (see Sect. 2.6.3) in NO_2 containing vessels (e.g. absorption cuvettes) unless they are kept very dry. Besides lowering the amount of NO_2, the highly structured HONO spectrum (see Sect. B.11) may contaminate the NO_2 spectrum.

B.3 Ammonia, NH_3

While the ground state of NH_3 is tetrahedral with C_{3v} symmetry, the electronically excited states are planar with D_{3h} symmetry. There are extended vibrational progressions of the $A^1A_2 \leftarrow X^1A_1$ and $A^3A_2 \leftarrow X^1A_1$ transitions. The $A^1A_2 \leftarrow X^1A_1$ transition gives rise to a band system from 216.8 nm to about 170 nm, the longer wavelength bands of which are used for observation of ammonia by DOAS (e.g. Edner et al., 1990; Mellqvist et al., 1992; Mellqvist and Rosén, 1996; Mount et al., 2002). Note that the strong atmospheric absorption and scattering does not allow light paths longer than about 1 km in this short UV wavelength range, as in the case of NO (Sect B.1).

B.4 Formaldehyde, HCHO

The HCHO molecule undergoes a change in configuration upon transition from the electronic ground state (X^1A_1 plane molecule with C_{2v} symmetry) to the exited state ($X^2A_1C_s$ symmetry). As a molecule with $n = 6$ atoms, formaldehyde has $3n - 6 = 6$ different vibrational modes. Absorptions due to formaldehyde ($A^1A_2 \leftarrow X^1A_1$ transition) are found in the near UV spectral range ca. 260–360 nm (Cantrell et al., 1990; Meller and Moortgart, 2000; Pope et al., 2005a,b). This structure has been used by numerous atmospheric studies to determine the atmospheric formaldehyde concentration by active (e.g. Platt et al., 1979; Platt and Perner, 1980; Notholt et al., 1992; Cardenas et al., 2000; Hak et al., 2005) and passive (Heckel, 2003; Palmer et al., 2003; Martin et al., 2004; Heckel et al., 2005; Wittrock et al., 2006] DOAS techniques.

B.5 Glyoxal, CHOCHO

The glyoxal (CHO-CHO) molecule exhibits a structured $A^1A_u - X^1A_g$ and $a^3A_u - X^1A_g$ band system in the blue spectral region (about 400–480 nm) and the UV region (250–350 nm), respectively, and a diffuse B–X transition. The absorption spectrum of the molecule was studied by several authors (e.g. Plum et al., 1983; Zhu et al., 1996; Orlando and Tyndall, 2001; Horowitz et al., 2001; Volkamer, 2001; Chen and Zhu, 2003). High-resolution absorption cross-sections were reported by Volkamer et al. (2005a). Up to now, only a few measurements using the bands in the blue spectral region have been reported (Volkamer et al., 2001, 2005b; Wittrock et al., 2006; Sinreich et al., 2007). The spectrum exhibits very narrow band spectral structures, which are not normally resolved by DOAS instruments; thus, apparent deviations from Lambert–Beer's law can occur, as demonstrated in Fig. B.1. Many of the previous determinations (e.g. Plum et al., 1983; Zhu et al., 1996; Orlando and Tyndall, 2001) did not fully resolve the spectral structures.

B.6 Sulphur Dioxide, SO_2

The sulphur dioxide molecule exhibits three band systems with narrow, discrete vibrational bands.

1. The strongest system in the range of 180 to 235 nm is due to transitions from the ground state to several electronic states, in particular the $c^1B_2 \leftarrow X^1A_1$ transition.
2. A relatively strong system from 260 to 340 nm is due to the $A^1B_1 \leftarrow X^1A_1$ transition.
3. A weak system in the spectral range from 340 to 390 nm corresponds to the sprin-forbidden $a^3B_2 \leftarrow X^1A_1$ transition.

DOAS measurements make use of the spectral range from about 290 to 310, where the strongest bands of the $A^1B_2 \leftarrow X^1A_1$ transition are located (Brand and Srikameswaran, 1972; Brand et al., 1972; Hamada and Merer, 1974; Brand and Nanes, 1973; Marx et al., 1980; Wu and Judge, 1981; Brassington, 1981; Brassington et al., 1984; Thomsen 1990). In polluted air, SO_2 dominates the absorption structure in this spectral range. Note that the structures of the O_3 Huggins bands (see Sect. B.8) are similar to those of SO_2 in spectral width (though about two orders of magnitude weaker), and thus may cause interferences in clean air if not taken into account, particularly at low spectral resolution.

The SO_2 absorption structures around 230 nm (Gall et al., 1991; Mellqvist and Rosén, 1996), but mostly near 300 nm, have been used in a large number of studies of its atmospheric abundance, both by active (e.g. Platt, 1977, 1978; Platt and Perner, 1980; Brauers et al., 1990; Coe et al., 1997) and passive (e.g. Lee et al., 2002; Bobrowski et al., 2003, 2006; McGonigle et al., 2005; Khokhar et al., 2005) DOAS techniques.

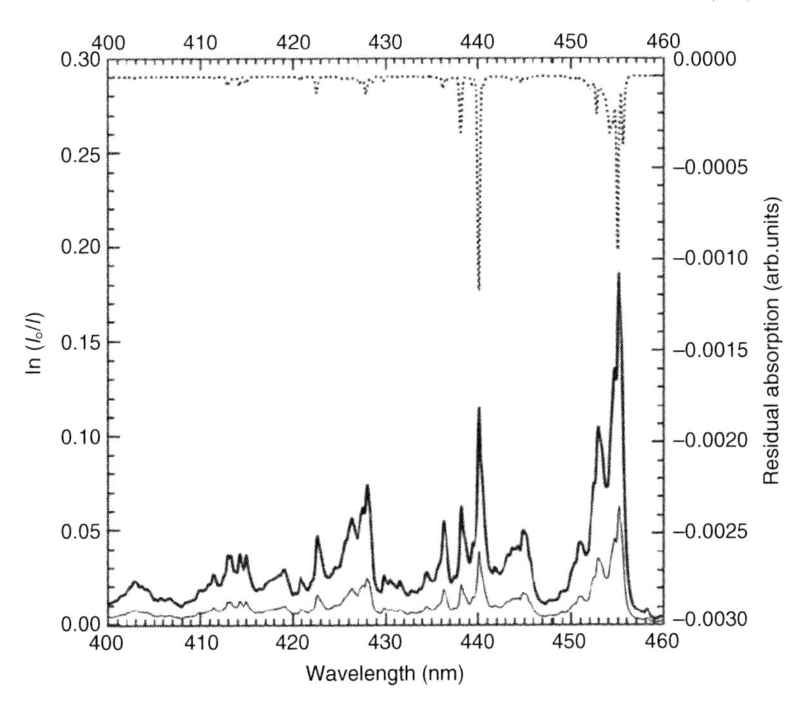

Fig. B. 1. Apparent deviation from Lambert–Beer's law for glyoxal. Two high-resolution spectra corresponding to column densities of 10^{17} molec./cm^2 (*thin line*) and 3×10^{17} molec./cm^2 (*thick line*) were convoluted to 0.25 nm spectral resolution (FWHM); residual structure (*dotted line*) after the two spectra were scaled to match column density and subtracted. This distortion of the spectrum can be understood from the considerable ro-vibrational structure, which, when observed at low spectral resolution, causes the apparent absorption cross-section to become lower at higher column densities (from Volkamer et al., 2005a, reproduced by permission of The Royal Society of Chemistry (RSC) on behalf of the European Society for Photobiology and the European Photochemistry Association)

B.7 Carbon Disulfide, CS_2

The CS_2 molecule exhibits two strongly structured absorption systems:

1. Between 190 and 210 nm,
2. In the near UV between 290 and about 360 nm.

Determinations of the cross-sections in both systems were reported by Ahmed and Kumar (1992); more recent data can be found on the internet (Keller-Rudek and Moortgat, 2005).

B.8 Ozone, O_3

Although ozone is similar in structure to SO_2, it exhibits only diffuse bands. There are four band systems in the UV-visible spectral range:

Table B.2. The relevant states and some transitions of the O_3 molecule

Electronic state	Energy of electronic state, T_0 cm^{-1}	Vibrational energies, cm^{-1}			Denomination of transition from X^1A_1 ground state
		ν_1	ν_2	ν_3	
D	33000		300		Hartley
C	28447	636	351		Huggins
B^1B_1	16625	1099			Chappuis
A^3B_1	10000		567		
X^1A_1	0	1110	705	1042	

1. The Hartley system from 220 to 300 nm.
2. The Huggins system from 300 to 374 nm.
3. The Chappuis system from 550 to 610 nm.
4. In addition, there is a further system of transitions in the 400 to 1000 nm region (Bass and Paur, 1985; Anderson and Mauersberger, 1992; Burkholder and Talukdar, 1994; Burrows et al., 1999; Voigt et al., 2001; Orphal, 2003).

Table B.2 summarises the relevant states and some transitions of the O_3 molecule. DOAS measurements of O_3 with artificial light sources and from satellites largely make use of the structured absorption of the Hartley and Huggins bands (e.g. Axelsson et al., 1990b; Edner et al., 1994; Hausmann and Platt, 1994), while stratospheric measurements with ground-based ZSL-DOAS instruments frequently use the structures of the Chappuis band (e.g. Hoiskar et al., 1997; Sarkissian et al., 1997; Vaughan et al., 1997; van Roozendael et al., 1998; Roscoe et al., 2001). It must be kept in mind that the structured absorption of all bands are very weak as compared to the total absorption; this is illustrated in Fig. 2 showing the total- and differential absorption cross-section spectra of ozone in the Huggins band.

B.9 Monocyclic Aromatic Hydrocarbons

The derivatives of benzene exhibit a strong band system in the vacuum–UV, and narrow, distinct bands between 240 and 280 nm. In addition, there is a very weak system of benzene in the 300 to 340 nm range. Since the vacuum–UV transition is in a range where the atmosphere is not sufficiently transparent and the $a^3B_{1u} \leftarrow X^1A_{1g}$ transition is too weak for DOAS purposes, the $A^1B_{2u} \leftarrow X^1A_{1g}$ transition is used (see Table B.3). Table B.4 gives an overview of absorption cross sections and DOAS detection limits of a number of aromatics.

As discussed by Volkamer et al. (1997), care has to be taken that spectrometers with sufficient spectral resolution are used, the overlap of the Hertzberg

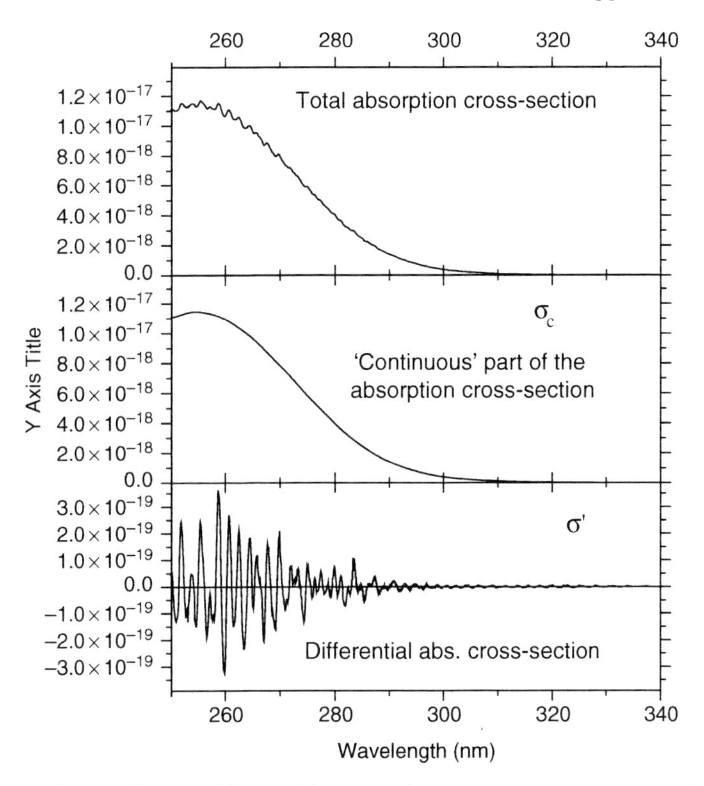

Fig. B. 2. The total- and differential absorption cross-section spectra of ozone in the Huggins band. Note that this is one of the few cases where there is a large difference between σ and σ' (from Geyer, 2000)

bands of molecular oxygen is taken into account, and correct absorption cross-sections for the aromatic compounds are taken. Unfortunately, in most of the literature on atmospheric DOAS measurements of aromatic compounds, no information is given about the absorption cross-sections used. Only Axelsson

Table B.3. Electronic terms and transitions of the benzene molecule (see e.g. Hollas)

Electronic term of C_6H_6	Spectral range of the transition from X^1A_{1g} ground state (nm)	Electric dipole transition probability
A^1B_{2u}	227–267	orbital-forbidden, spin-allowed
B^1B_{1u}	185–205	orbital-forbidden, spin-allowed
C^1E_{1u}	170–180	allowed
a^3B_{1u}	300–340	orbital- and spin-forbidden

Table B.4. Summary of absorption cross-sections and typical DOAS detection limits for a series of monocyclic aromatic hydrocarbons (adapted from Etzkorn et al., 1999). The second column shows the differential cross-section (σ' in units of 10^{-18} cm^2 at a spectral resolution (FWHM) of 0.146 nm) after high pass filtering for the strongest band together with the corresponding wavelength. The fourth column gives typical DOAS detection limits (assuming a light path of 480 m and a minimal detectable optical density of 0.001 at 25 ± 3°C)

Aromatic species (stated purity)	Differential cross-section of the strongest band (wavelength) 10^{-18} cm^2 (nm)	Uncertainty of the differential cross-section %	DOAS detection limit ppb	Uncertainty of the absolute cross-section 10^{-18} cm^2
Benzene (>99.8%)	2.19 (252.9)	7.2	0.39	0.21
Toluene (99.8%)	1.28 (266.6)	2.6	0.66	0.24
p-xylene (99%)	2.03 (272.2)	5.6	0.42	0.17
m-xylene (99%)	0.66 (270.6)	3.5	1.28	0.08
o-xylene (>97%)	0.21 (269.0)	5.1	3.99	0.05
1,3,5-trimethylbenzene (99%)	0.34 (270.7)	2.9	2.51	0.11
1,2,4-trimethylbenzene (98%)	0.50 (274.3)	4.9	1.71	0.08
Styrene (>99%)	1.55 (287.7)	8.3	0.55	0.17
Ethylbenzene[1] (99%)	0.38 (266.5)	3.2	2.25	0.10
Benzaldehyde (>99%)	4.40 (284.1)	4.7	0.19	1.82
o-tolulaldehyde (97%)	1.31 (292.2)	3.9	0.65	0.34
m-tolulaldehyde (97%)	1.26 (292.4)	3.5	0.67	0.44
p-tolulaldehyde (97%)	1.84 (284.5)	5.8	0.46	0.26
2,4,6-trimethylphenol (97%)	0.74 (283.0)	4.7	1.14	0.30
Phenol(>99.5%)	19.84 (275.1)	2.6	0.04	1.63
p-cresol (99%)	8.72 (283.0)	5.1	0.10	0.84
o-cresol (>99%)	2.01 (275.8)	2.8	0.42	0.60
m-cresol (99%)	3.18 (277.9)	4.1	0.27	0.49
2,3-dimethylphenol (99%)	2.16 (277.3)	3.7	0.39	0.41
2,4-dimethylphenol (98%)	1.27 (283.3)	9.2	0.66	0.32
2,5-dimethylphenol (>99%)	1.13 (263.6)	7.9	0.75	0.32
2,6-dimethylphenol (>99.5%)	0.94 (275.8)	7.6	0.90	0.30
3,4-dimethylphenol (99%)	1.24 (283.3)	4.6	0.68	0.34
3,5-dimethylphenol (>99%)	2.85 (279.1)	8.0	0.30	0.80

et al. (1995), Trost et al. (1997), Volkamer et al. (1997), and Ackermann et al. [2000] presented their measured absorption cross-sections or gave references. When comparing recent determinations of absorption cross-sections of aromatic compounds [Pantos et al. (1978): benzene; Milton et al (1992): benzene and toluene; Suto et al. (1992): benzene and the isomers of xylene;

Table B.5. Comparison of literature reports of differential UV absorption cross-sections of a series of monocyclic aromatic compounds. For each compound, the differential cross-section of the strongest band was divided by the differential cross-section obtained by Etzkorn et al. (1999). The different spectral resolutions have been taken into account (from Etzkorn et al., 1999)

Compound	Trost et al. 1997	Suto et al. 1992	Axelsson 1995	Milton et al. 1992	Pantos et al. 1978
Benzene	0.89	1.05	0.56	0.59	0.90
Toluene	0.85		0.98	0.46	
o-xylene	1.12	1.06	1.18		
m-xylene	1.06	1.13	1.22		
p-xylene	1.12	1.02	2.80		
Ethylbenzene			1.93		
1,2,4-trimethylbenzene			0.37		
1,3,5-trimethylbenzene			0.93		
Phenol	1.13				
Benzaldehyde	1.12				
o-cresol	0.44				
m-cresol	0.63				
p-cresol	0.60				

Axelsson et al. (1995): benzene, toluene, the isomers of xylene, ethylbenzene, 1,2,4-trimethylbenzene, and 1,3,5-trimethylbenzene; Trost et al. (1997): benzene, toluene, the isomers of xylene, phenol, benzaldehyde, and the isomers of cresol], differences of up to a factor of two occur between reported values (see Table B.5).

B.10 Polycyclic Aromatic Hydrocarbons

Polycyclic aromatic hydrocarbons also show structured absorption spectra in the near UV, which allow sensitive determination by DOAS: an important example is naphthalene and its derivatives. Besides the medium resolution structures (of the order of 0.5 nm width) usually used by DOAS, there are also narrow-band (0.001 nm) spectral structures, which allow detection of these species by high-resolution DOAS (George and Morris, 1968; Majewski and Meerts, 1984; Neuroth et al., 1991).

B.11 Nitrous Acid, HONO

Nitrous acid, HONO (HNO_2), has strong absorption bands in the 300–400 nm spectral range. It has been extensively studied by DOAS, as described in Sect. 2.6.3. There have been many determinations of the HONO absorption

Table B.6. Comparison of the differential and absolute absorption cross-sections of nitrous acid (HONO) with literature data (adapted from Stutz et al., 2000)

Author	Spectral resolution nm	Total absorption cross-section σ at 354 nm $10^{-20}\,cm^2$	σ at 354 nm with a resolution of 1 nm $10^{-20}\,cm^2$	Difference to Stutz et al. 2000 %	Difference to the σ' of Stutz et al. 2000 %
Johnston and Graham, 1974	0.87	13.6 ± 2.7	13.0 ± 2.7	−70	-70 ± 0.1
Cox and Derwent, 1976, 1977	< 0.1	67 ± 7	58.8 ± 6	26	9.7 ± 1.2
Stockwell and Calvert, 1978	< 1	49.6 ± 4.9	48.9 ± 4.9	13	30 ± 1
Perner and Platt, 1979	0.6	–	–	–	13
Vasudev, 1990	–	49.6 ± 4.9	43.6 ± 4.3	0.8	0.4 ± 0.8
Bongartz et al., 1994	0.1	54.9 ± 2	46.44 ± 1.7	7.4	4.8 ± 0.1
Febo et al., 1996		–	–	–	?
Pagsberg et al., 1997	–	50.2 ± 5	44.29 ± 5	2.4	-12 ± 0.5
Brust et al., 1999	0.5	38.9 ± 6	34.7 ± 5	−25	-26.8 ± 0.2
Stutz et al., 2000	0.08	51.9 ± 0.1	43.24 ± 0.1	–	–

cross-section; a summary is given in Table B.6. It can be seen that most of the data are in good agreement when the differences in resolution are considered. In addition, Table B.7 gives the spectral positions of the different HONO bands and the assignment to the cis and trans isomers of the molecule.

B.12 Halogen Monoxides

The halogen monoxide radicals have strongly banded spectra of the $A \leftarrow X^2\Pi$ transition in the near UV (ClO, BrO) and blue (IO) spectral regions. The peak cross-sections increase from just below $10^{-17}\,cm^2$ for ClO to several times $10^{-17}\,cm^2$ for IO, thus allowing ppt sensitivities for DOAS measurements of these species. The halogen monoxide molecules partly exhibit high spectral resolution structures (e.g. Neuroth et al., 1991), which can give rise to apparent non-linearity effects if not resolved (see also B.5). Moreover, in passive DOAS measurements, interference of the high resolution absorption structures with solar Fraunhofer lines can lead to interferences, the so called 'solar I_0 Effect' (see e.g. Mohammed-Tahrin, 2001).

Table B.7. Comparison of the spectral positions (in cm^{-1}) of the different HONO bands of the cis and trans isomers of the molecule table, but (adapted from Stutz et al., 2000)

Band	Bongartz et al., 1994	King and Moule 1962	Stutz et al. 2000 ($\pm 2\sigma$)
trans HONO			
0	26035	26034 ± 10	26036.3 ± 1.4
1	27153	27150 ± 2	27153.7 ± 1.5
2	28220	28225 ± 2	28227.8 ± 1.6
3	29249	29518 ± 2	29257.2 ± 3.4
4	30227	30231 ± 2	30230.3 ± 3.7
5	31158	31175 ± 10	31170.0 ± 3.9
6	32076	32075 ± 30	32078.2 ± 4.1
Cis HONO			
0	26312		
1	27430	27420 ± 50	27441.4 ± 3.0
2	28496	28495 ± 2	28495.0 ± 3.2
3	29513	29518 ± 2	29523.3 ± 3.5
4	30502	30505 ± 2	30510.8 ± 3.7
5	31456	31461 ± 10	31453.5 ± 4.0
6	32344	32350 ± 30	32359.6 ± 4.2

B.12.1 Chlorine Monoxide, ClO

The ClO molecule has a strong, structured absorption system ($A^2\Pi - X^2\Pi$) in the spectral range 270–310 nm. It has been subject to numerous studies (e.g. Johnston et al., 1969; Rigaud et al., 1977; Mandelman and Nicholls, 1977; Jourdain et al., 1978; Sander and Friedl, 1989; Simon et al., 1990; Trolier et al., 1990; Howie et al., 1999). There are few measurements of ClO in the atmosphere; in fact, to date only tropospheric ClO measurements by DOAS have been reported (Tuckermann, 1996; Tuckermann et al., 1997; Stutz et al., 2002b; Bobrowski, 2005; Bobrowski et al., 2007; Lee et al., 2005).

B.12.2 Bromine Monoxide, BrO

Strong bands of the $A(^2\Pi_{3/2}) \leftarrow X(^2\Pi_{3/2})$ transition of BrO are in the spectral region 310–350 nm. Quantitative determination of spectra was reported by several authors (e.g. Cox et al., 1982; Wahner et al., 1988; Wilmouth et al., 1999; Fleischmann et al., 2004). These spectra were employed by numerous researchers to study the abundance of BrO, both in the stratosphere (Carroll et al., 1989; Wahner et al., 1990b; Arpag et al., 1994; Platt and Hausmann, 1994; Fish et al., 1995; Pundt et al., 1996; Eisinger et al., 1997; Kreher et al., 1997; Aliwell et al., 1997, 2002; Solomon et al., 1989b,d; Richter et al., 1999, 2001; Ferlemann et al. 1998; Harder et al., 1998, 2000; Fitzenberger et al., 2000; Tørnkvist et al., 2002; Weidner et al., 2005; Schofield et al., 2006; Dorf et al. 2006) and in the troposphere (Hausmann et al., 1992, 1994; Tuckermann et al., 1997; Wagner and Platt, 1998; Hegels et al., 1998; Richter et al., 1998; Hebestreit et al., 1999; Martinez et al., 1999; McElroy et al., 1999; Frieß et al., 1999, 2004, 2005; Matveev et al., 2001; Wagner et al., 2001b; Hönninger and Platt, 2002; Stutz et al., 2002; Bobrowski et al., 2003; Leser et al., 2003; Van Roozendael et al., 2003; Hönninger et al., 2004a,b,c; Hollwedel et al., 2004; Saiz-Lopez et al., 2004a, 2005; Lee et al., 2005), as also outlined in Chap. 10 and 11.

B.12.3 Iodine Monoxide, IO

The standard IO spectral retrieval uses the wavelength interval 425–465 nm, where five strong bands of the 'cold' (4-0, 3-0, 2-0, 1-0, 0-0) and two weaker bands of the 'hot' (3-1, 2-1) vibrational absorption of the IO $A^2\Pi \rightarrow X^2\Pi$ electronic transitions are located. The total absorption cross-section of the IO molecule was measured by a number of authors (Lazlo et al., 1995; Himmelmann et al., 1996; Harwood et al., 1997; Hönninger, 1999; Cox et al., 1999; Dillon et al., 2005). These spectra were employed by several researchers to study the abundance of IO in the troposphere (e.g. Brauers et al., 1990; Platt. and Janssen, 1996; Alicke et al., 1999; Wittrock et al., 2000; Allan et al., 2000; Frieß et al., 2001; Saiz-Lopez et al., 2005; Zingler and Platt, 2005), while in the stratosphere only upper limits have been reported (Pundt et al., 1998; Bösch et al., 2003).

B.13 Halogen Dioxides

The symmetric halogen dioxide molecules exhibit strongly strutured absorption spectra in the near UV (OClO) and visible (OBrO, OIO) spectral ranges, which have been used to make sensitive measurements of OClO and OIO in the atmosphere.

B.13.1 Chlorine Dioxide, OClO

Studies of the UV absorption cross-section of the OClO $A^2A_2 \leftarrow X^2B_1$ transition in the wavelength range 300–440 nm were performed by Wahner et al. (1987) and Kromminga et al. (2003) at spectral resolutions better than 0.1 nm. These bands have been used in numerous studies of stratospheric OClO as described in Chap. 11.

B.13.2 Bromine Dioxide, OBrO

Studies of the UV absorption cross-section of the OBrO $C^2A_2 \leftarrow X^2B_1$ transition in the wavelength range 390–620 nm were reported by Rattigan et al. (1996), Knight et al. (2000), and Fleischmann and Burrows (2002).

B.13.3 Iodine Dioxide, OIO

There are few absolute determinations of the structured absorption cross-section of OIO in the spectral range between 467–667 nm (Himmelmann et al., 1996; Cox et al., 1999; Ashworth et al., 2002; Joseph et al., 2005; Tucceri et al., 2006). The peak absorption cross-section at 548.6 nm is estimated at $2.7 \times 10^{-17}\,\mathrm{cm^2/molec}$.

B.14 Molecular Iodine (I_2)

The electronic transition of the $B^3\Pi\,(O_u^+) \leftarrow X^1\Sigma_g^X$ electronic transition exhibits a distinct, regular vibrational structure in the range 500–630 nm, which can be used for DOAS evaluations. Absolute determinations of the cross-section were reported by Goy and Pritchard (1964), Ogryzlo and Thomas (1965), Tellinghuisen (1973), Saiz-Lopez et al. (2004c), and Spietz et al. (2006). The dissociation energy of I_2 corresponds to a threshold wavelength of 533 nm (Saiz-Lopez et al., 2004c). Atmospheric determinations of I_2 using the above transition were reported by Saiz-Lopez et al. (2004b) and Peters et al. (2005).

B.15 Water Vapour, H_2O

Water vapour has a series of bands in the visible and near IR spectral regions, which are suitable for DOAS measurements. Data on line strengths are found in Rothman et al. (1992, 1998), with several recent corrections (e.g. Belmiloud et al., 2000; Coheur et al., 2002; Schermaul et al., 2002; Mérienne et al., 2003). The bands correspond to the excitations 4ν, $3\nu + \delta$, 3ν, and $2\nu + \delta$, (located around 725, 820, 943, and 1136 nm, respectively), where ν denotes a quantum number of the stretching modes and δ of the bending mode; details of the line assignment can be found at Tolchenov et al. (2005).

The water molecule partly exhibits high spectral resolution structures, which can give rise to apparent non-linearity effects if not resolved (see also B.5) or corrected for. At high water vapour column densities ($S \approx 4 \times 10^{23}\,\mathrm{cm^{-2}}$), the individual H_2O lines have optical densities in excess of unity. Considering the variation of the water vapour column amount, a model study was carried out to investigate the effects of this non-linear dependence of H_2O absorption bands on the results of the NO_3 evaluation. Different optical densities of water vapour can be simulated by applying Lambert–Beer's law to the high-resolution water absorption cross-section (Mandin et al., 1986) and then smoothing the calculated spectrum to simulate the instrumental resolution. This simulation demonstrates that the column amount-dependent absorption structure of H_2O can be interpolated by simultaneously fitting two water reference spectra, which bracket the optical density of the atmospheric spectrum (see Geyer et al., 2000).

B.16 Nitrate Radical, NO_3

The absorption cross-section $\sigma(\lambda)$ of the nitrate radical shows strong features in the red spectral region, with prominent bands at wavelengths of 662 and 623 nm (Fig. 10.24). Various studies of its wavelength and temperature dependency have been performed in recent years (see Table B.8). Most of the studies report similar relative shapes of the spectrum; however, there are discrepancies concerning the absolute value of $\sigma(\lambda)$. The value of the cross-section at 298 K and 662 nm published by different authors varies from as low as $1.21 \times 10^{-17}\,\mathrm{cm}^2$ (Mitchell et al., 1984) up to $2.49 \times 10^{-17}\,\mathrm{cm}^2$ (Magnotta and Johnston, 1980). Recent publications reported $\sigma(622\,\mathrm{nm}) = (2.23 \pm 0.22)10^{-17}\,\mathrm{cm}^2$ (Yokelson et al., 1994). In addition, the temperature dependence of the NO_3 absorption cross-section is not clear. While Sander (1986), Ravishankara and Mauldin (1986), and Yokelson et al. (1994) determined an increase in $\sigma(\lambda)$ towards lower temperatures (Table B.8), Cantrell et al. (1987) did not find a temperature effect. In recent DOAS observations (e.g. Geyer et al. 2001a), the temperature-independent cross-section $\sigma(662\,\mathrm{nm}) = (2.1 \pm 0.2) \times 10^{-17}\,\mathrm{cm}^2$ recommended by Wayne et al. (1991) was, therefore, used as reference spectrum for the NO_3 evaluation. This cross-section was derived by averaging the data of four recent studies of Ravishankara and Mauldin (1986), Sander (1986), Cantrell et al. (1987), and Canosa-Mas et al. (1987).

B.17 OH Radicals

Several active DOAS studies of atmospheric OH have been performed using the rotational lines of the $A\,^2\Sigma^+(v' = 0) \leftarrow X^2\Pi_{3/2}(v'' = 0)$ and $A\,^2\Sigma^+(v' = 0) \leftarrow X^2\Pi_{1/2}(v'' = 0)$ transitions near 308 nm. Due to the relatively small reduced mass and corresponding small momentum of inertia of the OH radical, the spacing of the individual rotational lines is relatively large (of the order of 0.15 nm). Under ambient conditions (1 atmosphere, 300K), the strongest rotational lines are $Q_1(2)$ (307.9951 nm), $Q_1(3)$ (308.1541 nm), and $P_1(1)$ (308.1665 nm), with line widths (FWHM) around 0.0025 nm (Dorn et al. 1995); Doppler broadening and pressure broadening contribute approximately equally to the line width. Spectroscopic parameters for the OH radical have been measured by many authors (e.g. Dieke and Crosswhite, 1961; German 1975; Dimpfl and Kinsey, 1979; Goldman and Gillis, 1981). Recent measurements of the UV absorption cross-section of the OH radical were reported by Dorn et al. (1995). In contrast to most other DOAS applications, the specific and sensitive detection of atmospheric OH requires instruments that are able resolve the individual rotational lines. The narrow absorption lines dictate the use of a light source with high spectral intensity. For example, OH can be detected using a laser with an emission bandwidth broader than the OH lines in combination with a high-resolution spectrometer (Hübler et al., 1984; Perner

Table B.8. Results of studies of the NO_3 absorption cross-section since 1978. Reported absolute values of the peak absorption cross-section at 662 nm vary from (1.21 to 2.48) $\times 10^{-17}$ cm^2. The temperature dependence of the cross-section is not understood at present (adapted from Geyer et al., 2000)

Studies	Technique	Temp. [K]	σ (662 nm) [10^{-17} cm^2]	Temperature dependence of σ (662 nm) [10^{-17} cm^2]
Graham and Johnston, 1978	Modulated photolysis	298	1.71	not studied
Magnotta and Johnston, 1980	Calibration of quantum yield	298	2.48	not studied
Mitchell et al., 1980	Differential photomultiplier	294	1.21	not studied
Marinelli et al., 1982	Tuneable dye laser	298	1.90	not studied
Ravishankara and Wine, 1983	Discharge flow, tuneable dye laser	298	1.78	not studied
Cox et al., 1984	Modulated photolysis, diode array spectrometer	298	1.63 ± 0.15	not studied
Burrows et al., 1985	Modulated photolysis	298	1.85 ± 0.56	not studied
Ravishankara and Mauldin, 1986	Discharge flow, tuneable dye laser	298	1.90 ± 0.22	4.65–0.00932 T
Sander, 1986	Flash photolysis, diode array spectrometer	298	2.28 ± 0.34	4.19–0.0064 T
Cantrell et al., 1987	Fourier transform spectrometer	298	2.02 ± 0.2	not found
Canosa-Mas et al., 1987	Differential photomultiplier	296	2.23 ± 0.35	not studied
Wayne et al., 1991	Averaged value	–	2.10 ± 0.20	–
Yokelson et al., 1994	Diode array spectrometer	298	2.23 ± 0.22	4.56–0.00788 T
Orphal et al., 2003	Fourier transform spectroscopy	294		

et al., 1987; Platt et al., 1987, 1988; Dorn et al., 1988; Hofzumahaus et al., 1991; Mount 1992), or a narrow-band laser scanning the OH line (Zellner and Hägele, 1985; Amerding et al., 1990, 1992). Figure 10.21 shows a portion of the OH absorption spectrum. Several species, e.g. SO_2 and CH_2O, absorb at similar wavelengths to OH (Hübler et al., 1984; Neuroth et al., 1991).

B.18 Oxygen, O_2

The oxygen molecule exhibits several absorption systems (see Table B.9), which are relevant for DOAS applications:

1. The $B^3\Sigma_g^- \leftarrow X^3\Sigma_g^-$ transition known as Schuman–Runge bands in the short-wavelength UV makes the atmosphere essentially opaque below about 200 nm, preventing DOAS measurements at shorter wavelengths.
2. The $A^3\Sigma_g^+ \leftarrow X^3\Sigma_g^-$ transition of the Herzberg I system in the spectral range 242–286 nm is overlapping the absorption features of most monocyclic aromatic species (see Sect. B.9 and Volkamer et al., 1998).
3. The near IR and visible spectral range has been used to determine optical path lengths in the atmosphere as described in Sect. 11.9.2 and for cloud detection by satellite sensors.
4. An additional forbidden system gives rise to absorption in the near IR (Smith and Newnham, 1999; Newman et al.; 1999, 2000; Smith et al. 2001).

Table B.9. The absorption bands of oxygen. The atmospheric bands are in the VIS and NIR spectral regions. The band strength is the sum of all line strengths of the band. The spectroscopic data is taken from the HITRAN96 database (from Gamache et al., 1998)

Electronic transition	Vibrational transition	Band centre (nm)	Band strength (cm/molec.)	Designation
$B^3\Sigma_g^- \leftarrow X^3\Sigma_g^-$	–	UV 175.9	–	Schumann–Runge
$A^3\Sigma_g^+ \leftarrow X^3\Sigma_g^-$	–	UV 242-286	–	Herzberg I
$b^1\Sigma_g^+ \leftarrow X^3\Sigma_g^-$	$(0 \leftarrow 0)$	762.19	2.24×10^{-22}	A band
$b^1\Sigma_g^+ \leftarrow X^3\Sigma_g^-$	$(1 \leftarrow 0)$	688.47	1.49×10^{-23}	B band
$b^1\Sigma_g^+ \leftarrow X^3\Sigma_g^-$	$(2 \leftarrow 0)$	628.85	4.63×10^{-25}	γ band
$b^1\Sigma_g^+ \leftarrow X^3\Sigma_g^-$	$(1 \leftarrow 1)$	771.07	9.53×10^{-26}	
$b^1\Sigma_g^+ \leftarrow X^3\Sigma_g^-$	$(0 \leftarrow 1)$	864.75	7.88×10^{-27}	
$a^1\Delta_g \leftarrow X^3\Sigma_g^-$	$(0 \leftarrow 0)$	1268.6	3.68×10^{-24}	
$a^1\Delta_g \leftarrow X^3\Sigma_g^-$	$(1 \leftarrow 0)$	1067.7	9.53×10^{-27}	
$a^1\Delta_g \leftarrow X^3\Sigma_g^-$	$(0 \leftarrow 1)$	1580.8	2.75×10^{-28}	

B.19 Oxygen Dimer, O_4 or $(O_2)_2$

The broadband absorptions in gaseous O_2 (Fig. B.3) were first described by Janssen et al. (1885, 1886), who also showed that the intensity of these bands varies with the square of the oxygen pressure (i.e. O_2 concentration). The absorption of oxygen molecules in the UV, vis, and near IR spectral ranges is due to different types of chemical bonds and transitions. Besides the discrete, structured ro-vibrational bands of the electronic transition of the O_2 monomer (see Sect. B.18) and the structured bands of the bound van der Waal's molecule O_4, oxygen also shows broad unstructured absorptions (Figure 3 due to the collision induced oxygen dimer $(O_2)_2$ (see Greenblatt et al., 1990; Solomon et al., 1998, and references therein). Since the original work of Janssen et al. (1885, 1886), several studies investigated the $(O_2)_2$ absorptions demonstrating that, under atmospheric conditions, these bands contain no fine structure (see Solomon et al., 1998; Naus and Ubachs, 1999, and ref-

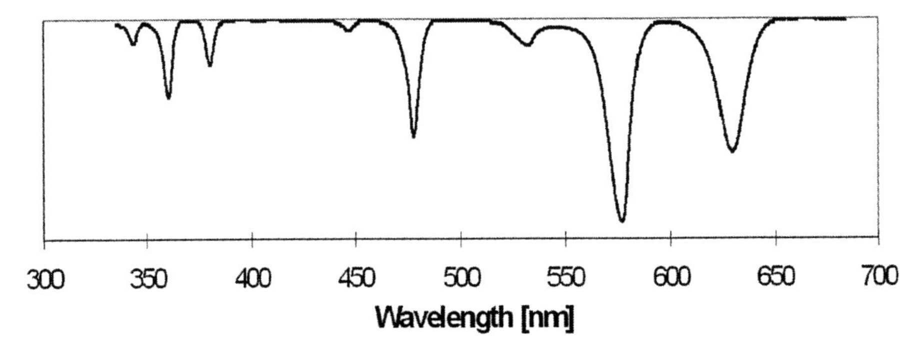

Fig. B. 3. Absorption spectrum of oxygen dimers (O_4) (from Greenblatt et al., 1990, Copyright by American Geophysical Union (AGU), reproduced by permission of AGU)

Table B.10. Wavelengths and transition assignments of the O_4 absorption bands in the UV-vis spectral region (adapted from Wagner et al., 2002b)

Wavelength (nm)	Upper state of transition (from $^3\Sigma_g^- +^3 \Sigma_g^-$ ground state)
343.4*	$^1\Sigma_g^+ +^1 \Sigma_g^+$ ($\nu = 2$)
360.5	$^1\Sigma_g^+ +^1 \Sigma_g^+$ ($\nu = 1$)
380.2	$^1\Sigma_g^+ +^1 \Sigma_g^+$ ($\nu = 0$)
446.7*	$^1\Sigma_g^+ +^1 \Delta_g$ ($\nu = 1$)
477.3	$^1\Sigma_g^+ +^1 \Delta_g$ ($\nu = 0$)
532.2	$^1\Delta_g +^1 \Delta_g$ ($\nu = 2$)
577.2	$^1\Delta_g +^1 \Delta_g$ ($\nu = 1$)
630.0	$^1\Delta_g +^1 \Delta_g$ ($\nu = 0$)

Table B.11. O_4 absorption cross-sections, as determined in the laboratory and the atmosphere from several studies (adapted from Wagner et al., 2002b)

O_4 band	Perner and Platt, 1980	Green-blatt et al., 1990	Green-blatt et al., 1990	Volkamer, 1996	Osterkamp et al., 1998		Newnham and Ballard, 1998		Naus and Ubachs, 1999	Wagner et al., 2002b*
Temp.	279 K	296 K	296 K	296 K	204 K	256 K	223 K	283 K	294 K	242 K
nm										
360	5.4 ± 1.5	4.1 ± 0.4	5.7 ± 0.6	5.42 ± 0.7						5.70 ± 0.5
380	< 1.4	2.4 ± 0.2	3.7 ± 0.4	2.4 ± 0.2						2.44 ± 0.4
477	5.9 ± 1.8	6.3 ± 0.6	7.6 ± 1.3	6.1 ± 0.3	7.9 ± 0.3	7.0 ± 0.3	6.99 ± 0.35	8.34 ± 0.83		7.80 ± 0.2
532				1.3 ± 0.3	1.4 ± 0.2	1.2 ± 0.2	1.31 ± 0.2	1.23 ± 0.38		1.74 ± 0.5
577	16 ± 6	11 ± 1	-	10.3 ± 0.3	12.2 ± 0.4	13.6 ± 0.4	12.61 ± 0.11	11.75 ± 0.2	11.41 ± 0.5	13.50 ± 0.4
630	-	7.2 ± 0.7	-	6.2 ± 0.6			8.8 ± 0.13	7.9 ± 0.15	7.55 ± 0.5	9.61 ± 0.3

* Direct moonlight measurements in the atmosphere.

erences therein). Greenblatt et al. (1990) concluded from the weak temperature dependence of the $(O_2)_2$ absorption that, under atmospheric conditions, they are most probably related to an oxygen collision complex rather than a bound dimer. The O_4 absorption bands in the UV-vis spectral regions (360–630 nm) belong to simultaneous transitions between two ground state oxygen molecules and electronically exited states of both oxygen molecules, as summarised in Table B.10. At larger wavelengths, transitions of only one exited oxygen molecule occur, and under atmospheric conditions N_2 can serve as a collision partner (see Solomon et al., 1998; Smith and Newnham, 2000, and references therein). Direct moonlight measurements by Wagner et al. (2002b) yielded improved absorption cross-sections of O_4. The investigated O_4 bands were found to show an increase of the peak absorption with decreasing temperature ranging from $\approx$13%/100 K at 477.3 and 532.2 nm, $\approx$20% at 360.5 and 577.2 nm to $\approx$33% at 380.2 and 630.0 nm. Moreover (except for the band at 380.2 nm), the O_4 absorption cross-sections were found to be somewhat larger than previous measurements (see Table B.11).

Atmospheric absorptions of O_4 were first reported by Perner and Platt (1980), and are important in atmospheric radiative transfer modelling.

Index